FUNGI
Applications and
Management Strategies

Series: Progress in Mycological Research

- Fungi from Different Environments (2009)
- Systematics and Evolution of Fungi (2011)
- Fungi from Different Substrates (2015)
- Fungi: Applications and Management Strategies (2016)

FUNGI

Applications and Management Strategies

Editors

S.K. Deshmukh
Biotechnology and Management of Bioresources Division
The Energy and Resources Institute (TERI)
New Delhi, India

J.K. Misra
Botany Department
Saroj Lalji Mehrotra Bhartiya Vidya Bhavan Girls Degree College
Lucknow
India

J.P. Tewari
Akron, Ohio
USA

Tamás Papp
University of Szeged
Szeged, Hungary

CRC Press
Taylor & Francis Group
Boca Raton London New York

CRC Press is an imprint of the
Taylor & Francis Group, an **informa** business

A SCIENCE PUBLISHERS BOOK

CRC Press
Taylor & Francis Group
6000 Broken Sound Parkway NW, Suite 300
Boca Raton, FL 33487-2742

First issued in paperback 2021

Version Date: 20160115

ISBN-13: 978-0-367-78309-9 (pbk)
ISBN-13: 978-1-4987-2491-3 (hbk)

Library of Congress Cataloging-in-Publication Data

Names: Deshmukh, S. K. (Sunil K.), editor. | Misra, J. K., editor.
Title: Fungi / editors: Sunil Kumar Deshmukh and J.K. Misra.
Description: Boca Raton, FL : CRC Press, Taylor & Francis Group, 2016. |
Includes bibliographical references and index.
Identifiers: LCCN 2015047485 | ISBN 9781498724913 (hardcover : alk. paper)
Subjects: LCSH: Fungi.
Classification: LCC QK603 .F958 2016 | DDC 579.5--dc23
LC record available at http://lccn.loc.gov/2015047485

Visit the Taylor & Francis Web site at
http://www.taylorandfrancis.com

and the CRC Press Web site at
http://www.crcpress.com

Preface

Fungi, the second largest group of organisms after insects, are estimated to be about 1.5 million in number including those that are in various culture collections. Fungi grow and proliferate in soil, water (both fresh water and marine), and also in mangrove environments. They also grow on various other substrates and in different environments and, therefore, fungi are said to be of ubiquitous and cosmopolitan in nature. Some fungi also grow in or on the plants and animals as parasites, symbionts or as mutualists. These organisms have been in association with human beings since times immemorial. They have both good and bad effects on our lives. Fungi serve as a rich source of food, are a source of a variety of useful chemicals including many antibiotics, are used to make bread, variety of beverages and so on. With the advancement and development of science we have got many things like penicillin and cyclosporine from these fascinating but bewildering organisms. On the other hand they cause many diseases in plants, animals and human beings.

This book presents some aspects of fungi which have helped us to use them for human welfare. The book contains 17 Chapters covering some of the important new applications of these organisms in medicine, agriculture, industry, etc. Endophytic fungi produce secondary metabolites that are proving to be of great value for pharmaceutical applications directly or indirectly, hence the endophyes and their uses have been dealt with in the book. The importance of secondary metabolites as source of antifungal, antibacterial, antiviral, and antimycobacterial compounds cannot be ignored. Therefore, wider and deeper search for fungi for such compounds is needed. This aspect has been elaborated in different chapters. More so, there are still many silent genes or gene clusters in fungi that can be harnessed for the production of secondary metabolites, hence genetic engineering is now playing a greater role to get compounds from fungi which are not yet known to us. Statins, the cholesterol-lowering agents that are saving millions of lives worldwide, have been briefly dealt in one of the chapters. How thermophilic fungi are being used for getting variety of useful compounds and how their application can help us in solving energy crisis have been discussed. Some secondary metabolites of fungi are quite toxic—the mycotoxins. These are being given very serious attention worldwide to study their effect on the commodity and the consumers. This aspect is covered in two separate chapters that deal with mycotoxins. Besides describing the nutritive value of some important mushrooms and wood-inhabiting fungi, and their applied aspects, strategies for the management of pathogenic fungi of rice and soybean have also been dealt with in two separate chapters. And, how the volatile organic compounds of fungi can be applied to arrest the other undesired fungal forms has been elaborated. In recent years, the nano-particles and their applications have attracted wider attention; therefore, a chapter is devoted to describe the green synthesis of nano-particles using fungal system.

The editors are indebted to all the contributors for their efforts in preparing and providing the chapters for the book.

Dr. S.K. Deshmukh is grateful to Dr. R.K. Pachauri, Director General and to Dr. Alok Adholeya, Senior Director, The Energy and Resources Institute, New Delhi for their encouragement. He also acknowledges the academic support of Dr. B.N. Ganguli, Emeritus Scientist (CSIR) and Chair Professor of the Agharkar Research Institute, Pune, India. Dr. J.K. Misra expresses his gratefulness to Justice V.K. Mehrotra, the Chairman and to Brig. R.N. Misra, the Secretary of the Management of the college for providing all encouragements and support. Dr. Rachna Mishra, Principal of the college also deserves thanks for all the help she rendered to Dr. Misra. Dr. Tamás Papp is grateful for the grants TÉT_12_DE-1-2013-0009 and OTKA NN 106394.

S.K. Deshmukh
J.K. Misra
J.P. Tewari
Tamás Papp

Contents

About The Book

The book comprising seventeen chapters deals with the application of fungi and the strategic management of some plant pathogens. The book has covered the fungal bioactive metabolites, with emphasis on those secondary metabolites that are produced by various endophytes, their pharmaceutical and agricultural uses, regulation of the metabolites, mycotoxins, nutritional value of mushrooms, prospecting of thermophilic and wood-rotting fungi, and fungi as myconano factories. Strategies for the management of some plant pathogenic fungi of rice and soybean have also been dealt with. Updated information, for all these aspects have been presented and discussed in different chapters.

CHAPTER 1

The Promise of Endophytic Fungi as Sustainable Resource of Biologically Relevant Pro-drugs: A Focus on Cameroon

Souvik Kusari* and Michael Spiteller*

ABSTRACT

Fungal endophytes constitute an extraordinarily multifarious cluster of polyphyletic fungi pervasive in plants, and retain an indiscernible vibrant relationship with their associated host plants for at least a part of their life cycle. The potential of 'novel' endophytes capable of biosynthesizing bioactive natural products has undoubtedly been acknowledged. However, it is disappointing that there is still no known breakthrough in the commercial production of these bioactive secondary metabolites using endophytes. Thus, vigilant bio-prospecting strategies should be devised, in selecting the best possible plants thriving in propitious ecological niches, for isolating endophytes with potent pharmaceutical value. Cameroon (Africa) is such a largely unexplored biodiversity "hotspot", where wild plants from ecologically distinct areas are intensively used by the local population in traditional medicine. Plants are a rich source of bioactive compounds and are vibrantly interacting with other organisms, including endophytes, for thriving in their distinct 'natural' ecological landscapes.

Institute of Environmental Research (INFU), Department of Chemistry and Chemical Biology, TU Dortmund, Otto-Hahn-Str. 6, D-44221 Dortmund, Germany.
Email: Souvik.Kusari@infu.tu-dortmund.de
Email: m.spiteller@infu.tu-dortmund.de
* Corresponding authors

In this chapter, we briefly highlight an interesting bio-prospecting strategy and provide a perspective of using the African traditional knowledge directed towards the study of plants and endophytic fungi harboring them in order to find new lead compounds with therapeutically relevant activities.

Introduction

The combined approach, using the knowledge from ecosystem analysis, phylogenetic background and also knowledge acquired from traditional medicinal applications (via traditional healers), is an efficient and promising route to discover new lead compounds, both from plants and the associated endophytic microorganisms. Endophytic organisms live together with plants in mutualistic association and produce plethora of secondary metabolites which, in case they have desired biological activities, are of high interest for pharmaceutical applications (Kusari et al., 2014). Even if the metabolites produced by the endophytes do not have the preferred biological function, they are of inestimable value as occasionally these natural products might have new lead structures (or act as pro-drugs), from which a number of new compounds can be derived by derivatization and/or chemical synthesis. Several reports have deliberated the products of endophytic microbes and their competence for usage in medicine, agriculture and industry (Kusari and Spiteller, 2011, 2012; Kusari et al., 2012, 2013a). In this chapter, we highlight and discuss about a particular tropical biodiversity hotspot of the world, namely Cameroon (Africa), in order to provide the readers an exhilarating perspective of bio-prospecting plants and endophytes from tropical rainforest microbes, which have unique and remarkable promise for utility in the pharmaceutical industry.

We started our endeavor of focusing on the tropical ecological niches in Africa, especially in Cameroon, within the purview of the recent "Welcome to Africa" initiative of the German Federal Ministry of Education and Research (BMBF) and the German Academic Exchange Service (DAAD), Germany. Throughout Cameroon, particularly in the south region, plants are intensively used by the local population in traditional medicine. Cameroonian plants have been found to be a rich source of bioactive compounds. Our program is directed towards the study of these plants and endophytic fungi harboring them in order to find new lead compounds with potent biological activity, mainly antibacterial and anti-inflammatory.

Introduction to Microbial Endophytes

Plants (or green plants) are a unique group of cellulose-containing multi-cellular organisms capable of performing photosynthesis. Beyond this generalized understanding of plants, however, is a much more complex veracity of comprehending them. This includes commanding the network of associations of plants with other organisms, various biotic and abiotic selection pressures, an assortment of cost-benefit mutualisms, and interaction-directed co-evolution of attack-defense-counterdefense mechanisms (Kusari et al., 2012, 2013a). One central 'partner' within these strata of natural acquaintances is a class of remarkably diverse group of microorganisms called endophytic microorganisms (typically known as endophytes). They inhabit living, internal tissues of plants, and retain a discreet association with their associated hosts for at least a part of their life (Bacon and White, 2000; Porras-Alfaro and Bayman, 2011; Kusari and Spiteller, 2012). This category of connotation is symptomless and established exclusively inside the living host plant tissues. "Endophytism", therefore, epitomizes a specific form of plant-microbe mutualism.

The establishment of microorganisms, both fungi and bacteria, as endophytes occurs commonly by their coincidental encounters with potential host plants over an evolutionary period wherein a number of factors play an important role, viz., the ecological niche of the plant, plant population (wild and/or agriculture), plant tissue type, and site (localized and/or systemic manner) (Hyde and Soytong, 2008). Consequently, even a microbe that is pathogenic in one ecological niche can be endophytic to plant hosts in another ecosystem (Kusari et al., 2013b). Especially for fungi, it has been demonstrated that endophytic-pathogenic lifestyles are transposable and are due to a number of environmental, chemical and/or molecular triggers (Eaton et al., 2011; Hyde and Soytong, 2008; Schulz et al., 1999). Endophytes have been reported to function as plant growth and defense promoters by synthesizing phytohormones, producing biosurfactants, enzymes or precursors for secondary plant metabolites. They also fix atmospheric nitrogen and carbon dioxide, control plant diseases, and help in plant tolerance towards environmental stress like drought and salinity (Stone et al., 2000; Redman et al., 2002; Arnold et al., 2003; Rodriguez et al., 2004, 2008; Waller et al., 2005; Marquez et al., 2007; Rodriguez and Redman, 2008; Porras-Alfaro and Bayman, 2011). The competence of endophytes to yield a plethora of bioactive natural products has also been decisively established. Sporadically, endophytes can also produce the bioactive host plant secondary metabolites or precursors. Some notable examples include the production of paclitaxel, podophyllotoxin, deoxypodophyllotoxin, camptothecin and structural analogues, hypericin, and emodin by endophytic fungi (Kusari and Spiteller, 2011; Kusari et al., 2012). In addition to plants producing volatile compounds induced by parasitic or pathogenic fungi (Piel et al., 1997, 1998), endophytes themselves are also capable of producing volatile organic compounds (Strobel et al., 2007).

Present-Day Situation

The production of bioactive compounds by endophytic fungi, including those exclusive to their host plants, is striking both from the molecular and biochemical viewpoint, and the even from the ecological standpoint. The production of secondary metabolites by endophytes, which have pharmaceutically-relevant pro-drug-like biological function, fosters anticipations of exploiting them as substitute and viable sources of these compounds. However, the commercial implication of production of desirable compounds by endophytic fungi still remains an impending intention (Kusari and Spiteller, 2011, 2012). A foremost deterrent foiling the pharmaceutical application of endophytes is the bewildering problem of diminution of secondary metabolite production on repetitive sub-culturing under axenic monoculture conditions. In addition to a persistent quest of discovering competent endophytes with promise for pharmaceutical use, it is crucial to follow-up these discoveries with cutting-edge research to establish, restore and sustain the *in vitro* biosynthetic competence of endophytes. This can be realized by a multifaceted tactic encompassing the comprehensive interpretation of the dynamic endophyte-endophyte, endophyte-host, endophyte-pathogen, endophyte-herbivore, host-pathogen, and host-herbivore multipartite interactions concerning their biological, biochemical and genetic contexts.

The individual mutualistic suites, namely the plant-endophyte and plant-insect mutualisms, have been the major focus of research in the last decades (Kusari et al., 2012, 2013a). However, the interaction between these two highly diverse and almost entirely different mutualisms yet remains to be explored (Fig. 1). While the reciprocal benefits of many plant-fungal and plant-herbivore mutualisms have been sporadically studied independent of each other, less

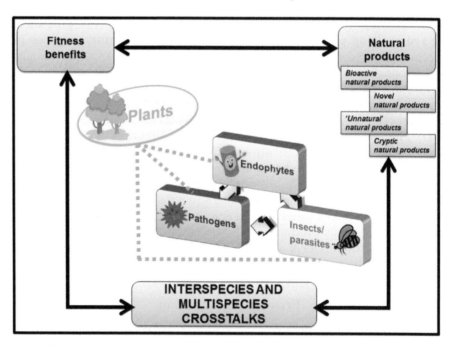

Figure 1. The interspecies and multispecies crosstalk between endophytes (bacteria and fungi), pathogens, plants and feeders.

understood are the interactions among the component members of these different symbioses. Moreover, even though endophytic interaction with other organisms has been somewhat exploited over the last decades, the role of such relationships on the endophytic production of host plant compounds is poorly known and only gained attention recently. It is highly desirable to investigate the various links between endophytes and plants, such as other coexisting endophytes and insects, with regard to their function in triggering chemical defense reactions. It has, therefore, become clear that it is vital to consider all the different types of interactions that endophytes have with other co-existing endophytes, host plants, insects and specific herbivores, and all their interconnectivity at the organ, species, population and community level will enable the comprehensive understanding of the plant-endophyte-insect networking.

Can Plant Selection Strategy Dictate Paramount Probabilities of Isolating Endophytes with Extraordinary Biosynthetic Potential?

Exhilarating prospects for exploring and utilizing the potential of endophytes have been realized by employing suitable plant bio-prospecting strategies (Kusari and Spiteller, 2012). However, selecting the most appropriate plants for isolating endophytes with pharmaceutical value can be a daunting task. Therefore, various bio-prospecting strategies, each based on specific rationale, might be affianced in order to discover potent endophytes with appropriate traits (Fig. 2). These include, but are not limited to, the following:

1. *Random plant sampling*: In this approach, sampling of different plants from diverse populations is randomly undertaken, followed by isolation of the associated endophytes, and their characterization.
2. *Traditional medicinal plants*: This is one of the frequently employed strategies in which medicinal plants used by indigenous people in traditional medicinal preparations, or plants having a historical record of ethno-botanical use are bio-prospected for endophytes, especially for the ones capable of producing one or more of the bioactive secondary metabolites mimetic to the associated the host plants.
3. *Phylogenetically related plants*: This is a very interesting strategy where plants are selected based on their phylogenetic relatedness to well-known, well-investigated, or invaluable traditional medicinal plants, with or without knowledge of the phyto-chemical constituents of the selected plants.
4. *Plants from unique ecosystems*: This is an alternate methodology wherein the evolutionary relatedness among groups of plants at a particular sampling site, correlating to species, genus, and populations, through morphological data matrices and molecular sequencing is evaluated followed by isolation of endophytes from the desired plants. A preliminary assessment of a given ecosystem with regard to its features correlating to its natural

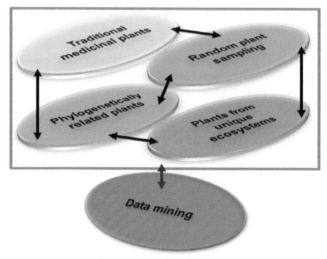

Figure 2. Various bio-prospecting strategies that might be exploited in order to discover novel or competent endophytes with desirable traits.

population of plant species, their association with the environment, soil composition, and biogeochemical cycles is often taken into account when employing this selection strategy.

5. *Data mining approaches*: This strategy can be used in combination and complementary to all the above four approaches. The valuable information acquired using the different bio-prospecting schemes can thus be assembled together, comparatively evaluated, and stored for further use applying suitable data mining approaches.

Dja Rainforest in Cameroon as a Natural Resource for Bioactive Compounds

The Dja Biosphere Reserve, located in the south of Cameroon, includes different geographic and climatic zones which contribute to its high biodiversity (Gartlan and Leakey, 1998). The area comprises mainly dense evergreen rainforest and is known to have a wide range of primate species (Betti, 2004). As the exploitation of timber resources increases, the Dja forest remains an important refuge for many plants and animal species that make the area attractive for many tourists (Motte, 1980). The place covers an area of 526000 hectares with more than 1500 plant species recorded (Baniakina et al., 1995; Betti, 2001, 2002; www.ecofac.org).

The harvest of plants for pharmaceutical use is a very important source of income (Magilu et al., 1996). For example, *Baillonella toxisperma* is the most important plant in the treatment of lumbago (low back pain) and *Nauclea diderrichii* is largely used for treating malaria and fever (Betti, 2004). The local population claims that *Picralima nitida* is at least twice as strong as chloroquine against malaria, and *Chenopodium ambrosioides* has a similarly stronger potency than "vermox" against intestinal worms (Betti, 2004).

During our recent research visits to Cameroon, we were able to identify with the aid of traditional healers and by discussions with some native pygmies (Fig. 3), along with an experienced botanist (Mr. Victor Nana of the National Herbarium of Yaoundé), some very interesting and potential plant species. Previous research on the phytochemistry and pharmacology of some species of different genera charted in the biosphere reserve has led to increasing positive results in other countries (Ohigashi et al., 1989; Okoli and Iroegbu, 2004). However, most plants inhabiting this reserve have not yet or have only partially been studied chemically as well as microbiologically. The species already studied have been found to be rich source of alkaloids, iridoids, terpenoids, flavonoids, and ergostane derivatives. For instance, secoiridoid, angustine-type alkaloids, and triterpenic glycosides have been isolated from the stems of *Nauclea diderrichii* (Adeoye and Waigh, 1983). Furthermore, monoterpene derivatives have been reported from aerial parts of *Chenopodium ambrosioides* (Ahmed, 2000; Kiuchi et al., 2002). *Entandrophragma* species have been found

Figure 3. (a) A traditional healer of Foumban in Cameroon, Elhadji Nji Nsangou Issah, at his medicine compound. (b) Healer shop showing the medicine containers used by him. (c) A pygmy-village in Kribi in Cameroon; a pygmy (left) with African traditional medicinal plants and the present 'king' of the village (right) is pictured; insert shows the extracts of the pictured plants used by the pygmies to treat rheumatism.

to produce especially ergostane and triterpene derivatives (Hotellier et al., 1975). While *Alstonia boonei* L., *Picralima nitida* (Apocynaceae) and *Pauridiantha callicarpoides* (Rubiaceae) were found to be rich source of indole alkaloids (Ojewole, 1984; Ama-asamoah et al., 1990; Pousset et al., 1971), *Detarium microcarpum* have been found to produce alkaloids and diterpenes, respectively (Cavin et al., 2006). A survey of the literature has shown alkaloids and iridoids as well as terpenoids to be good antimicrobials (Eberlin, 1994; WHO, 2003). These examples show that plants largely used by the local population have often effective chemical substances, which explains the impact of forest logging on the Baka's life and the need of developing alternative medicines from local knowledge.

Although several sources of natural products are known to date, plants continue to play an important role in the search for bioactive substances (Harvey, 2008). The research based on plants may also lead to new leads which can serve as models for chemists involved in drug discovery to synthesize analogues with even better bioactivities (Lee, 2004). Our research is focused on the search for bioactive molecules from unexplored Cameroonian medicinal plants and the endophytes harbored within them. The Dja Biosphere Reserve of Cameroon, which contains many unexplored plants, is currently being used in the framework of this research, as the source of the plants (and their endophytic counterparts) to be studied for identifying and isolating novel lead compounds.

Mount Cameroon as a Natural Resource for Bioactive Compounds

We are also bio-prospecting plants and their endophytic microflora from a second location (Mount Cameroon), having a very different climatic, soil and environmental conditions than the Dja forest. Mount Cameroon is located in the south-west region of Cameroon, on the coastal belt of the Gulf of Guinea (Biafra) between 3°57′–4°27′N and 8°58′–9°24′E. It has a humid tropical climate and the climatic pattern is sharply modified by the influence of the topography. The main annual rainfall of the area varies between 2085 mm near Ekona to 9086 mm at Debundscha. The mean monthly temperature, at sea level, varies from 19 to 30°C with the maximum in March-April. The mean annual temperature is about 25°C and this decreases by 0.6°C per 100 m of ascent, to 4°C at the summit. The soil of Mount Cameroon is mainly of volcanic origin and very fertile. The vegetation and plant species diversity make it a unique biotope to find new bioactive compounds. Therefore, within this project, we are also exploiting the biosynthetic potential of the endophytic fungi harbored within the plants of Mount Cameroon as a renewable and sustainable resource of bioactive pro-drugs, especially the compounds which can directly or indirectly have commercial implications.

Bioprospecting Endophytes from Cameroonian 'Hotspots'

In continuation of our quest for isolating and characterizing bioactive metabolites from Cameroonian medicinal plants and their endophytes, we have recently started working on a special medicinal plant called *Entandrophragma cylindricum* Sprague (Meliaceae), locally known as "Sapele" or "Sappelli", which has long been used in the traditional medicinal sector in Cameroon by the 'Bantu' and 'Baka' tribes in the eastern region of Cameroon, for the treatment of rheumatism. We recently characterized, both chemically and biologically (against inflammation), four acyclic triterpene derivatives named sapelenins G–J, along with eight known compounds, sapelenins A–D, ekeberin D2, (+)-catechin and epicatechin, and anderolide G from the bark of this plant (Kouam et al., 2012). We have further isolated a plethora of endophytic fungi from this plant and are presently evaluating their biosynthetic and plant-protective potential by producing host plant or host mimetic compounds (such as sapelenins) and/or other bioactive metabolites (with primary focus on anti-inflammatory compounds). The strategy for isolation and characterization of endophytic fungi from Cameroonian medicinal plants and further elucidation of desired bioactive compounds and/or their desired biological functions in the host plant is elaborated (Fig. 4).

Given the fact that malaria is presently a devastating menace in Africa (including and especially in Cameroon which is designated a malarial 'red zone'), our group recently started investigating the Cameroonian medicinal plants for endophytes particularly producing anti-malarial lead compounds. One example of our recent success story is the plant *Enantiachlorantha* Oliv., which was prospected from the Barumbi camp forest in Kumba (south-west region of Cameroon) for the harboring endophytes. A plethora of endophytes were isolated and characterized. One endophytic isolate, identified as *Preussia* sp. (strain CAD4), was capable of producing the novel dibenzofurans, preussiafurans A–B, having good anti-plasmodial activity against erythrocytic stages of chloroquine-resistant *Plasmodium falciparum* NF54 and even moderate cytotoxicity on L6 rat skeletal myoblasts cell lines (Talontsi et al., 2014). This endophyte was also capable of producing six other known compounds cissetin, asterric acid, methyl asterrate, alternariol, alternariol-5-O-methyl ether, and ergosterol. In fact, the 2015 Nobel Prize in Physiology or Medicine attests the importance of natural products (such as artemisinin that can cure Malaria) isolated from a plethora of remarkable natural resources. Further investigation of other Cameroonian medicinal plants for their endophytic microflora is underway.

Outlook

We are presently engaged in exploiting the virtually inexhaustible assortment of endophytic microflora from the fascinating and almost unexplored tropical niches of Cameroon. Our target remains in establishing endophytes as a

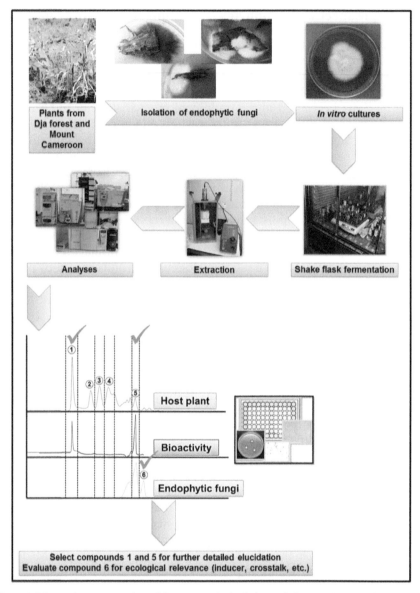

Figure 4. Schematic representation of the strategy for isolation and characterization of endophytic fungi from Cameroonian medicinal plants and further elucidation of desired bioactive compounds.

novel and sustainable resource of gene- and pro-drug 'pools' with immense implications not only in the pharmaceutical and medicinal sectors, but also in the agricultural areas. We presently stand at an interesting juncture of endophyte research where we have just begun to understand the intricacies of plant-endophyte engagements on different molecular levels. The following open questions still need to be addressed (in addition to several other related fundamental questions), on a case-by-case basis, in order to deliver on the promise of industrial utilization of endophytes:

1. Which plants have to be selected for maximum chance of plant-endophyte interaction leading to a desired endophytic trait (production of natural products, fitness benefits to host plants)?
2. Why and how do endophytes produce bioactive compounds? Why and how are some capable of producing the host plant compounds?
3. What are the origins and evolutions of biosynthetic pathways in endophytes and their associated host plants?
4. Why and how does interspecies and multispecies crosstalk drive the biosynthesis of natural products in endophytes and associated host plants?
5. How are the different mutualistic suites (such as endophyte-endophyte, endophyte-host, endophyte-pathogens, endophyte-herbivores, host-pathogens, host-herbivores) connected in nature? What are the attack-defense-counterdefense strategies linking these organisms?
6. Who is the original source organism for the production of a natural product, endophyte or host, and why?
7. How can the biosynthesis of bioactive or 'cryptic' natural products be triggered and enhanced in endophytes?
8. How can the biosynthetic potential of endophytes in producing a desirable compound (from above point) be maintained, optimized, and scaled-up under *in vitro* laboratory conditions for possible pharmaceutical applications?

Acknowledgements

Research at the Institute of Environmental Research (INFU), Department of Chemistry and Chemical Biology is supported by the International Bureau (IB) of the German Federal Ministry of Education and Research (BMBF/DLR), Germany, the Ministry of Innovation, Science, Research and Technology of the State of North Rhine-Westphalia, Germany, the German Academic Exchange Service (DAAD/BMBF "Welcome to Africa" initiative), and the German Research Foundation (Deutsche Forschungs Gemeinschaft, DFG). The authors express special thanks to Dr. Ferdinand Mouafo Talontsi, Dr. Marc Lamshoeft, and Dr. Sebastian Zuehlke (INFU, TU Dortmund) and to Prof. Dr. Simeon Fogue Kouam (Higher Teachers' Training College, University of Yaoundé I, Cameroon, Africa) for their invaluable contributions to the DAAD/BMBF "Welcome to Africa" initiative, valuable discussions and fantastic collaboration. The work and concepts highlighted in this chapter was funded by the DAAD/BMBF "Welcome to Africa" initiative, and is only a glimpse of the entire project. For further details, the readers are invited to visit the authors' websites.

References

Adeoye, A.O. and Waigh, R.D. 1983. Secoiridoid and triterpenic acids from the stems of *Nauclea diderrichii*. Phytochemistry, 22: 975–978.

Ahmed, A.A. 2000. Highly oxygenated monoterpenes from *Chenopodium ambrosioides*. J. Nat. Prod., 63: 989–991.

Ama-Asamoah, R., Kapadia, G.J., Lloyd, H.A. and Sokoloski, E.A. 1990. Picratidine, a new indole alkaloid from *Picralima nitida* seeds. J. Nat. Prod., 53: 975–977.

Arnold, A.E., Mejia, L.C., Kyllo, D., Rojas, E.I., Maynard, Z. and Robbins, N. 2003. Fungal endophytes limit pathogen damage in a tropical tree. Proc. Natl. Acad. Sci. USA, 100: 15649–15654.

Bacon, C.W. and White, J.F. 2000. Microbial Endophytes. Marcel DekerInc New York.

Baniakina, J., Eyme, J. and Adjanohoun, E. 1995. Recherches sur les s tructuresmorphologiquesetanatomiques des plantesmédicinales de la famille des sterculiaceae et des Bombacaceae. Revue Méd. Trad. Pharm. Afr., 9: 49–62.

Betti, J.L. 2001. Usages traditionnelsetvulnérabilité des plantesmédicinalesdans la Réserve de Biosphère du Dja et dans les marchés de Yaoundé, Cameroun. Thèse Doc. Sci. Agro., ULB, Bruxelles.

Betti, J.L. 2002. Usages traditionnels des plantesmédicinalesettraitement des maux de dos dans la Réserve de biosphère du Dja, Cameroun. pp. 117–154. *In*: Guerci, A. and Consiglière, S. (eds.). Proceedings of The Second International Congress on Anthropology, History of Health and Diseases: Living and "Curing" Old Age in The World/Old Age in the World, Genoa/Italy (3).

Betti, J.L. 2004. Impact of Forest Logging in the Dja Biosphere Reserve, Ministry of Environment and Forestry/PSRF, Yaoundé/Cameroon, pp. 13.

Cavin, A.L., Hay, A.E., Marston, A., Stoeckli-Evans, H., Scopelliti, R., Diallo, D. and Hostettmann, K. 2006. Bioactive diterpenes from the fruits of *Detarium microcarpum*. J. Nat. Prod., 69: 768–773.

Eaton, C.J., Cox, M.P. and Scott, B. 2011. What triggers grass endophytes to switch from mutualism to pathogenism? Plant Sci., 180: 190–195.

Eberlin, T. 1994. Les Antibiotiques, Ed. Nathan, Paris, pp. 123.

Gartlan, S. and Leakey, R. 1998. Conservation etutilisationrationnelle des écosystèmesforestiersen Afriquecentrale: Dossier d'exécution, annexe 1. UICN/FED, pp. 32.

Harvey, A.L. 2008. Natural products in drug discovery. Drug Discov. Today, 13: 894–901.

Hotellier, F., Delaveau, P. and Pousset, J.L. 1975. Nauclefineetnaucletinedeux nouveaux alcaloides de type indoloquinolizidineisoles du *Nauclea latifolia*. Phytochemistry, 14: 1407–1409.

Hyde, K.D. and Soytong, K. 2008. The fungal endophyte dilemma. Fungal Divers., 33: 163–173.

Kiuchi, F., Itano, Y., Uchiyama, N., Honda, G., Tsubouchi, A., Nakajima-Shimada, J. and Aoki, T. 2002. Monoterpene hydroperoxides with trypanocidal activity from *Chenopodium ambrosioides*. J. Nat. Prod., 65: 509–512.

Kouam, S.F., Kusari, S., Lamshöft, M., Tatuedom, O.K. and Spiteller, M. 2012. Sapelenins G-J, acyclic triterpenoids with strong anti-inflammatory activities from the bark of the Cameroonian medicinal plant *Entandrophragma cylindricum*. Phytochemistry, 83: 79–86.

Kusari, P., Kusari, S., Spiteller, M. and Kayser, O. 2013b. Endophytic fungi harbored in *Cannabis sativa* L.: Diversity and potential as biocontrol agents against host plant-specific phytopathogens. Fungal Divers., 60: 137–151.

Kusari, S., Hertweck, C. and Spiteller, M. 2012. Chemical ecology of endophytic fungi: origins of secondary metabolites. Chem. Biol., 19: 792–798.

Kusari, S., Pandey, S.P. and Spiteller, M. 2013a. Untapped mutualistic paradigms linking host plant and endophytic fungal production of similar bioactive secondary metabolites. Phytochemistry, 91: 81–87.

Kusari, S., Singh, S. and Jayabaskaran, C. 2014. Biotechnological potential of plant-associated endophytic fungi: hope versus hype. Trends Biotechnol., 32: 297–303.

Kusari, S. and Spiteller, M. 2011. Are we ready for industrial production of bioactive plant secondary metabolites utilizing endophytes? Nat. Prod. Rep., 28: 1203–1207.

Kusari, S. and Spiteller, M. 2012. Metabolomics of endophytic fungi producing associated plant secondary metabolites: progress, challenges and opportunities. pp. 241–266. *In*: Roessner, U. (ed.). Metabolomics, InTech ISBN 978-953-51-0046-1.

Lee, K.H. 2004. Current developments in the discovery and design of new drug candidates from plant natural product leads. J. Nat. Prod., 67: 273–283.

Magilu, M., Mbuyi, M. and Ndjélé, M.B. 1996. Plantesmédicinalesutilisées par les pygmées (Mbute) pour combattre le paludismedans la zone de Mambasa, Ituri, Zaïre. pp. 741–746. *In*: van der Maesen, L.J.G., van der Burgt, X.M. and van Medenbach de Rooy, J.M. (eds.). The Biodiversity of African Plants, Kluwer Academic Publishers. The Netherlands.

Márquez, L.M., Redman, R.S., Rodriguez, R.J. and Roossinck, M.J. 2007. A virus in a fungus in a plant: three-way symbiosis required for thermal tolerance. Science, 315: 513–515.

Motte, F. 1980. Les plantes chez les pygmées Aka et les Mozombo de la Lobaye (Centrafrique). Soc. Et. Ling. et Anthrop. de France, Etudes pygmées V, pp. 573.

Ohigashi, H., Kaji, M., Sakaki, M. and Koshimizu, K. 1989. 3-hydroxyuridine, an allelopathic factor of an African tree, *Baillonella toxisperma*. Phytochemistry, 28: 1365–1368.

Ojewole, J.A.O. 1984. Studies on the pharmacology of echitamine, an alkaloid from the stem bark of *Alstonia boonei* L. (Apocynaceae). Pharma. Biol., 22: 121–143.

Okoli, A.S. and Iroegbu, C.U. 2004. Evaluation of extracts of *Anthocleista djalonensis*, *Nauclea latifolia* and *Uvaria afzalii* for activity against bacterial isolates from cases of non-gonococcal urethritis. J. Ethnopharmacol., 92: 135–144.

Piel, J., Atzorn, R., Gaebler, R., Kuehnemann, F. and Boland, W. 1997. Cellulysin from the plant parasitic fungus *Trichoderma viride* elicits volatile biosynthesis in higher plants via the octadecanoid signaling cascade. FEBS Lett., 416: 143–148.

Piel, J., Donath, J., Bandemer, K. and Boland, W. 1998. Mevalonate-independent biosynthesis of terpenoid volatiles in plants: induced and constitutive emission of volatiles. Angew. Chem. Int. Ed., 37: 2478–2481.

Porras-Alfaro, A. and Bayman, P. 2011. Hidden fungi, emergent properties: Endophytes and microbiomes. Annu. Rev. Phytopathol., 49: 291–315.

Pousset, J.L., Bouquet, A., Cave, A. and et Paris, R.R. 1971. Structure de deux nouveaux alcaloïdesisolés de *Pauridianthacallicarpoides* Brem. (Rubiacées). C. R. Acad. Sci. Paris, 272: 665–667.

Redman, R.S., Sheehan, K.B., Stout, R.G., Rodriguez, R.J. and Henson, J.M. 2002. Thermotolerance conferred to plant host and fungal endophyte during mutualistic symbiosis. Science, 298: 1581.

Rodriguez, R. and Redman, R. 2008. More than 400 million years of evolution and some plants still can't make it on their own: plant stress tolerance via fungal symbiosis. J. Exp. Bot., 59: 1109–11014.

Rodriguez, R.J., Henson, J., Van Volkenburgh, E., Hoy, M., Wright, L., Beckwith, F., Kim, Y.O. and Redman, R.S. 2008. Stress tolerance in plants via habitat-adapted symbiosis. ISME J., 2: 404–416.

Rodriguez, R.J., Redman, R.S. and Henson, J.M. 2004. The role of fungal symbioses in the adaptation of plants to high stress environments. Mitig. Adapt. Strat. Global Change, 9: 261–272.

Schulz, B., Roemmert, A.K., Dammann, U., Aust, H.J. and Strack, D. 1999. The endophyte-host interaction: a balanced antagonism. Mycol. Res., 103: 1275–1283.

Stone, J.K., Bacon, C.W. and White, J.F. 2000. An overview of endophytic microbes: endophytism. pp. 3–30. *In*: Bacon, C.W. and White, J.F. (eds.). Microbial Endophytes, New York Marcel Dekker Inc.

Strobel, S.A., Kluck, K., Hess, W.M., Sears, J., Ezra, D. and Vargas, P.N. 2007. *Muscodor albus* E-6, an endophyte of *Guazuma ulmifolia* making volatile antibiotics: isolation, characterization and experimental establishment in the host plant. Microbiol., 153: 2613–2620.

Talontsi, M.F., Lamshöft, M., Douanla-Meli, C., Kouam, S.F. and Spiteller, M. 2014. Antiplasmodial and cytotoxic dibenzofurans from *Preussia* sp. harboured in *Enantia chlorantha* Oliv. Fitoterapia, 93: 233–238.

Waller, F., Achatz, B., Baltruschat, H., Fodor, J., Becker, K., Fischer, M., Heier, T., Hückelhoven, R., Neumann, C., von Wettstein, D., Franken, P. and Kogel, K.H. 2005. The endophytic fungus *Piriformospora indica* reprograms barley to salt-stress tolerance, disease resistance and higher yield. Proc. Natl. Acad. Sci. USA, 102: 13386–13391.

World Health Organisation (WHO). 2003. Manual for the laboratory identification and antimicrobial susceptibility testing of bacterial pathogens of Public Health Importance in the developing world. WHO/CdS/RMD/2003.6.(s).

CHAPTER 2

Endophytic Fungi and Their Bioprospection

A. González-Coloma,[1],* A. Cosoveanu,[2] R. Cabrera,[2] C. Giménez[2] and Nutan Kaushik[3]

ABSTRACT

Fungal endophytes have received special attention due to their potential to produce metabolites of pharmacological interest and the role that they play in regulating the ecological systems at different trophic levels. The study of the endophyte-host plant interaction has resulted in a better understanding of the role that the endophytes play in plant protection. This aspect is being intensively investigated for its potentials to offer alternative methods of pest control.

Therefore, the isolation, identification and conservation of endophytic species along with the study of their metabolites is of great importance, but a proper prospection must be carried out depending on the objectives pursued.

Introduction

The interest in endophytic fungi has grown significantly in the last decade, being promoted by the discovery of new properties of this group of microorganisms. An updated revision of the current researches concerning the bioprospection of endophytic fungi has been presented in the following pages.

[1] ICA-CSIC. Serrano 115 dpdo. 28006.Madrid.
[2] Dept. Botany, Ecology and Plant Physiology, Astrofísico Fco. Sánchez, s/n.38206. Univ La Laguna. Spain.
[3] The Energy and Resources Institute (TERI), India Habitat Center, Lodhi Road, New Delhi 110003, India.
* Corresponding author: azu@ica.csic.es

Definition

The term endophyte refers to a living organism which resides in a plant (endon = inside; phyton = plant). In the 19th century, De Bary referred to fungi living inside the tissue of plants as endophytes. From the taxonomic point of view, endophytic fungi have been organized into two main groups, based on host plant evolution, ecology and taxonomy: clavicipitaceous (C), which are able to infect only some grass species and non clavicipitaceous (NC), which are recovered from asymptomatic tissues of bryophytes, ferns, gymnosperms, and angiosperms (Rosa et al., 2010). Except for a few Basidiomycetes and Zygomycetes, most known taxa of grass endophytes belong to Ascomycetes and Deuteromycetes. Another significant group of all endophytic isolates is *Mycelia sterilia*, fungi which do not produce fruiting structures. Despite the systematic classification, Schulz and Boyle (2005) grouped the fungal endophytes into three ecological groups: mycorrizal, balansiaceous or pasture and nonpasture endophytes.

Transmission

The transmission of endophytes can be vertical or horizontal, the latter being the most common and has been previously revised (Giménez et al., 2007). Vertical transmission takes place when the seed is infected by the endophyte and then transmitted to the plant. This is the transmission mechanism of pasture endophytes such as *Neotyphodium* resulting in infected populations which are clustered in the field.

Horizontal transmission requires the production of external spores and their airborne dispersion to infect other plants. It occurs in the less specific species, generally those associated with woody plants and localized in certain organs of the host plant. Occasionally, the dispersion requires other organisms, like vectors (insects) which contribute to spore fecundation and dispersion as in *Epichloe festuca,* where the presence of *Botanophila* spp. (Insecta: Antomiidae) is associated with this fungus and contributes to its sexual reproduction and dispersion. Moreover, endophyte spores from *Alternaria, Cladosporium, Penicillium* and other non-specific taxa are pervasive in the atmosphere, sporulating when the host tissue senesces and dies (Sánchez Marquez et al., 2012).

Endophyte-host Interactions

Endophyte-host plant interactions are species dependent. For instance, *Rhytismataceae* endophytes are parasitic on certain species of *Pinus* and endophytes in others, thus raising the question about latent pathogenicity or specificity. Disease symptoms might express miscommunication with the host rather than active pathogenicity, leading to the hypothesis that plants participate or initiate the disease processes (Rodríguez and Redman, 2008; Aly et al., 2011).

The effects of the presence of endophyte in hosts may be observed during the growth of the plant, in the reduction of damage by fungal pathogens and pests and in the production of a wide array of secondary metabolites with different effects (Giménez et al., 2007; Quesada-Moraga et al., 2009; Vega et al., 2009; Dudeja et al., 2012; Sánchez Márquez et al., 2012). On the other hand, endophytes can decrease the host plant photosynthetic capacity, for example in *Musa* and *Zea* (Davis and Shaw, 2008).

As a result of the relationships between endophytes and their hosts, two lines of work have received special attention based on their potential practical application: a) endophytes for plant protection and b) endophytes as "biofactories" of natural products.

Endophytes for Plant Protection

The symbiosis that exists between the endophyte and its host is considered "defensive mutualism" where the resistance of the host plant to pathogens, phytophagous insects and environmental conditions increases. Secondary metabolites are involved in most cases. The biochemical and physiological mechanisms that trigger and regulate these relations is a subject of intense research.

The advantage achieved by the plant from its endophytic fungi in terms of resistance to pest and pathogens has been demonstrated in several studies. The benefits offered by endophytes may be exploited in agriculture and plant pest control (Kaul et al., 2012). A protective role for endophytes has been proposed against sucking, chewing and leaf mining insect pests (Bing and Lewis, 1991; Latch, 1993; Cherry et al., 2004; Jallow et al., 2008; Vega, 2008). In conifers, it has been reported that the endophyte *Phialocephala scopiformis* which lives in *Picea glauca* (Pinaceae), produces the toxin rugulosin which has an antifeeding activity against the spruce budworm *Choristoneura fumiferana* (Lep. Tortricidae) (Rohlfs and Churchill, 2011; Larkin et al., 2012).

The production of antioxidant compounds by plants, like phenolic acids and their derivatives (Huang et al., 2007b), isobenzofuranones (Strobel et al., 2002), isobenzofurans (Harper et al., 2003), as well as mannitol and other carbohydrates (Richardson et al., 1992), is attributed to the presence of reactive oxygen species (ROS) generated by endophytes. This antioxidant production increases the stress tolerance of the plant (Huang et al., 2007a; White and Torres, 2010; Aly et al., 2011) including extreme temperatures (up to 65°C) as in the case of *Curvularia protuberate* found in cotton panic grass, *Dichanthelium lanuginosum* (Poaceae), from geothermal soils of Yellowstone National Park (Redman et al., 2002; Loro et al., 2012). Indole-3-acetic acid (IAA) producing endophytic fungi, like *Colletotrichum* sp., may enhance the growth of rice plant under salinity, drought and temperature stress. *Fusarium* sp. E5 extract acts as an auxin (Dudeja et al., 2012). Fungal strains producing gibberellic acid (GA) (*Penicillium funiculosum* and *Aspergillus fumigatus*) allow soybean plant growth under moderate and high salinity stress.

Endophytic fungi might control plant hormone levels by stomatal behavior and osmotic adjustment leading to a maintained turgescence. In saline conditions osmo-protectants like proline, accumulate to provide an energy source for plant growth and survival by preventing ionic and osmotic imbalances as in the case of *Paecilomyces formosus* (Aly et al., 2011; Loro et al., 2012). Moreover, the genus *Curvularia* confers drought tolerance to rice, wheat, tomato and watermelon plants (Rodríguez et al., 2004, 2008).

The plant receives tolerance to heavy metals in soil when its endophytes have degradation, packaging, metal sequestration or chelation systems (Weyens et al., 2009), and in return the endophyte receives substrate to develop, essential compounds for its growth or self-defence (Metz et al., 2000; Strobel, 2002), protection from dessication, nutrients and, in the case of vertical transmission, dissemination to the next generation of hosts (Saikkonen et al., 1998; Faeth and Fagan, 2002; Rudgers et al., 2004).

In some cases, the plant produces lignin and other cell-wall deposits to limit the endophytic colonization. The fungus uses mechanisms to penetrate and reside in the host tissue such as exo enzymes; therefore, the fungal development might take place in the apoplastic fluid of the host (Schulz et al., 2002; Chandra et al., 2012). As a consequence, the cell wall becomes re-inforced after endophytic penetration, generating a defensive barrier against pathogens (Dudeja et al., 2012).

The endophyte-plant interaction may influence their relationships with other organisms. In tomato, the endophyte confers heat tolerance when it is infected with a virus (Loro et al., 2012) illustrating a mutualistic system network. The correlation between *Fusarium verticilloides* as endophyte, *Zea mays* (Poaceae) as host and *Ustilago maydis* as pathogen showed that each fungal species survival strategy involved catabolic activities producing compounds that either allowed or restricted the growth of their antagonist. In the presence of endophytes, *F. verticilloides* lowered the production of indol by *Ustilago maydis*, and produced fusaric acid (limitant to the development of *U. maydis*) and break down plant defensive compounds (Rodríguez Estrada et al., 2011, 2012).

A non-pathogenic *Fusarium oxysporum* (Fo162) isolate inoculated to the rhizosphere of squash or melon and pepper seedlings significantly reduced the reproductive rate of *Aphis gossypii* (Aphididae) and *Myzus persicae* (Aphididae) (Akello and Sikora, 2012). When aphids feed on tall fescue plants treated with high ergot alkaloid producing isolates, the reproduction was lower than with ergot alkaloid free endophytic isolates (Bultman and Bell, 2003; Bultman et al., 2004; Akello and Sikora, 2012). Clavicipitaceous endophyte inside the shoots of perennial ryegrass seedlings negatively affected the Russian wheat aphid *Rhopalosiphum padi* (Aphididae) population increment. Besides, entomopathogenic fungi have been found as endophytes, including species of *Acremonium, Beauveria, Cladosporium, Clonostachys, Isaria* (Vega, 2008), *Lecanicillium* and *Trichoderma* (Ownley et al., 2010). Feeding of *A. gossypii* on cotton leaves colonized by either *Beauveria bassiana* or *Lecanicillium lecanii* results in a slow aphid reproduction (Gurulingappa et al., 2010; Akello and Sikora, 2012).

The presence of alkaloids in plants affects phytophagous insects. Loline is present only in the endophyte-host plant interaction, others being produced by the endophytes in synthetic culture media as well. The possibility of inducing alkaloid production in plants by artificial inoculation of endophytes has been investigated as an opportunity to reduce the pest damages. For instance, *Lolium perenne* (Poaceae) pasture infected by *Acremonium lolii* in New Zealand is more resistant to the curculionid *Listronotus bonariensi*, the Argentine stem weevil, than the endophyte-free plants. This protection is caused by the toxicity of lolitrem B and the antifeedant effect of peramine (Rowan et al., 1990). *Lolium perenne* cultivars infected with a peramine producing *Neotipodium lolii* strain without toxicity to grazing animals are now commercially available (Shiba and Sugawara, 2005).

A strain of *Neotyphodium uncinatum*, the most common *Neotyphodium* endophyte in meadow fescue, *Festuca pratensis* (Poaceaea), has been isolated and inoculated into meadow type cultivars of *Lolium multiflorum* (Poaceaea), which is also a reproductive host of *Trigonotylus caelestialium* (Hemiptera: Miridae), one of the most serious insect pests in rice. The infected plants showed powerful insect feeding deterrent effects (Kasai et al., 2004; Shiba and Sugawara, 2005) that depended on the alkaloid composition (Hunt and Newman, 2005), the endophyte species and the interaction plant/endophyte/genotype/environment (Faeth and Fagan, 2002).

Some strains of entomopathogenic fungi can live as endophytes in the host plant and be used as a pest control method (Gomez-Vidal et al., 2006). The inoculation techniques depend on the natural dispersion of the fungus. In the case of *Beauveria bassiana* the inoculation should take place through the vegetative part of the plant. After applications of *B. bassiana* foliar spray on coffee leaves, around 25% isolates have shown a low ability to colonize, due to the main components of leaf cuticle, waxes and cutin, which might have a detrimental effect on *B. bassiana* conidia germination and the lack of stomata on the adaxial side. Endophytic *Beauveria bassiana* caused no mortality to *Coleomegilla maculata* (Coleoptera: Cocinellidae), a predator of *Ostrinia nubilalis* (Lepidoptera: Crambidae) eggs and larvae (Pingel and Lewis, 1996).

Secondary metabolites produced by the fungal endophyte, *Neotyphodium* spp., are able to inactivate and kill the plant-parasitic nematode *Meloidogyne incognita* (Nematoda: Meloidogynidae) (Hallmann and Sikora, 1996). Populations of phytopathogenic nematode *Meloydogine marylandi* are reduced in the rhizosphere of tall fescue infected with the endophyte *Neotyphodium coenophialum* (Elmi et al., 2000), probably due to the induction of high levels of quitinase in plants inoculated with the endophytes (Roberts et al., 1992).

In banana plantations, endophytic isolates of *Fusarium* spp. secrete metabolites toxic to the nematodes *Radopholus similis*, *Meloidogyne incognita* and *Pratylenchus zeae* (Nematoda: Pratylenchidae). Nematode populations are reduced and also the transmission of *Fusarium oxysporum* f. sp. *cubense* (E.F. Sm.) W.C. Snyder & H.N. Hans, responsible of 'Panama disease', associated to injuries in roots caused by nematode infestation (Pocasangre et al., 2000).

Endophyte nematode egg parasitizing fungi, like *Pochonia* spp., have higher probability to parasite nematode eggs inside the roots, and subsequently decrease spread and root infection by the second generation of juveniles. Some structures, similar to trapping devices, have been observed in epidermal cells colonized by *Arthrobotrys oligospora*, which trap newly hatched juveniles escaping the roots (Moosavi et al., 2010). Therefore, the presence of endophytes can reduce the amount of nematicidal and fungicidal agents needed.

Products of Pharmacological and Industrial Interest

There are many reviews on the topic, emphasizing the importance of its current applications in medicine, industry, and plant protection (Strobel and Daisy, 2003; Gunatilaka, 2006; Giménez et al., 2007; Yu et al., 2010; Chowdhary et al., 2012; Kharwar et al., 2012; Kumar and Kaushik, 2012; Radic and Strukelj, 2012).

Fungal endophytes are a source of novel drugs and their metabolites belong to various chemical groups, like phenols, steroids, flavonoids, quinones, terpenoids, xanthones, peptides, alkaloids, aliphatic compounds, phenylpropanoids, isocoumarins, benzopyranones, tetralones, cytochalasines and enniatines (Schulz et al., 2002; Schulz and Boyle, 2005; Aly et al., 2010; Rocha et al., 2011). Some of the metabolites are xanthine oxidase inhibitors, eosinophil inhibitors, acetylcholinesterase inhibitors, b-glucuronidase inhibitors, anti-inflammatory agents, insulin receptor activators and immunosuppressants (Gunatilaka, 2006; Zhang et al., 2006; Verma et al., 2009; Aly et al., 2010).

Cytochalasins are a group of toxic fungal metabolites which exhibited a broad spectrum of biological activities including antibiotic, antitumor activity and inhibition of HIV-1 protease (Hazalina et al., 2012). Among this group of compounds, cytochalasin is produced by a strain of *Phomopsis* isolated from the leaves of the mangrove species, *Kandelia candel* (Li et al., 2010). Other compounds of endophyte origin include lactones (Jiménez-Romero et al., 2008) and enalin derivatives (Hormazabal et al., 2005), colletotrichic acid (Zou et al., 2000), myrocin A and apiosporic acid (Klemke et al., 2004), phomopsilactone (Silva et al., 2005), cyclopentanoids, and (+)-ascochin and (+)-ascodiketone (Krohn et al., 2007), chaetocyclinones (Losgen et al., 2007), isofusidienols (Losgen et al., 2008) and naphthoquinones (Macías-Rubalcava et al., 2008). Anticancer compounds such as taxol (Stierle et al., 1993), Hsp 90 inhibitors (Turbyville et al., 2006), sequoiatones A and B (Stierle et al., 1999) and camptothecin (Amna et al., 2006) are also endophyte metabolites of pharmacological value.

Sometimes, the same compound is also found in the plant extract and in the endophyte as well (Wu et al., 2011). Similarly, taxol is produced by *Taxus brevifolia, Taxus cuspidata* and *Taxus baccata* (Nadeem et al., 2002; Aly et al., 2011) and it is also found in the endophytes living in *Taxus* and other genera. *Taxomyces andreanae* was the first taxol producing endophyte isolated from *Taxus brevifolia* (Stierle et al., 1993, 1995). *Seimatoantlerium tepuiense, Seimatoantlerium nepalense* (Bashyal et al., 1999), *Tubercularia* sp. strain TF5 (Wang et al., 2000), *Metarhizium anisopliae* (Liu et al., 2009), *Colletotrichum gloeosporioides* (Gangadevi and Muthumary, 2008a,b) or *Cladosporium oxysporum* (Gokul Raj et al., 2015), also produce this compound. Moreover, Vincristine, another anticancer drug originally obtained from *Catharanthus roseus* (Apocynaceae), was detected in cultures of an endophytic *Fusarium oxysporum* isolated from the same plant (Lingqi et al., 2000). Podophyllotoxin, another anticancer compound, has been reported in *Trametes hirsute* and *Phialocephala fortinii*, which are endophytes isolated from *Podophyllum hexandrum* and *Podophyllum peltatum* (Berberidaceae) (Eyberger et al., 2006; Puri et al., 2006). *Fusarium oxysporum* endophytic in the medicinal plant *Juniperus recurva* (Cupressaceae) and *Aspergillus fumigatus* endophytic in *Juniperus communis* (Cupressaceae) produce a deoxy analog of podophyllotoxin. Furthermore, Camptothecin, a natural topoisomerase I inhibitor has been reported in the plant families Cornaceae, Icacinaceae, Rubiaceae, Apocynaceae and Gelsemiaceae (Wink, 2008). Fungi like *Entrophospora infrequens* and *Fusarium solani* have also been isolated from the host plants producing the aforesaid compound (Kusari et al., 2009a,b). These findings suggest that the distribution of these compounds in the plants might be due to the infection of endophyte and subsequent gene transfer (Wink, 2008; Aly et al., 2011).

Endophytes can carry out stereoselective biotransformations of chemicals, thus aiding in drug modifications (Borges et al., 2007, 2008). A new alkaloid, 16α-hydroxy-5N-acetylardeemin was found in the fermentation broth of endophyte *Aspergillus terreus*, isolated from *Artemisia annua*. It presents an inhibitory activity against acetylcholinesterase (Ge et al., 2010). Endophytes producing VOC (volatile compounds) might have potential as fuel producers or fuel additives. There is an endophytic group known as Mycodiesel. *Hypoxylon* sp. produces the above mentioned compounds (Tomsheck et al., 2010).

The great diversity of compounds isolated from endophytic fungi and their activities justify the wider and deeper interest that these organisms have created in recent decades.

Prospection

Given the above considerations, some issues involved in the prospection of endophytes must be evaluated. The isolation of the different endophytes, the study of their role and interaction with the host and their bioactive metabolites is not a simple task. A taxonomic criterion selecting bioactive compound-producing species should be applied (Higginbotham et al., 2013). However, there are endophytes that only produce metabolites of interest in the presence of other microorganisms, therefore, they require co-cultivation techniques (Bertrand et al., 2014; Chen et al., 2015).

Studies including species diversity in the host plant, the endophyte-endophyte and endophyte-host interactions, or the variability of endophyte communities with locations or seasons are also quite complex. Also the effectiveness of endophyte isolation and identification methods depend on their biology (varying specificity for target organ, or environmental conditions), leading to underestimation of the existing populations.

Distribution and Specificity

The endophyte density in a host plant is related to its genotype, developmental stage and environmental conditions (Dudeja et al., 2012). The age of the plant organs can also influence the diversity of the endophyte community, with colonization rates increasing with age (Nascimento et al., 2015). Furthermore, growth conditions may also affect the diversity of endophytes. For example, the roots of *Anemopsis californica* (Saururaceae) from wild populations had higher diversity than roots from cultured ones (Bussey et al., 2015). The duration of the growing season may also influence the endophyte diversity (Petrini and Carroll, 1981; U'Ren et al., 2012).

The relation between the host plant and the endophyte is dynamic, thus the changing lifestyle of the fungi may represent evolutionary transitions resulting in a larger bio-geographic distribution (Rodríguez and Redman, 2008). Endophytic diversity can be found at the sub specific level. For instance, studies conducted on the leaves of *Quercus* sp. (Fagaceae), *Pinus ponderosa* (Pinaceae), *Cupressus arizonica* (Cupressaceae), and *Platycladus orientalis* (Cupressaceae) showed that species of four genera of endophytic fungi (*Botryosphaeria, Colletotrichum, Mycosphaerella*, and *Xylaria*) differed in 1–2% of their DNA sequences (obtained from the internal nuclear ribosomal transcription space region, ITS) depending on the plant from which they were isolated (Arnold et al., 2007). This variability could also explain the fact that sometimes the isolate is an endophyte and in other situation a phytopathogen. The difference between pathogenic and non pathogenic forms can be explained by one gene (Redman et al., 1999, 2001).

Numerous studies have shown that the distribution of endophytic fungi in the host plant is not homogenous and specificity for certain organs and tissues has been observed. For example, Dark Septate Endophytes (DSE), a group with dark melanized septa which forms intracellular hyphae and develops microsclerotia are found in plant roots (Xing et al., 2010). Mycobiota isolated from various locations has been greater in the aerial plant parts than in the underground organs (Sánchez Márquez et al., 2012). In the aerial plant parts a preference for different organs and tissues has been observed. For example, 2003 fungal isolates were obtained from 750 needles of *Pinus monticola* (Pinaceae) while only 16 endophytic isolates were recovered from 800 seeds. Furthermore, the needle isolates belonging to *Rhytismataceae*, did not occur in seeds. In *Musa acuminata* (Musaseae) some *Xylariaceous* seem to be specific to leaf tissue, *Pyriculariopsis parasitica* and *Dactylaria* sp. were most common in pseudostem while *Colletotrichum musae* and *C. gloeosporioides* infected mostly midribs and petioles (Sun et al., 2011). In *Pinus wallichiana* (Pinaceae), the fungal diversity is higher in the needles than in the branches, with 17 species being specific to the needles, 11 for the branches and 10 common to both (Qadri et al., 2014). In *Cupressus arizonica* (Cupressaceae) 15 of the isolated genotypes were found in twigs or foliage, whereas three species were recovered from both tissue types (Arnold, 2007). In the mangrove tree, *Kandelia candel* (Rhizophoraceae), a similar endophyte collection has been found in bark and stems (647 isolates) but a different one in leaves (63 isolates) (Aly et al., 2011).

Therefore, fungal biodiversity studies should include a screening of the different plant tissues in addition to screening different plant species and similar tissues (Hyde and Soytong, 2008). When the host plants are related, more similarities can be found between the isolated endophytes. An explanation might be the endophyte plant specificity; microclimate and ecosystem influence (Susan, 2004; Arnold, 2007; Higgins et al., 2007; Saikkonen, 2007; Sieber, 2007; Hoffman and Arnold, 2008). However, this is not a general rule since various examples show heterogenous colonization inside host plants. For instance, some species of *Phomopsis, Phoma, Colletotrichum*, and *Phyllosticta* have a wide host range and colonize several taxonomically unrelated plant hosts (Pandey et al., 2003; Murali et al., 2006; Sieber, 2007; Aly et al., 2011).

Geographical Distribution

The wide distribution of endophytic fungi in different geographical areas might be related to the interactions established with their hosts, which are controlled by a complex genetic system. The coevolution of both organisms (endophyte-host plant) and the mutual adaptation influenced by biotic and abiotic factors is present in the genome of both co-habitants (Herre et al., 1999; Moricca and Ragazzi, 2008). Endophytes have adapted to the host microenvironment by genetic variation, including the uptake of some plant DNA into their genome (Stierle et al., 1993; Germaine et al., 2004; Aly et al., 2011). Genes producing enzymes that make possible the infection of the host are expressed by the endophyte only in the presence of the plant. Moreover, they may regulate the production of secondary metabolites with antibiotic activity (Bailey et al., 2006; Aly et al., 2011). The synthesis of the same bioactive metabolites by the host and its endophyte supports this idea (Stierle et al., 1993; Lingqi et al., 2000; Puri et al., 2005, 2006; Amna et al., 2006; Eyberger et al., 2006; Kusari et al., 2009b).

Endophytic presence and diversity correlates with latitude. Fungal species collected from the lowland tropical forest (Barro Colorado Island, Panama) to the northern boreal forest (Schefferville, Québec, Canada) showed lower species diversity in high latitudes, represented by several classes of Ascomycota, while tropical sites had a larger species diversity but a small number of classes (Arnold and Lutzoni, 2007).

In desert areas, a low fungal diversity in plants may be due to the scarce vegetation and the climatic conditions, since the horizontal spreading of fungi requires humidity for sporulation and infection. Another explanation is that, desert endophytes have a wide area of distribution as opportunistic beneficiaries of the infected plant, the endophytes overcoming extreme conditions like intense heat, UV radiation and desiccation (Sun et al., 2011).

In the Antarctic area, the diversity of endophytes is low. The endophyte taxa found in Antartic plant, *Colobanthus quitensis* (Caryophyllaceae) showed the presence of species with different survival mechanisms such as *Microdochium* spp., *Fusarium* spp., and *Mycocentrospora* spp., *Geomyces pannorum*, *Aspergillus reptans* (Mercantini et al., 1989). Antarctic endophytes are of cosmopolitan, indigenous or endemic origins but they lack diversity, with lower index than temperate and tropical areas (Rosa et al., 2010; U'ren et al., 2012).

The temperate grasses are dominated by *Alternaria, Acremonium, Cladosporium, Penicillium, Epicoccum* and *Aureobasidium* spp., with low specificity and many temperate forest endophytes are restricted to a host family or genus. Specific interactions between the symbiont and its host are unlikely, due to their low frequency of occurrence.

Tropical plant communities are rich in biodiversity including endophytes. However, tropical endophytes are less host-specific than those from temperate areas. Tropical woody plant endophytes are dominated by *Xylariaceae* and *Ascomycetes* of low host specificity. *Acremonium, Fusarium, Phoma* and *Pleospora* are also isolated from tropical and temperate hosts while others are more frequent in tropical areas (*Colletotrichum, Guignardia, Phyllosticta* and *Pestalotiopsis*) but being also found in plant species of the Macaronesian laurel forest (a temperate relict flora) (Giménez, 2006).

Isolation and Identification of Endophytes

The characterization of an endophyte requires a correct morphological and/or molecular identification. The identification of the host plant is also needed. Additional data like geographical coordinates, state and age of the plant, collection date, local use and pictures should be registered. Samples of all plant organs can be used in the process of isolation. The plant samples must be preserved at low temperature (4°C) and should be placed on the culture media as soon as possible. A surface sterilization is needed (washing the plant fragments with ethanol, sodium hypochlorite, oxygenated water or flaming the sample). Surface sterilized small plant fragments are placed on to Petri dishes with PDA or other specific culture media. The number of the isolated endophytes depends on the isolation methods, sample manipulation, host plant condition, pH of the substrate, population density, etc. (Kowalski and Andruch, 2012).

The number and the size of the tissue samples influence the number of isolated colonies (Gamboa et al., 2002). For example, when the leaf is cut into small fragments, the number of isolates increase (Arnold, 2007; Sun et al., 2011). Although, increasing the number of fragments do increase the number of isolates, but not the number of different species (Bezerra et al., 2013; Qadri et al., 2014; Wu et al., 2014; Vaz et al., 2014; Ferreira et al., 2015; Nascimento et al., 2015).

For morphological identification, the colony morphology and color must be considered (Sreekanth et al., 2011). For instance, *Gliocladium* spp., a frequently isolated endophyte grows in OMA, followed by CMA, PDA and V-8 juice medium. As a result, the fungus sporulates better on PDA and OMA, while on V-8 it does not produce spores. Similarly, the use of different culture media can increase the number of species isolated (Wu et al., 2014). Additionally, non-competitive or slow growing species are difficult to isolate (Sun et al., 2011).

Study of the fungal DNA extracted directly from the plant tissue made possible the identification of endophytes which did not grow in synthetic media. For example, fourteen different species were identified from *Magnolia liliifera* (Magnoliaceae) which were not able to grow *in vitro* (Duong et al., 2006), and similarly eleven species have been detected in the leaves of *Rhododendron tomentosum* (Ericaceae) (Tejesvi et al., 2011).

The molecular identification process involves DNA extraction directly from the sterilized tissue or from the colonies grown in artificial media, PCR amplification, cloning, fingerprint and sequenciation. When studies of fungal DNA have been performed directly on the plant tissue show higher diversity in isolates than sequencing DNA extracted from the isolates grown on synthetic medium (Arnold and Lutzoni, 2007; Arnold et al., 2007; Unterseher and Schnittler, 2010; Tejesvi et al., 2011). Therefore, the diversity of endophytes is underestimated when the colonies are isolated. The DNA sequences are identified through BLAST bioinformatics program in the Genbank database (Duong et al., 2006; Tejesvi et al., 2011). This method is a convenient alternative to morphological identification (Gao et al., 2005; Rakotoniriana et al., 2008) however, erroneous identification may occur or the sequence may not be public. Therefore, a matrix should be built with morphological characterization data and DNA identification. Isolate identification by sequence comparison with public data bases express a percentage of similarity. As a consequence, in these data bases the same sequence has been assigned to different species. Some studies estimated that 86% of the published *Colletotrichum gloeosporioides* sequences and preserved in GenBank, do not correspond to the epitype (Cai et al., 2009).

The cryo-preservation of the identified endophyte fungi involves freezing at –70 or –80°C in glycerol or on a sterile substrate (like barley seeds) dried and freezed. For each species, it is recommended to check the maximum period to which the fungus can grow and develop normally. Once a strain has been identified and preserved, its continuous disponibility is guaranteed.

Conclusions

The endophytic fungi are biofactories of potentially active natural products. Some of the research fields in which attention should be focussed might be:

- Studies of composition and structure of endophytic communities.
- Studies of natural succession of endophytic communities in wild and their role in ecosystem composition.
- Identification of taxa with economical potential.
- Optimization methods for sampling, isolation and for *in vitro* cultivation and up-scaled production.

References

Akello, J. and Sikora, R. 2012. Systemic acropedal influence of endophyte seed treatment on *Acyrthosiphon pisum* and *Aphis fabae* offspring development and reproductive fitness. Biol. Control, 61: 215–221.

Aly, A.H., Debbab, A., Kjer, J. and Proksch, P. 2010. Fungal endophytes from higher plants: a prolific source of phytochemicals and other bioactive natural products. Fungal Divers., 41: 1–16.

Aly, A.H., Debbab, A. and Proksch, P. 2011. Fungal endophytes: unique plant inhabitants with great promises. Appl. Microbiol. Biot., 90: 1829–1845.

Amna, T., Puri, S.C., Verma, V., Sharma, J.P., Khajuria, R.K., Musarrat, J., Spiteller, M. and Qazi, G.N. 2006. Bioreactor studies on the endophytic fungus *Entrophospora infrequens* for the production of an anticancer alkaloid camptothecin. Can. J. Microbiol., 52: 189–196.

Arnold, A.E. 2007. Understanding the diversity of foliar endophytic fungi: progress, challenges, and frontiers. Fungal Biol. Revs., 21: 51–66.

Arnold, A.E., Henk, D.A., Eells, R.L., Lutzoni, F. and Vilgalys, R. 2007. Diversity and phylogenetic affinities of foliar fungal endophytes in loblolly pine inferred by culturing and environmental PCR. Mycologia, 99: 185–206.

Arnold, A.E. and Lutzoni, F. 2007. Diversity and host range of foliar fungal endophytes: are tropical leaves biodiversity hotspots? Ecology, 88: 541–549.

Bailey, B.A., Bae, H., Strem, M.D., Roberts, D.P., Thomas, S.E., Crozier, J., Samuels, G.J., Choi, I.Y. and Holmes, K.A. 2006. Fungal and plant gene expression during the colonization of cacao seedlings by endophytic isolates of four *Trichoderma* species. Planta, 224: 1449–1464.

Bashyal, B., Li, J.Y., Strobel, G.A. and Hess, W.M. 1999. *Seimatoantlerium nepalense*, an endophytic taxol producing coelomycete from Himalayan yew (*Taxus wallachiana*). Mycotaxon, 72: 33–42.

Bertrand, S., Bohni, N., Schnee, S., Schumpp, O., Gindro, K. and Wolfender, J. 2014. Metabolite induction via microorganism co-culture: A potential way to enhance chemical diversity for drug discovery. Biotechnol. Adv., 32(6): 1180–1204.

Bezerra, J.D.P., Santos, M.G.S., Barbosa, R.N., Svedese, V.M., Lima, D.M.M., Fernandes, M.J.S. and Souza-Motta, C.M. 2013. Fungal endophytes from cactus *Cereus jamacaru* in brazilian tropical dry forest: A first study. Symbiosis, 60(2): 53–63.

Bing, L.A. and Lewis, L.C. 1991. Suppression of *Ostrinia nubilalis* (Hübner) (Lepidoptera: Pyralidae) by endophytic *Beauveria bassiana* (Balsamo) Vuillemin. Environ. Entomol., 20: 1207–1211.

Borges, K.B., Borges, W.D.S., Pupo, M.T. and Bonato, P.S. 2008. Stereoselective analysis of thioridazine-2-sulfoxide and thioridazine-5-sulfoxide: an investigation of rac-thioridazine biotransformation by some endophytic fungi. J. Pharma. Biomed., 46: 945–952.

Borges, K.B., Borges, W.S., Pupo, M.T. and Bonato, P.S. 2007. Endophytic fungi as models for the stereoselective biotransformation of thioridazine. Appl. Microbiol. Biotech., 77(3): 669–674.

Bultman, T.L., Bell, G. and Martin, W.D. 2004. A fungal endophyte mediates reversal of wound-induced resistance and constrains tolerance in a grass. Ecology, 85(3): 679–685.

Bultman, T.L. and Bell, G.D. 2003. Interaction between fungal endophytes and environmental stressors influences plant resistance to insects. Oikos, 103(1): 182–190.

Bussey, R.O., Kaur, A., Todd, D.A., Egan, J.M., El-Elimat, T., Graf, T.N. and Cech, N.B. 2015. Comparison of the chemistry and diversity of endophytes isolated from wild-harvested and greenhouse-cultivated yerba mansa (*Anemopsis californica*). Phytochem. Lett., 11: 202–208.

Cai, L., Hyde, K.D., Taylor, P.W.J., Weir, B.S., Waller, J.M., Abang, M.M., Zhang, J.Z., Yang, Y.L., Phoulivong, S., Liu, Z.Y., Prihastuti, H., Shivas, R.G., McKenzie, E.H.C. and Johnston, P.R. 2009. A polyphasic approach for studying *Colletotrichum*. Fungal Divers., 39: 183–204.

Chandra, S. 2012. Endophytic fungi: novel sources of anticancer lead molecules. Appl. Microbiol. Biotech., 95: 47–59.

Chen, H., Daletos, G., Abdel-Aziz, M.S., Thomy, D., Dai, H., Brötz-Oesterhelt, H. and Proksch, P. 2015. Inducing secondary metabolite production by the soil-dwelling fungus *Aspergillus terreus* through bacterial co-culture. Phytochem. Lett., 12: 35–41.

Cherry, A.J., Banito, A., Djegui, D. and Lomer, C. 2004. Suppression of the stem borer *Sesamiacalamistis* (Lepidoptera: Noctuidae) in maize following seed dressing, topical application and stem injection with African isolates of *Beauveria bassiana*. Agric. Entom., 7: 171–181.

Chowdhary, K., Kaushik, N., Gonzalez Coloma, A. and Cabrera, R. 2012. Endophytic fungi and their metabolites isolated from Indian medicinal plants. Phytochem. Rev., 11(4): 467–485.

Davis, E.C. and Shaw, A.J. 2008. Biogeographic and phylogenetic patterns in diversity of liverwort-associated endophytes. Am. J. Bot., 95(8): 914–924.

Dudeja, S.S., Giril, R., Saini1, R., Suneja-Madan, P. and Kothe, E. 2012. Interaction of endophytic microbes with legumes. J. Basic Microb., 52: 248–260.

Duong, L.M., Jeewon, R., Lumyong, S. and Hyde, K.D. 2006. DGGE coupled with ribosomal DNA gene phylogenies reveal uncharacterized fungal phylotypes. Fungal Divers., 23: 121–138.

Elmi, A.A., West, C.P., Robbins, R.T. and Kirkpatrick, T.L. 2000. Endophyte effects on reproduction of a root-knot nematode (*Meloidogyne marylandi*) and osmotic adjustment in tall fescue. Grass Forage Sci., 55(2): 166–172.

Eyberger, A.L., Dondapati, R. and Porter, J.R. 2006. Endophyte fungal isolates from *Podophyllum peltatum* produce podophyllotoxin. J. Nat. Prod., 69: 1121–1124.

Faeth, S.H. and Fagan, W.F. 2002. Fungal endophytes: common host plant symbionts but uncommon mutualists. Integr. Comp. Biol., 42: 360–368.

Ferreira, M.C., Vieira, M.d.L.A., Zani, C.L., Alves, T.M.d.A., Junior, P.A.S., Murta, S.M.F. and Rosa, L.H. 2015. Molecular phylogeny, diversity, symbiosis and discovery of bioactive compounds of endophytic fungi associated with the medicinal amazonian plant *Carapa guianensis* aublet (meliaceae). Biochem. Syst. Ecol., 59: 36–44.

Gamboa, M.A., Laureano, S. and Bayman, P. 2002. Measuring diversity of endophytic fungi in leaf fragments: does size matter? Mycopathologia, 15: 41–45.

Gangadevi, V. and Muthumary, J. 2008a. Taxol, an anticancer drug produced by an endophytic fungus Bartalinia robillardoides Tassi, isolated from a medicinal plant, *Aegle marmelos* Correa ex Roxb. World J. Microb. Biot., 24: 717–724.

Gangadevi, V. and Muthumary, J. 2008b. Isolation of *Colletotrichum gloeosporioides*, a novel endophytic taxol-producing fungus from the leaves of a medicinal plant, *Justicia gendarussa*. Mycologia Balcanica, 5: 1–4.

Gao, X.X., Zhou, H., Xu, D.Y., Yu, C.H., Chen, Y.Q. and Qu, L.H. 2005. High diversity of endophytic fungi from the pharmaceutical plant, *Heterosmilax japonica* Kunth revealed by cultivation independent approach. FEMS Microbiol. Lett., 249: 255–266.

Ge, H.M., Peng, H., Guo, Z.K., Cui, J.T., Song, Y.C. and Tan, R.X. 2010. Bioactive alkaloids from the plant endophytic fungus *Aspergillus terreus*. Plant Med., 76: 822–824.

Germaine, K., Keogh, E., Garcia-Cabellos, G., Borreman, B., Lelie, D., Barac, T., Oeyen, L., Vangronsveld, J., Moore, F.P., Moore, E.R.B., Campbel, C.D., Ryan, D. and Dowling, D.N.

2004. Colonisation of poplar trees by gfp expressing bacterial endophytes. FEMS Microbiol. Ecol., 48: 109–118.

Giménez, C. 2006. Productos bioactivos de plantas de Canarias y sus hongos endofitos: deteccion de actividad y utilizacion en el control de plagas y enfermedades agrícolas. Ph.D. Thesis. University of La Laguna, Tenerife.

Giménez, C., Cabrera, R., Reina, M. and González-Coloma, A. 2007. Fungal endophytes and their role in plant protection. Curr. Org. Chem., 11: 707–720.

Gokul Raj, K., Manikandan, R., Arulvasu, C. and Pandi, M. 2015. Anti-proliferative effect of fungal taxol extracted from *Cladosporium oxysporum* against human pathogenic bacteria and human colon cancer cell line HCT 15. Spectrochimica Acta Part A: Mol. Biomol. Spectro., 138: 667–674.

Gomez-Vidal, S., Lopez-Llorca, L.V., Jansson, B.H. and Salinas, J. 2006. Endophytic colonization of Date Palm (*Phoenix dactylifera* L.) leaves by entomopathogenic Fungi. Micron., 37(7): 624–632.

Gunatilaka, A.A.L. 2006. Natural products from plant-associated microorganisms: Distribution, structural diversity, bioactivity, and implications of their occurrence. J. Nat. Prod., 69: 509–526.

Gurulingappa, P., Sword, G.A., Murdoch, G. and Mcgee, P.A. 2010. Colonization of crop plants by fungal entomopathogens and their effects on two insect pests when in planta. Biol. Control, 55(1): 34–41.

Hallmann, J. and Sikora, R.A. 1996. Toxicity of fungal endophyte secondary metabolites to plant parasitic nematodes and soil-borne plant pathogenic fungi. Europ. J. Plant Pathol., 102: 155–162.

Harper, J.K., Arif, A.M., Ford, E.J., Strobel, G.A., Porco, J.A., Tomer, D.P., Oneill, K.L., Heider, E.M. and Grant, D.M. 2003. Pestacin: a 1,3-dihydro isobenzofuran from *Pestalotiopsis microspora* possessing antioxidant and antimycotic activities. Tetrahedron, 59: 2471–2476.

Hazalina, N.A.M., Ramasamya, K., Lima, S.M., Cole, A.L.J. and Majeed, A.B.A. 2012. Induction of apoptosis against cancer cell lines by four ascomycetes (endophytes) from Malaysian rainforest. Phytomed., 19: 609–617.

Herre, E.A., Knowlton, N., Mueller, U.G. and Rehner, S.A. 1999. The evolution of mutualisms: exploring the paths between conflict and cooperation. Trends Ecol. Evol., 14: 49–53.

Higginbotham, S.J., Arnold, A.E., Ibañez, A., Spadafora, C., Coley, P.D. and Kursar, T.A. 2013. Bioactivity of fungal endophytes as a function of endophyte taxonomy and the taxonomy and distribution of their host plants. PLoS ONE, 8(9): e73192.

Higgins, K.L., Arnold, A.E., Miadlikowska, J., Sarvate, S.D. and Lutzoni, F. 2007. Phylogenetic relationships, host affinity, and geographic structure of boreal and arctic endophytes from three major plant lineages. Mol. Phylogenet. Evol., 42: 543–555.

Hoffman, M.T. and Arnold, A.E. 2008. Geographic locality and host identity shape fungi endophyte communities in cupressaceous trees. Mycol. Res., 112: 331–344.

Hormazabal, E., Schmeda-Hirschmann, G., Astudillo, L., Rodríguez, J. and Theoduloz, C. 2005. Metabolites from *Microsphaeropsis olivacea*, an endophytic fungus of *Pilgerodendron uviferum* (D. Don) Florin. Z. Naturfors. C., 60: 11–21.

Huang, W.Y., Cai, Y.Z., Hyde, K.D., Corke, H. and Sun, M. 2007a. Endophytic fungi from *Nerium oleander* L. (Apocynaceae): main constituents and antioxidant activity. World J. Microb. Biot., 23: 1253–1263.

Huang, W.Y., Cai, Y.Z., Xing, J., Corke, H. and Sun, M. 2007b. A potential antioxidant resource: endophytic fungi from medicinal plants. Econ. Bot., 61: 14–30.

Hunt, M.G. and Newman, J.A. 2005. Reduced herbivore resistance from a novel grass-endophyte association. J. Appl. Ecol., 42: 762–769.

Hyde, K.D. and Soytong, K. 2008. The fungal endophyte dilemma. Fungal Divers., 33: 163–173.

Jallow, M.F.A., Dugassa-Gobena, D. and Vidal, S. 2008. Influence of an endophytic fungus on host plant selection by a polyphagous moth via volatile spectrum changes. Arthropod–Plant Integration, 2: 53–62.

Jiménez-Romero, C., Ortega-Barria, E., Arnold, A.E. and Cubilla-Rios, L. 2008. Activity against *Plasmodium falciparum* of lactones isolated from the endophytic fungus *Xylaria* sp. Pharm. Biol., 46: 1–4.

Kasai, E., Sasaki, T. and Okazaki, H. 2004. Artificial infection of italian ryegrass (*Lolium multiflorum* Lam.) with *Neotyphodium uncinatum*. Grassland Sci., 50: 180–186.

Kaul, S., Gupta, S., Ahmed, M. and Dhar, M. 2012. Endophytic fungi from medicinal plants: A treasure hunt for bioactive metabolites. Phytochem. Rev., 11: 487–505.

Kharwar, R.N., Sharma, V., Mishra, A., Verma, S., Gond, S. and Kumar, A. 2012. *Azadirachta indica* (Neem) and its endophytes: Amazing repertoires for biologically active natural compounds. Phytochem. Rev., 11(4): 487–505.

Klemke, C., Kehraus, S., Wright, A.D. and Konig, G.M. 2004. New secondary metabolites from the marine endophytic fungus *Apiospora montagnei*. J. Nat. Prod., 67: 1058–1063.

Kowalski, T. and Andruch, K. 2012. Mycobiota in needles of *Abies alba* with and without symptoms of Herpotrichia needle browning. Forest Pathol., 42: 183–190.

Krohn, K., Kock, I., Elsisser, B., Florke, U., Schulz, B., Draeger, S., Pescitelli, G., Antus, S. and Kurtan, T. 2007. Bioactive natural products from the endophytic fungus *Ascochyta* sp. from *Melilotus dentatus*–configurational assignment by Solid-State CD and TDDFT calculations. Eur. J. Org. Chem., 7: 1223–1229.

Kumar, S. and Kaushik, N. 2012. Metabolites of endophytic fungi as novel source of biofungicide: A review. Phytochem. Rev., 11(4): 507–522.

Kusari, S., Lamshöft, M. and Spiteller, M. 2009a. *Aspergillus fumigatus* Fresenius, an endophytic fungus from *Juniperus communis* L. Horstmann as a novel source of the anticancer pro-drug deoxypodophyllotoxin. J. Appl. Microbiol., 107: 1019–1030.

Kusari, S., Zühlke, S. and Spiteller, M. 2009b. An endophytic fungus from *Camptotheca acuminata* that produces camptothecin and analogues. J. Nat. Prod., 72: 2–7.

Larkin, B.G., Hunt, L.S. and Ramsey, P.W. 2012. Foliar nutrients shape fungal endophyte communities in Western white pine (*Pinus monticola*) with implications for white-tailed deer herbivory. Fungal Ecol., 5(2): 252–260.

Latch, G.C.M. 1993. Physiological interactions of endophytic fungi and their hosts: biotic stress tolerance imparted to grasses by endophytes. Agric. Ecosys. Environ., 44: 143–156.

Li, Y.Y., Wang, M.Z., Huang, Y.J. and Shen, Y.M. 2010. Secondary metabolites from *Phomopsis* sp. A123. Mycology, 1: 254–261.

Lingqi, Z., Bo, G., Haiyan, L., Songrong, Z., Hua, S., Su, G. and Rongcheng, W. 2000. Preliminary study on the isolation of endophytic fungus of *Catharanthus roseus* and its fermentation to produce products of therapeutic value. Zhong Cao Yao, 31: 805–807.

Liu, K., Ding, X., Deng, B. and Chen, W. 2009. Isolation and characterization of endophytic taxol-producing fungi from *Taxus chinensis*. J. Ind. Microbiol. Biot., 36: 1171–1177.

Loro, M., Valero-Jiménez, C.A., Nozawa, S. and Márquez, L.M. 2012. Diversity and composition of fungal endophytes in semiarid Northwest Venezuela. J. Arid Environ., 85: 46–55.

Losgen, S., Magull, J., Schulz, B., Draeger, S. and Zeeck, A. 2008. Isofusidienols: novel chromone-3-oxepines produced by the endophytic fungus *Chalara* sp. J. Org. Chem., 4: 698–703.

Losgen, S., Schlorke, O., Meindl, K., Herbst-Irmer, R. and Zeeck, A. 2007. Structure and biosynthesis of chatocyclinones, new polyketides produced by an endosymbiotic fungus. Eurp. J. Org. Chem., (13): 2191–2196.

Macías-Rubalcava, M.L., Hernández-Bautista, B.E., Jiménez-Estrada, M., González, M.C., Glenn, A.E., Hanlin, R.T., Hernández-Ortega, S., Saucedo-García, A., Muria-González, J.M. and Anaya, A.L. 2008. Naphthoquinone spiroketal with allelochemical activity from the newly discovered endophytic fungus *Edenia gomezpompae*. Phytochem., 69: 1185–1196.

Mercantini, R., Marsella, R. and Cervellati, M.C. 1989. Keratinophilic fungi isolated from antarctic soil. Mycopathologia, 106(1): 47–52.

Metz, A., Haddad, A., Worapong, J., Long, D., Ford, E., Hess, W.M. and Strobel, G.A. 2000. Induction of the sexual stage of *Pestalotiopsis microspora*, a taxol producing fungus. Microbiol., 146: 2079–2089.

Moosavi, M.R., Zare, R., Zamanizadeh, H.R. and Fatemy, S. 2010. Pathogenicity of *Pochonia* species on eggs of *Meloidogyne javanica*. J. invert. Pathol., 104: 125–133.

Moricca, S. and Ragazzi, A. 2008. Fungal endophytes in Mediterranean oak forests: a lesson from Discula quercina. Phytopathol., 98: 380–386.

Murali, T.S., Suryanarayanan, T.S. and Geeta, R. 2006. Endophytic *Phomopsis* species: host range and implications for diversity estimates. Can. J. Microbiol., 52: 673–680.

Nadeem, M., Rikhari, H.C., Kumar, A., Palni, L.M.S. and Nandi, S.K. 2002. Taxol content in the bark of Himalayan Yew in relation to tree age and sex. Phytochem., 60: 627–631.

Nascimento, T.L., Oki, Y., Lima, D.M.M., Almeida-Cortez, J.S., Fernandes, G.W. and Souza-Motta, C.M. 2015. Biodiversity of endophytic fungi in different leaf ages of *Calotropis procera* and their antimicrobial activity. Fungal Ecol., 14: 79–86.

Ownley, B.H., Gwinn, K.D. and Vega, F.E. 2010. Endophytic fungal entomopathogens with activity against plant pathogens: Ecology and evolution. BioControl, 55: 113–128.

Pandey, A.K., Reddy, M.S. and Suryanarayanan, T.S. 2003. ITS-RFLP and ITS sequence analysis of a foliar endophytic *Phyllosticta* from different tropical trees. Mycol. Res., 107: 439–444.

Petrini, O. and Carroll, G. 1981. Endophytic fungi in foliage of some Cupressaceae in Oregon. Can. J. Bot., 59: 629–636.

Pingel, R.L. and Lewis, L.C. 1996. The Fungus *Beauveria bassiana* (Balsamo) Vuill. in a corn ecosystem: It's effect on the insect predator *Coleomegilla maculata* De Geer. Biol. Control, 6: 137–141.

Pocasangre, L., Sikora, R.A., Vilich, V. and Schuster, R.P. 2000. Encuesta Sobre los Hongos Endofíticos del Banano América Central y el Cribado Para el Control Biológico del Nematodo Barrenador (Radopholus similis). INFOMUSA, 9(1): 3–5.

Puri, S.C., Nazir, A., Chawla, R., Arora, R., Riyaz-Ul-Hasan, S., Amna, T., Ahmed, B., Verma, V., Singh, S., Sagar, R., Sharma, A., Kumar, R., Sharma, R.K. and Qazi, G.N. 2006. The endophytic fungus *Trametes hirsuta* as a novel alternative source of podophyllotoxin and related aryl tetralin lignans. J. Biotechnol., 122: 494–510.

Puri, S.C., Verma, V., Amna, T., Qazi, G.N. and Spiteller, M. 2005. An endophytic fungus from *Nothapodytes foetida* that produces camptothecin. J. Nat. Prod., 68: 1717–1719.

Qadri, M., Rajput, R., Abdin, M.Z., Vishwakarma, R.A. and Riyaz-Ul-Hassan, S. 2014. Diversity, molecular phylogeny and bioactive potential of fungal endophytes associated with the himalayan blue pine (*Pinus wallichiana*). Microb. Ecol., 67(4): 877–887.

Quesada-Moraga, E., Muñoz-Ledesma, F.J. and Santiago-Alvarez, C. 2009. Systemic protection of *Papaver somniferum* L. against *Iraella luteipes* (Hymenoptera: Cynipidae) by an endophytic strain of *Beauveria bassiana* (Ascomycota: Hypocreales). Environ. Entomol., 38(3): 723–730.

Radic, N. and Strukelj, B. 2012. Endophytic fungi—The treasure chest of antibacterial substances. Phytomedicine, 19(14): 1270–1284.

Rakotoniriana, E.F., Munaut, F., Decock, C., Randriamampionona, D., Andriambololoniaina, M., Rakotomalala, T., Rakotonirina, E.J., Rabemanantsoa, C., Cheuk, K., Ratsimamanga, S.U., Mahillon, J., El-Jazir, M., Quetin-Leclercq, J. and Corbisier, A.M. 2008. Endophytic fungi from leaves of *Centella asiatica*: occurrence and potential interactions within leaves. Anton. Leeuwen., 93: 27–36.

Redman, R.S., Sheehan, K.B., Stout, T.G., Rodríguez, R.J. and Henson, J.M. 2002. Thermotolerance generated by plant/fungal symbiosis. Science, 298: 1581.

Redman, R.S., Dunigan, D.D. and Rodríguez, R.J. 2001. Fungal symbiosis from mutualism to parasitism: Who controls the outcome, host or invader? New Phytol., 151(3): 705–716.

Redman, R.S., Ranson, J.C. and Rodríguez, R.J. 1999. Conversion of the pathogenic fungus *Colletotrichum magna* to a nonpathogenic, endophytic mutualist by gene disruption. Mol. Plant Microb. Interac., 12(11): 969–975.

Richardson, M.D., Chapman, G.W., Hoveland, C.S. and Bacon, C.W. 1992. Sugar alcohols in endophyte-infected tall fescue. Crop Sci., 32: 1060–1061.

Roberts, C.A., Marek, S.M., Niblack, T.L. and Karr, A.L. 1992. Parasitic *Meloidogyne* and mutualistic *Acremonium* increase chitinase in Tall Fescue. J. Chem. Ecol., 18(7): 1107–1116.

Rocha, A.C.S., Garcia, D., Uetanabaro, A.P.T., Carneiro, R.T.O., Araújo, I.S., Mattos, C.R.R. and Góes-Neto, A. 2011. Foliar endophytic fungi from *Hevea brasiliensis* and their antagonism on *Microcyclus ulei*. Fungal Divers., 47: 75–84.

Rodríguez Estrada, A.E., Jonkers, W., Corby Kistler, H. and May, G. 2012. Interactions between *Fusarium verticillioides, Ustilago maydis*, and *Zea mays*: An endophyte, a pathogen, and their shared plant host. Fungal Genet. Biol., 49(7): 578–587.

Rodríguez Estrada, A.E., Hegeman, A., Corby Kistler, H. and May, G. 2011. *In vitro* interactions between *Fusarium verticillioides* and *Ustilago maydis* through real-time PCR and metabolic profiling. Fungal Genet. Biol., 48(9): 874–885.

Rodríguez, R.J., Henson, J., Van Volkenburgh, E., Hoy, M., Wright, L., Beckwith, F., Kim, Y. and Redman, R.S. 2008. Stress tolerance in plants via habitat-adapted symbiosis. ISME J., 2(4): 404–416.

Rodríguez, R. and Redman, R. 2008. More than 400 million years of evolution and some plants still can't make it on their own: plant stress tolerance via fungal symbiosis. J. Exp. Bot., 59: 1109–1114.

Rodríguez, R.J., Redman, R.S. and Henson, J.M. 2004. The role of fungal symbioses in the adaptation of plants to high stress environments. Mitigation and Adaptation Strategies for Global Change, 9: 261–272.

Rohlfs, M. and Churchill, A.C.L. 2011. Fungal secondary metabolites as modulators of interactions with insects and other arthropods. Fungal Genet. Biol., 48: 23–34.

Rosa, L.H., Almeida Vieira, M.L., Furtado Santiago, I. and Rosa, C.A. 2010. Endophytic fungi community associated with the dicotyledonous plant *Colobanthus quitensis* (Kunth) Bartl. (Caryophyllaceae) in Antarctica. FEMS Microbiol. Ecol., 73(1): 178–189.

Rowan, D.D., Dymock, J.J. and Brimble, M.A. 1990. Effect of fungal metabolite Peramine and analogs on feeding and development of Argentine stem weevil (*Listronotus bonariensis*). J. Chem. Ecol., 16(5): 1683–1695.

Rudgers, J.A., Koslow, J.M. and Clay, K. 2004. Endophytic fungi alter relationships between diversity and ecosystem properties. Ecol. Lett., 7: 42–51.

Saikkonen, K. 2007. Forest structure and fungal endophytes. Fungal Biol. Rev., 21: 67–74.

Saikkonen, K., Faeth, S.H., Helander, M. and Sullivan, T.J. 1998. Fungal endophytes: a continuum of interactions with host plants. Annu. Rev. Ecol. Syst., 29: 319–343.

Sánchez Márquez, S., Bills, G.F., Herrero, N. and Zabalgogeazcoa, I. 2012. Non-systemic fungal endophytes of grasses. Fungal Ecol., 5: 289–297.

Schulz, B., Boyle, C., Draeger, S., Römmert, A.K. and Krohn, K. 2002. Endophytic fungi: a source of novel biologically active secondary metabolites. Mycol. Res., 106: 996–1004.

Schulz, B. and Boyle, C. 2005. The endophytic continuum. Mycol. Res., 109: 661–686.

Shiba, T. and Sugawara, K. 2005. Resistance to the rice leaf bug, *Trigonotylus caelestialium*, is conferred by *Neotyphodium* endophyte infection of perennial ryegrass, *Lolium perenne*. Entomol. Exp. Appl., 115: 387–392.

Sieber, T.N. 2007. Endophytic fungi in forest trees: are they mutualists? Fungal Biol. Rev., 21: 75–89.

Silva, G.H., Teles, H.L., Trevisan, H.C., Bolzani, V.S., Young, M.C.M., Pfenning, L.H., Eberlin, M.N., Haddad, R., Costa-Neto, C.M. and Araujo, R. 2005. New bioactive metabolites produced by *Phomopsis cassiae*, an endophytic fungus in *Cassia spectabilis*. J. Brazil Chem. Soc., 16: 1463–1466.

Sreekanth, D., Sushim, G.K., Syed, A., Khan, B.M. and Ahmad, A. 2011. Molecular and morphological characterization of a taxol-producing endophytic fungus, *Gliocladium* sp., from *Taxus baccata*. Mycobiol., 39(3): 151–157.

Stierle, A., Strobel, G. and Stierle, D. 1993. Taxol and taxane production by *Taxomyces andreanae*, an endophytic fungus of Pacific yew. Science, 260: 214–216.

Stierle, A., Strobel, G., Stierle, D., Grothaus, P. and Bignami, G. 1995. The search for a taxol-producing microorganism among the endophytic fungi of the pacific yew, *Taxus brevifolia*. J. Nat. Prod., 58: 1315–1324.

Stierle, A.A., Stierle, D.B. and Bugni, T. 1999. Sequoiatones A and B: novel antitumour metabolites isolated from a redwood endophyte. J. Org. Chem., 64: 5479–5484.

Strobel, G.A. 2002. Rainforest endophytes and bioactive products. Criti. Rev. Biotechnol., 22(4): 315–333.

Strobel, G. and Daisy, B. 2003. Bioprospecting for microbial endophytes and their natural products. Microbiol. Mol. Biol. R., 67(4): 491–502.

Strobel, G., Ford, E., Worapong, J., Harper, J.K., Arif, A.M., Grant, D.M., Fung, P.C.W. and Chau, R.M.W. 2002. Isopestacin, an isobenzofuranone from *Pestalotiopsis microspora*, possessing antifungal and antioxidant activities. Phytochem., 60: 179–183.

Sun, Y., Wang, Q., Lu, X., Okane, I. and Kakishima, M. 2011. Endophytic fungal community in stems and leaves of plants from desert areas in China. Mycol. Prog., 11(3): 781–790.

Susan, D.C. 2004. Endophytic-host selectivity of *Discula umbrinella* on *Quercus alba* and *Quercus rubra* characterized by infection, pathogenicity and mycelial compatibility. Eurp. J. Plant Pathol., 110: 713–721.

Tejesvi, M.V., Kajula, M., Mattila, S. and Pirttilä, A.M. 2011. Bioactivity and genetic diversity of endophytic fungi in *Rhododendron tomentosum* Harmaja. Fungal Divers., 47(1): 97–107.

Tomsheck, A.R., Strobel, G.A., Booth, E., Geary, B., Spakowicz, D., Knighton, B., Floerchinger, C., Sears, J., Liarzi, O. and Ezra, D. 2010. *Hypoxylon* sp., an endophyte of *Persea indica*, producing 1, 8-Cineole and other bioactive volatiles with fuel potential. Microb. Ecol., 60: 903–914.

Turbyville, T.J., Wijeratne, E.M.K., Liu, M.X., Burns, A.M., Seliga, C.J., Luevano, L.A., David, C.L., Faeth, S.H., Whitesell, L. and Gunatilaka, A.A.L. 2006. Search for HSP 90 inhibitors with potential anticancer activity: isolation and SAR studies of radicicol and monocillin I from two plant-associated fungi of the Sonoran desert. J. Nat. Prod., 69: 178–184.

U'ren, J.M., Lutzoni, F., Miadlikowska, J., Laetsc, A.D. and Arnold, A.E. 2012. Host and geographic structure of endophytic and endolichenic fungi at a continental scale. Am. J. Bot., 99(5): 898–914.

Unterseher, M. and Schnittler, M. 2010. Species richness analysis and ITS rDNA phylogeny revealed the majority of cultivable foliar endophytes from beech (*Fagus sylvatica*). Fungal Ecol., 3: 366–378.

Vaz, A.B.M., da Costa, A.G.F.C., Raad, L.V.V. and Góes-Neto, A. 2014. Fungal endophytes associated with three south american myrtae (myrtaceae) exhibit preferences in the colonization at leaf level. Fungal Biol., 118(3): 277–286.

Vega, F.E. 2008. Insect pathology and fungal endophytes. J. Inverteb. Pathol., 98(3): 277–279.

Vega, F.E., Goettel, M.S., Blackwell, M., Chandler, D., Jackson, M.A., Keller, S., Koike, M., Maniania, N.K., Monzón, A., Ownley, B.H., Pell, J.K., Rangel, D.E.N. and Roy, H.E. 2009. Fungal entomopathogens: new insights on their ecology. Fungal Ecol., 2(4): 149–159.

Verma, V.C., Kharwar, R.N. and Strobel, G.A. 2009. Chemical and functional diversity of natural products from plant associated endophytic fungi. Nat. Prod. Commun., 4(11): 1511–1532.

Wang, J., Li, G., Lu, H., Zheng, Z., Huang, Y. and Su, W. 2000. Taxol from *Tubercularia* sp. strain TF5, an endophytic fungus of *Taxus mairei*. FEMS Microbiol. Lett., 193: 249–253.

Weyens, N., Van Der Lelie, D., Artois, T., Smeets, K., Taghavi, S., Newman, L., Carleer, R. and Vangronsveld, J. 2009. Bioaugmentation with engineered endophytic bacteria improves contaminant fate in phytoremediation. Environ. Sci. Technol., 43(24): 9413–9418.

White, J.F., Jr. and Torres, M.S. 2010. Is plant endophyte-mediated defensive mutualism the result of oxidative stress protection? Physiol. Planta., 138(4): 440–446.

Wink, M. 2008. Plant secondary metabolism: diversity, function and its evolution. Nat. Prod. Commun., 3: 1205–1216.

Wu, S.H., Chen, Y.W. and Miao, C.P. 2011. Secondary metabolites of endophytic fungus *Xylaria* sp. YC-10 of *Azadirachta indica*. Chem. Nat. Compd., 47(5): 858–861.

Wu, Z., Yan, S., Zhou, S. and Chen, S. 2014. Diversity of endophytic mycobiota in *Fortunearia sinensis*. Acta Ecologica Sinica, 34(3): 160–164.

Xing, X., Guo, S. and Fu, J. 2010. Biodiversity and distribution of endophytic fungi associated with *Panax quinquefolium* L. cultivated in a forest reserve. Symbiosis, 51: 161–166.

Yu, H., Zhang, L., Li, L., Zheng, C., Guo, L., Li, W., Sun, P. and Qin, L. 2010. Recent developments and future prospects of antimicrobial metabolites produced by endophytes. Microbiol. Res., 165: 437–449.

Zhang, H.W., Song, Y.C. and Tan, R.X. 2006. Biology and chemistry of endophytes. Nat. Prod. Rep., 23: 753–771.

Zou, W.X., Meng, J.C., Lu, H., Chen, G.X., Shi, G.X., Zhang, T.Y. and Tan, R.X. 2000. Metabolites of *Colletotrichum gloeosporioides*, an endophytic fungus in *Artemisia mongolica*. J. Nat. Prod., 63: 1529–1530.

CHAPTER 3

Antimycobacterials from Fungi

Sunil Kumar Deshmukh,[1,] Shilpa Amit Verekar[2]
and B.N. Ganguli[3]*

ABSTRACT

Tuberculosis is an endemic disease of the poverty ridden, undernourished and over populated countries of the world. It is also a systemic disease that is extremely dependent on the physiology of the system it invades and thus varies significantly from person to person. New developments in the treatment of this disease have rarely percolated down to the larger sections of the under privileged in our societies. The need for highly active, long acting, yet less expensive drugs against Multi-Drug Resistant (MDR) *Mycobacterium tuberculosis* still exists. Research initiative on endophytic fungi as a source of such biotherapeutics is an important step that could help to tackle the need. Complete eradication of tuberculosis is certainly possible by integration of research results and public health programs. However, such initiatives have been hindered by the lack of effective communication lines in many countries of the world. Language is just one of the several hurdles! Nationalistic jingoism is another!!

A major initiative could be to investigate the effects of the mixtures of compounds already known to have activities against different strains of *M. tuberculosis*. Such as, those reported in the local knowledge forums of Ayurvedics in villages of India and in allopathic medical publications. We must have a "United Front to Combat Tuberculosis" (UFCT)—A Worldwide Effort.

---

[1] The Energy and Resources Institute, Darbari Seth Block, IHC Complex, Lodhi Road, New Delhi 110 003, India.

[2] Department of Microbiology, St. Xavier's College – Autonomous, 5, Mahapalika Marg, Mumbai, 400 001 India.
Email: shilfa1@rediffmail.com

[3] Emeritus Scientist of the CSIR, India, Chair Professor of the Agharkar Research Institute, Pune, India; (Residence-702/12 TulsidhamKalyani, Majiwade Thane, 400607).
Email: bganguli@mail.airtelmail.in

* Corresponding author: sunil.deshmukh@teri.res.in

Introduction

In the many countries of the world, the hunt for new anti-mycobacterial compounds is going on. In most of them, a marker MIC level is set so that both synthetic and natural (plants, fungal) compounds can be selected that have better therapeutic potential especially against clinically relevant Multi Drug Resistant strains of *Mycobacteria*. The marker compounds used are Isoniazid with a Minimum Inhibitory Concentration (MIC) of 0.04–0.09 µg/mL and Kanamycin sulphate with an MIC of 2.0–5.0 µg/mL usually. In the opinion of the authors, the use of selective mixtures of anti-tubercular compounds could be better, so that development of resistance is slowed down if not totally prevented. Choice of several different mixtures of compounds could be of advantage after extensive evaluation. Serum binding may not be a disadvantage if a slow but continuous release is observed and measured over time. What needs also to be borne in mind is that this chronic disease usually affects the poorer populations of the world where public health efforts are negligible, if not totally absent, and communications extremely difficult.

Scrutiny of the available literature of the years from 2002 to 2013 clearly indicated that research initiatives against *M. tuberculosis* are discouraging with the publication of 1–2 papers on anti-mycobacterial per year. Moreso, to our utter surprise, during the year 2006–2007 there was no report in the journals we reviewed. But in the year 2010 the largest number of publications appeared. This poses a million dollar question. What made the researches to jump on it too heavily and suddenly? But this momentum is a welcome move.

The World Health Organization (WHO) estimated that currently ca. 50 million people were infected and 1500 people dieper hour from Tuberculosis worldwide. After the detection of strains of *Mycobacterium tuberculosis* resistant to multiple drugs (MDRTB), the search for new antimycobacterials has been intensified (WHO, 2008). The world recognizes medicinal plants as repositories of fungal endophytes that produce metabolites with novel molecular structures that are active against various human diseases. For example, extracts of endophytic fungi isolated from Thailand's Garcinia plant species inhibit *M. tuberculosis* (Wiyakrutta et al., 2004). Several compounds reported from fungi with anti-mycobacterial activities are shown in Table 1.

Antimycobacterials from Fungi

From Ascomycetes

3-Nitropropionic acid (3-NPA) **(1)** (Fig. 1) is found in the extracts of several strains of endophytic genus *Phomopsis* sp. It is highly active against *M. tuberculosis* H37Ra with an MIC of 3.3 µM, but no *in vitro* cytotoxicity was seen in a number of cell lines. Endophytes produce high levels of 3-NPA which accumulates in certain plants and could, therefore be a marker

Table 1. Antimycobacterial from fungi.

Sr. No.	Fungus	Source	Compounds Isolated	Biological Activity*	Reference
1.	*Phomopsis* sp.	*Urobotrya siamensis, Grewia* sp., *Mesua ferrea, Rhododendron lyi, Tadehagi* sp., *Gmelina elliptica*	3-Nitropropionic acid (3-NPA) **(1)**	Compound **(1)** inhibits *Mycobacterium tuberculosis* H37Ra strain (MIC of 3.3 µM). Inhibits the isocitrate lyase (ICL), the enzyme involved in fatty acid catabolism and virulence in *M. tuberculosis*	Chomcheon et al., 2005 Munoz-Elias et al., 2005
2.	*Nodulisporium* sp.	Marine derived fungus	Vermelhotin **(3)**, Aspergillusidone D **(4)**	Vermelhotin **(3)** inhibits five reference strains of *M. tuberculosis* with MICs of 3.1–6.2 µg/mL. Aspergillusidone D **(4)** has an MIC value of 50.0 µg/mL	Kasettrathat et al., 2008; Ganihigama et al., 2015
3.	*Phomopsis* sp. BCC 1323	Leaf of *Tectona grandis* L.	Phomoxanthones A **(5)** and B **(6)**	Compounds **(5)** and **(6)**, *M. tuberculosis* H37Ra strain (MIC of 0.5 and 6.25 µg/mL respectively)	Isaka et al., 2001
4.	*Phomopsis* sp. PSU-D15	*Garcinia dulcis*	Phomoenamide **(7)**	Compound **(7)** *M. tuberculosis* (MIC of 6.25 mg/mL)	Rukachaisirikul et al., 2008
5.	*Diaporthe* sp. BCC 6140		Diaportheins A **(8)** and B **(9)**	Compound **(8)** inhibits *M. tuberculosis* (MIC 200 µg/mL) and Compound **(9)** at (MIC 3.1 µg/mL).	Dettrakul et al., 2003
6.	*Phoma* sp. NRRL 46751	*Saurauia scaberrinae*	Phomapyrrolidones B-C **(10-11)**	Compound **(10)** and **(11)** inhibits *M. tuberculosis* H37Pv (weak *in vitro* anti-tubercular activity when tested using the microplate Alamar Blue assay (MABA) for replicating cultures with MIC of 5.9 and 5.2 µg/ml respectively In the low oxygen recovery assay (LORA) with MIC 15.4 and 13.4 µg/ml respectively for non-replicating	Wijeratne et al., 2013

7.	Fruit hull of *Garcinia mangostana* L.	α-Mangostin (12)	-	Arunrattiyakorn et al., 2011
	Compound (12) Incubation with *C. gloeosporioides* (EYL131)	Mangostin 3-sulfate (13), Mangostanin 6-sulfate (14), 17,18-Dihydroxymangostanin 6-sulfate (15), Isomangostanin 3-sulfate (16)	Compound (12-13) inhibits *M. tuberculosis* (MICs 15.24 and 6.75 µM respectively) Compound 14-16 (MIC > 50 µg/mL)	Arunrattiyakorn et al., 2011
	Compound (12) Incubation with *N. spathulata* (EYR042).	Mangostin 3-sulfate (13)	Compound (13) inhibits *M. tuberculosis* (MIC 6.75 µM)	Arunrattiyakorn et al., 2011
8.	*Cynodon dactylon*	Chaetoglocins A-B (17-18)	Compounds (17-18) inhibits *B. subtilis*, *Streptococcus pyogens*, *Mirococcus luteus* and *M. smegmatis* with MICs between 8 and 32 µg/mL	Ge et al., 2011
9.	*Chaetomium globosum* KMITL-N0802	Echinuline (19) and Chaetomanone (20)	Compounds (19-20) inhibits *M. tuberculosis* with MIC of 169.92 and 216.62 µM respectively	Kanokmedhakul et al., 2002
10.	*Chaetomium brasiliense*	Mollicellins K (21)	Compound (21) inhibits *M. tuberculosis* MIC 12.5 µg/mL	Khumkomkhet et al., 2009
11.	*Chaetomium cochliodes* VTh01 and *C. cochliodes* CTh05	Cochliodone C (22), Chaetoviridine E and F (23-24), Chaetochalasin A (25), 24(R)-5α,8α-Epidioxyergosta-6-22-diene-3β-ol (26)	Compounds (22-26) inhibits *M. tuberculosis* (MICs 200, 50, 100, 100, and 200 µg/mL, respectively)	Phonkerd et al., 2008
12.	*Trichoderma* sp.	Trichoderins A (27), A1 (28), and B (29)	Compounds (27-29) active against both dormant and multiplying *M. tuberculosis* strain H37Rv. *M. smegmatis, Mycobacterium bovis* BCG, and *M. tuberculosis* H37Rv with MIC values in the range of 0.02–2.0 µg/mL	Pruksakorn et al., 2010

Marine sponge-derived fungus appears as the source for row 12.

Table 1. contd....

Table 1. contd....

Sr. No.	Fungus	Source	Compounds Isolated	Biological Activity*	Reference
13.	*Coniothyrium cereale*	Marine green alga *Enteromorpha* sp.	(–)-Trypethelone (30)	Compound (30) is active against *M. phlei*, *S. aureus*, and *E. coli*, at 20 µg/disk with inhibition zones of 18, 14, and 12 mm, respectively	Elsebai et al., 2011
14.	*Biscogniauxia formosana* BCRC 33718	*Cinnamomum* sp.	Biscogniazaphilones A (31) and B (32), N-trans-feruloy-3-O-methyldopamine (33), 5-Hydroxy-3,7,4-trimethoxyflavone (34), 4-Methoxycinnamaldehyde (35), Methyl 3,4-methylenedioxycinnamate (36), 4-Methoxy-trans-cinnamic acid (37)	Compounds (31) and (32) inhibits *M. tuberculosis* strain H37Rv with MIC of ≤ 5.12 and ≤ 2.52 µg/mL, respectively Compounds (33–37) inhibits *M. tuberculosis* strain H37Rv with MIC of 12.5, 25.0, 42.1, 58.2 and 50.0 µg/mL, respectively	Cheng et al., 2012
15.	*Fusarium* sp. BCC14842	Bamboo leaf	Javanicin (38), 3-O-Methylfusarubin (39), a diastereomer of Dihydronaphthalenone (40) and 5-Hydroxy-3-methoxydihydrofusarubin A (41)	Compounds (38) and (40), anti-mycobacterial activity (MICs of 25 µg/mL) Compounds (39)and (41), anti-mycobacterial activity(MICs of 50 µg/mL)	Kornsakulkarn et al., 2011
16.	*Fusarium* sp.	Mangrove plant	Cadmium (42) and copper (43) metal complexes of Fusaric acid	Compounds (42-43) *M. bovis* BCG (MIC 4 µg/mL) and the *M. tuberculosis* H37Rv strain (MIC 10 µg/mL)	Pan et al., 2011
17.	*Fusarium* spp. PSU-F14	Sea fan-derived fungi	9α-hydroxyhalorosellinia A (44), Nigrosporin B (45) anhydrofusarubin (46)	Compounds (44-46) inhibits *M. tuberculosis* H37Ra (MIC of 39, 41 and 87 µM respectively)	Trisuwan et al., 2010

No.	Fungus	Source	Compounds	Activity	Reference
18.	*Microsphaeropsis* sp. BCC 3050	Lichenicolous fungus isolated from *Dirinaria applanata*	3'-O-Demethylpreussomerin I (47), Preussomerins E-I (48–52), Deoxypreussomerin A (53) and Bipendensin (Palmarumycin C11) (54)	Compounds (47–54) inhibits *M. tuberculosis* H37Ra (MIC 25, 3.12, 3.12–6.25, 6.25, 12.5, 25, 1.56–3.12, 50 µg/mL, respectively)	Seephonkai et al., 2002
19.	*Phaeosphaeria* sp.		(3S,4R)-4,8-Dihydroxy-3-methoxy-3,4-dihydronaphthalen-1(2H)-one (55), (4S)-3,4,8-Trihydroxy-6-methoxy-3,4-dihydronaphthalen-1(2H)-one (56), (S)-4,6,8-Trihydroxy-3,4-dihydronaphthalen-1(2H)-one (57), 1-(1-Hydroxy-3,6-dimethoxy-5,8-dioxo-5,8-dihydronaphthalen-2-yl) ethyl acetate (58), 2,5,7-Trihydroxy-3-(1-hydroxy-3,6-dimethoxy-5,8-dioxo-5,8-dihydronaphthalen-2-yl)ethyl]naphthalene-1,4-dione (59), 6-Ethyl-5-hydroxy-2,7-dimethoxynaphthalene-1,4-dione (60)	Compounds (55-56) (*M. tuberculosis*) MICs 12.50 µg/mL, Compound (58) MIC of 12.50 µg/mL, compound (59), MIC 0.39 µg/mL. Compound (60) MIC of 6.25 µg/mL, compound (57) MIC of 25 µg/mL.	Pittayakhajonwut et al., 2008
20.	*Dothideomycete* sp. LRUB20	*Leea rubra*	2-hydroxymethyl-3-methylcyclopent-2-enone (61), Asterric acid (62), and hydrazone derivative of cis-2-hydroxymethyl-3-methylcyclopentanone (63)	Compounds (61–63), have mild anti-mycobacterial activities with MIC values of 200 µg/mL	Chomcheon et al., 2006
21.	*Penicillium dipodomyicola* - HN4-3A	Stem of the mangrove plant *Acanthus ilicifolius*	Peniphenone B (64), Peniphenone C (65)	Compounds (64), (65), inhibited Mptp B with IC_{50} values of 0.16 ± 0.02 and 1.37 ± 0.05 µM, respectively	Li et al., 2014

Table 1. contd....

Table 1. contd....

Sr. No.	Fungus	Source	Compounds Isolated	Biological Activity*	Reference
22.	*Geotrichum* sp.	*Crassocephalum crepidoides*	7-butyl-6,8-dihydroxy-3(R)-pent-11-enylisochroman-1-one **(66)**, 7-but-15-enyl-6,8-dihydroxy-3(R)-pent-11-enylisochroman-1-one **(67)** and 7-butyl-6,8-dihydroxy-3(R)-pentylisochroman-1-one **(68)**	Compound **(66–68)** inhibits *M. tuberculosis* H27Ra. MICs are respectively, 25 µg/mL, 50 µg/mL, and inactive	Kongsaeree et al., 2003
23.	*Verticillium hemipterigenum*	Pathogenic fungus	Enniatins H **(69)**, I **(70)**, B **(71)**, and B4 **(72)**	Compound **(69–72)** inhibits *M. tuberculosis* H37Ra (MICs 3.12–6.25 µg/mL)	Nilanonta et al., 2003
			Analogues H, I and MK1688 **(73–75)**	Compound **(73–75)** (MICs 3.12–6.25 µg/mL)	Vongvilai et al., 2004
24.	Unidentified fungus		Enniatins L **(76)**, M1 **(77)**, M2 **(78)** and N **(79)**	Compound **(76–79)** MICs of 6.25–12.5 µg/ml	Vongvilai et al., 2004
25.	*Nigrospora* sp.	Mangrove endophyte	4-Deoxybostrycin **(80)** and Nigrosporin **(45)**	Compound **(80 and 45)** In the Kirby-Bauer disk diffusion susceptibility test, both had inhibition zone sizes of over 25 mm against *M. tuberculosis*	Wang et al., 2013
26.	*Hirsutella kobayasii* BCC 1660	Entomopathogenic fungus	Hirsutellide A **(81)**	Compound **(81)** *M. tuberculosis* H37Ra using the microplate Alamar Blue Assay (MABA) showed a MIC with 6–12 µg/mL	Vongvanich et al., 2002
27.	*Hirsutella nivea* BCC 2594	Pathogenic fungus	Hirsutellones A–D **(82–85)**	The compounds **(82–85)**, inhibits *M. tuberculosis* H37Ra (MIC 0.78, 3.125, 0.78, 0.78 µg/mL)	Isaka et al., 2005

No.	Source	Compound	Activity	Reference
28.	*Trichoderma* sp. BCC 7579	Hirsutellone F (86), Hirsutellones A, B, and C	Compound (86) inhibits *M. tuberculosis* H37Ra (MIC 3.12 μg/mL)	Isaka et al., 2006
29.	*Periconia* sp. / *Piper longum*	Piperine (87)	Compound (87) inhibits *M. tuberculosis* and *M. smegmetis* with MIC of 1.74 and 2.62 μg/ml respectively	Verma et al., 2011
30.	*Aschersonia tubulata* BCC 1785 / Insect pathogenic fungus	Dustanin (88), 3 beta-acetoxy-15 alpha, 22-dihydroxyhopane (89)	Compounds (88), and (89), exhibited anti-mycobacterial activity with the MIC of 12.5 μg/ml	Boonphong et al., 2001
31.	*Aspergillus* sp.	Physcion (90)	Compound (90) exhibited mycobacterial detoxification enzyme mycothiol-S-conjugate amidase (MAC), with IC_{50} value of 50 μM against *M. smegmatis*	Nicholas et al., 2003
32.	Lichen	Usnic acid (91)	Compound (91) *M. tuberculosis* (MIC 2.5–5 μg/mL)	Ingólfsdóttir, 2002
33.	*Menisporopsis theobromae* / Seed fungus	Menisporopsin A (92)	Compound (92) MIC of 50 μg/ml against *M. tuberculosis* H37Ra	Chinworrungsee et al., 2004
34.	fungal strain WZ-4-11 of *Aspergillus carbonarius*	8'-O-Demethylnigerone (93) and 8'-O-demethylisonigerone (94)	Compounds (93) and (94) inhibits *M. tuberculosis* H37Rv with MIC values of 43.0 and 21.5 μM, respectively	Zhang et al., 2008
35.	*Cordyceps* sp. BCC 1861 / Insect pathogenic fungus from *Homoptera cicada* nymph	Cordyol A (95)	Compound (95) showed anti-mycobacterial with a MIC value of 100 μg/mL	Bunyapaiboonsri et al., 2007
36.	*Ophiocordyceps communis* BCC 16475 / Insect pathogenic Fungus	Cordycommunin (96)	Cordycommunin (96) inhibits *M. tuberculosis* H37Ra with an MIC value of 15 μM	Haritakun et al., 2010

Table 1. contd....

Table 1. contd....

Sr. No.	Fungus	Source	Compounds Isolated	Biological Activity*	Reference
37.	*Emericella variecolor*	Marine-derived fungus	Ophiobolin K (97), 6-epi-ophiobolin K (98) and 6-epi-ophiobolin G (99)	Ophiobolins (97–99) inhibited biofilm formation of *M. smegmatis* with MICs of 4.1–65 mM	Arai et al., 2013
				Ophiobolin K (97) was also effective against the biofilm formation of *M. bovis* BCG and was able to restore the antimicrobial activity of isoniazid against *M. smegmatis* by inhibiting biofilm formation	
38.	*Emericella rugulosa*		Bicyclol[3.3.1]nona-2,6-diene derivative, rugulosone (100)	Compound (100), anti-mycobacterial active	Moosophon et al., 2009
39.	*Conoideocrella tenuis* BCC 18627	Insect pathogenic fungus	Hopan-27-al-6β,11r,22-triol (101), Hopane-6β,11r,22,27-tetraol (102), Hopane-6β,7β,22-triol (103), (atropisomer of ES-242-2) (104), Compound (105)	Compounds (101–105) active against *M. tuberculosis* H37Ra with MIC of > 105, 52, > 107, > 75, > 75 μM/ml, respectively	Isaka et al., 2011
40.	*Scleroderma citrinum*	Thai mushroom	4,4'-dimethoxyvulpinic acid (106), dibromo derivative of (106) 3,3'-dibromo-4,4'-dimethoxyvulpinic acid (107) acetyl 4,4'-dimethoxyvulpinate (108)	Compounds (106–108) inhibits *M. tuberculosis* H37Ra with MIC of 25, 100 and 100 μg/ml	Kanokmedhakul et al., 2003
41.	*Astraeus pteridis*	Truffle-mimiking mushroom	3-epi-astrahygrol (109), astrahygrone (110) and 3-epi-astrapteridiol (111)	Compounds (109–111) inhibits *M. tuberculosis* with MIC values of 58.0, 64.0 and 34.0 μg/mL, respectively	Stanikunaite et al., 2008

No.	Source	Type	Compound	Description	Reference
42.	*Astraeus odoratus*	Edible mushroom	Astraodoric acids A (112) and B (113)	Compounds (112-113) inhibits *M. tuberculosis* H37Ra with MICs of 50 and 25 µg/mL	Arpha et al., 2012
43.	*Kionochaeta ramifera* BCC 7585	The coral mushroom	Ramiferin (114)	Compound (114) anti-tubercular MIC 12.7 µM	Bunyapaiboonsri et al., 2008
44.	*Ramaria cystidiophora*	The coral mushroom	Ramariolide (115)	Compound (115) *M. smegmatis* and *M. tuberculosis* active	Centko et al., 2012
45.	*Mycena* sp. (F205435)	Basidiomycetes	Gliotoxin (116), and S,S dimethyl gliotoxin (117)	Compounds (116-117) exhibit mycobacterial detoxification enzyme mycothiol-S-conjugate amidase (MAC), with IC_{50} values of 50 and 70 µM against *M. tuberculosis*. Gliotoxin inhibits MAC. Its IC_{50} value is 50 µM against *M. smegmatis*	Nicholas et al., 2003
46.	*Ganoderma orbiforme* BCC 22324	Reishi mushroom	Ganoderic acid T (118), and the C-3 epimer of Ganoderic acid T (119)	Compounds (118-119), *M. tuberculosis* H37Ra with MIC of 10.0 and 1.3 µM respectively	Isaka et al., 2013
47.	Endophytic fungi PSU-N24	*Garcinia nigrolineata*	9α-Hydroxyhalorosellinia A (120)	Compound (120) inhibits *M. tuberculosis* with the MIC value of 12.50 µg/ml	Sommart et al., 2008
48.	Nonsporulating filamentous fungus, F7524		Agonodepside A (121) and B (122)	Inhibit the 2-trans-enoyl-acyl-reductase involved in mycolic acid biosynthesis. Agonodepside A (121) has IC_{50} value of 75 µM, B (122) is not active at 100 µM	Cao et al., 2002
49.	*Mortierella alpina* FKI-4905		Calpinactam (123)	Calpinactam (123) inhibits *M. smegmatis* and *M. tuberculosis* with MICs of 0.78 and 12.5 µg/ml, respectively	Koyama et al., 2010

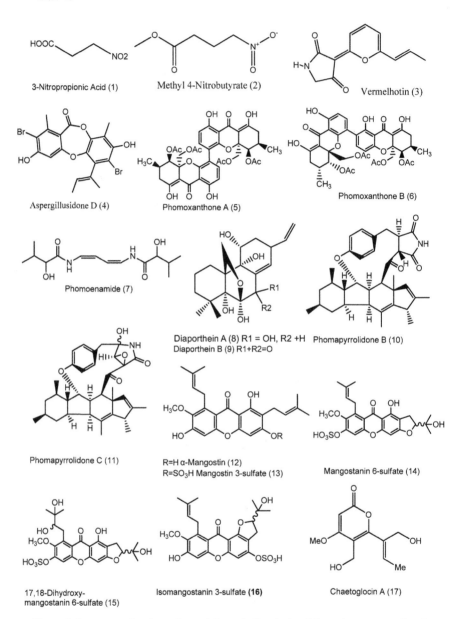

Figure 1. Structures of antimycobacterial metabolites isolated from Ascomycetes **(1-17)**.

for endophytic fungi (Chomcheon et al., 2005). 3-NPA inhibits Isocitrate Lyase (ICL), the enzyme involved in fatty acid catabolism and virulence in *M. tuberculosis* (Munoz-Elias et al., 2005). 3-NPA has MIC values of 12.5 and 50.0 µg/mL against the MTB H37Rv and H37Ra strains, respectively. Out of three derivatives of 3-Nitropropionic acid, Methyl 4-Nitrobutyrate **(2)** is active with MIC values of 12.5 and 25.0 µg/mL against H37Ra and H37Rv strains, respectively (Ganihigama et al., 2015).

Vermelhotin (3) and Aspergillusidone D (4) (Fig. 1), were isolated from the marine derived fungus, a *Nodulisporium* sp. (Kasettrathat et al., 2008). Vermelhotin (3) is active against five reference strains of *M. tuberculosis* with MICs of 3.1–6.2 µg/mL. Aspergillusidone D (4) has an MIC value of 50.0 µg/mL in comparative assays (Ganihigama et al., 2015).

Phomoxanthone A (5) and B (6) (Fig. 1) were obtained from *Phomopsis* sp. BCC 1323, collected from the leaves of *Tectonagrandis* from the Mee Rim district of Chaingmai Province, Northern Thailand. These compounds show moderate *in vitro* activities with MICs of 0.5 and 6.25 µg/mL, respectively against *M. tuberculosis* H37Ra strain, as compared to Isoniazid and Kanamycin sulphate (MICs of 0.050 and 2.5 µg/mL, respectively) (Isaka et al., 2001). Phomoenamide (7) isolated from the endophyte *Phomopsis* sp. PSU-D15 of *Garcinia dulcis* has an MIC of 6.25 µg/mL against *M. tuberculosis* (Rukachaisirikul et al., 2008).

The pimaranediterpenes Diaporthein A (8) and B (9) (Fig. 1), were isolated from *Diaporthe* sp. BCC 6140. Diaporthein B strongly inhibits *M. tuberculosis* with a MIC 3.1 µg/mL, while A is less active (MIC 200 µg/mL). As compared to Isoniazid, MIC 0.04–0.09 µg/mL and Kanamycin sulfate, MIC of 2.0–5.0 µg/mL (Dettrakul et al., 2003). The results suggest that the carbonyl function C-7 of Diaporthein B is essential for its anti-TB activity (Asres et al., 2001).

Phomapyrrolidone B-C (10-11) (Fig. 1), were isolated from the endophyte *Phoma* sp. NRRL 46751, of the plant *Saurauiasca berrinae*. Phomapyrrolidone B (10) and C (11) have weak *in vitro* anti-tubercular activities when tested in the microplate Alamar Blue assay (MABA) for replicating cultures with MICs of 5.9 and 5.2 µg/ml, respectively and the low oxygen recovery assay (LORA) with MICs of 15.4 and 13.4 µg/ml, respectively for non-replicating *M. tuberculosis* H37Pv (Wijeratne et al., 2013).

α-Mangostin (12) (Fig. 1), a prenylatedxanthone from the fruit hull of *Garcinia mangostana*, was individually metabolized by two fungi, *Colletotrichum gloeosporioides* (EYL131) and *Neosartorya spathulata* (EYR042), respectively. Incubation of compound (12) with *C. gloeosporioides* (EYL131) gave four metabolites identified as Mangostin 3-sulfate (13), Mangostanin 6-sulfate (14), 17,18-Dihydroxymangostanin 6-sulfate (15) and Isomangostanin 3-sulfate (16) (Fig. 1). Compound (13) was also formed by incubation with *N. spathulata* (EYR042). Compounds (12) and (13) are active against *M. tuberculosis* (MICs 15.24 and 6.75 µM for 12 and 13, respectively). In contrast, 14–16 showed very week activity (MIC > 50 µg/mL) (Arunrattiyakorn et al., 2011).

Chaetoglocin A (17) (Fig. 1) Chaetoglocin B (18) (Fig. 2) isolated from *Chaetomium globosum* strain IFB-E036, an endophyte of *Cynodon dactylon* are active against *B. subtilis*, *Streptococcus pyogenes*, *Mirococcus luteus* and *M. smegmatis* with MICs between 8 and 32 µg/mL (Ge et al., 2011). Echinuline (19) and Chaetomanone (20) (Fig. 2) were isolated from *Chaetomium globosum* KMITL-N0802 isolated from a Thai soil. Chaetomanone and Echinuline have week activities against *M. tuberculosis* with MICs of 169.92 and 216.62

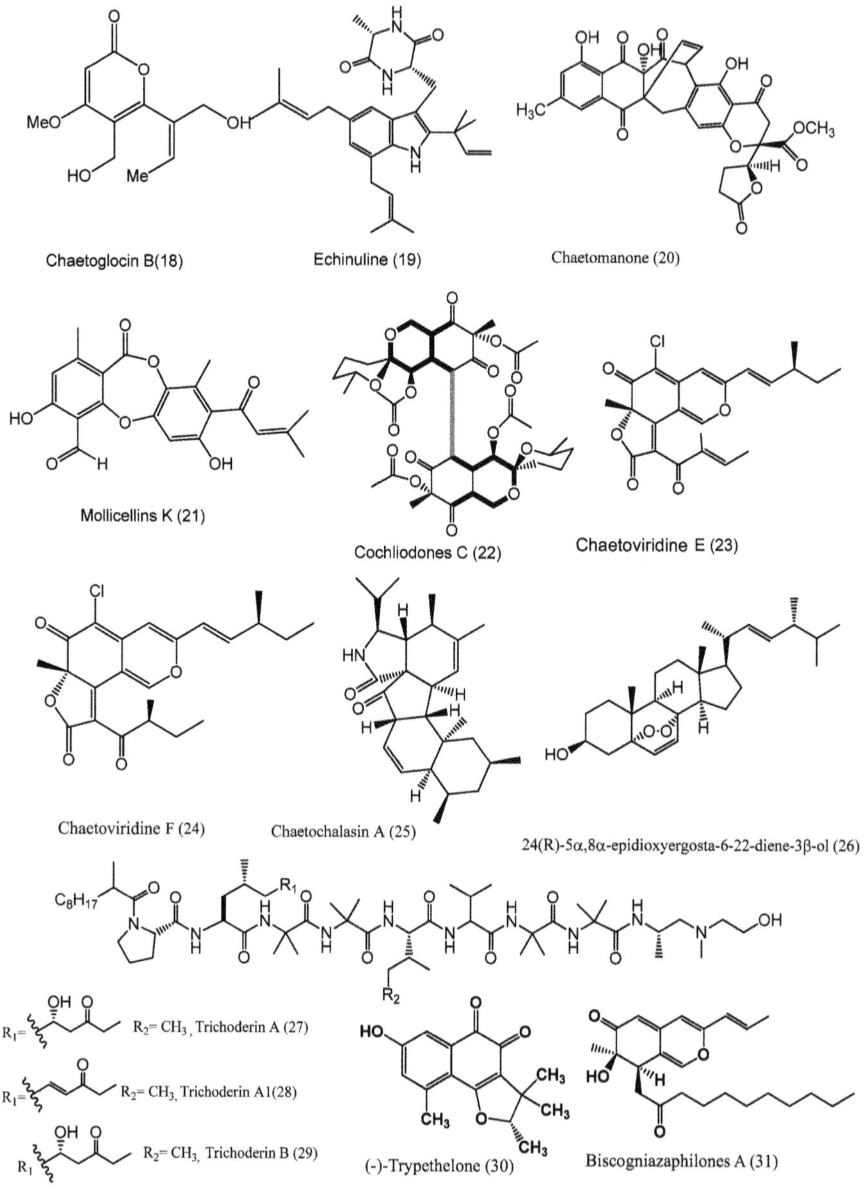

Chaetoglocin B(18)   Echinuline (19)   Chaetomanone (20)

Mollicellins K (21)

Cochliodones C (22)   Chaetoviridine E (23)

Chaetoviridine F (24)   Chaetochalasin A (25)

24(R)-5α,8α-epidioxyergosta-6-22-diene-3β-ol (26)

$R_2= CH_3$, Trichoderin A (27)

$R_2= CH_3$, Trichoderin A1(28)

$R_2= CH_3$, Trichoderin B (29)

(-)-Trypethelone (30)   Biscogniazaphilones A (31)

Figure 2. Structures of antimycobacterial metabolites isolated from Ascomycetes **(18-31)**.

μM, respectively (Kanokmedhakul et al., 2002). Mollicellin K **(21)** (Fig. 2) was isolated from the fungus *Chaetomium brasiliense* showed activity against *M. tuberculosis* (MIC 12.5 μg/ml) (Khumkomkhet et al., 2009).

Cochliodone C **(22)**, Chaetoviridine E and F **(23-24)**, Chaetochalasin A **(25)**, 24(R)-5α,8α-epidioxyergosta-6-22-diene-3β-ol **(26)** (Fig. 2) were isolated from the fungi *Chaetomium cochliodes* VTh01 and *C. cochliodes* CTh05. Compounds **(22–26)** are active against *M. tuberculosis* with MIC values of 200, 50, 100, 100, and 200 µg/mL, respectively (Phonkerd et al., 2008).

Trichoderins A **(27)**, A1 **(28)**, and B **(29)** (Fig. 2), aminolipopeptides from a *Trichoderma* sp., a marine sponge-derived fungus, are reported to be active against both dormant and multiplying *M. tuberculosis* strain H37Rv. Trichoderins are highly active against *M. smegmatis*, *M. bovis* BCG, and *M. tuberculosis* H37Rv with MIC values in the range of 0.02–2.0 µg/mL (Pruksakorn et al., 2010).

(–)-Trypethelone **(30)** (Fig. 2), isolated from the endophyte *Coniothyrium cereale* of the marine green alga *Enteromorpha* sp. is active against *M. phlei*, *S. aureus*, and *E. coli*, at 20 µg/disk/6 mm with inhibition zones of 18, 14, and 12 mm, respectively (Elsebai et al., 2011).

Biscogniazaphilone A **(31)** (Fig. 2) and B **(32)**, N-trans-feruloy-3-O-methyldopamine **(33)**, 5-Hydroxy-3,7,4-trimethoxyflavone **(34)**, 4-Methoxycinnamaldehyde **(35)**, Methyl 3,4-methylenedioxycinnamate **(36)**, 4-Methoxy-trans-cinnamic acid **(37)** (Fig. 3), were all isolated from the endophyte *Biscogniauxia formosana* BCRC 33718, of a *Cinnamomum* sp. Compounds **(31)** and **(32)** are active against *M. tuberculosis* strain H37Rv *in vitro* with MIC values of ≤ 5.12 and ≤ 2.52 µg/mL, respectively, as compared to the clinical drug Ethambutol (MIC 6.25 µg/mL). Compounds **(33–37)** have either moderate or weak anti-mycobacterial activities, MICs of 12.5, 25.0, 42.1, 58.2 and 50.0 µg/mL, respectively (Cheng et al., 2012).

Javanicin **(38)**, 3-O-methylfusarubin **(39)**, a diastereomer of Dihydronaphthalenone **(40)** and 5-Hydroxy-3-methoxydihydrofusarubin A **(41)** (Fig. 3) were isolated from the endophyte, a *Fusarium* sp. BCC 14842 of the Bamboo leaf, collected from a forest of Nam Nao National Park, Phetchabun Province, Thailand. Compounds **(38)** and **(40)**, have moderate activities (MICs of 25 µg/mL), while 3-O-methylfusarubin **(39)**, and 5-hydroxy-3-methoxydihydrofusarubin A **(41)**, have weak antimycobacterial activities (MICs of 50 µg/mL) (Kornsakulkarn et al., 2011).

Fusaric acid was isolated from a *Fusarium* sp., an endophyte of a mangrove plant. Cadmium and Copper complexes were prepared. The Cadmium **(42)** and Copper **(43)** (Fig. 3), complexes showed potent activities against *M. bovis* BCG (MIC 4 µg/mL) and *M. tuberculosis* H37Rv (MIC 10 µg/mL) (Pan et al., 2011).

9α-Hydroxyhalorosellinia A **(44)**, Nigrosporin **(45)** Anhydrofusarubin **(46)** (Fig. 3), were isolated from the sea fan-derived fungi *Fusarium* spp. PSU-F14. Compounds **(44–46)** were found active against *M. tuberculosis* H37Ra, with MICs of 39, 41 and 87 µM, respectively (Trisuwan et al., 2010).

3'-O-Demethylpreussomerin I **(47)**, Preussomerin E **(48)**, (Fig. 3), Preussomerins F–I **(49–52)** (Fig. 4), Deoxypreussomerin A **(53)** (Fig. 4) and Bipendensin (Palmarumycin C11) **(54)** (Fig. 4), were isolated from *Microsphaeropsis* sp. BCC 3050, a lichenicolous fungus of *Dirinaria applanata* collected from Phu Tee-Suan-Sai forest in Loei province, Northeastern

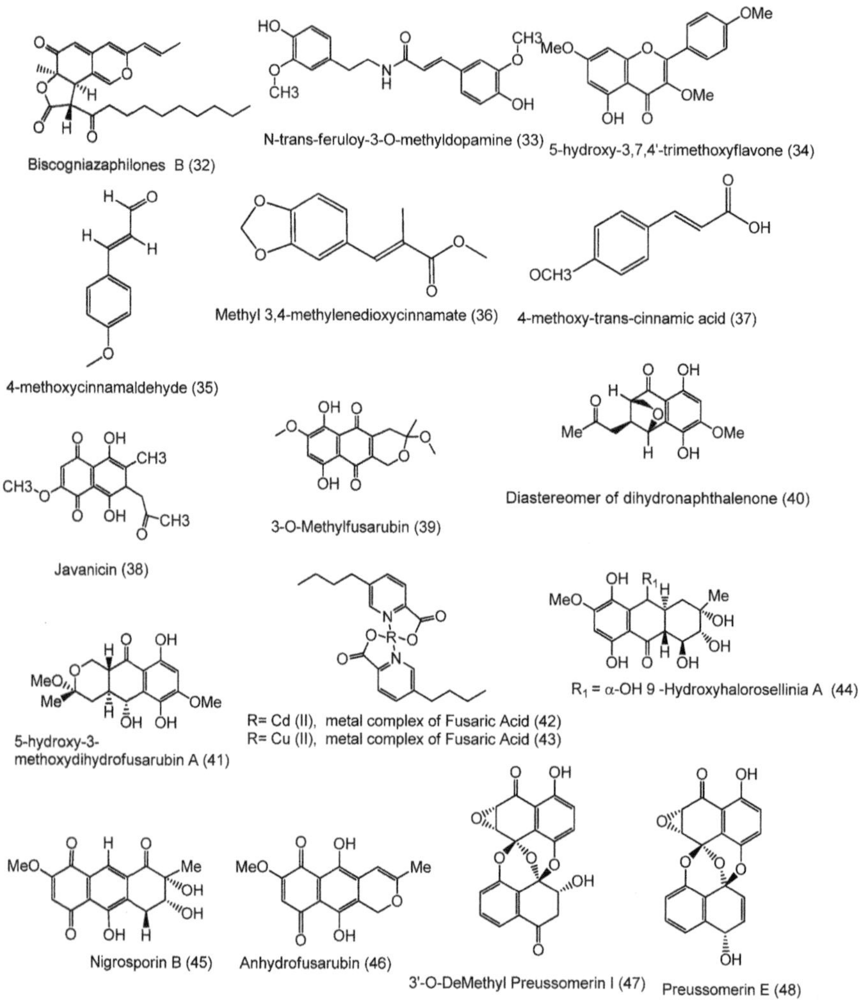

Figure 3. Structures of antimycobacterial metabolites isolated from Ascomycetes **(32-48)**.

Thailand. These compounds **(47–54)** are active against *M. tuberculosis* H37Ra (MICs 25, 3.12, 3.12–6.25, 6.25, 12.5, 25, 1.56–3.12, 50 µg/mL, respectively) (Seephonkai et al., 2002).

(3S,4 R)-4,8-Dihydroxy-3-methoxy-3,4-dihydro-1(2 H)-naphthalenone **(55)**, (S)-4,6,8-Trihydroxy-3,4-dihydro-1(2H)-naphthalenone **(56)**, (3S,4S)-3,4,8-Trihydroxy-6-methoxy-3,4-dihydro-1(2 H)-naphthalenone **(57)**, 6-Ethyl-5-hydroxy-2,7-dimethoxynaphthoquinone **(58)**, 6-(1-Acetoxyethyl)-5-hydroxy-2,7-dimethoxynaphthoquinone **(59)**, Deacetylkirschsteinin **(60)** (Fig. 4) were isolated from a *Phaeosphaeria* sp. Compounds **(55)** and **(56)** have good anti-mycobacterial activity with MICs of 12.50 µg/mL. Compound **(58)**

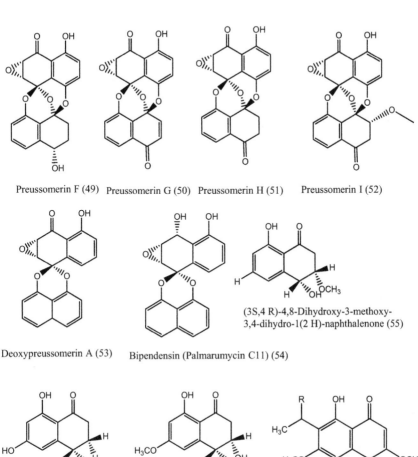

Preussomerin F (49)   Preussomerin G (50)   Preussomerin H (51)   Preussomerin I (52)

(3S,4 R)-4,8-Dihydroxy-3-methoxy-
3,4-dihydro-1(2 H)-naphthalenone (55)

Deoxypreussomerin A (53)   Bipendensin (Palmarumycin C11) (54)

(S)-4,6,8-Trihydroxy-3,4-
dihydro-1(2H)-
naphthalenone (56)

(3S,4S)-3,4,8-Trihydroxy
-6-methoxy -3,4-dihydro-1(2 H)
-naphthalenone (57)

R=H ; 6-Ethyl-5-hydroxy-2,7-
dimethoxynaphthoquinone (58)

R= OCOCH3; 6-(1-Acetoxyethyl)
-5-hydroxy-2,7-
dimethoxynaphthoquinone
(59)

Deacetylkirschsteinin (60)

Figure 4. Structures of antimycobacterials metabolites isolated from Ascomycetes **(49-60)**.

exhibited anti-TB activity with MIC of 12.50 µg/mL, while its acetyl derivative, compound **(59)**, has excellent anti-TB activity, MIC 0.39 µg/mL. Compound **(60)** has an MIC value of 6.25 µg/mL, while compound **(57)** has an MIC of 25 µg/mL as compared to MIC values of isoniazid and kanamycin sulphate that were 0.05 and 2.5 µg/mL, respectively (Pittayakhajonwut et al., 2008).

2-hydroxymethyl-3-methylcyclopent-2-enone **(61)**, Asterric acid **(62)** and hydrazone derivative of cis-2-hydroxymethyl-3-methylcyclopentanone **(63)** (Fig. 4), were all isolated from a *Dothideomycete* sp. LRUB20, an endophyte of the stem of a medicinal plant *Leearubra* in Thai. Compounds **(61-63)** have low anti-mycobacterial activities with MIC values of 200 µg/mL (Chomcheon et al., 2006).

Peniphenone B **(64)** and C **(65)** (Fig. 5), were isolated from *Penicillium dipodomyicola* - HN4-3A of the stem of the mangrove plant *Acanthusilicifolius* collected from the South China Sea in Hainan Province, China. Both B and C exhibited strong inhibitory activity against protein tyrosine phosphatase B (MptpB) with IC50 values of 0.16 ± 0.02 and 1.37 ± 0.05 µM, respectively (Li et al., 2014).

7-butyl-6,8-dihydroxy-3(R)-pent-11-enylisochroman-1-one **(66)**, 7-but-15-enyl-6,8-dihydroxy-3(R)-pent-11-enylisochroman-1-one **(67)** and 7-butyl-6,8-dihydroxy-3(R)-pentylisochroman-1-one **(68)** (Fig. 5) novel Dihydroisocoumarins were isolated from a *Geotrichum* sp., an endophyte of *Crassocephalum crepidoides*. The MICs of compounds **(66–68)** were, 25 µg/mL, 50 µg/mL, and inactive against *M. tuberculosis* H37Ra respectively. This suggests that the double bond C11-C12 and the aliphatic group at C14-C17 are important for the biological activities (Kongsaeree et al., 2003).

Four cyclic peptides, namely, Enniatins B **(69)**, B4 **(70)**, G **(71)**, C **(72)** (Fig. 5) were isolated from a pathogenic fungus *Verticillium hemipterigenum*. Analogues H **(73)**, I **(74)** and MK1688 **(75)** (Fig. 5), were prepared by feeding the substrate analogs L-leucine and L-isoleucine to the fermentation. Compounds **(69–75)**, inhibited *M. tuberculosis* H37Ra (MIC 3.12, 3.12, 6.25, 6.25, 6.25, 6.25 and 1.56 µg/mL, respectively) (Nilanonta et al., 2003). Fermentation of an unidentified Thai fungus led to the isolation of new hydroxyl analogs Enniatins L **(76)**, M1 **(77)**, M2 **(78)** and N **(79)** (Fig. 5) with MICs of 6.25–12.5 µg/ml (Vongvilai et al., 2004).

4-Deoxybostrycin **(80)** (Fig. 5) and Nigrosporin **(45)** (Fig. 3) were isolated from the mangrove endophyte, a *Nigrospora* sp. collected from the South China Sea. In the Kirby-Bauer disk diffusion susceptibility test, both showed zones of over 25 mm against *M. tuberculosis*. Compound **(80)** has activity against multidrug-resistant (MDR) *M. tuberculosis* strains with MICs of < 5–39.0 µg/ml. The gene expression profile of *M. tuberculosis* H37Rv after treatment with 4-Deoxybostrycin was compared with that of the untreated bacteria. One hundred and nineteen out of 3,875 genes were significantly different in *M. tuberculosis* exposed to 4-deoxybostrycin from that of the control. There are 46 functionally known genes involved in metabolism, information storage and processing, and cellular processes. The differential expressions of six

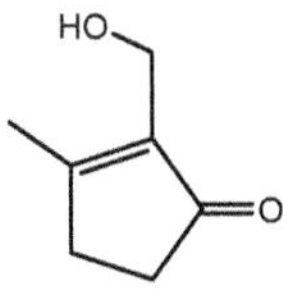

2-hydroxymethyl-
3-methylcyclopent
-2-enone (61)

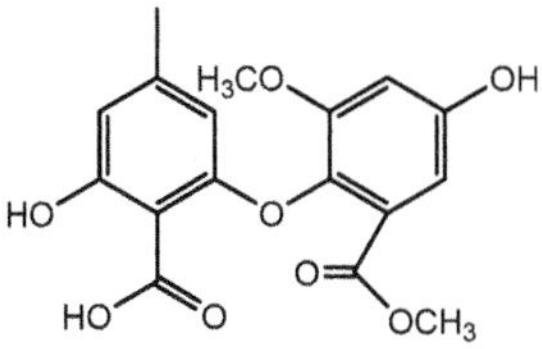

2-hydroxymethyl
-3-metasterric acid (62)

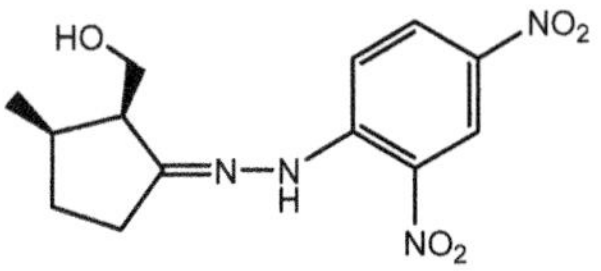

hydrazone derivative of cis-2-
hydroxymethyl
-3-methylcyclopentanone (63)

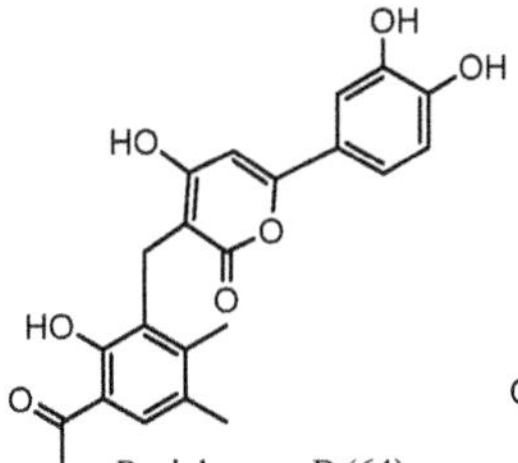

Peniphenone B (64)

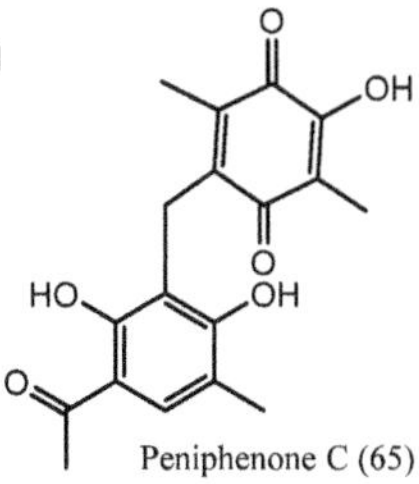

Peniphenone C (65)

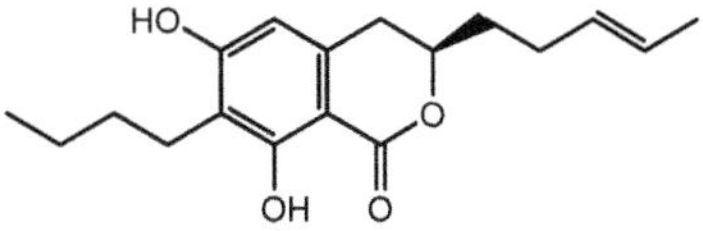

7-butyl-6,8-dihydroxy-
3(R)-pent-11-enylisochroman-1-one (66)

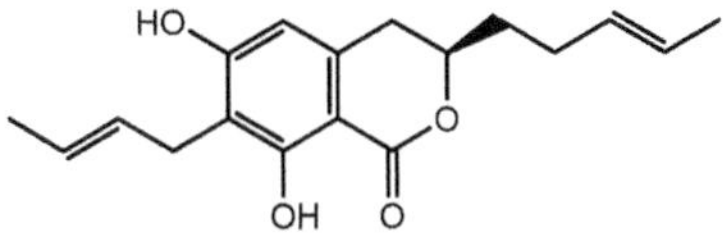

7-but-15-enyl-6,8-dihydroxy-3(R)-
pent-11-enylisochroman-1-one (67)

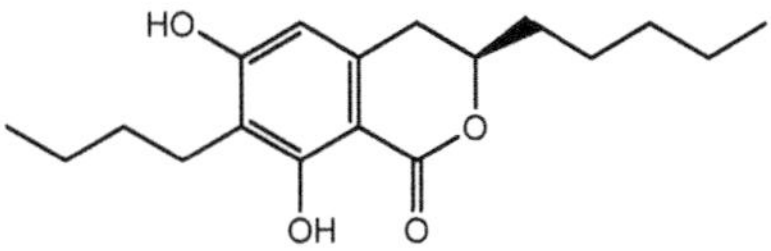

7-butyl-6,8-dihydroxy-3(R)-pentylisochroman-1-one (68)

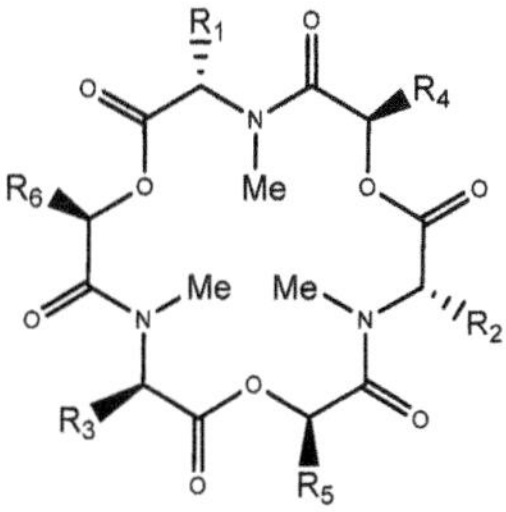

R1= R2= R3= R4=R5=R6 = i-Pr Enniatin B (69)

R1 =i-Bu, R2, R3, R4,R5,R6 = i-Pr Enniatin B$_4$ (70)

R1, R2, =i-Bu, R3, R4,R5,R6 = i-Pr Enniatin G (71)

R1 R2, R3=i-Bu, R4,R5,R6 = i-Pr Enniatin C (72)

R1,R2, R3, = iPr, R4=s-Bu,R5,R6 = i-Pr Enniatin H (73)

R1,R2,R3 =i-Pr, R4, R5 = s-Bu, R6 = i-Pr Enniatin I (74)

R1, R2, R3 = i-Pr, R4,R5,R6 = s-Bu, MK1688 (75)

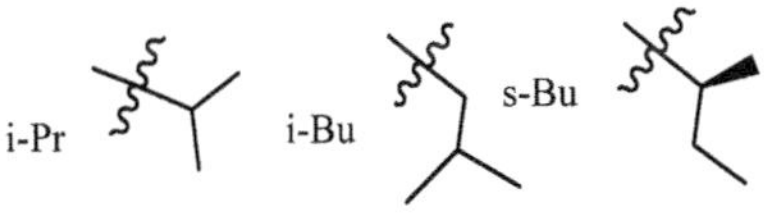

i-Pr   i-Bu   s-Bu

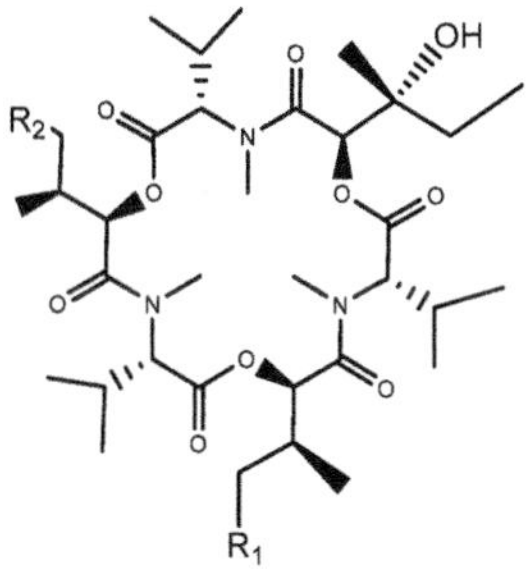

R1=R2 =H Enniatin L (76)

R1=Me, R2 =H Enniatin M1 (77)

R1= H, R2 =Me Enniatin M2 (78)

R1=R2 =Me Enniatin N (79)

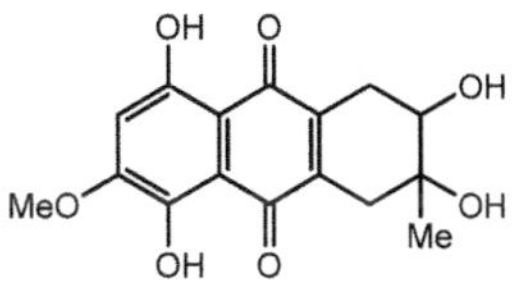

4-deoxybostrycin (80)

Figure 5. Structures of antimycobacterial metabolites isolated from Ascomycetes **(61-80)**.

genes were confirmed by quantitative real-time polymerase chain reaction (qRT-PCR) (Wang et al., 2013).

Hirsutellide A **(81)** (Fig. 6), was isolated from the entomopathogenic fungus *Hirsutella kobayasii* BCC 1660. It was active against *M. tuberculosis* H37Ra in

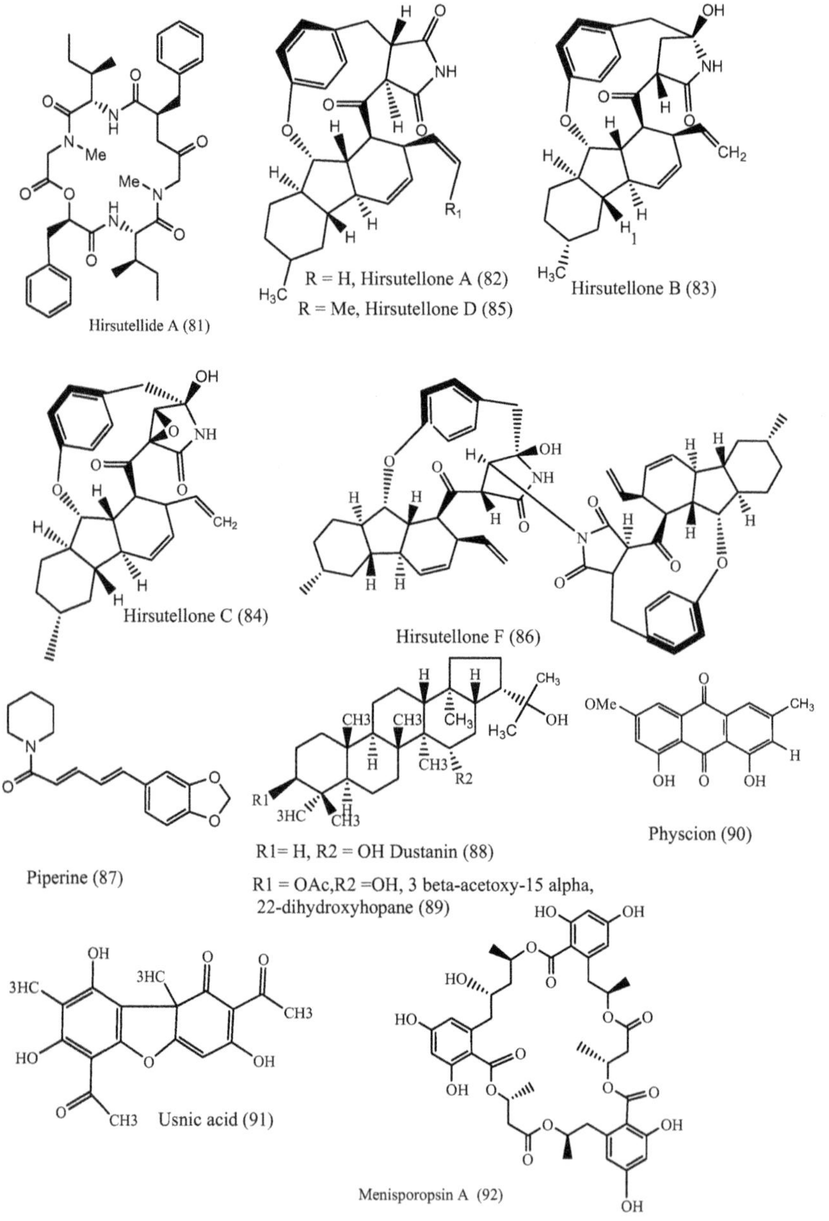

Figure 6. Structures of antimycobacterial metabolites isolated from Ascomycetes **(81-92)**.

Microplate Alamar Blue Assay (MABA) with an MIC with 6–12 µg/mL and no cytotoxicity against Vero cells at 50 µg/mL (Vongvanich et al., 2002).

Hirsutellones A–D **(82–85)** (Fig. 6), of the pathogenic fungus *Hirsutella nivea* BCC 2594 from Thailand, inhibited *M. tuberculosis* H37Ra (MIC 0.78, 3.125, 0.78, 0.78 µg/mL, respectively) (Isaka et al., 2005). Hirsutellone F **(86)** (Fig. 6), a new dimer alkaloid along with the known Hirsutellones A, B, and C, from the spores of the fungus *Trichoderma* sp. BCC 7579 showed a weaker activity against *M. tuberculosis* H37Ra (MIC 3.12 µg/mL) than the Hirsutellones A, B, and C (Isaka et al., 2006).

Piperine **(87)** (Fig. 5), is obtained from an endophytic *Periconia* sp. of *Piper longum*. Piperine has very good anti-mycobacterial activity against *M. tuberculosis* and *M. smegmetis* with MIC of 1.74 and 2.62 µg/ml, respectively (Verma et al., 2011).

Dustanin **(88)** and 3 beta-acetoxy-15 alpha, 22-dihydroxyhopane **(89)** (Fig. 6), were isolated from the insect pathogenic fungus *Aschersonia tubulata* BCC 1785. Compounds **(88)**, and **(89)**, have anti-mycobacterial activities with MICs of 12.5 µg/ml (Boonphong et al., 2001).

Physcion **(90)** (Fig. 6), isolated from an *Aspergillus* sp. inhibited the mycobacterial detoxification enzyme, mycothiol-S-conjugate amidase (MAC) with IC_{50} of 50 µM against *M. smegmatis* (Nicholas et al., 2003). The dibenzofuran derivative, Usnic acid **(91)** (Fig. 6), a secondary metabolite of lichen, inhibits *M. tuberculosis*, MIC 2.5–5 µg/mL (König and Wright, 1999; Ingólfsdóttir, 2002).

A phenolic macrocyclicpolylactone, Menisporopsin A **(92)** (Fig. 6), reported from the seed fungus *Menisporopsis theobromae* has weak activity with MIC of 50 µg/ml against *M. tuberculosis* H37Ra (Chinworrungsee et al., 2004).

8'-O-Demethylnigerone **(93)** and 8'-O-Demethylisonigerone **(94)** (Fig. 7), dimericnaphtho-gamma-pyrones, were isolated from strain WZ-4-11 of *Aspergillus carbonarius*. Compounds **(93)** and **(94)** have weak activities against *M. tuberculosis* H37Rv (MICs of 43.0 and 21.5 µM, respectively) (Zhang et al., 2008).

Cordyol A **(95)** (Fig. 7), was isolated from *Cordyceps* sp. BCC 1861 of *Homoptera cicada* nymph of the KhaoLaem National Park, Kanchanaburi Province, Thailand. Cordyol A has weak anti-mycobacterial activity with MIC 100 µg/mL (Bunyapaiboonsri et al., 2007).

A novel cyclodepsipeptide, Cordycommunin **(96)** (Fig. 7), isolated from the insect pathogenic fungus *Ophiocordyceps communis* BCC 16475 inhibits *M. tuberculosis* H37Ra, MIC 15 µM. This compound has weak cytotoxic effect on KB cell line with an IC_{50} of 45 µM but inactive against BC, NCI-H187 and Vero cell lines at 88 µM (50 µg/mL) (Haritakun et al., 2010).

Ophiobolin K **(97)**, 6-epi-ophiobolin K **(98)** and 6-epi-ophiobolin G **(99)** (Fig. 7), were isolated from the marine-derived fungus *Emericella variecolor*. Ophiobolins **(97–99)** inhibited biofilm formation of *M. smegmatis* at MICs of 4.1–65 mM, whereas these compounds do not show anti-microbial activity at the concentrations that show anti-biofilm formation. Ophiobolin K **(97)** is

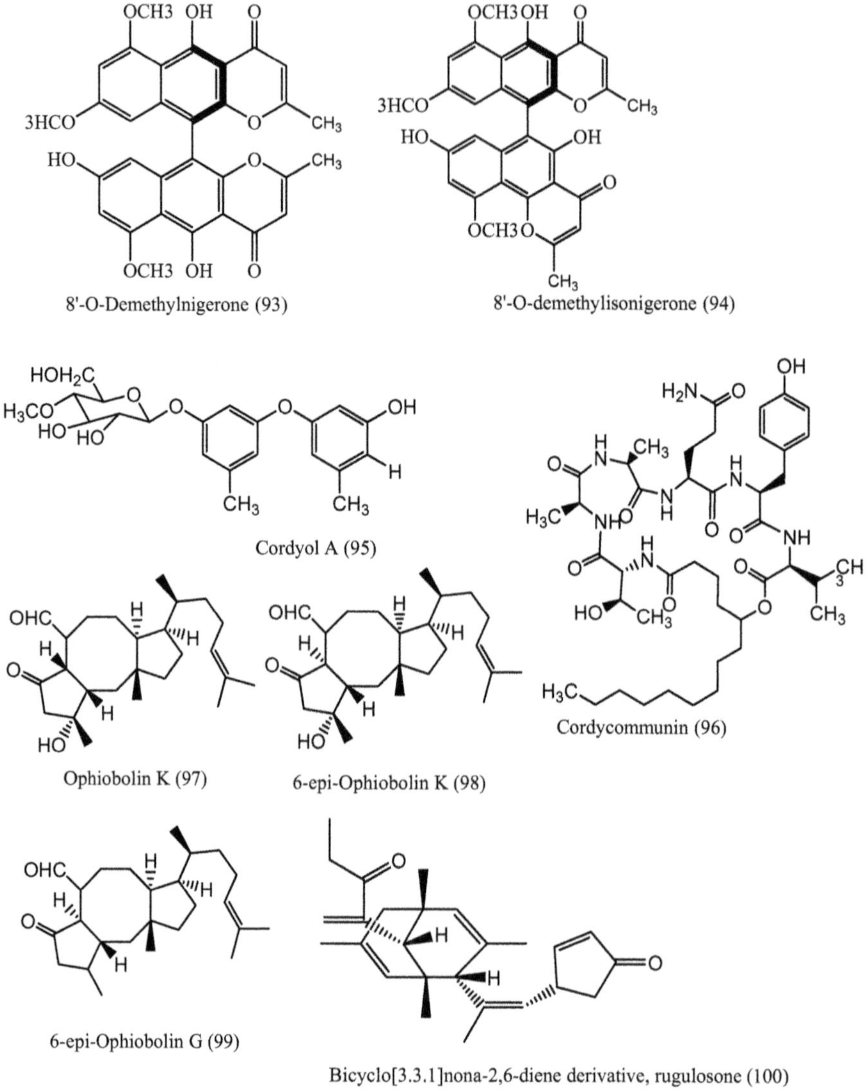

8'-O-Demethylnigerone (93)

8'-O-demethylisonigerone (94)

Cordyol A (95)

Cordycommunin (96)

Ophiobolin K (97)

6-epi-Ophiobolin K (98)

6-epi-Ophiobolin G (99)

Bicyclo[3.3.1]nona-2,6-diene derivative, rugulosone (100)

Figure 7. Structures of antimycobacterial metabolites isolated from Ascomycetes **(93-100)**.

also effective against the biofilm formation of *M. bovis* BCG and is thus able to restore the anti-microbial activity of isoniazid against *M. smegmatis* (Arai et al., 2013).

The Bicyclo[3.3.1]nona-2,6-diene derivative, Rugulosone **(100)** (Fig. 7), was isolated from *Emericella rugulosa*. It showed anti-malarial and anti-mycobacterial activities, as well as cytotoxicity against three cancer cell lines (Moosophon et al., 2009).

Hopan-27-al-6β,11α,22-triol **(101)**, Hopane-6β,11r,22,27-tetraol **(102)**, Hopane-6β,7β,22-triol **(103)**, Compound **(104)** (atropisomer of ES-242-2) and Compound **(105)** (Fig. 8), were isolated from the scale insect pathogenic fungus *Conoideocrella tenuis* BCC 18627. Compounds **(101–105)** are active against *M. tuberculosis* H37Ra, MIC of > 105, 52, > 107, > 75, > 75 µM/ml, respectively. The MIC values of standard anti-TB drug Isoniazid were 0.17–0.34 µM (Isaka et al., 2011).

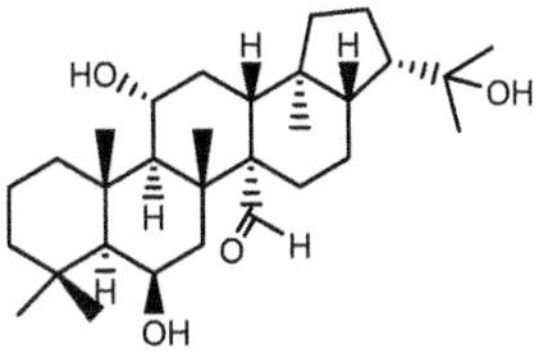

Hopan-27-al-6β,11r,22-triol (101) 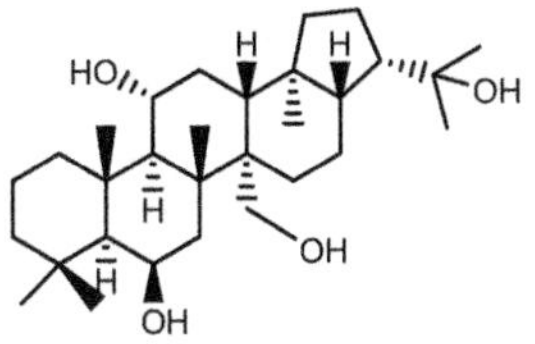   Hopane-6β,11r,22,27-tetraol (102) 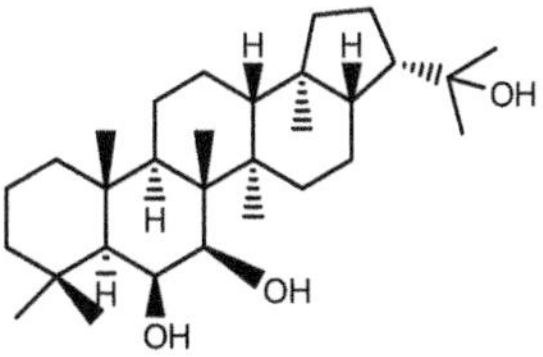   Hopane-6β,7β,22-triol (103)

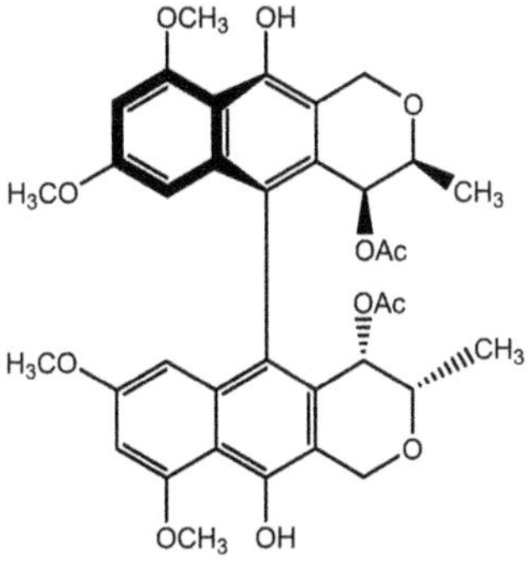

Compound (104) (Atropisomer of ES-242-2)

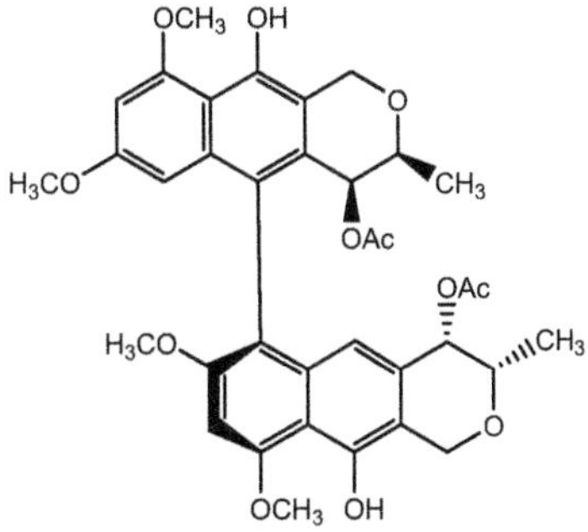

Compound (105)

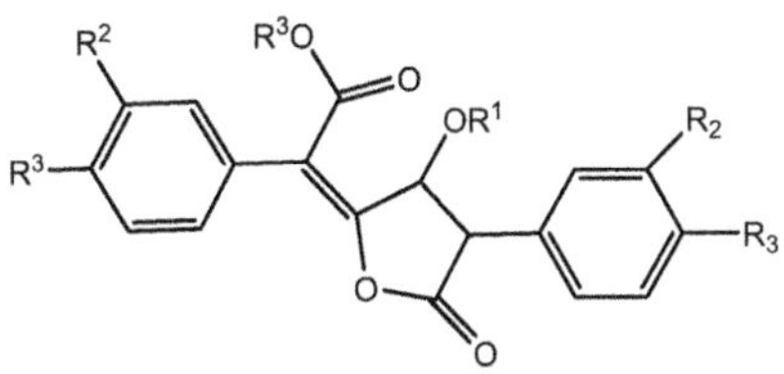

R1=R2= H, R3 =Me
4,4'-dimethoxyvulpinic acid (106)
R1= H, R2=Br, R3 =Me
3,3'-dibromo- 4,4'-dimethoxyvulpinic acid (107)
R1= Ac, R2= H, R3 =Me
Acetyl 4,4'-dimethoxyvulpinate (108)

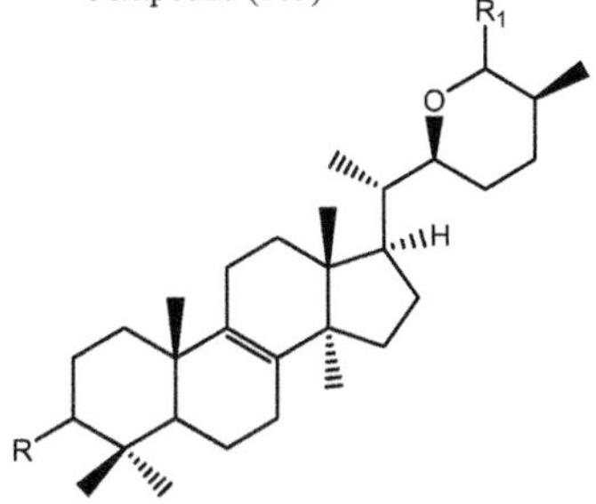

R= α-OH, R1=O 3-epi-astrahygrol (109)
R = R1 =O Astrahygrone (110)
R=R1= α-OH 3-epi-astrapteridiol (111)

Figure 8. Structures of antimycobacterial metabolites isolated from Ascomycetes **(101-105)** and Basidiomycetes **(106-111)**.

From Basidiomycetes

4,4′-dimethoxyvulpinic acid **(106)** (Fig. 8), was isolated from the Thai mushroom *Scleroderma citrinum*. In addition, the dibromo derivative of **(106)** 3,3′-dibromo-4,4′-dimethoxyvulpinic acid **(107)** and the acetate derivative acetyl 4,4′-dimethoxyvulpinate **(108)** were also prepared. All the compounds are active against *M. tuberculosis* H37Ra with MICs 25, 100 and 100 µg/ml, respectively (Kanokmedhakul et al., 2003).

3-Epi-astrahygrol **(109)**, Astrahygrone **(110)** and 3-epi-astrapteridiol **(111)** (Fig. 8), were isolated from, the truffle-mimicking mushroom, *Astraeus pteridis*. Compounds **(111) (109)** and **(110)** showed moderate activity against *M. tuberculosis* with MIC values of 34.0, 58.0, and 64.0 µg/mL, respectively (Stanikunaite et al., 2008).

Lanostanetriterpenes, Astraodoric acids A **(112)** and B **(113)** (Fig. 9), were isolated from, an edible mushroom, *Astraeus odoratus*. Compounds **(112)** and **(113)** exhibited moderate activities against *M. tuberculosis* $H_{37}Ra$ (MICs of 50 and 25 µg/mL) and cytotoxic activities (IC_{50}) values of 34.69 and 18.57 µg/mL against KB cancer cells lines and 19.99 and 48.35 µg/mL against NCI-H187 cancer cells lines, respectively (Arpha et al., 2012).

A new bisphenol-sesquiterpene, Ramiferin **(114)** (Fig. 9), isolated from the fungus *Kionochaeta ramifera* BCC 7585 has anti-tubercular activity, MIC 12.7 µM. It is toxic against three cancer cell lines (BC, KB and NCI-H187) and nonmalignant Vero cells with IC_{50} values of 9.1, 12.6, 13.0, and 9.7 µM, respectively (Bunyapaiboonsri et al., 2008).

Ramariolides A **(115)** (Fig. 9), a Butenolides was isolated from the fruiting bodies of a coral mushroom *Ramaria cystidiophora*. Ramariolide A has an unusual spirooxiranebutenolide moiety and shows *in vitro* activity against *M. smegmatis* and *M. tuberculosis* (Centko et al., 2012).

Gliotoxin **(116)**, and S,S dimethyl gliotoxin **(117)** (Fig. 9), isolated from *Mycena* sp. (F205435) inhibited the mycobacterial detoxification enzyme mycothiol-S-conjugate amidase (MAC) of *M. tuberculosis* with IC_{50} of 50 and 70 µM. Both compounds inhibited MAC of *M. smegmatis* with IC_{50} value of 50 µM each (Nicholas et al., 2003).

Ganoderic acid T **(118)**, and the C-3 epimer of Ganoderic acid T **(119)** (Fig. 9), were isolated from *Ganoderma orbiforme* BCC 22324. Compounds **(118-119)**, are active against *M. tuberculosis* H37Ra with MICs of 10.0 and 1.3 µM, respectively (Isaka et al., 2013).

From Unidentified Fungus

9α-hydroxyhalorosellinia A **(120)** (Fig. 9), was isolated from an endophytic fungus PSU-N24 from *Garcinia nigrolineata*, collected from the Ton Nga Chang wildlife sanctuary, Songkhla province, Southern Thailand. It is active against *M. tuberculosis*, MIC 12.50 µg/ml (Sommart et al., 2008).

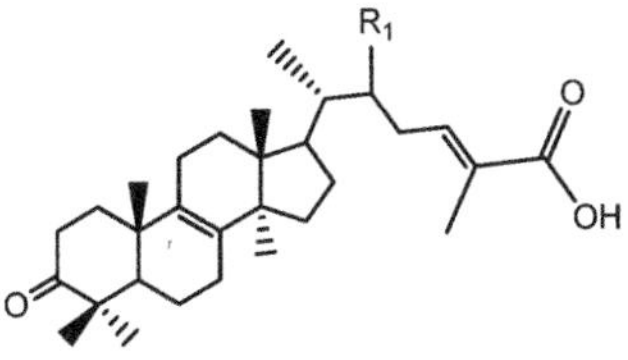

R1=α–OAc Astraodoric acid A (112)

R1=α–OH Astraodoric acid B (113)

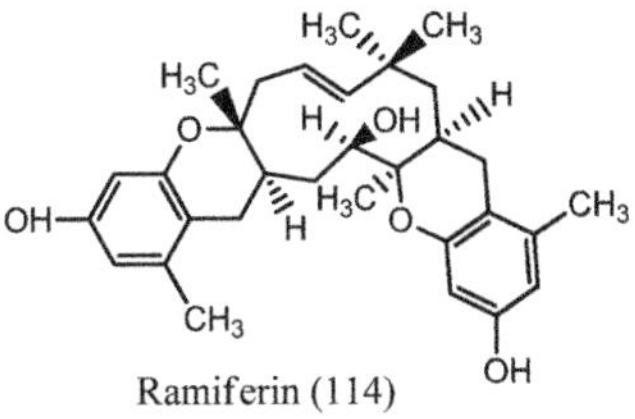

Ramiferin (114)

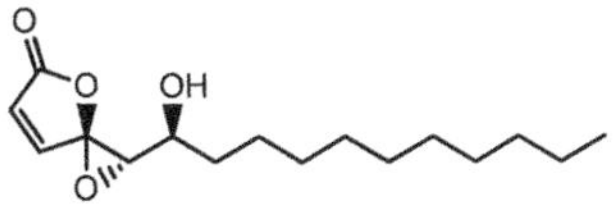

Ramariolide A (115)

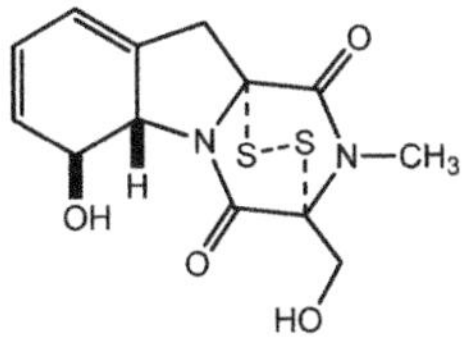

Gliotoxin (116)

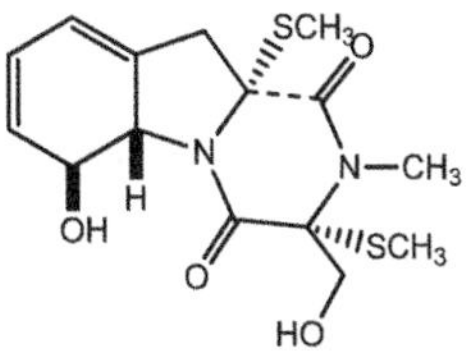

S,S dimethyl gliotoxin (117)

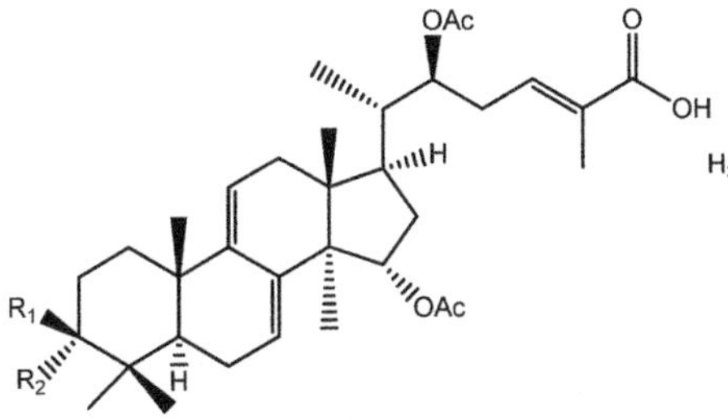

R1=H, R2=OAc, Ganoderic acid T (118)

R1= OAc, R2=H, C-3 epimer of ganoderic acid T (119)

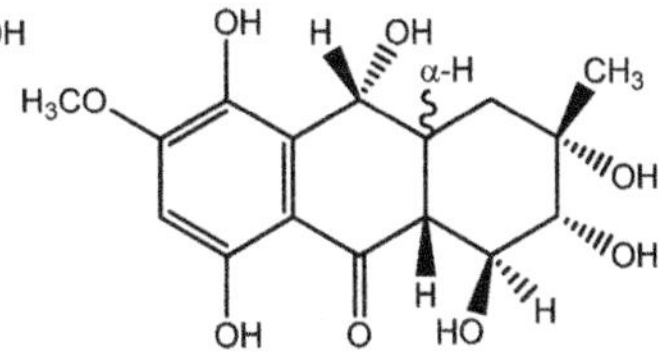

9α -hydroxyhalorosellinia A (120)

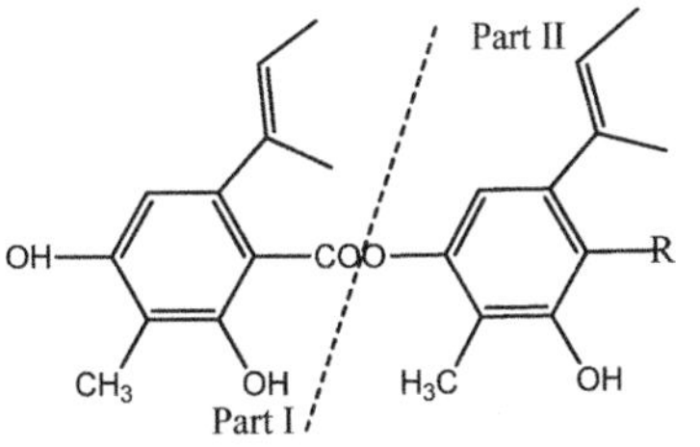

Part II

Part I

R=H, Agonodepside A (121)
R=COOH, Agonodepside B (122)

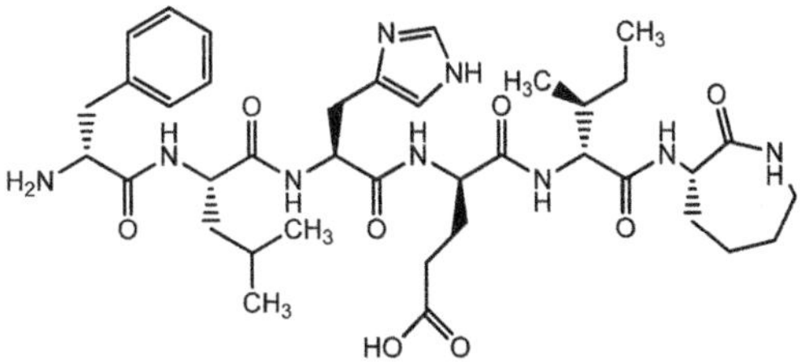

Calpinactam (123)

Figure 9. Structures of antimycobacterials metabolites isolated from Basidiomycetes **(107-119)**, Unidentified fungus **(120-122)** and Zygomycetes **(123)**.

Agonodepside A **(121)** and B **(122)** (Fig. 9), were isolated from a non-sporulating filamentous fungus, F7524. They inhibited the mycobacterial InhA enzyme, a 2-trans-enoyl-acyl-reductase involved in Mycolic acid biosynthesis, which is a major lipid of the mycobacterial envelope.

Agonodepside A had moderate activity, with an IC$_{50}$ of 75 μM, while Agonodepside B is not active at 100 μM (Cao et al., 2002).

From Zygomycetes

Calpinactam **(123)** (Fig. 9) was isolated from *Mortierella alpina* FKI-4905. Calpinactam inhibits *M. smegmatis* and *M. tuberculosis* with MIC values of 0.78 and 12.5 μg/ml, respectively (Koyama et al., 2010).

Volatile Organic Compounds (VOCs) as Antimycobacterials

A stain of *Muscodor* namely, *Muscodor crispans* of *Ananas ananassoides* (wild pineapple) growing in the Bolivian Amazon Basin produces VOCs namely, Propanoic acid, 2-methyl-, 1-butanol, 3-methyl-1-butanol, 3-methyl-, acetate propanoic acid, 2-methyl-, 2-methylbutyl ester, and ethanol. The VOCs of this fungus are effective against *Xanthomonas axonopodis* pv. *citri*, a citrus pathogen and also on several human pathogens, including *Yersinia pestis*, *M. tuberculosis* and *Staphylococcus aureus*. *Muscodor crispans* is only effective against the vegetative cells of *Bacillus anthracis* and not against its spores. Artificial mixtures of the fungal VOCs were both inhibitory and lethal to a number of human and plant pathogens, including three drug-resistant strains of *M. tuberculosis* (Mitchell et al., 2010). The mechanism of action of the VOCs of *Muscodor* spp. on target bacteria is unknown. A microarray study of the transcriptional response analysis of *B. subtilis* cells exposed to *M. albus* VOCs show that the expression of genes involved in DNA repair and replication increased, suggesting that VOCs induce some type of DNA damage in cells, possibly through the effect of one of their naphthalene derivatives (Mitchell et al., 2010).

Outlook /Conclusion/Suggestions

In the poorer countries of the world and particularly those of the Asian Subcontinent, *M. tuberculosis* remains a persistent problem with very few solutions in sight. This is of course due to the extreme poverty of the populations in such third world countries. Typically Nepal, Tibet, North Eastern India (such as Assam), where communications are very weak both due to the inaccessibility of many of the remote area and language problem. The extreme poverty leads to very poor nutrition. Inadequate medical facilities, some time totally missing in many parts of North India, Nepal and Tibet exits even today. Distribution of effective of effective medicine is a huge and difficult task. Affordable medicine? Follow up? There is no light of the end of this tunnel of disease!! Unless there is a 'United Front To Combat Tuberculosis' worldwide This front must be supported by a world with consortium of countries such as UN WHO plus the other advanced countries of the world!!

Will it happen? The mindset of the nations of the world should change from "what can we suggest" to a "what can we do" to solve such great a problem! Do!

Consider the use of complex mixture of Ayurvedic and allopathic compounds, already been used. Variations in regimens of treatment may also be the part such new initiatives.

References

Arai, M., Niikawa, H. and Kobayashi, M. 2013. Marine-derived fungal sesterterpenes, ophiobolins, inhibit biofilm formation of *Mycobacterium* species. J. Nat. Med., 67(2): 271–275.

Arpha, K., Phosri, C., Suwannasai, N., Monkolthanaruk, W. and Sodngam, S. 2012. Astraodoric acids A–D: New lanostantetriterpenes from edible mushroom *Astraeus odoratus* and their anti-*Mycobacterium tuberculosis* H37Ra and cytotoxic activity. J. Agr. Food. Chem., 60: 9834–9841.

Arunrattiyakorn, P., Suksamrarn, S., Suwannasai, N. and Kanzaki, H. 2011. Microbial metabolism of a-mangostin isolated from *Garcinia mangostana* L. Phytochemistry, 72(8): 730–734.

Asres, K., Bucar, F., Edelsbrunner, S., Kartnig, T., Hoger, G. and Thiel, W. 2001. Investigations on antimycobacterial activity of some Ethiopian medicinal plants. Phytother. Res., 15: 323–326.

Boonphong, S., Kittakoop, P., Isaka, M., Palittapongarnpim, P., Jaturapat, A., Danwisetkanjana, K., Tanticharoen, M. and Thebtaranonth, Y. 2001. A new antimycobacterial, 3 beta-acetoxy-15 alpha,22-dihydroxyhopane, from the insect pathogenic fungus *Aschersonia tubulata*. Planta Med., 67(3): 279–281.

Bunyapaiboonsri, T., Yoiprommarat, S., Intereya, K. and Kocharin, K. 2007. New diphenyl ethers from the insect pathogenic fungus *Cordyceps* sp. BCC 1861. Chemical & Pharmaceutical Bulletin, 55(2): 304–307.

Bunyapaiboonsri, T., Veeranondha, S., Boonruangprapa, T. and Somrithipol, S. 2008. Ramiferin, abisphenol-sesquiterpene from the fungus *Kionochaeta ramifera* BCC 7585. Phytochemistry Letters, 1: 204–206.

Cao, S., Lee, A.S., Huang, Y., Flotow, H., Ng, S., Butler, M.S. and Buss, A.D. 2002. Agonodepsides A and B: two new depsides from a filamentous fungus F7524. J. Nat. Prod., 65(7): 1037–1038.

Cheng, M.J., Wu, M.D., Yanai, H., Su, Y.S., Chen, I.S., Yuan, G.F., Hsieh, S.Y. and Chen, J.J. 2012. Secondary metabolites from the endophytic fungus *Biscogniauxia formosana* and their antimycobacterial activity. Phytochemistry Letters, 5(3): 467–472.

Centko, R.M., Ramón-García, S., Taylor, T., Patrick, B.O., Thompson, C.J., Miao, V.P. et al. 2012. Ramariolides A–D, antimycobacterial butenolides isolated from the mushroom *Ramaria cystidiophora*. Journal of Natural Products, 75: 2178–2182.

Chinworrungsee, M., Kittakoop, P., Isaka, M., Maithip, P., Supothina, S. and Thebtaranonth, Y. 2004. Isolation and structure elucidation of a novel antimalarial macrocyclic polylactone, menisporopsin A, from the fungus *Menisporopsis theobromae*. J. Nat. Prod., 67(4): 689–692.

Chomcheon, P., Wiyakrutta, S., Sriubolmas, N., Ngamrojanavanich, N., Isarangkul, D. and Kittakoop, P. 2005. 3-Nitropropionic acid (3-NPA), a potent antimycobacterial agent from endophytic fungi: is 3-NPA in some plants produced by endophytes? J. Nat. Prod., 68(7): 1103–1105.

Chomcheon, P., Sriubolmas, N., Wiyakrutta, S., Ngamrojanavanich, N., Chaichit, N., Mahidol, C., Ruchirawat, S. and Kittakoop, P. 2006. Cyclopentenones, scaffolds for organic syntheses produced by the endophytic fungus mitosporic *Dothideomycete* sp. LRUB20. J. Nat. Prod., 69(9): 1351–1353.

Dettrakul, S., Kittakoop, P., Isaka, M., Nopichai, S., Suyarnsestakorn, C., Tanticharoen, M. and Thebtaranonth, Y. 2003. Antimycobacterial pimarane diterpenes from the fungus *Diaporthe* sp. Bioorg. Med. Chem. Lett., 13(7): 1253–1255.

Elsebai, M.F., Natesan, L., Kehraus, S., Mohamed, I.E., Schnakenburg, G., Sasse, F., Shaaban, S., Gutschow, M. and Konig, G.M. 2011. HLE-inhibitory alkaloids with a polyketide skeleton from the marine-derived fungus *Coniothyrium cereal*. J. Nat. Prod., 74(10): 2282–2285.

Ganihigama, D.U., Sureram, S., Sangher, S., Hongmanee, P., Aree, T., Mahidol, C., Ruchirawat, S. and Kittakoop, P. 2015. Antimycobacterial activity of natural products and synthetic agents: Pyrrolodiquinolines and vermelhotin as anti-tubercular leads against clinical multidrug resistant isolates of *Mycobacterium tuberculosis*. Eur. J. Med. Chem., 89: 1–12.

Haritakun, R., Sappan, M., Suvannakad, R., Tasanathai, K. and Isaka, M. 2010. An antimycobacterial cyclodepsipeptide from the entomopathogenic fungus *Ophiocordyceps communis* BCC 16475. J. Nat. Prod., 73(1): 75–78.

Ingólfsdóttir, K. 2002. Usnic acid. Phytochemistry, 61(7): 729–736.

Isaka, M., Jaturapat, A., Rukseree, K., Danwisetkanjana, K., Tanticharoen, M. and Thebtaranonth, Y. 2001. Phomoxanthones A and B, novel xanthone dimers from the endophytic fungus *Phomopsis* species. J. Nat. Prod., 64(8): 1015–1018.

Isaka, M., Rugseree, N., Maithip, P., Kongsaeree, P., Prabpai, S. and Thebtaranonth, Y. 2005. Hirsutellones A–E, antimycobacterial alkaloids from the insect pathogenic fungus *Hirsutellanivea* BCC 2594. Tetrahedron, 61: 5577–5583.

Isaka, M., Prathumpai, W., Wongsa, P. and Tanticharoen, M. 2006. Hirsutellone F, a dimer of antitubercular alkaloids from the seed fungus *Trichoderma* species BCC 7579. Org. Lett., 8(13): 2815–2817.

Isaka, M., Palasarn, S., Supothina, S., Komwijit, S. and Luangsaard, J.J. 2011. Bioactive compounds from the scale insect pathogenic fungus *Conoideocrella tenuis* BCC 18627. J. Nat. Prod., 74: 782–789.

Isaka, M., Chinthanom, P., Kongthong, S., Srichomthong, K. and Choeyklin, R. 2013. Lanostane triterpenes from cultures of the Basidiomycete *Ganoderma orbiforme* BCC 22324. Phytochemistry, 87: 133–139.

Kanokmedhakul, S., Kanokmedhakul, K., Phonkerd, N., Soytong, K., Kongsaeree, P. and Suksamrarn, A. 2002. Antimycobacterial anthraquinone-chromanone compound and diketopiperazine alkaloid from the fungus *Chaetomium globosum* KMITL-N0802. Planta Med., 68(9): 834–836.

Kanokmedhakul, S., Kanokmedhakul, K., Prajuabsuk, T., Soytong, K., Kongsaeree, P. and Suksamrarn, A. 2003. A bioactive triterpenoid and vulpinic acid derivatives from the mushroom *Scleroderma citrinum*. Planta Med., 69(6): 568–571.

Kasettrathat, C., Ngamrojanavanich, N., Wiyakrutta, S., Mahidol, C., Ruchirawat, S. and Kittakoop, P. 2008. Cytotoxic and antiplasmodial substances from marine-derived fungi, *Nodulisporium* sp. and CRI247-01. Phytochemistry, 69(14): 2621–2626.

Khumkomkhet, P., Kanokmedhakul, S., Kanokmedhakul, K., Hahnvajanawong, C. and Soy-tong, K. 2009. Antimalarial and cytotoxic depsidones from the fungus *Chaetomium brasiliense*. J. Nat. Prod., 72: 1487–1491.

Kongsaeree, P., Prabpai, S., Sriubolmas, N., Vongvein, C. and Wiyakrutta, S. 2003. Antimalarial dihydroisocoumarins produced by *Geotrichum* sp., an endophytic fungus of *Crassocephalum crepidioides*. J. Nat. Prod., 66: 709.

König, G.M. and Wright, A.D. 1999. ^{1}H and ^{13}C NMR and biological activity investigations of four LIchen derived compounds. Phytochem. Anal., 10: 279–284.

Kornsakulkarn, J., Dolsophon, K., Boonyuen, N., Boonruangprapa, T., Rachtawee, P., Prabpai, S., Kongsaeree, P. and Thongpanchang, C. 2011. Dihydronaphthalenones from endophytic fungus *Fusarium* sp. BCC14842. Tetrahedron, 67(39): 7540–7547.

Koyama, N., Kojima, S., Nonaka, K., Masuma, R., Matsumoto, M., Omura, S. et al. 2010. Calpinactam, a new anti-mycobacterial agent, produced by *Mortierella alpina* FKI-4905. Journal of Antibiotics, 63: 183–186.

Li, H., Jiang, J., Liu, Z., Lin, S., Xia, G., Xia, X., Ding, B., He, L., Lu, Y. and She, Z. 2014. Peniphenones A–D from the mangrove fungus *Penicillium dipodomyicola* HN4-3A as inhibitors of *Mycobacterium tuberculosis* phosphatase MptpB. J. Nat. Prod., 77(4): 800–806.

Mitchell, A.M., Strobel, G.A., Moore, E., Robison, R. and Sears, J. 2010. Volatile antimicrobials from *Muscodor crispans*, a novel endophytic fungus. Microbiology, 156(1): 270–277.

Moosophon, P., Kanokmedhakul, S., Kanokmedhakul, K. and Soytong, K. 2009. Prenylxanthones and a bicyclo[3.3.1]nona-2,6-diene derivative from the fungus *Emericella rugulosa*. J. Nat. Prod., 72(8): 1442–1446.

Munoz-Elias, E.J., Munoz Elias, E.J. and McKinney, J.D. 2005. *Mycobacterium tuberculosis* Isocitratelyases 1 and 2 are jointly required for *in vivo* growth and virulence. Nat. Med., 11(6): 638–644.

Nicholas, G.M., Eckman, L.L., Newton, G.L., Fahey, R.C., Ray, S. and Bewley, C.A. 2003. Inhibition and kinetics of *Mycobacterium tuberculosis* and *Mycobacterium smegmatis* Mycothiol-S-conjugate amidase by natural product inhibitors. Bioorg. Med. Chem., 11(4): 601–608.

Nilanonta, C., Isaka, M., Chanphen, R., Thong-orn, N., Tanticharoen, M. and Thebtaranonth, Y. 2003. Unusual enniatins produced by the insect pathogenic fungus *Verticillium hemipterigenum*: Isolation and studies on precursor-directed biosynthesis. Tetrahedron, 59: 1015–1020.

Pan, J.H., Chen, Y., Huang, Y.H., Tao, Y.W., Wang, J., Li, Y., Peng, Y., Dong, T., Lai, X.M. and Lin, Y.C. 2011. Antimycobacterial activity of fusaric acid from a mangrove endophyte and its metal complexes. Arch. Pharm. Res., 34(7): 1177–1181.

Phonkerd, N., Kanokmedhakul, S., Kanokmedhakul, K., Soytong, K., Prabpai, S. and Kongsaeree, P. 2008. Bis-spiro-azaphilones and azaphilones from the fungi *Chaetomium cochliodes* VTh01 and *C. cochliodes* CTh05. Tetrahedron, 64: 9636–9645.

Pittayakhajonwut, P., Sohsomboon, P., Dramae, A., Suvannakad, R., Lapanun, S. and Tantichareon, M. 2008. Antimycobacterial substances from *Phaeosphaeria* sp. BCC 8292. Planta Med., 74: 281–286.

Pruksakorn, P., Arai, M., Kotoku, N., Vilchèze, C., Baughn, A.D., Moodley, P., Jacobs, W.R., Jr. and Kobayashi, M. 2010. Trichoderins, novel aminolipopeptides from a marine sponge-derived *Trichoderma* sp., are active against dormant mycobacteria. Bioorg. Med. Chem. Lett., 20(12): 3658–3663.

Rukachaisirikul, V., Sommart, U., Phongpaichit, S., Sakayaroj, J. and Kirtikara, K. 2008. Metabolites from the endophytic fungus *Phomopsis* sp. PSU-D15. Phytochemistry, 69(3): 783–787.

Salomon, C.E. and Schmidt, L.E. 2012. Natural products as leads for tuberculosis drug development. Current Topics in Medicinal Chemistry, 12: 735–765.

Seephonkai, P., Isaka, M., Kittakoop, P., Palittapongarnpim, P., Kamchonwongpaisan, S., Tanticharoen, M. and Thebtaranonth, Y. 2002. Evaluation of antimycobacterial, antiplasmodial and cytotoxic activities of preussomerins isolated from the lichenicolous fungus *Microsphaeropsis* sp. BCC 3050. Planta Med., 68(1): 45–48.

Sommart, U., Rukachaisirikul, V., Sukpondma, Y., Phongpaichit, S., Sakayaroj, J. and Kirtikara, K. 2008. Hydronaphthalenones and a dihydroramulosin from the endophytic fungus PSU-N24. Chem. Pharm. Bull. (Tokyo), 56(12): 1687–1690.

Stanikunaite, R., Radwan, M.M., Trappe, J.M., Fronczek, F. and Ross, S.A. 2008. Lanostane-type triterpenes from the mushroom *Astraeus pteridis* with antituberculosis activity. Journal of Natural Products, 71: 2077–2079.

Suwanborirux, K., Charupant, K., Amnuoypol, S., Pummangura, S., Kubo, A. and Saito, N. 2002. Ecteinascidins 770 and 786 from the Thai tunicate *Ecteinascidia thurstoni*. J. Nat. Prod., 65(6): 935–937.

Trisuwan, K., Khamthong, N., Rukachaisirikul, V., Phongpaichit, S., Preedanon, S. and Sakayaroj, J. 2010. Anthraquinone, cyclopentanone, and naphthoquinone derivatives from the seafan-derived fungi *Fusarium* spp. PSU-F14 and PSU-F135. J. Nat. Prod., 73(9): 1507–1511.

Verma, V.C., Lobkovsky, E., Gange, A.C., Singh, S.K. and Prakash, S. 2011. Piperine production by endophytic fungus Periconia sp. isolated from *Piper longum* L. J. Antibiot. (Tokyo), 64(6): 427–31.

Vongvanich, N., Kittakoop, P., Isaka, M., Trakulnaleamsai, S., Vimuttipong, S., Tanticharoen, M. and Thebtaranonth, Y. 2002. Hirsutellide A, a new antimycobacterial cyclohexadepsipeptide from the entomopathogenic fungus *Hirsutella kobayasii*. J. Nat. Prod., 65(9): 1346–1348.

Vongvilai, P., Isaka, M., Kittakoop, P., Srikitikulchai, P., Kongsaeree, P., Prabpai, S. and Thebtaranonth, Y. 2004. Isolation and structure elucidation of enniatins L, M1, M2, and N: novel hydroxy analogs. Helv. Chim. Acta, 87: 2066–2073.

Wang, C., Wang, J., Huang, Y., Chen, H., Li, Y., Zhong, L., Chen, Y., Chen, S., Wang, J., Kang, J., Peng, Y., Yang, B., Lin, Y., She, Z. and Lai, X. 2013. Anti-mycobacterial activity of marine fungus-derived 4-deoxybostrycin and nigrosporin. Molecules, 18: 1728–1740.

Wijeratne, E.M., He, H., Franzblau, S.G., Hoffman, A.M. and Gunatilaka, A.A. 2013. Phomapyrrolidones A–C, antitubercular alkaloids from the endophytic fungus *Phoma* sp. NRRL 46751. J. Nat. Prod., 76(10): 1860–5.

Wiyakrutta, S., Sriubolmas, N., Panphut, W., Thongon, N., Danwisetkanjana, K., Ruangrungsi, N. and Meevootisom, V. 2004. Endophytic fungi with anti-microbial, anti-cancer and antimalarial activities isolated from Thai medicinal plants. World J. Microbiol. Biotechnol., 20: 256–272.

Zhang, Y., Ling, S., Fang, Y., Zhu, T., Gu, Q. and Zhu, W.-M. 2008. Isolation, structure elucidation, and antimycobacterial properties of dimericnaphtho-γ-pyrones from the marine-derived fungus *Aspergillus carbonarius*. Chem. Biodivers., 5: 93–100.

CHAPTER 4

Antiphytopathogenic Metabolites Derived from Endophytic Fungi

Kanika Chowdhary[1] and *Nutan Kaushik*[2,]*

ABSTRACT

This review analysis the antiphytopathogenic secondary metabolites isolated from endophytic fungi. Among many others attributes of endophytic fungi, they are also a promising source of novel compounds exhibiting antagonistic activity against phytopathogens. Endophytes represent a dependable source of specific secondary metabolites and can be manipulated both physiochemically and genetically to increase yields of defined metabolites and to produce novel analogues of bioactive metabolites.

Introduction

Besides the several accomplishments of modern agriculture, certain cultural practices (such as use of genetically similar crop plants, plant cultivars susceptible to pathogens and the use of nitrogenous fertilizers in higher concentrations that enhance disease susceptibility and so on) have actually enhanced the destructive potential of diseases. Moreso, greater use of agrochemicals causes major problems, like soil and water pollution and

[1] TERI University, 10th Institutional Area, Vasant Kunj, New Delhi-110070, India.
[2] The Energy and Resources Institute (TERI), India Habitat Center, Lodhi Road, New Delhi 110003, India.
* Corresponding author: kaushikn@teri.res.in

negatively impact the biodiversity because of their ability to also affect non-target species (Janusauskaite et al., 2012). Recent epidemiological studies pertain to the effects of pyrethroids on male fertility and prenatal development. The main metabolites of pyrethroids have frequently been detected in urine samples from the general population, confirming widespread exposure of children and adults to one or more pyrethroids (Saillenfait et al., 2015).

Pathogenic fungi inflict severe damages during both developmental and post-harvest stages causing decline in aesthetic characteristics, nutritive values and shelf life of the end product (Agrios, 2004). Phytopathogenic fungi are largely controlled by synthetic fungicides; however, usage has been restricted due to heightened awareness amongst consumers regarding their ill-effects on plant, human health and environment (Harris et al., 2001). Therefore, there is a dire need to search a novel source of biologically and environmentally safe antiphytopathogenic metabolites.

Endophytic fungi, recently known reservoir of structurally and biologically novel secondary metabolites, are capable of fulfilling above mentioned requirements of effective, efficient and environmentally safe antiphytopathogenic products as deduced from a vast body of literature.

Fungal endophytes are categorized as highly diverse, polyphyletic group of primarily ascomycetous fungi, capable of colonizing tissues of plants asymptomatically (Aly et al., 2011). The term endophyte came into existence in 1866 when De Bary introduced it to the world as any organisms occurring within plant tissues (De Bary, 1866) and from last two decades it has been making critical waves in scientific community. Endophytes have been isolated from all plants studied to date. 420,000 plant species exist in nature and only a few have been completely studied relative to their endophytic biology (Vourela, 2004).

Being able to reside asymptomatically inside plant tissues both intracellularly and intercellularly, fungal endophytes are constantly in a state of "metabolic aggressiveness", thereby synthesizing inimitable array of metabolites (Tejesvi et al., 2007). These metabolites have exhibited a plethora of biological activities such as antimicrobial, antineoplastic, immunosuppressive and cytotoxic and so on as indicated in the literature. Recent literature cited that 51% of bioactive substances isolated from endophytic fungi were previously unknown as compared to 38% from soil fungi (Strobel, 2003; Kharwar et al., 2011).

Antiphytopathogenic Metabolites Screened from Endophytic Fungi

Endophytes are believed to carry out a resistance mechanism to overcome pathogenic invasion by producing secondary metabolites. Numerous antifungal metabolites have been reported from endophytic fungi belonging to different structural classes such as alkaloids, peptides, steroids, terpenoids, phenols, quinines and flavonoids. Extensive section of literature on endophytic fungi has reported antifungal metabolites exhibiting antiphytopathogenic activity (Table 1). Such antiphytopathogenic metabolites have enormous potential to be developed into an agrochemical product.

For instance, *Pestalotiopsis* namely, *P. jesteri*, an endophytic fungal species isolated from the inner bark of small limbs of a *Fragraea bodenii* located in Papua New Guinea, produces the highly functionalized cyclohexenone epoxides, Jesterone **(1)** and Hydroxyjesterone **(2)**. Jesterone was found active against *Pythium ultimum, Aphanomyces* sp., *Phytophthora citrophthora, Phytophthora cinnamomi, Sclerotinia sclerotiorum, Rhizoctonia solani, Geotrichum candidum* and *Pyricularia oryzae* with Minimum Inhibitory Concentration (MIC) of 25, 6.5, 25, 6.5, 100, 25, > 100 and 25 mcg/ml. Hydroxyjesterone was also found active against *Aphanomyces* sp. and *Phytophthora cinnamomi* with MIC of 125 and 62.5 µg/ml (Li and Strobel, 2001).

Excelsional **(3)**, 9-hydroxyphomopsidin **(4)**, Excelsione **(5)**, Phomopsidin **(6)**, alternariol **(7)**, alternariol-5-O-methyl ether **(8)**, the hitherto undescribed 5′-hydroxyalternariol **(9)**, altenusin **(10)** were isolated from endophyte *Phomopsis* sp. of *Endodesmia calophylloides*. All the metabolites displayed motility inhibition and lytic activities against zoospores of *Plasmospora viticola* in dose and time dependent manner from 1 to 10 µg/ml (Talontsi et al., 2012).

12β-hydroxy-13α-methoxyverruculogen TR-2 **(11)**, fumitremorgin B **(12)**, verruculogen **(13)**, and helvolic acid **(14)**, were isolated from endophytic fungus *Aspergillus fumigatus* of *Melia azedarach*. All the isolated compounds exhibited antifungal activity against *Botrytis cinerea, Alternaria solani, Alternaria alternata, Colletotrichum gloeosporioides, Fusarium solani, Fusarium oxysporum* f. sp. *niveum, Fusarium oxysporum* f. sp. *vasinfectum*, and *Gibberella saubinettii* with MIC values of 6.25–50 µg/ml as compared to positive controls—Carbendazim and Hymexazol (Li et al., 2012). Helvolic acid **(14)** was also reported from *Pichia guilliermondii* having strong inhibitory activity on spore germination of *Magnaporthe oryzae* with an IC_{50} value of 7.20 µg/ml (Zhao et al., 2010).

Phomoxanthone A **(15)** dimeric xanthone was isolated from an endophytic fungus *Phomopsis* sp., from the stem of *Costus* sp. (Costaceae) growing in the rain forest of Costa Rica. It showed moderate inhibition of *Ustilago violacea* at a concentration of 10 mg/ml (Elsaesser et al., 2005).

Two new metabolites, Ethyl 2, 4-dihydroxy-5,6-dimethylbenzoate **(16)** and Phomopsilactone **(17)** were isolated from *Phomopsis cassiae*, an endophytic fungus in *Cassia spectabilis*. Both the compounds displayed strong antifungal activity against the phytopathogenic fungi *Cladosporium cladosporioides* and *C. sphaerospermum* and the detection limit for both the compounds was 1 µg the same as for the positive control Nystatin (Silva et al., 2005).

Phomodione **(18)** was isolated from a *Phoma* sp., an endophyte on a Guinea plant (*Saurauia scaberrinae*). Phomodione exhibited antifungal activity against *Pythium ultimum, Sclerotinia sclerotiorum* and *Rhizoctonia solani*, with MIC between 3 and 8 µg/ml (Hoffmann et al., 2008).

Table 1. Novel antiphytopathogenic metabolites screened and characterised from endophytic fungi.

S. No.	Endophytic fungi	Host plant	Metabolites identified	Antiphytopathogenicity against phytopathogens	References
1	Pestalotiopsis jesteri	Fragraea bodenii	Jesterone (1), Hydroxyjesterone (2)	Pythium ultimum, Aphanomyces sp., Phytophthora citrophthora, Phytophthora cinnamomi, Sclerotinia sclerotiorum, Rhizoctonia solani, Geotrichum candidum and Pyricularia oryzae	Li et al., 2001
2	Phompopsis sp.	Endodesmia calophylloides	Excelsional (3) 9-hydroxyphomopsidin (4), Excelsione (5), Phomopsidin (6), alternariol (7), alternariol-5-O-methyl ether (8), 5'-hydroxyalternariol (9), altenusin (10)	Zoospores of Plasmopara viticola	Talontsi et al., 2012
3	Aspergillus fumigatus	Melia azedarach	12β-hydroxy-13α-methoxyverruculogen TR-2 (11), fumitremorgin B (12), verruculogen (13), and helvolic acid (14)	Botrytis cinerea, Fusarium oxysporum, Colletotrichum gloeosporioides, Alternaria solani, Alternaria alternata, Fusarium solani, Giberrella saubinetti	Li et al., 2012
4	Pichia guilliermondii	Paris polyphylla	Helvolic acid (14)	Magnaporthe oryzae	Zhao et al., 2010
5	Phomopsis sp.	Costus sp.	Phomoxanthone A (15)	Ustilago violacea	Elsaesser et al., 2005
6	Phomopsis cassiae	Cassia spectabilis	Ethyl 2, 4-dihydroxy-5,6-dimethylbenzoate (16), Phomopsilactone (17)	Cladosporium cladosporioides and C. sphaerospermum	Silva et al., 2005
7	Phoma sp.	Saurauia scaberrinae	Phomodione (18)	Pythium ultimum, Sclerotinia sclerotiorum and Rhizoctonia solani	Hoffmann et al., 2008
8	Phomopsis sp.	Laurus azorica	Cycloepoxylactone (19)	Microbotryum violaceum	Hussain et al., 2009a

9	*Phomopsis* sp.	*Gossypium hirustum*	Epoxycytochalasin H (20), Cytochalasin N (21), and cytochalasin H (22)	*Botrytis cinerea, Sclerotinia sclerotiorum, Rhizoctonia cerealis, Fusarium oxysporum*	Fu et al., 2011
10	*Pestalotiopsis adjusta*	Unidentified tree (China)	Pestalachlorides A–C (23–25)	*Fusarium culmorum, Gibberella zeae* and *Verticillium alboatrum*	Li et al., 2008
11	*Verticillium* sp.	*Rehmannia glutinosa*	Ergosterol peroxide (26), 2,6-dihydroxy-2-methyl-7-(prop-1E-enyl)-1-benzofuran3(2H)-one (27)	*Verticillium* sp., *Rhizoctonia* sp., *Fusarium* sp.	You et al., 2009
12	*Trichothecium roseum*	*Maytenus hookeri*	Trichothecene (28)	*Phytophthora infestans, Alternaria solani, phyriculatia oryzae*	Zhang, 2010
13	*Coniothyrium* sp.	*Sideritis chamaedryfolia*	1-hydroxy-5-methoxy-2-nitronapthalene (29), 1,5-dimethoxy-4-nitronapthalene (30), 1-5-dimethoxy-2,4-dinitronapthalene (31) and 1,5-dimethoxy-4,8-dinitronapthalene (32)	*Microbotrytum violaceum*	Krohn et al., 2008
14	*Colletotrichum gloeosporioides*	*Artemisia mongolica*	Colletotric acid (33)	*Helminthosporium sativum*	Zou et al., 2000
15	*Microdochium bolleyi*	*Fagonia cretica*	(12R)-12- hydroxymonocerin (34), (12S)-12-hydroxymonocerin (35) (3R, 4R, 10R)-4 (2-4) (36) and Monocerin (37)	*Microbotrytum violaceum*	Zhang et al., 2008
16	*Cryptosporiopsis quercina*	*Tripterigeum wilfordii*	Cryptocin (38)	*Pyricularia oryzae*	Li et al., 2000
17	*Penicillium* sp.	*Alibertia macrophylla*	Orcinol (39) and 4-hydroxymellein (40), 8-methoxymellein (41)	*Cladosporium cladosporioides, C. sphaerospermum*	Oliveira et al., 2009
18	Ascomycete unidentified	*Meliotus denatus*	5-Methoxy-7-hydroxyphthalide (42) and (3R, 4R)-cis-4-hydroxymellein (43)	*Microbotrytum violaceum*	Hussain et al., 2009b

Table 1. contd....

Table 1. contd....

S. No.	Endophytic fungi	Host plant	Metabolites identified	Antiphytopathogenicity against phytopathogens	References
19	Nigrospora sp. YB-141	Azadirachta indica	Solanapyrone C, N and O (**44–46**) and Phomalactone (**47**)	Aspergillus niger, Botrytis cinerea, Fusarium avenaceum, Fusarium moniliforme, Helminthosporium maydis	Wu et al., 2008a
20	Phomopsis sp. YM 311483	Azadirachta indica	8α-Acetoxy-5α-hydroxy-7oxodecan-10-olide (**48**), 7α,8α-Dihydoxy-3,50decadien-10-olide (**49**), 7α-Acetoxymultiplolide A (**50**), 8α-Acetoxymultiplolide A (**51**) and Multiplolide (**52**)	Aspergillus niger, Botrytis cinerea, Fusarium avenaceum, F. moniliforme, Helminthosporium maydis, Penicillium islandicum and Ophiostoma minus	Wu et al., 2008b
21	Botryosphaeriarhodina	Bidenspilosa	Botryorhodines A, D (**53, 54**)	Fusarium oxysporum	Abdou et al., 2010
22	Phoma sp.	Cinnamomum mollissimum	5-hydroxyramulosin (**55**)	Aspergillus niger	Santiago et al., 2012
23	Cordyceps dipterigena	--	Cordycepsidone A (**56**) and cordycepsidone B (**57**)	Gibberella fujikuroi	Varughese et al., 2012
24	Epicoccum sp.	Theobroma cacao	Epicolactone (**58**), epicoccolides A (**59**) and B (**60**)	Pythium ultimum, Aphanomyces cochlioides, Rhizoctonia solani	Talontsi et al., 2013
25	Cladosporium cladosporioides	--	Cladosporin (**61**), Isocladosporin (**62**)	Colletotrichum acutatum, Colletotrichum fragariae, Colletotrichum gloeosporioides, and Plasmospora viticola	Wang et al., 2013
26	Physalospora sp.	--	Cytochalasins E (**63**) and K (**64**)	E. repens and M. microspora	Hussain et al., 2014a
27	Trichoderma brevicompactum	Allium sativum	4β-acetoxy-12, 13-epoxy-Δ^9-trichothecene (**65**) (Trichodermin)	Rhizoctonia solani	Shentu et al., 2014

28	*Bipolaris* sp.	Bipolamide B (66)	*Cladosporium cladosporioides* FERMS-9, *Cladosporium cucumerinum* NBRC 6370, *Saccharomyces cerevisiae* ATCC 9804, *Aspergillus niger* ATCC 6275 and *Rhizopus oryzae*	Siriwach et al., 2014
29	*Botryosphaeria dothidea* KJ-1	Stemphyperylenol (67)	*Alternaria solani*	Xiao et al., 2014a
30	*Aspergillus* sp.	Asperpyrone A (68), (R)-3-hydroxybutanonitrile (69)	*Gibberella saubinetti, Magnaporthe grisea, Botrytis cinerea, Colletotrichum gloeosporioides* and *Alternaria solani*	Xiao et al., 2014b
31	Unidentified Ascomycete	Cis-4-Acetoxyoxymellein (70) and 8-deoxy-6-hydroxy-cis-4-acetoxyoxymellein (71)	*Microbotryum violaceum* and *Botrytis cinerea*	Hussain et al., 2014b

Note: column 2 species are *Gynura hispida* (28), *Melia azedarach* (29), *Melia azedarach* (30), *Meliotus dentatus* (31).

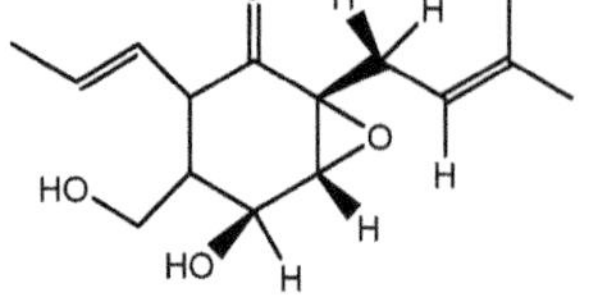

Jesterone (1)

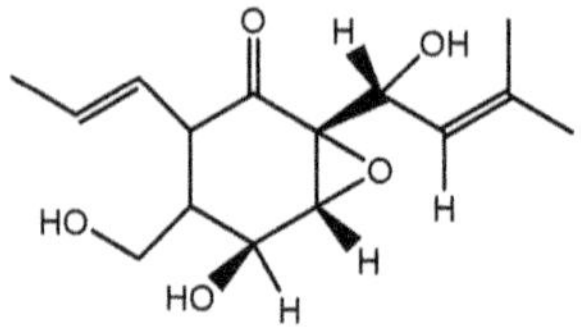

Hydroxyjesterone (2)

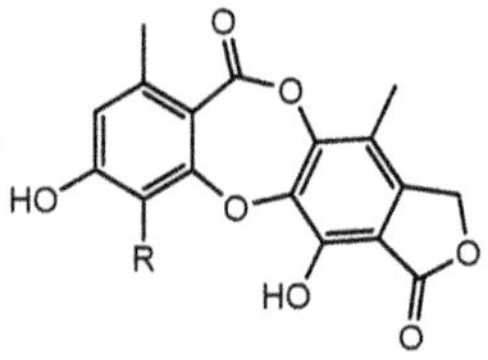

R=CHO, Excelsional (3)
R= CH₂OH Excelsione (5)

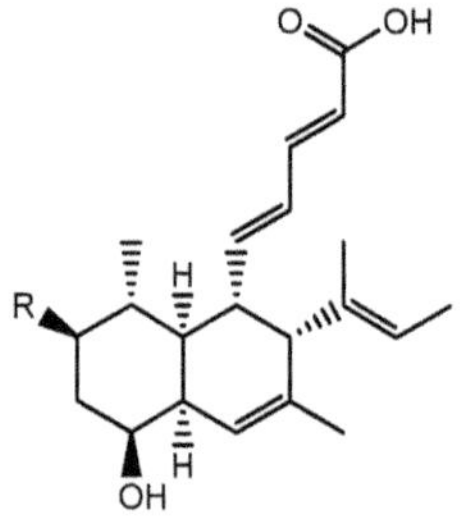

R= OH 9-Hydroxyphomopsidin (4)
R=H, Phomopsidin (6)

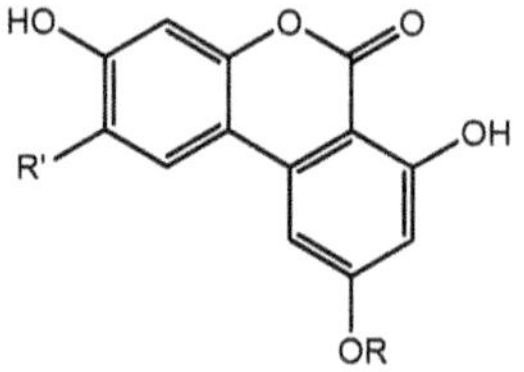

3a R=R'=H Alternariol (7)
3b R=CH3 R'=H Alternariol-5-O-methyl ether (8)
3c R = H, R'=OH 5'-hydroxyalternariol (9)

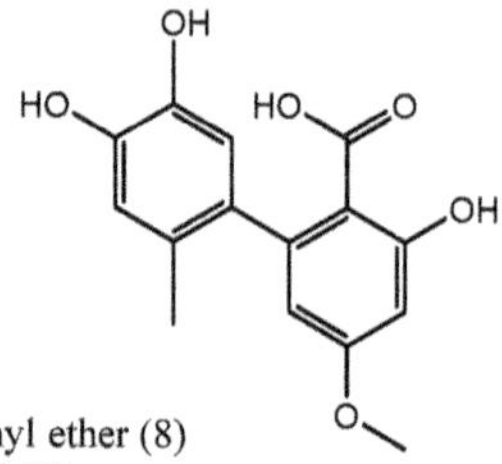

Altenusin (10)

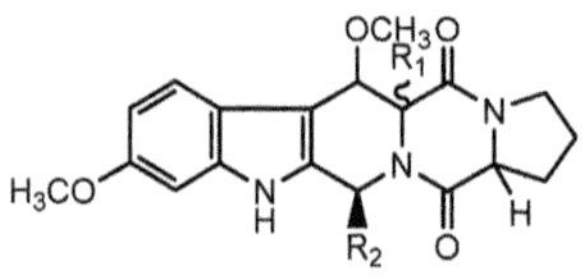

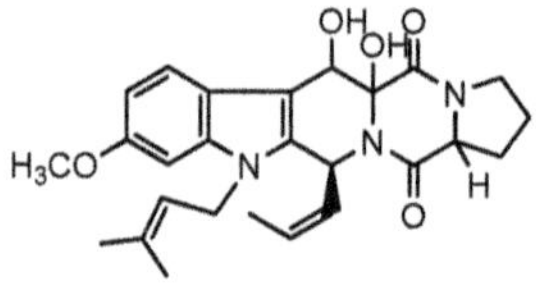

Fumitremorgin B (12)

Verruculogen (13)

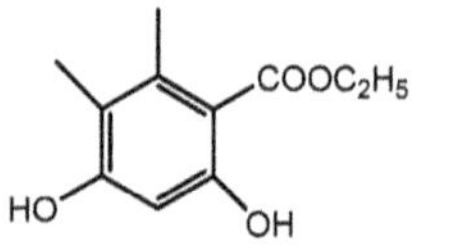

R1 =β OH R2 =   OH

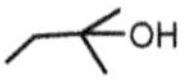

12β-hydroxy-13α-
methoxyverruculogen TR-2 (11),

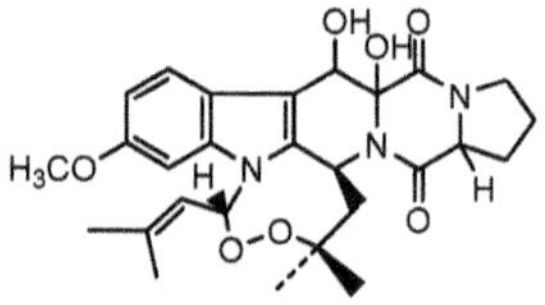

Ethyl 2, 4-dihydroxy- 5,
6-dimethylbenzoate (16)

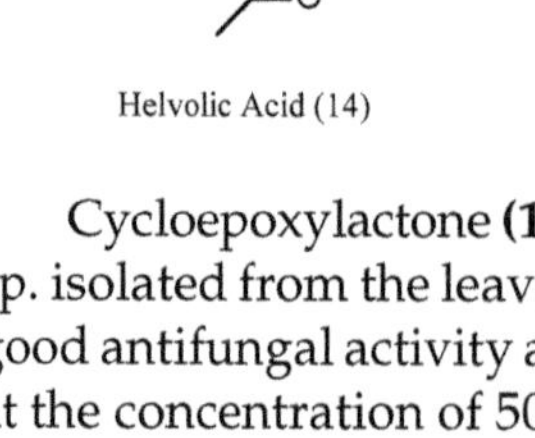

Helvolic Acid (14)

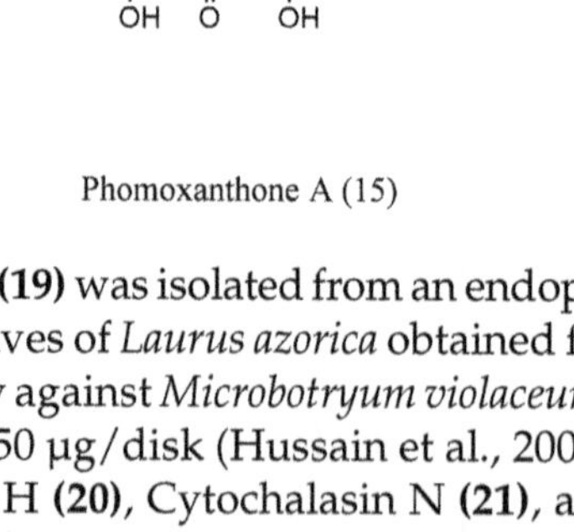

Phomoxanthone A (15)

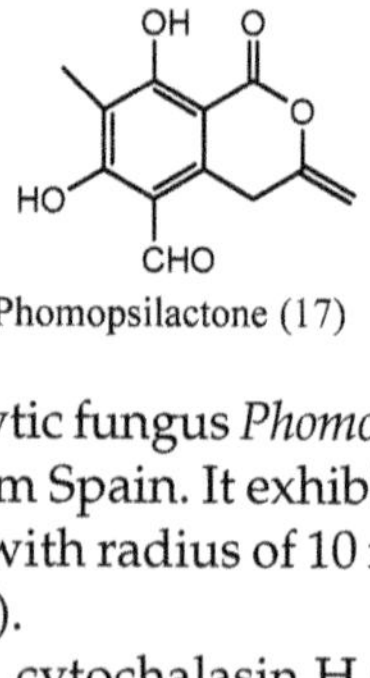

Phomopsilactone (17)

Cycloepoxylactone **(19)** was isolated from an endophytic fungus *Phomopsis* sp. isolated from the leaves of *Laurus azorica* obtained from Spain. It exhibited good antifungal activity against *Microbotryum violaceum* with radius of 10 mm at the concentration of 50 µg/disk (Hussain et al., 2009a).

Epoxycytochalasin H **(20)**, Cytochalasin N **(21)**, and cytochalasin H **(22)** demonstrated antifungal activity against four phytopathogens, namely *Botrytis cinerea, Sclerotinia sclerotiorum, Rhizoctonia cerealis, Fusarium oxysporum* with IC$_{50}$ ranging from 0.1 to 50 µg/ml (Fu et al., 2011).

Three chlorinated benzophenone derivatives, Pestalachlorides A–C **(23–25)** were isolated from endophytic fungus *Pestalotiopsis adjusta* recovered from the stem of an unidentified tree in Xinglong, China. Compounds **23** and **25** displayed significant antifungal activity against three plant pathogenic fungi namely, *Fusarium culmorum*, *Gibberella zeae* and *Verticillium aibo-atrum*. Pestalachloride A displayed potent antifungal activity against *F. culmorum*, with an IC_{50} value of 0.89 micromolar, while Pestalachloride B exhibited remarkable activity against *G. zeae*, with an IC_{50} value of 1.1 µm, whereas Pestalachloride C did not show any noticeable *in-vitro* antifungal activities against *F. culmorum*, *G. zeae* and *V. aibo-atrum* ($IC_{50} > 100$ micromolar) (Li et al., 2008).

Ergosterol peroxide **(26)** and 2,6-dihydroxy-2-methyl-7-(prop-1E-enyl)-1-benzofuran-3(2H)-one **(27)** were isolated from *Verticillium* sp., an endophyte of roots of wild *Rehmannia glutinosa*. Both the compounds were found active against *Fusarium*, *Verticillium*, *Septoria* and *Rhizoctonia* sp. when grown in PDB medium. Minimum morphological deformation concentration (MMDC) of compounds **(26)** and **(27)** towards *Pyricularia oryzae* P-2b was 1.95 and 7.8 mcg/ml, respectively. Both the Compounds clearly inhibited biomass accumulations at a low concentration (0.97 mcg/ml) in liquid culture (You et al., 2009).

Trichothecene **(28)** was isolated from *Trichothecium roseum* the endophyte isolated from *Maytems hookeri*. Compound **(28)** exhibited antifungal activity against *Typhula incarnate* (MIC 50 µg/ml), *Gaeumannomyces graminis* (MIC 30 µg/ml), *Phytophthora infestans* (MIC 30 µg/ml), *Alternaria solani* (MIC 5 µg/ml), *Pyricularia oryzae* (MIC 20 µg/ml) (Zhang et al., 2010).

Four natural nitro napthalenes derivatives, 1-hydroxy-5-methoxy-2-nitronapthalene **(29)**, 1,5-dimethoxy-4-nitronapthalene **(30)**, 1-5-dimethoxy-2,4-dinitronapthalene **(31)** and 1,5-dimethoxy-4,8-dinitronapthalene **(32)** were isolated from *Coniothyrium* sp., an endophyte of the shrub *Sideritis chamaedryfolia* from an arid habitat near Spain. All the compounds displayed strong antifungal activity against *Microbotryum violaceum* at 1 µg/µl (Krohn et al., 2008).

Colletotric acid **(33)** was isolated from *Colletotrichum gloeosporioides*, an endophytic fungus colonized inside the stem of *Artemisia mongolica* and inhibited the growth of the pathogenic fungus *Helminthosporium sativum* with MIC value of 50 mcg/ml (Zou et al., 2000).

Isocoumarin derivatives, (12R)-12-hydroxymonocerin **(34)**, (12S)-12-hydroxymonocerin **(35)**, (3R, 4R, 10R)-4 (2-4) **(36)** and Monocerin **(37)** were isolated from *Microdochium bolleyi*, an endophytic fungus from *Fagonia cretica*, from the semi-arid coastal regions of Gomera. Compounds **(35)**, **(36)** and **(37)** exhibited good antifungal activity while compound **(34)** exhibited moderate antifungal activity against *Microbotryum violaceum* (Zhang et al., 2008).

Cryptocin **(38)** a tetramic acid is an antifungal compound also obtained from *Cryptosporiopsis quercina* isolated from the inner bark of the stem of *Tripterigeum wilfordii*. This unusual compound possesses potent activity

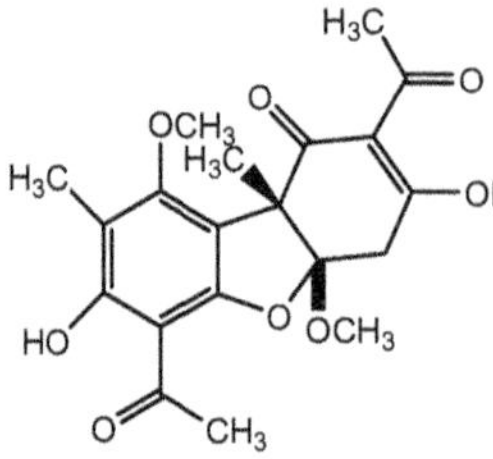

Phomodione (18)

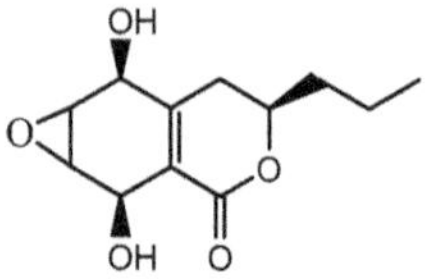

Cycloepoxylactone (19)

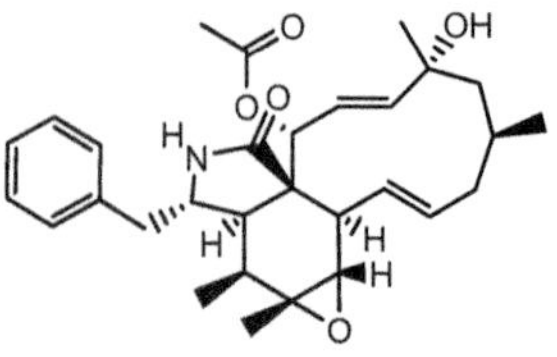

Epoxycytochalasin H (20),

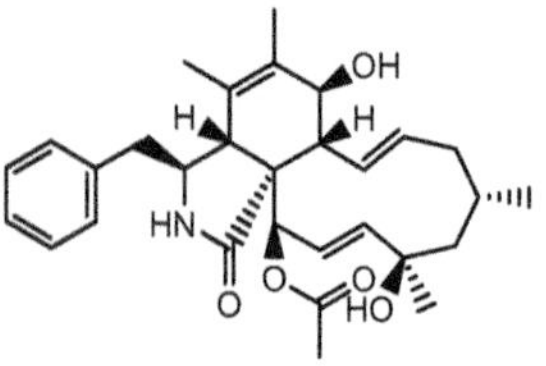

Cytochalasin N (21),

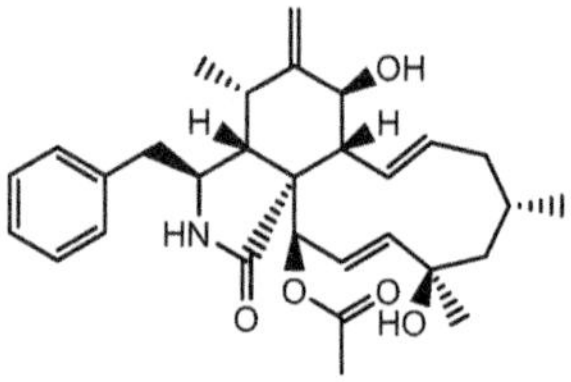

Cytochalasin H (22)

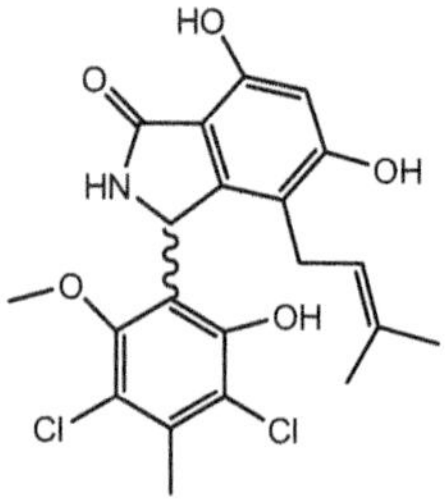

Pestalachlorides A (23)

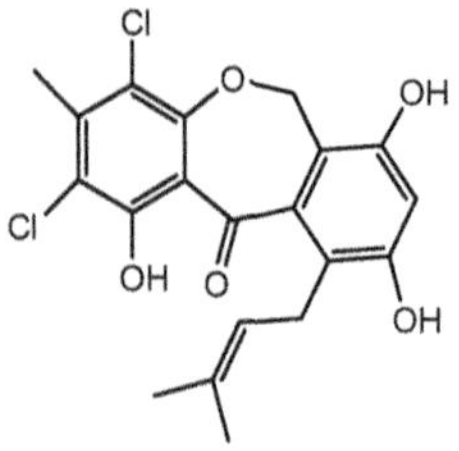

Pestalachlorides B (24)

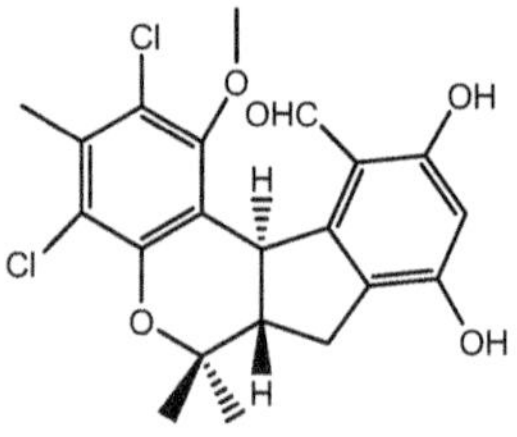

Pestalachlorides C (25)

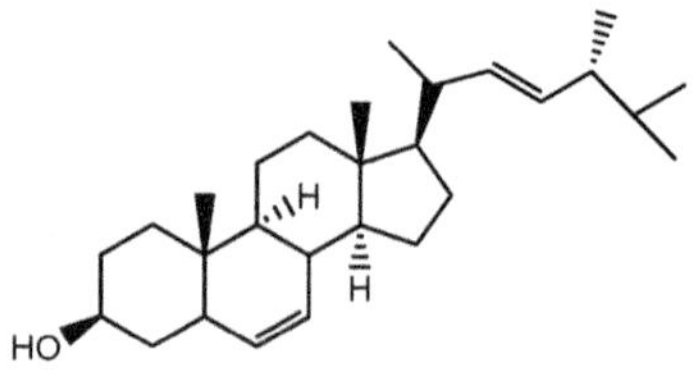

Ergosterol peroxide (26)

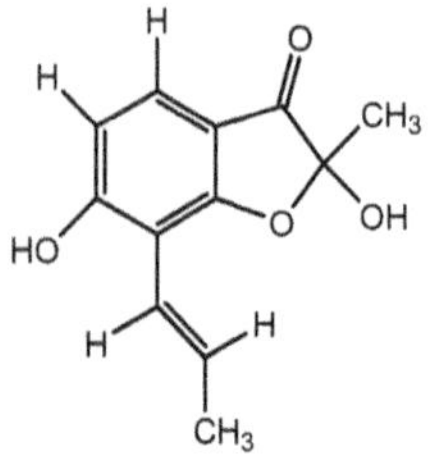

2,6-dihydroxy-2-methyl-7-(prop-1E-enyl)
-1-benzofuran-3(2H)-one (27)

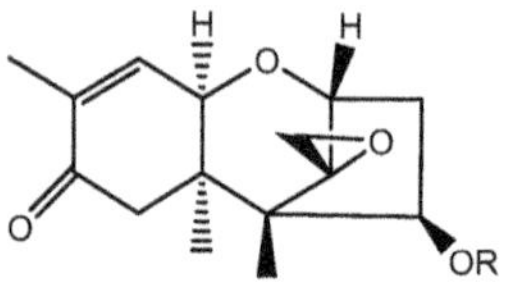

R = COCHCHCH₃

Trichothecene (28)

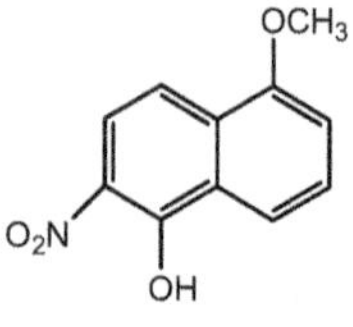

1-hydroxy-5-methoxy
-2-nitronapthalene (29)

against *Pyricularia oryzae*, a causal agent of rice blast disease with MICs of 0.39 mcg/ml. This compound also has potent activity for a number of other plant pathogenic fungi (Li et al., 2000).

Orcinol (**39**) and 4-hydroxymellein (**40**), 8-methoxymellein (**41**) were isolated from *Penicillium*, an endophyte of the leaves of *Alibertia macrophylla* (Rubiaceae). Compounds (**39**) and (**40**) Orcinol (**39**) and 4-hydroxymellein (**40**), showed a potent effect towards the yeasts, exhibiting a detection limit of 5.00 and 10.0 µg against *C. cladosporioides* and *C. sphaerospermum*, respectively. Compound (**41**) exhibited moderate fungi toxicity towards *C. cladosporioides* and *C. sphaerospermum*, showing a detection limit of 10.0 and 25.0 µg, respectively (Oliveira et al., 2009).

Two polyketide metabolites namely, 5-Methoxy-7-hydroxyphthalide (**42**) and (3*R*, 4*R*)-*cis*-4-hydroxymellein (**43**) were isolated from an unidentified Ascomycete, **endophytes** of *Meliotus dentatus*. Compound **42** and **43** exhibited 7 mm and 8 mm as radius of zone of inhibition against *Microbotryum violaceum* (Hussain et al., 2009b).

The solanapyrone analogues, Solanapyrone C, N and O (**44–46**) and Phomalactone (**47**) were isolated from *Nigrospora* sp. YB-141, an endophytic

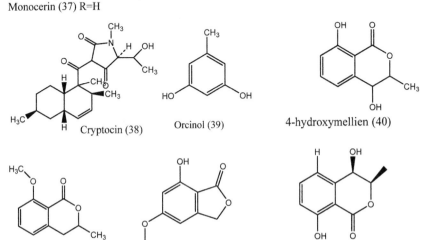

1,5-dimethoxy-4 -nitronapthalene (30)   1-5-dimethoxy-2,4 -dinitronapthalene (31)   1,5-dimethoxy-4,8 -dinitronapthalene (32)   Colletotric acid (33)

(12R)-12- hydroxymonocerin (34) R= α-OH   (3R, 4R, 10R)-4 (2-4) (36) R1=OH, R2=α-OH
(12S)-12-hydroxymonocerin (35) R=β-OH
Monocerin (37) R=H

Cryptocin (38)   Orcinol (39)   4-hydroxymellien (40)

8-methoxymellein (5)(41)   5-Methoxy-7-hydroxyphthalide (42)   (3R, 4R)-cis-4-hydroxymellein (43)

fungus isolated from *Azadirachta indica*. All the compounds were tested for their antifungal properties against seven phytopathogenic fungal strains namely, *Aspergillus niger, Botrytis cinerea, Fusarium avenaceum, Fusarium moniliforme, Helminthosporium maydis, Ophiostoma minus* and *Penicillium islandicum*. All the compounds showed antifungal activities against *B. cinerea* with MIC values in the range of 31.25–250 mcg/ml (Wu et al., 2008a).

Five 10-membered lactones: 8α-Acetoxy-5α-hydroxy-7oxodecan-10-olide (48), 7α,8α-Dihydoxy-3,50decadien-10-olide (49), 7α-Acetoxymultiplolide A (50), 8α-Acetoxymultiplolide A (51) and Multilplolide (52) were isolated from *Phomopsis* sp. YM 311483, obtained from the stem of *Azadirachta indica* growing in Yuanjiang Country, a tropical region in Yunnan province, People's Republic of China. These lactones were evaluated for their antifungal activity against seven plant pathogens namely, *Aspergillus niger, Botrytis cinerea, Fusarium avenaceum, F. moniliforme, Helminthosporium maydis, Penicillium islandicum* and *Ophiostoma minus*, using the dose dependent paper-disk diffusion method. All five compounds showed weak antifungal activities. Compound (51) was the most potent, with MIC values in the range of 31.25–500 mcg/ml, interestingly compound (51) was more active than compound (50), even though their structures differed only in the position of the acetoxy substituent (Wu et al., 2008b).

Botryorhodines A and D (53, 54) have been isolated from *Botryosphaeria rhodina*, an endophytic fungus of the medicinal plant *Bidens pilosa* and they have antifungal activity against pathogenic strain of *Fusarium oxysporum* with MIC values 191.60 μM for Botryorhodine A and 238.80 μM for B (Abdou et al., 2010).

A polyketide compound identified as 5-hydroxyramulosin (55) was reported as a bioactive constituent obtained from endophytic fungus *Phoma* sp. harbored inside *Cinnamomum mollissimum*. This compound inhibited *Aspergillus niger* at IC_{50} 1.56 μg/mL (Santiago et al., 2012).

Two new depsidone metabolites, cordycepsidone A (56) and cordycepsidone B (57), were isolated from the PDA culture extract of *Cordyceps dipterigena* and identified as being responsible for the antifungal activity against plant pathogenic fungus *Gibberella fujikuroi* at MIC 8.3 μg/ml (Varughese et al., 2012).

Epicolactone (58) and epicoccolides A (59) and B (60) were isolated from an endophytic fungus, *Epicoccum* sp. CAFTBO, of *Theobroma cacao*. Epicolactone (58) and epicoccolides A (59) and B (60) showed potent antimicrobial activities and significant inhibitory effects on the mycelial growth of two peronosporomycete phytopathogens, *Pythium ultimum* and *Aphanomyces cochlioides*, and the basidiomycetous fungus *Rhizoctonia solani* (Talontsi et al., 2013).

Cladosporin (61), Isocladosporin (62), were isolated by bioassay guided fractionation of *Cladosporium cladosporioides*. At 30 μM, compound 61 exhibited 92.7, 90.1, 95.4, and 79.9% growth inhibition against *Colletotrichum acutatum, Colletotrichum fragariae, Colletotrichum gloeosporioides*, and *P. viticola*,

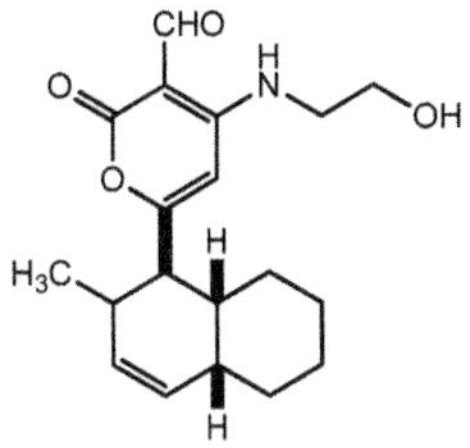

Solanapyrone C (44)

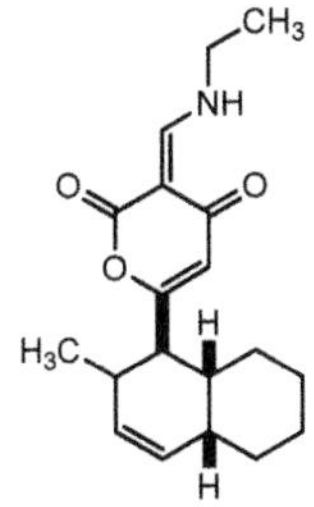

Solanapyrone N (45)

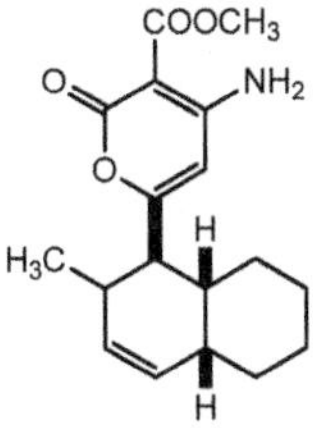

Solanapyrone O (46) 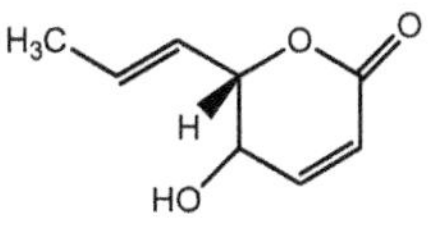 Phomalactone (47)

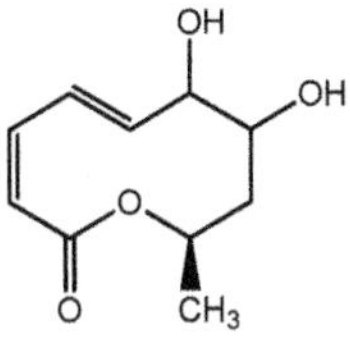

8α-Acetoxy-5α-hydroxy-7oxodecan-10-olide (48)

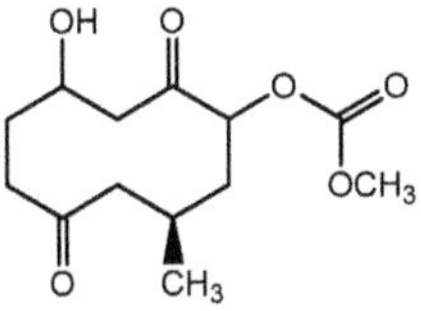

7α,8α-Dihydoxy-3,5 decadien-10-olide (49)

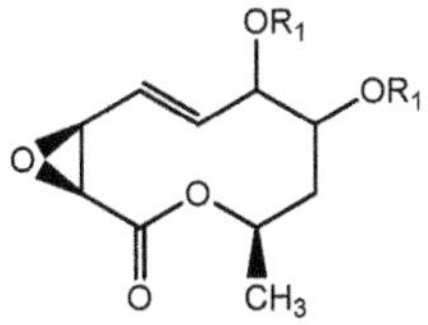

7α-Acetoxymultiplolide A (50)
R1=Ac, R2=H
8α-Acetoxymultiplolide A (51)
R1=H, R2=Ac
Multilplolide (52) R1, R2=H

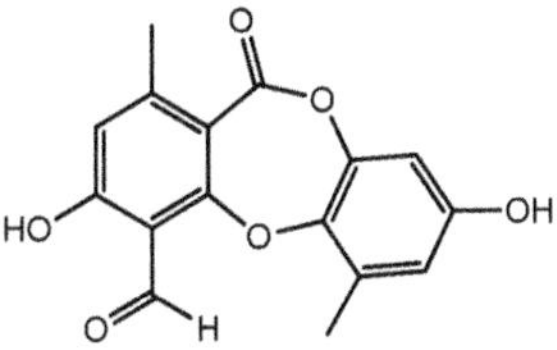

Botryorhodines A (53)

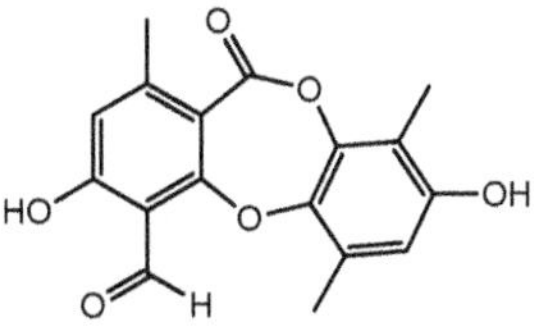

Botryorhodines D (54)

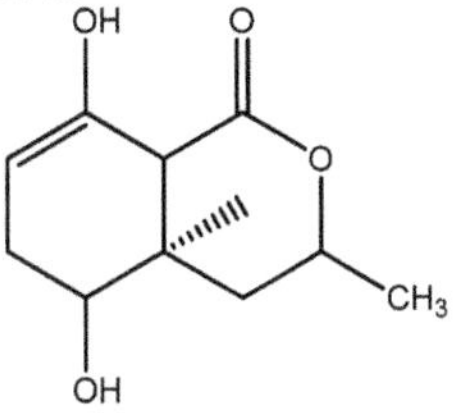

5-hydroxyramulosin (55)

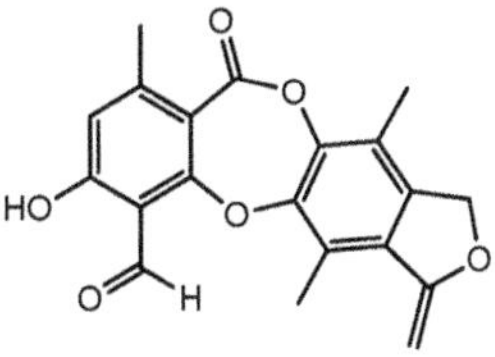

Cordycepsidone A (56)

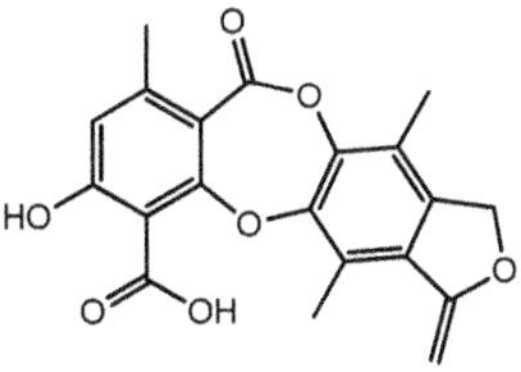

Cordycepsidone B (57)

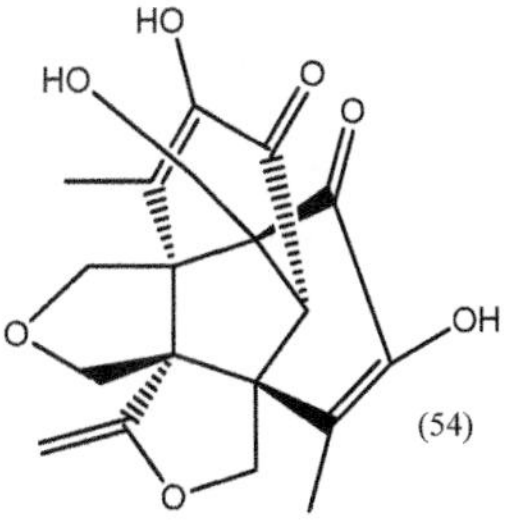

(54)

Epicolactone (58)

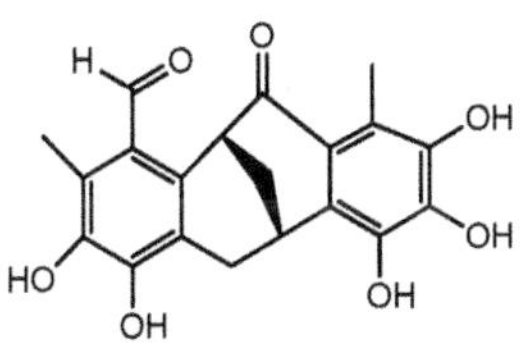

Epicoccolide A (59)

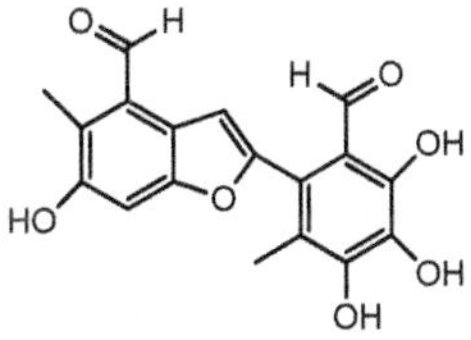

Epicoccolide B (60)

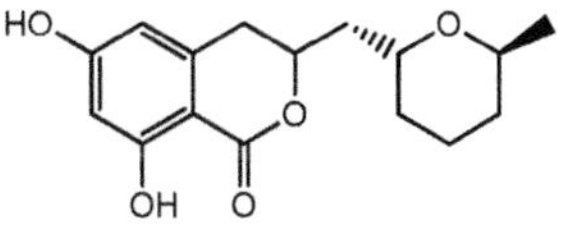

Cladosporin (61)

respectively. Compound **62** showed 50.4, 60.2, and 83.0% growth inhibition at 30 µM against *Co. fragariae, Co. gloeosporioides,* and *P. viticola,* respectively (Wang et al., 2013).

Cytochalasins E **(63)**, and K **(64)** isolated from endophytes *Physalospora* sp., exhibited antifungal activity against *E. repens* and *M. microspore* (Hussain et al., 2014a).

The endophytic fungus strain 0248, isolated from garlic, was identified as *Trichoderma brevicompactum*. The bioactive compound 4β-acetoxy-12, 13-epoxy-Δ^9-trichothecene **(65)** (trichodermin) has a marked inhibitory activity on *Rhizoctonia solani*, with an EC_{50} of 0.25 µg mL^{-1}. Trichodermin showed inhibitory activity against *Rhizoctonia solani* (EC50 of 0.25 µg mL^{-1}, *Botrytis cinerea* (EC50 of 2.02 µg mL^{-1}), and *Co. lindemuthianum* (EC50 = 25.60 µg mL^{-1}) (Shentu et al., 2014).

Bipolamide B **(66)** was isolated from *Bipolaris* sp., an endophyte of leaves of *G. hispida* Thwaites collected from the botanical garden in Mahidol University, Bangkok, Thailand. Compound **(66)** is active against *Cladosporium cladosporioides* FERMS-9, *Cladosporium cucumerinum* NBRC 6370, *Saccharomyces cerevisiae* ATCC 9804, *Aspergillus niger* ATCC 6275 and *Rhizopus oryzae* ATCC 10404, with minimum inhibitory concentration (MIC) values of 16, 32, 32, 64 and 64 µg/ml, respectively (Siriwach et al., 2014).

Stemphyperylenol **(67)** displayed a potent antifungal activity against the plant pathogen *Alternaria solani* with MIC of 1.57 µM comparable to the commonly used fungicide Carbendazim. It was isolated from the solid culture of the endophytic fungus *Botryosphaeria dothidea* KJ-1, collected from the stems of *Melia azedarach* (Xiao et al., 2014a).

Asperpyrone A (3) **(68)** and (R)-3-hydroxybutanonitrile (7) **(69)**, were isolated from the culture of *Aspergillus* sp. KJ-9, a fungal endophyte isolated from *Melia azedarach*. Both the compounds were found active against the phytopathogenic fungi (*Gibberella saubinetti, Magnaporthe grisea, Botrytis cinerea, Colletotrichum gloeosporioides* and *Alternaria solani*) with minimum inhibitory concentration (MIC) in the range of 6.25–50 µM (Xiao et al., 2014b).

Cis-4-Acetoxyoxymellein **(70)** and 8-deoxy-6-hydroxy-cis-4-acetoxyoxymellein **(71)** displayed significant antifungal activity (0.05 mg) towards *Microbotryum violaceum, Botrytis cinerea* and *Septoria tritici,* isolated from an unidentified Ascomycete, isolated from *Meliotus dentatus* (Hussain et al., 2014b).

Endophytic Fungi: Source of Volatile Organic Compound

Pioneer Endophytologist Prof. Strobel has documented several endophytic fungi as a myriad source of Volatile Organic Compounds (VOCs) having potential of "bio-fuels" in future. For example, *Muscodar* sp., a novel fungal genus, produces extremely bioactive volatile organic compounds. This fungal isolate was initially discovered as an endophyte in *Cinnamomum zeylanicum*

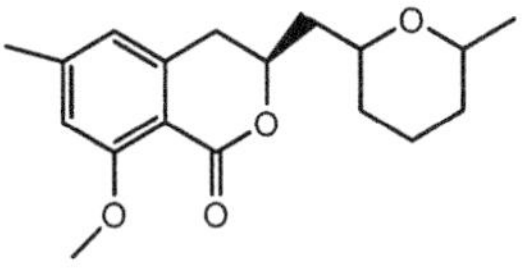

Isocladosporin (62)

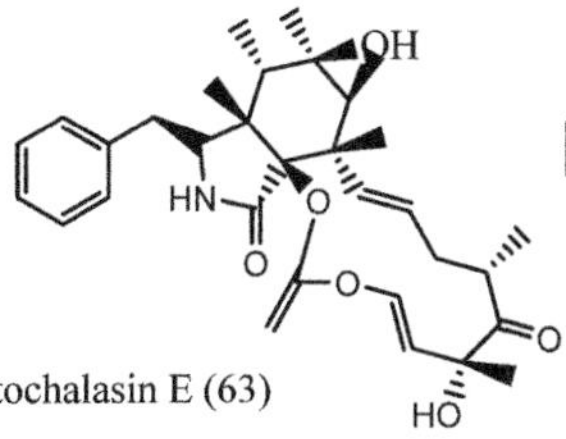

Cytochalasin E (63)

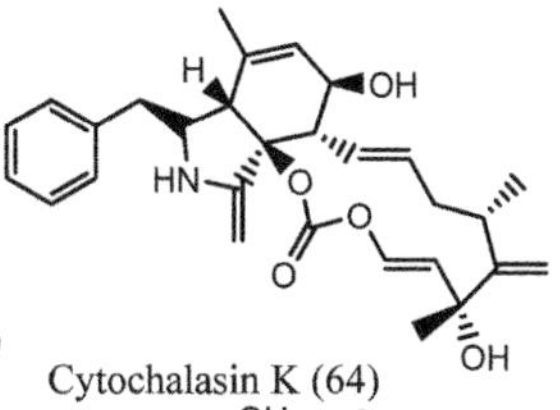

Cytochalasin K (64)

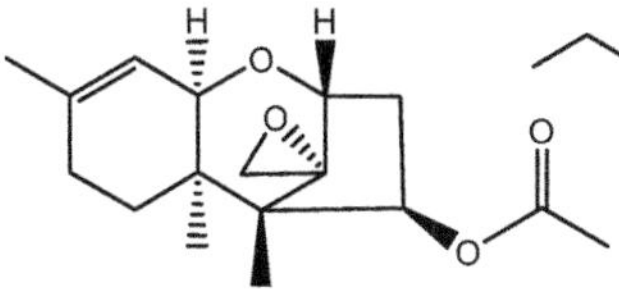

4β-acetoxy-12, 13-epoxy-9
Δ-trichothecene (65)
(Trichodermin)

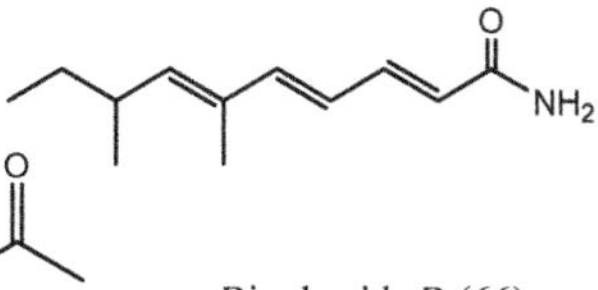

Bipolamide B (66)

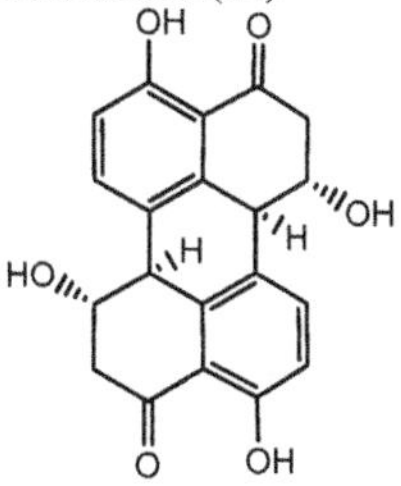

Stemphyperylenol (67)

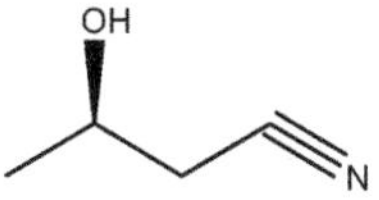

(R)-3-hydroxybutanonitrile (69)

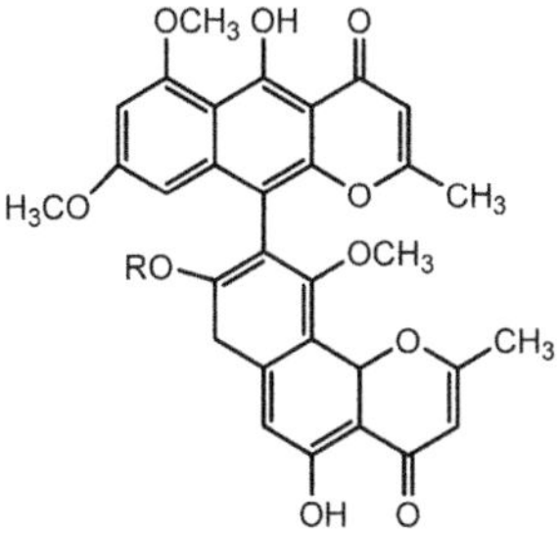

Asperpyrone A (68), R=H

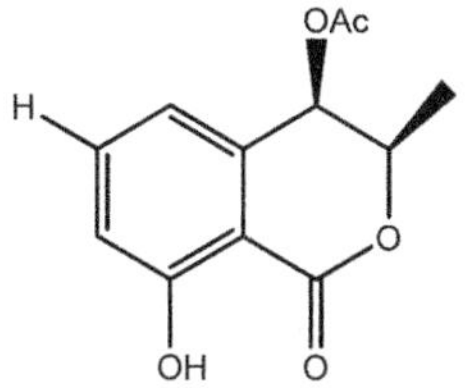

Cis-4-Acetoxyoxymellein (70)

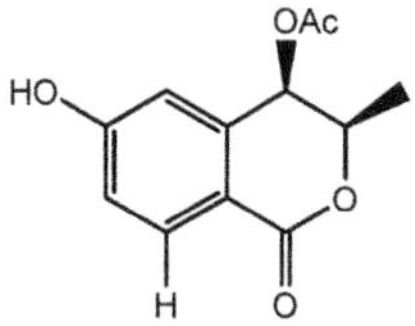

8-deoxy-6-hydroxy-cis-4-
acetoxyoxymellein (71)

in a botanical garden in Honduras. *Muscodor albus*, produces a mixture of VOCs that are lethal to a wide variety of plant and human pathogens—fungi and bacteria. It is also effective against nematodes and certain insects. The mixture of VOCs has been analyzed using GC/MS and are reported to consist of various alcohols, acids, esters, ketones, and lipids (Strobel, 2001). The term "mycofumigation" has been applied to the practical applications of this fungus. Fumigation of adult codling moth with VOC's of *M. albus* for 3 days resulted in 81% mortality (Lacey, 2009).

Another endophytic fungus, *Gliocladiun roseum*, produced a series of volatile hydrocarbons and hydrocarbon derivatives on an oatmeal-based agar under microaerophilic conditions as analysed by solid-phase micro-extraction (SPME)-GC/MS. An extract of the host plant, *Eucryphia cordifolia* supported the growth and hydrocarbon production of this fungus. Hydrocarbons that were produced by this organism, included undecane, 2,6-dimethyl; decane, 3,3,5-trimethyl; cyclohexene, 4-methyl; decane, 3,3,6-trimethyl; and undecane, 4,4-dimethyl (Ezra, 2004).

Hypoxylon sp. inhabiting *Peresa indica* produces an impressive spectrum of VOCs, e.g., 1,8-Cineole, 1-methyl-1,4-cyclohexadiene, (+)-alpha-methylene-alpha-fenchocamphorone, among many others. Six-day old cultures displayed maximal VOC-antifungal activity against *Botrytis cinerea, Sclerotinia sclerotiorum* and *Phytophthora cinnamomi* with 100%, 90.4% and 100% growth inhibition, respectively. Production of 1,8-Cineole by a fungal source seems very promising. 1,8-cineole also known as Eucalyptol oil is the active component of essential oil of eucalyptus and thus can be utilized in pharmaceuticals, fragrances and degreasing detergents. Furthermore, 1,8-cineole improves octane number and, therefore, can be used as fuel additive (Tomsheck et al., 2010).

Endophytic fungi inhabiting *Paris polyphylla* namely, *Pichia guilliermondii* and *Gliomastix murorum* yielded (E)-9-octadecenoic acid and (7Z, 10Z)-7,10-hexadecadienoic acid respectively, depicted antifungal activity against *Magnaporthe oryzae* having IC_{50} value of 1.56 mg/mL (Zhao et al., 2009).

Phoma sp. was isolated and characterized as an endophytic and as a pathogen of *Larrea tridentata* (creosote bush) growing in the desert region of southern Utah, USA. This fungus produces a unique mixture of volatile organic compounds (VOCs), including a series of sesquiterpenoids, some alcohols and several reduced naphthalene derivatives. Trans-caryophyllene, a product in the fungal VOCs, was also noted in the VOCs of this pungent plant. The gases of *Phoma* sp. possess antifungal properties and are markedly similar to that of a methanolic extract of the host plant. Some of the test organisms with the greatest sensitivity to the *Phoma* sp. VOCs were *Verticillium, Ceratocystis, Cercospora* and *Sclerotinia* while those being the least sensitive were *Trichoderma, Colletotrichum* and *Aspergillus* (Strobel, 2011).

Conclusions

Diverse array of secondary metabolites (phenols, alkaloids, terpenoids, steroids, flavonoids, etc.) isolated from different species of fungi (*Phomopsis, Phoma, Pestalotiopsis,* etc.) have exhibited outstanding antiphytopathogenic activity. Studies and their results discussed have been conducted *in-vivo* so far. Efficacy of potential metabolites should further be explored in field conditions also. It is imperative to know, before huge promises are made, that what are the characteristic variables between laboratory conditions and land conditions and how they affect the effectiveness of end product. However, selection of host plant for isolation of endophytic fungi is very crucial. Medicinal plants collected from rich biodiversity have shown to have fungal endophytes with diverse bioactivities and chemical structure. Also, substrate and pH optimisation studies of culture conditions to obtain desired metabolite are required to be done diligently. Finally, gene insertion and other molecular techniques such as genetic transformation are relatively uncomplicated in microbes, and can be used to up-regulate production of a specific compound or to generate analogues of a potent metabolite. Discovery and evaluation of natural product fungicides is largely dependent upon the availability of miniaturized antifungal bioassays. Essentials for natural product bioassays include sensitivity to microgram quantities, selectivity to determine optimum target pathogens, and adaptability to complex mixtures. Experimental accuracy and precision must be stable between assays over time. These assays should be relevant to potential pathogen target sites in the natural infection process of the host and applicable to the agrochemical industry. To maximize the detection of natural products, high-throughput bioassay techniques must target important agricultural pathogens, include relevant commercial pesticide standards, and adhere to sound statistical principles. Ultimately, the discovery of natural products with low mammalian and environmental toxicity for the control of pests and diseases in agriculture will provide for a safe and dependable food supply. The use of natural product-based agrochemicals provides an opportunity for better natural resource management by reducing dependence on synthetic chemicals.

References

Abdou, R., Scherlach, K., Dahse, H.M., Sattler, I. and Hertweck, C. 2010. Botryorhodines A-D, antifungal and cytotoxic depsidones from *Botryosphaeria rhodina*, an endophyte of the medicinal plant *Bidens pilosa*. Phytochem., 71: 110–116.

Agrios, G.N. 2004. Plant Pathology. Elsevier, Oxford, UK.

Aly, A.H., Debbab, A. and Proksch, P. 2011. Fungal endophytes: unique plant inhabitants with great promises. Appl. Microbiol. Biotechnol., 90(6): 1829–1845.

DeBary 1866. Morphologie und PhysiologiePilze, Flechten, und myxomyceten. Hofmeister's Handbook of Physiological Botany Vol. 2. Leipzig.

Elsaesser, B., Krohn, K., Floerke, U., Root, N., Aust, H.J., Draeger, S., Schulz, B., Antus, S. and Kurtan, T. 2005. X-ray structure determination, absolute configuration and biological activity of phomoxanthone A. Eur. J. Org. Chem., 21: 4563–4570.

Ezra, D., Hess, W.M. and Strobel, G.A. 2004. New endophytic isolate of *Muscodar albus*, volatile antibiotic producing fungi. Microbiol., 150: 4023–4031.

Fu, J., Zhou, Y., Li, H.F., Ye, Y.H. and Guo, J.H. 2011. Antifungal metabolites from *Phomopsis* sp. By254, an endophytic fungus in *Gossypium hirsutum*. Afr. J. Microbiol. Res., 5(10): 1231–1236.

Harris, C.A., Renfrew, M.J. and Woolridge, M.W. 2001. Assessing the risk of pesticide residues to consumers: recent and future developments. Food Addit. Contam., 18: 1124–1129.

Hoffman, A.M., Mayer, S.G., Strobel, G.A., Hess, W.M., Sovocool, G.W., Grange, A.H., Harper, J.K., Arif, A.M., Grant, D.M. and Kelley-Swift, E.G. 2008. Purification, identification and activity of phomodione, a furandione from an endophytic *Phoma* species. Phytochem., 69(4): 1049–1056.

Hussain, H., Akhtar, N., Draeger, S., Schulz, B., Pescitelli, G., Salvadori, P., Antus, S., Kurtan, T. and Krohn, K. 2009a. New bioactive 2,3-epoxycyclohexenes and isocoumarins from the endophytic fungus *Phomopsis* sp. from *Laurus azorica*. Eur. J. Org. Chem., 5: 749–756.

Hussain, H., Krohn, K., Draeger, S., Meier, K. and Schulz, B. 2009b. Bioactive chemical constituents of a sterile endophytic fungus from *Meliotus denatus*. Rec. Nat. Prod., 3(2): 114–117.

Hussain, H., Kliche-Spory, C., Al-Harrasi, A., Al-Rawahi, A., Abbas, G., Green, I.R., Schulz, B., Krohn, K. and Shah, A. 2014a. Antimicrobial constituents from three endophytic fungi. Asian Pacific J. Trop. Med., 7(S1): 224–227.

Hussain, H., Jabeen, F., Krohn, K., Al-Harrasi, A., Ahmad, M., Mabood, F., Shah, A., Badshah, A., Rehman, N.U., Green, I.R., Ali, I., Draeger, S. and Schulz, B. 2014b. Antimicrobial activity of two mellein derivatives isolated from an endophytic fungus. Med. Chem. Res., DOI 10.1007/s00044-014-1250-3.

Janusauskaitė, D., Česnu levičienė, R. and Gaurilcikiene, I. 2012. Non-target effects of fungicidal pea (*Pisum sativum* L.) seed treatment on soil microorganisms. Žemdirbystė = Agriculture, 99(4): 387–392.

Kharwar, R.N., Mishra, A., Gond, S.K., Stierle, A. and Stierle, D. 2011. Anticancer compounds derived from fungal endophytes: Their importance and future challenges. Nat. Prod. Rep., 28(7): 1208–1228.

Krohn, K., Kouam, S.F., Brandt, S.C., Draeger, S. and Schulz, B. 2008. Bioactive nitronaphthalenes from an endophytic fungus, *Coniothyrium* sp., and their chemical synthesis. Eur. J. Org. Chem., 3615–3618.

Lacey, L.A., Horton, D.R., Jones, D.C., Headrick, H.L. and Neven, L.G. 2009. Efficacy of the biofumigant fungus *Muscodor albus* (Ascomycota: Xylariales) for control of codling moth (Lepidoptera: Tortricidae) in simulated storage conditions. J. Econ. Entomol., 102(1): 43–49.

Li, E., Jiang, L., Guo, L., Zhang, H. and Che, Y. 2008. Pestalachlorides A-C, antifungal metabolites from the plant endophytic fungus *Pestalotiopsis adusta*. Bioorg. Med. Chem., 16(17): 7894–7899.

Li, J.Y. and Strobel, G.A. 2001. Jesterone and hydroxy-jesterone anti-oomycete cyclohexenenone epoxides from the endophytic fungus *Pestalotiopsis jesteri*. Phytochem., 57: 261–265.

Li, J.Y., Strobel, G.A., Harper, J., Lobkovsky, E. and Clardy, J. 2000. Cryptocin, a potent tetramic acid antimycotic from the endophytic fungus *Cryptosporiopsis* cf. quercina. Organic Lett., 767–770.

Li, X.J., Zhang, Q., Zhang, A.L. and Gao, J.M. 2012. Metabolites from *Aspergillus fumigatus*, an endophytic fungus associated with *Melia azedarach*, and their antifungal, antifeedant and toxic activities. J. Agric. Food Chem., 60(13): 3424–3431.

Oliveira, C.M., Silva, G.H., Regasini, L.O., Zanardi, L.M., Evangelista, A.H., Young, M.C., Bolzani, V.S. and Araujo, A.R. 2009. Bioactive metabolites produced by *Penicillium* sp.1 and sp.2, two endophytes associated with *Alibertia macrophylla*. Z. Naturforsch C., 64(11-12): 824–830.

Saillenfait, A.M., Ndiaye, D. and Sabaté, J.P. 2015. Pyrethroids: Exposure and health effects—An update. International Journal of Hygiene and Environmental Health, 218: 281–292.

Santiago, C., Fitchett, C., Munro, M.H.G., Jalil, J. and Santhanam, J. 2012. Cytotoxic and antifungal activities of 5-hydroxyramulosin, a compound produced by an endophytic fungus isolated from *Cinnamomum mollisimum*. Evidence-based Complemen. Altern. Med., 2012: 689310.

Shentu, X., Zhan, X., Ma, Z., Yu, X. and Zhang, C. 2014. Antifungal activity of metabolites of the endophytic fungus *Trichoderma brevicompactum* from garlic. Brazilian J. Microbiol., 45(1): 248–254.

Silva, G.H., Teles, H.L., Trevisan, H.C., Bolzani Vanderlan, da S., Young, M.C.M., Pfenning, L.H., Eberlin, M.N., Haddad, R., Costa-Neto, C.M. and Araujo, A.R. 2005. New bioactive metabolites produced by *Phomopsis cassiae*, an endophytic fungus in *Cassia spectabilis*. J. Brazilian Chem. Soc., 16(6B): 1463–1466.

Siriwach, R., Kinoshita, H., Kitani, S., Igarashi, Y., Pansuksan, K., Panbangred, W. and Nihira, T. 2014. Bipolamides A and B, triene amides isolated from the endophytic fungus *Bipolaris* sp. MU34. J. Antibiotics, 67(2): 167–170.

Strobel, G.A. 2003. Endophytes as sources of bioactive products. Microbes Infect., 5: 535–544.

Strobel, G.A., D'irkse, E., Sears, J. and Marksworth, C. 2001. Volatile antimicrobials from *Muscodar albus*, a novel endophytic fungi. Microbiol., 147: 2943–50.

Strobel, G., Singh, S.K., Hassan, R.U., Mitchell, A.M., Geary, B. and Sears, J. 2011. An endophytic/ pathogenic *Phoma* sp. from creosote bush producing biologically active volatile compounds having fuel potential. FEMS Microbiol. Lett., 320: 87–94.

Talontsi, F.M., Dittrich, B., Schüffler, A., Sun, H. and Laatsch, H. 2013. Epicoccolides: Antimicrobial and antifungal polyketides from an endophytic fungus *Epicoccum* sp. associated with *Theobroma cacao*. Eur. J. Org. Chem., 2013(15): 3174–3180.

Talontsi, F.M., Islam, M.T., Facey, P., Douanla Meli, C., Von Tiedemann, A. and Laatsch, H. 2012. Depsidones and other constituents from *Phomopsis* sp. and its host plant *Endodesmia calophylloides* with potent inhibitory effect on motility of zoospores of grapevine pathogen *Plasmopara viticola*. Phytochem. Lett., 5(3): 657–664.

Tejesvi, M.V., Kini, K.R., Prakash, H.S., Subbiah, V. and Shetty, H.S. 2007. Genetic diversity and antifungal activity of species of *Pestalotiopsis* isolated as endophytes from medicinal plants. Fungal Divers., 24: 37–54.

Tomsheck, A.R., Strobel, G.A., Booth, E., Geary, B., Spakowicz, D., Knighton, B., Floerchinger, C., Sears, J., Liarzi, O. and Ezra, D. 2010. *Hypoxylon* sp., an endophyte of *Persea indica*, producing 1,8-Cineole and other bioactive volatiles with fuel potential. Microbial Ecol., 60(4): 903–914.

Varughese, T., Rios, N., Higginbotham, S., Arnold, A.E., Coley, P.D., Kursar, T.A., Gerwick, W.H. and Rios, L.C. 2012. Antifungal depsidone metabolites from *Cordyceps dipterigena*, an endophytic fungus antagonistic to the phytopathogen *Gibberella fujikuroi*. Tetrahedron Lett., 53(13): 1624–1626.

Vourela, H. 2004. Natural products in the process of finding new drug candidates. Curr. Med. Chem., 11: 1375–1389.

Wang, X., Radwan, M.M., Taráwneh, A.H., Gao, J., Wedge, D.E., Rosa, L.H., Cutler, H.G. and Cutler, S.J. 2013. Antifungal activity against plant pathogens of metabolites from the endophytic fungus *Cladosporium cladosporioides*. J. Agric. Food Chem., 61(19): 4551–4555.

Wu, S.H., Chen, Y.W., Shao, S.C., Wang, L.D., Yu, Y., Li, Z.Y., Yang, L.Y., Li, S.L. and Huang, R. 2008a. Two new solanapyrone analogues from the endophytic fungus *Nigrospora* sp. YB-141 of *Azadirachta indica*. Chem. Biodivers., 6(1): 79–85.

Wu, S.H., Chen, Y.W., Shao, H.C., Wang, L.D., Li, Z.Y., Yang, L.Y., Li, S.L. and Huang, R. 2008b. Ten-membered lactones from *Phomopsis* sp., an endophytic fungus of *Azadirachta indica*. J. Nat. Prod., 71: 731–734.

Xiao, J., Zhang, Q., Gao, Y.Q., Shi, X.W. and Gao, J.M. 2014a. Secondary metabolites from the endophytic *Botryosphaeria dothidea* of *Melia azedarach* and their antifungal, antibacterial, antioxidant, and cytotoxic activities. J. Agric. Food Chem., 62(16): 3584–3590.

Xiao, J., Zhang, Q., Gao, Y.Q., Shi, X.W. and Gao, J.M. 2014b. Antifungal and antibacterial metabolites from an endophytic *Aspergillus* sp. associated with *Melia azedarach*. Nat. Prod. Res., 28(17): 1388–1392.

You, F., Han, T., Wu, J.Z., Huang, B.K. and Qin, L.P. 2009. Antifungal secondary metabolites from *Verticillium* sp. Biochem. Syst. Ecol., 37: 162–165.

Zhang, W., Krohn, K., Draeger, S. and Schulz, B. 2008. Bioactive isocoumarins isolated from the endophytic fungus *Microdochium bolleyi*. J. Nat. Prod., 71(6): 1078–1081.

Zhang, X.M., Li, G., Ma, J., Zeng, Y., Ma, W. and Zhao, P. 2010. Endophytic fungus *Trichothecium roseum* LZ93 antagonising pathogenic fungi *in vitro* and its secondary metabolites. J. Microbiol., 48: 784–790.

Zhao, J., Mou, Y., Shan, T., Li, Y., Zhou, L., Wang, M. and Wang, J. 2010. Antimicrobial metabolites from the endophytic fungus *Pichia guilliermondii* isolated from *Paris polyphylla* var. *yunnanensis*. Molecules, 15: 7961–7970.

Zhao, J., Shan, T., Huang, Y., Liu, X., Gao, X., Wang, M., Jiang, W. and Zhou, L. 2009. Chemical composition and *in vitro* antimicrobial activity of the volatile oils from *Gliomastix murorum* and *Pichia guilliermondii*, two endophytic fungi in *Paris polyphylla* var. *yunnanensis*. Nat. Prod. Commun., 4(11): 1491–1496.

Zou, W.X., Meng, J.C., Lu, H., Chen, G.X., Shi, G.X., Zhang, T.Y. and Tan, R.X. 2000. Metabolites of *Colletotrichum gloeosporioides*, an endophytic fungus in *Artemisia mongolica*. J. Nat. Prod., 63(11): 1529–1530.

CHAPTER 5

Fungal Secondary Metabolites as Source of Antifungal Compounds

Francisca Vicente, * *Fernando Reyes* and *Olga Genilloud*

ABSTRACT

Natural products are composed of a reservoir of privileged chemical scaffolds which have been naturally selected by microbes to specifically interact with a diversity of biological targets present in the environment, thereby providing a fitness advantage to the producing organism. The vast number and variety of chemotherapeutic agents isolated from microbial natural products–approximately 80% of all available clinically used antibiotics are directly or indirectly derived from natural products and used to treat bacterial infection have greatly contributed to the improvement of human health over the past century.

However, the mortality rate of invasive fungal infections is alarming and has increased significantly over the past decades, and early and accurate diagnosis being increasingly difficult. Furthermore, most antifungal drugs in use today are not completely effective due to the development of increasing resistance and undesirable side effects which limit their use. In this scenario, the development of new antifungal agents, preferably with novel mechanisms of action, is an urgent medical need. This chapter summarizes a selection of antifungal agents produced by microorganisms classified according to their mechanisms of action and provides a general description of the producing organisms.

Fundación MEDINA, Parque Tecnológico de Ciencias de la Salud, Avda. del Conocimiento 34, 18016 Granada, Spain.
* Corresponding author: francisca.vicente@medinaandalucia.es

Introduction

Microbial natural products, as sources of therapeutically useful compounds, have a much shorter and lesser-known history than the use of plants and plant extracts in human medicine. Microorganisms may have evolved the ability to produce biologically active compounds as a consequence of the selection advantages conferred from the interactions of these molecules with specific receptors in other organisms (Demain, 1983; Omura, 1986). Whereas more than 20,000 microbial metabolites and approximately 100,000 plant products have been described so far, still secondary metabolites appear to be an inexhaustible source of lead structures for new antimicrobials, antivirals, and antitumor drugs, agricultural and pharmacological agents. Natural products (NPs) are the basis of a vast majority of anti-infective therapies in current clinical use. Indeed, approximately 80% of all the available and clinically used antibiotics are directly (or indirectly) derived from NPs (Newman et al., 2003; Newman and Cragg, 2012). This success of NPs exploits the optimized characteristics of microbially produced small molecules, namely their immense chemical diversity, intrinsic cell permeability, and target specificity versus that typically reflected in synthetic or combinatorial libraries. NPs reflect a reservoir of privileged chemical scaffolds which have been naturally selected by microbes to specifically interact with a diversity of biological targets in the environment, thereby providing a fitness advantage to the producing organism (Genilloud et al., 2010; Genilloud and Vicente, 2012). Compounds isolated from natural sources (variety of chemotherapeutic agents) provide a wealth of bioactive molecules that in some cases have been directly used as drugs or as lead for the development of potent inhibitors useful for the characterization of enzymes of interest, and to design future therapeutic drugs (Singh and Pelaez, 2008; Fischbach and Walsh, 2009). These agents have greatly contributed to the improvement of human health over the past century.

Despite the state-of-the-art antifungal therapy, the mortality rates for invasive infections from the three most common species of human fungal pathogens (*Candida albicans*, *Aspergillus fumigatus* and *Cryptococcus neoformans*) are between 40–90% (Lai et al., 2008; Park et al., 2009). Unfortunately, researches on antifungal agents is limited as compared to the number of agents available for bacterial infections. In fact, it took 20 years for the newest class of antifungal drugs, the echinocandins, to progress from bench-to-bedside. Until the 1970s, fungal infections were considered largely treatable and the demand for new medicines to treat them was very limited. Before this period, antifungal chemotherapy included only two kinds of compounds: potassium iodide, effective in the treatment of sporotrichosis; and the polyenes: amphotericin B, nystatin and pimaricin, which were introduced in the 1950s; amphotericin B for the treatment of severe systemic mycoses and nystatin and pimaricin for cutaneous, vaginal and intestinal candidiasis. Except for the development of flucytosine (first synthesized in 1957 but its antifungal properties were discovered in 1964), there was little progress until the development of the azole drugs in the early 1970s. Therefore, only a limited number of antifungal agents (polyenes and azoles plus the recently introduced echinocandins) are currently available for the treatment of life-threatening fungal infections. However, these antifungal agents show some limitations, such as the significant nephrotoxicity of amphotericin B and emerging resistance to the azoles (Georgopapadakou and Walsh, 1994), despite several recent improvements such as lipid formulations of polyenes with lower toxicity and new triazoles (voriconazole, rovuconazole and pasaconazole) with a wider spectrum of action, including activity against some azole-resistant isolates (Granier, 2000).

This chapter reviews a selection of novel agents produced by microorganisms with potential against fungal infections. The compounds can be broadly categorized based on the target they address and on the compound class. They are specifically classified according to their essential processes for the pathogen's biology such as cell wall components (glucan, chitin and mannoproteins), sphingolipid synthesis (inositol phosphoceramide synthase, serine palmitoyltransferase, ceramide synthase, and fatty acid elongation), and protein synthesis or according to genome-wide chemical genetic profiling. In addition, some considerations related to the producing organisms are also discussed.

Compounds Targeting Cell Wall Synthesis

The fungal cell wall protects the cell from the environment; it is a dynamic structure that is continually modified by enzymes to facilitate growth (Latge, 2001). Since fungal cell wall is different from that of mammals, its components represent good targets under active investigation for the development of specific and highly selective novel antifungal agents. Fungal cell wall composition varies among species, but it is generally made up of glucan (alpha and beta glucan), chitin and mannoproteins. The subcellular mechanisms of their synthesis and assembly have been used as potential targets to identify new antifungals and numerous natural products have been identified as inhibitors acting at these levels (Calugi et al., 2011).

Glucan Synthase Inhibitors

β-glucan, a polysaccharide composed of glucose monomers linked by (1,3)-β or (1,6)-β bonds, is widespread among fungi and forms an essential amorphous, insoluble component of the cell wall, guaranteeing many of its properties and its physical integrity (Latge, 2007; Fleet, 1991). At present, glucan synthesis is the only component of cell wall synthesis machinery that has successfully led to the development of a new drug in the market. The development of echinocandins, the first class of antifungals to target the fungal cell wall, was a milestone achievement in antifungal chemotherapy. Echinocandins were discovered as fermentation metabolites with antifungal activity during screening programs for new antibiotics (Hector, 1993). They have inherent selective activity *in vitro* and *in vivo* animal models. Echinocandins are cyclic hexapeptides N-acylated with an aliphatic chain of different length and different substituents in the hexapeptide ring (Fig. 1) (Morrison, 2005). Since the first echinocandin was discovered in the early 1970s (Nyfeler and Keller, 1974), many members of the family have been described as produced by diverse fungi (Table 1).

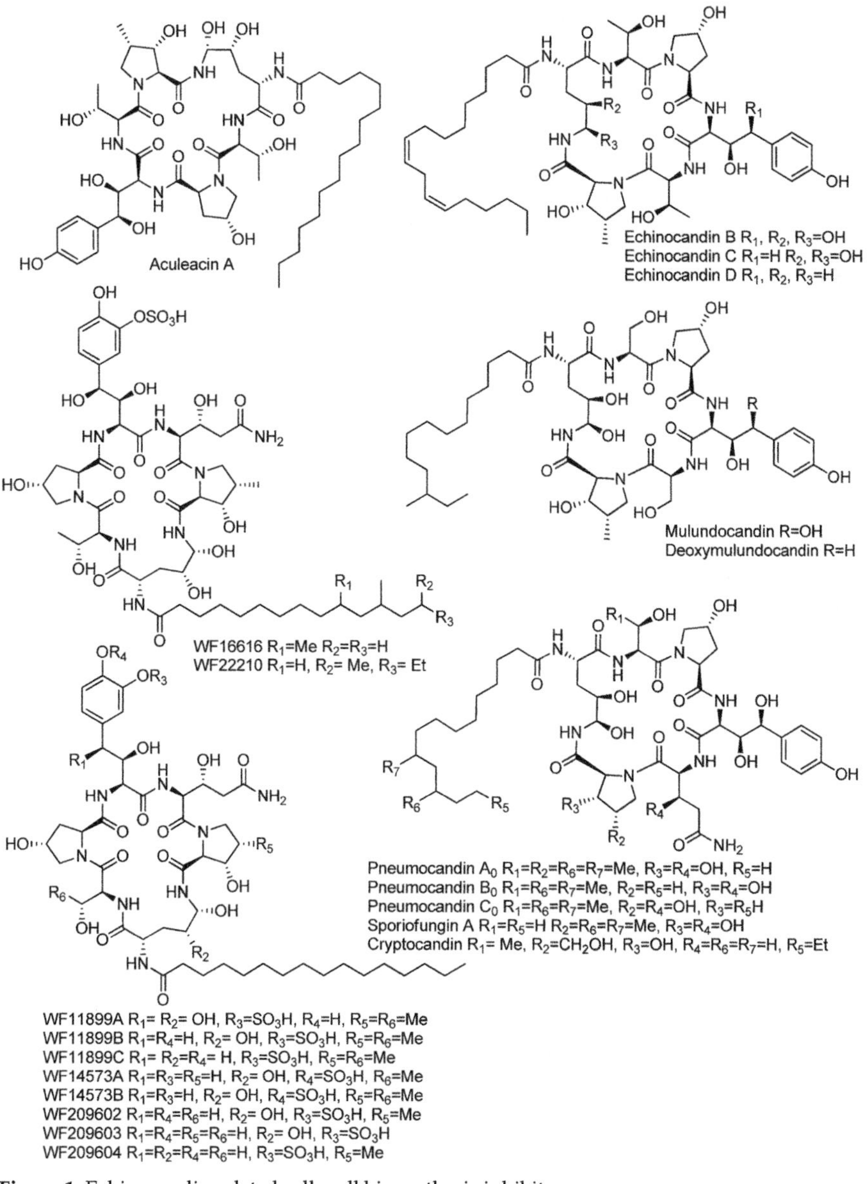

Figure 1. Echinocandin related cell wall biosynthesis inhibitors.

Table 1. Fungi as producer of metabolite inhibitors of fungal cell wall.

Fungi	Metabolite(s)	Reference
Arthrinium arundinis, A. phaeospermum, Leotiales anamorphs, *Coelomycete* undet	Arundifungin	Cabello et al., 2001
Ascotricha amphitricha, Mycoleptodiscus atromaculans	Ascoteroside	Onishi et al., 2000
Aspergillus aculeatus, Eupenillin sp., *Aspergillus japonicus* var. *aculeatus*	Aculeacin A	Mizuno et al., 1977
Aspergillus nidulans var. *echinolatus, A. rugulosus, Emericella rugulosa,* others	Echinocandin B-D	Nyfeler and Keller, 1974
Aspergillus Sydowii, Eupenicillium sp.	Mulundocandin Deoxymulundocandin	Roy et al., 1987; Mukhopadhyay et al., 1992
Clavariopsis aquatica	Clavariopsins	Kaida et al., 2001
Coleophoma empetri, Coleophoma crateriformis, Tolypocladium parasiticum, Chalara sp.	WF11899 and related sulfate-derivatives	Fujie, 2007
Coryneum modonium	Corynecandin	Gunawardana et al., 1997
Cryptosporiopsis quercina	Cryptocandin	Strobel et al., 1999
Dictyochaeta simplex	L-687781	VanMiddlesworth et al., 1991
Fusarium sambucinum	Fusacandin	Yeung et al., 1996
Gilmaniella sp.	BU-4794F	Aoki et al., 2000
Glarea lozoyensis, Pezicula sp., *Cryptosporiopsis* sp.	Pneumocandin A_0	Schwartz et al., 1989, 1992; Bills et al., 1999, 2009a; Noble et al., 1991
Glarea lozoyensis	Other pneumocandins	Schwartz et al., 1992
Hormonema carpetanum	Enfumafungin	Pelaez et al., 2000; Onishi et al., 2000
Monocillium sp., *Triocothecin* sp.	Furanocandin	Magome et al., 1996
Papularia sphaerosperma	Papulacandins	Traxler et al., 1977
Penicillium arenicola, Cryptosporiopsis sp.	Sporiofungins	Dreyfuss et al., 1986
Phialophora cyclaminis	Mer-WF3010	Kaneto et al., 1993
Taifanglania inflate Pecilomices inflatus	YW3548, BE-49385	Sutterlin et al., 1997; Wang et al., 1991
Trichoderma longibrachiatum, T. koningii, T. viride	Ergokonin A	Vicente et al., 2001
Unidentified fungus	FR901469	Fujie et al., 2000
Unidentified fungus	Arborcandins	Ohyama et al., 2000
Coleophoma empetri	FR901379 (WF11899A), WF11899B-C	Hino et al., 2001
Coleophoma empetri	FR220897 (WF14573B) FR220899 (WF14573A)	Kanasaki et al., 2006b; Hino et al., 2001
Tolypocladium parasiticum	FR190293 (WF16616)	Kanasaki et al., 2006a; Hino et al., 2001
Chalara sp.	FR227673 (WF22210)	Kanasaki et al., 2006a; Hino et al., 2001
Coleophoma crateriformis	FR209602-4 (WF738A-C)	Kanasaki et al., 2006c,d; Hino et al., 2001

The class of pneumocandin and echinocandin has been successfully used to develop CANCIDAS™ (caspofungin acetate), the first of the three antifungal agents currently approved for treating serious fungal infections by *Candida* and *Aspergillus* species (Fig. 2) (Morrison, 2006). Pneumocandins (Fig. 1) were originally discovered from *Glarea lozoyensis* a fungal strain isolated from a water sample from Spain (Bills et al., 1999; Schwartz et al., 1992; Schwartz et al., 1989). A minor component from *G. lozoyensis* fermentation, pneumocandin B_0, was the basis used to develop CANCIDAS™ (Bartizal et al., 1993; Conners and Pollard, 2005). This semisynthetic pneumocandin, caspofungin acetate, is an aza-substituted derivative of pneumocandin B_0 (Bouffard et al., 1996). The introduction of additional amino groups in the peptide ring of pneumocandin B_0 has doubly increased the solubility of the molecule and the potency against fungal pathogens (Bouffard, 1994). The compound has shown its effectiveness in animal models for disseminated candidiasis, aspergillosis, coccidiomycosis

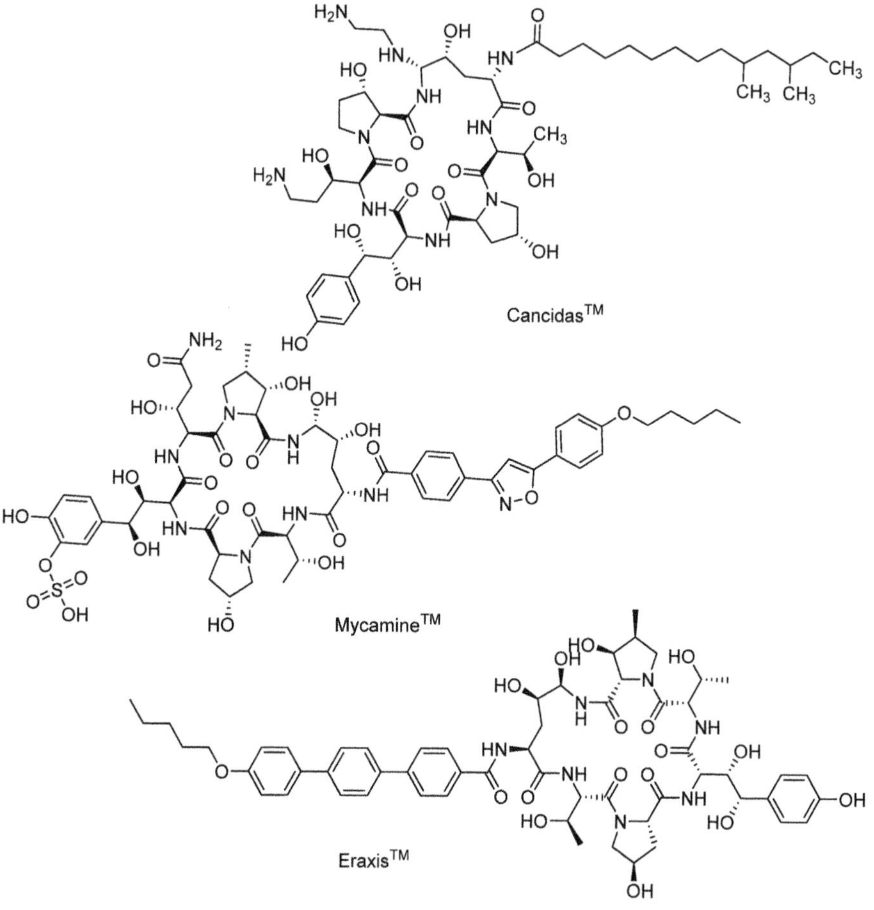

Figure 2. Marketed echinocandin derivatives.

and pneumonia caused by *Pneumocystis carinii* (Abruzzo et al., 1997; Smith et al., 1996; Powles et al., 1998; González et al., 2001). Good tolerance of the compound and its efficacy in the treatment of oropharyngeal and esophageal candidiasis, as well as in invasive aspergillosis, has been noticed in its clinical trial (Arathoon et al., 1998; Maertens et al., 2000). CANCIDAS™ was initially approved by the FDA for its use against invasive aspergillosis, refractory to or intolerant of other therapies. Later it was also considered for treating systemic candidiasis.

Other echinocandin-like antifungal agents were modified to improve solubility, spectrum of activity and pharmacokinetic characteristics (Debono and Gordee, 1994). These have been developed for clinical use and include FK463, MYCAMINE™ (micafungin) (Fig. 2), an echinocandin derivative with a sulfate ester moiety in the hexapeptide nucleus (Ikeda et al., 2000; Akihiko, 2007) and LY303366, ERAXIS™ (V-echinocandin, anidulafungin) (Fig. 2), which has a terphenyl head group and a C5 chain (Petraitis et al., 2001). All three echinocandins are structurally similar cyclic hexapeptide antibiotics with a modified N-linked acyl lipid side, which plays a role in anchoring the hexapeptide nucleus to the fungal cell membrane where the drug interacts with the target enzyme complex involved in cell wall synthesis (Denning, 2003). Experience with this antifungal class suggests that it is among the best tolerated and safest class of antifungals available.

Other echinocandin derivatives have also been reported from Fujisawa (Tojo et al., 2000), Roche (aerothricins) (Aoki et al., 2000) and Aventis (mulundocandins, deoxymulundocandin) (Roy et al., 1987; Mukhopadhyay et al., 1992) (Fig. 1, Table 1). The success of the lipopeptide class of glucan synthesis inhibitors has prompted interest from the industry in the search for other structural types with improved features over the echinocandins, especially for their lack of oral absorption.

Additional pneumocandin-producing isolates of *G. lozoyensis* were obtained from plant litter collected in southern Argentina and New Hampshire, USA (Peláez et al., 2011). Moreover, it has been observed that the production of pneumocandins and cryptocandins is widespread across many *Pezicula* species and their anamorphic *Cryptosporiopsis* states (Bills et al., 2009b; Peláez et al., 2011; Roemer et al., 2011). *Cryptosporiopsis* or *Pezicula* strains have been reported in the literature producing pneumocandin A_0 (Noble et al., 1991) and the serine analog of pneumocandin A_0, named sporiofungin (Dreyfuss, 1986), as well as cryptocandin (Strobel et al., 1999) (Fig. 1).

Bills and his co-workers mapped the distribution of fungi producing similar members of the echinocandin class, characterized by a sulphate group at the para or meta position of the homotyrosine in the hexapeptide ring (Bills et al., 2009a; Peláez et al., 2011; Roemer et al., 2011). Astellas Pharmaceuticals reported the sulphated echinocandins (Table 1), using WF11899A (Fig. 1) as the starting point for the commercial antifungal MYCAMINE™ (Fujie, 2007). Interestingly, the production of another echinocandin containing a sulphate group in the homotyrosine residue (compound FR227673 or WF22210)

(Fig. 1) has been reported from a strain assigned to the anamorphic genus *Chalara* (Kanasaki et al., 2006a; Hino et al., 2001).

Besides echinocandins and the like, other cyclic peptides that are glucan synthesis inhibitors have been identified (Table 1). Arborcandins (Fig. 3) are recently described antifungal agents, containing a 10-aminoacid ring and two lipophilic tails (Ohyama et al., 2000). Likewise, the compound named as FR901469 (Fig. 3) is a macrocyclic lipopeptidolactone composed of 12 aminoacids and a 3-hydroxypalmitoyl moiety (Fujie et al., 2000). Clavariopsins (Fig. 3), cyclic depsipeptides lacking a long lipophilic radical, have also been suggested to act as inhibitors of glucan synthesis (Kaida et al., 2001). However, to date, besides cyclic lipopeptides, only two other types of glucan synthesis inhibitors are known: the papulacandins and related compounds, and the acidic triterpenes.

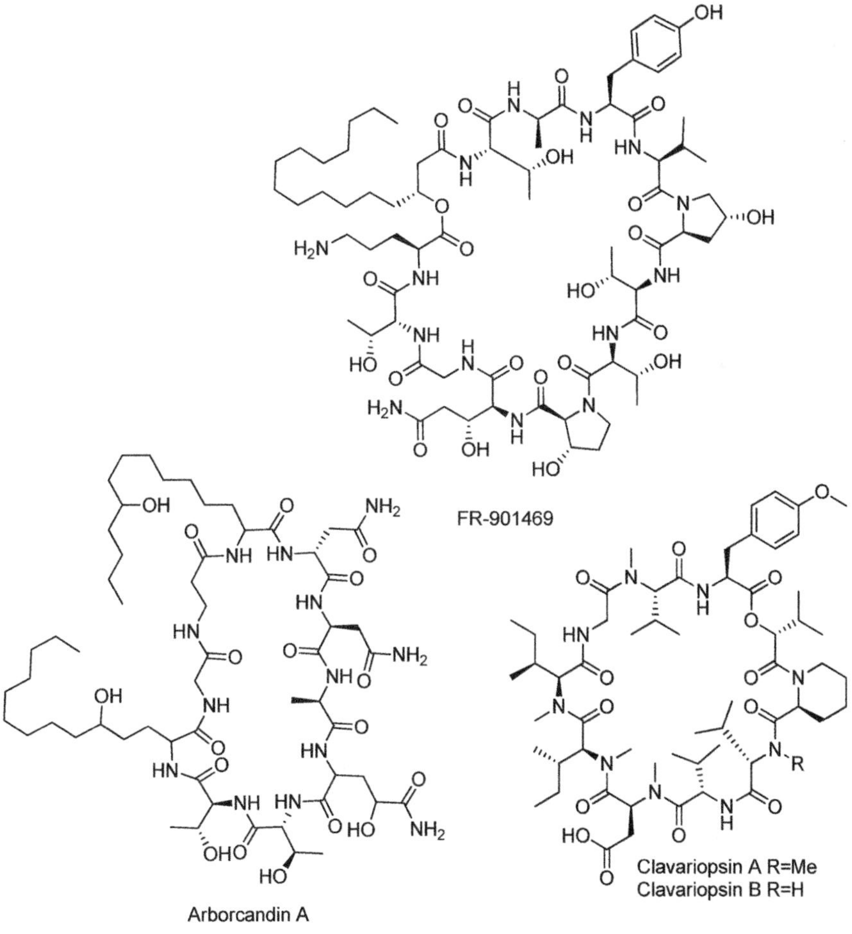

Figure 3. Other peptidic cell wall biosynthesis inhibitors.

Furthermore, aculeacin A (Fig. 1), is another antifungal antibiotic with a structure related to echinocandin B isolated from the mycelial cake of *Aspergillus aculeatus* M-4214 that differs in the long fatty acid chain. Upon acid hydrolysis aculeacin A generates palmitic acid and five ninhydrin-positive products including theonine and hydroxyproline. The antibiotic showed potent activity against molds and yeasts, but exhibited no antibacterial activity and has relatively low toxicity in mice (Mizuno et al., 1977).

The papulacandins (Fig. 4) are glycolipids, discovered in the late 1970s (Traxler et al., 1977). A series of related compounds has been discovered over the years, all of them are produced by fungi (Table 1, Fig. 4). Despite the efforts of medicinal chemists, neither papulacandins nor any of their relatives have been developed as drugs, primarily due to their limited potency in animal models (Georgopapadakou and Tkacz, 1995; Yeung et al., 1996).

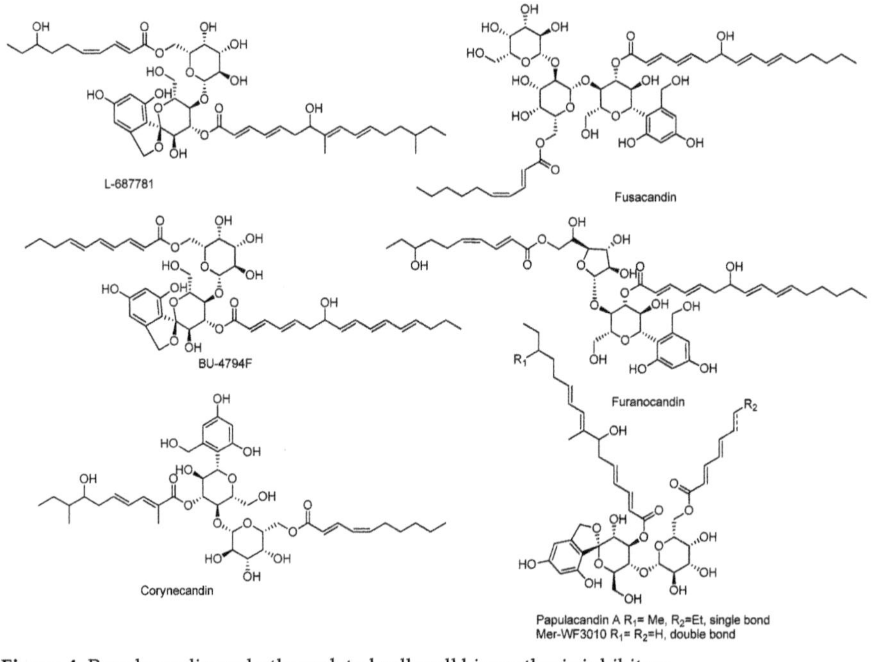

Figure 4. Papulacandin and other related cell wall biosynthesis inhibitors.

The most recently discovered class of β-glucan synthase inhibitors are triterpenoids containing a polar moiety (Onishi et al., 2000). This polar moiety can be a glycoside (as in enfumafungin and ascosteroside), a succinate (as in arundifungin) or a sulfate-derivative amino acid (as in ergokonin A) (Fig. 5). Fungi that produce these compounds are listed in Table 1 (Cabello et al., 2001; Onishi et al., 2000; Peláez et al., 2000; Vicente et al., 2001). Although these compounds were inactive or only weakly active *in vivo* (mouse) models, collectively they represent a new molecular platform for antifungal drugs

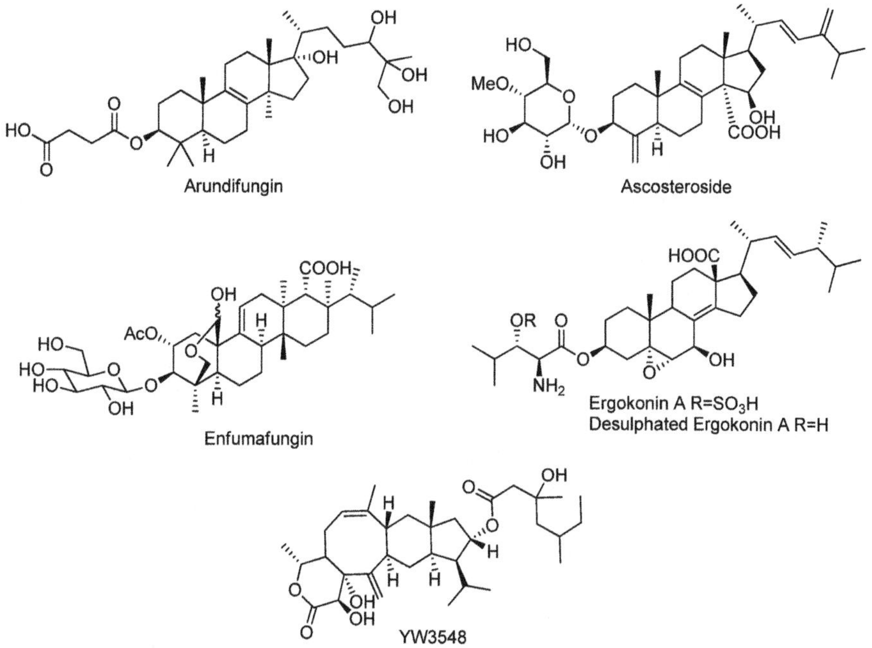

Figure 5. Polar triterpenoids inhibiting cell wall biosynthesis.

acting on the fungal cell wall. The most exciting acidic terpenoid is the endophyte metabolite, enfumafungin, an acidic triterpene glycoside that potently inhibits fungal cell wall biosynthesis. Enfumafungin was originally isolated from a strain identified as *Hormonema* sp. Subsequently, the fungus was recovered from various substrata, including living and decaying leaves of *Juniperus* spp. and other plants, and even from rock surfaces in Spain and was named *Hormonema carpetanum* (Bills et al., 2004). Extensive modifications to the chemistry of fungal natural product (enfumafungin) have resulted into a semi-synthetic derivative orally available β-1,3-glucan synthase inhibitor (e.g., MK-3118/SCY-078) that recently entered Phase I Clinical Trial (Motyl et al., 2010).

The MK-3118/SCY-078, exhibited *in vitro* antifungal activity against non-*Aspergillus* molds comparable to that of echinocandins with one important exception. It was the only agent tested with activity against *Scedosporium prolificans*, a notoriously pan resistant mold for which there are no good treatment options (MEC90 4 µg/ml). This discovery warrants further investigation into the potential role of SCY-078 for treating *S. prolificans* infections. Also from a clinical perspective, the good activity of SCY-078 against *Paecilomyces variotii* is of particular interest. *P. variotii* has been associated with disseminated infections including fungemia and peritonitis in patients undergoing peritoneal dialysis (Lamoth and Alexander, 2015).

The conclusion that these compounds are acting as inhibitors of glucan synthesis is based on several lines of evidence: first, the spectrum of activity, being active against *Aspergillus* and *Candida* species, but inactive against *Cryptococcus*; second, they induced the same alterations in the micro-morphology of *Aspergillus fumigatus* as other inhibitors of glucan synthesis (Cabello et al., 2001; Peláez et al., 2000; Vicente et al., 2001); and third, direct measurement of the effect of these compounds on the synthesis of cell wall macromolecules indicates that glucan is the only polymer whose synthesis is significantly altered upon treatment with these agents (Onishi et al., 2000). Although these compounds were inactive or only weakly active in the *in vivo* mouse model (Cabello et al., 2001; Peláez et al., 2000; Vicente et al., 2001), they represent a new paradigm in the search for antifungal compounds with this mode of action, and hence they could be useful as a lead for the development of improved drugs.

Clearly, a weak point in all the glucan synthesis inhibitors discovered or developed to date is their lack of activity against *C. neoformans*. The reasons for this lack of activity are unclear. The hypothesis that echinocandins do not inhibit β-(1,6) glucan synthesis, which seems to be the main glucan in *C. neoformans* cell wall (Georgopapadakou and Tkacz, 1995), has been suggested recently. Moreover, the FKS1 homologue gene, coding the catalytic subunit of β-(1,3) glucan synthase, has been shown to be essential in *C. neoformans*. However, the enzyme could be relatively resistant to the action of echinocandins and the other glucan synthesis inhibitors (Feldmesser et al., 2000). Other cell wall inhibitors produced by fungi such as: corynecandin, BU-4794F, furanocandin, etc., are also summarized (Table 1).

Chitin Synthesis Inhibitors

The fungal cell wall protects the cell from the environment and is a dynamic structure that is continually modified by enzymes to facilitate growth (Latge, 2001). The cell wall of all pathogenic fungi contains chitin, as a major and essential component, that is broken down by chitinases during cell wall remodeling. Disrupting this process is expected to result in a decrease of fungal viability and/or virulence.

Chitin is a homopolymer constructed from units of β-(1,4)glycoside linked with N-acetyl-D-glucosamine bond. This biological polymer is one of the structural microfibrillar components of the fungal cell wall structure that maintains the morphologic shape of the cells and plays an essential role in fungal morphogenesis (Ruiz-Herrera et al., 1992). Chitin, which is synthesized by chitin synthases (chitin synthase I, involved in a repair function at the time of cytokinesis; chitin synthase II, an essential enzyme for primary septum formation between mother and daughter cells; and chitin synthase III, which synthesizes lateral chitin in the cell wall (Shaw et al., 1991), accounts for 1% of the cell wall and is distributed differently from glucan. Chitin links covalently to the cellular glucan, thereby strengthening the wall. From the three chitin synthases present in *Candida albicans*, chitin synthases I and III are the main targets for the development of new cell wall biogenesis inhibitors (Ruiz-Herrera and San-Blas, 2003). The classical inhibitors of chitin synthesis are nikkomycins and polyoxins (Gabib, 1991). Both belong to a family of peptide-nucleoside antimycotic agents that are substrate analogs of UDP-N-acetylglucosamine, the essential building block for chitin biosynthesis, and they were isolated from two different *Streptomyces* species. Nikkomycins and polyoxins are currently used exclusively as agricultural fungicides, due to their modest activity against human pathogens (Cohen, 1993).

Recently, as part of the continuing screening for new chitin synthase inhibitors, three novel antifungal compounds were found from fungal isolates: phellinsin A, arthrichitin and argifin. Phellinsin A, a phenolic compound isolated from the cultured broth of fungus PL3 identified as *Phellinus* sp., selectively inhibited chitin synthase I and II of *S. cerevisiae*. In addition, this compound exhibited antifungal activity against human pathogens such as *Trichophyton mentagrophytes* and *Aspergillus fumigatus* and exhibited very weak activity against other human pathogens such as *Cryptococcus neoformans* and *Coccidioides immitis*. It selectively inhibited chitin synthase I and II of *Saccharomyces cerevisae* with IC_{50} values of 76 and 28 μg/ml, respectively. However, it showed no activity against *Candida albicans*, *Candida lusitaniae*, *Candida krusei*, *Candida tropicalis* and *Fusarium oxysporum* (Hwang et al., 2000). Arthrichitin is a cyclic depsipeptide which was isolated from *Arthrinium phaeospermum* and (as LL156256g) from the marine fungus *Hypoxylon oceanicum* (Vijayakumar et al., 1996; Schlingmann et al., 1998). This compound showed activity against *Candida* spp., *Trychophyton* spp. and several phytopathogens. Although it's *in vitro* potency is too low for its use in the clinic, it has been suggested that analogs with improved activity could be developed (Vijayakumar et al., 1996). Argifin is another natural product that was first isolated from *Gliocladium* fungal cultures from a sample collected in Micronesia (Omura et al., 2000). The structure of argifin was shown to be an unusual arginine-containing cyclopentapeptide (Arai et al., 2000). A fragment identified from the natural product argifin, dimethylguanylurea, was identified as the minimal fragment necessary for competitive inhibition of the bacterial-type chitinase (Andersen et al., 2008). This argifin derivative has inhibitory activity against the *Aspergillus fumigatus* (Rush et al., 2010).

Mannosyl Transferase Inhibitors

Mannoproteins represent the third most important component of the fungal cell wall, as they form the outer layer of the cell wall with radially extending fibrillae outside the cell wall. The majority of the cell wall mannoproteins are anchored by β-(1,6) and β-(1,3)-glucan (Van der Vaart et al., 1996) and play several roles in the function of fungal membranes. Mannosylation of these proteins is catalyzed by a family of mannosyl transferases (Protein O-mannosyltransferases, PMT). These have been reported in *S. cerevisiae*, *C. albicans*, *A. fumigatus* and other fungal species (Arroyo-Flores et al., 2000). Therefore, they have been considered as another potential target in fungal membranes for antimycotic agents. To date, no mannoprotein associated with the cell wall has been identified as essential, by evaluating cell viability after gene disruption experiments. The chemicals of pradimicin/benanomycin family inhibit the function of mannoproteins. Their chemical structure possesses a benzo[a]naphthacenequinone skeleton (Oki et al., 1988; Takeuchi et al., 1988). The free carboxyl group of these compounds interacts with the saccharide portion of cell-surface mannoproteins, which is followed by disruption of the plasma membrane and leakage of intracellular potassium. These antifungal (and antiviral) agents produced by *Actinomadura* species are without significant acute toxicities. Pradimicin/benanomycin and analogues were effective in experimental animal models. However, phase I clinical trials suggested drug-related toxicities and the development was, therefore, discontinued (Fromtling, 1998).

Compounds Targeting Cell Membrane Synthesis

Sphingolipids are ubiquitous and essential structural components of eukaryotic membranes as well as some prokaryotic organisms and viruses (Merrill and Sandhoff, 2002). They are predominantly found in outside, or outer leaflet of the plasma membrane bilayer (Gurr et al., 2002). Essential functions of sphingolipids coupled with the divergence of the biosynthetic pathway between mammals and eukaryotic pathogens have resulted in the investigation of the biosynthetic enzymes as possible drug targets for antifungals and antiprotozoals. While many steps in the human and fungal sphingolipid biosynthetic pathway are similar, there are several enzymes uniquely found in fungi that make sphingolipid synthesis an attractive potential target for antifungal therapy. Thus, three key enzymes of this pathway have been targeted—inositol phosphorylceramide (IPC) synthase, an enzyme without mammalian counterpart, serine palmitoyltransferase and ceramide synthase. Natural products have proven to be a rich source of structurally diverse inhibitors of sphingolipid biosynthesis and hence inhibitors of many of the steps in sphingolipid biosynthesis have been described (Mina et al., 2009).

Sterols, essential lipids in most eukaryotic cells, ensure important structural and signaling functions. The selection pressure that has led to different dominant sterols in the three eukaryotic kingdoms remains unknown. Ergosterol is a major constituent of the fungal plasma membrane hence the ergosterol biosynthesis pathway is important (Paltauf et al., 1992; Parks and Casey, 1995). This pathway and plasma membranes of other organisms are composed predominantly of other types of sterols. However, the pathway is fungal-specific and not universally present in all fungi. The ergosterol biosynthesis pathway has been the subject of intensive investigation as a target for antifungal drugs (Lupetti et al., 2002).

Inositol Phosphorylceramide (IPC) Synthase Inhibitors

The post-ceramide divergence, represented by the absence of inositol phosphorylceramide (IPC) synthase and inositol-based sphingolipids in mammalian cells, highlights the therapeutic potential of inhibitors targeting fungal IPC synthases. These inhibitors could result in selective antifungal drugs with minimal host toxicity. To date, only the natural compounds like aureobasidin A, khafrefungin and rustmicin have been reported as potent inhibitors of the fungal IPC synthase (Table 2). Aureobasidins A to R (produced by *Aureobasidium pullulans*) are cyclic depsipeptides described as novel antifungal agents with potent *in vitro* antifungal activity, especially against *C. albicans*. Of these, aureobasidins A, B, C and E are the most potent (Takesako et al., 1991). Aureobasidin A (Fig. 6) inhibits the IPC synthase of *S. cerevisiae* (Nagiec et al., 1997) and has an inhibitory effect against kinetoplastid enzyme orthologues (Harris et al., 1998).

There are several potent inhibitors of the sphingolipid synthesis at the IPC synthase level produced by fungi and actinomycetes, which are specific inhibitors of this enzyme such as khafrefungin (Fig. 6), a fungal compound consisting of a novel 22-carbon linear polyketide acid esterified with an aldonic acid. This exhibits broad spectrum antifungal activity with *C. albicans* (Mandala et al., 1997a). Rustmicin (also known as galbonolide A) is a macrolide antifungal agent produced by a strain of *Micromonospora* sp. (Fig. 6) with potent activity against wheat stem rust fungus *Puccinia graminis* (Takatsu et al., 1985), *Botrytis cinerea* and several other phytopathogens (Fauth et al., 1986). Rustmicin also has extraordinarily potent antifungal activity against several human pathogens, especially *C. neoformans*. Its antifungal activity is due to the inhibition of sphingolipid synthesis at the IPC synthase level at picomolar to low nanomolar concentrations (Mandala et al., 1998). The rustmicin-related macrolide galbonolide B (Fig. 6) was also reported to inhibit IPC synthase but with lesser potency than galbonolide A (Harris et al., 1998).

Khafrefungin and rustmicin did not have any detectable effect on lipid synthesis in mammalian cells (Mandala et al., 1997a). Rustmicin and aureobasidin A have been nontoxic in animal studies (Mandala et al., 1998; Takesako et al., 1993), supporting the hypothesis that inositol phosphoceramide synthase is a fungal specific target, and hence are the preferred targets for further research to look for novel antifungals, blocking sphingolipid biosynthesis.

Further development of all the three inhibitors (aureobasidin A, khafrefungin and rustmicin) of IPC synthase described above has been stalled either due to the lack of physical properties required for an acceptable

Table 2. Natural products inhibitors of sphingolipid biosynthesis, protein synthesis and ergosterol synthesis.

Fungi	Metabolite(s)	Reference
Sphingolipid biosynthesis		
Aspergillus fumigatus, Paecilomyces variotii	Sphingofungins	Zweerink et al., 1992; Horn et al., 1992; Mandala et al., 1994
Trichoderma viride	Viridiofungins	Mandala et al., 1997a,b
Isaria sinclairii	Myriocin	Miyake et al., 1995
Fusarioum moniliforme	Fumonisin B1	Wang et al., 1991
Sporormiella australis	Australifungin	Mandala et al., 1995
Aureobasidium pullulans	Aureobasidin A	Nagiec et al., 1997
Unidentified sterile fungus	Khafrefungin	Mandala et al., 1997a,b
Sporomiella minimoides	Minimoidin	Mandala and Harris, 2001
Phoma spp.	Phomafungin	Herath et al., 2009
Protein synthesis		
Sordaria araneosa	Sordarin	Sigg et al., 1969
Zopfiella marina	Zofimarin	Ogita et al., 1987
Penicillium minioluteum	BE31405	Okada et al., 1998
Unidentified sterile fungus *Xylaria* spp.	SCH57404 Xylarin	Coval et al., 1995 Schneider et al., 1995
Hypoxylon croceum	Hypoxysordarin	Daferner et al., 1999
Graphium putredinis	GR135402	Kinsman et al., 1998
Aspergillus spp.	Aspirochlorine	Monti et al., 1999 Wang et al., 2012
Morinia pestalozzioides	Moriniafungin	Basilio et al., 2006
Fusarium larvarum	Parnafungins	Bills et al., 2009b Parish et al., 2008
Ergosterol synthesis		
Petriella spp. *Pseudallescheria* spp.	Tyroscherin	Hayakawa, 2004
Penicillium spp. *Neosartorya* spp.	Restricticin	Schwartz et al., 1991
Crytosporiopsis spp.	Hymeglusin (Antibiotic 1233A)	Omura et al., 1987
Phoma spp. *Penicillium* spp.	Antibiotic PF1163A	Nose et al., 2000

pharmacokinetic profile/poor bioavailability or because of their highly complex structures that render chemical synthesis more challenging (Young et al., 2012).

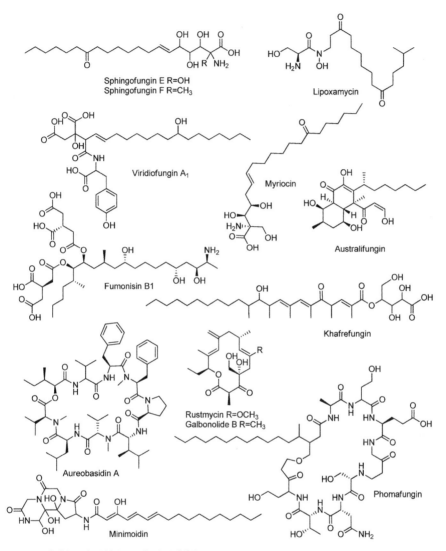

Figure 6. Sphingolipid biosynthesis inhibitors.

Ceramide Synthase Inhibitors

The sphingolipid biosynthesis pathway before the synthesis of ceramide/phytoceramide is largely conserved across evolution. Therefore, all the inhibitors identified as targeting the fungal enzymes as ceramide inhibitors are nonselective because they also inhibit the mammalian orthologues (Mina et al., 2009), which prevent their clinical application as antifungal agents. The fumonisins, mycotoxins, initially isolated as tumor promoting agents are associated with severe toxicological effects in animals. They inhibit *de novo* sphingolipid biosynthesis blocking the reaction catalyzed by ceramide synthase in rat hepatocytes, and their toxicity and carcinogenicity have been attributed to the inhibition of ceramide synthesis and the concomitant accumulation of sphingoid bases (Wang et al., 1991). Although fumonisin B_1 produced by *Fusarium moniliforme* (Fig. 6) does inhibit fungal ceramide synthase *in vitro* (Wu et al., 1995), the fumonisins have very poor activity against whole cell fungal sphingolipid synthesis or growth and their limited penetration could account for their poor antifungal activity.

Australifungin (Fig. 6) was the first non-sphingosine-based inhibitor identified for the sphingolipid biosynthetic pathway. It shows a unique combination of α-diketone and β-ketoaldehyde functional groups and shows a broad spectrum antifungal activity. It inhibits ceramide synthase *in vitro* at nanomolar concentrations and the enzyme inhibition accounts for the arrest of sphingolipid synthesis. An analog isolated from the same fungus, australifunginol, also blocked the enzyme converting sphinganine to ceramide, but it was at least 50-fold less potent. Australifungin had particularly good activity against *Candida pseudotropicalis*, *C. tropicalis*, and *Cryptococcus neoformans*. Much weaker activity was detected for australifunginol (Mandala et al., 1995). The main limitation to the therapeutical use of australifungin for the treatment of fungal infections is its activity against ceramide synthesis in HepG2 (Mandala et al., 1995).

Serine Palmitoyltransferase Inhibitors

The first step involved in sphingolipid biosynthesis is the condensation of serine and palmitoyl CoA, a reaction catalyzed by serine palmitoyltransferase (SPT), a key enzyme of sphingolipid metabolism to produce 3-ketodihydrosphingosine (KDS). SPT is suggested to be a key enzyme for the regulation of sphingolipid levels in cells. Its regulation of sphingolipid synthesis, at the SPT step, prevents a harmful accumulation of metabolic sphingolipid-intermediates including sphingoid bases and ceramide (Hanada, 2003). There are several fungal inhibitors of this enzyme. Sphingofungins A through F (Horn et al., 1992; Zweerink et al., 1992) were isolated from thermotolerant species of *Aspergillus fumigatus* and *Paecilomyces variotti*. These constitute a family of novel compounds whose chemical structures resemble the long chain base intermediates in the sphingolipid pathway (Mandala et al., 1994). They were initially identified as broad spectrum antifungal agents, inhibiting the growth of various *Candida* species and showing an especially potent activity against *Cryptococcus neoformans*, but they are inactive against filamentous fungi (Horn et al., 1992; VanMiddlesworth et al., 1991). Viridiofungins A (Fig. 6), B and C are members of a novel family of amino alkyl citrates isolated from the fungus *Trichoderma viride* that have potent, broad spectrum antifungal activity, inhibiting the growth of different pathogenic fungi such as *C. neoformans*, *Candida* species and *A. fumigatus* (Onishi et al., 1997). They are not as specific for serine palmitoyltransferase inhibition as the other inhibitors of the enzyme. They also inhibit squalene synthase and other enzymes that are sensitive to dicarboxylic acids, although at higher concentrations than required for serine palmitoyltransferase inhibition (Mandala et al., 1997b).

It has also been demonstrated by a variety of biological and biochemical means that sphingofungins and viridiofungins are specific inhibitors of serine palmitoyltransferase in fungi at nanomolar concentrations (Zweerink et al., 1992).

Lipoxamycin, a bacterial inhibitor (Fig. 6) and hydroxylipoxamycin, an analog co-produced in the fermentation by strains of *Streptomyces* have a long alkyl chain and an amino-containing polar head group, but otherwise do not resemble the sphingoid bases as closely as the sphingofungins do. These compounds have antifungal activity against many human pathogenic fungi with better potency against some of the *C. neoformans* and *Candida* species. *A. fumigatus* was not inhibited in broth dilution assays, but other filamentous fungi were sensitive to the lipoxamycins in disk diffusion assays (Mandala et al., 1994).

The serine palmitoyltransferase inhibitors described above also have potent activity against the mammalian enzyme (Zweerink et al., 1992; Horn et al., 1992; Mandala et al., 1997b). But the toxicity may be mechanism based, since studies with a CHO cell mutant have shown that this enzyme is essential in mammalian cells. Another natural product isolated from the fungus *Isaria sinclairii*, the potent immunosuppressant ISP-1/myriocin (Fig. 6), has also been reported to inhibit serine palmitoyltransferase at picomolar concentrations and the proliferation of an IL-2 dependent mouse cytotoxic T cell line, CTLL-2, at nanomole concentrations (Miyake et al., 1995).

Fingolimod (FTY720/Gilenya; Novartis) is a derivative of ISP-1 (myriocin; 2-amino-2-[2-(4-octylphenyl)ethyl] propane-1,3-diol) and is phosphorylated *in vivo* by SphK to produce FTY720 phosphate (FTY720P; Kumlesh et al., 2008). Fingolimod is the first oral treatment for multiple sclerosis. It is the first-in-class sphingosine 1-phosphate receptor modulator that binds to sphingosine 1-phophate receptors on lymphocytes and via downregulation of the receptor prevents lymphocyte egress from lymphoid tissues into the circulation. Two large phase III studies with fingolimod have shown superior efficacy of the drug in two dosages compared to placebo and to weekly intramuscular injections of Interferonbeta-1a. Among the possible side effects of the drug is a transient bradycardia after the first dose of fingolimod including possible AV blockade and, therefore, monitoring of pulse rate and blood pressure for 6 hours following the first application is needed. It is noted that most of the deaths and cardiovascular problems had occurred in patients with a history of cardiovascular problems or taking other medicines.

However, the data, reviewed by the EMA agency on the safety issue, were not conclusive as to whether Gilenya was the cause of the deaths. In fact, the EMA was of the opinion that the possible risk of heart problems in patients taking Gilenya could be minimized by further strengthening the existing warnings on the cardiovascular effects of the medicine and ensuring close monitoring of all patients (Fazekas, 2012).

Fatty Acid Elongation Inhibitors

A compound that indirectly inhibits sphingolipid synthesis by blocking the fatty acid elongation pathway is minimoidin (Fig. 6), thus depriving the substrate for ceramide synthase (Mandala and Harris, 2001). Minimoidin blocks radioactivity uptake into C26 fatty acids and inhibits the incorporation of ^{14}C-malonyl-CoA into long chain fatty acids.

Ergosterol Inhibitors

The major target of azole antifungal drugs is lanosterol 14-alpha demethylase, a member of the cytochrome P450 family known as Erg11 protein in many fungal species. Squalene epoxidase (Erg1p in *S. cerevisiae*) is the specific target of allylamine drugs such as terbinafine (Leber et al., 2003). The polyenes are the oldest class of the antifungal agents having been introduced for the first time in the 1950s. These compounds are produced by different actinomycetes: amphotericin B (*Streptomyces nodosus*), nystatin (different *Streptomyces* spp.) and natamycin (commercial poultry feed) (*S. natalensis, S. chatanoogensis*), bind ergosterol, the principal sterol in fungal cytoplasmatic membrane, with their hydrophobic moiety.

In addition, there are several ergosterol inhibitors which are produced by fungi (Table 2 and Fig. 7). Tyroscherin was isolated from *Petriella* sp. and *Pseudallescheria* sp. (Hayakawa, 2004) and was originally reported to inhibit insulin-like growth factor-1 (IGF-1)-dependent growth of MCF-7 human breast cancer cells with an IC_{50} values of 30 nM-6.2 µM. Different approaches have been adopted by different groups to obtain the total synthesis or analogs of tyroscherin (Yoon et al., 2013). Tae et al. (2010) accomplished a concise total synthesis of the proposed and revised structure of tyroscherin in 12 steps from commercially available materials, however, their findings contradict the reported IGF-specificity of growth inhibition, previously reported for tyroscherin. Studies to identify tyroscherin's *in vivo* biological targets and elucidate its mode of action are underway (Tae et al., 2010), restricticin is a naturally-occurring antifungal agent which contains triene, pyran and glycine ester functionalities and is unrelated to any previously known family of natural products. This compound has been produced and isolated from both, solid and liquid fermentations, of *Penicillium restrictum* and *Neosartorya* sp. The antifungal spectrum of restricticin was similar to ketoconazole, as was the significant increase in potency observed in the complex Kimmig agar vs. the defined yeast nitrogen base glucose agar. The results of this evaluation indicated that the antifungal spectrum of this compound is broad with activity against both, yeast and filamentous fungi

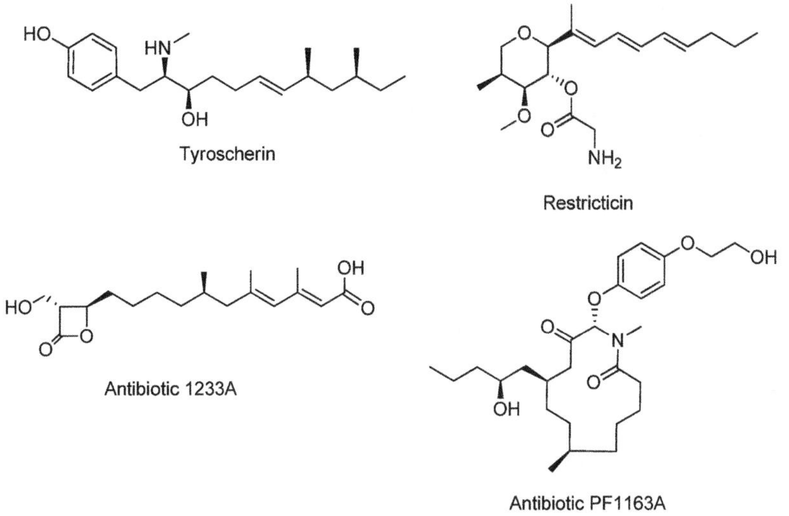

Figure 7. Ergosterol synthesis inhibitors.

(Schwartz et al., 1991). The fungal beta-lactone, hymeglusin (Antibiotic 1233A) isolated from *Crytosporiopsis* spp. (Omura et al., 1987) is a specific inhibitor of mevalonate biosynthesis (specifically inhibits 3-hydroxy-3-methylglutaryl-CoA (HMG-CoA) synthase) and found to be an inhibitor of Vero cell growth in MEM medium containing 2% calf serum, but not in the above medium supplemented with 1 mM mevalonate. The absolute configuration was determined by Chiang et al. (1988). The total synthesis of hymeglusin has been reported by several groups. The first total synthesis was reported by Mori et al. (1991). Tomoda and his group showed that hymeglusin inhibits hamster cytosolic HMG-CoA synthase by covalently modifying the active Cys 129 residue of the enzyme (Tomoda et al., 2004). Two novel antifungal antibiotics PF1163A and B produced by *Phoma* spp. and *Penicillum* spp. showed potent growth inhibitory activity against the pathogenic fungal strain *Candida albicans* but did not show cytotoxic activity against mammalian cells (Nose et al., 2000). The total synthesis of a novel antifungal antibiotic PF1163A is reported utilising Keck asymmetric allylation, sharpless kinetic resolution, regioselective epoxide-ring opening, esterification and ring-closing metathesis as the key reactions (Krishna and Srinivas, 2012).

Compounds Targeting Protein Biosynthesis

Protein synthesis is a basic and universal biological process that employs significantly different components and displays substantially different mechanisms in bacterial, archaeal and eukaryotic cells. Protein synthesis has always been considered as one of the most attractive targets in the development of antimicrobial agents (Hall et al., 1992).

Fungal Elongation Factor 2 (EF2) Inhibitors

Two soluble elongation factors show some fungal-specificity: EF3, a factor that is required by fungal ribosomes only (Kamath et al., 1989; Skogerson and Engelhardt, 1977), and EF2, which has been demonstrated to possess at least one functional distinction from its mammalian counterpart (Dominguez et al., 1998; Justice et al., 1998). The latter protein has been identified as a target as it is exclusive to fungal cells, and the sordarins were identified as compounds that block the interaction of fungal elongation factor 2 (EF2) at the ribosomal P-protein stalk (Shastry et al., 2001). Although the protein EF2 is conserved across all eukaryotic organisms, sordarins exclusively inhibit fungal protein synthesis and do not affect orthologous gene products in mammalian or plant cells. Their biology and chemistry have been reviewed extensively (Liang, 2008; Dominguez and Martin, 2005).

Sordarin (Fig. 8) was isolated by scientists at Sandoz from fermentations of the fungus *Sordaria araneosa* (Table 2). The compound was patented in 1969

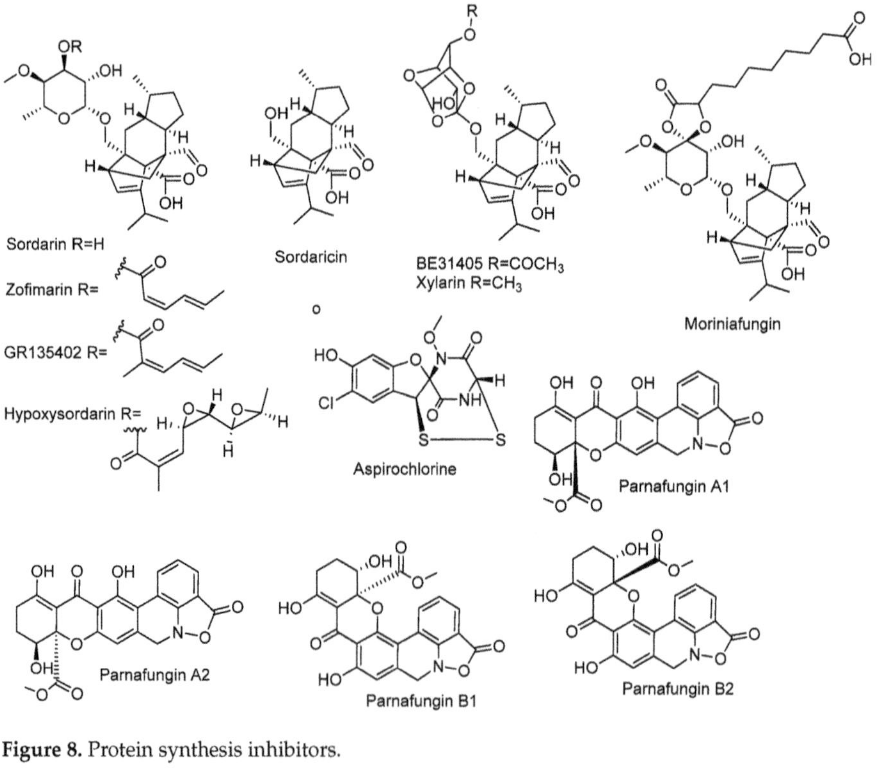

Figure 8. Protein synthesis inhibitors.

under the name SL (Sigg and Binningen, 1969). Its purification and degradation by acid hydrolysis to a diterpenic aglycon, sordaricin (Fig. 8), and a novel sugar, 6-desoxy-4-O-methyl-D-altrose, were described in 1971 (Hauser and Sigg, 1971). Its full structure was reported shortly afterwards (Thaddaus Vasella, 1972). Recent publications from Merck and Glaxo Wellcome report that the sordarins are potent inhibitors of translation in fungi with an extremely high level of selectivity. They act via a specific interaction with EF2, by stabilizing the fungal EF2-ribosome complex, despite the high degree of amino acid sequence homology exhibited by EF2 from various eukaryotes. This fungal specificity of the sordarins makes EF2 an attractive antifungal target (Justice et al., 1998; Dominguez et al., 1998). The lack of activity of the sordarins against *C. krusei, C. glabrata* and *C. parapsilosis*, in comparison with their extremely high levels of potency against *C. albicans*, suggests that these compounds have a highly specific binding site, which may also be the basis for the greater selectivity of these compounds in inhibiting the fungal, but not the mammalian protein synthesis.

After the discovery of sordarin, several compounds structurally related, sharing the common aglycone of sordarin, sordaricin, were isolated from diverse fungal species. Those compounds and the microorganisms from which they were isolated are given in Fig. 8 and Table 2.

Table 2 also lists other inhibitors of the protein synthesis, together with the microorganism that produce them and the corresponding references. Zofimarin was patented by Sankyo in 1987 (Ogita et al., 1987) and like sordarin, it is active against *C. albicans, S. cervisiae* and *C. neoformans* but with additional activity against *Aspergillus* at higher concentrations. No further development appears to have been undertaken with zofimarin.

Other examples of inhibitors of protein synthesis produced by strains of fungi such as SCH57404 and BE31405 are described in Table 2. SCH57404 was isolated from an unidentified sterile fungus by Schering-Plough (Coval et al., 1995). It differs from BE31405 isolated from *P. minioluteum* by the presence of a methoxy in place of the acetate at C-4 of the tricyclic sugar moiety. It has a narrow *in vitro* spectrum, with antifungal activity against *C. albicans,* but is poorly active against dermatophytes and *Aspergillus*. Another group identified the same compound from a *Xylaria* sp. The isolated compound, named as xylarin, exhibited potent activity against yeasts and filamentous fungi and as expected was only weakly cytotoxic towards mammalian cells (Schneider et al., 1995).

BE31405 was isolated from the culture broth of a fungal strain, *Penicillium minioluteum*, which had been obtained from a soil sample collected in Saitama prefecture, Japan. BE-31405 was isolated by absorption on high porous polymer resin (Diaion HP-20), and successively purified by solvent extraction and finally by crystallization. BE31405 contains a unique sugar with an unusual tricyclic structure. It showed broad spectrum activity against pathogenic fungal strains such as *Candida albicans, Candida glabrata* and *Cryptococcus neoformans* with more potent antifungal activity than sordarin, but did not show cytotoxic activity against mammalian cells such as P388 mouse leukemia. The mechanism studies indicated that BE-31405 inhibited the protein synthesis of *C. albicans* but not of mammalian cells (Okada et al., 1998).

In addition, a new natural sordarin derivative, hypoxysordarin, has been isolated from cultures of *Hypoxylon croceum*. Like the parent compound sordarin, this new derivative presented potent antifungal activities against yeasts and several filamentous fungi (Daferner et al., 1999).

Glaxo Wellcome carried out several screening programs to identify compounds that inhibit fungal protein synthesis; the most promising compound that emerged was a natural product showing a close similarity to zofimarin, GR135402. The spectrum of activity included *C. albicans, C. tropicalis* and *C. neoformans*, but not some other *Candida* species or *Aspergillus* species. GR135402 showed a therapeutic effect in mice with systemic candidiasis (Kinsman et al., 1998). A synthetic chemical program at Glaxo Wellcome was initiated to synthesize novel analogs. Several sordarin derivatives with a broad spectrum of activity and remarkable potencies *in vivo* were synthesized by modifications of the basic sordarin molecule sugar moiety (different types of fused rings at position C3' to C4'). Four of these sordarin derivatives (GM222712, GM237354, GM193633 and GM211676) have demonstrated significant *in vitro* and *in vivo* antifungal activity against most *Candida* species, including azole-resistant isolates, *C. neoformans, Pneumocystis carinii*, important dimorphic endemic fungal pathogens and significant *in vitro* activity against yeast-like fungi (Gargallo-Viola, 1999).

Azasordarins, a new series of derivatives, are compounds characterized by the presence of a 6-methylmorpholin-2-yl group with different N-4' substituents at position 8a of the sordaricin indacene ring system instead of the 4' sugar moiety present in sordarin. They have the additional advantage of easier chemical synthesis. These compounds displayed significant activities against *Candida* species, *P. carinii, Rhizopus arrhizus* and *Blastoschizomyces capitatus*. The *in vitro* selectivity of azasordarins was investigated by cytotoxicity studies performed with five cell lines and primary hepatocytes. The cytotoxicity values of some of the azasordarins (GW 471552 and GW 471558) were > 100 µg/ml for all cell lines tested. Tox50s on hepatocytes were in the following order for these azasordarins: GW 471558 > GW 471552 > GW 570009 > GW 587270 > GW 515716 > GW 479821, with values ranging from higher than 100 µg/ml to 23 µg/ml. The cytotoxicity results obtained with fully metabolizing rat hepatocytes were in total agreement with those obtained with cell lines. These results justify additional studies to determine the clinical usefulness of azasordarins (Herreros et al., 2001).

Merck worked on sordarin production originally centered on *Rosellinia subiculata*, a wood associated fungus that can cause endophytic infections of plants (Bills et al., 2002). Moriniafungin (Fig. 8) is one of the most potent natural sordarins found to date which shows a much broader antifungal spectrum (Basilio et al., 2006; Collado et al., 2006). The compound was produced by an endophytic isolate of *Morinia pestalozzioides* (*Amphisphaeriaceae*) from Almeria. Moriniafungin's 2-hydroxysebacic acid residue linked to C-30 of the sordarose residue of sordarin through a 1,3-dioxolan-4-one ring is unique and is the first example of a fungal natural product containing a 1,3-dioxolan-4-one ring. Based on years of systematically searching for new sordarins with greater potency and broader antifungal spectrum, Vicente et al. (2009) have published a preliminary phylogenetic map of the distribution of the sordarins across the fungal kingdom mapping sordarin production among the Xylariales, Microascales, and Sordariales from different habitats, including soil, coprophilous and endophytic isolates. The potent broad-spectrum *in vitro* activity and the fact that some of the sordarins have shown oral efficacy in animal models are significant advantages of these compounds. These are sufficient to justify an ongoing effort into developing the full clinical potential of the sordarin class.

Genome-Wide Chemical Genetic Profiling in *Candida albicans*

Starting in 2004, Merck's approach to antifungal discovery shifted from single targets (EF2, 1,3-β-D-glucan synthase) to chemical genetic strategies applied to the human pathogen *C. albicans* (Jiang et al., 2008). The approach relies on a phenomenon known as chemically-induced haplo-insufficiency, where deletion of one allele of the target gene of the diploid *C. albicans* renders the heterozygous deletion strain hypersensitive to cognate inhibitors of the depleted drug target (Roemer et al., 2003). Chemically-induced growth phenotypes of individual mutants within a defined library of deletion mutants reveal important clues as to the drug target and, therefore, the mechanism of action of bioactive compounds. Central to this approach is the ability to screen bioactive molecules or mixtures across a comprehensive set of potential drug targets, thereby, maximizing the likelihood of linking growth inhibitory compounds to their cognate target and circumventing the inefficiencies of single, 'target centric' screening, which narrows the available antimicrobial chemical space to a single chosen target of interest (Rodriguez-Suarez et al., 2007).

To comprehensively analyse across broad arrays of drug targets while using limited amounts of compound or extracts, strain-specific barcodes were introduced to uniquely identify individual mutants and facilitate their parallel screening in a multiplex culture. Challenged with a sublethal concentration of a single growth impairing compound or fermentation extract mixture, individual strains displaying an altered fitness (either reduced or enhanced) within this genetically defined population are identified by microarray analyses. Fitness Test (FT) based screening on a genome-wide scale provides multiple advantages over traditional target-based screenings. Applying this fitness test approach, essential drug targets can be readily linked to their cognate inhibitors when screened against a comprehensive heterozygote deletion mutant strain collection, which could yield additional mechanistic information regarding drug uptake, metabolism, off-target activities and/or export (Xu et al., 2007).

Based on this knowledge, Merck Research Laboratories implemented a large-scale study involving multidisciplinary teams and demonstrated the potential of the *C. albicans* Fitness Test (FT) screening to mechanistically annotate microbial-derived crude extracts containing bioactive components and identified the parnafungins (Fig. 8), a new fungal metabolite structural class (Parish et al., 2008). Genetic and biochemical data supported the FT-based prediction of their MOA as inhibitors of poly(A) polymerase-mediated mRNA processing. Parnafungins exhibited potent broad spectrum microbiological activity across medically relevant *Candida* and *Aspergillus* pathogens without displaying, obvious *in vitro*, cytotoxicity or adverse *in vivo* effects (Bills et al., 2009b). Thus, parnafungins clearly validated the chemical genetic screening approach to identify new and efficacious microbial metabolites with an unanticipated MOA and a novel antifungal lead molecule.

Roemer and co-workers have described the identification and mechanistic characterization of a number of novel and known antifungal natural products produced by fungi. These compounds target a wide range of cellular processes including: cell wall, ergosterol, fatty acid biosynthesis, protein synthesis, proteosome function, mRNA processing, etc. These include virgineone (*Lachnum virgineum*), calmodulin and heat shock response (Ondeyka et al., 2009), phomafungin (*Phoma* sp.) (Fig. 6), sphingolipid biosynthesis (Herath et al., 2009), yefafungin (a new uridine analog), aspirochlorine (*Aspergillus* spp.) (Fig. 8), elongation factor 3 (Monti, 1999; Wang et al., 2012), a novel pentaene metabolite predicted to inhibit adenylate cyclase activity and/or cAMP regulation, campafungin, a new 1,3-glucan synthase inhibitor analog of the ergokonin and papulacandin structural classes, and two new proteasome inhibitor (*Metulocladosporiella* sp.), fellutamides C and D (Roemer et al., 2011; Xu et al., 2011; Wang et al., 2012). In summary a full chemical-genetic profile of over 1800 bioactive fermentation extracts discerned biases toward prominent targets and pathways revealed to be 'druggable' after interrogation by more than 70 different microbial metabolites or metabolite classes (Roemer et al., 2011; Wang et al., 2012).

Other Fungal Targets

There are other targets that have also been explored for discovering antifungal drug discovery and some inhibitors of these targets are produced by fungal strains such as: TCA cycle (conocandin, *Microsphaeropsis* sp.) (Muller et al., 1976), microtubule assembly, 12-deoxo-hamigerone (the species are not shown because they have not been published yet) (Roemer et al., 2011; Wang et al., 2012), ATP leakage, potentiates caspofungin (phaeofungin, *Phaeosphaeria* sp.) (Singh et al., 2013), etc.

Other promising therapeutic strategies for targeting fungal strains which are produced by microorganisms are the secreted aspartyl proteases (SAPs inhibitors), such as peptides acting as pepstatin A and antimicrobial peptides such as CAP37 (Calugi et al., 2011), intracellular acidification via inhibition of H^+-extrusion was also identified as possible mechanism for cell death of *Candida* species when these strains where treated with the polyphenolic compound curcumin produced by *Curcuma longa*. The study of curcumin against 14 strains of *Candida* including (ATCC strains and clinical isolates) showed that curcumin is a potent fungicide compound with MIC values range from 250 to 2000 μg/mL. In another study, anti-Candida activity of curcumin was demonstrated against *Candida* including some fluconazole resistant strains and clinical isolates of *C. albicans, C. glabrata, C. krusei, C. tropicalis,* and *C. guilliermondii*. The MIC_{90} values for sensitive and resistant strains were 250–650 and 250–500 μg/mL, respectively. Curcumin also showed inhibitory effect on *Cryptococcus neoformans* and *C. dubliniensis* with MIC value of 32 mg/L. This strong antifungal activity and its low side effect were one of the main reasons to evaluate its synergistic effect with known fungicides (Moghadamtousi et al., 2014). This polyphenol compound is now used as supplement in several countries (Gupta et al., 2013).

In addition, it has been identified and mechanistically characterized the Mcd4-specific natural product inhibitor, M743, from natural product extracts. M743 (previously named BE-049385A as well as YM3548) is a terpenoid lactone ring-based natural product previously demonstrated to inhibit GPI biosynthesis by blocking Mcd4 ethanolamine phosphotransferase activity (EtNP) transferase activity of mannose1 (Man1) of the GPI precursor. Mcd4 inhibitor M720 is a semisynthetic analogue of M743, which is efficacious in a murine infection model of systemic candidiasis. Mcd4 is considered as a promising antifungal target and the GPI cell wall anchor synthesis pathway is also a promising antifungal target area by demonstrating that effects of inhibiting it are more general than previously recognized (Mann et al., 2015).

Perspective or Future Challenges

The development of new antifungal agents that may pose threat to emerging pathogens, preferably naturally-occurring with novel mechanisms of action, is still an urgent medical need (Ostrosky-Zeichner et al., 2010). It is utmost desire of the researchers of the area that the new antifungal drugs should possess extended spectrum of activity against fungi, including rare but medically important moulds such as *Scedosporium* spp., *Fusarium* spp. or Mucorales; should have improved pharmacokinetic profile so as to reduce dosing frequency; have fungicidal activity; have possibility of oral and parenteral administration and should show low adverse effects and few drug-drug interactions. In addition, they should act by new and selective mechanisms of action (Moriyama et al., 2014).

Among the new prospects for identifying new antifungal drugs, the concept of "re-purposing" established medications to treat new diseases has emerged as an approach to expediting drug development in general. Non-antifungal drugs have been recognized as having antifungal activity. Therefore, these drugs have been evaluated as single potential agents or in combination with established antifungal drugs.

Regarding new ways of looking for novel antifungal screening strategies, natural products are the basis for the vast majority of anti-infective therapies in clinical use. Perhaps there is a more limited biosynthetic potential to produce novel natural products than that of microbiota. This question could certainly be addressed (in part) by considering plant or marine extracts. Metagenomics and/or novel approaches to culturing previously unculturable organisms could be other alternatives for increasing NP chemical diversity (Lewis et al., 2010).

However, new approaches need to emerge to improve the success of antifungal screening, although the former approaches have been very successful in both academic and industrial settings. Some of the new screening alternatives could be chemical genetic-based screenings or application of non-growth-based assays (Arielle and Krysan, 2012).

Another future challenge will be to identify new antifungal targets considering that many fungal target proteins have orthologs in human cells. Therefore, it is essential to exploit differences in protein structure between host and pathogen in order to identify molecules with selectivity for the fungal protein and acceptable toxicity for the host. The advances in structural biology and medicinal chemistry have made this possible.

Moreover, the field of nanomedicine is the most innovative in coming up with new products as modern antifungal therapeutics, and the use of nanoparticles is ever increasing in this context. In addition, efforts in discovering new therapeutic approaches and combined strategies using antibodies may lead to innovative therapeutic alternatives in the future.

Lastly, it is important to emphasize that new developments in drug discovery should not replace older approaches but rather supplement them. In particular, advances in the generation of therapeutic agents with fungus-specific mechanisms of action are of the highest priority. Thus, among the new compounds patented, those targeting the β-1,3-D-glucan synthase are of great interest since cell wall appears to be the most selective target for antifungal drugs due to its absence in human beings. The new analogs of enfumafungin possess a broad antifungal spectrum and have the important advantage over commercial echinocandins that they can be prepared in oral and inhalation formulations. Among inhibitors of chitin synthases (CHS), new berberrubine derivatives with C-9 and C-13 substitution were patented. This modification led to an improvement in their antifungal and CHS inhibitory activities when compared with conventional berberine (Castelli et al., 2014).

As mentioned by the experts in the review of Castelli et al., progress has been made in developing new structures with a broad spectrum of activity against clinically important fungi, with a more favorable pharmacokinetic profile, good bioavailability and low adverse effects. However, there is still much to be done regarding the *in vivo* activity, generation of therapeutic agents with fungus-specific mechanisms of action, drug-drug interactions and other aspects that lead a compound to be a good antifungal agent (Castelli et al., 2014).

Acknowledgements

All our colleagues and collaborators in the field of natural products research who contributed to the development of some of the most important antifungal discoveries during the last decades are gratefully acknowledged.

References

Abruzzo, G.K., Flattery, A.M., Gill, C.J., Kong, L., Smith, J.G., Pikounis, V.B., Balkovec, J.M., Bouffard, A.F., Dropinski, J.F., Rosen, H., Kropp, H. and Bartizal, K. 1997. Evaluation of the echinocandin antifungal MK-0991 (L-743,872): efficacies in mouse models of disseminated aspergillosis, candidiasis, and cryptococcosis. Antimicrob. Agents Chemother., 41(11): 2333–2338.

Akihiko, F. 2007. Discovery of micafungin (FK463): A novel antifungal drug derived from a natural product lead. Pure Appl. Chem., 79(4): 603–614.

Andersen, O.A., Nathubhai, A., Dixon, M.J., Eggleston, I.M. and van Aalten, D.M. 2008. Structure-based dissection of the natural product cyclopentapeptide chitinase inhibitor argifin. Chem. Biol., 15(3): 295–301.

Aoki, M., Kohchi, M., Masubuchi, K., Mizuguchi, E., Murata, T., Ohkuma, H., Okada, T., Sakaitani, M., Shimma, N., Watanabe, T., Yanagisawa, M. and Yasuda, Y. 2000. Aerothricin analogs, their preparation and use. International patent, WO 0005251.

Arai, N., Shiomi, K., Iwai, Y. and Omura, S. 2000. Argifin, a new chitinase inhibitor, produced by *Gliocladium* sp. FTD-0668. II. Isolation, physico-chemical properties, and structure elucidation. J. Antibiot., 53(6): 609–614.

Arathoon, A., Gotuzzo, E., Noriega, L., Andrade, J., Kim, Y.S., Sable, C.A., DeStefano, M., Stone, J., Berman, R. and Chodakewitz, J.A. 1998. A randomized, double-blind, multicenter trial of MK-0991, an echinocandin antifungal agent, vs. amphotericin B for the treatment (Tx) of oropharyngeal (OPC) and esophageal (EC) candidiasis in adults. Clin. Infect. Dis., 27: 939–949.

Arielle, B. and Krysan, J.D. 2012. Antifungal drug discovery: Something old and something new. PLOS Pathogens, 8(9): e1002870.

Arroyo-Flores, B.L., Rodríguez-Bonilla, J., Villagómez-Castro, J.C., Calvo-Méndez, C., Flores-Carreón, A. and López-Romero, E. 2000. Biosynthesis of glycoproteins in Candida albicans: activity of mannosyl and glucosyl transferases. Fungal Genet. Biol., 30(2): 127–133.

Bartizal, K., Abruzzo, G.K. and Schmatz, D.M. 1993. Cutaneous antifungal agents: Selected the compounds in clinical practice. pp. 421–58. *In*: Rippon, J.W. and Fromtling, R.A. (eds.). Marcel Dekker, New York.

Basilio, A., Justice, M., Harris, G., Bills, G., Collado, J., de la Cruz, M., Diez, M.T., Hernandez, P., Liberator, P., Nielsen kahn, J., Pelaez, F., Platas, G., Schmatz, D., Shastry, M., Tormo, J.R., Andersen, G.R. and Vicente, F. 2006. The discovery of moriniafungin, a novel sordarin derivative produced by *Morinia pestalozzioides*. Biorganic & Medicinal Chem., 14: 560–566.

Bills, G.F., Platas, G., Pelaez, F. and Masurekar, P. 1999. Reclassification of a pneumocandin-producing anamorph, Glarea lozoyensis gen. et sp. nov., previosly identified as *Zalerion arboricola*. Mycol. Res., 103: 179–192.

Bills, G.F., Dombrowski, A.W., Horn, W.S., Jansson, R.K., Rattray, M., Schmatz, D. and Schwartz, R.E. 2002. *Rosellinia subiculata* ATCC 74386 and fungus ATCC 74387 for producing sordarin compounds for fungi control. US Patent., US6436395 B1.

Bills, G.F., Collado, J., Ruibal, C., Peláez, F. and Platas, G. 2004. *Hormonema carpetanum* sp. nov., a new lineage of dothideaceous black yeasts from Spain. Studies in Mycology, 50: 149–157.

Bills, G., Martin, J., Collado, J., Platas, G., Overy, D., Tormo, J.R., Vicente, F., Verkleij, G.J.M., Crous, P.W., Bills, G.F., Martín, J., Collado, J., Platas, G., Overy, D., Tormo, J.R., Vicente, F., Verkleij, G. and Crous, P. 2009a. Measuring the distribution and diversity of antibiosis and secondary metabolites in the filamentous fungi. Society of Industrial Microbiol. News, 59: 133–147.

Bills, G.F., Platas, G., Overy, D.P., Collado, J., Fillola, A., Jiménez, M.R., Martín, J., del Val, A.G., Vicente, F., Tormo, J.R., Peláez, F., Calati, K., Harris, G., Parish, C., Xu, D. and Roemer, T. 2009b. Discovery of the parnafungins, antifungal metabolites that inhibit mRNA poly-adenylation, from the *Fusarium larvarum* complex and other hypocrealean fungi. Mycologia, 101: 445–469.

Bouffard, F.A., Zambias, R.A., Dropinski, J.F., Balkovec, J.M., Hammond, M.L., Abruzzo, G.K., Bartizal, K.F., Marrinan, J.A., Kurtz, M.B., McFadden, D.C., Nollstadt, K.H., Powles, M.A. and Schmatz, D.M. 1994. Synthesis and antifungal activity of novel cationic pneumocandin Bo derivatives. J. Med. Chem., 37: 222–225.

Bouffard, F.A., Dropinski, J.F., Balkovec, J.M., Black, R.M., Hammond, M.L., Nollstadt, K.H. and Dreikorn, S. 1996. L-743,872, a novel antifungal lipopeptide: synthesis and structure-activity relationships of new aza-substituted pneumocandins [Abstract F27]. *In*: 36th Interscience Conference on Antimicrob. Agents and Chemo., New Orleans.

Cabello, M.A., Platas, G., Collado, J., Díez, M.T., Martín, I., Vicente, F., Meinz, M., Onishi, J.C., Douglas, C., Thompson, J., Kurtz, M.B., Schwartz, R.E., Bills, G.F., Giacobbe, R.A., Abruzzo, G.K., Flattery, A.M., Kong, L. and Peláez, F. 2001. The discovery of arundifungin, a novel antifungal compound produced by fungi. Biological activity and taxonomy of the producing organisms. Int. Microbiol., 4: 93–102.

Calugi, C., Trabocchi, A. and Guarna, A. 2011. Novel small molecules for the treatment of infections caused by *Candida albicans*: a patent review (2002–2010). Expert Opin. Ther. Patents, 21(3): 381–397.

Castelli, M.V., Butassi, E., Monteiro, M.C., Svetaz, L.A., Vicente, F., Susana, A. and Zacchino, S.A. 2014. Novel antifungal agents: a patent review (2011–present). Expert Opin. Ther. Patents, 24(3): 1–16.

Chiang, P.Y., Chiang, M.N., Chang, S.S.Y., Chabala, J.C. and Heck, J.V. 1988. Absolute configuration of L-659,699, a novel inhibitor of cholesterol biosynthesis. J. Org. Chem., 53: 4599–4603.

Cohen, E. 1993. Chitin synthesis and degradation as targets for pesticide action. Arch. Insect Biochem. Physiol., 22: 245–261.

Collado, J., Platas, G., Bills, G.F., Basilio, A., Vicente, F., Tormo, J.R., Hernández, P., Díez, M.T. and Peláez, F. 2006. Studies on *Morinia*. Recognition of *Morinia longiappendiculata* sp. nov. as a new endophytic fungus, and a new circumscription of *Morinia pestalozzioides*. Mycologia, 98: 616–627.

Conners, N. and Pollard, D. 2005. Handbook of Industrial Mycology. *In*: An, Z. (ed.). pp. 515–538. Marcel Dekker, New York.

Coval, S.J., Puar, M.S., Phife, D.W., Terracciano, J.S. and Patel, M. 1995. SCH 57404 an antifungal agent possessing the rare sordaricin skeleton and a tricyclic sugar moiety. J. Antibiot., 48: 1171–1172.

Daferner, M., Mensch, S., Anke, T. and Sterner, O. 1999. Hypoxysordarin, a new sordarin derivative from Hypoxylon croceum Z. Naturforsch, 54c: 474–480.

Debono, M. and Gordee, R.S. 1994. Antibiotics that inhibit fungal cell wall development. Annu. Rev. Microbiol., 48: 471–497.

Demain, A.L. 1983. New applications of microbial products. Science, 219: 709–714.

Denning, D.W. 2003. Echinocandin antifungal drugs. Lancet, 4, 362(9390): 1142–1151.

Dominguez, J.M. and Martin, J.J. 1998. Identification of elongation Factor 2 as the essential protein targeted by sordarins in *Candida albicans*. Antimicrob. Agents Chemother., 42: 2279–2283.

Domínguez, J.M. and Martín, J.J. 2005. *In*: Handbook of Industrial Mycology (An, Z., Ed.), pp. 335–53, Marcel Dekker, New York, USA.

Domínguez, J.M., Kelly, V.A., Kinsman, O.S., Marriott, M.S., Gómez de las Heras, F. and Martín, J.J. 1998. Sordarins: A new class of antifungal with selective inhibition of the protein synthesis elongation cycle in yeasts. Antimicrob. Agents Chemother., 42: 2274–2278.

Dreyfuss, M.M. 1986. Neue Erkenntnisse aus einem pharmakologischen Pliz-screening. Sydowia., 39: 22–36.

Fauth, U., Zahner, H., Muhlenfeld, A. and Achenbach, H. 1986. Galbonolides A and B- two non glycosidic antifungal macrolides. J. Antibiot., 39: 1760–1764.

Fazekas, F., 2012. Fingolimod in the treatment algorithm of relapsing remitting multiple sclerosis: a statement of the Central and East European (CEE) MS Expert Group. Wien Med. Wochenschr., 162: 354–366.

Feldmesser, M., Kress, Y., Mednik, A. and Casadevall, A. 2000. The effect of the echinocandin analogue caspofungin on cell wall glucan synthesis by *Cryptococcus neoformans*. J. Infect. Dis., 182: 1791–1795.

Fischbach, M.A. and Walsh, C.T. 2009. Antibiotics for emerging pathogens. Science, 325(5944): 1089–1093.

Fleet, G.H. 1991. Cell walls. pp. 199–277. *In*: Rose, A.H. and Harrison, J.S. (eds.). The Yeasts. London: Academic Press.

Fromtling, R.A. 1998. Human mycoses and current antifungal therapy. Drug News and Perspect., 11: 185–191.

Fujie, A., Iwamoto, T., Muramatsu, H., Okudaira, T., Nitta, K., Nakanishi, T., Sakamoto, K., Hori, Y., Hino, M., Hashimoto, S. and Okuhara, M. 2000. FR901469, a novel antifungal antibiotic from an unidentified fungus No. 11243 I. Taxonomy, fermentation, isolation, physicochemical properties and biological properties. J. Antibiot., 53: 912–919.

Fujie, A. 2007. Discovery of micafungin (FK463): A novel antifungal drug derived from a natural 'product lead. Pure Appl. Chem., 79: 603–614.

Gabib, E. 1991. Differential inhibition of chitin synthetases 1 and 2 from *Saccharomyces cerevisiae* by polyoxin D and nikkomycins. Antimicrob. Agents Chemother., 35: 170–173.

Gargallo-Viola, D. 1999. Sordarins as antifungal compounds. Current opinion in anti-infective investigational drugs, 1: 297–305.

Genilloud, O., González, I., Salzaar, O., Martín, J., Tormo, J.R. and Vicente, F. 2010. Current approaches to exploit actinomycetes as source of novel natural products. Journal of Industrial Microbiol. and Biotech., 38: 375–389.

Genilloud, O. and Vicente, F. 2012. Novel approaches to exploit natural products from microbial resources. *In*: Genilloud, O. and Vicente, F. (eds.). Drug Discovery from Natural Products. RSC series. Ch., 11: 221–248.

Georgopapadakou, N.H. and Tkacz, J.S. 1995. The fungal cell wall as a drug target. Trends in Microbiol., 3: 98–104.

Georgopapadakou, N.H. and Walsh, T.J. 1994. Human mycoses: drugs and targets for emerging pathogens. Science, 264: 371–373.

González, G.M., Tijerina, R., Najvar, L.K., Bocanegra, R., Luther, M., Rinaldi, M.G. and Graybill, J.R. 2001. Correlation between antifungal susceptibilities of *Coccidioides immitis in vitro* and antifungal treatment with caspofungin in a mouse model. Antimicrob. Agents Chemother., 45: 1854–1859.

Granier, F. 2000. Invasive fungal infections. Epidemiology and new therapies. Presse Med., 29: 2051–2056.

Gunawardana, G., Rasmussen, R.R., Scherr, M., Frost, D., Brandt, K.D., Choi, W., Jackson, M., Karwowski, J.P., Sunga, G., Malmberg, L.H., West, P., Chen, R.H., Kadam, S., Clement, J.J. and McAlpine, J.B. 1997. Corynecandin: a novel antifungal glycolipid from *Coryneum modonium*. J. Antibiot., 50(10): 884–886.

Gupta, S.C., Kismali, G. and Aggarwal, B.B. 2013. Curcumin, a component of turmeric: from farm to pharmacy. Biofactors, 39: 2–13.

Gurr, M.I., Harwood, J.L. and Frayn, K.N. 2002. Lipid Biochemistry: An Introduction, Blackwell Science, 5th edition.

Hall, C.C., Bertasso, A.B., Watkins, J.D. and Georgopapadakou, N.H. 1992. Screening assays for protein synthesis inhibitors. J. Antibiot., 45: 1697–1699.

Hanada, K. 2003. Serine palmitoyltransferase, a key enzyme of sphingolipid metabolism. Biochim. et Biophy. Acta, 1632: 16–30.

Harris, G.H., Shafiee, A., Cabello, M.A., Curotto, J.E., Genilloud, O., Göklen, K.E., Kurtz, M.B., Rosenbach, M., Salmon, P.M., Thornton, R.A., Zink, D.L. and Mandala, S.M. 1998. Inhibition of fungal sphingolipid biosynthesis by rustmicin, galbonolide B and their new 21-hydroxy analogs. J. Antibiot., 51(9): 837–844.

Hauser, D. and Sigg, H.P. 1971. Isolierung und abbau von sordarin. Helv. Chim. Acta, 54: 1178–1190.

Hayakawa, Y., Yamashita, T., Mori, T., Nagai, K., Shin-Ya, K. and Watanabe, H. 2004. Structure of tyroscherin, an antitumor antibiotic against IGF-1-dependent cells from *Pseudallescheria* sp. J. Antibiot., 57(10): 634–638.

Hector, R.F. 1993. Compounds active against cell walls of medically important fungi. Clin. Microbiol. Rev., 6(1): 1–21.

Herath, K., Harris, G., Jayasuriya, H., Zink, D., Smith, S., Vicente, F., Bills, G., Collado, J., González, A., Jiang, B., Kahn, J.N., Galuska, S., Giacobbe, R., Abruzzo, G., Hickey, E., Liberator, P., Xu, D., Roemer, T. and Singh, S.B. 2009. Isolation, structure and biological activity of phomafungin, a cyclic lipodepsipeptide from a widespread tropical *Phoma* sp. Bioorganic & Medicinal Chem., 17: 1361–1369.

Herreros, E., Almela, M.J., Lozano, S., Gomez de las Heras, F. and Gargallo-Viola, D. 2001. Antifungal activities and cytotoxicity studies of six new azasordarins. Antimicrob. Agents Chemother., 45: 3132–3139.

Hino, M., Fujie, A., Iwamoto, T., Hori, Y., Hashimoto, M., Tsurumi, Y., Sakamoto, K., Takase, S. and Hashimoto, S. 2001. Chemical diversity in lipopeptide antifungal antibiotics. J. Ind. Microbiol. Biotechnol., 27(3): 157–162.

Horn, W.S., Smith, J.L., Bills, G.F., Raghoobar, S.L., Helms, G.L., Kurtz, M.B., Marrinan, J.A., Frommer, B.R., Thornton, R.A. and Mandala, S.M. 1992. Sphingofungins E and F. Novel serine palmitoyltransferase inhibitors from *Paecilomyces variotii*. J. Antibiot., 47: 376–379.

Hwang, E.I., Yun, B.S., Kim, Y.K., Kwon, B.M., Kim, H.G., Lee, H.B., Jeong, W.J. and Kim, S.U. 2000. Phellinsin A, A novel chitin synthases inhibitor produced by *Phellinus* sp. PL3. J. Antibiot., 53(9): 903–911.

Ikeda, F., Wakai, Y., Matsumoto, S., Maki, K., Watabe, E., Tawara, S., Goto, T., Watanabe, Y., Matsumoto, F. and Kuwahara, S. 2000. Efficacy of FK463, a new lipopeptide antifungal agent, in mouse models of disseminated candidiasis and aspergillosis. Antimicrob. Agents Chemother., 44: 614–618.

Jiang, B., Xu, D., Allocco, J., Parish, C., Davison, J., Veillette, K., Sillaots, S., Hu, W., Rodriguez-Suarez, R., Trosok, S., Zhang, L., Li, Y., Rahkhoodaee, F., Ransom, T., Martel, N., Wang, H., Gauvin, D., Wiltsie, J., Wisniewski, D., Salowe, S., Kahn, J.N., Hsu, M.J., Giacobbe, R., Abruzzo, G., Flattery, A., Gill, C., Youngman, P., Wilson, K., Bills, G., Platas, G., Pelaez, F., Diez, M.T., Kauffman, S., Becker, J., Harris, G., Liberator, P. and Roemer, T. 2008. PAP inhibitor with *in vivo* efficacy identified by *Candida albicans* genetic profiling of natural products. Chemistry & Biology, 15: 363–374.

Justice, M.C., Hsu, M.J., Tse, B., Ku, T., Balkovec, J., Schmatz, D. and Nielsen, J. 1998. Elongation factor 2 as a novel target for selective inhibition of fungal protein synthesis. J. Biol. Chem., 273: 3148–3151.

Kaida, K., Fudou, R., Kameyama, T., Tubaki, K., Suzuki, Y., Ojika, M. and Sakagami, Y. 2001. New cyclic depsipeptide antibiotics, clavariopsins A and B, produced by an aquatic hyphomycete, *Clavariopsis aquatica*. I. Taxonomy, fermentation, isolation and biological properties. J. Antibiot., 54: 17–21.

Kamath, A. and Chakraburtty, K. 1989. Role of yeast elongation factor 3 in the elongation cycle. J. Biol. Chem., 264: 15423–15428.

Kanasaki, R., Kobayashi, M., Fujine, K., Sato, I., Hashimoto, M., Takase, S., Tsurumi, Y., Fujie, A., Hino, M., Hashimoto, S. and Hori, Y. 2006a. FR227673 and FR190293, novel antifungal lipopeptides from *Chalara* sp. No. 22210 and *Tolypocladium parasiticum* No. 16616. The J. Antibiot., 59: 158–167.

Kanasaki, R., Abe, F., Kobayashi, M., Katsuoka, M., Hashimoto, M., Takase, S., Tsurumi, Y., Fujie, A., Hino, M., Hashimoto, S. and Hori, Y. 2006b. FR220897 and FR220899, novel antifungal lipopeptides from Coleophoma empetri no. 14573. J. Antibiot., 59(3): 149–57.

Kanasaki, R., Abe, F., Furukawa, S., Yoshikawa, K., Fujie, A., Hino, M., Hashimoto, S. and Hori, Y. 2006c. FR209602 and related compounds, novel antifungal lipopeptides from *coleophoma crateriformis* no. 738. II. *In vitro* and *in vivo* antifungal activity. J. Antibiot., 59(3): 145–148.

Kanasaki, R., Sakamoto, K., Hashimoto, M., Takase, S., Tsurumi, Y., Fujie, A., Hino, M., Hashimoto, S. and Hori, Y. 2006d. FR209602 and related compounds, novel antifungal lipopeptides from *Coleophoma crateriformis* no. 738. I. Taxonomy, fermentation, isolation and physico-chemical properties. J. Antibiot., 59(3): 137–144.

Kaneto, R., Chiba, H., Agematu, H., Shibamoto, N., Yoshioka, T., Nishida, H. and Okamoto, R. 1993. Mer-WF3010, a new member of the papulacandin family. I. Fermentation, isolation and characterization. J. Antibiot., 46(2): 247–250.

Kinsman, O.S., Chalk, P.A., Jackson, H.C., Middlehon, R.F., Shuttleworth, A., Rudd Bam, J.C.A., Noble, H.M., Wildman, H.G., Dawson, M.J., Lynn, S. and Hayes, M.V. 1998. Isolation and characterisation of an antifungal antibiotic (GR 135402) with protein synthesis inhibition. J. Antibiot., 51: 41–49.

Krishna, P.R. and Srinivas, P. 2012. Total synthesis of the antifungal antibiotic PF1163A. Tetrahedron: Asymmetry, 23: 769–774.

Kumlesh, K., Dev., Mullershausen, F., Mattes, H., Kuhn, R.R., Bilbe, G., Hoyer, D. and Anis Mir, A. 2008. Brain sphingosine-1-phosphate receptors: Implication for FTY720 in the treatment of multiple sclerosis. Pharmacology & Therapeutics, 117: 77–93.

Lai, C.C., Tan, C.K., Huang, Y.T., Shao, P.L. and Hsueh, P.R. 2008. Current challenges in the management of invasive fungal infections. J. Infect. Chemother., 14: 77–85.

Lamoth, F. and Alexander, B.D. 2015. Antifungal activity of SCY-078 (MK-3118) and standard antifungal agents against clinical non-aspergillus mold isolates. Antimicrob. Agents Chemother., 59(7): 4308–4311.

Latge, J.P. 2001. The pathobiology of *Aspergillus fumigatus*. Trends Microbiol., 9: 382–389.

Latge, J.P. 2007. The cell wall: a carbohydrate armour for the fungal cell. Molecular Microbiol., 66: 279–290.

Leber, R., Fuchsbichler, S., Klobucnikova, V., Schweighofer, N., Pitters, E., Wohlfarter, K., Lederer, M., Landl, K., Ruckenstuhl, C., Hapala, I. and Turnowsky, F. 2003. Molecular mechanism of terbinafine resistance in *Saccharomyces cerevisiae*. Antimicrob. Agents Chemother., 47(12): 3890–3900.

Lewis, K., Epstein, S., D'Onofrio, A. and Ling, L.L. 2010. Uncultured microorganisms as a source of secondary metabolites. J. Antibiot., 63: 468–476.

Liang, H. 2008. Sordarin, an antifungal agent with a unique mode of action. Beilstein J. Organic Chemistry 4, doi: 10.3762/bjoc.4.31.

Lupetti, A., Danesi, R., Campa, M., Del Tacca, M. and Kelly, S. 2002. Molecular basis of resistance to azole antifungals. Trends Mol. Med., 8(2): 76–81.

Maertens, J., Raad, I., Sable, C.A., Ngai, A., Berman, R., Patterson, T.F., Denning, D.W. and Walsh, T. 2000. Multicenter, noncomparative study to evaluate safety and efficacy of caspofungin (CAS) in adults with invasive aspergillosis (IA) refractory R or Intolerant (I) to amphotericin B (AMB), AMB lipid formulations (lipid AMB), or azoles [abstract J-1103]. *In*: Proceedings of the 40th Interscience Conference Antimicrobial Agents and Chemotherapy, Toronto. American Society for Microbiology: Washington, DC. 371.

Magome, E., Harimaya, K., Gomi, S., Koyama, M., Chiba, N., Ota, K. and Mikawa, T. 1996. Structure of furanocandin, a new antifungal antibiotic from *Tricothecium* sp. J. Antibiot., 49(6): 599–602.

Mandala, S.M., Frommer, B.R., Thornton, R.A., Kurtz, M.B., Young, N.M., Cabello, M.A., Genilloud, O., Liesch, J.M., Smith, J.L. and Horn, W.S. 1994. Inhibition of serine palmitoyltransferase activity by lipoxamycin. J. Antibiot., 47: 376–379.

Mandala, S.M., Thornton, R.A., Frommer, B.R., Curotto, J.E., Rozdilsky, W., Kurtz, M.B., Giacobbe, R.A., Bills, G.F., Cabello, M.A. and Martín, I. 1995. The discovery of australifungin, a novel inhibitor of sphinganine N-acyltransferase from Sporormiella australis. Producing organism, fermentation, isolation and biological activity. J. Antibiot., 48: 349–356.

Mandala, S.M., Thornton, R.A., Rosenbach, M., Milligan, J., Garcia-Calvo, M., Bull, H.G. and Kurtz, M.B. 1997a. Khafrefungin, a novel inhibitor of sphingolipid synthesis. J. Biol. Chem., 272(51): 32709–32714.

Mandala, S.M., Thornton, R.A., Frommer, B.R., Dreikorn, S. and Kurtz, M.B. 1997b. Viridiofungins, novel inhibitors of sphingolipid synthesis. J. Antibiot., 50: 339–343.

Mandala, S.M., Thornton, R.A., Milligan, J., Rosenbach, M., Garcia-Calvo, M., Bull, H.G., Harris, G., Abruzzo, G.K., Flattery, A.M., Gill, C.J., Bartizal, K., Dreikorn, S. and Kurtz, M.B. 1998. Rustmicin, a potent antifungal agent, inhibits sphingolipid synthesis at inositol phosphoceramide synthase. J. Biol. Chem., 273(24): 14942–14949.

Mandala, S.M. and Harris, G.H. 2001. Isolation and characterization of novel inhibitors of sphingolipid synthesis: australifungin, viridiofungins, rustmicin, and khafrefungin. Methods Enzymol., 311: 335–348.

Mann, P.A., McLellan, C.A., Koseoglu, S., Si, Q., Kuzmin, E., Flattery, A., Harris, G., Sher, X., Murgolo, N., Wang, H., Devito, K., De Pedro, N., Genilloud, O., Nielsen Kahn, J., Jiang, B., Costanzo, M., Boone, C., Garlisi, C.G., Lindquist, S. and Roemer, T. 2015. Chemical genomics-based antifungal drug discovery: Targeting glycosylphosphatidylinositol (GPI) precursor biosynthesis. ACS Infect. Dis., 1: 59–72.

Merrill, Jr., A.H. and Sandhoff, K. 2002. Sphingolopids: metabolism and cell signaling. Vance, D.E. and Vance, J.E. (eds.). Biochemistry of Lipids, Lipoproteins and Membranes (4th Edt.). Elserviser Science B. V.

Mina, J.G., Pan, S.Y., Wansadhipathi, N.K., Bruce, C.R., Shams-Eldin, H., Schwarz, R.T., Steel, P.G. and Denny, P.W. 2009. The Trypanosoma brucei sphingolipid synthase, an essential enzyme and drug target. Mol. Biochem. Parasitol., 168: 16–23.

Miyake, Y., Kozutsumi, Y., Nakamura, S., Fujita, T. and Kawasaki, T. 1995. Serine palmitoyltransferase is the primary target of a sphingosine like immunosuppressant, ISP.1/myriocin. Biochem. Biophys. Res. Commun., 211: 396–403.

Mizuno, K., Yagi, A., Satoi, S., Takada, M. and Hayashi, M. 1977. Studies on aculeacin. I. Isolation and characterization of aculeacin A. J. Antibiot., 30(4): 297–302.

Moghadamtousi, S.Z., Kadir, H.A., Hassandarvish, P., Tajik, H., Abubakar, S. and Zandi, K. 2014. A review on antibacterial, antiviral, and antifungal activity of curcumin. Biomed. Res. Int., 2014. doi: 10.1155/2014/186864.

Monti, F., Ripamonti, F., Hawser, S.P. and Islam, K. 1999. Aspirochlorine: A highly selective and potent inhibitor of fungal protein synthesis. J. Antibiot., 52(3): 311–318.

Mori, K. and Takahashi, Y. 1991. Synthetic microbial chemistry, XXIV. Synthesis of antibiotic 1233A, an inhibitor of cholesterol biosynthesis. Liebigs Annalen der Chemie., 10: 1057–1065.

Morrison, V.A. 2005. Caspofungin: an overview. Expert Review of Anti-Infective Therarpy, 3: 697–705.

Morrison, V.A. 2006. Echinocandin antifungals: review and update. Expert Review of Anti-Infective Therarpy, 4: 325–342.

Moriyama, B., Gordon, L.A., McCarthy, M., Henning, S.A., Walsh, T.J. and Penzak, S.R. 2014. Emerging drugs and vaccines for Candidemia. Mycoses, 57: 718–733.

Motyl, M.R., Tan, C., Liberator, P., Giacobbe, R., Racine, F., Hsu, M., Nielsen-Kahn, J., Douglan, C., Bowman, J., Hammond, M., Balkovec, J., Greenlee, M., Meng, D., Parker, D., Peel, M., Fan, W., Mamai, A., Hong, J., Orr, M., Ouvray, G., Perrey, D., Liua, H., Jones, M., Nelson, K., Ogbu, C., Lee, S., Li, K., Kirwan, R., Note, A., Sligar, J. and Martensen, P. 2010. MK-3118, an orally active enfumafungin with potent *in vitro* anti-fungal activity *in* "50th Interscience Conference on Antimicrob. Agents and Chemother.", pp. F1–847.

Mukhopadhyay, T., Roy, K., Bhat, R.G., Sawant, S.N., Blumbach, J., Ganguli, B.N., Fehlhaber, H.W. and Kogler, H. 1992. Deoxymulundocandin-a new echinocandin type antifungal antibiotic. J. Antibiot., 45(5): 618–623.

Muller, J.M., Fuhrer, H., Gruner, J. and Voser, W. 1976. Stoffwechselprodukte von Mikroorganismen. 160. Mitteilung159. Mitteilung s. [1]. Conocandin, ein fungistatisches Antibiotikum aus Hormococcus conorum. Helvetica Chimica Acta, 59: 2506–2514.

Nagiec, M.M., Nagiec, E.E., Baltisberger, J.A., Wells, G.B., Lester, R.L. and Dickson, R.C. 1997. Sphingolipid synthesis as a target for antifungal drugs. Complementation of the inositol phosphorylceramide synthase defect in a mutant strain of *Saccharomyces cerevisiae* by the *AUR1* gene. J. Biol. Chem., 272: 9809–9817.

Newman, D.J. and Cragg, G.M. 2012. Natural products as sources of new drugs over the 30 years from 1981 to 2010. J. Nat. Prod., 75(3): 311–335.

Newman, D.J., Cragg, G.M. and Snader, K.M. 2003. Natural products as sources of new drugs over the period 1981–2002. J. Nat. Prod., 66: 1022–1037.

Noble, H.M., Langley, D., Sidebottom, P.J., Lane, S.J. and Fisher, P.J. 1991. An echinocandin from an endophytic *Cryptosporiopsis* sp. and *Pezicula* sp. in *Pinussylvestris* and *Fagus sylvatica*. Mycol. Res., 95: 1439–1440.

Nose, H., Seki, A., Yaguchi, T., Hosoya, A., Sasaki, T., Hoshiko, S. and Shomura, T. 2000. PF1163A and B, new antifungal antibiotics produced by *Penicillium* sp. I. Taxonomy of producing strain, fermentation, isolation and biological activities. J. Antibiot., 53(1): 33–37.

Nyfeler, R. and Keller, S.W. 1974. Metabolites of microorganisms. 143. Echinocandin B, a novel polypeptide-antibiotic from *Aspergillus nidulans* var. *echinulatus*: isolation and structural components. Helv. Chim. Acta, 57: 2459–2477.

Ogita, T., Hayashi, A., Sato, S. and Furaya, K. 1987. Antibiotic zofimarin. Jpn Kokai Tokkyo Koho JP 6240292.

Ohyama, T., Kurihara, Y., Ono, Y., Ishikawa, T., Miyakoshi, S., Hamano, K., Arai, M., Suzuki, T., Igari, H., Suzuki, Y. and Inukai, M. 2000. Arborcandins A, B, C, D, E and F, novel 1,3-β-glucan synthase inhibitors: production and biological activity. J. Antibiot., 53: 1108–1116.

Okada, H., Kamiya, S., Shiina, Y., Suma, H., Nagashima, M., Nakajima, S., Shimokawa, H., Sugiyama, E., Kondo, H., Kojiri, K. and Suda, H. 1998. BE-31405, a new antifungal antibiotic produced by *Penicillium minioluteum*. J. Antibiot., 51: 1081–1086.

Oki, T., Konishi, M., Tomatsu, K., Tomita, K., Saitoh, K., Tsunakawa, M., Nishio, M., Miyaki, T. and Kawaguchi, H. 1988. Pradimicin, a novel class of potent antifungal antibiotics. J. Antibiot., 41(11): 1701–1704.

Omura, S. 1986. Philosophy of new drug discovery. Microbiol. Rev., 50: 259–279.

Omura, S., Arai, N., Yamaguchi, Y., Masuma, R., Iwai, Y., Namikoshi, M., Turberg, A., Kölbl, H. and Shiomi, K. 2000. Argifin, a new chitinase inhibitor, produced by Gliocladium sp. FTD-0668. I. Taxonomy, fermentation, and biological activities. J. Antibiot. (Tokyo), 53(6): 603–608.

Omura, S., Tomoda, H., Kumagai, H., Greenspan, M.D., Yodkovitz, J.B., Chen, J.S., Alberts, A.W., Martin, I., Mochales, S., Monaghan, R.L., Chabala, J.C., Schwartz, R.E. and Patchett, A.A. 1987. Potent inhibitory effect of antibiotic 1233A on cholesterol biosynthesis which specifically blocks 3-Hydroxy-3-Methylglutaryl coenzyme A synthase. J. Antibiot., 40: 1356–1357.

Ondeyka, J., Harris, G., Zink, D., Basilio, A., Vicente, F., Bills, G., Platas, G., Collado, J., González, A., de la Cruz, J., Martin, M., Kahn, J.N., Galuska, S., Giacobbe, R., Abruzzo, G., Hickey, E., Liberator, P., Jiang, B., Xu, D.M., Roemer, T. and Singh, S.B. 2009. Isolation, structure elucidation, and biological activity of virgineone from Lachnum virgineum using the genome-wide *Candida albicans* fitness fest. J. Nat. Prod., 72: 136–141.

Onishi, J.C., Milligan, J.A., Basilio, A., Bergstrom, J., Curotto, J., Huang, L., Meinz, M., Nallin-Omstead, M., Pelaez, F., Rew, D., Salvatore, M., Thompson, J., Vicente, F. and Kurtz, M.B. 1997. Antimicrobial activity of viridiofungins. J. Antibiot., 50: 334–338.

Onishi, J., Meinz, M., Thompson, J., Curotto, J., Dreikorn, S., Rosenbach, M., Douglas, C., Abruzzo, G., Flattery, A., Kong, L., Cabello, A., Vicente, F., Pelaez, F., Diez, M.T., Martin, I., Bills, G., Giacobbe, R., Dombrowski, A., Schwartz, R., Morris, S., Harris, G., Tsipouras, A., Wilson, K. and Kurtz, M.B. 2000. Discovery of novel antifungal beta (1,3)-glucan synthase inhibitors. Antimicrob. Agents Chemother., 44: 368–377.

Ostrosky-Zeichner, L., Casadevall, A., Galgiani, J.N., Odds, F.C. and Rex, J.H. 2010. An insight into the antifungal pipeline: selected new molecules and beyond. Nat. Rev. Drug Discov., 9: 719–727.

Paltauf, F., Kohlwein, S. and Henry, S.A. 1992. Regulation and compartmentalization of lipid synthesis in yeast. The Molecular and Cellular Biology of the yeast *Saccharomyces*: Gene Expression, 2: 415–500.

Parish, C.A., Smith, S.K., Calati, K., Zink, D., Wilson, K., Roemer, T., Jiang, B., Xu, D.M., Bills, G., Platas, G., Peláez, F., Díez, M.T., Tsou, N., McKeown, A.E., Ball, R.G., Powles, M.A., Yeung, L., Liberator, P. and Harris, G. 2008. Isolation and structure elucidation of parnafungins, antifungal natural products that inhibit mRNA polyadenylation. J. American Chem. Society, 130: 7060–7066.

Parks, L.W. and Casey, W.M. 1995. Physiological implications of sterol biosynthesis in yeast. Annu. Rev. Microbiol., 49: 95–116.

Park, B.J., Wannemuehler, K.A., Marston, B.J., Govender, N., Pappas, P.G. and Chiller, T.M. 2009. Estimation of the current global burden of cryptococcal meningitis among persons living with HIV/AIDS. AIDS, 23: 525–530.

Peláez, F., Cabello, A., Platas, G., Díez, M.T., González del Val, A., Basilio, A., Martán, I., Vicente, F., Bills, G.E., Giacobbe, R.A., Schwartz, R.E., Onish, J.C., Meinz, M.S., Abruzzo, G.K., Flattery, A.M., Kong, L. and Kurtz, M.B. 2000. The discovery of enfumafungin, a novel antifungal compound produced by an endophytic *Hormonema* species biological activity and taxonomy of the producing organisms. System. and App. Microbiol., 23: 333–343.

Peláez, F., Collado, J., Platas, G., Overy, D.P., Martín, J., Vicente, F., González del Val, A., Basilio, A., de la Cruz, M., Tormo, J.R., Fillola, A., Arenal, F., Villareal, M., Rubio, V., Baral, H.O., Galán, R. and Bills, G.F. 2011. The phylogeny and intercontinental distribution of the pneumocandin-producing anamorphic fungus *Glarea lozoyensis*. Mycology, 2: 1–17.

Petraitis, V., Petraitiene, R., Groll, A.H., Sein, T., Schaufele, R.L., Lyman, C.A., Francesconi, A., Bacher, J., Piscitelli, S.C. and Walsh, T.J. 2001. Dosage-dependent antifungal efficacy of V-echinocandin (LY303366) against experimental fluconazol-resistant oropharyngeal and esophageal candidiasis. Antimicrob. Agents Chemother., 45: 471–479.

Powles, M.A., Liberator, P., Anderson, J., Karkhanis, Y., Dropinski, J.F., Bouffard, F.A., Balkovec, J.M., Fujioka, H., Aikawa, M., McFadden, D. and Schmatz, D. 1998. Efficacy of MK-991 (L-743,872), a semisynthetic, pneumocandin, in murine models of *Pneumocystis carinii*. Antimicrob. Agents Chemother., 42: 1985–1989.

Rodriguez-Suarez, R., Xu, D., Veillette, K., Davison, J., Sillaots, S., Kauffman, S., Hu, W., Bowman, J., Martel, N., Trosok, S., Wang, H., Zhang, L., Huang, L.Y., Li, Y., Rahkhoodaee, F., Ransom, T., Gauvin, D., Douglas, C., Youngman, P., Becker, J., Jiang, B. and Roemer, T. 2007. Mechanism-of-action determination of GMP synthase inhibitors and target validation in *Candida albicans* and *Aspergillus fumigatus*. Chem. & Biol., 14: 1163–1175.

Roemer, T., Jiang, B., Davison, J., Ketela, T., Veillette, K., Breton, A., Tandia, F., Linteau, A., Sillaots, S., Marta, C., Martel, N., Veronneau, S., Lemieux, S., Kauffman, S., Becker, J., Storms, R., Boone, C. and Bussey, H. 2003. Large-scale essential gene identification in *Candida albicans* and applications to antifungal drug discovery. Molec. Microbiol., 50: 167–181.

Roemer, T., Xu, D., Singh, S.B., Parish, C.A., Harris, G., Wang, H., Davies, J.E. and Bills, G.F. 2011. Confronting the challenge of natural product-based antifungal discovery. Chem. & Biol., 18: 148–164.

Roy, K., Mukhopadhyay, T., Reddy, G.C., Desikan, K.R. and Ganguli, B.N. 1987. Mulundocandin, a new lipopentide antibiotic. I. Taxonomy, fermentation, isolation and characterization. J. Antibiot., 40: 275–280.

Ruiz-Herrera, J. and San-Blas, G. 2003. Chitin synthesis as target for antifungal drugs. Curr. Drug Targets Infect. Disord., 3: 77–91.

Ruiz-Herrera, J., Sentandreu, R. and Martinez, J.P. 1992. Chitin biosynthesis in fungi. *In*: Arora, D.K., Elander, R.P. and Mukerji, K.G. (eds.). Handbook of Applied Mycology, New York Marcel Dekker, 4: 281–312.

Rush, C.L., Schüttelkopf, A.W., Hurtado-Guerrero, R., Blair, D.E., Ibrahim, A.F., Desvergnes, S., Eggleston, I.M. and van Aalten, D.M. 2010. Natural product-guided discovery of a fungal chitinase inhibitor. Chem. Biol., 17(12): 1275–1281.

Schlingmann, G., Milne, L., Williams, D.R. and Carter, G.T. 1998. Cell wall active antifungal compounds produced by the marine fungus *Hypoxylon oceanium* LL-15G256 II. Isolation and structure determination. J. Antibiot., 51: 303–316.

Schneider, G., Anke, H. and Sterner, O. 1995. Xylarin, an antifungal *Xylaria* metabolite with an unusual tricyclic uronic acid moiety. Nat. Prod. Letters, 7: 309–316.

Schwartz, R.E., Sesin, D.F., Joshua, H., Wilson, K.E., Kempf, A.J., Goklen, K.A., Kuehner, D., Gailliot, P., Gleason, C., White, R., Inamine, E., Bills, G., Salmon, P. and Zitano, L. 1992. Pneumocandins from *Zalerion arboricola*. 1. Discovery and isolation. J. Antibiot., 45: 1853–1866.

Schwartz, R.E., Giacobbe, R.A., Bland, J.A. and Monaghan, R.L. 1989. L-671,329, a new antifungal agent. 1. Fermentation and isolation. J. Antibiot., 42: 163–67.

Schwartz, R.E., Dufresne, C., Flor, J.E., Kempf, A.J., Wilson, K.E., Lam, T., Onishi, J., Milligan, J., Fromtling, R.A., Abruzzo, G.K., Jenkins, R., Glazomitsky, K., Bills, G., Zitano, L., Del Val, S.M. and Omstead, M.N. 1991. Restricticin, a novel glycine-containing antifungal agent. J. Antibiot., 44(5): 463–471.

Shastry, M., Nielsen, J., Ku, T., Hsu, M.J., Liberator, P., Anderson, J., Schmatz, D. and Justice, M.C. 2001. Species-specific inhibition of fungal protein synthesis by sordarin: identification of a sordarin-specificity region in eukaryotic elongation factor 2. Microbiol., 147: 383–390.

Shaw, J.A., Mol, P.C., Bowers, B., Silverman, S.J., Valdivieso, M.H., Durán, A. and Cabib, E. 1991. The function of chitin synthases 2 and 3 in the *Saccharomyces cerevisiae* cell cycle. J. Cell Biol., 114: 111–123.

Sigg, H.P. and Binningen, S.C. 1969. Antibiotic SL 2266. United States patent, US 3432598.

Singh, S.B. and Pelaez, F. 2008. Biodiversity, chemical diversity and drug discovery. Prog. Drug Res., 65: 143–147.

Singh, S.B., Ondeyka, J., Harris, G., Herath, K., Zink, D., Vicente, F., Bills, G., Collado, J., Platas, G., González del Val, A., Martin, J., Reyes, F., Wang, H., Kahn, J.N., Galuska, S., Giacobbe, R., Abruzzo, G., Roemer, T. and Xu, D. 2013. Isolation, structure, and biological activity of phaeofungin, a cyclic lipodepsipeptide from a *Phaeosphaeria* sp. using the genome-wide *Candida albicans* fitness test. J. Nat. Prod., 76: 334–345.

Skogerson, L. and Engelhardt, D. 1977. Dissimilarity in chain elongation factor requirements between yeasts and rat liver ribosomes. J. Biol. Chem., 252: 1471–1475.

Smith, J.G., Abruzzo, G.K., Gill, C.J., Flattery, A.M., Kong, L., Rosen, H., Kropp, H. and Bartizal, K. 1996. Evaluation of pneumocandin L-743,872 in neutropenic mouse models of disseminated candidiasis and aspergillosis [abstract F41]. In: Program and Abstracts of the 36th Interscience Conference on Antimicrobial Agents and Chemotherapy. Washington, DC, 1996.

Strobel, G.A., Miller, R.V., Martinez-Miller, C., Condron, M.M., Teplow, D.B. and Hess, W.M. 1999. Cryptocandin, a potent antimycotic from the endophytic fungus *Cryptosporiopsis* cf. *quercina*. Microbiol., 145: 1919–1926.

Sütterlin, C., Horvath, A., Gerold, P., Schwarz, R.T., Wang, Y., Dreyfuss, M. and Riezman, H. 1997. Identification of a species-specific inhibitor of glycosylphosphatidylinositol synthesis. EMBO J., 16(21): 6374–6383.

Tae, H.S., Hines, J., Schneekloth, A.R. and Crews, C.M. 2010. Total synthesis and biological evaluation of tyroscherin. Org. Lett., 12(19): 4308–4311.

Takatsu, T., Nakayama, H., Shimazu, A., Furihata, K., Ikeda, K., Furihata, K., Seto, H. and Otake, N. 1985. Rustmicin, a new macrolide antibiotic active against wheat stem rust fungus. J. Antibiot., 38: 1806–1809.

Takesako, K., Ikai, K., Haruna, F., Endo, M., Shimanaka, K., Sono, E., Nakamura, T., Kato, I. and Yamaguchi, H. 1991. Aureobasidins, new antifungal antibiotics. Taxonomy, fermentation, isolation, and properties. J. Antibiot., 44: 919–924.

Takesako, K., Kuroda, H., Inoue, T., Haruna, F., Yoshikawa, Y., Kato, I., Uchida, K., Hiratani, T. and Yamaguchi, H. 1993. Biological properties of aureobasidin A, a cyclic depsipeptide antifungal antibiotic. J. Antibiot., 46: 1414–1420.

Takeuchi, T., Hara, T., Naganawa, H., Okada, M., Hamada, M., Umezawa, H., Gomi, S., Sezaki, M. and Kondo, S. 1988. New antifungal antibiotics benanomicins A and B from an Actinomycete. J. Antbiot., 41: 807–810.

Thaddaus Vasella, A. 1972. Uber ein neuartiges diterpen aus Sordaria Araneosa Cain. Ph.D. thesis. Eidgenössischen Technischen Hochschule, Zürich.

Tojo, T., Ohki, H., Shiraishi, N., Matsuya, T., Matsuda, H., Murano, K., Barrett, D., Ogino, T., Matsuda, K., Ichihara, M., Hashimoto, N., Kanda, A. and Ohigashi, A. 2000. Cyclic hexapeptides having antibiotic activity. International Patent, WO 0064927.

Tomoda, H., Ohbayashi, N., Morikawa, Y., Kumagai, H. and Satoshi Omura, S. 2004. Binding site for fungal β-lactone hymeglusin on cytosolic3-hydroxy-3-methylglutaryl coenzyme A synthase. Biochim. et Biophys. Acta, 1636: 22–28.

Traxler, P., Gruner, J. and Auden, A. 1977. Papulacandins, a new family of antibiotics with antifungal activity. I. Fermentation, isolation, chemical and biological characterization of papulacandins A, B, C, D and E. J. Antibiot., 30: 289–296.

Van Der Vaart, J.M., Van Schagen, F.A. and Mooren, A.T. 1996. The retention mechanism of cell wall proteins in *Sccharomyces cerevisiae*. Wall bound Cwp2p is b1,6-glucosy-lated. Biochim. et Biophys. Acta, 1291: 206–214.

VanMiddlesworth, F.M., Giacobbe, R.A., Lopez, M., Garrity, G., Bland, J.A., Bartizal, K., Fromtling, R.A., Polishook, J., Zweerink, M., Edison, A.M., Rozdilsky, W., Wilson, K.E. and Monaghan, R.L. 1991. Sphingofungins A, B, C, and D; a new family of antifungal agents. I. Fermentation, isolation and biological activity. J. Antibiot., 45: 861–867.

Vicente, M.F., Cabello, A., Platas, G., Basilio, A., Díez, M.T., Dreikorn, S., Giacobbe, R.A., Onishi, J.C., Meinz, M., Kurtz, M.B., Rosenbach, M., Thompson, J., Abruzzo, G., Flattery, A., Kong, L., Tsipouras, A., Wilson, K.E. and Peláez, F. 2001. Antimicrobial activity of ergokonin A from *Trichoderma longibrachiatum*. J. Appl. Microbiol., 91: 806–813.

Vicente, F., Basilio, A., Platas, G., Collado, J., Bills, G.F., González del Val, A., Martín, J., Tormo, J.R., Harris, G.H., Zink, D.L., Justice, M., Kahn, J.N. and Peláez, F. 2009. Distribution of the antifungal agent's sordarins across filamentous fungi. Mycol. Resear., 113: 754–770.

Vijayakumar, E.K.S., Roy, K., Chatterjee, S., Deshmukh, S.K., Ganguli, B.N., Kogler, H. and Fehlhaber, H.W. 1996. Arthrichitin. A new cell wall active metabolite from Arthrinium phaeospermum. J. Org. Chem., 61: 6591–6593.

Wang, E., Norred, W.P., Bacon, C.W., Riley, R.T. and Merrill, A.H., Jr. 1991. Inhibition of sphingolipid biosynthesis by fumonisins. Implications for diseases associated with *Fusarium moniliforme*. J. Biol. Chem., 266: 14486–14490.

Wang, H., Parish, G.A., Xu, D. and Roemer, T. 2012. Coupling chemical genomics and natural products: The discovery of parnafungins and novel antifungal leads. pp. 278–306. *In*: Genilloud, O. and Vicente, F. (eds.). Drug Discovery from Natural Products. Chapter 13, RSC Drug Discovery Series N° 25.

Wu, W.I., McDonough, V.M., Nickels, J.T., Jr., Ko, J., Fischl, A.S., Vales, T.R., Merrill, A.H., Jr. and Carman, G.M. 1995. Regulation of lipid biosynthesis in *Saccharomyces cerevisiae* by fumonisin B_1. J. Biol. Chem., 270: 13171–13178.

Xu, D., Jiang, B., Ketela, T., Lemieux, S., Veillette, K., Martel, N., Davison, J., Sillaots, S., Trosok, S., Bachewich, C., Bussey, H., Youngman, P. and Roemer, T. 2007. Genome-wide fitness test and mechanism-of-action studies of inhibitory compounds in *Candida albicans*. PLoS Pathog., 3: e92.

Xu, D., Ondeyka, J., Harris, G.H., Zink, D., Nielsen-Kahn, J., Wang, H., Bills, G., Platas, G., Wang, W., Szewczak, A.A., Liberator, P., Roemer, T. and Singh, S.B. 2011. Isolation, structure and biological activities of fellutamides C and D from an undescribed *Metulocladosporiella* (Chaetothyriales) using the genome-wide *Candida albicans* fitness test. J. Nat. Prod., 74: 1721–1730.

Yeung, C.M., Klein, L.L. and Lartey, P.A. 1996. Preparation and antifungal activity of fusacandin analogs: C-6' sidechain esters. Bioorg. Med. Chem. Lett., 6: 819–822.

Yoon, D.-H., Ji, M.-K., Ha, H.-J., Park, J., Philjun Kang, P. and Lee, W.K. 2013. Synthesis and biological activities of tyroscherin analogs. Bull. Korean Chem. Soc., 34(6): 1899–1902.

Young, S.A., Mina, J.G., Denny, P.W. and Smith, T.K. 2012. Sphingolipid and ceramide homeostasis: Potential therapeutic targets. Biochem. Resear. Internat., Review article ID 248135: 1–12.

Zweerink, M.M., Edison, A.M., Wells, G.B., Pinto, W. and Lester, R.L. 1992. Characterization of a novel, potent, and specific inhibitor of serine palmitoyltransferase. J. Biol. Chem., 267: 25032–25038.

CHAPTER 6

Modulation of Fungal Secondary Metabolites Biosynthesis by Chemical Epigenetics

Jacqueline Aparecida Takahashi,[1,*]
Dhionne Corrêia Gomes,[2] *Fernanda Henrique Lyra*[1,a] and
Gabriel Franco dos Santos[1,b]

ABSTRACT

Secondary metabolites produced by fungi are an exceptionally rich source of drugs for the treatment of human diseases. These compounds have also been used as structural leads for synthesizing innovative drugs. The regulation of biosynthesis of secondary metabolites is due to the action of structural genes that control the transcription factors. The production of these metabolites can also be controlled by transcription factors encoded by genes unrelated to the known biosynthesis. Conventional conditions of fungal cultivation used in laboratories are poor to mimic natural habitats of fungi, and this can partially explain why most of the genes responsible for production of unknown metabolites are transcriptionally silenced in

[1] Departamento de Química, UFMG, Av. Antônio Carlos, 6627, Belo Horizonte, MG, Brazil, CEP 31270-901.
[a] Email: fernanda_lyra@yahoo.com.br
[b] Email: gfsantos@ymail.com
[2] Faculdade de Farmácia, UFMG, Av. Antônio Carlos, 6627, Belo Horizonte, MG, Brazil, CEP 31270-901.
   Email: dhionnegomes@hotmail.com
* Corresponding author: jat@qui.ufmg.br

the artificial environment. This chapter embodies several examples as to how the epigenetic tools, like modifications in the histones and in the DNA, can be used to modulate the biosynthetic pathways for the production of secondary metabolites by fungi.

Introduction

Fungi are eukaryotic organisms, capable of producing a large number of secondary metabolites biosynthetically that are chemically diverse and pharmacologically active. Due to these biological properties, the secondary metabolites produced by fungi are exceptionally rich source of drugs for the treatment of human diseases, and also as structural leads for synthesizing innovative drugs. Several chemotherapeuticals are product of fungi, and many fungal metabolites are under evaluation at different clinical stages. Of more than 20,000 bioactive products reported as antifungal, antitumor, immunosuppressants for organ transplants, enzyme inhibitors, cholesterol reducers, anti-parasite agents, anti-inflammatory drugs and antibiotics, among others, 42% are produced by fungi (Williams et al., 2008; Chen et al., 2013; Chung et al., 2013).

Secondary metabolites of fungi are generally of low molecular weight and are structurally very diverse. They are mostly synthesized to assist the fungi in the competition with other microorganisms in their natural habitats as many of them are toxic or inhibitors of other organisms (Takahashi et al., 2013). While, in most of the eukaryotic cells, the genes responsible for the production of substances are scattered throughout the genome, in fungi, the genes responsible for the production of secondary metabolites are often grouped in clusters on a single chromosome (Shwab and Keller, 2008).

The regulation of secondary metabolites biosynthesis occurs due to the action of structural genes that control the transcription factors, distinguishing themselves from the genes related to specific metabolic pathways. The production of these metabolites can also be controlled by transcription factors encoded by genes unrelated to the known biosynthesis. These genes mediate the action of factors involved in environmental signals, such as pH, carbon and nitrogen sources, temperature and presence of light (Fox and Howlett, 2008).

Research related to fungal genome sequencing revealed the presence of a large number of genes clusters not expressed or silent but biosynthesis-related. The number of such genes exceeds the number of genes linked to the synthesis of compounds so far isolated from the studied species. Suppression of transcription serves as a defense for the fungi, protecting them from possible autotoxic effects of secondary metabolites that accumulate.

Conventional conditions of fungal cultivation used in laboratories, unfortunately are poor to mimic natural habitats of fungi, and this can partially explain why most of the genes responsible for the production of unknown metabolites are transcriptionally quiet in the artificial media. Thus, since the biosynthetic pathways for the production of secondary metabolites expressed by fungi are always the same, it seemed that the biosynthetic potential of these organisms with regards to the prospects for the discovery of new drug was limited (Williams et al., 2008; Asai et al., 2011; Chen et al., 2013; Chung et al., 2013).

The recent genetic mapping techniques in fungi indicated that these organisms possess a large unexplored capacity for metabolites production. Studies on *Aspergillus niger* showed that several of the clusters of genes responsible for production of polyketides (PKS), non-ribosomal peptides (NRPS) and PKS-NRPS hybrids (HPN) are suppressed under laboratory conditions. Another example is the fungus *Aspergillus nidulans*, for which 28 gene clusters producers of PKS and 24 producers of NRPS were identified, enabling this fungus to express at least 52 different metabolites. However, the number of metabolites produced by species so far studied can be even greater, since there may be communication between these clusters of genes (VanderMolen et al., 2014).

The vast majority of fungi that have had their genetic sequencing performed, with the exception of those belonging to the sub-phyla *Saccharomycotina* and *Taphrinomycotina*, indicated the presence of clusters of genes responsible for producing many metabolites that are still not identified (Fisch et al., 2009). Therefore, genetic engineering can be used to activate these genes (Nützmann et al., 2011) and this activation of silent gene expression is a promising way to stimulate the production of new bioactive compounds (Chung et al., 2013).

Epigenetics

The term epigenetic refers to reversible and potentially inheritable changes in functional genome, which cause alterations in gene expression, without changing the DNA nucleotides sequence. While epigenetic modifications may be transmitted by cell division, these modifications are not usually transmitted to subsequent generations through sexual reproduction, though some modifications may persist in future generations (Rapp and Wendel, 2005; Pang et al., 2013).

Recent studies have revealed the use of epigenetic modifiers as an alternative to the activation of silent biosynthetic pathways, allowing access to a myriad of hidden natural products. This approach can also be used to increase the production of already known metabolites. These modulators act by modifying specific proteins required for chromatin remodeling (Chung et al., 2013), changing transcription rate of some genes. Stimulation of gene expression therefore has an important role in the production of secondary metabolites not produced under usual laboratory growing conditions (Tang et al., 2015).

The main mechanisms used by these modulators involve the methylation of cytosines in DNA and post-translational histones changes, which are related to changes in chromatin structure, leading to activation or repression of gene transcription (Asai et al., 2012). The development of such as approach represents a breakthrough for researchers in the area of natural products for the control of the expression of biosynthetic pathways and latent opportunities for discovery of new bioactive molecules and drug prototypes (Williams et al., 2008).

Histone Modifications

The nucleosome is the fundamental unit of chromatin, characterized by about 150 DNA base pairs, arranged around a central octamer of histone proteins (H2A, H2B, H3, and H4). It also holds a connection histone H1, which binds externally to the DNA that surrounds the octamer, stabilizing the chromatin. The histones are very abundant small proteins, whose main function is to condense the DNA, being support for the formation of its basic skeleton. They directly interfere in the control of DNA transcription.

The N-terminal terminations of histones H3, H4, H2A, and H2B extend to the nucleosome surface and can suffer post-translational modifications that alter chromatin structure (Verdone et al., 2006; Kouzarides, 2007; Cichewicz, 2010; Gonzalo, 2010). These modifications occur mostly in the N-terminal domain of flexible histones H3 and H4. Depending on a higher or smaller DNA interaction with histones proteins that comprise the nucleosome, gene expression can be altered (Graessle et al., 2001; Eissenberg and Shilatifard, 2010).

Histones contain a globular domain, in which the DNA filament wraps around a long and flexible N-terminal tail domain, which is exposed. Histone's affinity for DNA occurs because they present a high proportion of amino acid residues, particularly lysine and arginine, which are positively charged at physiological pH and helps the histone tail to interact and connect with the negatively charged phosphate groups present in the main structure of DNA (Kouzarides, 2007; Cichewicz, 2010). This portion can be used as a substrate for a variety of modifications related to fundamental changes in chromatin structure through different mechanisms, leading to the activation or repression of gene transcription (Cichewicz, 2010). Depending on the level of DNA condensation, chromatin can be found in two main forms: euchromatin and heterochromatin. The first form has low density of nucleosome, being less compressed, while heterochromatin has high density, and large amount of packed nucleosomes, being therefore, more compressed (Fig. 1) (Palmer and Keller, 2010).

The least compressed regions are more accessible to transcription, while the more compacted ones are associated with repression and gene silencing. When post-translational changes of histones favour greater compaction of chromatin, the binding sites of transcription factors become inaccessible. In this way, the same gene sequence can be well expressed or transcriptionally silent, depending on the compacting level (euchromatin or heterochromatin) (Bender, 2004).

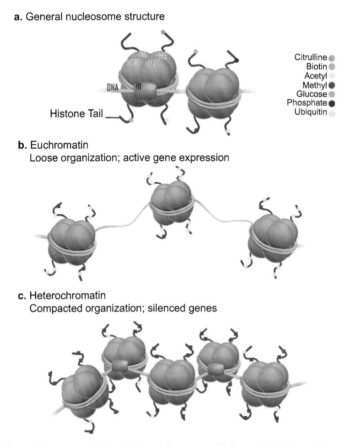

a. General nucleosome structure

Histone Tail

Citrulline
Biotin
Acetyl
Methyl
Glucose
Phosphate
Ubiquitin

b. Euchromatin
Loose organization; active gene expression

c. Heterochromatin
Compacted organization; silenced genes

Figure 1. Illustrative representation of the nucleosome and chromatin (Adapted from Cichewicz, 2010 and Gonzalo, 2010).

Since changes in chromatin depend on the degree of DNA compaction and consequent production of metabolites, the participation of histones in this process can be understood. Modifications in histones can modulate and cause a higher or lower genes expression (Cichewicz, 2010). The histones are subject to a rich variety of post-translational modifications, resulting in a true molecular language, known as "histone code", which is interpreted by different cellular factors. Modifications in histones involve the acetylation of lysines, phosphorylation of serines and treonines, methylation of lysines and arginines, the ubiquitylation of lysines, ADP-ribosylation of lysines and arginines, as well as the glycosylation, carbonylation and biotinylation of different N-terminal tail residues (Figs. 1, 2).

These modifications occur in different combinations and alter the electrostatic and biochemical properties of histones, resulting in different interactions of these with the DNA which in turn affect chromatin architecture and transcriptional activity. Once the histones are modified, they implement

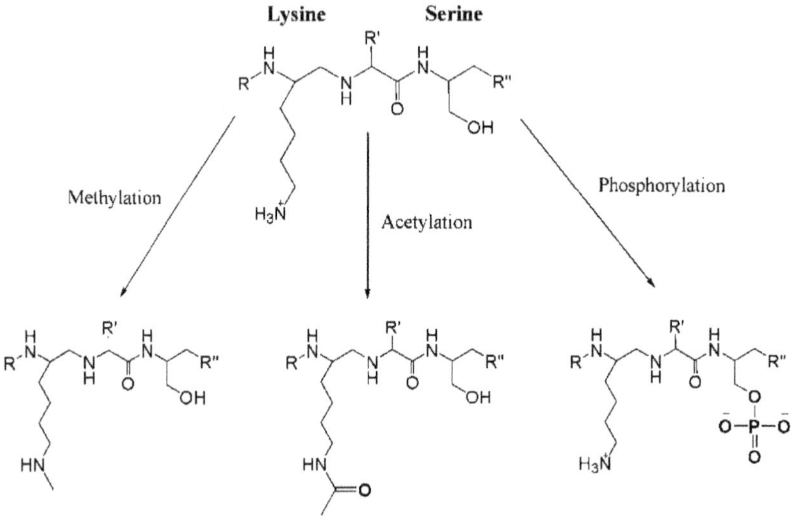

Figure 2. Main modifications in histones.

regulatory functions in different genetic events, with either activation or transcriptional silencing (Brosch et al., 2008; Mai et al., 2008; Gonzalo, 2010; Strauss et al., 2012).

Epigenetic Modifications in Histones

Among the post-translational histones modifications, the most widely studied and better understood is the acetylation. Generally, in transcriptionally active sites, there is a high degree of acetylation at lysine 4 of histone H3 (H3K4), while in heterochromatin, where there is little transcription, there is a low rate of acetylation of histone H3 lysine 9 (H3K9) (Palmer and Keller, 2010). Acetylation of histone H4 is also associated with the regulation of DNA transcription (Bender, 2004).

The mechanism of histone acetylation occurs mainly in the N-terminal tails of lysine residues by neutralization of the positively charged ammonium group. Acetylation decreases intensity of electrostatic interaction with the DNA, which is negatively charged. It also relaxes the structure of chromatin. The acetylation of lysine residues in histones increases the spatial distance between the DNA strands and the core protein, allowing access to complexes of transcription factors, leading to increased transcriptional activity (Cichewicz, 2010).

Acetylated histone forms are dynamic and controlled by the opposing activities of histone acetyltransferases (HAT) and histone deacetylases (HDACs). The HAT is responsible for adding an acetyl group from acetyl-coenzyme A to the lysine residues. This brings about a relaxation in chromatin structure, facilitates the process of transcription and consequently, the gene expression. On the other hand, the HDAC removes the acetyl group in histones, making the structure more compact. In this way, the system HAT-HDAC is important in the regulation of the gene expression (Strahl and Allis, 2000; Boutillier et al., 2003; Kazantsev and Thompson, 2008).

HDA2, one of the major codifier genes of HDACs, suppresses the production of secondary metabolites in several species of *Aspergillus nidulans*. On the other hand, the HDAC inhibition functions as an inducer of metabolites biosynthesis, illustrating how epigenetic regulation, by chromatin remodeling via chemical modifications in DNA and histones, can change the secondary metabolism in fungi (Asai et al., 2011). HDAC inhibitors include the natural product trichostatin A, besides suberoylanilidehydroxamic acid (SAHA) and bis-suberoyl-hydroxamic acid (SBHA), which have been successfully used (Fig. 3) (Fisch et al., 2009).

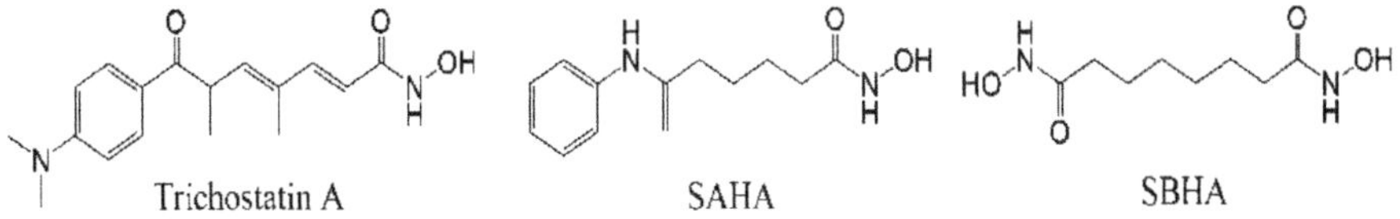

Figure 3. Molecular structures of the HDAC main inhibitors.

SAHA is able to radically restructure the secondary metabolome encoded by silent pathways in *Aspergillus niger* (Henrikson et al., 2009). The fungus *A. niger*, in the presence of SAHA, was able to produce a new fungal metabolite, Nygerone (Fig. 4), with an interesting 1-phenylpyridin-4(1H)-one skeleton, which had not been previously reported from any natural source until then.

SBHA also led to significant changes in the secondary metabolism of entomopathogenic fungus *Torrubiella luteorostrata*, inducing the production of three new prenylated tryptophan analogs (Luteorides A, B and C), along with Terezine D and a depsipeptide (Fig. 5). These structures are characterized by the presence of an (*E*)-oxime, a rare functional group in natural products, and a 3-methylbuta-1,3-dienilunit (Asai et al., 2011).

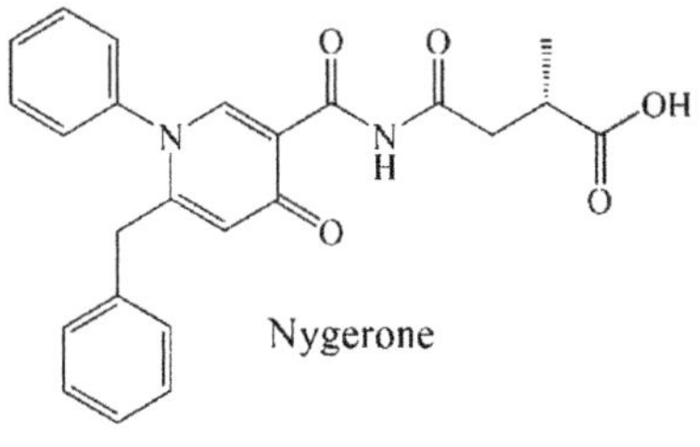

Nygerone

Figure 4. Molecular structure of Nygerone, a secondary metabolite produced by *A. niger* upon SAHA induction.

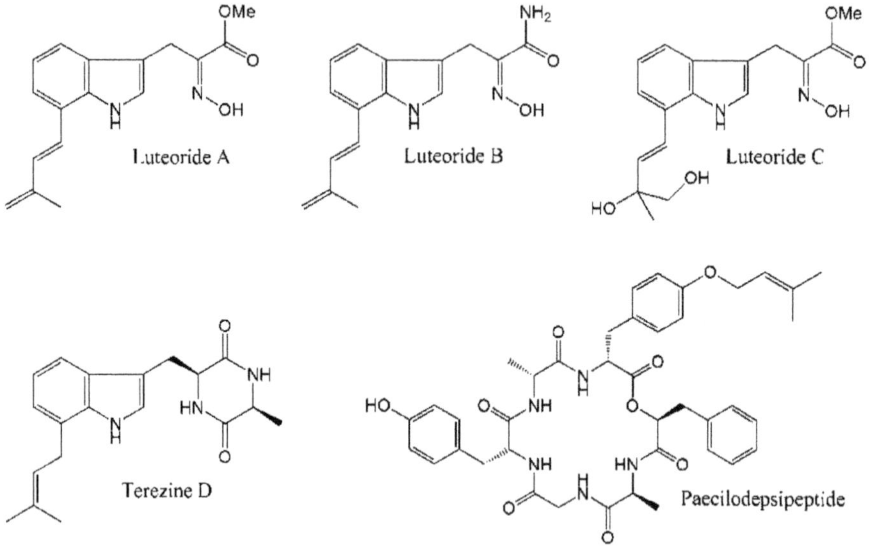

Figure 5. Molecular structure of metabolites produced by *Torrubiella luteorostrata* cultivated with SBHA.

These HDAC inhibitors can also lead to production of mycotoxins such as alternariol, tenuazonic acid and altertoxin II, produced by strains of an endophytic fungus obtained from *Datura stramonium* L., after addition of SBHA to the fermentation. Another species of fungus isolated from the root of *D. stramonium* L., identified as *Fusarium oxysporum* f. sp. *conglutinans,* when grown in the presence of SBHA, produced two new compounds, which had not been detected in the absence of epigenetic modulators. These new secondary metabolites were characterized as fusaric acid derivatives, 5-butyl-6-oxo-1,6-dihydropyridine-2-carboxylic acid and 5-(but-9-enyl)-6-oxo-1,6-dihydropyridine-2-carboxylic acid (Fig. 6) (Chen et al., 2013).

Epigenetic agents are usually added in small quantities to the growth medium, at the concentrations in the range of 0.5 mM to 1 mM for SBHA, and at the concentrations of 10 μM for SAHA (Williams et al., 2008; Henrikson et al., 2009; Asai et al., 2011; Asai et al., 2012; Chen et al., 2013). Generally the epigenetic agents are added in the log phase of growth (24 h after microorganism inoculation) and the modifiers are left in contact with the fungus for a variable

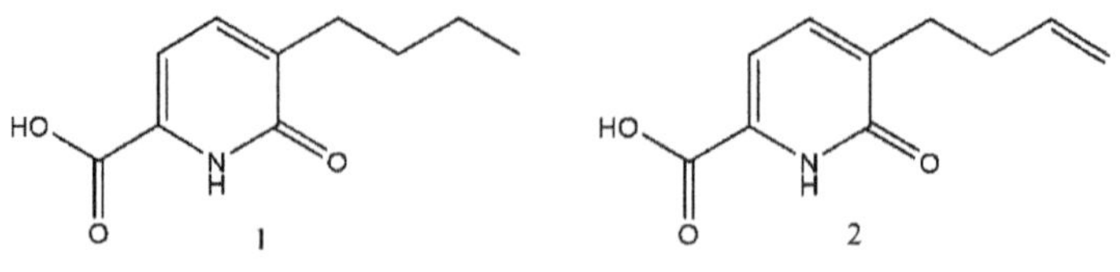

Figure 6. Molecular structure of fusaric acid derivatives. (1) 5-butyl-6-oxo-1,6-dihydropyridine-2-carboxylic acid; (2) 5-(but-9-enyl)-6-oxo-1,6-dihydropyridine-2-carboxylic acid.

period of time prior to extraction (Fisch et al., 2009). The use of these agents is not simple, since their effects on fungal metabolism are much more extensive than just the inhibition of specific proteins. In *Saccharomyces cerevisiae*, for example, many other effects such as interference in cell cycle progression and also in the metabolism of carbohydrates and amino acids were observed when using HDAC inhibitors-Trichostatin A, for example (Chung et al., 2013). These effects were observed in less than 15 min after addition indicating that the profile of action of this modulator includes other modifications, besides HDAC inhibition, at least in *S. cerevisiae* (Fisch et al., 2009).

Besides acetylation, the other post-translational modification of histones which has been extensively studied is methylation. Methylation of histones involves the transfer of a methyl group from *S*-adenosyl-*L*-methionine (SAM), and is promoted by methyltransferases that target certain residues of arginine and lysine. The sites of methylation are particularly present in the histones H3 where one, two or three methyl groups can be added (Brosch et al., 2008).

While the acetylation of lysine usually leads to inhibition of transcription, the effects of methylation in the genes are variable. It may be activation or even silencing (Cichewicz, 2010). Arginine methylation on histones is related to the activation of transcription, while the methylation of lysine is related to the repression of transcription. Histone methylation pattern also influences DNA methylation, contributing to a smaller or greater chromatin compaction (Brosch et al., 2008). A low index of lysine 9 of histone H3 (H3K9) methylation is related to a decrease of DNA methylation in fungi (Aoyama et al., 2008). Arginine residues of histones can be methylated by protein arginine methyltransferases (PRMT). The PRMTs catalyze the transfer of methyl groups of SAM to arginine. This modification occurs within the N-terminal tails of H3 and H4. Arginine can be mono or dimethylated, the latter in a symmetrical or asymmetrical shape (Trojer et al., 2004).

Several lysine residues, including lysines 4, 9, 14, 27, 36 and 72 of histone H3, lysines 20 and 59 of histone H4, and lysine 26 of H1 are sites of methylation. Each lysine residue can be mono, di or trimethylated by histone methyltransferase (HKMT). The trimethylation of lysine 4 on histone H3 (H3K4me3) is found in euchromatic structures which are open for transcription. However, the trimethylation of lysines 9 or 27 (H3K9me3 and H3K27me3, respectively) is accumulated in heterochromatic regions (Brosch et al., 2008; Franz et al., 2009). Methylation of lysines residues on histones is considered an important regulatory mechanism as it can activate or repress transcription (Peters et al., 2002; Brosch et al., 2008).

Some studies have also shown that phosphorylation of serine 10 of the histone H3 N-terminal portion plays a crucial role in chromatin remodeling and in the progression of the cell cycle during mitosis and meiosis. Furthermore, this modification is important during interphase, as it allows the transcription of an increasing number of genes (Nowak and Corces, 2004; Rossetto et al., 2012).

DNA Methylation

Among the epigenetic modifications, DNA methylation is the most studied and its well established methods are available in the literature. DNA methylation is an epigenetic reversible modification which plays an important role on cellular differentiation. The mechanism of DNA methylation occurs when the DNA (cytosine-5)-methyltransferase (C5 MTase) catalyzes the transfer of a methyl group from SAM to the position 5 of a cytosine on DNA-specific positions (Fig. 7) (Gabbara and Bhagwat, 1995; Volodymyr et al., 2005; Pang et al., 2013).

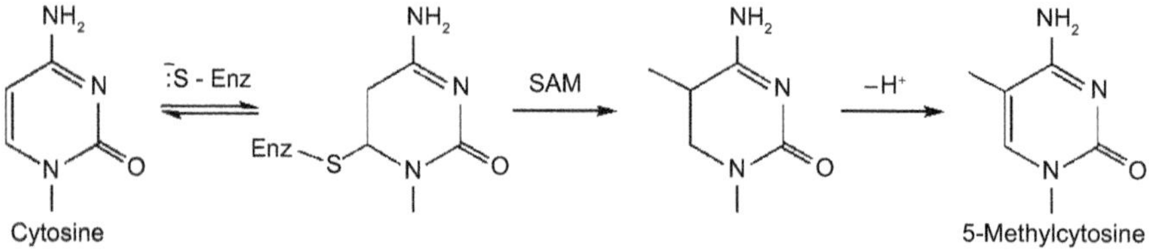

Figure 7. Formation of 5-methylcytosine by methylation of cytosine.

Cytosine is a DNA base derived from pyrimidine and its methylation leads to the formation of 5-methylcytosine (Meyer, 2011). Methylation of cytosine occurs in cooperation with the modifications in histones to regulate DNA transcription and genome stability. Thus, the 5-methylcytosine (5mC) is considered as an epigenetic marker and can be quantified to indicate regions with high or low gene transcription activity (Booth et al., 2012; Schübeler, 2015).

Two groups of enzymes are involved in the process of DNA methylation. The first group contains the DNA methyltransferase enzyme and is involved in the methylation of carbon 5 of cytosine. The second group comprises the enzymes that remove the methylated cytosines from the DNA, replacing them by non-methylated cytosines.

The process of DNA methylation, which occurs mainly in the region CG (cytosine-guanine), also known as "CpG islands", and less frequently in the region CHG and CHH (where H is adenine, cytosine or thymine), is extremely important for the maintenance of genome integrity and stability (Paszkowski and Scheid, 1998; Deaton and Bird, 2013). The hypomethylation, that occurs due to inhibition of methylation by the mechanism of DNA methyltransferase, can lead to total or partial gene modulation. This process can activate the production of new enzymes and consequently the activation of new metabolic pathways (Fisch et al., 2009; Supokawej et al., 2013).

Epigenetic Modifications in DNA

Among the main epigenetic modulators, 5-azacytidine (5-AZA) (Fig. 8) is the most widely used DNA methyltransferase inhibitor. The 5-AZA interacts with the methyltransferases resulting in hypomethylation of DNA and consequently in chromatin restructuring (Paszkowski and Scheid, 1998; Fisch et al., 2009; Supokawej et al., 2013). Administering 5-AZA to a fungal culture, DNA hypomethylation and chromatin restructuring lead to silencing and/or activation of genes, allowing the fungus to enable new pathways or even interrupting the biosynthesis of secondary metabolites that are naturally produced (González et al., 2013; Klironomos et al., 2013). Among the various studies on secondary metabolites produced by fungi upon induction by 5-AZA, some methodologies suggest the use of 5-AZA at concentrations of 0.05–1 mM (Kiselev et al., 2013; Lin et al., 2013).

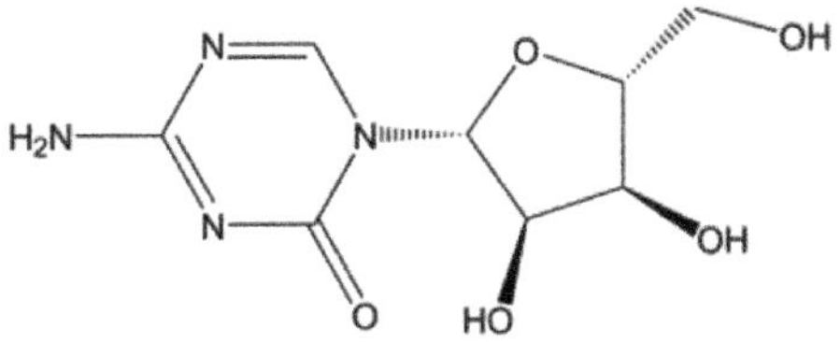

Figure 8. Molecular structure of epigenetic agent 5-azacytidine (5-AZA).

An interesting example showed that 5-AZA is able to silence the gene encoding *O*-methyltransferase A production. This enzyme is responsible for the conversion of sterigmatocystin to methylsterigmatocystin and dihydro-methylsterigmatocystin into its methyl derivative, hydrosterigmatocystin. By action of 5-AZA, there is inhibition of the biosynthesis of aflatoxins (Fig. 9), potent mycotoxins produced by fungi like *Aspergillus flavus* and *Aspergillus parasiticus* (Lin et al., 2013).

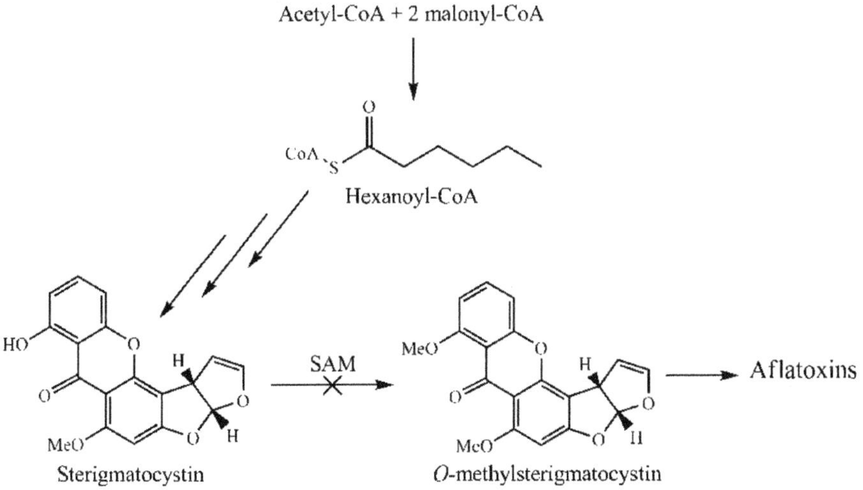

Figure 9. Stage of methylation silenced in the biosynthesis of aflatoxins.

Chemical epigenetic manipulation of *Penicillium citreonigrum* also led to significant changes in the secondary metabolites profile. The metabolites obtained from the fungus cultivated with 50 μM of 5-AZA includes six azaphilones, sclerotiorin, sclerotioramine, ochrephilone, dechloroisochromophilone III, dechloroisochromophilone IV, and 6-((3*E*,5*E*)-5,7-dimethyl-2-methylenone-3,5-dienyl)-2,4-dihydroxy-3-methylbenzaldehyde, and two new meroterpenes (Atlantinones A and B) (Fig. 10). When grown without epigenetic agents, the fungus *P. citreonigrum* did not produce any substance of azaphilones and meroterpenes classes, proving that new biosynthetic routes have been activated in the fungus as a result of 5-AZA addition (Wang et al., 2010).

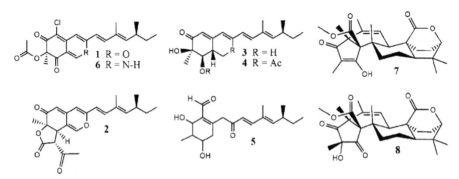

Figure 10. Secondary metabolites produced by *P. citreonigrum* cultivated in presence of 5-AZA. (1) Sclerotiorin; (2) Ochrephilone; (3) Dechloroisochromophilone III; (4) Dechloroisochromophilone IV; (5) 6-((3E, 5E)-5,7-dimethyl-2-methylenone-3,5-dienyl)-2,4-dihydroxy-3-methylbenzaldehyde; (6) Sclerotioramine; (7) Atlantinones A and (8) Atlantinones B.

The addition of 5-AZA to a culture of *Aspergillus sydowii* stimulated the production of new sesquiterpenoids of bisabolane type, some of them described for the first time in the literature and showed high anti-inflammatory and anti-diabetic activity (Chung et al., 2013).

Another example of production of new substances within a previously known biosynthetic route is the isolation of a new lipid, ustilagic acid, by the fungus *Ustilago maydis*. This fungus is known for producing a variety of glycolipids, a class of substances that has several properties that are of interest to the pharmaceutical, cosmetics and food industry. It has been proved that some metabolites, such as ustilagic acids A and B, are only produced upon addition of 5-AZA in the culture medium (Fig. 11) (Yang et al., 2013).

The 5-AZA is the most widely used substance for DNA demethylation and one of the first to be identified, however, other substances that have the same ability are also known. Other DNA methyltransferase inhibitors utilized are 5-aza-2'-deoxycytidine (5-azacytidine derived), zebularine and *N*-Phthalyl-*L*-Tryptophan (Fig. 12). Recently, a large number of studies have been focusing

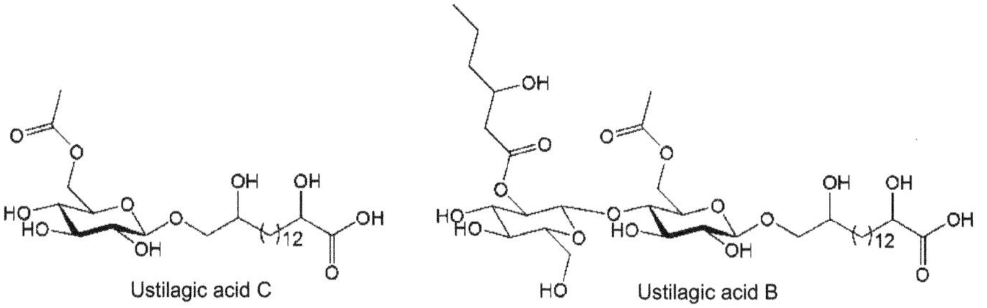

Figure 11. Molecular structures of new lipids produced by *Ustilago maydis* cultivated in the presence of 5-AZA.

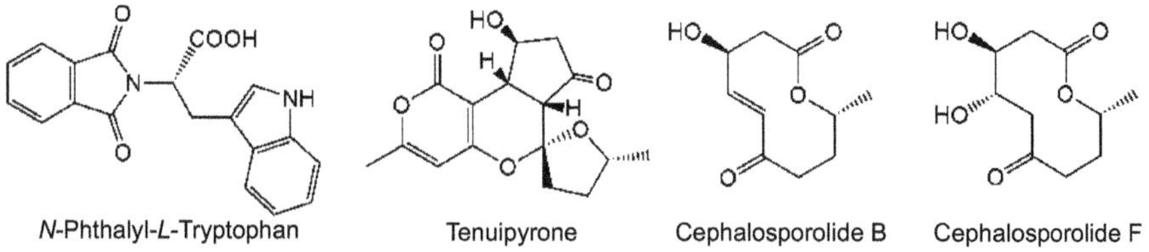

Figure 12. Molecular structure of N-Phthalyl-L-Tryptophan and metabolites obtained from *Isaria tenuipes*.

not only on the application of these epigenetic modulators for the modification of secondary fungal metabolites but also in the study of mechanism of action and discovery of new epigenetic agents (Arase et al., 2012; Gertych et al., 2013).

Final Remarks

Epigenetic modifiers, which are histone deacetylase inhibitors (HDAC), or DNA methyltransferase inhibitors, have recently come into use and are quite promising for the modulating the fungal metabolism. These chemicals alter the pattern of gene expression in fungi which results in the production of new secondary metabolites hitherto unreported (Williams et al., 2008; Chen et al., 2013).

HDAC inhibitors can be used in conjunction with DNA methyltransferase inhibitors with the aim to activate silent biosynthetic routes in filamentous fungi (Chung et al., 2013). Significant changes in metabolic profile of an entomopathogenic fungus *Isaria tenuipes* in the presence of two inhibitors, SBHA and N-Phthalyl-L-Tryptophan (Fig. 12) were observed. The metabolism was increased and a new polyketide with a tetracyclic ring (tenuipyrone), along with cephalosporolide B, a plausible biosynthetic precursor of tenuipyrone and cephalosporolide F were isolated (Fig. 12) (Asai et al., 2012).

On the other hand, in some cases, combinations are inefficient. The simultaneous treatment of fungal cultures with an inhibitor of DNA methyltransferase and a histone deacetylase inhibitor may be only moderately effective, due to significant growth constraint and/or the generation of metabolites dominated by the effects of only one of the components of the mixture (Williams et al., 2008).

Interactions with other microorganisms can also trigger epigenetic changes, as described for *Aspergillus nidulans*, when grown in presence of the bacterium *Streptomyces rapamycinicus*. This bacterium is able to lead to the activation of a cluster of polyketide synthase producing gene in *A. nidulans*. The changes caused by this bacterium in the fungus metabolism occur through the catalyst complex of histone acetyltransferase Saga/Ada. This complex is known to be responsible for the production of penicillin, sterigmatocystin and terrequinone in *A. nidulans*. Simultaneous cultivation was able to activate also the production of the polyketide or sellinic acid, lecanoric acid and two inhibitors of cathepsin K (Nützmann et al., 2011).

Miscellaneous fungi genera such as *Aspergillus, Cladosporium, Penicillium, Diatrype, Clonostachys*, among others, when cultivated in different culture media, were able to provide metabolic diversification, though the recent genomic data suggest that there is an even greater number of substances which can be produced upon induction by epigenetic agents. The same fungi, when treated with antibiotics such as amphotericin B, cycloheximide and 5-fluorouracil were not able to increase the production of metabolites or to activate the routes of production of new secondary metabolites, showing that the epigenetic modifiers do not work in fungus through a cytotoxic mechanism (Williams et al., 2008).

Epigenetic agents are also being studied for use in treatments of different autoimmune diseases such as lupus, and their use to understand the mechanism of action of different lymphomas, carcinomas and several cancerous cells have been reported (Gertych et al., 2013; Hewagama et al., 2013; Matteucci et al., 2013; Sung et al., 2013; Johnson et al., 2015; Mottamal et al., 2015). Some epigenetic agents studied both in humans, plants or microorganisms are genistein and lycopene (Arase et al., 2012; Gertych et al., 2013).

Through a classical view, fungi with the same genetics should always produce the same phenotype. However, when looking through an epigenetic view, it is possible that species bearing the same genetics can express different phenotypes, showing diversified appearance and producing different secondary metabolites in different percentages (Rapp and Wendel, 2005).

Acknowledgements

We acknowledge FAPEMIG, CNPq and CAPES (Brazilian Funding Agencies) for grants and scholarships. MatheusThomazNogueira Silva Lima (UFMG, Brazil) is acknowledged for creating the illustrative representation of the nucleosome and chromatin (Fig. 1) and for suggestions on the text.

References

Aoyama, T., Okamoto, T., Kohno, Y., Fukiage, K., Otsuka, S., Furu, M., Ito, K., Jin, Y., Nagayama, S., Nakayama, T., Nakamura, T. and Toguchida, J. 2008. Cell-specific epigenetic regulation of ChM-I gene expression: Crosstalk between DNA methylation and histone acetylation. Biochem. Biophys. Res. Commun., 365: 124–130.

Arase, S., Kasai, M. and Kanazawa, A. 2012. *In planta* assays involving epigenetically silenced genes reveal inhibition of cytosine methylation by genistein. Plant Methods, 8: 1–10.

Asai, T., Chung, Y.-M., Sakurai, H., Ozeki, T., Chang, F.-R., Yamashita, K. and Oshima, Y. 2012. Tenuipyrone, a novel skeletal polyketide from the entomopathogenic fungus, *Isaria tenuipes*, cultivated in the presence of epigenetic modifiers. Org. Lett., 14: 513–515.

Asai, T., Yamamoto, T. and Oshima, Y. 2011. Histone deacetylase inhibitor induced the production of three novel prenylated tryptophan analogs in the entomopathogenic fungus, *Torrubiella luteorostrata*. Tetrahedron Lett., 52: 7042–7045.

Bender, J. 2004. DNA methylation and epigenetics. Annu. Rev. Plant Biol., 55: 41–68.

Booth, M.J., Branco, M.R., Ficz, G., Oxley, D., Krueger, F., Relk, W. and Balasubramanian, S. 2012. Quantitative sequencing of 5-methylcytosine and 5-hydroxymethylcytosine at single-base resolution. Science, 336: 934–937.

Boutillier, A.L., Trinh, E. and Loeffler, J.P. 2003. Selective E2F-dependent gene transcription is controlled by histone deacetylase activity during neuronal apoptosis. J. Neurosci., 84: 814–828.

Brosch, G., Loidl, P. and Graessle, S. 2008. Histone modifications and chromatin dynamics: a focus on filamentous fungi. FEMS Microbiol. Rev., 32: 409–439.

Chen, H.-J., Awakawa, T., Sun, J.-Y., Wakimoto, T. and Abe, I. 2013. Epigenetic modifier-induced biosynthesis of novel fusaric acid derivatives in endophytic fungi from *Datura stramonium* L. Nat. Prod. Bioprospect., 3: 20–23.

Chung, Y.M., Wei, C.K., Chuang, D.W., El-Shazly, M., Hsieh, C.T., Asai, T., Oshima, Y., Hsieh, T.-J., Hwang, T.L., Wu, Y.C. and Chang, F.R. 2013. An epigenetic modifier enhances the production of anti-diabetic and anti-inflammatory sesquiterpenoids from *Aspergillus sydowii*. Bioorg. Med. Chem. Lett., 21: 3866–3872.

Cichewicz, R.H. 2010. Epigenome manipulation as a pathway to new natural product scaffolds and their congeners. Nat. Prod. Rep., 27: 11–22.

Deaton, A.M. and Bird, A. 2013. CpG islands and the regulation of transcription. Genes Dev., 25: 1010–1025.

Eissenberg, J.C. and Shilatifard, A. 2010. Histone H3 lysine 4 (H3K4) methylation in development and differentiation. Dev. Biol., 339: 240–249.

Fisch, K.M., Gillaspy, A.F., Gipson, M., Henrikson, J.C., Hoover, A.R., Jackson, L., Najar, F.Z., Wägele, H. and Cichewicz, R.H. 2009. Chemical induction of silent pathway transcription in *Aspergillus niger*. J. Ind. Microbiol. Biotechnol., 36: 1199–1213.

Fox, E.M. and Howlett, B.J. 2008. Secondary metabolism, regulation and role in fungal biology. Curr. Opin. Microbiol., 11: 481–487.

Franz, H., Mosch, K., Soeroes, S., Urlaub, H. and Fischle, W. 2009. Multimerization and H3K9me3 binding are required for CDYL1b heterochromatin association. J. Biol. Chem., 284: 35049–35059.

Gabbara, S. and Bhagwat, A.S. 1995. The mechanism of inhibition of DNA (cytosine-5-)-methyltransferases by 5-azacytosine is likely to involve methyl transfer to the inhibitor. Biochem. J., 307: 87–92.

Gertych, A., Oh, J.H., Wawrowsky, K.A., Weisenberger, D.J. and Tajbakhsh, J. 2013. 3-D DNA methylation phenotypes correlate with cytotoxicity levels in prostate and liver cancer cell models. BMC Pharmacol. Toxicol., 14: 1–21.

González, A.I., Sáiz, A., Acedo, A., Ruiz, M.L. and Polanco, C. 2013. Analysis of genomic DNA methylation patterns in regenerated and control plants of rye (Secalecereale L.). Plant Growth Regul., 70: 227–236.

Gonzalo, S. 2010. Epigenetic alterations in aging. J. Appl. Physiol., 109: 586–597.

Graessle, S., Loidl, P. and Brosch, G. 2001. Histone acetylation: plants and fungi as model systems for the investigation of histone deacetylases. Cell. Mol. Life Sci., 58: 704–720.

Henrikson, J.C., Hoover, A.R., Joyner, P.M. and Cichewicz, R.H. 2009. A chemical epigenetics approach for engineering the *in situ* biosynthesis of a cryptic natural product from *Aspergillus niger*. Org. Biomol. Chem., 7: 435–438.

Hewagama, A., Gorelik, G., Patel, D., Liyanarachchi, P., McCune, W.J., Somers, E., Gonzalez-Rivera, T., Cohort, T.M.L., Strickland, F. and Richardson, B. 2013. Overexpression of X-Linked genes in T cells from women with lupus. J. Autoimmun., 41: 60–71.

Johnson, C., Warmoes, M.O., Shen, X. and Locasale, J.W. 2015. Epigenetics and cancer metabolism. Cancer Letters, 356: 309–314.

Kazantsev, A.G. and Thompson, L.M. 2008. Therapeutic application of histone deacetylase inhibitors for central nervous system disorders. Nat. Rev. Drug Discovery, 10: 854–868.

Kiselev, K.V., Tyunin, A.P. and Zhuravlev, Y.N. 2013. Involvement of DNA methylation in the regulation of STS10 gene *Vitisa murensis*. Planta, 237: 933–941.

Klironomos, F.D., Berg, J. and Collins, S. 2013. How epigenetic mutations can affect genetic evolution: Model and mechanism. Bioessays, 35: 571–578.

Kouzarides, T. 2007. Chromatin modifications and their function. Cell, 128: 693–705.

Lin, J.-Q., Zhao, X.-X., Wang, C.-C., Xie, Y., Li, G.-H. and He, Z.-M. 2013. 5-Azacytidine inhibits aflatoxin biosynthesis in *Aspergillus flavus*. Ann. Microbiol., 63: 763–769.

Mai, A., Cheng, D., Bedford, M.T., Valente, S., Nebbioso, A., Perrone, A., Brosch, G., Sbardella, G., Bellis, F., Miceli, M. and Altucci, L. 2008. Epigenetic multiple ligands: Mixed histone/protein methyltransferase, acetyltransferase, and class III deacetylase (Sirtuin) inhibitors. J. Med. Chem., 51: 2279–2290.

Matteucci, E., Maroni, P., Bendinelli, P., Locatelli, A. and Desiderio, M.A. 2013. Epigenetic control of endothelin-1 axis affects invasiveness of breast carcinoma cells with bone tropism. Exp. Cell Res., 319: 1865–1874.

Meyer, P. 2011. DNA methylation systems and targets in plants. FEBS Lett., 585: 2008–2015.

Mottamal, M., Zheng, S., Huang, T.L. and Wang, G. 2015. Histone deacetylase inhibitors in clinical studies as templates for new anticancer agents. Molecules, 20: 3898–3941.

Nowak, S.J. and Corces, V.G. 2004. Phosphorylation of histone H3: a balancing act between chromosome condensation and transcriptional activation. Trends Genet., 20: 214–220.

Nützmann, H.-W., Reyes-Dominguez, Y., Scherlach, K., Schroeckh, V., Horn, f., Gacek, A., Schümann, J., Hertweck, C., Strauss, J. and Brakhage, A.A. 2011. Bacteria-induced natural product formation in the fungus *Aspergillus nidulans* requires Saga/Ada-mediated histone acetylation. Proc. Natl. Acad. Sci. USA, 108: 14282–14287.

Palmer, J.M. and Keller, N.P. 2010. Secondary metabolism in fungi: does chromosomal location matter? Curr. Opin. Microbiol., 13: 431–436.

Pang, J., Dong, M., Li, N., Zhao, Y. and Li, B. 2013. Functional characterization of a rice *de novo* DNA methyltransferase, OsDRM2, expressed in *Escherichia coli* and yeast. Biochem. Biophys. Res. Commun., 432: 157–162.

Paszkowski, J. and Scheid, O.M. 1998. Plant genes: The genetics of epigenetics. Curr. Biol., 8: 206–208.

Peters, A.H.F.M., Mermoud, J.E., Carroll, D., Pagani, M., Schweizer, D., Brockdorff, N. and Jenuwein, T. 2002. Histone H3 lysine 9 methylation is an epigenetic imprint of facultative heterochromatin. Nat. Genet., 30: 77–80.

Rapp, R.A. and Wendel, J.F. 2005. Epigenetics and plant evolution. New Phytol., 168: 81–91.

Rossetto, D., Avvakumov, N. and Côté, J. 2012. Histone phosphorylation: a chromatin modification involved in diverse nuclear events. Epigenetics, 10: 1098–108.

Schübeler, D. 2015. Function and information content of DNA methylation. Nature, 517: 321–326.

Shwab, E.K. and Keller, N.P. 2008. Regulation of secondary metabolite production in filamentous ascomycetes. Mycol. Res., 112: 225–230.

Strahl, B.D. and Alls, C.D. 2000. The language of covalent histone modifications. Nature, 403: 41–45.

Strauss, J. and Reyes-Dominguez, Y. 2012. Regulation of secondary metabolism by chromatin structure and epigenetic codes. Fungal Genet. Biol., 48: 62–69.

Sung, C.K., Li, D., Andrews, E., Drapkin, R. and Benjamin, T. 2013. Promoter methylation of the SALL2 tumor suppressor gene in ovarian cancers. Mol. Oncol., 7: 419–427.

Supokawej, A., Kheolamai, P., Nartprayut, K., U-pratya, Y., Manochantr, S., Chayosumrit, M. and Issaragrisil, M. 2013. Cardiogenic and myogenic gene expression in mesenchymal stem cells after 5-azacytidine treatment. Turk. J. Hematol., 30: 115–121.

Takahashi, J.A., Teles, A.P.C., Bracarense, A.A.P. and Gomes, D.C. 2013. Classical and epigenetic approaches to metabolite diversification in filamentous fungi. Phytochem. Rev., 12: 773–789.

Tang, H.-Y., Zhang, Q., Gao, Y.-Q., Zhang, A.-L. and Gao, J.-M. 2015. Miniolins A–C, novel isomeric furanones induced by epigenetic manipulation of *Penicillium minioluteum*. Royal Society of Chemistry, 5: 2185–2190.

Trojer, P., Dangl, M., Bauer, I., Graessle, S., Loidl, P. and Brosch, G. 2004. Histone methyltransferases in *Aspergillus nidulans*: Evidence for a novel enzyme with a unique substrate specificity. Biochemistry, 43: 10834–10843.

Verdone, L., Agricola, E., Caserta, M. and Mauro, E. 2006. Histone acetylation in gene regulation. Brief. Funct. Genom. Proteom., 5: 209–221.

Volodymyr, V.R., Nese, S., Ruslana, I.R., Ulrich, W. and Winfriede, W. 2005. The methylation cycle and its possible functions in barley endosperm development. Plant Mol. Biol., 59: 289–307.

Wang, X., SenaFilho, J.G., Hoover, A.R., King, J.B., Ellis, T.K., Powell, D.R. and Cichewicz, R.H. 2010. Chemical epigenetics alters the secondary metabolite composition of guttate excreted by an Atlantic-Forest-soil-derived *Penicillium citreonigrum*. J. Nat. Prod., 73: 942–948.

Williams, R.B., Henrikson, J.C., Hoover, A.R., Lee, A.E. and Cichewicz, R.H. 2008. Epigenetic remodeling of the fungal secondary metabolome. Org. Biomol. Chem., 6: 1895–1897.

Yang, X.-L., Awakawa, T., Wakimoto, T. and Abe, I. 2013. Induced production of the novel glycolipid ustilagic acid C in the plant pathogen *Ustilago maydis*. Tetrahedron Lett., 54: 3655–3657.

CHAPTER 7

Volatile Organic Compounds from Fungi: Impact and Exploitation

Sanjai Saxena

ABSTRACT

Fungi are ubiquitous in nature and they compete with other organisms for their survival. These produce carbon based compounds as mixtures in gaseous phase called the volatile organic compounds (VOC) which diffuses into the atmosphere from the substrate, they grow upon. These fungal VOC's serve as infochemicals and play an important role of signaling for fungi. Apart from this, they are also indicators of bio-deterioration and serve as taxonomic markers. The fungal VOCs are increasingly finding their applications in biological control and mycofumigation. There is a possibility that they may used as next generation biofuels. Thus, bioprospecting of fungi for the study of these VOCs is a new paradigm for the discovery of new products for human use apart from providing a fundamental knowledge of their production.

Introduction

Volatile organic compounds are organic chemicals which have a vapour pressure at room temperature (20–25°C) as a result of low boiling point and, therefore, sublimate at room temperature from liquid or solid forms (Pagans et al., 2006). Microorganisms produce a variety of primary and secondary

Department of Biotechnology, Thapar University, Patiala, Punjab, India 147004.
  Email: sanjaibiotech@yahoo.com; ssaxena@thapar.edu

metabolites which play a critical role in organismal and organism-environment interactions. VOCs produced by microorganisms, are derived from primary and secondary metabolism and include a variety of aroma compounds. In the primary metabolism the fungi extract nutrients from the natural substrate by breaking that down for their growth and maintenance and, during the process they produce characteristic volatile compounds. The production of fungal volatile organic compounds (FVOCs) as a result of secondary metabolism is broadly dependent upon the competition for the resources in a nutrient poor environment. The FVOC generally belong to different chemical classes' viz., monoterpenes, sesquiterpenes, alcohols and aldehydes. *Aspergillus niger*, *Aspergillus flavus* and *Penicillium roquefortii* are able to produce 1-octen-3-ol (Chittara et al., 2004). *Aspergillus flavus* is known to produce 2-octen-1-ol that has been described to possess a strong musty, oily odour (Kaminski et al., 1972). The grassy odor of *Botrytis cinerea* is due to benzyl cyanide production and its emission (Korpe et al., 1998).

Volatile organic compounds emitted by bacteria and fungi have been studied in context of their roles in different aspects such as chemotaxonomic markers or as infochemicals, i.e., molecules taking part in intra and inter cellular communication apart from their role in growth and development, deterioration/decay of fruits, vegetables and buildings. In this chapter the role and impact of FVOC's in fungal development, as taxonomic and deterioration markers and their applications in biological control, mycofumigation and as biofuels have been discussed (Fig. 1).

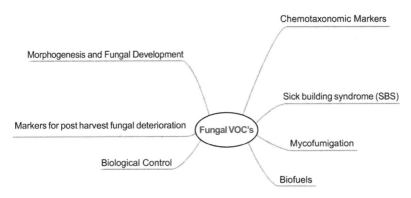

Figure 1. Intervention Areas of fungal volatile organic compounds (FVOC's).

FVOCs as Chemotaxonomic Markers

Chemotaxonomy has supplemented the traditional morphologically based taxonomy of fungi. Chemotaxonomy of filamentous fungi has largely been based on classification and identification of fungi on their profiles of secondary metabolites. The prominent precursor in FVOC biosynthesis is acetate present in acetyl coenzyme A in the cells. Acetyl CoA is basically derived from the glycolytic pathway from pyruvate and is a precursor of fatty acids and mevalonate which play a very important role in terpenoid metabolism. Volatile Terpenes which are produced during secondary metabolism have been largely profiled as taxonomic markers and have been used for the classification of plants, lichens and bacteria (Harbone and Turner, 1984; Barberio and Twibell, 1991; Adams, 1994; Labows et al., 1980; Jollivet et al., 1992). Forty seven taxa of *Penicillium* have been profiled for their volatile signatures for their use in chemosystematics. A clear separation between taxa was observed when Jaccard distance coefficient was employed on a binary matrix. *Penicillium discolor* and *Penicillium echinulatum* always appeared as separate clusters far from each other, strongly suggesting their separate taxonomic status. The dynamics of microbial groups viz., bacteria and fungi in ecosystems have been established by using VOC patterns as ecological indicators of microbial community composition (McNeal and Herbert, 2009). Most of the FVOC focus on emissions by single species. Thus, FVOC's can be used effectively as non-destructive markers to differentiate and identify a fungal species from unknown VOC mixture in nature. Chemotyping, based on VOC emissions, could also be an efficient method of identifying pathogens in early infection stage. There are some common types of VOCs emitted by all types of fungi, viz., ectomycorrhizal, plant pathogenic and saprophytic. Linalool is a monoterpene which is released by *Gymnopilus penetrans*, *Piloderma olivaceum* (ectomycorrhizal fungi), *Coprinus cinereus* and several *Penicillium* species and yeast *Kluveromyces lactis* (Drawert and Barton, 1978; Wihlborg et al., 2008).

The other common sesquiterpenes emitted by all fungi are trans-β-farnesene, β-bisabolene, β-himachalene, α-amorphene, δ-cadinene, and nerolidol. Other VOC's like 2-methyl-1-butanol, 1-octen-3-ol, 3-octanone, 2-pentylfuran are released by truffle species (Zeppa et al., 2004; Splivallo et al., 2011). *Muscodor albus* also emits 2-methyl-1-butanol and 1-octen-3-ol which has generally been referred to as mushroom alcohol. Similarly, caryophyllene, a sesquiterpene present is in *P. involutus*, endophyte of *Phialocephala fortinii*, saprophyte, *Fusarium oxysporum*, a pathogen *Phialophora fastigiata*, a mold *Penicillium caseifulvum* and an endophyte *Muscodor albus*. Caryophyllene has been found to be emitted from green plant and is also found to play an important role in plant-microbe interactions apart from possessing anti-microbial properties. Longipenene is also a sesquiterpenes which is reported from plant pathogenic fungus *Armillaria mallea*, *Paecilomyces variotii*, saprophyte *Fomitopsis pinicola* and truffle *Tuber magnatum* (Gioacchini et al., 2008; Kramer and Abraham, 2012). Similarly, Geosmin is a potential indicator of fungal activity which is generally identified as earth's odor due to its musty character (Magan and Evans, 2000).

Interestingly, a very specific mixture of sesquiterpenes viz., β-silenene; α-amorphene; α-bisabolene; caryophyllene and valencene have shown to be lethal for plant and human pathogenic fungi and bacteria (Strobel et al., 2001) and thus, could be used as defence agents. Members of xylariaceae have been prolific producers of volatile organic compounds, although these have been reported by some species of Ascomycota. Genus *Muscodor* has been largely identified based on their volatile signatures. It would not be out of place to mention that *Muscodor* is one of the best studied volatile antibiotic producing endophyte. *Muscodor* species generally possesses white hyphae which are ropy and sterile in nature on all media tested as they do not form any fruiting structures.

Till date fourteen *Muscodor* species have been described: *Muscodor albus*, *M. roseus* (Worapong et al., 2001, 2002), *M. vitigenus* (Daisy et al., 2002), *M. crispans* (Mitchell et al., 2008), *M. yucatanensis* (Gonzalez et al., 2009), *M. fengyangensis* (Zhang et al., 2010), *M. cinnanomi* (Suwannarach et al., 2010), *M. sutura* (Kudalkar et al., 2012), *M. musae, M. oryzae, M. equiseti, M. suthepensis* (Suwannarach et al., 2013) and quite recently *M. kashayum* (Meshram et al., 2013) and *M. tigerii* (Saxena et al., 2015). Geographically, *Muscodor* species have been reported from South America/USA, and South East Asia comprising of Thailand and China and very recently from India (Meshram et al., 2013; Banerjee et al., 2014; Saxena et al., 2014, 2015). A critical volatile organic analysis of these species have indicated that 2-methylpropanoic acid is a major compound emitted by *M. albus, M. crispans,* and *M. sutura* while the volatile emitted by *Muscodor yucatanensis, Muscodor roseus,* and *Muscodor vitigenus* are altogether unique. Similarly, the volatile profiles of *Muscodor* species reported from Thailand (South East Asia) are also derivatives of propanoic acid, while the one from China, *M. fengyangensis* produces a β-phellandrene. The Indian species, *Muscodor kashayum* and *Muscodor tigerii*, produce β-bisabolol and 4-octadecylmorpholine, respectively (Table 1). Further, the ITS 1-5.8-ITS 2 sequence based phylogenetic analysis suggests *Muscodor* being monophyletic with the presence of different haplotypes (Meshram et al., 2013; Saxena et al., 2014).

The work of Muller et al. (2013) has clearly indicated that FVOCs can be highly predictable for distinguishing ectomycorrhizal fungi, saprophytes and pathogens independently of their phylogenetic and familial relationships. Thus, the possibilities of using fungal volatiles as chemotaxonomic markers are on the increase and at the same time newer technologies of their detection and analysis are also coming up.

FVOCs in Fungal Morphogenesis and Development

The fungal volatiles probably are infochemicals or signal molecules which have a role in fungal growth and development. Phenylethanolis an *in situ* indicator of *Candida albicans* and *Candida dubliniensis* during planktonic and biofilm growth. This VOC is also responsible for the morphological transitions from yeasts to filamentous forms (Martins et al., 2007). It has been recently found that the induction of conidiation in *Trichoderma* species, apart from being controlled by circadian cycle, is also regulated by a volatile organic compound produced by the fungus. GC/MS profiling of the VOC spectrum produced

Table 1. Volatile characteristics and bioactivity comparison of *Muscodor* species.

Taxon	Host	Geographic location	Major VOC compounds	Bioactivity	References
M. albus	*Cinnamomum zeylanicum*	Central America	2-methyl propanoic acid, naphthalene and azulene derivatives	Antifungal Antibacterial	Worapong et al., 2001
M. vitigenus	*Paullinia paullinioides*	South America	Naphthalene	Anti-insect	Daisy et al., 2002
M. roseus	*Grevillea pteridifolia*	North Australia	Ethyl-2-butenoate and 1,2,4-trimethylbenzene	Antifungal	Worapong et al., 2002
M. crispans	*Ananas ananassoides*	South America	2-methyl propanoic acid	Antifungal Antibacterial	Mitchell et al., 2008
M. yucatanensis	*Bursera simaruba*	South America	1R, 4S, 7S, 11R-2,2,4,8-tetramethyltricyclo [5.3.1.0 (4,11)-undec-8-ene	Phytoinhibitory	Gonzalez et al., 2009
M. fengyangensis	*Actinidia chimensis*	China	2-methyl propanoic acid	Antifungal Antibacterial	Zhang et al., 2010
M. sutura	*Prestonia trifidi*	South America	2-methyl propanoic acid	Antifungal	Kudalkar et al., 2012
M. cinnamomi	*Cinnamomum bejolghota*	Thailand	Ethyl-2-methylpropanoate	Antifungal Antibacterial	Suwannarach et al., 2013
M. equisetii	*Equisetum debile*	Thailand	2-methyl propanoic acid	Antifungal Antibacterial	Suwannarach et al., 2013
M. musae	*Musa acuminata*	Thailand	2-methyl propanoic acid	Antifungal Antibacterial	Suwannarach et al., 2013
M. oryzae	*Oryza rufipogon*	Thailand	3-methylbutan-1-ol	Antifungal Antibacterial	Suwannarach et al., 2013

M. suthepensis	Cinnamomum bejolghota	Thailand	2-methyl propanoic acid	Antifungal Antibacterial	Suwannarach et al., 2013
M. kashayum	Aegle marmelos	India	β-Bisabolol and 2,6-Bis(1,1-dimethylethyl)-4-(1-oxopropyl)phenol	Antifungal Anticandidal	Meshram et al., 2013
M. tigerii	Cinnamomum camphora	India	4-octadecylmorpholine	Anticandidal Antibacterial	Saxena et al., 2015
M. darjeelingensis	Cinnamomum camphora	India	2, 6-Bis (1,1-dimethylethyl) -4-(1-oxopropyl) phenol; 4-octadecylmorpholine	Antifungal Anticandidal	Saxena et al., 2014

by non-conidiated and conidiated colonies have indicated production of 8-carbon VOC compounds-1-octen-3-ol, 3-octanol and 3-octanone specifically during conidiation (Nemcovic et al., 2008). These three compounds induced conidiation on exposure to the colonies placed in the dark in very minute concentrations only. 0.1 μM of 1-octen-3-ol suppress conidiation and inhibition of growth beyond 500 μM suggesting its putative fungi static and fungicidal activities (Chitarra et al., 2004, 2005). The volatile organic compounds influence the developmental process such as conidiation in fungi. It has been demonstrated in *Aspergillus parasiticus* that volatile organic compounds regulate developmental processes in strains in which the VeA gene is deleted. VeA is a global regulator of conidiation and sclerotia formation, besides secondary metabolism (Roze et al., 2010). 1-octen-3-ol has been recognized as the principal VOC which inhibits the germination in high conidial densities of *Aspergillus nidulans* while 3-octanone is the most active promoter of conidiogenesis (Gracia et al., 2011). Some oxylipins other than 1-octen-3-ol have been found to serve as factors which help in a critical balance between sexual and asexual reproduction in *Aspergillus nidulans*. These are referred to as precocious sexual induction factors (psi). Three psi factors viz., psiAα, psiBα and psiCα have been recognized. psi-Aα induces conidiation and counteracts sexual production while psiBα, psiCα stimulated ascospore formation (Champe and el Zayat, 1989). psi-Aα is a lactone of 5(S), 8(R)-dihydroxy-9(Z),12(Z)-octadecadienoic acid while psiBα is 8(R)-hydroxy-9(Z), 12(Z)-octadecenoic acid and psiCα is 5(S), 8(R)-dihydroxy-9(Z), 12(Z)-octadecadienoic acid (Calvo et al., 2001; Tsitsigiannis et al., 2005). Aromadendrene has been detected as a VOC in immature fruit bodies of *Tuber borchii* and also in mycelium which is grown in the presence of *Tilia platyphyllos* suggesting that the terpenoid signal molecule which is released during the plant-fungus interaction (Zeppa et al., 2004). It is probably associated with the breaking of the fruit body.

A deeper understanding of the mechanisms of the control of VOC production by fungi may provide insights on their role during germination, growth and reproduction and formation of morphological variants.

FVOCs and Sick Building Syndrome

Damp indoor environment, that causes microbial growth, is responsible for frequent health complaints in people inhabiting such environment commercial as well as residential buildings leading to an increased risk of respiratory symptoms, respiratory problems and exacerbation of asthma (WHO, 2009). Besides these, many other symptoms have also been noticed among the inhabitants of damp indoor environment that include, fatigue, headache, dermatological symptoms and gastrointestinal problems. The term mold related illness or sick building syndrome is generally used for ill-defined clinical conditions mentioned above. Trichothecenes, a mycotoxin had been a focus on building related sicknesses and sick building syndrome

(Jarvis and Miller, 2005). The VOCs emitted by molds have received less attention than mycotoxins. The symptoms of VOC exposure include lethargy, headache as well as irritation of eyes, mucous membranes of nose and throat (Araki et al., 2010). The indoor mold species, such as *Aspergillus, Penicillium, Stachybotrys* which grow on building materials, have been profiled for their VOC signatures under controlled laboratory conditions. The VOCs generated by these species vary with the length of incubation and other environmental parameters. 3-methylbutan-1-ol, hexan-2-one, heptan-2-one and octan-3-ol are considered reliable indicators of the occurrence of mold in damp houses and buildings. Occurrence of *Aspergillus* species has been correlated with volatiles viz., 3-methylbutan-1-ol, hexan-2-one, heptan-2-one, octan-3-ol and α-terpineol, while *Eurotium* sp. is associated with high indoor concentration of 3-methylbutan-1-ol, 3-methylbutan-2-ol, heptan-2-one, hexan-2-one, octan-3-ol and thujopsene (Elke et al., 1999; Pallotino et al., 2012) in sick buildings.

1-octen-3-ol has been reported to be in the problematic buildings at a concentration of 0.16 ppm (Morey et al., 1997). *Trichoderma viridae* produces alcohol like 2-methyl-1-propanol, 3-methyl-1-butanol, 2-methyl-1-butanol and 1-pentanol. These cause histamine release from human bronchoalveolar cells (Larsen et al., 1998). 1-octen-3-ol at a concentration of 10 mg/m^3 causes minor eyes, nose, and throat irritation in humans after 2 hours of exposure. It has also been reported to be neurotoxic to *Drosophila melanogaster* as well as human embryonic stem cells (Inamdar et al., 2010, 2011). The VOC seems to be interesting indicators of presence of fungi in buildings as markers of building health and indoor air quality. There is growing need for developing global standards of VOC's detection for indoor air quality monitoring, defining their toxicological limits based on *in vitro* toxicity studies, to protect humans form their adverse effects due to chronic exposure.

FVOCs as Markers of Post-Harvest Fungal Deterioration

Humans differentiate contaminated and non-contaminated food based on their odours and flavours. Fruits and vegetables are consumed raw or cooked based on their palatability, taste, flavour and odour. The selection of fruits and vegetables from vendors and outlet is also based on their flavour and odour.

The fungal spoilage is often manifested by different sensory signals such as off-colours, odours, and flavours. The deteriorated fruits become soft and vegetables appear slimy. However, the fungi start spoiling the aforesaid items long before it eventually becomes evident. Thus, early detection of spoilage of any materials by fungi is immensely important in preventing food losses and consequently the health hazard. Volatile profiles of infected and uninfected fruits and vegetables could be a better option for sensitive and advanced perception of possible fungal infestation. For this purpose, an electronic nose could be a suitable tool that would be able to detect non-destructively the presence of contaminating fungi. Electronic nose, an olfactory sensor device has been effectively used in biomedical research (Gardner et al., 2000); for food analysis (Di Natalie et al., 1997); to assess the deterioration of grain (Magan and Evans, 2000), for fruit maturity assessment (Athamneh et al., 2008) and to find out the meat and fish freshness (Winquist et al., 1993). The metabolic alterations due to fungal pathologies can be effectively detected by olfactometric analysis. This allows detection of specific volatile biomarkers and, therefore, could be used as an effective tool for post-harvest quality control and maintenance of fruits and vegetables (Riva et al., 2001). Spoilage of stored grain by fungi can be detected through the dominant volatiles that they produce—3-methyl-1-butanol, 1-octen-3-ol and 3-octanone (Tuma et al., 1989; Borjesson et al., 1990).

Potatoes infected with *Phytophthora infestans* and *Fusarium coeruleum* exhibited different VOC profiles when compared to uninfected potatoes. Of these VOCs, the most abundant were benzothiazole, 2-ethyl-1-hexanol, hexanal, 2-methylpropanoic acid-2, 2-dimethyl-1-(2-hydroxy-1-methylethyl)-propyl ester, 2-methylpropanoic acid-3-hydroxy-2, 4, 4-trimethyl-pentyl ester and phenol. These were common for the above mentioned fungal pathogens. Thus, the VOCs provide key information for sensor based detection and monitoring system of post-harvest diseases in stored potato tubers (de Lacy Costello et al., 2001).

A similar study has been carried out for moldy diseases in apples to develop an indicator method for storage houses. The apples were inoculated with *Penicillium expansum*. It was observed that there is a change in concentration of Terpenes and esters during spoilage of the apples. Pentanoic acid, ethyl ester and propanoic acid, and hexyl ester were found to decrease while concentration of butanoic acid, methyl-3,3-dimethyl ester, hexanoic acid ethyl ester and 1-methoxy-3-methyl benzene was increased during the spoilage process differentiating between healthy and *Penicillium expansum* inoculated apples (Sagi-Kiss and Fodor, 2011).

The major VOCs identified around oranges infected with *Penicillium digitatum* have been identified as terpenes specifically, limonene, β-myrcene, α-pinene, together with acetaldehyde, ethanol, carbon dioxide and ethylene. These VOCs are common and indicate the maturity of the fruit. Production of these is accelerated due to the presence of *Penicillium* (Eckert and Ratnayake, 1994; Pallottino et al., 2012).

There has been a recent upsurge in developing methods for identifying deterioration of fruits and vegetables by spiking them with possible pathogen flora and subsequently monitoring and carrying out a comparative analysis to define the indicators of spoilage during post-harvest storage conditions.

Thus, FVOCs offer novel methods of monitoring deterioration and spoilage of fruits and vegetables after harvest and during storage.

FVOCs as Biological Control Agents and Mycofumigants

Mycofumigation is a term which has been used for exploitation of VOC's elaborated by *M. albus* for the control of post-harvest plant diseases (Strobel, 2006).

Fungal VOCs possess potential to be exploited as biological control agents since they exhibit antagonistic effects towards fungi, bacteria, nematodes and insects. They, therefore, could be designated as volatile biopesticides as they can permeate into the air filled pores of soil and traverse long distance. An endophytic *Gliocladium* species from *Eucryphia cordifolia* was found to produce a mixture of volatile organic compounds which were lethal to plant pathogens—*Pythium ultimum* and *Verticillium dahlia*. The primary volatile compound produced by the species of *Gliocladium* is annulene or 1, 3, 5, 7-cyclooctatetraene which itself inhibits the fungal growth (Stinson et al., 2003). The genus *Muscodor* is one of the best studied endophytic fungus which produces synergistic mixture of VOCs having lethal effects against a wide variety of plant and human pathogenic fungi, nematodes, bacteria as certain insects (Strobel et al., 2001; Daisy et al., 2002; Grimme et al., 2007; Strobel et al., 2008). The volatile of *Muscodor* species has been used to replace methyl bromide (MeBr), a traditional soil fumigant which has been globally banned as it causes depletion of ozone layer. *M. albus* effectively controls soil borne plant pathogenic fungi like *Rhizoctonia solani* (which causes damping off of broccoli), *Phytophthora capsici* (which causes root rot of bell pepper) and *Tilletia caries* (causing bunt of wheat) (Camp et al., 2008; Mercier and Jimenenz, 2009; Worapong and Strobel, 2009; Goates and Mercier, 2011). *Mycosphaerella fijiensis*, the black sigatoka pathogen of bananas and *Xanthomonas axonopodis* pv *citri*, responsible for serious bacterial disease of citrus, were inhibited by the synergistic mixture of VOC produced by *Muscodor crispans* (Mitchell et al., 2010). Lethal effects of VOCs of *Muscodor yucatanensis* were observed against *Guignardia mangifera*, *Phomopsis* species, *Colletotrichum* sp., *Alternaria solani*, *Phytophthora capsici* but that did not affect *Fusarium oxysporum* (Macias-Rubalcava et al., 2010). *Muscodor cinnamomi* CMU-cib461 has been found to control *Rhizoctonia solani* AG-2 infections which include leaf blight, leaf spots, damping off and root rot (Suwannarach et al., 2012). *Muscodor kashayum* volatiles inhibited the growth of *Cerecospora beticola*, *Colletrotichum gloesporioides*, *Mycosphaerella fijiensis*, *Chaetomium heterosperum* and *Fusarium oxysporum* (Meshram et al., 2013). The growth of *Alternaria alternata* and *Cercospora bataticola* were completely suppressed *in vitro* by the volatiles of *Muscodor tigerii* (Saxena et al., 2015).

Volatiles from other endophytic fungi have also been proved to be effective mycofumigants in controlling fungal diseases of fruits and vegetables in farms as well as in storage houses. *Oxysporus latemarginatus* EF069 is a volatile producing endophytic fungi which was isolated from healthy tissues of pepper plant

(*Capsicum annum*) (Kim et al., 2007). The VOCs produced by *O. latemarginatus* were inhibitory action against *A. alternata, B. cineria, C. gloeosporioides, F. oxysporum* f. sp. *lycopersici, Magnaporthe grisea* and *Rhizoctonia solani*. The major antifungal VOC produced by *O. latemarginatus* EF069 is 5-pentyl-2-furaldehyde (PTF) (Lee et al., 2008). Similarly, volatiles emitted by *Nodulisporium* sp. CF016, isolated from the stem of *Cinnamomum lourcinii*, kill a wide range of plant and storage pathogens like *Pythium ultimum, Rhizoctonia solani, F. oxysporum, Phytophthora capsici, Colletotichum coccodes, Magnaporthe oryzae, Botrytis cineria* and *Penicillium expansum*. The most abundant volatile compound produced by *Nodulisporium* sp. CF016 is β-elemene followed by 1-methyl-1, 4-cyclohexadiene, β-selinene and α-selinene. Thus, *Nodulisporium* sp. CF016 could also be used as prospective mycofumigant in controlling the post-harvest diseases of various fruits (Park et al., 2010).

The fungal volatile compounds could also be effectively used as insecticidal and nematicidal agents. *Phthorimaea operculella* or the Potato tuber moth (PTM) is a serious pest of stored potato in potatoes producing countries of the world. *Bacillus thuringiensis* and granulovirus of *P. operculella* have also been used as biocontrol agents for the protection of tubers from PTM (Hamilton and Macdonald, 1990; Kroschel et al., 1996). Naphthalene is produced by *Muscodor vitigenus* which was previously used in moth balls as an effective insect repellent (Daisy et al., 2002). Insecticidal potential of *M. albus* volatiles were evaluated as rye grain cultures in 15 and 30 g formulations against PTM that were exposed as neonate larvae. A highly significant mortality, as well as repression of pupal stage, was observed due to the exposure of VOCs produced by *Muscodor albus*. This suggests their applications as volatile bioinsecticides with least traces of residual toxicity as compared to the traditional insecticides (Lacey and Neven, 2006). The nematicidal/nematostatic potentials of fungal volatiles elaborated by *Muscodor albus* were also studied against four parasitic nematodes *Meloidogyne chitwoodi, Meloidogyne haple, Paratrichodorus allius* and *Pratylenchus penetrans* which were exposed to VOC for 72 hrs (Riga et al., 2008). *Trichoderma* species has been a promising biocontrol agent against nematodes comprising *Meliodogyne* species, *Bursaphelenchus xylophilus, Panagrellus redivivus* and *Caenorhabditis elegans* (Maehara and Futai, 1997; Yang et al., 2010; Khan and Haque, 2011). FVOCs have been effectively used to control fungi as well as bacteria, however there are limited reports on their nematicidal properties. Different secondary metabolites produced by *Trichoderma* spp. viz., Trichodermin, Acetic acid and Gliotoxin have also been found to possess nematicidal activity (Djian et al., 1991; Anitha and Murugesan, 2005; Yang et al., 2010) although none of these are volatiles. But recently, a strain of *Trichoderma* sp. YMF 1.00416 is reported to produce volatile compounds that kill nematodes, and a nematicidal VOC 6-pentyl-2H-pyran-2-one has been isolated and identified (Yang et al., 2012). That FVOCs provide ample opportunities for the development of volatile biocontrol agents cannot be ignored.

FVOCs as Biofuels

Currently, the whole world has eyes for alternative energy sources to meet or makeup the diminishing resources of fossil fuels. Volatile high energy compound producing microorganisms can probably be the potential candidates to be given attention to meet partly our future energy need.

Fungi are prolific producers of bioactive natural products and have also been studied for the production of volatile secondary metabolites or VOCs. A common VOC, like 3-methyl-1-butanol, is currently being targeted as biofuel alternative. Endophytic fungi that are abundant within the living tissues of approximately 99% of the higher plants, produce fuel related substances such as octane, 1-octene and low molecular mass hydrocarbons and so on (McAfee and Taylor, 1999). The production of FVOC may be based on the utilization of biologically based resources and thereby converting the plant wastes directly into diesel (Strobel et al., 2011). The Patagonian fungal endophyte NRRL 50072, now identified as *Ascocoryne sarcoides*, is reported to produce a variety of medium chain and highly branched volatile organic compounds (VOCs) like octane; 1-octene; heptane, 2-methyl; hexadecane; undecane, 4-methyl; nonane, 3-methyl; and benzene, 1, 3-dimethyl. These compounds have been highlighted for their potential as a fuel as well as a fuel additive and termed as "Mycodiesel". The most comprehensive fermentation work on *A. sarcoides*, using cellulose, glucose and potato dextrose broth has indicated that fuel-range organics are produced by this organism on each substrate (Griffin et al., 2010).

An endophytic *Hypoxylon* CI-4A, isolated from *Perseaindica*, an evergreen tree native to the Canary Islands, when grown on PDA-Petri plates, the VOCs were produced by the fungus and that were primarily: 1, 8-cineole, 1-methyl-1, 4-cyclohexadiene, and (+)-α-methylene-α-fenchocamphorone. 1, 8-cineole production, as a fungal VOC, has generated a significant interest because this compound has not been reported from any other biological source except plant tissue. Further, this compound also prevents the phase separation when used as an additive in ethanol-gasoline fuel blends (Tomsheck et al., 2010). Fuels comprising of gasoline–eucalyptus oil mixture (having 80% or more 1, 8-cineole) improved the octane number and reduced the carbon monoxide levels in the exhaust. Hence 1, 8-cineole is a worthy target molecule for replacing fossil based hydrocarbons as fuel additive. VOCs that have both, fuel and biological potentials were isolated from *Daldinia* sp. (EC-12), an endophyte of *Myroxylon balsamum* from the Ecuadorian Amazon. The organism produced 1, 4-cyclohexadiene, 1-methyl-, 1-4 pentadiene and cyclohexene, 1-methyl-4-(1-methylethenyl)-, along with some alcohols and terpenoids of interest as potential fuels (Mends et al., 2012). *Phomopsis* sp. was isolated as endophyte of *Odontoglossum* sp. (Orchidaceae), associated with a cloud forest in Northern Ecuador. It produces a unique mixture of volatile organic compounds(VOCs) comprising of 1-butanol, 3-methyl; benzeneethanol; 1-propanol, 2-methyl; 2-propanone and sabinene (Singh et al., 2011). More recently an endophytic *Nigrograna mackinnonii* has been reported to produce volatiles including terpenes and odd chains polyenes which may find applications for use as biotechnologically derived fuels (Shaw et al., 2015).

Thus, FVOCs clearly indicate that fungi are excellent resource for hydrocarbon biofuels or biofuel precursors. It is to be remembered that no individual fungus makes all of the compounds found in diesel, but some do make VOCs that are representative of the four diesel classes. However, there is a need to address some critical bottlenecks for commercial applications of these FVOCs as fuels or fuel additives.

Conclusion

Bioprospecting for new microbes from soil for novel bioactive compounds that have applications in agrochemical and pharmaceutical industries have remained a thrust area over six decades. Microbes are ubiquitous; their interactions in different ecological niche play a predominant role in their evolution both at physiological as well as molecular levels. Thus, these microbes have better or different metabolomic profiles which could be explored and exploited. One such group of microorganisms thrives within the vascular plants (endophytes) (Bacon and White, 2000). Fungal endophytes, have not only been a source of bioactive secondary metabolites but also of bioactive VOCs which could be exploited as biocontrol agents, mycofumigants, information molecules, aromas, as markers of post-harvest deterioration, and sources of hydrocarbons. The present studies are just a beginning in harnessing the potential of FVOCs. However, roadblocks remain until we completely harness their potential. The FVOCs are produced in minute quantities. Therefore, their detection, isolation and characterization are a difficult task. There is not information or knowledge of their biosynthesis and regulatory aspects, although genomic, transcriptomic and metabolomic studies are underway to correlate gene expression to volatile production (Gianoulis et al., 2012). Understanding the genetic pathways of FVOC production is of paramount importance. The genes responsible for the FVOC expression could be over-expressed in the FVOC producing species or else transferred to an industrially tractable heterologous host for the large scale production of the FVOC of human interest. Thus, FVOCs represent a new frontier in bioprospecting of fungi.

References

Adams, R.P. 1994. Geographic variations in the volatile terpenoids of *Juniperus monospermae* and *Juniperus osteosperma*. Biochem. Syst. Ecol., 22: 65–71.

Anitha, R. and Murugesan, K. 2005. Production of gliotoxin on natural substrates by *Trichoderma virens*. J. Basic Microbiol., 45: 12–19.

Araki, A., Kawai, T., Eitaki, Y., Kanazawa, A., Morimoto, K., Nakayama, K., Shibata, E., Tanaka, M., Takigawa, T., Yoshimura, T., Chikara, H., Saijo, Y. and Kishi, R. 2010. Relationship between selected indoor volatile organic compounds, so-called microbial VOC, and the prevalence of mucous membrane symptoms in single family homes. Sci. Total Environ., 408: 2208–2215.

Athamneh, A.C., Zoecklein, E.W. and Mallikarjun, K. 2008. Electronic nose in evaluation of Cabernet Sauvignon fruit maturity. J. Wine Research, 19: 69–80.

Bacon, C.W. and White, J.W. 2000. Microbial Endophytes. Marcel Dekker, New York.

Banerjee, D., Pandey, A., Jana, M. and Strobel, G. 2014. *Muscodor albus* MOW 12 an endophyte of *Piper nigrum* (Piperaceae) collected from North East India produces volatile antimicrobials. Indian J. Microbiol., 54(1): 27–32.

Barberio, J. and Twibell, J. 1991. Chemotaxonomy of plant species using headspace sampling, thermal desorption and capillary GC. Journal of High Resolution Chromatography, 14: 637–639.

Borjesson, T., Stollman, U. and Schnurer, J. 1990. Volatile metabolites and other indicators of *Penicillium aurantiogriseum* growth on different substrates. Appl. Environ. Microbiol., 56: 3705–3710.

Calvo, A.M., Gardner, H.W. and Keller, N.P. 2001. Genetic connection between fatty acid metabolism and fungal development. Microbiol. Mol. Biol. Rev., 66: 447–459.

Camp, A.R., Dillard, H.R. and Smart, C.D. 2008. Efficacy of *Muscodor albus* for the control of *Phytophthora* blight of sweet pepper and butternut squash. Plant Dis., 92: 1488–1492.

Champe, S.P. and el Zayat, A.A. 1989. Isolation of a sexual sporulation hormone from *Aspergillus nidulans*. Microbiology, 151: 1803–1821.

Chitarra, G.S., Abee, T., Rombouts, F.M., Posthumas, M.A. and Dijksterhuis, J. 2004. Germination of *Penicillium paneum* conidia is regulated by 1-octen-3-ol, a volatile self-inhibitor. Appl. Environ. Microbiol., 70(5): 2823–2829.

Chitarra, G.S., Abee, T., Rombouts, F.M. and Dijksterhuis, J. 2005. 1-octen-3-ol inhibits conidia germination of *Penicillium paenum* despite mild effects on membrane permeability, respiration, intracellular pH and changes the protein composition. FEMS Microbiol. Ecol., 54(1): 67–75.

Daisy, B.H., Gary, A., Strobel, G.A., Castillo, U., Sears, J., Weaver, D.K. and Runyon, J.B. 2002. Naphthalene, an insect repellent is produced by *Muscodor vitigenus*, a novel endophytic fungus. Microbiology, 148: 3737–3741.

De Lacy Costello, B.P.J., Evans, P., Gunson, H.E., Ratcliffe, N.M. and Spencer-Phillips, T.N. 2001. Gas chromatography-mass spectrometry analysis of volatile organic compounds from potato tubers inoculated with *Phytophthora infestans* or *Fusarium coeruleum*. Plant Pathol., 50: 489–496.

DiNatalie, C., McAgnano, A., Davide, F., D'Amico, A., Paolesse, R., Boschi, T. et al. 1997. An electronic nose for food analysis. Sensor Actuators B, 44(1-3): 521–526.

Djian, C., Pijarouvski, L., Ponchet, M. and Arpin, N. 1991. Acetic acid, a selective nematicidal metabolite from culture filtrate of *Paecilomyces lilacinus* (Thom) Samsan and *Trichoderma longibrachiatum* Rifai. Nematologica, 37: 101–102.

Drawert, F. and Barton, H. 1978. Biosynthesis of flavor compounds by microorganisms. 3. Production of monoterpenes by the yeast *Kluyveromyces lactis*. J. Agric. Food Chem., 26: 765–766.

Eckert, J.W. and Ratnayake, M. 1994. Role of volatile compounds from wounded oranges in induction of germination of *Penicillium digitatum* conidia. Phytopathol., 84: 746–750.

Elke, K., Begerow, J., Oppermann, H., Kramer, U., Jarmann, E. and Dunema, L. 1999. Determination of selected microbial volatile organic compounds by diffusive sampling and dual column capillary GC-FID a new feasible approach for the detection of an exposure to indoor mold fungi? J. Environ. Monitor, 5: 445–452.

Gardner, J.W., Shin, H.W. and Hines, E.L. 2000. An electronic nose system to diagnose disease. Sensor Actuators B, 70: 19–24.

Gianoulis, T.A., Griffin, M.A., Spakowicz, D.J., Dunican, B.F., Alpha, C.J., Sboner, A., Sismour, A.M., Kodira, C., Egholm, M., Church, G.M., Gerstein, M.B. and Strobel, S.A. 2012. Genomic analysis of the hydrocarbon-producing, cellulolytic, endophytic fungus *Ascocoryne sarcoides*. PLoS Genet., 8: e1002558.

Gioacchini, A.M., Menotta, M., Guescini, M., Saltarelli, R., Ceccaroli, P., Amicucci, A., Barbieri, E., Giomaro, G. and Stocchi, V. 2008. Geographical traceability of Italian white truffle (*Tuber magnatum* Pico) by the analysis of volatile organic compounds. Rapid Commun. Mass Spectrom., 22: 3147–3153.

Goates, B.J. and Mercier, J. 2011. Control of common bunt of wheat under field conditions with the biofumigant fungus *Muscodor albus*. Eur. J. Plant. Pathol., 131(3): 403–407.

González, M.C., Anaya, A.L., Glenn, A.E., Macías-Rubalcava, M.L., Hernández-Bautista, B.E. and Hanlin, R.T. 2009. *Muscodor yucatanensis*, a new endophytic ascomycete from Mexican chakah, *Bursera simaruba*. Mycotaxon, 110: 363–372.

Gracia, E.H., Garzia, A., Cordobes, S., Espeso, E.A. and Ugalde, U. 2011. 8-Carbon oxylipins inhibit germination and growth, and stimulate aerial conidiation in *Aspergillus nidulans*. Fungal Biol., 115(4-5): 393–400.

Griffin, M.A., Spackowicz, D.J., Gianoulis, I.A. and Strobel, S.A. 2010. Volatile organic compounds production by organisms of the genus *Ascocoryne* and a re-evaluation of mycodiesel production by NRRL 50072. Microbiology, 156: 3814–3829.

Grimme, E., Zidack, N.K., Sikora, R.A., Strobel, G.A. and Jacobsen, B.J. 2007. Comparison of *Muscodor albus* volatiles with a biorational mixture for control of seedling diseases of sugar beet and root-knot nematode on tomato. Plant Dis., 91: 220–225.

Hamilton, J.T. and Macdonald, J.A. 1990. Control of potato moth, *Phthorimaea operculella* (Zeller) in stored seed potatoes. Gen. Appl. Entomol., 22: 3–6.

Harbone, J.B. and Turner, B.L. 1984. Plant Chemosystematics. Academic Press, London.

Inamdar, A.A., Masurekar, P. and Bennett, J.W. 2010. Neurotoxicity of fungal volatile organic compounds in *Drosophila melanogaster*. Toxicol. Sci., 117: 418–426.

Inamdar, A.A., Moore, J.C., Cohen, R.I. and Bennett, J.W. 2011. A model to evaluate the cytotoxicity of the fungal volatile organic compound 1-octen-3-ol in human embryonic stem cells. Mycopathologia, 173: 13–20.

Jarvis, B.B. and Miller, J.D. 2005. Mycotoxins as harmful indoor air contaminants. Appl. Microbiol. Biotechnol., 66: 367–372.

Jollivet, N., Bezenger, M.C., Vayssier, Y. and Belin, J.M. 1992. Production of volatile compounds in liquid cultures by six strains of *Coryneform* bacteria. Appl. Microbiol. Biotechnol., 3: 790–794.

Kaminski, E., Libbay, E., Stawicki, S. and Wasowicz, E. 1972. Identification of predominant volatile compounds produced by *Aspergillus niger*. Applied Microbiology, 24(5): 721–726.

Khan, M.R. and Haque, Z. 2011. Soil application of *Pseudomonas fluorescens* and *Trichoderma harzianum* reduces root-knot nematode, *Meloidogyne incognita*, on tobacco. Phytopathol. Mediterr., 50: 257–266.

Kim, H.Y., Choi, G.J., Lee, H.B., Lee, S.W., Lim, H.K., Jang, K.S., Son, S.W., Lee, S.O., Cho, K.Y., Sung, N.D. and Kim, J.C. 2007. Some fungal endophytes from vegetable crops and their anti-oomycete activities against tomato late blight. Lett. Appl. Microbiol., 44(3): 332–337.

Korpe, A., Pasanen, A.L. and Pasanen, P. 1998. Volatile compounds originating from mixed microbial cultures on building materials under various humidity conditions. Appl. Environ. Microbiol., 64(8): 2914–2919.

Kramer, R. and Abraham, W.R. 2012. Volatile sesquiterpenes from fungi: what are they good for Phytochem. Rev., 11: 15–37.

Kroschel, J., Kaack, H.J., Fritsch, E. and Huber, J. 1996. Biological control of the potato tuber moth (*Phthorimaea operculella* Zeller) in the Republic of Yemen using granulosis virus: propagation and effectiveness of the virus in Weld trials. Biocontrol Sci. Technol., 6: 217–226.

Kudalkar, P., Strobel, G., Ul Hassan, S.R., Geary, B. and Sears, J. 2012. *Muscodor sutura*, a novel endophytic fungus with volatile antibiotic activities. Mycoscience, 53: 319–325.

Labows, J.N., McGinley, K.J., Webster, G.F. and Leyden, J.J. 1980. Headspace analysis of volatile metabolites fungi based on diffusive sampling from Petri dishes. J. Microbiol. Meth., 19: 297–305.

Lacey, L.A. and Neven, L.G. 2006. The potential of the fungus, *Muscodor albus*, as a microbial control agent of potato tuber moth (Lepidoptera: Gelechiidae) in stored potatoes. J. Invertebr. Pathol., 91: 195–198.

Larsen, F.O., Clementsen, P., Hansen, M., Maltback, N., Ostenfeldt Larsen, T., Neilsen, K.F., Gravesen, S., Stahlskov, P. and Norn, S. 1998. Volatile organic compounds from indoor mold *Trichoderma viridae* cause histaminic release from human bronchoalveolar cells. Inflamm. Res., 47(1): 55–56.

Larsen, T.O. and Frisvad, J.C. 1995. Chemosystematics of *Penicillium* based on profile of volatile metabolites. Mycol. Res., 99: 1167–1174.

Lee, S.O., Kim, H.Y., Choi, G.J., Lee, H.B., Jang, K.S., Choi, Y.H. and Kim, J.C. 2008. Mycofumigation with *Oxyporus latemarginatus* EF069 for control of postharvest apple decay and *Rhizoctonia* root rot on moth orchid. J. Appl. Microbiol., 106: 1213–1219.

Macías-Rubalcava, M.L., Hernández-Bautista, B.E., Oropeza, F., Duarte, G., González, M.C., Glenn, A.E., Hanlin, R.T. and Anaya, A.L. 2010. Allelochemical effects of volatile compounds and organic extracts from *Muscodor yucatanensis*, a tropical endophytic fungus from *Burserasimaruba*. J. Chem. Ecol., 36: 1122–1131.

Maehara, N. and Futai, K. 1997. Effect of fungal interactions on the numbers of the pinewood nematode, *Bursaphelenchus xylophilus* (Nematoda: Aphelenchoididae), carried by the Japanese pine sawyer, *Monochamus alternatus* (Coleoptera: Cerambycidae). Fundam. Appl. Nematol., 20: 611–617.

Magan, N. and Evans, P. 2000. Volatiles as an indicator of fungal activity and differentiation between species, the potential use of electronic nose technology for early detection of grain spoilage. J. Stored Prod. Res., 36: 319–340.

Martins, M., Henriques, M., Azenedo, J., Rocha, S.M., Coimbra, H.A. and Oliveira, R. 2007. Morphogenesis control in *Candida albicans* and *Candida dublineinsis* through signaling molecules produced by planktonic and biofilm cells. Eukaryot Cell, 6: 2429–2436.

McAfee, B.J. and Taylor, A. 1999. A review of the volatile metabolites of fungi found on wood substrates. Nat. Toxins, 7: 283–303.

McNeal, K.S. and Herbert, B.E. 2009. Volatile organic metabolites as indicators of soil microbial activity and community composition shifts. Soil Sci. Soc. Am. J., 73: 579–588.

Mends, M.T., Yu, E., Strobel, G.A., Riyaz-Ul-Hassan, S., Booth, E., Geary, B., Sears, J., Taatjas, C.A. and Hadi, M.Z. 2012. An Endophytic *Nodulisporium* sp. producing volatile organic compounds having bioactivity and fuel potential. J. Pet. Environ. Biotechnol., 3: 117.

Mercier, J. and Jiménez, J.I. 2009. Demonstration of the biofumigation activity of *Muscodor albus* against *Rhizoctonia solani* in soil and potting mix. BioControl, 54: 797–805.

Meshram, V., Kapoor, N. and Saxena, S. 2013. *Muscodor kashayum* sp. nov.—a new volatile anti-microbial producing endophytic fungus. Mycology, 4(4): 196–204.

Mitchell, A.M., Strobel, G.A., Hess, W.M., Vargas, P.N. and Ezra, D. 2008. *Muscodor crispans*, a novel endophyte from *Ananas ananassoides* in the Bolivian Amazon. Fungal Divers., 31: 37–44.

Mitchell, A.M., Strobel, G.A., Moore, E., Robison, R. and Sears, J. 2010. Volatile antimicrobials from *Muscodor crispans*, a novel endophytic fungus. Microbiology, 156: 270–277.

Morey, P., Worthham, A., Weber, A., Horner, E., Black, M. and Muller, W. 1997. Microbial VOCs in moisture damage buildings. Healthy Buildings, 1: 245–250.

Muller, A., Faubert, P., Hagen, M., zu Castell, W., Polle, A., Peter-Schnitzler, J. and Rosenkranz, M. 2013. Volatile profiles of fungi- Chemotyping of species and ecological functions. Fungal Genet. Biol., 54: 25–33.

Nemčovič, M., Jakubíková, L., Víden, I. and Farkaš, V. 2008. Induction of conidiation by endogenous volatile compounds in *Trichoderma* spp. FEMS Microbiol. Lett., 284(2): 231–236.

Pagans, E., Font, X. and Sanchez, A. 2006. Emission of volatile organic compounds from composting of different solid wastes: abatement by biofiltration. J. Hazard Mater., 131: 179–186.

Pallotino, F., Costa, C., Antonucci, F., Strano, M.C., Calandra, M., Solanini, S. and Menesatti, P. 2012. Electronic nose application for determination of *Penicillium digitatum* in Valencia oranges. J. Sc. Food Agric., 92: 2008–2012.

Park, M.S., Ahn, J., Choi, G.J., Choi, Y.H., Jang, K.S. and Kim, J.C. 2010. Potential of the volatile producing fungus *Nodulisporium* sp. CF016 for the control of post-harvest diseases of Apple. Plant Pathology J., 26(3): 253–259.

Riga, E., Lacey, L.A. and Guerra, N. 2008. *Muscodor albus*, a potential biocontrol agent against plant parasitic nematodes and economically important vegetable crops of Washington State, USA. BioControl, 45: 380–385.

Riva, M., Benedeth, S. and Mannino, S. 2001. Shelf life of fresh cut vegetables as measured by electronic nose: a preliminary study. Ital. J. Food Sci., 13: 201–212.

Roze, L.V., Chanda, A., Laivenieks, M., Beaudry, R.M., Artymovich, K.A., Koptina, A.V., Awad, D.W., Valeeva, D., Jones, A.D. and Linz, J.E. 2010. BMC Biochemistry, 11: 33.

Sági-Kiss, V. and Fodor, P. 2011. Development of a SPME-GC-MS method for spoilage detection in case of plums inoculated with *Penicillium expansum*. Acta Alimentaria, 40: 188–197.

Saxena, S., Meshram, V. and Kapoor, N. 2014. *Muscodor darjeelingensis*, a new endophytic fungus of *Cinnamomum camphora* collected from North East Himalayas. Sydowia 66(1): 55–68.

Saxena, S., Meshram, V. and Kapoor, N. 2015. *Muscodor tigerii* sp. nov.-Volatile antibiotic producing endophytic fungus from the Northeastern Himalayas. Ann. Microbiol., 65: 47–57.

Shaw, J.J., Spackowicz, D.J., Dalal, R.S., Davis, J.H., Lehr, N.A., Dunican, B.F., Orellana, E.A., Trujillo, A.N. and Strobel, S.A. 2015. Biosynthesis and genomic analysis of medium-chain hydrocarbon production by the endophytic fungal isolate *Nigrogranamackinnonii* E5202H. Appl. Microbiol. Biotechnol., DOI 10.1007/s00253-014-6206-5.

Singh, S.K., Strobel, G.A., Knighton, B., Geary, B., Sears, J. and Ezra, D. 2011. An endophytic *Phomopsis* sp. possessing bioactivity and fuel potential with its volatile organic compounds. Microb. Ecol., 61: 729–739.

Splivallo, R., Ottonello, S., Mello, A. and Karlovsky, P. 2011. Truffle volatiles: from chemical ecology to aroma biosynthesis. New Phytol., 189: 688–699.

Stinson, M., Ezra, D., Hess, W.M., Sears, J. and Strobel, G.A. 2003. An endophyte *Gliocladium* sp. of *Eucryphia cordifolia* producing selective volatile antimicrobial compounds. Plant Science, 165(4): 913–922.

Strobel, G.A. 2006. Harnessing endophytes for industrial microbiology. Curr. Opin. Microbiol., 9: 240–244.

Strobel, G.A. 2011. *Muscodor* species-endophytes with biological promise. Phytochem. Rev., 10: 165–172.

Strobel, G.A., Dirkse, E., Sears, J. and Markworth, C. 2001. Volatile antimicrobials from *Muscodor albus* a novel endophytic fungus. Microbiology, 147: 2943–2950.

Strobel, G.A., Kinghton, B., Kluck, K., Ren, Y., Livinghouse, T., Griffin, M., Spakowicz, D. and Sears, J. 2008. The production of myco-diesel hydrocarbons and their derivatives by the endophytic fungus *Gliocladium roseum* (NRRL 50072). Microbiology, 154: 3319–3328.

Suwannarach, N., Bussaban, B., Hyde, K.D. and Lumyong, S. 2010. *Muscodor cinnamomi*, a new endophytic species from *Cinnamomum bejolghota*. Mycotaxon, 114: 15–23.

Suwannarach, N., Kumla, J., Bussaban, B., Hyde, K.D., Matsui, K. and Lumyong, S. 2013. Molecular and morphological evidence support four new species in the genus *Muscodor* from northern Thailand. Ann. Microbiol., 63(4): 1341–1351.

Suwannarach, N., Kumla, J., Bussaban, B. and Lumyong, S. 2012. Biocontrol of *Rhizoctonia solani* AG-2, the causal agent of damping-off by *Muscodor cinnamomi* CMU-Cib 461. World J. Microbiol. Biotechnol., 28(11): 3171–3177.

Tomsheck, A.R., Strobel, G.A., Booth, E., Geary, B., Spakowicz, D., Knighton, B., Floerchinger, C., Sears, J., Liarzi, O. and Ezra, D. 2010. *Hypoxylon* sp., an endophyte of *Persea indica*, producing 1,8-Cineole and other bioactive volatiles with fuel potential. Microb. Ecol., 60: 903–914.

Tsitsigiannis, D.I., Kowieski, T.M., Zarnowskii, R. and Keller, N.P. 2005. Three putative oxylipin biosynthetic genes integrate sexual and asexual development in *Aspergillus nidulans*. Microbiology, 151: 1809–1821.

Tuma, D., Sinha, R.N., Muir, W.E. and Abramson, D. 1989. Odor volatiles associated with microflora in damp ventilated and bin-stored bulk wheat. Int. J. Food Microbiol., 8: 103–119.

WHO. 2009. Guidelines for Indoor Air Quality: Dampness and Mold. Druckpartner Moser, Germany.

Wihlborg, R., Pippitt, D. and Marsili, R. 2008. Headspace sorptive extraction and GC-TOFMS for the identification of volatile fungal metabolites. J. Microbiol. Meth., 75: 244–250.

Winquist, F., Hornsten, E.G., Sundgren, H. and Lundstorm, J. 1993. Performance of an electronic nose for quality estimation of ground meat. J. Meat Sci. Technol., 4: 1493–1500.

Worapong, J. and Strobel, G.A. 2009. Biocontrol of a root rot of kale by *Muscodor albus* strain MFC2. BioControl, 54: 301–306.

Worapong, J., Strobel, G.A., Daisy, B., Castillo, Baird, G. and Hess, W.M. 2002. *Muscodor roseus* anna. nov. an endophyte from *Grevillea pteridifolia*. Mycotaxon, 81: 463–475.

Worapong, J., Strobel, G.A., Ford, E.J., Li, J.Y., Baird, G. and Hess, W.M. 2001. *Muscodor albus* anam. nov. an endophyte from *Cinnamomum zeylanicum*. Mycotaxon, 79: 67–79.

Yang, Z., Yu, Z., Lei, L., Xia, Z., Shao, L., Zhang, K. and Li, G. 2012. Nematicidal effects of volatiles produced by *Trichoderma* species. J. Asia Pacific Entomol., 15: 647–650.

Yang, Z.S., Li, G.H., Zhao, P.J., Zheng, X., Luo, S.L., Li, L., Niu, X.M. and Zhang, K.Q. 2010. Nematicidal activity of *Trichoderma* spp. and isolation of an active compound. World J. Microbiol. Biotechnol., 26: 2297–2302.

Zeppa, S., Gioacchini, A.M., Guidi, A., Guescini, M., Raffaella, P., Zambonelli, A. and Stocchi, V. 2004. Determination of specific volatile organic compounds synthesized during *Tuber borchii* fruit body development by solid-phase microextraction and gas chromatography/ mass spectrometry. Rapid Commun. Mass Spectrom., 18: 199–205.

Zhang, C., Wang, G., Mao, L., Komon-Zelazowska, M., Yuan, Z., Lin, F., Druzhinina, I.S. and Kubicek, C.P. 2010. *Muscodor fengyangensis* sp. nov. from southeast China: morphology, physiology and production of volatile compounds. Fungal Biol., 114: 797–808.

CHAPTER 8

Fungi and Statins

J.K. Misra

ABSTRACT

We have been using fungi for our benefit since the beginning of human civilization. Fungi are a rich protein source of food, and are used in food processing and beverage manufacture. They have been exploited commercially for the production of organic acids and other useful compounds such as antibiotics, enzymes, vitamins, steroids, etc. While, fungi also cause a number of diseases both in plants and animals, they also provide us many life-saving compounds such as statins that are potent inhibitors of *de novo* synthesis of cholesterol in the liver of the human body where approximately 70% of the blood cholesterol is synthesized. In the pages that follow, a brief history of the discovery of statins from fungi is presented. A need for cooperative research using all available modern tools and techniques is emphasized.

Introduction

Like other microbes, fungi produce a variety of compounds while growing in natural or artificial environments, but only a few have received attention so far, either as useful chemicals, toxins or medicines (Adrio and Demain, 2003; Demain and Zhang, 2005; Demain, 2006; Demain and Adrio, 2008; Goldstein and Brown, 2015). Of 23,000 active natural products from microbes, 42% are made by fungi (Demain, 2014) including statins. Statins include a group of fungal polyketides (compactin or mevastatin, and lovastatin) and their derivatives, which specifically inhibit cholesterologenesis in liver, and

Biological Research Unit, SLM Bhartiya Vidya Bhavan Girls Degree College, Gomti Nagar, Lucknow-226 010, India.
Email: Jitrachravi@gmail.com

represent best-selling drugs because their use has remarkably reduced the frequency of deaths due to heart attack and stroke. Thus, by preventing cardiovascular diseases, statins have provided longer and better lives to millions of people across the world (Veillard and Mach, 2002; Srinivasa Rao et al., 2011). Besides lowering plasma cholesterol (20–40%), the statins have also shown many other beneficial effects, such as improving endothelial function, and anti-inflammatory, anti-atherothrombosis, immunomodulation, and anti-migration activities (Youssef et al., 2002; Gullo and Demain, 2013). They are also useful in multiple sclerosis and cancer (Moon et al., 2014). Additional activities, such as stimulants for bone formation and antioxidants, are under study (Wrigley, 2004; Barrios-González and Miranda, 2010).

Cholesterol in the body arises by two mechanisms. One involves absorption by the body from the diet one eats (25%), and the other is synthesis by the liver (75%). Interestingly, in humans, cholesterol produced in the liver exceeds what is absorbed from the diet.

Statins are inhibitors of the enzyme hydroxymethylglutaryl coenzyme A (HMG-CoA) reductase that catalyzes the reduction of HMG-CoA to mevalonate during synthesis of cholesterol (Endo, 1992; Bobek et al., 1997). Natural statins have a similar molecular structure, i.e., a hexahydro-naphthalene system and a β-hydroxylactone, but with differences in side chains and the presence of a methyl group around the ring.

Here, I present a brief history of the discovery of statins from different fungi with short descriptions. A need for continuous and wider collaborative and a deeper search for naturally-occurring compound from fungi occurring in varying ecosystems are also emphasized.

Historical Background of Statin Discovery

Akira Endo, who has been credited with the discovery of compactin from a blue-green fungus, i.e., *Penicillium citrinum* Pen-51, was motivated for his discovery by noticing that very often, aged people in the borough of the Bronx in New York City, where he was living in the USA during his research pursuit, were often hospitalized because of heart attack, the main cause of death at that time due to hypercholesterolemia. He decided to search for a compound that could be a cholesterol-synthesis inhibitor. Although a few lipid-lowering agents like clofibrate, niacin and cholestylamine were available, they were not safe and effective. This fact further strengthened his determination to find a cholesterol-lowering drug. After finishing his studies in the USA, he returned to the Sankyo Company in Japan and started looking for the solution. His presumption was that microbes would produce an antibiotic in their defense against those microbes that require sterols or other mevalonate-derived isoprenoids for their growth. This led him to search for such compounds from fungi as one of their secondary metabolites. Endo's quest for such an assumed compound took two years as he examined thousands of microbes (Endo et al.,

1976a). He successfully isolated from the broth culture of *Penicillium citrinum*, the compound of his choice, i.e., compactin, which was similar in structure to that of mevalonate, the product of the HMG-CoA reductase reaction. Compactin turned out to be the potent competitive inhibitor that Endo had desired; he considered it to be a wonderful gift from nature (Endo, 2008).

After having compactin in hand, Endo and his associates started to access the efficacy of the compound using animals. Initially, experiments on rats, dogs, and monkeys were not too encouraging, but the results strengthened Endo's conviction that the compound would work. He came up with the possibility that it was the abnormal induction of HMG-CoA reductase in the liver of the rats that caused the lack of reduction of plasma cholesterol in the animals under experiment. Later, experiments indicated the efficacy of compactin in laying hens, dogs and monkeys (Tsujita et al., 1979; Endo, 1992).

Compactin was also obtained from *Penicillium brevicompactum* by British researchers, but as an antibiotic agent. They also expressed the view that compactin had no ability to reduce blood cholesterol in rats and mice, as observed by Endo in his experiments (Steinberg, 2006). However, the research continued and finally it was found that compactin has remarkable efficiency and is safe. In this venture, two companies, i.e., Sankyo and Merck joined hands. However, researchers at Merck independently discovered lovastatin from *Aspergillus terreus*, which differed from compactin by only one methyl group. Lovastatin showed properties similar to that of compactin (Alberts et al., 1980; Steinberg, 2006). Later on, compactin proved to be a very useful in treating patients with hypercholesterolemia (Mabuchi et al., 1981). Physicians in the USA also used lovastatin to treat cases of hypercholesterolemia and obtained results similar to that of Mabuchi and coworkers. Merck resumed clinical trials of lovastatin and the US Food and Drug Administration approved it in 1987. Thus, lovastatin became the first commercial statin (Steinberg, 2006). Other statins such as simvastatin, and pravastatin-compactin with an extra hydroxyl group are also known. Synthetic statins like fluvastatin, cerivastatin, atorvastatin, rosvastatin, and pitavastatin are also available. However, cerivastatin is not in use due to its side effects.

Many large-scale clinical trials under way since 1990 have shown decreases in LDL-cholesterol and thereby an appreciable decrease in coronary heart disease and general mortality. Also, incidence of stroke declined by 25–30%. According to an estimate, more than 30 million people are given a statin to prevent coronary heart disease and stroke. Thus, the story of the discovery of statins is extremely interesting (Endo, 2010).

Fungi that Produce Statins

A number of fungi produce statins (Manzoni and Rollini, 2002; Gullo and Demain, 2013), i.e., *Aspergillus terreus* (Buckland et al., 1989), *Aspergillus flavipes* (Valera et al., 2005), species of *Monoascus*, i.e., *M. ruber* (Endo, 1979;

Manzoni et al., 1999), *M. pupureus, M. vitreus, M. pilosus* (Miyake et al., 2006a,b), and *M. pubigerus* (Negishi et al., 1986), *Penicillium* species, i.e., *P. citrinum* (Endo et al., 1976a,b; Endo, 1985), *P. brevicompactum* (Brown et al., 1976), and *P. cyclopium* (Bazaraa et al., 1998). Also, *Doratomyces, Eupenicillium, Gymnoascus, Hypomyces* (*H. chrysospermum*) (Endo et al., 1986), *Paecilomyces, Phoma, Trichoderma* (*T. longibrachiatum* and *T. pseudokoningii*) (Endo et al., 1986) and *Pleurotus* (*P. ostreatus*) (Alarcon et al., 2003), *P. ostreatus, P. sapidus, P. saca* (Gunde-Cimerman et al., 1993), and *P. pulmonarus* (Gunde-Cimerman and Cimerman, 1995). The natural statins that come from fungi are compactin, lovastatin, and pravastatin.

Compactin

As mentioned earlier, compactin (Fig. 1) was initially found to be produced by *Penicillium citrinum* and then was discovered in other fungi as well. Its production details and other information can be found in the publication of Gullo and Demain (2013). Compactin has never been commercialized; rather it has been used in a modified form.

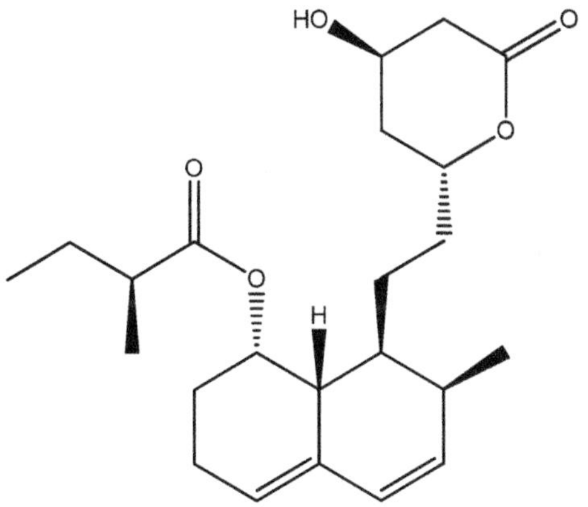

Figure 1. Structure of compactin.

Lovastatin

Monascus ruber and some of other *Monascus* species produced lovastatin (Fig. 2). It is more potent than compactin and is biosynthesized from acetate and methionine by a polyketide pathway (Moore et al., 1985). For more details, one can refer to Praveen and Savitha (2012) and Gullo and Demain (2013).

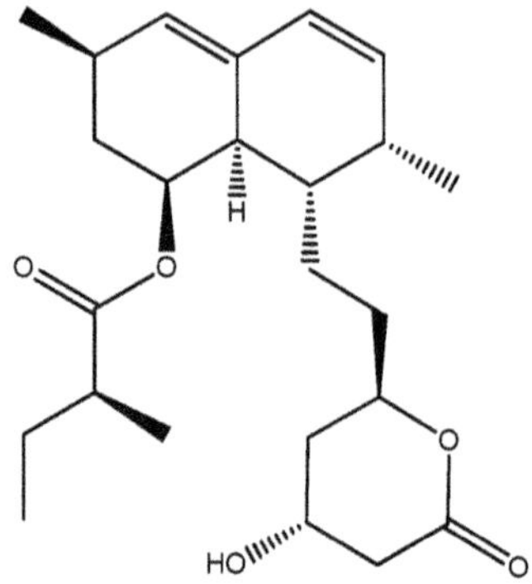

Figure 2. Structure of lovastatin.

Pravastatin

Pravastatin (Fig. 3) is the 3β-hydroxy derivative of compactin. It can be obtained by bioconversion and is more active than compactin in inhibiting HMG-CoA reductase (Hosobuchi et al., 1993a-d; see also McLean et al., 2015). Some *Aspergillus* and *Monascus* species are known to produce pravastatin (Manzoni et al., 1998, 1999).

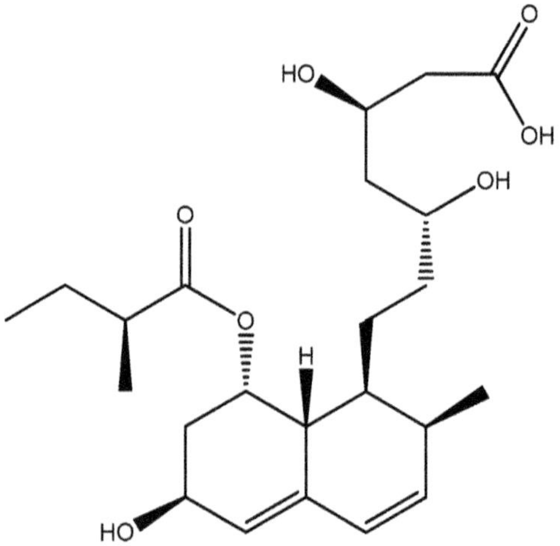

Figure 3. Structure of pravastatin.

It is worth mentioning here that available semi-synthetic and synthetic statins are almost as efficient as or even more efficient in lowering plasma cholesterol than the natural occurring ones. Simvastatin is a semi-synthetic statin produced by Merck. It is a more potent analog of lovastatin. Synthetic statins include atorvastatin, fluvastatin, and rosuvastatin. Atorvastatin is superior to those obtained naturally from fungi.

Atorvastatin   Fluvastatin   Rosuvastatin

Further Research

It is evident with the published work that out of a huge number of fungi occurring in a variety of natural environment, only a few genera (11) and their few species have been examined for production of important compounds such as statins among their secondary metabolites. It would be profitable for chemists engaged in medicinal chemistry to join hands with mycologists studying biodiversity to examine the secondary metabolites of such fungi that occur in unusual environments under stressed conditions. The reason is that fungi in such environments develop competitive ability to survive the stresses of the environment as well as the ability to compete with other microorganisms growing with them in the same niche. This environment might be helping them to produce unique molecules which we can use for human welfare. A cooperative/collaborative search for useful compounds, using all available modern tools and techniques, should be very fruitful.

Acknowledgement

The author is grateful to Dr. Arnold L. Demain of the Research Institute for Scientists Emeriti (R. I. S. E.), Drew University, Madison, NJ 07940 USA for kindly going through the manuscript and suggesting improvements.

References

Adrio, J.L. and Demain, A.L. 2003. Fungal biotechnology. Int. Microbiol., 6: 191–199. DOI 10.1007/s 10123-003-0133-0.

Alarcón, Julio, Anguila, Sergio, Arancibia-Avila, Patricia, Fuentes, Oscar, Zamorano-Ponce, Enrique and Hernández, Margarita. 2003. Production and purification of statins from *Pleurotus ostreatus* (Basidiomycetes) strains. Z. Naturforsch., 58c: 62–64.

Alberts, A.W., Chen, J., Kuron, G., Hunt, V., Huff, J., Hoffman, C., Rothrock, J., Lopez, M., Joshua, H., Harris, E., Patchett, A., Monaghan, R., Currie, S., Stapley, E., Albers-Schonberg, G., Hensens, O., Hirshfield, J., Hoogsteen, K., Liesch, J. and Springer, J. 1980. Mevinolin: a highly potent competitive inhibitor of hydroxymethylglutaryl-coenzyme A reductase and a cholesterol-lowering agent. Proc. Natl. Acad. Sci., USA, 77: 3957–3961.

Barrios-González, J. and Miranda, R.U. 2010. Biotechnological production and application of statins. Appl. Microbiol. Biotechnol., 85: 869–883. DOI 10.1007/s00253-009-2239-6.

Bazaraa, W.A., Hamdy, M.K. and Toledo, R. 1998. Bioreactor for continuous synthesis of compactin by *Penicillium cyclopium*. J. Ind. Microbiol. Biotechnol., 21: 192–202.

Bobek, P., Ozdin, L., Kuniak, L. and Hromadova, M. 1997. Regulation of cholesterol metabolism with dietary addition of oyster mushroom (*Pleurotus ostreatus*) in rats with hypercholesterolemia. Cas. Lek. Cesk., 136: 186–190.

Brown, A.G., Smale, T.C., King, T.J., Hasenkamp, R. and Thompson, R.H. 1976. Crystal and molecular structure of compactin, a new antifungal metabolite from *Penicillium brevicompactum*. J. Chem. Soc. Perkins. Trans., 1: 1165–1170.

Buckland, B., Gbewonyo, K., Hallada, T., Kaplan, L. and Masurekar, P. 1989. Production of lovastatin, an inhibitor of cholesterol accumulation in humans. pp. 161–169. *In*: Demain, A.L., Somkuti, G.A., Hunter-Cevera, J.C. and Rossmoore, H.W. (eds.). Novel Microbial Products for Medicine and Agriculture, Society for Industrial Microbiology. Elsevier Science Ltd., Amsterdam.

Demain, A.L. and Adrio, J.L. 2008. Contribution of microorganisms to industrial biology. Mol. Biotechnol., 38: 41–55.

Demain, A.L. and Zhang, L. 2005. Natural products and drug discovery. pp. 3–29. *In*: Zhang, L. and Demain, A.L. (eds.). Natural Products: Drug Discovery and Therapeutic Medicine. Humana Press, Totowa.

Demain, A.L. 2006. From natural products discovery to commercialization: a success story. J. Ind. Microbiol. Biotechnol., 33: 486–495.

Demain, A.L. 2014. Importance of microbial natural products and the need to revitalize their discovery. J. Ind. Microbiol. Biotechnol., 41: 185–201.

Endo, A. 1979. Monocolin K, a new hypocholesterolemic agent produced by a *Monoascus* species. J. Antibiot., 32: 852–854.

Endo, A. 1985. Compactin (ML-236B) and related compounds as potential cholesterol-lowering agents that inhibit HMG-CoA reductase. J. Med. Chem., 28: 401–405.

Endo, A. 1992. The discovery and development of HMG-CoA reductase inhibitors. J. Lipid Res., 33: 1569–1582.

Endo, A. 2008. A gift from nature: the birth of the statins. Nature Med., 14: 24–26.

Endo, A. 2010. A historical perspective on the discovery of statins. Proc. Jpn. Acad. Ser B, 86: 484–492.

Endo, A., Hasumi, K., Yamada, A., Shimoda, R. and Takeshima, H. 1986. The synthesis of compactin (ML-236B) and monocolin K in fungi. J. Antibiot., 39: 1609–1610.

Endo, A., Kuroda, M. and Tanzawa, K. 1976a. Competitive inhibition of 3-hydroxy-3-methylglutaryl coenzyme A reductase by ML-236A and ML-236B, fungal metabolites, having hypocholesteromic activity. FEBS Lett., 72: 323–326.

Endo, A., Kuroda, M. and Tsujita, Y. 1976b. ML-236A, ML-236B, and ML-236C, new inhibitors of cholesterogenesis produced by *Penicillium citrinum*. J. Antibiot., 29: 1346–1348.

Goldstein, J.L. and Brown, M.S. 2015. A centuary of cholesterol and coronaries: From plaques to genes to statins. Cell, 161: 161–172.

Gullo, V. and Demain, A.L. 2013. Statins: Fermentation products for cholesterol control in humans. pp. 435–444. *In*: Brahmachari, G. (ed.). Chemistry and Pharmacology of Naturally Occurring Bioactive Compounds. CRC Press, Boca Raton.

Gunde-Cimerman, N. and Cimerman, A. 1995. *Pleurotus* fruiting bodies contain the inhibitor of 3-hydroxy-3-methylglutaryl-Coenzyme A reductase-lovastatin. Exp. Mycol., 19: 1–6.

Gunde-Cimerman, N., Plemenitas, A. and Cimerman, A. 1993. *Pleurotus* fungi produce mevinolin, an inhibitor of HMG CoA reductase. FEMS Microbiol. Lett., 113: 333–338.

Hosobuchi, M., Fukui, F., Matsukawa, H., Suzuki, T. and Yoshikawa, H. 1993a. Morphology control of preculture during production of ML-236B, a precursor of pravastatin sodium, by *Penicillium citrinum*. J. Ferm. Bioeng., 76: 476–481.

Hosobuchi, M., Kurosawa, K. and Yoshikawa, H. 1993b. Application of computer to monitoring and control of fermentation process: microbial conversion of ML-236B Na to pravastatin. Biotechnol. Bioeng., 42: 815–820.

Hosobuchi, M., Ogawa, K. and Yoshikawa, H. 1993c. Morphology study in production of ML-236B, a precursor of pravastatin sodium by *Penicillium citrinum*. J. Ferm. Bioeng., 76: 470–475.

Hosobuchi, M., Shiori, T., Ohyama, J., Arai, M., Iwado, S. and Yoshikawa, H. 1993d. Production of ML-236B, an inhibitor of 3-hydroxy-3-methylglutaryl CoA reductase by *Penicillium citrinum*: Improvement of strain and culture conditions. Biosci. Biotech. Biochem., 57: 1414–1419.

Mabuchi, H., Haba, T., Tatami, R., Miyamoto, S., Sakai, Y., Wakasugi, T., Watanabe, A., Koizumi, J. and Takeda, R. 1981. Effect of an inhibitor of 3-hydroxy-3-methylglutaryl coenzyme A reductase on serum lipoproteins and ubiquinone-10-levels in patients with familial hypercholesterolemia. New Engl. J. Med., 305: 478–482.

Manzoni, M. and Rollini, M. 2002. Biosynthesis and biotechnological production of statins by filamentous fungi and application of these cholesterol-lowering drugs. Appl. Microbiol. Biotechnol., 58: 555–564.

Manzoni, M., Bergomi, S., Rollini, M. and Cavazzoni, V. 1999. Production of statins by filamentous fungi. Biotechnol. Lett., 21: 253–257.

Manzoni, M., Rollini, M., Bergomi, S. and Cavazzoni, V. 1998. Production and purification of statins from *Aspergillus terreus* strains. Biotechnol. Technol., 12: 529–532.

McLean, K., Hans, M., Meijrink, Ben, van Scheppingen, W.B., Vollebregt, A., Tee, K.L., van der Laan, Jan-Metske, Leys, D., Munro, A.W. and van den Berg, M.A. 2015. Single-step fermentative production of the cholesterol-lowering drug pravastatin via reprogramming of *Penicillium chrysogenum*. PNAS, 112: 2847–2852. www.pnas.org/cgi/doi/10.1073/pnas.1419028112.

Moon, Hyeongsun, Hill, Michelle M., Roberts, Matthew J., Gardiner, Robert A. and Brown, Andrew J. 2014. Statins: protectors or pretenders in prostate cancer? Trends Endocronol. Metab., 25: 188–196.

Moore, R.N., Chan, G.J.K., Hogg, A.M., Nakashima, T.T. and Vederas, J.C. 1985. Biosynthesis of hypocholesterolemic agent mevinolin by *Aspergillus terreus*. Determination of the origin of carbon, hydrogen, and oxygen atoms by 13C NMR and mass spectrometry. J. Am. Chem. Soc., 107: 3694–3701.

Negishi, S., Cai-Huang, Z., Husuni, K., Murakawa, S. and Endo, A. 1986. Productivity of monacolin K (mevinolin) in the genus *Monascus*. Hakkokogaku Kaishi, 64: 509–512.

Praveen, V.K. and Savitha, J. 2012. Solid state fermentation: An effective method for lovastatin production by fungi over submerged fermentation. J. Biotechnol. Pharmaceu. Res., 3: 15–21.

Srinivasa Rao, K., Prasad, T., Mohanta, G.P. and Manna, P.K. 2011. An overview of statins as hypolipidemic drugs. Intl. J. Pharmace. Sci. Drug Res., 3: 178–183.

Steinberg, D. 2006. An interpretative history of the cholesterol controversy. V. The discovery of the statins and the end of the controversy. J. Lipid Res., 47: 1339–1351.

Tsujita, Y., Kuroda, M., Tanzawa, K., Kitano, N. and Endo, A. 1979. Hypolipidemic effects in dogs of ML-2368, a competitive inhibitor of 3-hydroxy-methylglutaryl coenzyme A reductase. Atherosclerosis, 32: 307–313.

Valera, H.R., Gomes, J., Lakshmi, S., Guraraja, R., Suryanarayan, S. and Kumar, D. 2005. Lovastatin production by solid state fermentation using *Aspergillus flavipes*. Enzyme Micro. Technol., 37: 521–526.

Veillard, N.R. and Mach, F. 2002. Statins: the new aspirin? Cell. Mol. Life Sci., 59: 1771–1786.

Wrigley, S.K. 2004. Pharmacologically active agents of microbial origin. pp. 356–374. *In*: Bull, A.T. (ed.). Microbial Diversity and Bioprospecting. ASM Press, Washington, DC.

Youssef, S., Stueve, O., Patarroyo, J.C., Ruiz, P.J., Radosevich, J.L., Hur, E.M., Bravo, M., Mitchell, D.J., Sobel, R.A., Steinman, L. and Zamvil, S.S. 2002. The HMG-CoA reductase inhibitor, atorvastatin, promotes a Th2 bias and reverses paralysis in central nervous system autoimmune disease. Nature, 420: 78–84.

CHAPTER 9

Glycans of Higher Basidiomycetes Mushrooms with Antiphytoviral Properties: Isolation, Characterization, and Biological Activity[#]

O.G. Kovalenko[1] and S.P. Wasser[2,*]

ABSTRACT

The higher Basidiomycetes mushrooms produce a wide range of biologically active glycans, which have different chemical structures and different types of biological activity. The chapter focuses on the antiphytoviral properties of glycans of higher Basidiomycetes mushrooms on TMV-infection and their ability to induce virus resistance in plants. The different species of higher Basidiomycetes: *Ganoderma lucidum*, *G. applanatum*, *G. adspersum*, and *Tremella mesenterica* were screened for

[1] D.K. Zabolotny Institute of Microbiology and Virology, National Academy of Sciences of Ukraine, acad. Zabolotny St., Kyiv D 03680, Ukraine.
Email: udajko@ukr.net
[2] Institute of Evolution & Department of Evolutionary and Environmental Biology, Faculty of Natural Sciences, University of Haifa, Mount Carmel, Haifa 31905, Israel; N.G. Kholodny Institute of Botany, National Academy of Sciences of Ukraine, 2, Tereshchenkivska St., Kyiv 01601, Ukraine.
* Corresponding author: spwasser@research.haifa.ac.il
[#] The work is conducted with the support of Science & Technology Centre in Ukraine (project #4973).

production of antiviral polysaccharides and the prolific producer was selected. It was established that the polysaccharide extracted from the culture broths of *G. lucidum* and *G. applanatum* have moderate antiviral activity and do not induce expressed plant resistance for the infection of tobacco mosaic virus (TMV). Glucuronoxylomannan (GXM), obtained from culture broth of the *T. mesenterica*, inhibits TMV infection and induces high virus resistance in tobacco plants.

Three glucan fractions (water, acid, and alkaline) were isolated from *G. adspersum* mycelium. Glucans, obtained with acid and alkali agents, were homogeneous with molecular masses of 48 and 70 kDa, respectively, whereas the fraction extracted with water was heterogeneous and contained macromolecular components: 95, 40, and 20 kDa. The most active against the TMV-infection glucan preparation was extracted with water, although the preparations obtained with acid and alkali also inhibited viral infection. Using the transmission oxidation of polysaccharides, it was noticed that *G. adspersum* glucan contains 1-3-linked glucose in the main chain, and in side chains monosaccharides are linked with other types of bonds. The obtained glucan, like GXM, induced the plant resistance to viruses, which was partially inhibited by actinomycin D and α-methyl-D-mannoside and depended on RNA synthesis in the plant DNA matrix, on the one hand, and specific protein-carbohydrate interaction on the other. The possibility of using glycans of higher Basidiomycetes mushrooms to create complex plant protection agents against viruses and other infectious diseases is discussed.

Introduction

Higher Basidiomycetes, from ancient times, were used in many preparations against several diseases in folk medicine, especially in medieval Oriental countries. The records on the healing properties of Basidiomycetous fungi in Europe and North America belong to a later time, and their investigation as potential sources of pharmaceutical and medical products began only recently (Lindequist, 2005). The first studies that have shown that Basidiomycetous mushrooms are an inexhaustible source of substances that have medicinal properties for numerous diseases are of Danylyak and Reshetnikov (1996) and Ooi (2000). Glycans, peptidoglycans and glycoproteins, which have varying chemical structures and different types of biological activity, are primarily noteworthy among the medicinal substances from higher Basidiomycetes mushrooms (Mizuno, 1999; Polishchuk and Kovalenko, 2009; Wasser, 2002).

It was established that these biopolymers not only protect people from the negative impact of environmental factors (Andrianova et al., 2010; Harris, 2003), but also have a perspective as medical preparations because of their immunomodulatory, antiviral, antibacterial, and anticancer properties as well as their ability to maintain overall homeostasis of the organism (Mizuno et al., 1995; Wasser and Weis, 1999; Wasser and Didukh, 2005).

The chemical nature and structural peculiarities of carbohydrate-containing biologically active preparations of higher mushrooms depend on various factors including the producer species and strain, growth/environmental conditions, methods of extraction, etc. Higher Basidiomycetes, particularly the species of genus *Ganoderma*, produce biologically active branched β-glucans containing the main chain of (1→3)- and inside chains -(1→4)- and (1→6)-bound β-D-glucopyranosyl units (Mizuno et al., 1995; Mizuno, 1999; Wasser, 2002). It is the β-(1→3)-glycoside bond, molecular weight and degree of branching of polysaccharide chains, which is associated with antitumor and other biological properties of fungal β-glucan (Bohn and BeMiller, 1995; Wasser, 2010).

Glycans produced by some bacteria, microscopic fungi, and yeasts have antiphytoviral activity (Dyakov and Kovalenko, 1983; Kovalenko, 1987; Kovalenko et al., 1977, 1993). The mechanism of action of these biopolymers depends on their chemical structure. For example, neutral mannans, produced by yeasts of *Candida* genus, inhibit the formation of local lesions induced by tobacco mosaic virus (TMV), by activation of protective mechanisms that do not require new synthesis and are based on conformational protein-carbohydrate interaction (Kovalenko, 1993; Kovalenko et al., 2000).

In contrast, the antiviral action of sulfated mannans consists of the induction of resistance which is stipulated by the synthesis of protein components and antiviral factors *de novo* (Kovalenko et al., 1988, 1992, 1993). The mechanism of action of oligosaccharides is associated with the induction of chemical signals necessary for the initiation of supersensitive reactions (Dyakov and Kovalenko, 1983).

However, there are several reports on the action of the carbohydrate-containing biopolymers of higher mushrooms on plant viruses in the literature. In particular, it was shown that glycoprotein with lectin, isolated from mycothallus of *Agrocybe aegerita*, can suppress the development of viral infection through a breach of the penetration process of viral particles into a cell (Sun et al., 2003), and exocellular glucuronoxlylomannan of *Tremella mesenterica* (Kovalenko et al., 2009), in contrast to the total polysaccharide preparations from culture liquid *Ganoderma lucidum* and *G. applanatum* (Kovalenko et al., 2008), can induce the development of virus resistance in plants by activating cellular synthesis *de novo*. Since Basidiomycetes mushrooms are the source of glycans of different structures and have extremely useful biological properties and the activity of these biological properties on phytoviruses has not been studied sufficiently, it is reasonable to further explore the ability of these compounds to inhibit viral infection and induce protective mechanisms of action in plants. This chapter presents the data of research in this area.

Higher Basidiomycetes Species as Producers of Antiviral Glycans

Structural features and biological activity of carbohydrate-containing biopolymers of higher Basidiomycetes mushrooms depend on various factors including the species and strain of the producer, conditions of its cultivation, extraction methods, etc. Therefore, the screening aimed at identifying most active producers of inhibitors of phytoviral infections was conducted initially among the strains and species of mushrooms available in our collection. The criterion for the selection of producers was the ability of the obtained polysaccharide preparations to suppress the development of TMV infection in supersensitive plants, primarily in *Datura stramonium* L., which was the most sensitive species, according to our observations. For this purpose, polysaccharide preparations were obtained from a liquid culture or fungal mycelium by ethanol precipitation. The preparations so obtained were in the concentration of 1-2000 µg/ml, and were added to the purified suspension of TMV. Suspension so obtained was used to inoculate half leaf of the *D. stramonium*. The other half of leaves (used as the control) was inoculated by the same virus (without the addition of the preparation). Results were recorded after 5–7 days of inoculation. Antiviral activity was determined by the number of viral brown local lesions appearing on inoculated leaves in experimental and control samples.

Tremella mesenterica as Producers of Antiviral Glycans

T. mesenterica mushroom in culture media produce glucuronoxylomannan polysaccharide (GXM), which is a linear polymer of α-(1→3)-bound mannose, glycolized with β-(1→2)(1→4)-bound oligosaccharides of xylose and glucuronic acid, which gives it the polyanionic properties. Based on the chemical structure and properties of GXM (Vinogradov et al., 2004), polysaccharides can induce virus resistance in plants similar to the action of sulfated mannans and other polyanions (Kovalenko et al., 1988, 1992, 1993). Therefore, the total glucan preparation isolated from the culture liquid of *T. mesenterica*, and its components—neutral polysaccharide and acidic glucuronoxylomannan polysaccharide were used in further research. The preparations tested in different ways suppressed the development of local lesions induced by TMV in *D. stramonium* (Table 1).

Neutral glycans were more active according to this criterion. Their degree of inhibiting the development of local necrosis ranged from 80 to 99.4%. GXM was much less active, and the total preparation showed inhibiting ability in between. The results may indicate that the activity of the total preparation on TMV infectivity resulted largely by neutral glycans, rather than by acidic GXM.

This is consistent with earlier reports on higher efficiency of neutral polysaccharides in inhibiting virus infectivity in supersensitive plants, compared with their sulfated derivatives, which primarily induce virus resistance in plants *de novo* (Kovalenko et al., 1988, 1993). Since the aim of the research was to find out glycans, which may induce plant resistance to viruses. Therefore, only acidic GXM was used in further investigations. In general, we

Table 1. Effects of exocellular polysaccharides of *Tremella mesenterica* 623 on TMV infectivity in *Datura stramonium* plants.

Polysaccharide	Concentration, µg/ml	Number of local lesions per half leaf		Inhibition, %
		Experiment	Control	
Total	1000	2.5	28.1	$91 \pm 3^{+++}$
	100	15.9	27.2	$42 \pm 2^{+++}$
	10	19,3	27,4	$30 \pm 6^{+++}$
Neutral	1000	0,7	37.8	$98 \pm 4^{+++}$
	100	3,7	53.1	$93 \pm 5^{+++}$
	10	17,0	36,6	$54 \pm 12^{+}$
GXM	1000	3.9	28,1	$86 \pm 6^{++}$
	100	20.4	25.2	$11 \pm 7^{\circ}$
	10	16.0	16.9	$5 \pm 13^{\circ}$

Note: [+++]: $p \leq 0.1\%$; [++]: $0.1\% < p \leq 1\%$; [+]: $1\% < p \leq 5\%$; [°]: $p > 5\%$.

have extracted GXM preparations and studied their antiphytoviral properties from liquid culture of four more strains of *T. mesenterica*—15, 34, 41, and 45 HAI and they were less active GXM producers. Therefore, the strain 623, which produced GXM actively inducing plant resistance to TMV infection, was selected for further studies. This preparation has been a model for further research to find out the mechanism of action of acidic polysaccharides of higher Basidiomycetes as inducers of virus resistance in plants and as inactivators of viruses.

Species of Genus *Ganoderma* as Producers of Antiviral Glucans

Species of genus *Ganoderma* are the source of biologically active polysaccharides, widely used for medicinal purposes (Mizuno, 1999; Mizuno et al., 1995; Polishchuk and Kovalenko, 2009; Wasser and Weis, 1999; Wasser, 2002; Wasser and Didukh, 2005). However, unlike genus *Tremella*, the producers of acidic polysaccharides, genus *Ganoderma* mainly produces neutral glucans. The source of glucans can be culture medium (liquid), fruit body biomass, and biomass of mycelium. The process of extraction of these compounds from fruit body biomass and biomass of mycelium is quite labore intensive and pretty expensive. Therefore, initially liquid culture, from which biologically active glycans can be obtained more easily, was initially used. It is known that *Ganoderma lucidum* and *G. applanatum* produce extracellular glycoproteins of molecular weight approx. 2000 kDa (Hung et al., 2008; Smirnov et al., 2006). However, only a few strains could efficiently produce extracellular biologically active polysaccharides. Therefore, it was imperative to identify strains of mushrooms that could be inactive glycan producer having antiviral properties. During further studies *Ganoderma lucidum* and *G. applanatum* were cultured in liquid mineral media supplemented with glucose as a source of exopolysaccharides.

Ganoderma applanatum and *G. lucidum* as Producers of Antiviral Glucans

While choosing a nutrient medium for cultivation of *G. lucidum* and *G. applanatum* an important consideration was that there should not be any polymeric organic compounds, including polysaccharides in the medium, which can prevent the isolation and determination of exopolysaccharide activity produced by mushrooms. Therefore, the mineral medium (Litchfield, 1967) was used for the cultivation of fungi. Two percent of glucose was added to the medium as a source of carbon. After 14 days of cultivation on shakers at 28°C, the mycelium was separated; liquid culture was lyophilized and used as needed. If necessary, the aqueous solutions (1–1000 µg/ml) were prepared from the concentrates and tested against TMV, as mentioned above. In general, seven strains each of *G. lucidum* and *G. applanatum* were tested.

It was observed that all the strains of *G. lucidum* were able to produce substances that suppressed the development of TMV infection in *Datura stramonium* plants to some extent. However, the evident activity as a viral inhibitor producer was noted only for strain 900 of the fungus *G. applanatum* and strains 1900, 1788, 1683, and 1912 of the fungus *G. lucidum*. Their metabolic products suppressed the TMV infectivity at the concentrations of 1–1000 µg/ml on 46–95%, and products of strain 920 had a strong inhibitory effect (65%) even at low concentrations (Kovalenko et al., 2008). Thus, the biological activity of metabolites depends on the producer strain.

Ganoderma adspersum as Producer of Antiviral Glucans

Apart from extracellular glycans of *G. lucidum* and *G. applanatum*, we investigated endopolysaccharides isolated from the lyophilized mycelium of *G. adspersum*, grown on a synthetic medium of the following composition (g/l): glucose - 15, peptone - 2.5, yeast extract - 3; KH_2PO_4 – 0.5; $MgSO_4$ × $7H_2O$ – 0.3; Na_2HPO_4 – 0.2; pH 5.5. It is known that glycans of this mushroom have a lower molecular weight as compared to *G. lucidum* and *G. applanatum*. It can be assumed that low-molecular glycans are more actively adsorbed by plant tissues and thus more effectively inhibit viral infection in plants. It is known that yeast mannans with a molecular weight of 50–70 kDa (Kovalenko and Votzelko, 1977; Kovalenko et al., 1977), inhibit the TMV infection by or up to 90–100% and activate the virus resistance in plants (Kovalenko et al., 1999, 2000).

Glucan preparations were obtained by boiling the mycelium of *G. adspersum* in water, acid and alkaline solutions and purifying further from protein impurities (Kovalenko et al., 2010a,b) and were then used for further research. It was observed that the glycans isolated by different extraction methods showed antiphytoviral activity. Glycan preparation obtained by water extraction was the most active against TMV in all tested strains. The ability to inhibit 50–100% viral infection was observed at concentrations of 1–2000 µg/ml. The activity of the preparation increased proportionally with the increase of the dose, clearly demonstrating the accuracy of the results. Other preparations also had high levels of inhibitory activity, but it was slightly lower than that of the preparations obtained by aqueous extraction. Given the results obtained, a preparation obtained from mycelia of strain II of *G. adspersum* by water extraction was chosen for further research of the antiphytoviral activity of fungal glycans and their mechanisms of action. This preparation was the most active against TMV.

Physico-chemical and Antiviral Properties of Glycans

The main active components of metabolic products of higher Basidiomycetes mushrooms are polysaccharides, which are mainly represented by glycans with monomers of glucose and mannose (Marchessalt et al., 1977; Gorin and Barreto-Berger, 1983; Ben-Zion Zaidman et al., 2005). Glycoproteins can be found among the metabolites as well (Mizuno et al., 1995). Biological activity of glycans may depend not only on primary structure, molecular weight, degree of branching chains, solubility in water and other properties but also on their source, i.e., cultivation medium and fruiting bodies (Ovodov, 1997). As it was shown above, the antiviral properties of glycans differ depending on the producer and their source. However, it was necessary to establish chemical composition and structural elements of polysaccharide preparations, as well as the presence of impurities like proteins or nucleic acids before starting the study of the inducer and antiviral properties. In addition, it was important to establish the role played by individual structural elements (molecule charge, types of glycoside links, etc.) and the possible roles played by contaminants in antiviral activity of the preparations.

Glycans of Species of Genus *Ganoderma*

Glycans of *G. lucidum* and *G. applanatum*

A lyophilized liquid culture containing high-molecular weight polysaccharides and possibly other low-molecular weight metabolic products of the fungi including protein as contaminants, was used in the screening of the antiviral exopolysaccharides produced by *G. lucidum* and *G. applanatum*. Therefore, it was necessary to isolate all polymeric preparations and to investigate some physical characteristics of the active components, particularly, their thermolability and ability of penetration through the cellulose film during dialysis. For this purpose, dialysis and heating, on a water bath, at 100°C for

10 min of liquid culture was done and the dialyzates were then tested for their antiviral activity on plant *D. stramonium* and *Nicotiana tabacum*.

It was observed that there is a slight decrease in antiviral activity of the preparation from *G. applanatum* after dialysis of liquid culture whereas there was no significant change in *G. lucidum* preparation (Fig. 1A). This indicated that the antiviral activity of preparations of *G. lucidum* involves a factor (or factors), which is a polymer(s) with a molecular weight higher than 7–10 kDa (Keil, 1966). In preparations from *G. applanatum* the components with a lower molecular weight are apparently involved in virus inhibition.

Products isolated from the liquid culture of *G. lucidum* were resistant to high temperature, whereas heat treatment of *G. applanatum* liquid culture

A

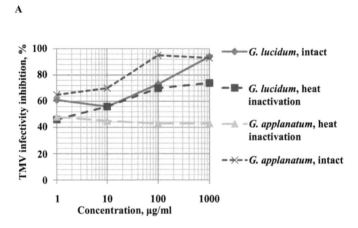

B

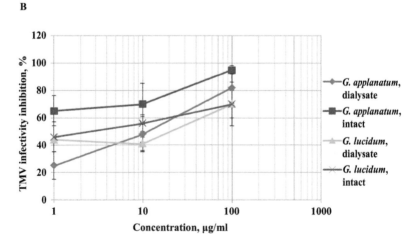

Figure 1. Effect of intact and heated (100°C, 10 min) (A) metabolic products of *G. lucidum* and *G. applanatum*, and their dialyzates (B) on TMV infectivity.

slightly reduced its activity (Fig. 1B). The results may indicate the presence of thermolabile proteins or glycoproteins with antiviral activity apart from polysaccharides in the preparations (Smirnov et al., 2006) observed that in the liquid culture medium *G. lucidum* produces glycoproteins in which the carbohydrate part was formed by the D-glucose residues bound by 1→3-glycoside bonds, and their protein content was up to 1.5–3.6%. This results indicate the thermostability and polymer nature (at least their molecular weight, which is higher than 10 kDa) of the extracted substances. It can be determined that they are mainly represented by glycopolymers.

Given the results of previous studies, partially purified polysaccharide preparations were obtained by the sedimentation of polymers from liquid culture by two volumes of alcohol. It was determined that the total polysaccharide preparation isolated from the liquid culture of *G. lucidum* did not contain protein impurities. In preparations obtained from liquid culture of *G. applanatum*, the protein content did not exceed 0.2%. In both preparations no nucleic acid impurities were found. At the same time, both preparations had antiviral activity. That is, exogenous products of higher basidiomycetous mushrooms *G. applanatum* and *G. lucidum* were most likely glycans of high-molecular weight (Kovalenko et al., 2008).

Since it is known that the antiviral activity of polysaccharides depends on the genotype of the host plants, hence future experiments were planned to study the influence of polysaccharide preparations on TMV infection on different plant indicators, which differ in the mechanisms of resistance (Kovalenko, 1987; Kovalenko et al., 1977; Kovalenko and Kyrychenko, 2004). To determine whether polysaccharides inhibit TMV infection through their influence on the mechanisms of resistance, the antiviral activity of preparations in *Nicotiana sanderae* Hort and *N. tabacum* L. was tested. These species as TMV hosts are different. *Nicotiana sanderae* Hort has no mechanism of viral infection localization, although primary lesions occur on inoculated plants, and resistance is induced after repeated infections (Kovalenko and Kyrychenko, 2004). In the second type—tobacco of variety Immunny 580, which inherited the hypersensitivity N-gene of *Nicotiana glutinosa* L. as a result of interspecific hybridization, mechanisms of localization of viral infection (LVI) and induced virus resistance (IVR) are implemented independently (Dyakov and Kovalenko, 1983).

It was established that in tobacco plants of var. Immunny 580 antiviral activity of both preparations was about the same as in the *Datura* plants (Table 2). The development of viral infection was suppressed by 40–44% by the tested preparations. However, the inhibition of TMV infection was also observed in *N. sanderae* plants, which were highly sensitive to TMV.

Results may indicate that one of the main mechanisms of antiviral activity of the studied polysaccharides together with virus resistance induction may be the inhibition of the initial stages of virus reproduction, for example, penetration of the virus into the cell, virion deproteinization, etc. (Kovalenko et al., 2008).

Table 2. Per cent inhibition of TMV infection by glycans isolated from the culture liquids, *G. lucidum* and *G. applanatum*, in plants of two tobacco species (number of local lesions per half leaf).

Producer*	Number of local lesions per half leaf		Inhibition, %
	Experiment	Control	
N. tabacum, var. Immuny 580			
G. applanatum	62.8	112.2	44.0[++]
G. lucidum	95.9	158.3	39.4[++]
N. sanderae			
G. applanatum	3.8	14.2	73.2[+++]
G. lucidum	19.2	39.6	51.3[+]

Note: *concentrations of preparations 2500 µg/ml; [+++]: $p \leq 0.1\%$; [++]: $0.1\% < p \leq 1\%$; [+]: $1\% < p \leq 5\%$.

Thus, *G. lucidum* and *G. applanatum* produced high-molecular weight glucans with moderate antiviral activity in culture medium; however, extracted polymers were able to inhibit viral infection in different indicator plants in different ways. One of the mechanisms of limiting viral infection may be the inhibition of the initial stages of virus reproduction.

Glycans of *Ganoderma adspersum*

As it is mentioned earlier that the activity of glycans depends on the source from which they were obtained, their chemical structure, and molecular weight. Therefore, antitumor activity of preparations extracted from mycothalluses was higher than that of mycelium or culture liquid (Mizuno, 1999; Mizuno et al., 1995). Both homopolysaccharides obtained primarily by water extraction and heteropolysaccharides, including xylose, mannose, galactose and uronic acids, are found among fungal polysaccharides. The latter are obtained from aqueous solution by alkali treatment and salt fractionation (Mizuno et al., 1995). To find out, how the methods of extraction of glycans effect their physicochemical properties and antiviral activity, different approaches to the extraction of polymers from biological material were further tested on a model fungus.

Isolation and Purification of *Ganoderma adspersum* Glycans

Extraction with water, acid and alkali, followed by purification of the resulted preparations from protein impurities with mixture of chloroform-amyl alcohol was performed to obtain glucan from the mycelium of *G. adspersum* (Kovalenko et al., 2010a,b). This method provides three polysaccharide fractions that have different properties.

The investigation of the selected preparations has shown that they, along with carbohydrates, contain proteins (Table 3). It is unknown what these substances are, whether are they the components of biopolymers-glycoproteins or accidental contaminants of glycans.

Table 3. Per cent yield and their content of proteins and carbohydrates using different types of extraction methods for *G. adspersum* mycelium.

Extraction	Preparation yield	Carbohydrates	Proteins
Water	8.4	44.0	2.0
Acid – 0.01 NH$_2$SO$_4$	4.4	61.3	3.3
Alkaline – 5% NaOH and 0.05% NaBH$_4$	11.5	31.5	1.4

Available data indicate that endopolysaccharides obtained by water extraction from mycothalluses of *G. lucidum* are exopolysaccharides, including peptidoglycans in which the protein part yields is 1.5%. Therefore, it can be concluded that the obtained preparations also are glycoproteins and contain no separate proteins or peptides as contaminants.

Monosaccharide Composition of *Ganoderma adspersum* Glycans

Determination of monosaccharide composition of the preparations using the HPLC-MS method has shown that the content of monomers in different preparations varied greatly. The main component of all polysaccharides was glucose (Table 4). Minor components were mannose, galactose and rhamnose, and in some of these—xylose, ribose, arabinose, etc. Particularly, in glucans

Table 4. Monosaccharide composition of glucans derived from the mycelia of *G. adspersum* by water, acid and alkaline extraction.

Monosaccharide	Per cent monosaccharide content in glucan preparations from *G. adspersum* mycelium		
	Water extraction	Acidic extraction	Alkaline extraction
Glucose	75.67	81.55	56.92
Mannose	8.96	7.31	18.31
Galactose	10.27	1.32	4.28
Ribose	1.46	-	1.72
Fucose	2.37	0.46	1.57
Arabinose	0.73	-	-
Rhamnose	0.54	0.99	0.84
Xylose	-	2.17	2.82
X1	-	3.34	7.43
X2	-	1.77	6.11
X3	-	0.75	-
X4	-	0.33	-

Note: "-" absence of monosaccharides, X1-X4 - not identified components.

obtained by acid extraction, additional components of a non-carbohydrate nature were found, but their total content is insignificant. Therefore, it can be assumed that they do not affect the antiviral properties of preparations.

Thus, our study confirmed the earlier reports on the glucan nature of the products derived from mycelium of *G. adspersum*, like other carbohydrate products of *Ganoderma* mushroom. Since the highest antiviral activity was observed by aqueous extract of the preparation, further research on antiviral properties of glycans produced by *G. adspersum* was conducted.

Gel-chromatography of *Ganoderma adspersum* Glucans

The monosaccharide composition and molecular characteristics of glycans from mycelium of *G. adspersum* was studied (Kovalenko et al., 2010b). Glucose as well as other minor components, including mannose and galactose (Table 4) was found. This may indicate that these drugs may contain other polysaccharide components with chemical composition other than glucan, which may vary by their molecular weight and antiviral activity. To investigate this possibility, studies of the component composition of glucan preparations obtained by water, acid, and alkali extraction were conducted.

Component identification based on molecular weight distribution was performed by column chromatography (1.7 × 41.0 cm) with Sephadex G-100 gel. The calibration of chromatographic column with a set of different molecular weight dextrans—17, 40 and 70 kDa (Biochemica)—was conducted for this purpose. A polysaccharide solution of 2 ml with a concentration of 15 mg/ml was applied to the column, and then eluted with 0.01 M pyridine-acetate buffer with a pH of 5.4. The fractions of 2 ml were collected using the fraction collector (LKB) and analyzed for carbohydrate content by qualitative reaction with phenol and concentrated H_2SO_4 (Varbanets et al., 2010). The activity of "water" preparation against TMV in *Datura stramonium* plants was also defined (as described in Kovalenko et al., 1977).

Glucan preparations obtained by acid and alkali extraction were chromatographically homogeneous and contained only one component that appears immediately by free volume (Fig. 2A,B). The presence of four components in preparation extracted with water one had the highest molecular weight and eluted most immediately by the free volume of column; others with lower molecular weights are delayed by gel and have a greater output volume when compared to the first component; the last, or fourth component, is apparently represented by oligosaccharides (Fig. 2C).

As a result of the comparison of output volume of the components eluted from the column, the acid extracted glucan had a molecular weight of 48 kDa and the alkali extracted glucan had—70 kDa. The first component of "water" glucan, which appeared immediately after the free column volume, was a polymer with a molecular weight of 95 kDa; molecular weight of the components that followed were 48 and 20 kDa, respectively (Fig. 3).

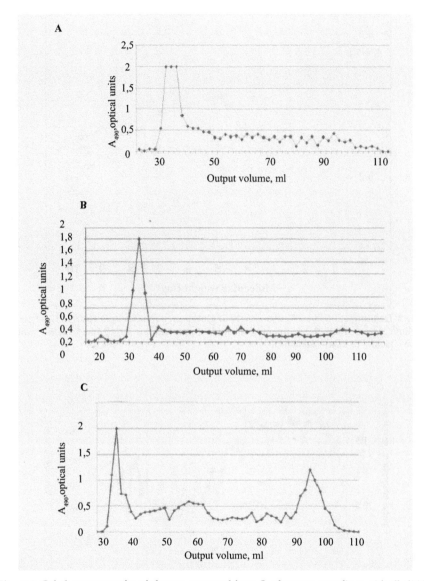

Figure 2. Gel-chromatography of glycans, extracted from *G. adspersum* mycelium with alkali (A), acid (B) and water (C) on Sephadex G-100 column.

Given the gel-filtration results, the differences of discovered chromatographic components of "water" preparation in ability to inhibit TMV infection were recorded. The tests of antiviral activity of "water" glucan factions have shown that they all have the ability to inhibit the virus infectivity, but the activity of high-molecular components was higher when compared with that of low-molecular components (Fig. 4).

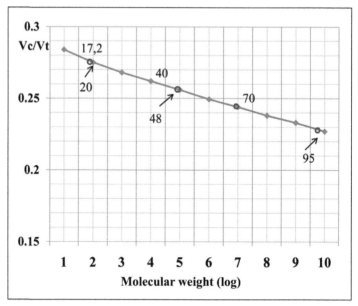

Figure 3. Determination of molecular weight of glucan preparations isolated from *G. adspersum* mycelium. Arrows mark the components of studied preparations: 20, 48 and 95 kDa - "water"; 48 kDa - "acid", 70 kDa - "alkaline".

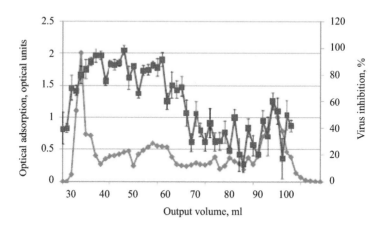

1. -------- – carbohydrate content in fractions

2. ------- decrease of TMV induced local lessions in Datura plants

Figure 4. Gel-chromatography of polysaccharide preparation, extracted with water from the mycelium of fungus *G. adspersum* on Sephadex G-100 column and antiviral activity of the obtained fractions. The 2 ml fractions were collected and tested *in vitro* against TMV via inoculation on *Datura stramonium* leaves. (1) carbohydrate content in fractions; (2) decrease of TMV induced local lesions in Datura plants.

Therefore, the antiviral activity of high-molecular components hardly differs. That is, the main contribution to the overall antiviral activity of glucan preparation extracted from the mycelium of *G. adspersum* with water was the high-molecular components.

Periodate Oxidation of *Ganoderma adspersum* Glucan

Basidiomycetes mushroom, in particular, the genus *Ganoderma*, produce biologically active branched β-glucans containing the main chain (1→3)- and/or β-(1→4)- and in the side chains -(1→4)- and (1→6)-bound β-D-glucopyranosyl units (Hung et al., 2008; Maeda et al., 1988; Ovodov, 1997; Yanaki et al., 1983). It is β-(1→3)-glycoside bonds in the main chain and (1→6)-bonds in side chains that are associated with antitumor and other biological properties of fungal β-glucan (Maeda et al., 1988; Mizuno, 1999). To clarify the role played by individual glycoside bonds, branching and polymerization of *G. adspersum* glucan in antiphytoviral activity, its periodate oxidation, borane reduction and partial hydrolysis with 0.05N H_2SO_4 were carried out. During the periodate oxidation, the decomposition of glucose bound by 1→2-, 1→4-, and 1→6-bonds occurs, while 1→3-bonds remain intact. As a result of oxidation, subsequent reduction and partial hydrolysis of the reduced derivative-end products are produced, which are represented: (1) in the case of 1→2-bond - glycerol and glycerol aldehyde, (2) in the case of 1→4-bond – glycol aldehyde and erythrite, (3) in the case 1→6-bond - glycol aldehyde and glycerol (Kochetkov et al., 1967).

The complete periodate oxidation of glucan under the applied conditions (+12°C) occurred within 72 hours. Therefore, only 24.6% of glucan is oxidized, while 76.4% remained intact. These data indicate that the studied glucan is represented by polymers, which are probably formed by glucose residues connected with 1→3 glycoside bonds, while other types of bonds (24.6%) belong to glucose residues present in the lateral branches of the main chain.

The examination of antiviral activity of oxidized, reduced, and partially hydrolyzed (with no side chains) glucan derivatives revealed a significant reduction of the ability of TMV inhibition in *Datura stramonium* plants after periodate oxidation, subsequent reduction, and hydrolysis when compared with native glucan activity (Fig. 5).

Thus, the results indicate the important role played by glycoside bonds decomposed by periodic acid in the antiviral activity of glucan extracted with water from *G. adspersum* mycelium. These are probably the 1→2-, 1→4- and/or 1→6-bonds. The fact that probably the 1→3-bound main chain, which remains intact after periodate oxidation, may also inhibit the TMV infection development, although to a lesser extent than native glucan, suggests its significant role in the implementation of biological (antiphytoviral) activity of polysaccharides.

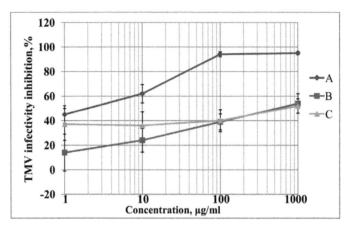

Figure 5. Influence of periodate oxidation on antiviral activity of *G. adspersum* glucan. (A) native glucan; (B) glucan A with reduction; (C) glucan B with partial hydrolysis.

Glucuronoxylomannan of *Tremella mesenterica*

A heteropolysaccharide, whose main chain is represented by α-(1→3)-bound D-mannose, glycosylated with xylose and glucuronic acid, was isolated by extraction with organic solvents from the culture liquid of *T. mesenterica* and purified by ion exchange chromatography (Vinogradov et al., 2004). A fractional precipitation of a polysaccharide with an organic solvent in the presence of 0.05M $CaCl_2$ was proposed to simplify the technique of its production (Kovalenko et al., 2009; Kovalenko and Polishchuk, 2009). This method allows the splitting of the acidic glucuronoxylomannan (GXM) and neutral polysaccharide present in the culture liquid without being time-consuming or an expensive distribution (ion exchange) chromatography.

The yield of glycans from *T. mesenterica* culture liquid, according to our long-term observations, is within 45–50% of the dry weight output. The share of neutral glycans in the total mass of preparation is 5–7%. A comparative antiviral activity of the total preparation as well as separated GXM and neutral polysaccharide is presented in a table (Table 1).

Polysaccharides with anionic properties, for instance, sulfated mannans, can induce virus resistance in plant, which is associated with the synthesis of antiviral factors (Kovalenko et al., 1993). On the contrary, neutral polysaccharides activate protective mechanisms in plants, based on a protein-carbohydrate interaction (Kovalenko, 1993; Kovalenko et al., 2000).

The data of previous experiments (Kovalenko and Polyshchuk, 2009) confirm a very high degree of inhibition of TMV infectivity by glucuronoxylomannan in *D. stramonium* (Table 1). As it is known that *D. stramonium* plants possess localized resistance to TMV (Loebesten, 1972; Dyakov and Kovalenko, 1983). On this basis, it could be assumed that GXM may involve in the mechanism of localization of viral infection (LVI). Therefore, its activity was examined in *N. sanderae* plants where LVI mechanism is non-effective (Kovalenko and Kyrychenko, 2004). Experiments with GXM on these plants were conducted by the same method as in the *Datura stramonium* plants, namely polysaccharide in different concentrations (1000–2500 µg/ml) was added to the inoculum before inoculation procedure. It was established that GXM can prevent the formation of viral lesions on the leaves of *N. sanderae* only in higher concentrations - 2500 pg/ml (Table 5). The reliable inhibition of this process at a concentration 1000 µg/ml was not observed in contrast to the results on the *D. stramonium* plants.

Thus, GXM inhibits the development of TMV infection in sensitive and hypersensitive host plants at the simultaneous introduction of the virus and polysaccharides. It proves the ability of GXM to induce virus resistance in plants similar to sulfated mannans (Kovalenko et al., 1988, 1993) and other anionic polymers (Gicherman and Loebenstein, 1968; Dyakov and Kovalenko, 1983; Gianinazzi, 1983).

Table 5. Glucuronoxylomannan (GXM) influence on TMV infection in *Nicotiana sanderae* plants.

GXM concentration, µg/ml	Number of local lesions per half leaf		Inhibition, %
	Experiment	Control	
1000	7.0	10.5	33°
2500	1.75	6.0	71+

Note: +: $1\% < p \le 5\%$; °: $p > 5\%$.

Thus, antiviral properties of exogenic and endogenic polysaccharides obtained from the studied strains: *G. adspersum, G. lucidum, G. applanatum,* and *T. mesenterica,* indicate that the given mushrooms produce biologically active substances capable of inhibiting the viral infection in plants. The degree of inhibition of TMV infection depends on the strain of the producer as well as on the chemical properties of the preparations. Thus, glycans from *G. adspersum, G. lucidum,* and *G. applanatum* with neutral properties are able to inhibit the local TMV infection to a higher extent when compared with GXM, which is negatively charged. Since the polysaccharide preparations had different influences on TMV infection in plants, depending on the presence of infection localization mechanism, it was important to study their ability to induce systemic and local resistance in plants to further inoculation.

Higher Basidiomycetes Glycans as Inducers of Plant Resistance to TMV Infection

The ability to induce systemic and local resistance in plants to further infection with the pathogen was already reported for many substances of natural and synthetic origin (Dyakov and Kovalenko, 1983). Publications on polysaccharides as inducers of plant resistance to virus infection are insufficient. The first research in this area was conducted by Gupta et al. (1974) demonstrating that polysaccharides obtained from culture liquid of microscopic fungi, *Trychothecium roseum,* can induce in tobacco leaves resistance to TMV infection. Therefore, the induction process depended on the cell's *de novo* genome activity since it appeared to be sensitive to the action of specific inhibitors of RNA transcription on cell DNA matrix. The *T-poly* preparation of a carbohydrate nature was for a long-time only known inducer of plant resistance to viruses.

Later it was found that active preparations of T-poly contain traces of double-stranded RNA of mycoviral origin infecting this fungus *in vitro* (George, 1981). This RNA, like a yeast RNA (Giccherman and Loebenstein, 1968) and double stranded synthetic polynucleotide complexes (Stein and Loebestein, 1970), is capable of inducing a resistance for TMV in hypersensitive plant. The ability of a virus infecting fungus *Penicillium funiculosum* and its RNA to induce resistance to TMV in hypersensitive varieties of tobacco was established previously (Bobyr et al., 1974). In relation to plant resistance, inducing properties of glycans to viruses remained unproven. (Modderman et al., 1985) had proven that the products of partial hydrolysis of plant cell walls can be the inducers of the plant resistance to virus. A similar feature was revealed a little later by Wieringa-Brants and Decker (1987) in glycoproteins, isolated from the intercellular liquid of TMV-infected supersensitive plants. However, the authors did not refer to the structure of polysaccharides and their charge. At the same time, it is known that inducers of virus resistance *de novo,* as is proved by many authors, may preferably be polyanions (Dyakov and Kovalenko, 1983). In addition, Dutch workers have not reported about the inhibitor of DNA-dependent RNA polymerase actinomycin D that is usually used for evidence host-based resistance induced be glycopolymers.

It was convincingly shown only on a model of neutral and sulfated (anionic) yeast polysaccharides that the "classic" induction of plant resistance to viruses *de novo* is inherent only to acid polysaccharides (Kovalenko et al., 1988, 1993). In contrast to that, neutral glycans can be involved in the realization of hypersensitive reactions by protein-carbohydrate interaction (Kovalenko, 1993; Kovalenko et al., 2000). The investigation of Basidiomycete glycans as inducers of plant resistance to viruses gives a unique opportunity to examine the validity of hypotheses on the multilevel character of exogenous glycopolymers influence on protective functions of plants and to study the dependence of its mechanism on the structural peculiarities of molecules.

Glucuronoxylomannan (GXM) of *Tremella mesenterica* as Inducers of Plant Resistance to TMV Infection

On the basis of polyanionic nature of GXMs produced by species of the genus *Tremella,* which is stipulated by the presence of glucuronic acid in side chains of polyglycoside (Yui et al., 1995; Vinogradov et al., 2004), it can be predicted that one of the mechanisms of *T. mesenterica* GXM antiviral activity is the activation of protective reactions of plants, which depends on the functioning of cell genome, particularly the *de novo* induction of viral resistance of plant tissues, similar to sulfated mannans (Kovalenko et al., 1988, 1992, 1993). To examine this hypothesis GXM in various concentrations (1000 and 2500 μg/ml) was introduced sub-epidermally into one half leaf of TMV supersensitive tobacco variety Immunny 580. The opposite halves served as the control were injected with sterile water instead of the preparation. The treated leaves were inoculated with purified virus suspension after 1, 3, 5, and 7 days. In 7 days after inoculation the number and/or size of local lesions on experimental and control leaf halves were counted. The reduction of plant sensitivity to the virus was determined by the ratio of the number of lesions in experimental and control samples, the sensitivity of the latter was considered as 100%. As a result, GXM in studied concentrations induced the development of resistance to TMV already on the first day after introduction of the inducer; the induced resistance remains on the same level for at least 7 days (Fig. 6).

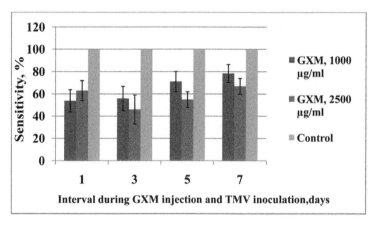

Figure 6. Effect of *T. mesenterica* GXM on the sensitivity of tobacco var. Immunny 580 to TMV infection. Right columns of each group show plants' sensitivity to the virus in control samples as 100%.

The level of the induced resistance was dependent on the concentration of glucan. IVR was reducing faster when glucan was used at a lower concentration.

GXM ability to induce plant resistance to TMV was also investigated in *N. sanderae* plants, in which, unlike *N. tabacum* variety Immunny 580, TMV infection is not localized; however, local necroses in inoculated leaves are formed and IVR develops against repeated virus inoculation (Kovalenko and Kyrychenko, 2004). As evident from the data (Fig. 7), GXM induces resistance to TMV of this species of tobacco. However, the IVR level, in this case, was

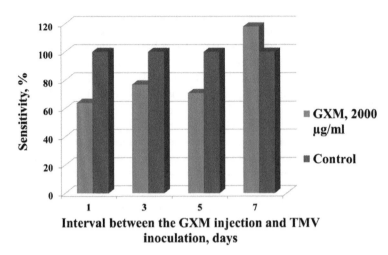

Figure 7. Sensitivity of *N. sanderae* plants to TMV infection in presence of *T. mesenterica* GXM.

lower than that of *N. tabacum*, and the stability was transient. This may indicate the dependence of induced resistance of host genotype, including the N-gene availability (Dyakov and Kovalenko, 1983).

It should be noted that virus-induced local lesions on leaf-halves of tobacco variety Immunny 580 treated with GXM and inoculated with TMV grew faster than controls on the 7th day after treatment with the inducer. The lesion size increased almost two-fold (Fig. 8). However, significant changes in the growth of lesions caused by TMV were found only at the concentrations of

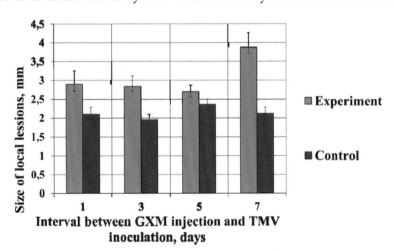

Figure 8. Growth of TMV induced local lesions in tobacco plants (var. Immunny 580) under the influence of *T. mesenterica* GXM (2500 µg/ml). The right column of each pair represents the size of necrotic blotches in controls (mm). Sizes of local lesions on experimental and control leaf-halves in 7 days after inoculation were measured.

2500 µg/ml of GXM. The GXM at the concentrations of 1000 µg/ml did not cause significant differences in the amount of necroses, both in the experimental and control samples.

GXM does not only induce plant tissues-resistance to viral infection, but it also interferes with the mechanism of localization of viral infection. A similar effect on the growth of TMV-induced local necroses in supersensitive tobacco plants is also caused by yeast RNA (Gicherman and Loebenstein, 1968). The authors interpret this phenomenon as the activation of virus-induced hypersensitivity reactions by exogenous inducer.

Thus, the results indicate the ability of GXM to induce plant resistance to TMV infection. Studied glycan causes not only reduction of quantity but also acceleration of the growth of local lesions. That is, GXM can be an inductor capable of activating the protective mechanisms of localization of viral infection, perhaps by activating the plant hypersensitivity mechanism and related IVR. However, the existence of other mechanisms of influence of glycans on the development of TMV infection in plants, namely, the inhibition of infection and reproduction processes of virus in plant cells, and even direct inactivation of virus particles *in vitro*—cannot be completely ignored.

Ganoderma lucidum and *G. applanatum* Glycans as Inducers of Plant Resistance to TMV Infection

The inducer properties of partially purified glycans obtained from the 14-day culture liquid of *G. lucidum* and *G. applanatum* were reported (Kovalenko et al., 2008). A transient increase (up to 3 days) of plant resistance to TMV infection was observed when introducing the preparation derived from the culture liquid of *G. applanatum* (1000 µg/ml) into the intercellular space of tobacco Immunny 580 leaves (Fig. 9A). On the first day, the level of induced resistance was 58%, while on the third day it was 43%. But no resistance in plants was found with virus inoculation on 5th and 7th days after the introduction of the inducer. At higher concentrations (2500 µg/ml) the activity of the preparation was not observed. However, the glycan preparation isolated from *G. lucidum* culture liquid had the ability to induce plant resistance to TMV infection only at high concentrations, i.e., 2500 µg/ml, while at 1000 µg/ml the inducer activity was not observed (Fig. 9B).

Therefore, exopolysaccharides of *G. lucidum* and *G. applanatum* are moderate inducers of virus resistance in tobacco plants. They caused a low level of virus resistance. Moreover, the latter was transient. They were completely unable to cause resistance to TMV infection in *N. sanderae* plants, which may support the assumption that the inhibition of viral infection in supersensitive plants by fungal glycans is due to the activation of protective mechanisms, like the action of neutral and sulfated yeast mannans (Kovalenko, 1993; Kovalenko et al., 1993, 2000).

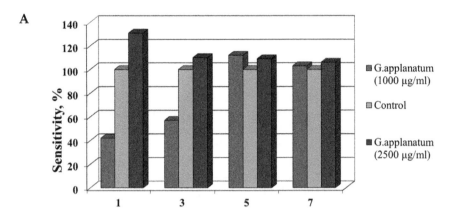

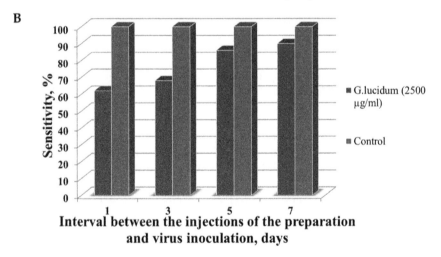

Figure 9. Influence of glycans isolated from culture liquid of *G. applanatum* (A) and *G. lucidum* (B) on sensitivity of tobacco var. Immunny 580 to TMV.

Ganoderma adspersum Glucan as Inducers of Plant Resistance to TMV Infection

Previous studies have shown that the preparations isolated from the mycelium of *G. adspersum* quite actively suppress TMV infection in experiments *in vitro*, particularly the water extracted preparation. Therefore, the inducer properties of this preparation were studied in detail (Kovalenko et al., 2010a,b). Preliminary experiments established that the optimal concentration of the preparation with no toxicity to plants, showing fairly high level of antiviral activity is 500 µg/ml. This glucan concentration induced a higher virus resistance level (65%) when it was introduced into intercellular space of leaves (tobacco leaves -var. Immunny 580) exactly 1 day before inoculation. This high level of resistance remained until day 3–5 and even on the 7th up to 40% (Fig. 10). However, similar treatment of *N. sanderae* plants with *G. adspersum* glucan in the same concentration did not result in the inhibition of virus-induced lesions formed on plant leaves (data not shown). Thus, it is evident either polysaccharide activates protective mechanisms present in healthy plants or to those that are activated by viral infection.

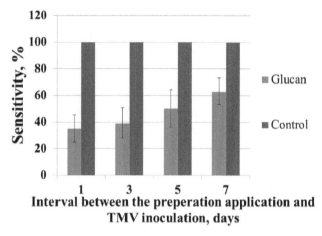

Figure 10. Virus resistance in Immunny 580 tobacco induced by treating the leaves with glucan isolated from a *G. adspersum* mycelium preparation (500 μg/ml).

Thus, as a result of the research, the ability of fungal glycans to interfere with the functioning of plant defense mechanisms acting against viral infections was determined. They can induce plant resistance to viral infection and activate the development of a virus-induced hypersensitivity response. The extent and type of the influence of polysaccharides on the protective mechanisms of plants depends on the chemical structure of polymers. Thus, polyanionic GXM activates the expression of hypersensitivity to viral infection while inducing the resistance of plants to viral infection. Antiviral activity of glucan extracted from species of genus *Ganoderma*, similar to the activity of neutral mannans (Kovalenko, 1993; Kovalenko et al., 2000), which influences the genetically stipulated protective mechanisms in plants and induces resistance in plants, is characterized by an effective LVI mechanism (Kovalenko and Kyrychenko, 2004). Both total polysaccharides of *G. lucidum*, *G. applanatum*, and *G. adspersum* glucan inhibit the formation of virus lesions in supersensitive tobacco plants of var. Immunny 580 and do not influence virus infection in *N. sanderae* plants, which is also evident for the possible impact of these biopolymers on the protective function of plants. However, it is possible that glucans of species of genus *Ganoderma* and *T. mesenterica* GXM can directly influence the reproduction of the virus in the cell or inactivate viral particles.

Higher Basidiomycetes Glycans as Inhibitors of TMV Reproduction

The results elaborated above clearly demonstrate the ability of higher *Basidiomycetes* glycans to influence on antiviral protective mechanisms of plants, stimulating or inducing their action *de novo*. However, we cannot exclude the possibility of direct (inhibitory or stimulatory) influence of glycans on virus reproduction in plants. The possibility of suppression of TMV reproduction by some species of higher Basidiomycetes (*T. mesenterica* GXM and *G. adspersum* glucan) glycans was tested on the model of N-gene-deficient mutant forms of tobacco of var. Immunny 580, which has no protective mechanisms, such as IVR and LVI, and responds to the inoculation with virus by systemic infection (Scherbatenko and Oleschenko, 1994). Isolated leaves or leaf discs were treated using the method mentioned above. Treatments were conducted with the time interval 30 or 60 min before or after inoculation. Leaf discs from opposite halves of tobacco leaves treated by water and inoculated TMV at the same time as experimental ones served as a control. Infectivity of the virus, which replicated in the treated leaf tissues, was examined after 72 h by inoculation of *Datura stramonium* plants with water extracts of experimental plants. The detailed data on these studies is already published (Kovalenko and Polishchuk, 2007, 2009; Polishchuk and Kovalenko, 2008).

GXM of *Tremella mesenterica* as Inhibitor of TMV Reproduction

GXM in concentrations of 1000 µg/ml can inhibit the reproduction of TMV in tobacco cells by systemic viral infection (Fig. 11). Moreover, such ability of glycans was observed when treating the plant tissues both before and after virus infection: there was no significant difference in the inhibition of the reproduction of the virus 60 minutes before or 60 min after treatment of tissues with preparation; in both cases the inhibition degree was 80% over the control.

Antiviral effect of GXM in systemic infection of plants may be caused by the influence on the early stages of TMV reproduction, probably, viral RNA deproteinization. It is also possible that the processes of cellular metabolism which support the viral RNA replication are disrupted in tissues treated by the given polyanion concentration.

Glucan of *Ganoderma adspersum* as Inhibitors of TMV Reproduction

Application of *G. adspersum* glucan for control of TMV infection in sensitive tobacco plants at concentrations of 500 µg/ml for 30 or 60 min before and after virus inoculation caused virus inhibition upto 80 and 97%, respectively (Fig. 12), i.e., this polysaccharide inhibits virus reproduction in prophylactic

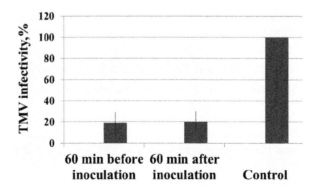

Figure 11. Inhibition of TMV reproduction under the influence of *T. mesenterica* GXM in tissues of sensitive tobacco plants (N-gene-deficient mutant of tobacco var. Immunny 580). Glucan concentration—1000 µg/ml.

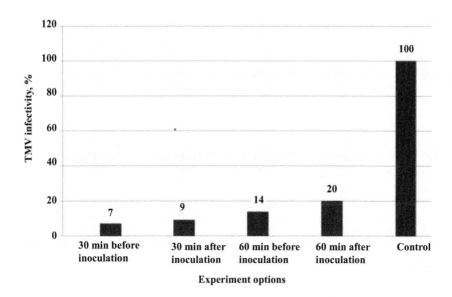

Figure 12. Influence of *G. adspersum* glucan (500 µg/ml) on the TMV reproduction in sensitive tobacco plants (N-gene deficient mutant of tobacco var. Immunny 580).

and therapeutic applications. Significant difference in the activity of the preparation in different ways and terms of application was not observed. Thus, one of the properties of *G. adspersum* glucan is the ability to inhibit TMV reproduction, which may be due to its influence both on the process of infection of sensitive cells and on early virus reproduction.

Thus, the investigated glycans of higher Basidiomycetes, which differ in their chemical structure, have approximately the same effect on TMV reproduction, suppressing it both before and after it infects system of the host tissue. The mechanism of such influence of glycans may be based on disruption of virion penetration into cells and/or early stages of reproduction, particularly viral RNA deproteinization.

Possible Mechanisms of Antiviral Action of Glycans

The investigation of antiviral properties of polysaccharides produced by higher Basidiomycetes has shown that they can suppress the development of TMV infection both when added to the TMV suspension *in vitro* and when introduced into the tissues using different methods and ways for the virus inoculation. A more detailed study of the antiviral activity on different plant-indicators allows distinguishing two of the most active producers that synthesize polysaccharides inducing plant resistance to viral infection and affect virus reproduction in plants with systemic viral infection— *T. mesenterica* and *G. adspersum*. However, the detection of antiviral activity by itself did not explain the mechanisms of their action. The mechanisms of action of polysaccharide may depend on their chemical structure and type of experimental system. Taking this into account, the biological, absorption, and competitive activity of given glycans was further studied to establish the mechanisms by which they can inhibit TMV infection *in vivo* and *in vitro*.

Actinomycin D Influence on Glycan-induced Virus Resistance

One of the approaches to investigate mechanisms of action of antiphytoviral preparations is the application of cellular metabolism inhibitors including antibiotic actinomycin D (AMD). AMD is an inhibitor of DNA-dependent RNA synthesis. Its mechanism of action is based on blocking the activity of cellular RNA polymerase (Dyakov and Kovalenko, 1983; Semal, 1967). However, under certain conditions of application and concentration of the antibiotic, the RNA-dependent synthesis of viral RNA is either not disturbed or only partially disturbed, which was shown particularly on the model TMV–tobacco (Matthews, 1981). Thus, when introducing the antibiotic immediately after TMV infection, viral RNA synthesis was inhibited, but when introducing it 12 hours after infection, the RNA synthesis, by contrast, increased. Treating plants with the inhibitor later than 12 hours after inoculation, the impact of AMD on virus reproduction was not observed. Similar data were obtained on a model of barley infected with brome mosaic virus (Hiruki and Kaesberg, 1965). The property of this selective inhibitor of cellular metabolism helps in studying the mechanisms of action of antiphytoviral preparations, including inducers of virus resistance and to differente those that induce resistance on a genetic level, i.e., involving cell synthesis of mRNA *de novo*, and also to those that do not require synthesis of new RNA and proteins (Dyakov and Kovalenko, 1983; Gianinazzi, 1983). AMD was successfully used for studying the mechanism of inducing activity of polysaccharides, including the ones isolated from yeasts

(Kovalenko, 1987; Kovalenko et al., 1986, 1988, 1993, 1994) and from higher mushrooms (Kovalenko and Polishchuk, 2007; Kovalenko et al., 2009, 2010a,b).

GXM of *Tremella mesenterica* as Inducer of Virus Resistance in Plants

Water solutions of AMD in concentrations of 10 and 20 μg/ml were used to study the GXM-induced virus resistance in plants. The antibiotic at these concentrations did not significantly affect the TMV reproduction, was not phytotoxic, and was used in studying the mechanism of resistance induced by this virus (Gicherman and Loebenstein, 1968; Loebenstein, 1972). AMD was introduced sub-epidermally simultaneously or over a period of time after the GXM introduction. Inoculation of the virus was performed 5 days after introduction of the inducer or AMD.

As a result, it was found that AMD inhibits virus resistance in GXM-induced plants regardless of the method of its introduction (simultaneously or within 2 days after the GXM introduction) (Fig. 13). IVR is completely inhibited when using AMD at the concentration of 20 μg/ml, and partially inhibited at the concentration of 10 μg/ml. Partial IVR inhibition by actinomycin D at the concentration of 20 μg/ml can be caused by some antibiotic toxicity to plant tissues, as evidenced by the reduction of necrosis in controls, where AMD was introduced into the leaf without GXM.

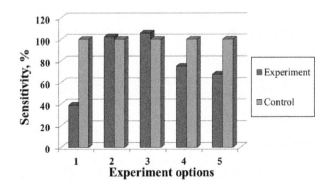

Figure 13. Influence of AMD on the resistance of Immunny 580 tobacco to TMV infection induced by GXM (1000 μg/ml): (1) GXM; (2) mixture of GXM and AMD (10 μg/ml); (3) AMD (10 μg/ml) injected 2 days after GXM; (4) AMD (20 μg/ml) injected 2 days after GXM; (5) AMD (10 μg/ml) injected without GXM.

Thus, the results generally indicate that AMD inhibits plant resistance induced by GXM. Since AMD inhibits RNA synthesis in cellular DNA matrix, it can be concluded that the GXM-induced resistance of plants is based on *de novo* cellular RNA synthesis. The results are consistent with data obtained in the study of sulfated mannan derivatives, which have anionic properties and similar mechanism of antiviral action (Kovalenko et al., 1988, 1993).

Ganoderma adspersum Glucan as Inducer of Virus Resistance of Plants

A concentration of 10 µg/ml of antibiotic was used to study the impact of AMD on glucan-induced resistance. The inoculation of plants was performed on the second day after the introduction of AMD or glucan (500 µg/ml) because during this period the resistance induced by this polysaccharide was highest.

The performed studies have shown that AMD inhibited glucan-induced resistance completely—regardless of the method of application, used—antibiotics alone, or in combination with the inducer (Fig. 14).

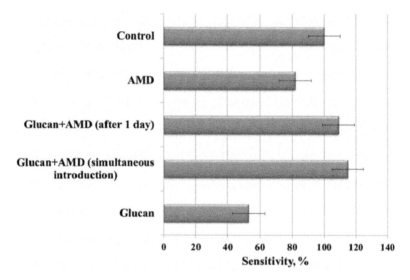

Figure 14. AMD (10 µg/ml) influence on the resistance development in Immunny 580 tobacco plants to TMV caused by *G. adspersum* glucan (500 µg/ml).

Thus, the conducted research gives evidence for the activation of mRNA synthesis, which causes the synthesis of new proteins and the development of plant resistance to TMV infection, as one of the mechanisms of antiviral action of *G. adspersum* glucan, is similar to GXM of *T. mesenterica*.

Thus, GXM of *T. mesenterica* and glucan of *G. adspersum* are classical inducers of *de novo* resistance. The activity of both polysaccharides was similar to previously studied sulfated polysaccharides (Kovalenko et al., 1988, 1993), yeast RNA (Gicherman and Loebenstein, 1968), and other polyanions (Dyakov and Kovalenko, 1983). However, it is possible that plant virus resistance induction in the presence of these polysaccharides may develop, particularly due to alosteric effects occurring based on protein-carbohydrate interactions (Kovalenko, 1993; Kovalenko et al., 2000).

Antiviral Activity of Glycans in Presence of Methyl Glycosides

It is known that substituted monosaccharides, in particular α-methyl-D-glucoside and α-methyl-D-mannoside, are active lectin haptens (Khomutovsky et al., 1986) and can inhibit the development of virus resistance in plants induced by several polysaccharides (Kovalenko et al., 1991, 2000). According to the hypothesis (Kovalenko, 1993), the activity of exogenous glycans as inducers of protective reactions of plants is based on their interaction with cellular receptors responsible for initiation of supersensitive reaction. Similar activity was determined for neutral yeast mannans (Kovalenko et al., 1991) and endogenous glycopolymers (Kovalenko et al., 2000), which, like the exogenous glycans, activate the protective mechanisms of plants based on protein-carbohydrate interactions and do not require new syntheses in cell (Kovalenko, 1993). In accordance with the published data it was assumed that together with the activation of *de novo* plant cell synthesis, fungal glycans have the ability to activate virus resistance mechanisms through specific interactions with cellular lectins that serve as supersensitive reaction triggers. In this case, lectin-specific haptens, particularly methylglycosides, such as α-methyl-D-glucoside and α-methyl-D-mannoside, may inhibit the antiviral activity of investigated glucans, as they inhibit the activity of yeast mannans (Kovalenko et al., 1991).

Influence of α-methyl-D-glycosides on Antiviral Activity of *Ganoderma adspersum* Glucan

Aqueous solutions of substituted monosaccharides (0.1 M) were added to TMV inoculum, containing *G. adspersum* glucan at the concentrations of 1–1000 µg/ml with the aim of investigating their effects on glucan-induced resistance. Control halves were inoculated with TMV without glucan, but with glycosides. Other variant plants were inoculated with the mixture of virus and glucan.

It was determined that α-methyl-D-mannoside significantly reduces the ability of glucan to inhibit TMV infection in plants and α-methyl-D-glucoside insignificantly inhibits it (Fig. 15A). It was revealed during the examination of the antiviral activity of monosaccharides that α-methyl-D-mannoside does not inhibit the TMV infectivity, while α-methyl-D-glucoside inhibits 96% TMV infectivity (data not shown).

Influence of α-methyl-D-glycosides on Antiviral Activity of *Tremella mesenterica* GXM

Only α-methyl-D-mannoside was used in experiments with GXM since α-methyl-D-glucoside has the ability to inhibit TMV, apparently due to toxic effects on host plant cells. α-methyl-D-mannoside solution was added to TMV

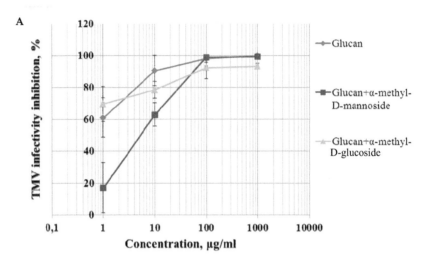

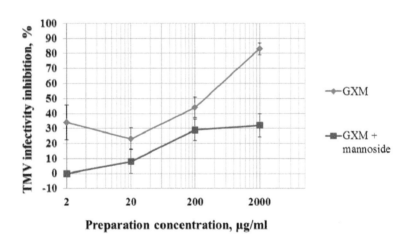

Figure 15. Influence of α-methyl-D-glucoside and α-methyl-D-mannoside on the antiviral activity of *G. adspersum* glucan (A) and *T. mesenterica* GXM (B) in *D. stramonium* plants.

inoculum (4 µg/ml), containing GXM at the concentrations of 2–2000 µg/ml, and *Datura stramonium* plants were inoculated with this mixture.

It was established that α-methyl-D-mannoside, like *G. adspersum* glucan, significantly inhibits GXM antiviral activity (Fig. 15B). The results confirm the assumption that the polysaccharide together with the activation of cellular synthesis *de novo* may induce protective mechanisms of plants based on protein-carbohydrate interactions. This property is most likely caused by the presence of free OH-groups in 1-2-bound mannopyranose at the 3rd and 4th carbon atoms (Kovalenko et al., 1995).

Thus, it was found that α-methyl-D-mannoside inhibits the antiviral activity of *T. mesenterica* GXM and *G. adspersum* glucan. The data confirm the possibility of the interaction of polysaccharides with pectin cell receptors and the activation of defense mechanisms due to protein-carbohydrate interactions. Such properties are particularly pronounced in yeast mannans, which have residues of 1-2-bound mannopyranose with free OH-groups at the 3rd and 4th carbon atoms in their side chain (Kovalenko et al., 1991, 1995, 2000).

Complex Formation of GXM with TMV Virions

Given the peculiarities of the *T. mesenterica* GXM chemical structure as a polyanion (Vinogradov et al., 2004), it can be assumed that this glycan may not only induce the resistance of supersensitive plants to *de novo* TMV infection (Kovalenko and Polishchuk, 2009; Kovalenko et al., 2009) and inhibit the reproduction of the virus in sensitive plants (Kovalenko and Polishchuk, 2007; Polishchuk and Kovalenko, 2008), but also form a complex with coat proteins *in vitro* or *in vivo*.

To test this hypothesis, a series of experiments was conducted (Boltovets et al., 2009). To determine whether the degree of TMV infectivity reduction depends on its incubation time with polysaccharides, TMV suspension (10 µg/ml) was mixed with GXM water solution (500 µg/ml); the mixture was incubated 0–60 seconds *in vitro* and then *D. stramonium* leaf-halves were inoculated. The opposite control halves were inoculated with TMV in the same concentration without polysaccharide. The calculation of the local lesions showed (Fig. 16) that TMV inactivation within the first 15 seconds significantly increased with increase in virus incubation time with GXM. These data may indicate the interaction of TMV virions with GXM *in vitro*, probably due to ionic bonds, resulting in the formation of non-infectious complexes or virion aggregation.

The virus suspension was mixed with a polysaccharide (5000 and 2000 µg/ml) mixture, was incubated for 30 min, then a series of fivefold dilutions was conducted to determine the strength of possible complexes

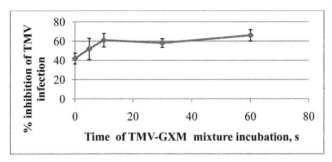

Figure 16. Change in TMV infection in *D. stramonium* plants depending on the time of incubation with GXM.

and also to check whether the TMV virion structure is disrupted by GXM. As in the previous experiment virus infection in diluted solutions was tested on *Datura stramonium* and compared with that of the control group, where GXM was absent. It was established that the dilution of mixture (reduction in the virus and GXM concentration), antiviral activity of polysaccharide was gradually falling.

Obviously, in experiments in aqueous medium, a dynamic equilibrium of reversible physico-chemical reaction between GXM and TMV is set:

$$A + B \leftrightarrow [AB]$$

In the presence of GXM excess, the reaction proceeds towards the formation of the complex. With decreasing GXM concentration, the complex probably falls apart. That is, the formation of the complex *in vitro*. Consequently, polyanion-caused inactivation of the virus is reversible. Perhaps in this case an ion interaction between the carboxyl groups of glucuronic acid residues and NH_2-groups of TMV protein shell takes place. The formation of TMV-GXM complex, due to shielding of the virion surface by glycan, is similar to glycoprotein isolated from *Agrocybe aegerita* mycothallus, which has lectin properties (Sun et al., 2003).

In any case, the process of cell infection by the virus is blocked. Thus, the obtained results generally indicate the possibility of formation of complex between GXM and TMV virions. However, this interaction does not significantly affect the structure and biological activity of virions. This is evidenced by the fact that infectivity of TMV does not change after mixture was diluted (Fig. 17). The method of surface plasmon resonance (SPR) was used to confirm the complexation of GXM with TMV *in vitro*.

The SPR method is based on the interaction of light with a thin metal film (e.g., Au, Ag), which is deposited on the surface of an insulator (glass). Light that passes through a glass prism on the edge of glass and thin metal film can

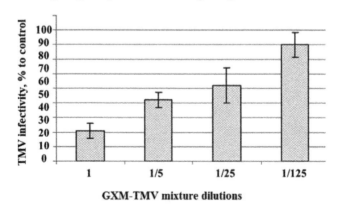

GXM-TMV mixture dilutions

Figure 17. Renewal in TMV infectivity at dilution of GXM + TMV mixture. Incubating during 30 min experimental (TMV + GXM) and control (TMV + water) mixtures by water and tested on the *D. stramonium* leaves.

excite surface polaritons. This effect is observed as a sharp decrease in the intensity of the reflected light with a minimum from a certain angle, called the SPR angle. The SPR angle's minimum shift occurs after the application of molecules on the surface of the sensor element, and, as a result, changes in the thickness of the reaction layer. Thus, measuring the magnitude of this angle, one can get the information on the processes of adsorption and desorption of molecules on the surface of metal films (Beketov et al., 1998). The SPR angle's minimum shift depends on the thickness of the reaction layer as well as on its density, resulting in a change of the refractive index.

With the aim of detecting the complex of virus and polysaccharide, it was assumed that GXM, interacting with TMV virions, blocks antibody binding sites on the capsid protein. Depending on the number of antibodies adsorbed on the virion, the SPR angle changes. With the aim of determining the GXM impact on further interaction of viral particles with antibodies, the virus was incubated with glucan for 30 min, then added to the mixture of antibodies, and again incubated for 30 min. In the control variants of the experiment, water was added instead of the polysaccharide in order to follow the appropriate ratio between the components of the reaction mixture. The incubation time was 30 min during the experiment (Fig. 18).

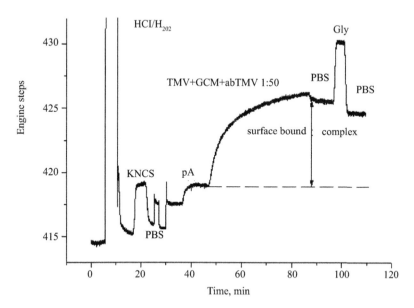

Figure 18. Course of the experiment in the study of TMV-GXM complexes. Solutions that consistently filled cell sensor are labeled with arrows: (1) water, (2) $HCl/H_2O_2/H_2O$, (3) KNCS, (4) PBS, (5) protein A, (6) studied complex of virus and TMV specific serum with or without polysaccharide, (7) glycin buffer pH 2.2.

As a result, it was established that the optimal serum dilution for detecting the interaction between virus and antibodies, with constant TMV concentration of 100 µg/ml, was 1:200. Antigen-antibody complexes formed the most dense structure on the surface with this concentration, and the signal naturally reached its maximum (Fig. 19). When an additional factor, such as anionic GXM, was added to the system, the value of the signal decreased.

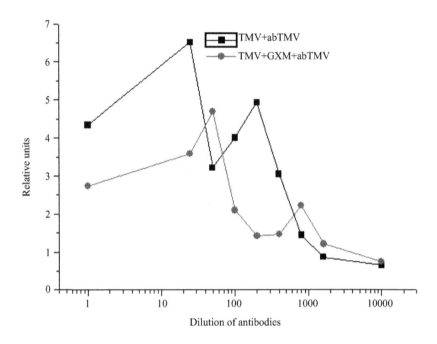

Figure 19. The dependence of the resonance angle on the concentration of antibodies to TMV in mixture with and without GXM. Relative units—the absolute value of the response to the virus/ absolute value of the response to buffer.

Thus, the interaction of polysaccharide with the surface of the virus resulted in the significant reduction of the number of TMV-bound antibodies, which leads to a decrease of the SPR signal (about twice at the maximum) and to a change of the position at this maximum, to 1:800. This is evident that GXM reduces the number of potential antigen binding sites of antibodies, on the one hand, and provides an additional negative charge to virus particle—on the other hand, as a result of the selective interaction with the virus shell protein.

Thus, the study of mechanisms of GXM-TMV interaction *in vitro* by the SPR method confirmed that GXM forms a complex with virions. This effect, obviously, can play a definite role in the antiviral activity of polysaccharides. As a result of GXM adsorption on virions, the interaction of the virus with cells and, consequently, the process of infection, are probably blocked.

The hypothetical mechanism, according to which the decrease of biosensor reaction virus-antibody in the presence of GXM occurs, is shown in figure (Fig. 20).

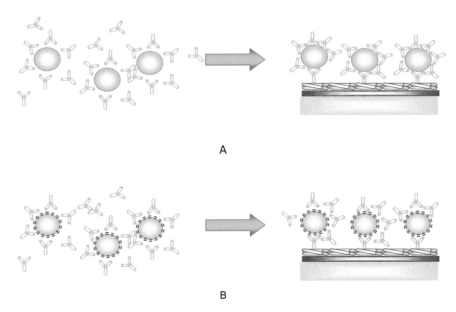

Figure 20. Scheme demonstrating the formation of the complex virus-antibody on the surface of the sensor in the absence (A) and in the presence (B) of GXM in the system. The diagram shows the virus as circles, antibodies—as triple asterisks, glycans—as dotted lines.

To study the possibility of GXM to form the complex with TMV, the method of centrifugation in the sucrose density gradient was used (Matthews, 1981). After centrifugation the virus from opalescent zone was isolated and investigated using an electron microscope. It was determined that in the control, separate TMV virions with coherent structure were observed in the field of view (Fig. 21a). In the experiment their agglomerations and aggregates

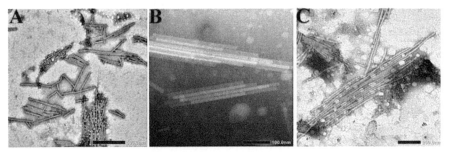

Figure 21. Electron microscope images of TMV obtained after centrifugation in sucrose density gradient: a - virus after centrifugation without GXM, b, c—the virus after centrifugation in the presence of GXM (Size range: (A) 500 nm, (B) 100 nm, (C) 200 nm).

side-to-side and end-to-end can be clearly observed (Fig. 21b,c), which may be caused by the presence of GXM in the virus suspension.

Thus, the experiments determined that GXM of *T. mesenterica* was able to form a complex with TMV *in vitro*. This property of the glucan, generally, can play an important complementary role in its biological activity as an inhibitor of viral infection and inducer of virus resistance in plants.

Conclusions

The ability of the glycans of higher Basidiomycetes mushrooms to inhibit the biological activity of phytoviruses and induce plant resistance against them was demonstrated on the model of tobacco mosaic virus (TMV) for the first time. The antiviral and inducing activities of polysaccharides was studied; the mechanisms by which they can influence the course of TMV infection in sensitive and supersensitive plants were differentiated. It was shown that *Tremella mesenterica, Ganoderma lucidum, G. applanatum,* and *G. adspersum* mushrooms produce substances that can inhibit the infection of TMV in tobacco and *Datura stramonium* plants. Polysaccharide preparations, isolated from culture liquid of *G. lucidum* and *G. applanatum,* do not influence the N-gene's determined protective mechanisms in tobacco plants, while acidic glucuronoxylomannan (GXM) of *T. mesenterica* and *G. adspersum* glucan can activate them. The inhibitor of DNA-dependent RNA polymerase actinomycin D suppresses the resistance induced by *G. adspersum* glucan and *T. mesenterica* GXM, which indicates its participation in the formation of cellular genomes, including *de novo* RNA synthesis on the cell DNA matrix. Antiviral activity of *T. mesenterica* GXM and *G. adspersum* glucan in *Datura stramonium* plants supersensitive to TMV is partially blocked by hapten of mannose-specific lectin -α-methyl-D-mannoside. It may indicate the ability of these glycopolymers to activate protective mechanisms based on conformational protein-carbohydrate interaction of molecules. *T. mesenterica* GXM and *G. adspersum* glucan suppress the early stages of TMV reproduction in sensitive tobacco plants in preventive and therapeutic treatments, indicating the possible influence of glycans on intracellular processes associated with virus reproduction. *T. mesenterica* GXM and *G. adspersum* glucan, in effective concentrations, have no toxic effects on plant tissues and protoplasts, and therefore, eliminates the phytotoxicity of glycans as one of the mechanisms of inhibition of TMV infection. GXM of *T. mesenterica* can form a very unstable complex with TMV virions *in vitro*, which breaks at a dilution of the interacting components. The aggregation of viral particles under the influence of polysaccharides was revealed using a transmission electron microscopy in several areas obtained by centrifugation of the mixture of GXM with TMV virions in sucrose density gradient. The data are generally consistent with the literature on the activity of acid and neutral polysaccharides as inhibitors of virus infection (Kovalenko, 1993; Kovalenko et al., 1988, 1991, 1992, 1993, 2000). These antiphytoviral properties of the investigated glycans may be lead to further study of these substances as means for plant protection and recovery from viral infections. Furthermore, their study as inhibitors of viral infections can be extended to other systems, including human and animal cells infected with viruses.

Acknowledgements

Authors would like to express gratitude to their colleagues: V.M. Vasilev Ph.D., Ph.D. students O.M. Polishchuk, Dipl. engineer V.O. Isakova for active participation in conduct of experiments. Thanks are due to O.Y. Karpenko, a Ph. D. student for translating the text into English and help in the organization of the paper.

References

Andrianova, D.A., Meychyk, N.R. and Nikolaeva, Y.I. 2010. Studying the composition of functional groups of cell wall of filamentous fungi. Immunopathology, Allergology, Infectology, 1: 14–15.

Beketov, G.V., Shirshov, Yu.M., Shynkarenko, O.V. and Chegel, V.I. 1998. Surface plasmon resonance spectroscopy: prospects of superstate refractive index variation for separate extraction of molecular layer parameters. Sensors and Actuators, 48: 432–438.

Ben-Zion Zaidman, M., Yassin, M., Mahajna, J.A. and Wasser, S.P. 2005. Medicinal mushroom modulators of molecular targets as cancer therapeutics. Applied Microbiology and Biotechnology, 67: 453–468.

Bobyr, A.D., Sadovsky, Yu.P. and Kovalenko, A.G. 1974. Induction of local and sysem resistance of tobacco to tobacco mosaoc virus with the help of virus and RNA of the virus affecting *Penicillium funiculosum*. Biol. Sci., Moskov, 10(130): 99–105.

Bohn, J.B. and BeMiller, J.N. 1995. (1→3)-β-D-Glucans as biological response modifiers: a review of structure – functional activity relationships. Carbohydrate Polymers, 28: 3–14.

Boltovets, P.M., Polishchuk, O.M., Kovalenko, O.G. and Snopok, B.A. 2009. Investigation of multi component virus contain – ING mixtures by the surface plasmon resonance, Program and Abstracts of the 2nd International Conference «Physical Methods in Ecology, Biology and Medicine», Vorokhta – Lviv, Ukraine, p. 54.

Danylyak, M.I. and Reshetnikov, S.V. 1996. Medicinal fungi. Medical application and biotechnology problems, Kiev, Ukraine, 64 p.

Dyakov, Ju.T. and Kovalenko, A.G. 1983. Mechanisms of plant resistance to viruses and fungi in plants. The Results of Science and Engineering. Series of Plant Protection Moscow, Russia, 171 p.

George, C.X., Gupta, B.M. and Paul Khurana, S.M. 1981. Antiviral activity in plants of a mycoviral double-stranded RNA from *Trichothecium roseum*. Acta Virologica, 25(6): 408–414.

Gianinazzi, S. 1983. Genetic and molecular aspects of resistance induced by infections or chemicals. *In*: Kosuqe, T. and Nester, E.W. (eds.). Plant – Microbe Interaction. Molecular and Genetic Perspectives, New York, Makmillan, 1: 321–342.

Gicherman, G. and Loebenstein, G. 1968. Competitive inhibition by foreign nucleic acids and induced interference by yeast-RNA with the infection of tobacco mosaic virus. Phytopathology, 58: 405–409.

Gorin, P.A. and Barreto-Berger, E. 1983. The chemistry of polysaccharides of fungi and lichens. The Polysaccharides, 2: 365–409.

Gupta, B.M., Chandra, K., Verma, H.N. and Verma, C.S. 1974. Induction of antiviral resistance in *Nicotiana glutinosa* plants by treatment with *Trichothecium* polysaccharide and its reversal by actinomycin D. Journal of General Virology, 24(2): 211–213.

Harris, B. 2003. Growing wild mushrooms a complete guide to cultivating edible and hallucinogenic mushrooms, Oakland, Roning Publishing, 96 p.

Hiruki, C. and Kaesberg, P. 1965. Double-stranded replicative form of viral RNA from barley plants infected with brome grass mosaic virus. Phytopathology, 55: 1061–1063.

Hung, W.-T., Wang, S.-H., Chen, Ch.-H. and Yang, Wen-Bin. 2008. Structure determination of β-glucans from *Ganoderma lucidum* with matrix-assisted laser desorption/ionization (MALDI) mass spectrometry. Molecules, 13: 1538–1550.

Keil, B. (ed.). 1966. Laboratory methods in organical chemistry. Moscow (Russia), Mir, 512 p.

Khomutovsky, O.A., Lutsyk, M.D. and Perederei, O.F. 1986. Electronic hystochemie of cell membrane receptors, Kiev, Naukova Dumka, Ukraine, 167.

Kochetkov, N.K., Bochkov, N.F. and Dmitriev, B.A. 1967. Chemistry of Carbohydrates, Moskow, Russia, 546.

Kovalenko, A.G. 1987. Antiphytovirale Eigehscnaften microbieler Polysaccharide – ein Überblick. Zentralblatt für Mikrobiologie, 142: 301–310.

Kovalenko, A.G. 1993. Protein-carbohydrate interaction in the realization of plant resistance to viruses. Mikrobiologichny zhurnal, 55(6): 74–91.

Kovalenko, O.G. and Kyrychenko, A.M. 2004. Localization of TMV-infection and induced resistance in *Nicotiana sanderae* Hort., *Datura stramoniun* L. and *D. metel* L. plants. Mikrobiologichny zhurnal, 66(4): 43–47.

Kovalenko, A.G. and Votzelko, S.K. 1977. Molecular weight and some structural features of yeast polysaccharides possessing antiviral properties. Applied Biochemistry and Microbiology, 13(1): 46–54.

Kovalenko, A.G., Barkalova, A.A. and Bobyr, A.D. 1986. Effect of actinomycin D on the antiviral activity of yeast mannan preparations in plants hypersensitive to TMV. Mikrobiologichny zhurnal, 48(4): 58–62.

Kovalenko, A.G., Bobyr, A.D., Votzelko, S.K., Menzel, G. and Barkalova, A.A. 1977. Untersuchungen über gegen pflanzenviren wirksame Hemmstofe aus Hefepilzen. Phytopathologische Zeitschrift, 88(3): 322–340.

Kovalenko, A.G., Grabina, T.D., Bobyr, A.D., Korbelainen, E.S., Vitovskaya, G.A. and Elinov, N.P. 1988. Mannansulfates as inducers of plant resistance to viral infection. Voprosy virusologii, 33: 732–737.

Kovalenko, A.G., Grabina, T.D., Kolesnik, L.V., Didenko, L.F., Oleschenko, L.T., Olevinskaya, Z.M. and Telegeeva, T.A. 1993. Virus resistance induced with mannan sulfates in hypersensitive host plants. Journal Phytopathology, 137(2): 133–147.

Kovalenko, A.G., Korobko, O.P., Korbelainen, E.S., Barkalova, A.A., Telegeeva, T.A. and Papp, V.T. 1992. Influence of *Candida maltose* mannan and its sulfated derivatives on the sensitivity of plants to viral and bacterial infections. Mikrobiologichny zhurnal, 54(2): 63–69.

Kovalenko, A.G., Kovalenko, E.A. and Grabina, T.D. 1991. Lectine-binding and antiviral activity of the yeast mannans in hypersensitive plants. Mikrobiologichny zhurnal, 53(2): 83–89.

Kovalenko, O.G. and Polishchuk, E.M. 2007. Antiviral activity glucuronoxylomannan *Tremella mesenterica* Ritz. Er. Modern Problems of Microbiology and Biotechnology. Book of Abstracts, Odesa, Astroprint, 186.

Kovalenko, O.G. and Polishchuk, O.M. 2009. Influence of *Tremella mesenterica* Ritz.: Fr. (Basidiomycota) glucuronoxylomannan on the plant resistance to tobacco mosaic virus. Mikrobiologichny zhurnal, 71(1): 50–56.

Kovalenko, O.G., Polishchuk, O.M. and Wasser, S.P. 2009. Virus resistance induced by glucuronoxylomannan isolated from submerged cultivated yeast-like cell biomass of medicinal yellow brain mushroom *Tremella mesenterica* Ritz.: Fr. (Heterobasidiomycetes) in the hypersensitive host plants. International Journal of Medicinal Mushrooms, 11(2): 199–205.

Kovalenko, O.G., Polishchuk, O.M., Krupodorova, T.A., Bisko, N.A. and Buchalo, A.S. 2008. Screening of metabolites produced by strains of *Ganoderma lucidum* (Curt.: Fr.) P. Karst and *Ganoderma applanatum* (Pirs.: Waller) Pat for their activity against tobacco mosaic virus// Visnyk Kiyvskogo Nazionalnogo Univresitrty, 51: 32–34.

Kovalenko, O.G., Polishchuk, O.M. and Wasser, S.P. 2010a. Glycans of higher Basidiomycetes mushroom *Ganoderma adspersum* (Schulzer) Donk: isolation and antiphytoviral activity. Biotechnology, 3(5): 83–91.

Kovalenko, O.G., Polishchuk, O.M., Vasyliev, V.M. and Wasser, S.P. 2010b. Glycans of *Ganoderma adspersum* (Schulzer) Donk: production, physico-chemical characteristics, antiviral activity. Abstracts of 6th International Conference "Bioresources and viruses", Kiev, Ukraine, 156–157.

Kovalenko, A.G., Scherbatenko, I.S. and Oleschenko, L.T. 1999. Virus resistance expression in isolated protoplasts of tobacco and the influence on it exogenous polysaccharides. Cytology and Genetics, 33(4): 16–24.

Kovalenko, O.G., Shashkov, O.S., Vasyliev, V.M. and Telegeeva, T.A. 1995. Structural features and biological activity of mannans of *Candida* spec. Biopolymers and cell, 11(3-4): 77–84.

Kovalenko, A.G., Telegeeva, T.A., Shtakun, A.M., Pgorila, Z.O. and Telegeeva, T.A. 2000. Influence of some mono- and polysaccharides on localization of viral infection and induced virus resistance in plants. Biopolymers and cell, 16(4): 53–59.

Lindequist, U., Niedermeyer, T.H.J. and Julich, W. 2005. The pharmacological potential of mushrooms. eCAM, 2: 285–299.

Litchfield, J.H. 1967. Submerged culture of mushroom mycelium. Microbial technology. Peppler, H.J. (ed.). Reinhold Publishing Corp., New York, pp. 107–144.

Loerenstein, G. 1972. Localization and induced resistance in virus-infected plants. Annual Review of Phytopathology, 10: 177–206.

Maeda, Y.Y., Watanabe, S.T., Chihara, C. and Rokutanda, M. 1988. Denaturation and renaturation of a β-1,6; 1,3-glucan, lentinan associated with expression of T-cell-mediated responses. Cancer Research, 48: 671–675.

Marchessalt, R.H., Deslandes, Y., Ogawa, K. and Sundararajan, P.R. 1977. X-ray diffraction data for β-D-glucan. Canadian Journal Chemistry, 55: 300–303.

Matthews, A.G. 1981. Plant Virology, 2-nd edit., Acad. Press, New York, London.

Mizuno, T. 1999. The extraction and development of antitumor-active polysaccharides from medicinal mushrooms in Japan (Review). International Journal of Medicinal Mushrooms, 1: 9–29.

Mizuno, T., Saito, H., Nishitoba, T. and Kawagishi, H. 1995. Antitumor-active substances from mushrooms. Food Reviews International, 11: 23–61.

Modderman, P.M., Schot, C.P., Klis, F.M. and Wieringa-Brants, P.M. 1985. Acquired resistance in hypersensitive tobacco against tobacco mosaic virus, induced by plant cell wall components. Phytopathologysche Zeitschrift, 113: 165–170.

Ooi, V.E.C. 2000. Medicinally important fungi. Science and Cultivation of Edible Fungi. Balkema Publishers, Rotterdam, 1: 41–51.

Ovodov, Ju.S. 1997. Polysaccharides of fungi, moss and lichen, structure and physiological activity. Papers of Komi Research Center of URO RAS, No. 156, 12 p.

Polishchuk, O.M. and Kovalenko, O.G. 2008. Effect of *Tremella mesenterica* glucuronoxylomannan on tobacco mosaic virus reproduction in isolated tobacco leaves. International scientific conference. Biotechnology in Kazakhstan: problems and prospects of innovative development, Alma-Aty, pp. 17–18.

Polishchuk, O.M. and Kovalenko, O.G. 2009. Biological activity of glycopolymers of basidiomycetous fungi. Biopolymers and cell, 25(3): 181–193.

Scherbatenko, I.S. and Oleschenko, L.T. 1994. Somaclonal variability of hypersensitivity reaction to TMV in tobacco protoclones Mikrobiologichny zhurnal, 56(6): 35–40.

Semal, J. 1967. Effects of actinomycin D in plant virology. Phytopathology, 59: 55–71.

Smirnov, D.A., Puchkova, T.A., Sherba, V.V. and Bisko, N.A. 2006. Physico-chemical properties of *Ganoderma lucidum* polysaccharides. Proceedings of international conference modern state and prospects of development of microbiology and biotechnology. Minsk, pp. 177–179.

Stein, A. and Loebenstein, G. 1970. Induction of resistance to tobacco mosaic virus by poly-I-poly-C in plants. Nature, 226, N 5243. 363–364.

Sun, H., Zhao, C.G., Tong, X. and Qi, Y.P. 2003. A lectin with mycelia differentiation and antiphytovirus activities from edible mushroom *Agrocibe aegerita*. Biochemistry and Molecular Biology, 36(2): 214–222.

Varbanets, L.D., Zdorovenko, G.M. and Knirel, Ju.A. 2010. Investigation of endotoxins, Naukova dumka, Kiev, Ukraine, 137 p.

Vinogradov, B.O., Petersen, J.O., Duus, J.E. and Wasser, S.P. 2004. The structure of the glucuronoxylomannan produced by culinary-medicinal yellow brain mushroom (*Tremella mesenterica* Ritz.: Fr., Heterobasidiomycetes) grown as one cell biomass in submerged culture. Carbohydrate Research, 339: 1483–1489.

Wasser, S.P. 2002. Medicinal mushrooms as a source of antitumor and immunomodulating polysaccharides. Applied Microbiology and Biotechnology, 60: 258–274.

Wasser, S.P. 2010. Medicinal mushroom science: history, current status, future trends, and unsolved problems. International Journal of Medicinal Mushrooms, 12: 1–16.

Wasser, S.P. and Weis, A.L. 1999. Medicinal properties of substances occurring in higher Basidiomycetes mushrooms: current perspectives. International Journal of Medicinal Mushrooms, 1: 31–62.

Wasser, S.P. and Didukh, M.Y. 2005. Mushroom polysaccharides in human healthcare. Biodiversity of fungi. Their role in human life. S.K. Deshmukh, M.K. Rai. (eds.). Science Publishers, Inc. Enfield USA, pp. 289–328.

Wieringa-Brants, D.H. and Decker, W.C. 1987. Induced resistance in hypersensitive tobacco against tobacco mosaic virus by injection of intracellular fluid from tobacco plants with systemic acquired resistance. Phytopathology, 118(2): 165–170.

Yanaki, T., Ito, W. and Tabata, K. 1983. Correlation between the antitumor activity of a polysaccharide schizophyllan and its triple-helical conformation in dilute aqueous solution. Biophysical Chemistry, 1: 337–342.

Yui, T., Ogawa, K., Kakuta, M. and Misaki, A. 1995. Chain conformation of a glucuronoxylomannan isolated from fruit body of *Tremella fuciforms* Berk. Journal of Carbohydrate Chemistry, 14: 255–263.

CHAPTER 10

Fungi: Myconanofactory, Mycoremediation and Medicine

Mahendra Rai,[1,2,]* *Avinash Ingle,*[1] *Swapnil Gaikwad,*[1]
Indarchand Gupta,[1] *Alka Yadav,*[1] *Aniket Gade*[1] and
Nelson Duran[2,3]

ABSTRACT

Fungi have been playing very significant roles in nature by way of decomposing organic matter, cycling the nutrients and exchanging them. They also have long been used as a direct source of food, such as mushrooms and truffles, as a leavening agent for bread, and in the fermentation of various food products. Besides these, fungi are also known to play role in remediation of toxic materials from the environment (mycoremediation). Recently, fungi have been enormously used for the synthesis of nanoparticles (myconanotechnology) which have their applications in different fields. Nanomaterials have many applications in different fields like electronics, agriculture, pharmaceuticals and medicine, etc. Nanoparticles are used as antimicrobials in the field of medicine. How fungi remove the heavy metals and other toxic compounds from the environment by virtue of their ability to make nanoparticles has also been dealt with. This area of research is now gaining importance and

[1] Department of Biotechnology, S.G.B. Amravati University, Amravati 444 602, Maharashtra, India.
[2] Institute of Chemistry, Biological Chemistry Laboratory, Universidade Estadual de Campinas, Campinas, SP, Brazil.
[3] Center of Natural and Human Sciences, Universidade Federal do ABC, Santo André, SP, Brazil.
* Corresponding author: mkrai123@rediffmail.com; pmkrai@hotmail.com

momentum. In this chapter the role of fungi in the field of nanotechnology has been discussed and the application potentials of this new and emerging branch of science have also been emphasized.

Introduction

Contamination of environment is an increasingly major issue now-a-days, of which heavy metal contamination is very important due to their high reactivity and are harmful effects to both humans and the environment. However, microorganisms are known to employ a number of substrates as energy source and play a major role in the process of natural degradation by breaking down these waste materials to their less toxic forms (Karigar and Rao, 2011). The interactions between metals and micro-organisms have been exploited for diverse biotechnological applications in the fields of bioremediation, bioleaching, biomineralization and biocorrosion (Narayanan and Sakthivel, 2010). Among the different microorganisms used for bioremediation, fungi are taking the centre stage due to their tolerance and metal bioaccumulation ability (Thakkar et al., 2010). Thus, mycoremediation is defined as degradation or sequester management of contaminants in the environment.

Rai et al. (2009) for the first time coined the term 'myconanotechnology' and drew the attention of researchers to the novel field of biosynthesis of nanomaterials by fungi. This new field of nanoscience is at the interference of nanotechnology and mycology, comprising an interesting new applied science with considerable potential owing to the wide range of diversity of fungi (Gade et al., 2010). Initially, Ingle et al. (2008) proposed the term 'mycosynthesis' which denotes to the synthesis of nanomaterials using fungi. Till date, many fungi have been successfully used for the synthesis of different metal nanoparticles and it was proved that fungi are the better systems for the biological synthesis of nanoparticles among all other biological agents like bacteria, algae, actinomycetes, plants, cyanobacteria, etc., due to their advantages over other biological systems. Since a large number of fungi have already proved their potential or superiority for the biological synthesis, these are considered to be factories for the synthesis of nanomaterials and hence fungi can be termed as 'myconanofactories'.

Microorganisms have developed defence system against toxicity of heavy metals by pumping out toxic metals or reducing toxic metals to ions with the help of enzymes (Sanghi and Verma, 2009). Thus, there might be reducing agents produced by the micro-organisms to reduce ions to metallic particles (Zhang, 2011). To accomplish this process the metal ions are trapped by the cells probably by two ways, firstly by the electrostatic interaction and/or secretion of substances that adhere to the metal ions. In the electrostatic interaction positively charged metal ions and negatively charged microbial cells interact electrostatically (Rai et al., 2010). After this, the metal ions are reduced by the electrons in the presence of certain enzymes like NADH-dependent reductase and oxido-reductase generated on the cell surface.

Generally, in microbes, detoxification of toxic metals is mediated through enzymes—oxidoreductase. Most fungi remove chlorinated phenolic compounds from the environment by the action of extracellular oxidoreductase enzymes, which are released from the mycelium. These enzymes also form a key factor for nanoparticle synthesis (Karigar and Rao, 2011). Oxygenases catalyze is the oxidation of reduced substrates from alkanes to steroids to fatty acids by moving oxygen from molecular oxygen (O_2) and utilizing NADH/NADPH as co-substrate (Karigar and Rao, 2011).

The process of bioremediation is very slow and only a few bacteria and fungi have proven to be useful for bioremediation (Thakkar et al., 2010). Fungal diversity and organic amendment in soil are some of the factors that contribute to making bioremediation effective (Tortella et al., 2005; Briceño et al., 2007).

Compared to the bacterial species, fungi are considered better candidates for bioremediation as they secrete large amounts of enzymes and are also simpler to handle/culture or grow under laboratory conditions (Jain et al., 2011). Fungi show high tolerance towards metals, wall-binding capacity and intracellular metal uptake (Hemanth et al., 2010; Jain et al., 2011). They also possess the capacity to reduce metal ions, both by intracellular and extracellular methods (Rai et al., 2009; Dhillon et al., 2012). A number of filamentous fungi has shown accumulation of heavy metals and its reduction to metal ions in the form of nanoparticles (Narayanan and Sakthivel, 2010; Zhang, 2011). Fungi such as *Verticillium* sp. (Mukherjee et al., 2001), *Fusarium oxysporum* (Duran et al., 2005; Kumar et al., 2007a,b; Namasivayam et al., 2011), *Phoma* sp., *Aspergillus flavus* (Vigneshwaran et al., 2007; Jain et al., 2011), *Coriolus versicolor* (Sanghi and Verma, 2009), *Pleurotus sajor-caju* (Nithya and Raghunathan, 2009), have been extensively used to reduce the metal ions to metal nanoparticles.

Myconanofactory and Mycoremediation

Microorganisms including fungi, reduce metal toxicity of the environment and consequently form nanoparticles and, therefore, are supposed to be nanofactories (Sanghi and Verma, 2010).

The word 'mycoremediation' was coined by Paul Stamets to denote bioremediation using fungi to degrade or sequester contaminants in the environment (Stamets, 1999). Microorganisms interact with metals and minerals in natural and synthetic environments and alter their physical and chemical properties (Gadd, 2010). These interactions have been well documented and the ability of microorganisms to extract and/or accumulate metals has already been employed in biotechnological processes such as bioleaching and bioremediation (Navazi et al., 2010). Microorganisms such as certain bacteria, yeast and fungi have been shown to play an important role in the remediation of toxic metals through reduction of metal ions (Vahabi et al., 2011). These environment-friendly microorganisms reduce the toxicity of metals in course of synthesis of nanoparticles and hence may be considered as benign nanofactories (Sanghi and Verma, 2010). However, a number of bacteria, like the *Lactobacillus* strains found in buttermilk, have been used in the remediation of toxic metals to nanoparticles for example, synthesis of gold, silver and gold-silver alloy crystals (Nair and Pradeep, 2002). *Desulfovibrio desulfuricans* and *Escherichia coli* were harnessed to recover precious metals by reducing Pd(II) to Pd(0) (Yong et al., 2009). But, fungi have recently been utilized in the remediation of toxic metal ions to nanoparticles.

Mukherjee et al. (2001) exposed the fungus, *Verticillium* sp. (AAT-TS-4), to aqueous Tetrachloroaurate(1-) ($AuCl_4$) ions which resulted in the reduction of the metal ions and formation of gold nanoparticles of around 20 nm diameter. The gold nanoparticles were formed on both the surfaces and within the fungal cells (on the cytoplasmic membrane) with negligible reduction of the metal ions in solution. Freeze-dried fungal biomass was also exploited for the synthesis of silver nanoparticles (Chen et al., 2003). Freeze-dried mycelium of *Phoma* sp. 3.2883 was treated with silver nitrate ($AgNO_3$) solution for 50 hours. It was found that up to 13.4 mg of silver nanoparticles were produced per gram of dry mycelium and TEM micrographs showed the presence of a large number of silver nanoparticles of around 70 nm within the fungal mycelium. The fungus *Aspergillus flavus*, when treated with aqueous silver ions, accumulation of silver nanoparticles on its cell wall was found (Vigneshwaran et al., 2007). The TEM micrographs showed the synthesis of monodisperse silver nanoparticles. The nanoparticles were protein stabilized which were assumed to be fungal proteins.

Nanoparticles of Platinum group of metals are highly significant in the industrial production of fuel cells. Riddin et al. (2006) harnessed *Fusarium oxysporum* f. sp. *lycopersici* for the intracellular synthesis of platinum nanoparticles. The formation of the nanoparticles was evident by a change in the colour of the fungal biomass from yellow to dark brown. The particles produced were of various shapes, ranging from spherical, hexagonal and pentagonal, square to rectangular, etc.

Heavy metals are viable options of recycling, and fungi help in the recovery of valuable metal nanoparticles from electronic waste of heavy metals. *Fusarium oxysporum* is used for synthesis of copper nanoparticles from a dumpsite of e-waste along with *Pseudomonas* sp. and *Lantana camara*. In shake flask studies, *Fusarium oxysporum* showed bioremediation of copper with a distinct colour change and the nanoparticles formed were in the range of 93–115 nm. Thus, mycoremediation of metal ions to nanoparticles is a vital part of natural biosphere processes and is beneficial for human society (Majumder, 2012). Therefore, a multidisciplinary approach is required to increase the use of this important area of microbiology and to exploit it for different biotechnological applications.

Silver Nanoparticles in Medicine: As Antimicrobials

Nowadays, the field of nanomaterials is one of the areas of intense research in materials science. Many applications of nanoparticles and nanomaterials are evolving rapidly. Nanocrystalline particles have tremendous applications in the field of high sensitivity biomolecular detection and diagnostics, antimicrobials and therapeutics, catalysis and microelectronics (Jain et al., 2009). The nanoscale materials are more beneficial than the original materials because of their high surface-area-to-volume ratio, which results in specific physiochemical characteristics and increased reactivity. Large scale synthesis of silver nanoparticles attracts researchers nowadays because these are being used in many consumer products—from medical devices to sport socks and washing machines, to prevent microbial growth (Jo et al., 2009). The use of nano-sized particles as antimicrobial agents has become common as technological advances make their production quite economical. It is well known that nanoparticles such as that of silver, copper, zinc (Horiguchi et al., 1980), iron (Lee et al., 2008), iron oxide (Tran et al., 2010), titanium, magnesium, gold, alginate and so on have strong antibacterial capabilities (Gu et al., 2003; Ahmad et al., 2005). Among these, silver nanoparticles have proved to be the most effective antimicrobial agent against bacteria (Gade et al., 2008; Ingle et al., 2008), viruses and other eukaryotic microorganisms (Gong et al., 2007).

The antimicrobial properties of silver in the form of metallic silver, silver nitrate, silver sulfadiazine have been exploited since ancient times for the treatment of burns, wounds and several bacterial infections (Rai et al., 2009). Compared to the bulk metal, metal nanoparticles, like that of silver, have large surface area, providing well contact with microorganism and show competent antimicrobial property. Feng et al. (2000) observed the effect of silver nanoparticles on *E. coli* and *S. aureus* and found that the free state of DNA changed to a condensed form in the center of the electron-light region in the cells; many electron-dense granules appeared surrounding the cell wall or electron-light region. And, X-ray microanalysis also showed the existence of silver in electron-dense granules, cytoplasm, and DNA molecules. These results suggest that due to denaturation effects of silver ions, DNA molecules become condensed and lose their replication abilities and silver ions interact with thiol groups in protein, which induce the inactivation of the bacterial proteins.

Preferably, the nanoparticles attack the respiratory chain and the cell division causing the cell to die. The nanoparticles release silver ions in the bacterial cells, which enhance the bactericidal activity (Sondi and Salopek-Sondi, 2004; Morones et al., 2005; Song et al., 2006).

According to Morones et al. (2005), Scanning Electron Microscopy reveals the presence of silver nanoparticles within the cell membrane and inside the bacteria, angled annular dark field images showed that the smaller nanoparticles have significant antibacterial activity. An important antibacterial activity of textile impregnated with biogenic silver nanoparticles has also been reported (Marcato et al., 2012a). Silver nanoparticles also show efficient antifungal activity against the fungi that are resistant to antibiotic similar to bacteria. According to Kim et al. (2008) *Trichophyton mentagrophytes, Trichophyton beigelii* and *Candida albicans* showed conclusive sensitivity to silver nanoparticles as compared to commercially available antifungal agents' amphotericin B and fluconazole. Also, synergistic effect of fluconazole and silver nanoparticles against *Phoma glomerata, Phoma herbarum, Fusarium semitectum, Trichoderma* sp. and *Candida albicans* has been studied by Gajbhiye et al. (2009). Biogenic silver nanoparticles and its antifungal activity as a new topical transungual drug delivery against *Trichophyton rubrum* was recently published (Marcato et al., 2012b).

Antiviral activity of silver nanoparticles was described by Elechiguerra et al. (2005) for the first time. They found that interaction of silver nanoparticles with HIV-1 is size dependent because only nanoparticles within the range of 1–10 nm were able to bind to the virus. They also discovered that physiochemical properties of nanoparticles may depend on the nanoparticle interactions with capping agent molecules, that's why they verified silver nanoparticles with three different surface agents namely, foamy carbon, poly (N-vinyl-2-pyrrolidone) (PVP), and bovine serum albumin (BSA). The antiviral (HIV-1 BaL) activity of silver nanoparticles (10 nm in size) towards post-infected Hut/CCR5 cells was investigated by Sun et al. (2005). Lara et al. (2010) reported that silver nanoparticles exert anti-HIV activity, at an early stage of viral replication binding. Silver nanoparticle binds to gp120 (Glycoprotein 120) to prevent CD-4 dependent virion binding, fusion and infectivity. Moreover, silver nanoparticles also inhibit post entry stage of the HIV-1 life cycle.

Other Nanoparticles as Antimicrobials

Metal oxide particles are being used for research in health related applications (Durán and Seabra, 2012). Inorganic nanocrystalline metal oxides are more significant and suitable for biological applications because they have particularly high surface area. Ionic metal oxides are interesting because of their wide variety of physical and chemical properties and antimicrobial activity. Inorganic nanoparticles show superior durability, less toxicity, heat resistance and greater selectivity (Berry and Curtis, 2003; Gupta and Gupta, 2005). Bactericidal effect of iron oxide nanoparticles against *S. aureus* was evaluated by Tran et al. (2010). Park et al. (2011) investigated the antimicrobial activity of peptide antibiotic polymyxin B (PMB). Their findings were that Au nanoparticles and CdTe quantum dots conjugated against *E. coli* and *S. aureus* and that the *E. coli* is more sensitive than *S. aureus*. Generally, Gram negative and Gram positive bacteria as well as fungi are present in the polluted water. Bacteria like *Pseudomonas fluorescens, B. subtilis* and a fungus, *Aspergillus versicolor,* can be inactivated using iron nanoparticles (Diao and Yao, 2009).

Babushkina et al. (2010) reported the concentration and time dependent antibacterial action of iron nanoparticles against *S. aureus* strains. They verified that 0.1 mg/ml and 1 mg/ml of iron nanoparticles decreased the microbial population from 3 to 34%. Nanoparticles of ZnO were found to be better in controlling the growth of various microorganisms and thus, can be used externally to control the spread of bacterial infections more efficaciously than the ZnO powder that has been in use for a long time as an active ingredient for dermatological applications in creams, lotions and ointments (Sawai, 2003). Also, the ZnO and Al_2O_3 nanoparticles showed significant growth inhibition of Gram positive and Gram negative bacteria (Jones et al., 2008). The mechanism of microbiocidal activity of TiO_2 nanoparticles has been studied that indicated that TiO_2 nanoparticles cause oxidative damage onto the cell wall when that gets in contact with the cell, photocatalytic reaction takes place and cell permeability increases, and photooxidation of intracellular components is also caused (Galvez et al., 2007). Desai and Kowshik (2009) also checked the susceptibility of organisms to TiO_2 nanoparticles in which they found *S. aureus* to be more susceptible than *Klebsiella pneumoniae, Pseudomonas aeruginosa* and *E. coli*. Antibacterial study of magnesium oxide nanoparticles against food-borne pathogens, viz., *E. coli* and *Salmonella stanley* was evaluated by Zhonglin and Yiping (2012). They concluded that magnesium oxide nanoparticles could be used directly in foods or incorporated in food packaging materials to improve microbiological food safety. Gold nanoparticles, do not act as antimicrobial agent. However, antibiotic coated with gold nanoparticles showed more hindrance activity against *Salmonella typhimurium* (Zawrah and Abd El-Moez, 2011). It means gold nanoparticles could be used as enhancer of commercial available antibiotics. Copper nanoparticles have expressed superior inhibitory effect on growth of clinical strains of golden *Staphylococcus* than iron nanoparticles (Babushkina et al., 2010). Antimicrobial activity of the copper nanoparticles was tested against many pathogenic bacteria such as *Salmonella, Klebsiella, Shigella* and *Pseudomonas* species (Mahapatra et al., 2008). In mechanistic action of copper ions, it may be reacting with negatively charged cell wall components such as peptidoglycan through electrostatic attraction. This reaction generates H_2O_2 that damages the cytoplasmic membrane (Stoh and Bagchi, 1995; Kim et al., 2000). Thus, the nanoparticle would be helpful to develop novel antimicrobial agents for those microorganism that have developed resistance to the antibiotics in use.

Mechanism Proposed for The Synthesis of Metal Nanoparticles Using Fungi

The fungal cell wall is a dynamic structure, which changes and modifies at different stages of its life cycle. Basically, it is composed of a microfibrillar component located to the inner side of the wall and usually embedded in an amorphous matrix material. The prime components of the fungal cell wall include β-linked glucans and chitin, while the matrix consists mainly of polysaccharides that are mostly water-soluble and many other enzymes (Alexopoulus et al., 1996). Considering these facts, many attempts have been made to elucidate the mechanism for the synthesis of metal nanoparticles by different fungi (Table 1). There are reports of the hypothetical mechanisms about the synthesis of metal nanoparticles however, exact mechanism is yet not clear.

The fungal cell wall plays an important role in the absorption of heavy metals (Dias et al., 2002). On the basis of these properties of fungi, Mukherjee et al. (2001) proposed a two-step mechanism for the intracellular synthesis of metal nanoparticles. In first step of bioreduction, trapping of metal ions takes place at the fungal cell surface. It is probably due to the electrostatic interaction of the positively charged groups in enzymes present on the cell wall (mycelia). In the second step, the metal ions are reduced by the enzymes within the cell wall, which leads to the aggregation of metal ions and formation of nanoparticles. The TEM micrographs of the fungus taken show the presence of nanoparticles on the cytoplasmic membrane as well as within the cytoplasm. This indicates the possibility that some Au^+ ions diffuse through the cell wall and get also reduced by the enzymes present on the cytoplasmic membrane and cytoplasm, while some of the smaller nanoparticles diffuse through the fungal cell wall and get trapped within the cytoplasm.

Ahmad et al. (2003) reported the extracellular synthesis of silver nanoparticles using the fungus, *Fusarium oxysporum* and for the first time proposed a mechanism involved in the synthesis. Their studies confirmed the presence of NADH dependent reductase enzyme in the fungal cell filtrate which was thought to be responsible for the reduction of Ag^+ ions and the subsequent formation of silver nanoparticles in the presence of NADH.

In another study, Durán et al. (2005) worked on the mechanistic aspects of biosynthesis of silver nanoparticles and reported that the synthesis of silver nanoparticles occurs in the presence of anthraquinone and the NADPH nitrate reductase, in this case the electron required to fulfill the deficiency of aqueous silver ions (Ag^+) and convert it into Ag neutral (Ag^0) was donated by both quinone and NADPH. Similarly, Anilkumar et al. (2007) reported that the process of silver nanoparticles formation requires the reduction of α-NADPH to α-NADP$^+$ and the hydroxyquinoline, Hydroxyquinoline probably acts as an electron shuttle transferring the electron generated during the reduction of nitrate to Ag^+ ions converting them to Ag^0. In accordance with the above studies Ingle et al. (2008) also confirmed the presence of NADH dependent nitrate reductase enzyme in fungal cell filtrate using commercially available nitrate reductase discs. The colour of the disc turned reddish from white when challenged with fungal filtrate signifying the presence of nitrate reductase. Thus, it can be concluded that the enzyme NADH dependent nitrate reductase may be associated with reduction of Ag^+ to Ag^0 in the case of fungus *F. acuminatum*.

Table 1. Synthesis of nanoparticles by different fungi.

Sr. No.	Name of fungi	Type of nanoparticle synthesized	Size (nm)	Shape	Reference
1	*Verticillium* sp.	Gold	12–28	Quasi hexagonal	Mukherjee et al., 2001
2	*Phoma* sp. 3.2883	Silver	71.06 ± 3.46	Spherical	Chen et al., 2003
3	*Colletotrichum species*	Gold	20–40	Rod, flat sheets, triangles	Shivshankar et al., 2003
4	*Fusarium oxysporum*	Silver	5–15	Spherical	Duran et al., 2005
5	*Trichothecium* sp.	Gold	Not mentioned	Spherical	Ahmad et al., 2005
6	*Fusarium oxysporum*	Gold-Silver alloy	8–14	Spherical, ellipsoidal	Senapati et al., 2005
7	*Fusarium oxysporum* f. sp. *lycopersici*	Platinum	10–100	Hexagons, Pentagons, circles, Square, rectangles	Riddin et al., 2006
8	*Aspergillus fumigatus*	Silver	5–25	Spherical, triangular	Bhainsa and D'souza, 2006
9	*Phaenerochaete chrysosporium*	Silver	50–200	Hexagonal, pyramidal	Vigneshwaran et al., 2006
10	*Aspergillus flavus*	Silver	8.92 ± 1.61	Isotropic	Vigneshwaran et al., 2007
11	*Fusarium oxysporum*	Zirconia	2–10	Spherical	Bansal et al., 2007
12	*Fusarium acuminatum*	Silver	5–40	Spherical	Ingle et al., 2008
13	*Fusarium semitectum*	Silver	10–60	Spherical	Basavaraja et al., 2008
14	*Penicillium* sp.	Silver	16–40	Not mentioned	Sadowski et al., 2008

Table 1. contd....

Table 1. contd....

Sr. No.	Name of fungi	Type of nanoparticle synthesized	Size (nm)	Shape	Reference
15	*Aspergillus niger*	Silver	~ 20	Spherical	Gade et al., 2008
16	*Alternaria* sp.	Silver	20–60	Spherical	Gajbhiye et al., 2009
17	*Cladosporium cladosporioides*	Silver	10–100	Spherical	Balaji et al., 2009
18	*Coriolus versicolor*	Cadmium sulphide	25–75	Spherical	Sanghi and Verma, 2009
19	*Phoma glomerata*	Silver	60–80	Spherical	Birla et al., 2009
20	*Fusarium solani*	Silver	5–35	Spherical	Ingle et al., 2009
21	*Penicillium fellutanum*	Silver	5–25	Spherical	Kathiresan et al., 2009
22	*Trichoderma viride*	Silver	2–4	Spherical	Fayaz et al., 2010
23	*Fusarium culmorum*	Silver	5–25	Spherical	Bawaskar et al., 2010
24	*Aspergillus niger*	Silver	3–30	Spherical	Jaidev and Narasimha, 2010
25	*Coriolus versicolor*	Gold	5–30	Spherical	Sanghi and Verma, 2010
26	*Rhizopus stolonifer*	Silver	5–30	Spherical	Rathod et al., 2011
27	*Trichoderma reesei*	Silver	5–50	Spherical	Vahabi et al., 2011
28	*Trichosporon beigelii*	Silver	50–100	Spherical	Ghodake et al., 2011
29	*Pestolotia* sp.	Silver	10–40	Spherical	Raheman et al., 2011
30	*Phoma sorghina*	Silver	120–160 X 30–40	Rods	Gade et al., 2011

31	*Phoma glomerata*	Silver	30–60	Iron oxide	Gudadhe et al., 2011
32	*Rhizopus stolonifer*	Silver	5–30	Spherical	Rathod et al., 2011
33	*Neurospora crassa*	Silver Gold Platinum	11 (avg.) 32 (avg.) 20–110	All are mostly spherical	Castro-Longoria et al., 2011, 2012
34	*Aspergillus terreus*	Silver	10–30	Spherical	Li et al., 2012
35	*Aspergillus flavus*	Silver	~7	Not mentioned	Moharrer et al., 2012
36	*Aspergillus niger*	Silver	Not mentioned	Not mentioned	Sangappa and Thiagarajan, 2012
37	*Trichophyton mentagrophytes* *Trichophyton rubrum* *Microsporum canis*	Silver Silver Silver	Less than 50 50–100 50–70	Spherical Spherical Spherical	Moazeni et al., 2012
38	*Cryphonectria* sp.	Silver	30–70	Spherical	Dar et al., 2013
39	*Trichoderma* sp.	Silver	8–60	Spherical	Devi et al., 2013
40	*Aspergillus fumigatus*	Silver	Not mentioned	Not mentioned	Bala and Arya, 2013

Gade et al. (2011) proposed a three-step mechanism for the synthesis of silver nanorods using the fungus *Phoma sorghina*. According to them, first step includes nucleation, which involves the role of proteins acting as capping agent to initiate the silver nanorod formation. Second step is the elongation, in which an anthraquinone derivative secreted by fungus acts as the electron shuttle, which takes up the electron donated by inorganic nitrate and transfers it to silver ions and, thereby, reducing them to form silver particles (Ag^0). Anthraquinone derivatives play a key role in the elongation of silver nanorod synthesis. The third and final step is the termination of the silver nanorod synthesis process. The process will be terminated once the anthraquinone molecule involved in the synthesis is either recruited by another nucleation center for the elongation or until the distance an anthraquinone can act as an electron shuttle has been achieved. Recently, Li et al. (2012) also reported that reduced nicotinamide adenine dinucleotide (NADH) was found to be an important reducing agent for the biosynthesis, and the formation of AgNPs might be an enzyme-mediated extracellular reaction process in *Aspergillus terreus*. A probable mechanism of the biosynthesis of silver nanoparticles using *Aspergillus flavus* involves two steps: first, the reduction of bulk silver ions into silver nanoparticles (32 kDa protein) and, second, is the capping of the synthesized nanoparticles (35 kDa proteins) (Jain et al., 2011).

As proposed in the aforementioned studies, NADH dependent nitrate reductase enzyme is required for the reduction of metal ions in presence of electron shuttle/carrier (Anilkumar et al., 2007; Ingle et al., 2008). These results are in agreement with the results of the study also performed by Duran et al. (2005). However, according to them, not only nitrate reductase is essentially required for the reduction of metal ions, but any electron shuttle/carrier is also equally important. Through their study, they reported that an anthraquinone pigment produced by *F. oxysporum* is responsible for the synthesis of silver nanoparticles because the pigment acts as electron shuttle/carrier in the presence of nitrate reductase enzyme (Duran et al., 2005). The findings of Gade et al. (2011) for the synthesis of silver nanorods using *Phoma sorghina* also agree with the results obtained by Duran et al. (2005). Further, they also opined that *F. moniliforme* was not able to synthesize the metal nanoparticles because, it produced only reductase enzyme and did not produce anthraquinones, of the electron shuttle/carrier.

Here, we propose a modification in the mechanism proposed by those of Duran et al. (2005) and Gade et al. (2011) that is a quinone derivative based mechanism for the reduction of metal (silver) ions (Fig. 1). In this, reduction of silver nanoparticles occurs either in the presence of reductase enzyme (Duran et al., 2005) or inorganic nitrate (Gade et al., 2011), but quinone derivatives (anthraquinone) which act as electron shuttle/carrier are common.

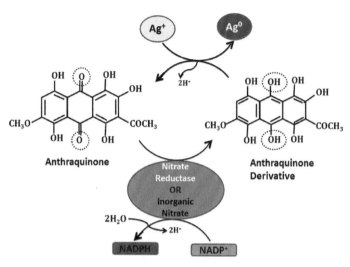

Figure 1. Schematic representation of general mechanism for the synthesis of metal nanoparticles.

Toxicity of Nanoparticles

Nanoparticles (NPs) are applicable in various fields like healthcare, agriculture and cosmetics. Therefore, in order to confirm the safe use of these materials, it is necessary to assess their consequences on human health and environment. In this respect, various researchers across the globe are evaluating the toxic effects of nanoparticles if used at high concentrations. Nanoparticles cause harmful effects on the cells and organs of living beings, particularly skin and disturb their normal functioning/metabolism. In this regard, Kiss et al. (2008) and Pan et al. (2009) demonstrated the cell type-dependent toxicity of titanium oxide (TiO_2) nanoparticles disturbing the normal cellular proliferation, differentiation, mobility, and apoptosis of human skin derived cells and dermal fibroblasts. Hussain et al. (2005) for the first time reported the toxicity of rat liver cells *in vitro*. According to them silver nanoparticles are highly toxic to those cells which mainly show the oxidative stress. Furthermore, they also studied toxic effect of silver nanoparticles on neural cell lines (Hussain et al., 2006). The silver nanoparticles also found to exert the toxic effect on mouse undifferentiated reproductive cells disturbing the mitochondrial function and causing the increased ion leakage though the cellular membrane (Braydich-Stolle et al., 2005). Additionally, Koike and Kobayashi (2006) also detected the oxidative stress in alveolar epithelial cells exposed to Carbon Black (CB) nanoparticles of variable size. Likewise, exposure to Fullerene and C-60 nanoparticles are reported to cause the lipid peroxidation and membrane damage on HDF, HepG2 and NHA cells (Sayes et al., 2005). Bai et al. (2009) has reported that Zinc oxide nanoparticles affect the natural process of Zebrafish embryo development which ultimately lead to their death. Recently, Jarvis et al. (2013) have also reported the toxicity of ZnO nanoparticles to copepod *Acartiatonsa* causing reduction in its survival and reproduction. Thus, in recent years, the understanding about interaction of nanoparticles with cell has increased considerably. A review with toxicological aspect of silver nanoparticles has been reviewed by De Lima et al. (2012).

The aforesaid reports on the toxicological action of nanoparticles clearly indicate that nanoparticles are useful due to their nanosize, but have the harmful effect to the human health and environment. These studies further suggest that while studying the application of nanoparticles, the risk associated with their use should also be carefully assessed.

Conclusion

Nanotechnology is a very promising field having many applications which can help the mankind. Among them, resolving the environmental problem is of utmost importance. Heavy metal contamination is an important problem, which causes various harmful effects to the living beings. For such a purpose, use of heavy metal utilizing/processing organism could be a better choice. Among all organisms fungi are preferred for such purposes because it has been proved that the fungi have the ability to convert the toxic metals to their respective nanoparticles.

The nanoparticles, synthesised by using fungi, have shown to be an antimicrobial agent. Since fungi secret enzymes in their surrounding environment, they can reduce the metal ions and generate the nanoparticles. But nanoparticles also show harmful effects after certain level of their concentration. Therefore, with the careful application of nanotechnology, we can enjoy its benefits for our welfare.

References

Ahmad, A., Senapati, S., Khan, M.I., Kumar, R., Ramani, R., Srinivas, V. and Sastry, M. 2003. Intracellular synthesis of gold nanoparticles by a novel alkalotolerant actinomycete, *Rhodococcus* species. Nanotechnol., 14: 824–828.

Ahmad, Z., Pandey, R., Sharma, S. and Khuller, G.K. 2005. Alginate nanoparticles as antituberculosis drug carriers: formulation development, pharmacokinetics and therapeutic potential. Indian J. Chest Dis. Allied Sci., 48: 171–176.

Ahmad, A., Senapati, S., Khan, M.I., Kumar, R. and Sastry, M. 2005. Extra-/Intracellular biosynthesis of gold nanoparticles by an alkalotolerant fungus, *Trichothecium* sp. J. Biomed. Nanotechnolo., 1(1): 47–53.

Alexopoulus, C.J., Mims, C.W. and Blackwell, M. 1996. Introductory Mycology. 4th Edition, John Wiley & Sons, Canada.

Anilkumar, S., Abyaneh, M.K., Gosavi, S.W., Kulkarni, S.K., Pasricha, R., Ahmad, A. and Khan, M.I. 2007. Nitrate reductase-mediated synthesis of silver nanoparticles from $AgNO_3$. Biotechnol. Lett., 29: 439–443.

Babushkina, I.V., Borodulin, V.B., Korshunov, G.V. and Puchinjan, D.M. 2010. Comparative study of antibacterial action of iron and copper nanoparticles on clinical *Staphylococcus aureus* strain. Saratov J. Med. Sci. Res., 6(1): 11–14.

Bai, W., Zhang, Z., Tian, W., He, X., Ma, Y., Zhao, Y. and Chai, Z. 2009. Toxicity of zinc oxide nanoparticles to zebrafish embryo: a physicochemical study of toxicity mechanism. J. Nanoparticle Res., 12(5): 1645–1654.

Bala, M. and Arya, V. 2013. Biological synthesis of silver nanoparticles from aqueous extract of endophytic fungus *Aspergillus fumigatus* and its antibacterial action. Intl. J. Nanomat. Biostruc., 3(2): 37–41.

Balaji, D.S., Basavaraja, S., Deshpande, R., Bedre, M.D., Prabhakar, B.K. and Venkataraman, A. 2009. Extracellular biosynthesis of functionalized silver nanoparticles by strains of *Cladosporium cladosporioides* fungus. Coll. Surf. B: Biointerf., 68(1): 88–92.

Bansal, V., Syed, A., Bhargava, S.K., Ahmad, A. and Sastry, M. 2007. Zirconia enrichment in zircons and by selective fungus-mediated bioleaching of silica. Langmuir, 23: 4993–5008.

Basavaraja, S., Balaji, S.D., Legashetty, A., Rasab, A.H. and Venkatraman, A. 2008. Extracellular biosynthesis of silver nanoparticles using the fungus *Fusarium semitactum*. Mat. Res. Bull., 43(5): 1164–1170.

Bawaskar, M., Gaikwad, S., Ingle, A., Rathod, D., Gade, A., Duran, N., Marcato, P. and Rai, M. 2010. A new report on mycosynthesis of silver nanoparticles by *Fusarium culmorum*. Curr. Nanosci., 6: 376–380.

Berry, C.C. and Curtis, A.S.G. 2003. Functionalization of magnetic nanoparticles for applications in biomedicine. J. Phys. D: Appl. Phys., 36: 198–206.

Bhainsa, K.C. and D'souza, S.K. 2006. Extracellular biosynthesis of silver nanoparticles using the fungus *Aspergillus fumigatus*. Coll. Surf. B: Biointerf., 47: 160–164.

Birla, S.S., Tiwari, V.V., Gade, A.K., Ingle, A.P., Yadav, A.P. and Rai, M.K. 2009. Fabrication of silver nanoparticles by *Phoma glomerata* and its combined effect against *Escherchia coli*, *Pseudomonas aeruginosa* and *Staphylococcus aureus*. Lett. Appl. Microbiol., 48: 173–179.

Braydich-Stolle, L., Hussain, S., Schlager, J.J. and Hofmann, M.C. 2005. *In vitro* cytotoxicity of nanoparticles in mammalian germline stem cells. Toxicol. Sci., 88(2): 412–419.

Briceño, G., Palma, G. and Durán, N. 2007. Influence of organic amendment on the biodegradation and movement of pesticides. Crit. Rev. Environ. Sci. Technol., 37: 233–271.

Castro-Longoria, E., Moreno-Velasquez, S.D., Vilchis-Nestor, A.R., Arenas-Berumen, E. and Avalos-Borja, M. 2012. Production of platinum nanoparticles and nanoaggregates using *Neurosporacrassa*. J. Microbiol. Biotechnol., 22(7): 1000–1004.

Castro-Longoria, E., Vilchis-Nestor, A.R. and Avalos-Borja, M. 2011. Biosynthesis of silver, gold and bimetallic nanoparticles using the filamentous fungus *Neurosporacrassa*. Coll. Surf. B: Biointerf., 83: 42–48.

Chen, J.C., Lin, Z.H. and Ma, X.X. 2003. Evidence of the production of silver nanoparticles via pretreatment of *Phoma* sp. 3.2883 with silver nitrate. Lett. Appl. Microbiol., 37: 105–108.

Dar, M.A., Ingle, A. and Rai, M. 2013. Enhanced antimicrobial activity of silver nanoparticles synthesized by *Cryphonectria* sp. evaluated singly and in combination with antibiotics. Nanomed. NBN, 9: 105–110.

De Lima, R., Seabra, A.B. and Durán, N. 2012. Silver nanoparticles: a brief review of cytotoxicity and genotoxicity of chemically and biogenically synthesized nanoparticles. J. Appl. Toxicol., 32: 867–879.

Desai, V.S. and Kowshik, M. 2009. Antimicrobial activity of titanium dioxide nanoparticles synthesized by sol gel method. Res. J. Microbiol., 4(3): 97–103.

Devi, T.P., Kulanthaivel, K.D., Borah, J.L., Prabhakaran, N. and Srinivasa, N. 2013. Biosynthesis of silver nanoparticles using *Trichoderma* sp. Indian J. Experi. Biol., 51: 543–547.

Dhillon, G.S., Brar, S.K., Kaur, S. and Verma, M. 2012. Green approach for nanoparticles biosynthesis by fungi: current trends and applications. Criti. Rev. Biotechnol., 32(1): 49–73.

Diao, M. and Yao, M. 2009. Use of zero valent iron nanoparticles in inactivating microbes. Water Res., 43: 5243–5251.

Dias, M.A., Lacerda, I.C.A., Pimentel, P.F., de Castro, H.F. and Ros, C.A. 2002. Removal of heavy metals by an *Aspergillus terreus* strain immobilized in a polyurethane matrix. Lett. Appl. Microbiol., 34: 46–50.

Duran, N., Marcato, P.D., Alves, O.L., DeSouza, G. and Esposito, E. 2005. Mechanistic aspects of biosynthesis of silver nanoparticles by several *Fusarium oxysporum* strains. J. Nanobiotechnol., 3: 1–8.

Duran, N. and Seabra, A.B. 2012. Metallic oxide nanoparticles: state of the art in biogenic syntheses and their mechanisms. Appl. Microbiol. Biotechnol., 95: 275–288.

Elechiguerra, J.L., Burt, J.L., Morones, J.R., Camacho-Bragado, A., Gao, X., Lara, H.H. and Yacaman, M.J. 2005. Interaction of silver nanoparticles with HIV-1. J. Nanobiotechnol., 29: 3–6.

Feng, Q.L., Wu, J., Chen, G.O., Cui, F.Z., Kim, T.N. and Kim, J.O. 2000. A mechanistic study of the antibacterial effect of silver ions on *Escherichia coli* and *Staphylococcus aureus*. J. Biomed. Mat. Res. Part A, 52: 662–668.

Fayaz., A.M., Balaji, K., Girilal, M., Yadav, R., Kalaichelvan, P.T. and Venketesan, R. 2010. Biogenic synthesis of silver nanoparticles and their synergistic effect with antibiotics: a study against gram-positive and gram-negative bacteria. Nanomed. NBM, 6: 103–109.

Gadd, G.M. 2010. Metals, minerals and microbe: Geomicrobiol Bioremidi. Microbiol., 156: 609–643.

Gade, A., Rai, M. and Kulkarni, S. 2011. *Phoma sorghina*, a phytopathogen mediated synthesis of unique silver rods. Intl. J. Green Nanotechnol., 3: 153–159.

Gade, A.K., Bonde, P., Ingle, A.P., Marcato, P.D., Duran, N. and Rai, M.K. 2008. Exploitation of *Aspergillus niger* for synthesis of silver nanoparticles. J. Biobased Mat. Bioener., 2(3): 243–247.

Gade, A., Ingle, A., Whiteley, C. and Rai, M. 2010. Mycogenic metal nanoparticles: progress and applications. Biotechnol. Lett., 32(5): 593–600.

Gajbhiye, M., Kesharwani, J., Ingle, A., Gade, A. and Rai, M. 2009. Fungus mediated synthesis of silver nanoparticles and their activity against pathogenic fungi in combination with fluconazole. Nanomed.: Nanotechnol. Biol. Med., 5: 382–386.

Galvez, J.B., Ibanez, P.F. and Sixton, M.R. 2007. Solar photocatalytic detoxification and disinfection of water: Recent overview. J. solar Ener. Engin., 129: 12–12.

Ghodake, V.P., Kininge, P.T., Magdum, S.P., Dive, A.S. and Pillai, M.M. 2011. Biosynthesis of silver nanoparticles using. *Trichosporon beigelii* NCIM 3326 and evaluation of their antimicrobial activity. J. Eng. Res. Stud., 3(1): 32–36.

Gong, P., Li, H., He, X., Wang, K., Hu, J., Tan, W., Tan, S. and Zhang, X.Y. 2007. Preparation and antibacterial activity of Fe3O4@Ag nanoparticles. Nanotechnol., 18: 604–611.

Gu, H., Ho, P.L., Tong, E., Wang, L. and Xu, B. 2003. Presenting vancomycin on nanoparticles to enhance antimicrobial activities. Nanotechnol. Lett., 3: 1261–1263.

Gudadhe, J.A., Bonde, S.R., Gaikwad, S.C., Gade, A.K. and Rai, M.K. 2011. *Phoma glomerata*: A novel agent for fabrication of iron oxide nanoparticles. J. Bionanosci., 5: 138–142.

Gupta, A.K. and Gupta, M. 2005. Synthesis and surface engineering of iron oxide nanoparticles for biomedical applications. Biomat., 26(18): 3995–4021.

Hemanth, N.K.S., Kumar, G., Karthik, L. and Bhaskar Rao, K.V. 2010. Extracellular biosynthesis of silver nanoparticles using the filamentous fungus *Penicillium* sp. Arch. Appl. Sci. Res., 2(6): 161–167.

Horiguchi, H. 1980. Chemistry of Antibacterial and Antimildew. Sankyo Press, Tokyo.

Hussain, S.M., Hess, K.L., Gearhart, J.M., Geiss, K.T. and Schlager, J.J. 2005. *In vitro* toxicity of nanoparticles in BRL 3A rat liver cells. Toxicol. *In Vitro*, 19: 975–983.

Hussain, S.M., Javorina, A.K., Schrand, A.M., Duhart, H.M., Ali, S.F. and Schlager, J.J. 2006. The interaction of manganese nanoparticles with PC-12 cells induces dopamine depletion. Toxicol. Sci., 92: 456–463.

Ingle, A., Gade, A., Pierrat, S., Sonnichsen, C. and Rai, M. 2008. Mycosynthesis of silver nanoparticles using the fungus *Fusarium acuminatum* and its activity against some human pathogenic bacteria. Curr. Nanosci., 4: 141–144.

Ingle, A., Gade, A., Bawaskar, M. and Rai, M. 2009. *Fusarium solani*: A novel biological agent for the extracellular synthesis of silver nanoparticles. J. Nanopart. Res., 11(8): 2079–2085.

Jaidev, L.R. and Narasimha, G. 2010. Fungal mediated biosynthesis of silver nanoparticles, characterization and antimicrobial activity. Coll. Surf. B: Biointerf., 81: 430–433.

Jain, D., Daima, H.K., Kachhwaha, S. and Kothari, S.L. 2009. Synthesis of Plant-mediated silver nanoparticles using papaya fruit extract and evaluation of their antimicrobial activities. J. Nanomat. Biostru., 4: 557–563.

Jain, N., Bhargava, A., Majumdar, S., Tarafdar, J.C. and Panwar, J. 2011. Extracellular biosynthesis and characterization of silver nanoparticles using *Aspergillus flavus* NJP08: A mechanism perspective. Nanoscale, 3: 635–641.

Jarvis, T.A., Miller, R.J., Lenihan, H.S. and Bielmyer, G.K. 2013. Toxicity of ZnO nanoparticles to the copepod *Acartiatonsa*, exposed through a phytoplankton diet. Environ. Toxicol. Chem., 32(6): 1264–1269.

Jo, Y.K., Kim, B.H. and Jung, G. 2009. Antifungal activity of silver ions and nanoparticles on phytopathogenic fungi. Plant Dis., 93: 1037–1043.

Jones, N., Ray, B., Ranjit, K.T. and Mannas, A.C. 2008. Antibacterial activity of ZnO nanoparticle suspension on a broad spectrum of microorganism. FEMS Microbiol. Lett., 279(1): 71–76.

Kathiresan, K., Manivannan, S., Nabeel, M.A. and Dhivya, B. 2009. Studies on silver nanoparticles synthesized by a marine fungus, *Penicillium fellutanum* isolated from coastal mangrove sediment. Coll. Surf. B: Biointerf., 71: 133–137.

Karigar, C.S. and Rao, S.S. 2011. Role of microbial enzymes in the bioremediation of pollutants: A review. Enzy. Res., 805187 (11 pages).

Kim, J.H., Cho, H., Ryu, S.E. and Choi, M.U. 2000. Effect of metal ions on the activity of protein tyrosine phosphatase VHR: Highly potent and reversible oxidative inactivation by Cu^{2+} ion. Arch. Biochem. Biophys., 382(1): 72–80.

Kim, K.J., Sung, W.S., Moon, S.K., Choi, J.S., Kim, J.G. and Lee, D.G. 2008. Antifungal effect of silver nanoparticles on dermatophytes. J. Microbiol. Biotechnol., 18(8): 1482–1484.

Kiss, B., Biro, T., Czifra, G., Tóth, B.I., Kertész, Z., Szikszai, Z., Kiss, A.Z., Juhász, I., Zouboulis, C.C. and Hunyadi, J. 2008. Investigation of micronized titanium dioxide penetration in human skin xenografts and its effect on cellular functions of human skin-derived cells. Experi. Dermatol., 17: 659–667.

Koike, E. and Kobayashi, T. 2006. Chemical and biological effects of carbon black nanoparticles. Chemosphere, 65: 946–951.

Kumar, A.S., Ansary, A.A., Ahmad, A. and Khan, M.I. 2007a. Extracellular biosynthesis of CdSe quantum dots by the fungus *Fusarium oxysporum*. J. Biomed. Nanotechnol., 3: 190–194.

Kumar, A.S., Abyaneh, M.K., Gosavi, S.W., Kulkarni, S.K., Pasricha, R., Ahmad, A. and Khan, M.I. 2007b. Nitrate reductase-mediated synthesis of silver nanoparticles from $AgNO_3$. Biotechnol. Lett., 29: 439–445.

Lara, H.H., Ayala-Nunez, N.V., Ixtepan-Turrent, L. and Rodriguez-Padilla. 2010. Mode of antiviral action of silver nanoparticles against HIV-1. J. Nanobiotechnol., 8(1): 1–10.

Lee, Z., Kim, J.Y., Lee, W.I.I., Nelson, K.L., Yoon, J. and Sedlak, D.L. 2008. Bactericidal effect of zero-valent. Iron nanoparticles on *Escherichia coli*. Environ. Sci. Technol., 42(13): 4927–4933.

Li, G., He, D., Qian, Y., Guan, B., Gao, S., Cui, Y., Yokoyama, K. and Wang, L. 2012. Fungus mediated green synthesis of silver nanoparticles using *Aspergillus terreus*. Intl. J. Mol. Sci., 13: 466–476.

Mahapatra, O., Bhagat, M., Gopalkrishnan, C. and Arunachalam, K.D. 2008. Ultrtafine dispersed CuO nanoparticles and their antibacterial activity. J. Exper. Nanosci., 3(3): 185–193.

Majumder, D.R. 2012. Bioremediation: Copper nanoparticles from electronic-waste. Intl. J. Engi. Sci. Technol., 4(10): 4380–4389.

Marcato, P.D., Nakasato, G., Brocchi, M., Melo, P.S., Huber, S.C., Ferreira, I.R., Alves, O.L. and Durán, N. 2012a. Biogenic silver nanoparticles: Antibacterial and cytotoxicity applied to textile fabrics. J. Nano Res., 20: 69–76.

Marcato, P.D., Durán, M., Huber, S., Rai, M., Melo, P.S., Alves, O.L. and Durán, N. 2012b. Biogenic silver nanoparticles and its antifungal activity as a new topical transungual drug delivery. J. Nano Res., 20: 99–107.

Moazeni, M., Rashidi, N., Shahverdi, A.R., Noorbakhsh, F. and Rezaie, S. 2012. Extracellular production of silver nanoparticles by using three common species of dermatophytes: *Trichophyton rubrum*, *Trichophyton mentagrophytes* and *Microsporumcanis*. Iranian Biomed. J., 16(1): 52–58.

Moharrer, S., Mohammadi, B., Gharamohammadi, R.A. and Yargoli, M. 2012. Biological synthesis of silver nanoparticles by *Aspergillus flavus*, isolated from soil of Ahar copper mine. Indian J. Sci. Technol., 5(S3): 2443–2444.

Morones, J.R., Elechiguerra, J.L., Camacho, A. and Ramirez, J.T. 2005. The bactericidal effect of silver nanoparticles. Nanotechnol., 5(16): 2346–2353.

Mukherjee, P., Ahmad, A., Mandal, D., Senapati, S., Sainkar, S.R., Khan, M.I., Ramani, R., Parischa, R., Ajayakumar, P.V., Alam, M., Sastry, M. and Kumar, R. 2001. Bioreduction of $AuCl_4^-$ ions by the fungus, *Verticillium* sp. and surface trapping of the gold nanoparticles formed. Angewandte Chemie Intl. Edition, 40(19): 3585–3588.

Nair, B. and Pradeep, T. 2002. Coalescence of nanoclusters and formation of submicron crytallites assisted by *Lactobacillus* strains. Crys.l Grow. Design, 2: 293–298.

Namasivayam, S.K.R., Ganesh, S. and Avimanyu. 2011. Evaluation of anti-bacterial activity of silver nanoparticles synthesized from *Candida glabrata* and *Fusarium oxysporum*. Intl. J. Medicobiol. Res., 1(3): 130–136.

Narayanan, K.B. and Sakthivel, N. 2010. Biological synthesis of metal nanoparticles by microbes. Adva. Coll. Interf. Sci., 156: 1–13.

Navazi, Z.R., Pazouki, M. and Halek, F.S. 2010. Investigation of culture conditions for biosynthesis of silver nanoparticles. Iranian J. Biotechnol., 8(1): 56–61.

Nithya, R. and Ragunathan, R. 2009. Synthesis of silver nanoparticles using *Pleurotus sajor-caju* and its antimicrobial study. J. Nanomate. Biostruct., 4(4): 623–629.

Pan, Z., Lee, W., Slutsky, L., Clark, R.A., Pernodet, N. and Rafailovich, M.H. 2009. Adverse effects of titanium dioxide nanoparticles on human dermal fibroblasts and how to protect cells. Small, 5: 511–520.

Park, S., Chibli, H., Wong, J. and Nadeau, J.L. 2011. Antimicrobial activity and cellular toxicity of nanoparticle–polymyxin B conjugates. Nanotechnol., 22: 185101. doi: 10.1088/0957-4484/22/18/185101.

Raheman, F., Deshmukh, S., Ingle, A., Gade, A. and Rai, M. 2011. Silver nanoparticles: Novel antimicrobial agent synthesized from a endophytic fungus *Pestalotia* sp. isolated from leaves of *Syzygiumcumini* (L.). Nano Biomed. Engin., 3(3): 174–178.

Rai, M., Yadav, A. and Gade, A. 2010. Mycofacbrication, mechanistic aspects and multifunctionality of metal nanoparticles- where are we? And where should we go? pp. 1343–1354. *In*: Mendez-Vilas, A. (ed.). Current Research, Technology and Education Topics in Applied Microbiology and Microbial Biotechnology. Formatex Research Center, Badajoz, Spain.

Rai, M., Yadav, A., Bridge, P. and Gade, A. 2009. Myconanotechnology: A new and emerging science. pp. 258–267. *In*: Mahendra Rai and Paul Bridge (eds.). Applied Mycology, Edi. 14, CAB International New York.

Rathod, V., Banu, A. and Ranganath, E. 2011. Biosynthesis of highly stabilized silver nanoparticles by *Rhizopusstolonifer* and their anti-fungal efficacy. Intl. J. Mol. Clin. Microbiol., 1: 65–70.

Riddin, T.L., Gericke, M. and Whiteley, C.G. 2006. Analysis of the inter- and extracellular formation of platinum nanoparticles by *Fusarium oxysporum* f. sp. *Lycopersici* using response surface methodology. Nanotechnol., 17: 3482–3489.

Sadowski, Z., Maliszewska, I.H., Grochowalska, B., Polowczyk, I. and Kozlecki, T. 2008. Synthesis of silver nanoparticles using microorganisms. Mater. Sci. Poland, 26: 419–425.

Sangappa, M. and Thiagarajan, P. 2012. Mycobiosynthesis and characterization of silver nanoparticles from *Aspergillus niger*: A soil fungal isolate. Int. J. Life Sc. Bt. Pharm. Res., 1(2): 282–289.

Sanghi, R. and Verma, P. 2009. Biomimetic synthesis and characterization of protein capped silver nanoparticles. Bioreso. Technol., 100: 502–504.

Sanghi, R. and Verma, P. 2010. pH dependent fungal proteins in the "green" synthesis of gold nanoparticles. Adv. Mat. Lett., 1(3): 193–199.

Sawai, J. 2003. Quantitative evaluation of antibacterial activities of metallic oxide powders (ZnO, MgO and CaO) by conductimetric assay. J. Microbiol. Metho., 54: 177–182.

Sayes, C.M., Gobin, A.M., Ausman, K.D., Mendez, J., West, J.L. and Colvin, V.L. 2005. Nano-C60 cytotoxicity is due to lipid peroxidation. Biomat., 26(36): 7587–7595.

Senapati, S., Ahmad, A., Khan, M.I., Sastry, M. and Kumar, R. 2005. Extracellular biosynthesis of bimetallic Au-Ag alloynanoparticles. Small, (1): 517–520.

Shivshankar, S., Ahmad, A., Pasricha, R. and Sastry, M. 2003. Bioreduction of chloroaurate ions by Geranium leaves and its endophytic fungus yields gold nanoparticles of different shapes. J. Mater. Chem., 13: 1822–1826.

Sondi, I. and Salopek-Sondi, B. 2004. Silver nanoparticles as antimicrobial agent: a case study on *E. coli* as a model for gram-negative bacteria. J. Coll. Interf., 275: 177–182.

Song, H.Y., Ko, K.K., Oh, L.H. and Lee, B.T. 2006. Fabrication of silver nanoparticles and their antimicrobial mechanisms. Europ. Cells Mat. J., 11: 58–63.

Stamets, P. 1999. Helping the Ecosystem through Mushroom Cultivation. Adapted from Stamets, P. 1998. Earth's Natural Internet. Whole Earth Magazine, Fall 1999.

Stoh, S.J. and Bagchi, D. 1995. Oxidative mechanisms in the toxicity of metal ions. Free Radical Biol. Med., 18: 321–336.

Sun, W.Y.R., Chen, R., Chung, N.P.Y., Ho, C.M., Lin, C.L.S. and Che, C.M. 2005. Silver nanoparticles fabricated in Hepes buffer exhibit cytoprotective activities towards HIV-1 infected cells. Chem. Commun., (camb): 5059–5061.

Thakkar, K.N., Mhatre, S.S. and Parikh, R.Y. 2010. Biological synthesis of metallic nanoparticles. Nanomed., 6(2): 257–262.

Tortella, G.R., Diez, M.C. and Durán, N. 2005. Fungal diversity and use in decomposition of environmental pollutants. Crit. Rev. Microbiol., 31: 197–212.

Tran, N., Mir, A., Malik, D., Sinha, A., Nayar, S. and Webster, J.T. 2010. Bactericidal effect of iron oxide nanoparticles on *S. aureus*. Intl. J. Nanomed., 5: 277–283.

Vahabi, K., Mansoori, G.A. and Karimi, S. 2011. Biosynthesis of silver nanoparticles by fungus *Trichoderma reesei*. Insci. J., 1(1): 65–79.

Vigneshwaran, N., Kathe, A.A., Varadarajan, P.V., Nachane, R.P. and Balasubramanya, R.H. 2006. Biomimetics of silver nanoparticles by white rot fungus, *Phanerochaete chrysosporium*. Coll. Surf. B: Biointer-faces, 53(1): 55–59.

Vigneshwaran, N., Ashtaputre, M., Nachane, R.P., Paralikar, K.M. and Balasubramanya, H. 2007. Biological synthesis of silver nanoparticles using the fungus *Aspergillus flavus*. Mat. Lett., 61(6): 1413–1418.

Yong, P., Mikheenko, I.P., Deplanche, K., Sargent, F. and Macaskie, L.E. 2009. Biorecovery of precious metals from wastes and conversion into fuel cell catalyst for electricity production. Adv. Mat. Res., 71-73: 729–732.

Yoon, K.Y., Byeon, J.H., Park, J.H. and Hwang, J. 2007. Susceptibility constant of *Escherichia coli* and *Bacillus subtilis* to silver and copper nanoparticles. Sci. Total Environ., 373: 572–575.

Zawrah, M.F. and Abd El-Moez, S.I. 2011. Antimicrobial activities of gold nanoparticles against major foodborne pathogens. Life Sci. J., 8(4): 37–44.

Zhang, X. 2011. Application of microorganisms in biosynthesis of nanomaterials—a review. Wei Sheng Wu XueBao, 51(3): 297–304.

Zhonglin, J. and Yiping, H. 2012. Antibacterial activities of magnesium oxide (MgO) nanoparticles against foodborne pathogens. J. Nanopart. Res., 13: 6877–6885.

CHAPTER 11

Potential Biotechnological Applications of Thermophilic Moulds

Bijender Singh[1] *and T. Satyanarayana*[2,*]

ABSTRACT

Thermophilic moulds are the eukaryotic microorganisms which are ubiquouts in their occurrence in both thermogenic and non-thermogenic natural and man-made environments. These moulds are able to degrade organic matter efficiently, produce an array of useful enzymes, antibiotics and nutritionally enriched feeds are suitable as agents in bioconversions. Their enzymes are also useful in the treatment of industrial wastes and effluents that are rich in oil, heavy metals, anti-nutritional factors (e.g., phytic acid) and other polysaccharides (cellulose, hemicellulose, chitin and pectin). The enzymes are thermostable, and therefore, find applications in different industrial process including mushroom composting. The utility of their enzymes in generating glucose, xylose and mannose from the hydrolysis of agro-residues and their fermentation to bioethanol is a major venture in the field of biotechnology.

Introduction

Microorganisms have originated on the earth about 4 billion years ago at a time when temperatures were likely to be in extremes, which has been proven by the discovery of microorganisms from geothermal areas all over the world.

[1] Department of Microbiology, Maharshi Dayanand University, Rohtak-124001, Haryana.
[2] Department of Microbiology, University of Delhi South Campus, Benito Juarez Road, New Delhi-110021, India.
* Corresponding author: tsnarayana@gmail.com

Prokaryotes have a wide range of temperature tolerance, while in case of eukaryotic organisms, only a few species of fungi can tolerate up to 62°C. Depending upon their ability to survive at different temperature, fungi are catagorised into psychrophiles, mesophiles and thermophiles. Psychrophilic fungi have T_{max} at 20°C and T_{min} at or below 10°C. Thermophilic fungi have T_{max} at or above 50°C and T_{min} at or above 20°C, while mesophilic fungi have T_{max} at 45°C and T_{min} at or below 20°C (Johri et al., 1999; Satyanarayana and Singh, 2004). In fungi, it is not as extreme as in eubacteria or archea, some species of which are able to grow up to 122°C in thermal springs and hydrothermal vents (Tansey and Brock, 1978; Takai et al., 2008).

Heaps of waste plant material, piles of agricultural and forestry products and other organic materials provide a suitable environment for the growth and development of thermophilic fungi (Johri et al., 1999; Satyanarayana and Singh, 2006). Cooney and Emerson (1964) initially provided a taxonomic description of 13 species known at that time and mentioned about their habitats and general biology. Now a large number of thermophilic fungi are known. Not only taxonomy, but their physiology, enzymes and their potential applications have well been studied and reviewed (Satyanarayana et al., 1988; Johri et al., 1999; Satyanarayana et al., 1992; Maheshwari et al., 2000; Archana and Satyanarayana, 2001; Singh and Satyanarayana, 2009b). In this chapter attempts have been made to highlight the biotechnological applications of thermophilic moulds in the production of an array of enzymes and in the management of environmental pollution.

Potential Applications of Thermophilic Moulds

Thermophilic fungi are well known to produce thermostable enzymes, which are useful in many industries (Singh, 2014). Berka et al. (2011) have recently reported comparative genome analysis of two thermophilic moulds, *Myceliophthora thermophila* and *Thielavia terrestris* and revealed the presence of a variety of genes encoding different enzymes responsible for the degradation of organic matter. *Rhizomucor pusillus* has been used in the biotransformation of antihelmintic drug albendazole to produce novel and active metabolite of commercial interest (Prasad et al., 2011). The enzymes produced by thermophilic moulds are mostly extracellular and obtained in substantial quantities in the culture filtrates. The research is focused mainly on the identification of suitable thermophilic fungus that can produce the desired enzymes. These moulds are also been used to manage environmental pollution by their application in decomposition of organic matter, effluent treatment and biosorption of heavy metals.

Enzymatic Machinery

The thermophilic microbes have been considered as a good source of thermostable enzymes with high catalytic activity, greater resistance to denaturing agents, and lower incidence of contamination (Singh, 2014). Thermophilic enzymes are receiving considerable attention because of their utility in high-temperature catalysis of various enzymatic industrial processes. The thermophilic moulds produce a large number of enzymes which are useful in various industries such as food, textile and detergent, dairy, pharmaceutical and others (Table 1). The important enzymes obtained from thermophilic moulds are briefly discussed below:

Phosphatases

The phosphoserine residues in caseins can be hydrolyzed by both alkaline and acid phosphatases. In molecular biology, alkaline phosphatases (EC 3.1.3.1) can be used for dephosphorylation of the 5' end of DNA or RNA. Bilai et al. (1985) screened 775 strains of thermophilic fungi for the production of acid phosphatases but only isolates of 15 species were found positive. *Rhizopus micromyces, Rhizomucor pusillus, Talaromyces thermophilus, Populaspora thermophila, Thermomyces lanuginosus, Acremonium thermophile, Thermoascus aurantiacus* and *Chrysosporium thermophilum* grew well on the medium (Emerson, YpSs), but only the latter two fungi produced acid phosphatase. Of the 13 thermophilic fungi tested by Satyanarayana et al. (1985) for the production of extracellular acid and alkaline phosphatases, *Acremonium alabamensis* and *Rhizopus rhizoidiformis* secreted only acid phosphatase where as the other fungi secreted both types of enzymes.

An extracellular (conidial) and an intracellular (mycelial) alkaline phosphatase from the thermophilic fungus *Scytalidium thermophilum* were purified by DEAE-cellulose and Concanavalin A-Sepharose chromatography (Guimarães et al., 2001). The molecular masses of the conidial and mycelial enzymes, estimated by gel filtration, were 162 and 132 kDa, respectively. Both proteins are glycoprotein and migrated on SDS-PAGE as a single polypeptide of 63 and 58.5 kDa, respectively, suggesting that these enzymes are dimers of identical subunits. The optimum pH for the conidial and mycelial alkaline phosphatases was 10.0 and 9.5 at 70–75°C respectively.

Phytases (*myo*-inositol hexakisphosphate phosphohydrolase) are the phosphatases, which catalyze the hydrolysis of phytic acid to inorganic phosphate and *myo*-inositol phosphate derivatives, while phosphatases are able to hydrolyze a wide variety of esters and anhydride phosphoric acids, releasing phosphate, and are also able to perform transphosphorylation reactions (Singh and Satyanarayana, 2011; Singh et al., 2011). The reduction of phytic acid content in the foods and feeds by enzymatic hydrolysis using phytase is desirable, since physical and chemical methods of phytate removal negatively affect their nutritional value. These enzymes, therefore,

Table 1. List of enzymes produced by thermophilic moulds.

Enzyme	Thermophilic mould	References
Phosphatases	*Thermoascus aurantiacus, Chrysosporium thermophilum*	Bilai et al. (1985)
	Acremonium alabamensis, Rhizopus rhizoidiformis	Satyanarayana et al. (1985)
	Scytalidium thermophilum	Guimaraes et al., 2001
Phytases	*Aspergillus fumigatus*	Pasamontes et al., 1997a
	Thermomyces lanuginosus	Berka et al., 1998
	Myceliophthora thermophila	Mitchell et al., 1997
	Rhizomucor pusillus	Chadha et al., 2004
	Thermoascus aurantiacus	Namapoothiri et al., 2004
	Talaromyces thermophilus	Pasamontes et al., 1997b
	Sporotrichum thermophile	Singh and Satyanarayana, 2006a,b; 2008a,b,c
Amylases	*Thermomyces lanuginosus*	Mishra and Maheshwari, 1996; Arnesen et al., 1998; Chadha et al., 1997; Petrova et al., 2000; Nguyena et al., 2002
	Scytalidium thermophilum	Roy et al., 2000; Aquino et al., 2001
	Thermomucor indicae-seudaticae	Kumar and Satyanarayana, 2003
	Malbranchea sulfurea	Gupta and Gautam, 1993
	Humicola grisea	Tosi et al., 1993; Campos and Felix, 1995
Lipases	*Humicola lanuginosa*	Arima et al., 1968; Omar et al., 1987
	Rhizomucor miehei	Huge-Jensen et al., 1989; Rao and Divaker, 2002
	Rhizopus arrhizus	Kumar et al., 1993
Cellulases	*Sporotrichum thermophile*	Bhat and Maheshwari, 1987
	Thermoascus aurantiacus	Gomes et al., 2000; Parry et al., 2001; Kalogeris et al., 2003; Hong et al., 2003
	Humicola insolens	Moriya et al., 2003
	Melanocarpus albomyces	Hirvonen and Papageorgiou, 2003
	Talaromyces emersonii	Murray et al., 2003
	Scytalidium thermophilum	Arifolua and Ögel, 2000
	Chaetomium thermophilum var. *coprophile*	Venturi et al., 2002
Xylanases	*Thermoascus aurantiacus*	Tong et al., 1980; Gomes et al., 1994; Kalogeris et al., 1998; dos Santos et al., 2003

Table 1. contd....

Table 1. contd.

Enzyme	Thermophilic mould	References
Xylanases	*Chaetomium thermophile* var. *coprophile*	Ganju et al., 1989
	Malbranchea pulchella var. *sulfurea*	Kvesitadze et al., 1998
	Melanocarpus albomyces	Prabhu and Maheshwari, 1999; Narang et al., 2001; Roy et al., 2003
	Sporotrichum thermophile	Katapodis et al., 2003
	Thermomyces lanuginosus	Haarhoff et al., 1999; Lin et al., 1999; Singh et al., 2000
	Chaetomium cellulolyticum	Baraznenoka et al., 1999
Pectinases	*Penicillium duponti, Humicola stellata, H. lanuginosa, H. insolens, Mucor pusillus*	Craveri et al., 1967
	Talaromyces thermophilus	Tong and Cole, 1975
	Sporotrichum thermophile	Adams and Deploey, 1978; Whitehead and Smith, 1989
	Chaetomium thermophile, Talaromyces emersonii CBS 814.70, *T. emersonii* UCG 208 and *Thermoascus aurantiacus*	Tuohy et al., 1989
Proteases	*Mucor pusillus*	Arima et al., 1967; Etoh et al., 1979; Aikawa et al., 1990
	M. miehei	Ottesen and Rickert, 1970; Etoh et al., 1979
	Malbranchea pulchella var. *sulfurea, Humicola lanuginosa*	Stevenson and Gaucher, 1975
	Humicola lanuginosa	Shenolikar and Stevenson, 1982
	Thermomyces lanuginosus	Jensen et al., 2002
	Scytalidium thermophilum	Ifrij and Ögel, 2002

have potential applications in food and feed industries for mitigating their phytic acid and other organic phosphorus compounds to liberate utilizable inorganic phosphate and for improving digestibility as a result of elimination of antinutrient characteristics.

Thermophilic moulds are known to secrete phytases in submerged as well as in solid state fermentations. First report came in 1997, when phytase from *Aspergillus fumigatus* was cloned and over-expressed (Pasamontes et al., 1997). The phytase from *Thermomyces lanuginosus* was cloned and overexpressed, which exhibited optimum activity at 65°C and pH of 6.0 (Berka et al., 1998). *Chaetomium thermophilum* ATCC 58420, *Rhizomucor miehi* ATCC22064, *Thermomucor indicae-seudaticae* ATCC28404, and *Myceliophthora thermophila* ATCC48102 are also known to produce phytases (Mitchell et al., 1997). *Rhizomucor pusillus* secreted phytase optimally at 50°C and pH 5.5 in SSF using wheat bran (Chadha et al., 2004). While *Thermoascus aurantiacus* produced phytase in semisynthetic medium using glucose, starch and wheat bran (Namapoothiri et al., 2004). *Sporotrichum thermophile* secreted phytase in both solid state (Singh and Satyanarayana, 2006a, 2008b) and submerged fermentations (Singh and Satyanarayana, 2006b, 2008a, 2009a). Phytase of *S. thermophile* was effective in the dephytinization of sesame oil cake, wheat flour, bread and soymilk with concomitant reduction in phytic acid content and liberation of utilizable inorganic phosphate (Singh and Satyanarayana, 2006a, 2008a, 2008b, 2008c). The enzyme hydrolyzed insoluble phytates to a varied extent. Furthermore, both enzyme as well as thermophilic mould promoted the growth of wheat plants (Singh and Satyanarayana, 2010).

The phytase and acid phosphatase enzymes can work in coordination, where phytase can split the molecule of phytate in a selective manner, while acid phosphatase can attack the inositol phosphate intermediates independently, and as a result accelerate the total dephosphorylation process.

Starch Hydrolyzing Enzymes

Amylolytic enzymes are produced by an extremely wide variety of microorganisms. α-Amylase and glucoamylase are the most widely reported in microorganisms, while β-amylase is generally of plant origin, and this has been reported in only a few microbes. There are many advantages in the use of thermostable enzymes in the starch processing industry, such as increased reaction rates and decreased contamination risk (Maheshwari et al., 2000). The culture of *Thermomyces lanuginosus* was found to produce maximally α-amylase and glucoamylase (18.4 and 11.2 U ml^{-1}), respectively when the medium contained rice flour (2% w/v) as carbon and corn steep liquor as nitrogen source with medium pH adjusted to 5.5 and incubated for 72 h under shaking conditions (150 rpm) at 50°C (Chadha et al., 1997). Glucoamylase produced by *Scytalidium thermophilum* was purified 80-fold by DEAE-cellulose, ultrafiltration and CM-cellulose chromatography. The enzyme is a glycoprotein containing 9.8% saccharide, pI of 8.3 and molar mass of 75 kDa (SDS-PAGE) or 60 kDa (Sepharose 6B) (Aquino et al., 2001). Optima of pH and temperature with starch or maltose as substrates were 5.5/70°C and 5.5/65°C, respectively.

Amylolytic enzymes (α-amylase and glucoamylase) from *Thermomyces lanuginosus* ATCC 34626 were purified to electrophoretic homogeneity (Nguyena et al., 2002). The molecular mass of purified α-amylase and glucoamylase were 61 and 75 kDa with pI values of 3.5–3.6 and 4.1–4.3, respectively. The amylolytic enzymes from *T. lanuginosus* exhibit pH optima in the range 4.4–6.6 at 70°C. The K_m and V_{max} of α-amylase on soluble starch were 0.68 mg/ml and 45.19 U mg^{-1}, respectively. The K_m values of glucoamylase on maltose, maltotriose, maltotetraose, maltopentose and soluble starch were 6.5, 3.5, 2.1, 1.1 mM and 0.8 mg ml^{-1}, respectively. The first 37 residues of N-terminal of the purified α-amylase of *T. lanuginosus* ATCC 34626 showed complete homology with the α-amylase from *Aspergillus oryzae* and *Emericella nidulans.* A glucoamylase from *Thermomucor indicae-seudaticae,* was purified to near homogeneity. It was a glycoprotein with a carbohydrate content of 9–10.5%, which is optimally active at 60°C and pH 7.0. It had a molecular mass of 42 kDa with a pI of 8.2. The enzyme hydrolyzed soluble starch at 50°C (K_m 0.50 mg mL^{-1} and V_{max} 109 μ mol mg^{-1} protein min^{-1}) and at 60°C (K_m 0.40 mg mL^{-1} and V_{max} 143 μ mol mg^{-1} protein min^{-1}). Its experimental activation energy was 43 KJ mol^{-1} with temperature quotient (Q_{10}) of 1.35 (Kumar and Satyanarayana, 2003). An ideal starch saccharification process was developed using enzymes (glucoamylase, amylopullulanase and α-amylase) from thermophiles (*Thermomucor indicae-seudaticae* and *Geobacillus thermoleovorans*) (Satyanarayana et al., 2004).

An α-amylase has been purified from the culture of *Scytalidium thermophilum* (Roy et al., 2000). A nine-fold purification was achieved in a single step using fluidized bed chromatography wherein alginate was used as the affinity matrix. There are at least two isoenzymes as shown by concanavalin A (Con A)–agarose column chromatography which slightly differ in their pH and temperature optima. The isoenzymes have similar molecular weights of around 45 kDa as shown by SDS–PAGE analysis.

The thermophilic fungus *Thermomyces lanuginosus* was cultivated in shake flasks for up to 120 h with low molecular weight dextran as carbon source supplemented with either Tween 80 or Triton X-100 (Arnesen et al., 1998). Addition of Tween 80 to the growth medium gave a 2.7-fold increase in maximum α-amylase activity as compared with the controls. Triton X-100 did not affect α-amylase production. *Malbranchea sulfurea* produce α-amylase extracellularly and α-glucosidase was present in cell bound fraction (Gupta and Gautam, 1993). This glucosidase was latter purified 31.6 fold with a yield of 11.68%. The molecular mass of the enzyme was 110 kDa, which showed optimally activity at pH 4.8 and at 60°C. Campos and Felix (1995) purified and characterized a glucoamylase from *Humicola grisea* having molecular mass and isoelectric point of 74 kDa and 8.4, respectively. It was a glycoprotein with 5% carbohydrate content and showed maximal activities at pH 6.0 and 60°C. The K_m value of soluble starch hydrolysis at 50°C and pH 6.0 was 0.14 mg ml^{-1}. An extracellular glucoamylase from *Humicola grisea* var. *thermoidea* was purified and characterized by Tosi et al. (1993). The molecular mass of the purified protein was estimated 63 kDa by SDS-PAGE and 65 kDa by gel filtration chromatography. It was a glycoprotein with 1.8% carbohydrate content and pH and temperature optima of 5.0 and 55°C, respectively. A *Thermomyces lanuginosus,* strain IISc 91, secreted one form each of α-amylase and glucoamylase during growth (Mishra and Maheshwari, 1996). Both enzymes were purified near to homogeneity. α-Amylase was considered to be a dimeric protein of ~42 kDa and contained 5% carbohydrate content with optimum activity at pH 5.6 and at 65°C. Its activation energy was 44 kJ mol^{-1} and K_m for soluble starch was 2.5 mg ml^{-1}. The glucoamylase is a monomeric protein with a molecular mass of ~45 kDa and 11% carbohydrate content. It was stable at 60°C for over 10 h with activation energy of 61 kJ mol^{-1}. Its apperent

K_m and V_{max} for soluble starch were 0.04 mg ml^{-1} and 660 µ mol glucose mg^{-1} protein min^{-1}.

Two α-amylases from the thermophilic fungus, *T. lanuginosus* ATCC 34626 (wild and mutant strains), were purified to homogeneity by a simple procedure including, consecutively, precipitation with ice-cold 2-propanol, anion-exchange and molecular-sieve chromatographic methods (Petrova et al., 2000). The molecular masses of the purified α-amylases (both with pI values of 3.0) were 58 kDa by SDS-PAGE. The optimal pH of α-amylase activity was 5.0 for the wild enzyme and 4.5 for the mutant one.

Han et al. (2013) studied a novel α-amylase from the thermophilic fungus, *Malbranchea cinnamomea*. The purified enzyme displayed optimal activity at pH 6.5 and 65°C with a molecular mass of 60.3 kDa. The enzyme was stable over a broad pH range (pH 5.0–10.0) and showed broad substrate specificity.

Lipases

Lipases (EC 3.1.1.3) catalyze the hydrolysis of triglycerols and the synthesis of esters from glycerols and long chains fatty acids. In detergents, the alkaline lipases (pH 10.0–11.0) with a temperature range of 30–60°C are preferred (Table 2). Thermophilic moulds are also reported to produce lipases. Arima et al. (1968) purified an extracellular lipase from *Humicola lanuginosa* strain Y-38, isolated from compost in Japan. The fungi produced lipase in a medium containing soybean oil, starch, corn steap liquor, and antifoaming agent. It was purified to homogeneity by ammonium sulphate precipitation, followed by dialysis, ion exchange and gel filteration chromatography, with 30% recovery. The protein was a monomeric 27.5 kDa protein with optimal at pH 8.0 and at a temperature of 60°C. A more thermostable enzyme was produced by a *Humicola lanuginosa* strain in a medium containing sorbitol, corn steap liquor, silicone oil as an antifoaming agent and whale or castor oil as an inducer (Omar

Table 2. List of industries using enzymes secreted by thermophilic moulds.

S. No.	Industry	Enzymes used
1.	Animal food and feed	Xylanase, Phytase, Cellulase
2.	Textile	Cellulase, Laccase
3.	Detergent	Protease, Cellulase, Lipase, Amylase
4.	Starch	α-Amylase, Glucoamylase, Glucose isomerase
5.	Pulp and paper	Xylanase, Phytase
6.	Fruit juice	Pectinase, Cellulase, Xylanase
7.	Baking	α-Amylase, Glucoamylase, Xylanase, Glucose oxidase
8.	Dairy	Rennin, Lactase, Protease
9.	Brewing	Glucanase, papain

et al., 1987). It was purified with acetone precipitation followed by successive chromatographic steps. This enzyme was stable at 60°C for 20 hours having an optimum neutral pH.

A lipase from *H. lanuginosa* was cloned and overexpressed in *Aspergillus oryzae* (Huge-Jensen et al., 1989). The recombinant enzyme was purified by a two step procedure involving hydrophobic interaction chromatography and ion exchange chromatography. A recombinant lipase of *R. miehei* was expressed in *A. oryzae* and heterologous expression did not affect the characteristics of the wild enzyme (Huge-Jensen et al., 1989). Thermophilic strain of *Rhizopus arrhizus* accumulates an acidic lipase in culture fluid when grown in a medium containing groundnut oil, milk powder and inorganic salts (Kumar et al., 1993). Addition of 2.0% groundnut oil yielded higher enzyme activity. Soyabean meal and arabinose were found to be the best nitrogen and carbon sources for enzyme production, respectively. Esterification of α-terpineol with acetic anhydride or propionic acid mediated by *R. miehei* lipase was subjected to a response surface study in order to optimize conditions for maximum esterification (Rao and Divaker, 2002). The variables were enzyme/substrate (acid) ratio, α-terpineol concentration and incubation period using lipase from *R. miehei*. Between acetic anhydride and propionic acid, the former showed better yields at lower enzyme/substrate ratios than the latter. Yields predicted by the models for α-terpinyl propionate and α-terpinyl acetate formation were found to be in agreement with the experimentally determined ones. Lipases, stable at pH 10–11 and 30–60°C obtained from *H. lanuginosa* and *Rhizomucor miehei* are used in detergents for the removing oil stains (Maheshwari, 2012).

Cellulolytic Enzymes

Cellulose, a polysaccharide of β-1, 4-linked D-glucosyl units, is the major component of plant cell walls and is one of the most abundant biopolymers in nature. Cellulases (cellobiohydrolases and endoglucanases) are enzymes that catalyse the hydrolysis of cellulose to smaller oligosaccharides, a process of paramount importance in biotechnology. Bhat and Maheshwari (1987) demonstrated that the β-glucosidase, endo- and exoglucanase activities in the culture filterate of their best strain of *S. thermophile* were lower than that in *Trichoderma reesei*. Despite these lower activities, *S. thermophile* degraded cellulose faster than *T. reesei*. Extracellular cellulolytic enzymes were produced under solid state cultivation by the thermophilic fungus *Thermoascus aurantiacus* (Kalogeris et al., 2003). Under optimal growth conditions endoglucanase and β-glucosidase activities of 1572 and 101.6 Units per g of carbon source were obtained, respectively. The zymogram indicated endoglucanase and β-glucosidase with pI values of 3.5 and 3.9, respectively. Major cellulase gene of *Humicola insolens* FERM BP-5977 showed high homology with other family 6 cellulases (Moriya et al., 2003). *Melanocarpus albomyces* produced a 20 kDa endoglucanase known as 20K-cellulase that has been found particularly useful in the textile industry (Hirvonen and Papageorgiou, 2003). A gene (cbh2) encoding cellobiohydrolase II was isolated from the fungus *Talaromyces emersonii* by rapid amplification of cDNA ends techniques and the equivalent genomic sequence was subsequently cloned (Murray et al., 2003). DNA sequencing revealed that cbh2 has an open reading frame of 1377 bp, which encodes a putative polypeptide of 459 amino acids, and is interrupted by seven introns. Expression of the *T. emersonii* cbh2 gene is induced by cellulose, xylan, xylose, and gentiobiose and clearly repressed by glucose. The culture medium composition was optimized, on a shake-flask scale, for simultaneous production of high activities of endoglucanase and β-glucosidase by *Thermoascus aurantiacus* using statistical factorial designs (Gomes et al., 2000). The optimized medium containing Solka Floc as the carbon source and soymeal as the organic nitrogen source yielded 1130 nkat ml^{-1} endoglucanase and 116 nkat ml^{-1} β-glucosidase activities after 264 h as shake cultures. In addition, good levels of β-xylanase (3479 nkat ml^{-1}) and low levels of filter-paper cellulase, β-xylosidase, β-L-arabinofuranosidase, β-mannanase, β-mannosidase, β-galactosidase and β-galactosidase were detected. Batch fermentation in a 5L laboratory fermentor using the optimized medium resulted in the production of 940 nkat ml^{-1} endoglucanase and 102 nkat ml^{-1} β-glucosidase after 192 h.

Scytalidium thermophilum type culture *Torula thermophila* was isolated from mushroom compost and the total cellulase, endoglucanase, Avicel-adsorbable endoglucanase activities, as well as the fungal biomass generation and cellulose utilisation were analyzed in shake flask cultures with Avicel as the carbon source (Arifolua and Ögel, 2000). The pH and temperature optima for endoglucanase activities were pH 6.0 and 65°C for *Torula thermophila*, and pH 6.5 and 60°C for *Humicola insolens*. The mould *T. thermophila* can grow and produce cellulases in the range of 35 to 55°C with optima at 45°C.

A gene encoding a thermo-stable endo-beta-1,4-glucanase was isolated from the thermophilic fungus, *Thermoascus aurantiacus* IFO9748, and designated as eg1 (Hong et al., 2003). Induction of this gene expression at 50°C was stronger than at 30°C. The deduced amino acid sequence encoded by eg1 showed that it belongs to the glycoside hydrolase family 5. The cloned gene was expressed in *Saccharomyces cerevisiae* and the gene product was purified and characterized. No significant activity loss was detected over 2 h at 70°C and the product was stable from pH 3–10. The enzyme was optimally active at pH 6.0 and 70°C.

An extracellular β-glucosidase from *Thermoascus aurantiacus* was purified to homogeneity by DEAE-Sepharose, Ultrogel AcA 44 and Mono-P column chromatography (Parry et al., 2001). The enzyme was a homotrimer, with a monomer molecular mass of 120 kDa; only the trimer was optimally active at 80°C and at pH 4.5. The enzyme had the lowest K_m towards p-nitrophenyl β-D-glucoside (0.1137 mM) and the highest k_{cat} towards cellobiose and β-trehalose (17052 min^{-1}). The thermophilic fungus *Chaetomium thermophilum* var. *coprophilum* produced large amounts of extracellular and intracellular β-glucosidase activity when grown on cellulose or cellobiose as carbon sources (Venturi et al., 2002). Charavgi et al. (2013) determined the structures of the thermophilic StGE2 esterase from *M. thermophila*, a member of the CE15 family, and its S213A mutant at 1.55 and 1.9 Å resolution, respectively. All of the three-dimensional protein structures have been shown to contain α/β-hydrolase fold with three-layer αβα-sandwich architecture and a Rossmann topology and comprise one molecule per asymmetric unit. The β-glucosidase gene (*bgl3a*) from *M. thermophila* was cloned and expressed in *P. pastoris* after the excision of one intron and the secreting signal peptide under the control of the strong alcohol oxidase promoter in the plasmid pPICZαC (Karnaouri et al., 2013). The recombinant enzyme had a molecular mass of 90 kDa and displayed its optimal activity at 5.0 and 70°C. Recombinant *P. pastoris* efficiently secreted high level of enzymatic activity (41 U ml^{-1}) after 192 h of growth, under methanol induction. The enzyme was able to hydrolyze low molecular weight substrates containing β-glucosidic residues. The K_m value was found to be 0.39 mM on p-β-NPG and 2.64 mM on cellobiose.

An endoglucanase from *M. thermophila* was functionally expressed in *P. pastoris* (Karnaouri et al., 2014). The purified recombinant enzyme showed a molecular mass of 65 kDa and exhibited high activity on substrates containing β-1, 4-glycosidic bonds (carboxymethyl cellulose, barley β-glucan, and cello-oligosaccharides) as well as xylan-containing substrates (arabinoxylan and oat spelt xylan).

A novel β-glucosidase from *P. thermophila* was cloned and expressed in *Pichia pastoris* by Yang et al. (2013) that showed significant similarity to other fungal β-glucosidases from glycoside hydrolase family 1. Purified recombinant enzyme exhibited broad substrate specificity. The K*m* values for pNP-β-D-glucopyranoside, cellobiose, gentiobiose and salicin were 0.55 mM, 1.0 mM, 1.74 mM and 6.85 mM, respectively.

Xylanolytic Enzymes

These are the xylan hydrolyzing enzymes. Xylan is the most abundant structural polysaccharide, next to cellulose in nature. Xylanases of thermophilic fungi are receiving considerable attention due to their application in biobleaching of pulp in the paper industry, where the enzymatic removal of xylal from lignin-carbohydrate complexes facilitates the leaching of lignin from the fibre cell wall, obviating the need of chlorine for pulp bleaching in the brightening process. They also have application in animal feed to improve its digestibility (Table 2). The majority of xylanases have pH and temperature optima between 4.5–6.5 and 55–65°C, respectively. Pure xylan and other xylan rich natural substrates like sawdust, wheat bran, corncob, and sugarcane bagasse, have been used for the induction of xylanases in microorganisms. Production of xylanase by *Thermoascus aurantiacus* in solid-state fermentation was enhanced by optimization of C and N sources, inoculum size, moisture level, and particle size of the C source (Kalogeris et al., 1998). Xylanases are known to be co-induced with cellulases as in *T. aurantiacus* (Tong et al., 1980), *Chaetomium thermophile* var. *Coprophile* (Ganju et al., 1989). *Malbranchea pulchella* var. *sulfurea* also produced an extracellular xylosidase, but it was periplasmic in case of *Humicola grisea* var. *thermoidea* and *Talaromyces emersonii*. *Allescheria terrestris, Chaetomium thermophile, Aspergillus wentii,* and *Aspergillus versicolor* produced thermostable endoxylanases during submerged fermentation at 40°C (Kvesitadze et al., 1998). *Melanocarpus albomyces* isolated from compost produced cellulase-free xylanase in culture medium containing bagasse (Prabhu and Maheshwari, 1999). Size-exclusion and anion-exchange chromatography separated four xylanases in the culture filtrate. The medium components for the production of extracellular xylanase by *Thermoascus aurantiacus* was optimized in shake-flask culture using the Box-Wilson method and a central composite design (Gomes et al., 1994). The optimized culture conditions produced 5347.4 U ml^{-1} of xylanase, 3.0 U ml^{-1} of β-xylosidase, 1.0 U ml^{-1} of acetyl esterase, 89.7 U ml^{-1} of acetyl xylan esterase and 3.5 U ml^{-1} of α-arabinosidase. The xylanase and β-xylosidase were optimally active at pH 5.0 and 75–80°C.

An endo-beta-1,4-xylanase of *Sporotrichum thermophile* ATCC 34628 was purified to homogeneity by Q-Sepharose and Sephacryl S-200 column chromatographies (Katapodis et al., 2003). The enzyme has a molecular mass of 25 kDa, an isoelectric point of 6.7, and is optimally active at pH 5 and at 70°C. Xylanase was produced by solid-state fermentation using *Thermoascus aurantiacus* (dos Santos et al., 2003). Maximum production (500 U g^{-1} bagasse) was achieved on the sixth day of cultivation on solid sugarcane bagasse medium supplemented with 15% (v/w) rice bran extract. Xylanase of *Melanocarpus albomyces* IIS 68 was immobilized on Eudragit L-100 (Roy et al., 2003). The K$_{m}$ of the enzyme increased from 5.9 mg ml^{-1} to 9.1 mg ml^{-1} upon immobilization. The V$_{max}$ of the immobilized enzyme showed an increase from 90.9 micro mol ml^{-1} min^{-1} (for the free enzyme) to 111.1 micro mol ml^{-1} min^{-1}. The immobilized enzyme could be reused up to ten times without impairment of the xylanolytic activity.

An extracellular xylanase was purified to homogeneity from the culture filtrate of *Thermomyces lanuginosus*-SSBP, and its biochemical characteristics were studied (Lin et al., 1999). A yield of 70–80% was achieved through the procedures of 80%-satd. ammonium sulphate precipitation, DEAE-Sephadex A25 and quaternary aminoethyl (QAE)-Sephadex A25 column chromatography. The molecular mass of the purified xylanase was 23.6 kDa with a pI value of 3.8.

The culture supernatant of *Thermomyces lanuginosus* strains MED 2D and MED 4B1 had high activities of xylanase with low inducible activities of β-xylosidase (Haarhoff et al., 1999). The crude xylanase was optimally active at 70°C and at pH 6.0 to 6.5. Properties of an endo-xylanase produced by a locally isolated *Thermomyces lanuginosus* strain SSBP was compared to seven other *T. lanuginosus* strains isolated from different geographical regions

(Singh et al., 2000). Strain SSBP produced the highest xylanase activity of 59600 nkat ml^{-1} when cultivated on corn cobs (maize) medium. The optimal temperature and pH for xylanase production by the strains was either 40 or 50°C and between pH 6 and 7, respectively. Optimal xylanase activity was observed at pH 6 or 6.5 and 70°C having a molecular mass of 24.7 kDa and pI 3.9.

   Chaetomium cellulolyticum produced three xylanases with molecular weights of 25, 47, and 57 kDa and pIs of 8.9, 8.4, and 5.0, respectively (Baraznenoka et al., 1999). According to their biochemical characteristics and specificity, the 25- and 47-kDa xylanases were related to the family of G and the 57-kDa xylanase to the family of F. These xylanases had a neutral pH optimum within the range of 6–7 and the 25-kDa xylanase maintained high activity at alkaline pH up to 10. An important characteristic of the 25-kDa xylanase was its high stability at pH 9.

   Xylanase production by the thermophilic fungus, *Melanocarpus albomyces* IIS 68, during solid state fermentation of wheat straw was studied and the effects of various variables were observed using response surface methodology (Narang et al., 2001). The optimum levels of the variables (600–850 m particle size, 43 h inoculum age, 1.37% Tween 80, 86% initial moisture content, 5.1% urea, 0.74% yeast extract and a harvest time of 96 h) resulted in xylanase activity of 7760 U g^{-1} initial dry substrate. Xylanase production by *M. thermophila* was studied with 1% xylan at pH 6.0 (Yadav and Jaitly, 2011). The mould showed the highest activity at 2% salt level. Two novel GH11 endo-xylanases (Xyl7 and Xyl8) from *M. thermophila* C1, were purified to study the effect of solubility and molecular structure of various xylans on their efficiency (Gool et al., 2013). The two GH11 xylanases released different products from the xylans due to the presence of a specific residue at position 163 in the amino acid sequence of Xyl8 as tyrosine and valine in Xyl7. Both xylanases were more efficient on self-associated xylan compared to C1 GH10 endo-xylanases and they released more small xylooligomers from these xylans.

   Endoxylanase of *M. thermophila* was used for XOS production by xylan hydrolysis at pH 7.0 and 45°C (Sadaf and Khare, 2014). The products analyzed by HPLC revealed the presence of xylobiose, xylotriose and xylotetraose. Two novel 12 xyloglucanase genes were cloned from *R. miehei* by (Song et al., 2013). The deduced amino acid sequences shared 68% identity with each other and less than 60% with other xyloglucanases. Both enzymes expressed in *E. coli* displayed very high specific activities toward tamarind xyloglucan, but no activity toward carboxymethylcellulose, Avicel, or p-nitrophenyl derivatives.

Pectinases

Pectin is an important constituent of plant cell wall, which is a heteropolysaccharide mainly comparising of polygalacturonic acid linked by α-1,4 linkages. It plays an important role by maintaining the integrity of the plant cell wall and middle lamella. The major pectin hydrolyzing enzyme, pectinase (also known as polygalacturonase) helps in the decomposition of vegetable matter (Table 2). Thermophilic fungi are also known to produce pectinases. Craveri et al. (1967) described pectinase from *Penicillium duponti, Humicola stellata, Humicola lanuginosa, Mucor pusillus* and *Humicola insolens.* The thermophilic fungus *Talaromyces thermophilus* was found to produce polygalacturonate hydrolase in stationary liquid culture (Tong and Cole, 1975). The enzyme production was highest with sodium polypectate as the substrate and it showed activity at 50°C but not at 25°C. Deploey (1976) screened ten thermophilic fungi for growth in stationary culture. With pectin as the sole carbon source, the growth was poor for all the fungi as compared with the growth on glucose. However moderate growth on pectin and formation of reducing substances during growth may indicate that pectinolytic enzymes were formed by all the fungi. This investigation was further carried out by Adams and Deploey (1978) who spotted pectinolytic enzymes when thermophilic fungi were grown on petridishes on agar with pectin. Polygalacturonase and pectate lyase were produced by *Sporotrichum thermophile* whereas, only the pectate lyase was found with *Mucor miehei, Populaspora thermophila, Talaromyces leycettanus* and *Thermoascus aurantiacus.*

Tuohy et al. (1989) studied pectinolytic activities from the solid state culture of 4 different strains of thermophilic fungi (*Chaetomium thermophile, Talaromyces emersonii* CBS 814.70, *Talaromyces emersonii* UCG 208 and *Thermoascus aurantiacus*) and found pectinase and poly-galacturonase with all strains, whereas pectin lyase was only found in *Chaetomium thermophile.* Whitehead and Smith (1989) studied the production of pectinase by *Sporotrichum thermophile* in both static as well as shake flask cultures. Static culture had a higher yield than shake flask culture. *Sporotrichum thermophile* is also known to produce pectinase in solid as well liquid state fermentation using citrus peal as substrate (Pandey, 2003). The mould produces 330 fold higher enzyme titres in SSF than in SmF (submerged fermentation). Optimization of enzyme production by response surface methodology enhanced enzyme secretion from 5039 U L^{-1} to 5120.32 U L^{-1} and lactose and pectin were found to be important parameters for attaining maximum enzyme secretion by the mould. The enzyme showed maximum activity at 60°C and pH 7.0.

Proteases

Proteases are protein degrading enzymes classified on the basis of a critical amino acid required for the catalytic function (e.g., serineprotease), the pH optimum of their activity (acidic, neutral, or alkaline protease), their site of cleavage (e.g., aminopeptidases, which act at the free N terminus of the polypeptide chain, or carboxypeptidases, which act at the C terminus of the polypeptidechain), or their requirement of a free thiol group (e.g., thiolproteinase).

Proteases have long been used in the food, dairy, and detergent industries and also for leather processing (Table 2). The need to overcome the limitation of obtaining chymosin, the milk-curdling enzyme from the stomach contents of milk-feeding calves, which is used in the industrial preparation of cheese, led to a search for substitutes. Arima et al. (1968) screened about 800 microorganisms and obtained a soil isolate of *Mucor pusillus* that produced an enzyme with a high ratio of milk-clotting to proteolytic activity, enabling the production of high yields of curds. The *Mucor rennins* were produced by growing the fungus on wheat bran, from which they were extracted with water (Arima et al., 1968). The crude extract was then purified and crystallized. Since the rennins hydrolyzed both casein and haemoglobin optimally at pH 3.7, they were classified as acid protease.

Malbranchea pulchella var. *sulfurea* and *Humicola lanuginosa* produced proteases during active growth in the presence of 2 and 8% (wt/vol) casein, respectively, suggesting that the enzyme was induced by external protein substrate (Stevenson and Gaucher, 1975). The production of protease by *M. pulchella* var. *sulfurea* was repressed by glucose, peptides, amino acids, or yeast extracts (Voordouw et al., 1974).

Shenolikar and Stevenson (1982) purified an alkaline protease of *Humicola lanuginosa* in one step based on its specific binding to an organomercury-Sepharose column, from which the enzyme was selectively eluted with a buffer containing mercuricchloride. Hasnain et al. (1992) purified a protease by hydrophobic affinity chromatography and shown to have a pH optimum of 8.0 and a molecular mass of 38 kDa. The production of extracellular enzymes by the *Thermomyces lanuginosus* was studied in chemostat cultures at a dilution rate of 0.08 h⁻¹ in relation to variation in the ammonium concentration in the feed medium (Jensen et al., 2002). *Scytalidium thermophilum* produced extracellular proteases on microcrystalline cellulose which were optimally active at pH 6.5–8 and 37–45°C (Ifrij and Ögel, 2002).

The full-length cDNA coding for serine proteases of *C. thermophilum* was generated using RACE-PCR (Li and Li, 2009). Recombinant enzyme expressed in *P. pastoris* was secreted into the culture medium. Purified recombinant enzyme showed optimal activity at pH 8.0 and 60°C. A serine protease from *Thermoascus aurantiacus* var. *levisporus* was cloned, sequenced, and expressed in *P. pastoris* (Li et al., 2011). Recombinant enzyme was expressed in *P. pastoris* with specific activity of 115.58 U mg⁻¹. The enzyme displayed broad substrate specificity and it was inhibited by PMSF, but not by DTT or EDTA.

Thermophilic Moulds and Management of Environmental Pollution

Environmental pollution is one of the most terrible dangers that deal with the man-kind today. The increased human activity has led to the expansion of industries at an accelerated rate which in turn has resulted in the deterioration of environment. Biological cleaning procedures make use of the fact that most organic chemicals are degraded by the action of microbes. Thermophilic moulds have been extensively employed in the disposal of organic matter and toxic chemicals from domestic and industrial wastes (Singh and Satyanarayana, 2009b). In the natural habitats, they colonize the organic matter and degrade their major components by the action of a large number of hydrolytic enzymes (Table 3). The hydrolyzed products are utilized by these moulds for their growth and development.

Table 3. Thermophilic moulds in environmental management.

Environmental aspect	Thermophilic mould	Substrate	Product	References
Solid waste management	*Chaetomium cellulolyticum*	Agricultural residue	Upgraded feed, SCP, enzymes	Satyanarayana et al., 1992
	Mucor miehei	--	Protease	Thakur et al., 1993
	Thermomucor indicae-seudaticae	Wheat bran	Glucoamylase	Kumar and Satyanarayana, 2004
	Sporotrichum thermophile	Sesame oil cake	Phytase	Singh and Satyanarayana, 2006a
	S. thermophile	Citrus peel	Xylanase, pectinase, cellulase	Kaur and Satyanarayana, 2004
	Humicola lanuginosa	Wheat bran	Xylanase	Kamra and Satyanarayana, 2004
	Chaetomium thermophilum, Chaetomium sp., Malbranchea sulfurea, Myriococcum thermophilum, Scytalidium thermophilum, Stilbella thermophila, Thielavia terrestris	Wheat straw, manure	Mushroom	Straatsma et al., 1994
Bioremediation	*Talaromyces emersonii*	Polluted water	Uranium	Bengtsson et al., 1995
	Mucor sp., Rhizopus sp.	Polluted water	Heavy metal	
Bioethanol	Thermophilic moulds	Agricultural residue	Sugars, ethanol	Hahn-Hagerdal et al., 2006

Solid Waste Management and Composting

A wide range of agricultural and forest residues are degraded by thermophilic moulds. Available substrates such as starch, cellulose, hemicellulose and lignin are utilized by these fungi (Sharma and Johri, 1992,

Table 3). Several attempts have been made to produce protein-enriched upgraded feeds, single cell protein (SCP) and some enzymes by solid state fermentation (SSF) using thermophilic moulds (Satyanarayana et al., 1992). *Chaetomium cellulolyticum* has been used in the conversion of forest and agricultural residues into protein enriched feed. The SCP products are nutritious, digestible and non toxic in animal feed protein rations. Thakur et al. (1993) produced fungal rennet using *Mucor miehei* in SSF on a large scale in trays. Kalogeris et al. (1998) reported the production of endoxylanase by solid state fermentation of wheat straw. *Thermomucor indicae-seudaticae* secreted 10-fold higher glucoamylase in SSF than SmF using wheat bran as a substrate (Kumar and Satyanarayana, 2004). *S. thermophile* secreted phytase in SSF using sesame oil cake as the substrate (Singh and Satyanarayana, 2006a). In solid-state fermentation, wheat bran in combination with citrus peel supported maximum xylanolytic, pectinolytic and cellulolytic enzyme secretion by *S. thermophile* (Kaur and Satyanarayana, 2004; Kaur et al., 2004). Among the lignocellulosic substrates tested, wheat bran supported a high xylanase secretion by *Humicola lanuginosa* in solid-state fermentation which was 23-fold higher than that in submerged fermentation (Kamra and Satyanarayana, 2004).

Mushroom composting is the conversion of solid organic residues into simpler substrate for mushroom cultivation by the action of microorganisms. The white button mushroom (*Agaricus bisporus*) is cultivated on compost prepared from the mixture of wheat straw, horse manure, chicken manure, and gypsum (Table 3). Thermophilic moulds play an important role in the mushroom composting due to their dense mycelial growth on the substrate. These fungi penetrate the dense and tight areas of composting substrate or into the balls of the compost (Straatsma et al., 1994). Of the 34 species of thermophilic fungi tested, 9 strains promoted the mycelial growth of *Agaricus bisporus* on sterilized compost.

Bioethanol from Lignocellulosic Biomass

The most common renewable fuel today is bioethanol produced from the sugar or grain which is not adequate to support the fuel requirement of world transportation. One of the greatest challenges for society in the 21st century would be to meet the growing demand for energy for transportation, heating and industrial processes, and to provide raw material for the industry in a sustainable way. Liquid biofuels from renewable resources such as lignocellulosic materials will have a substantial role in meeting these goals. Ethanol can be blended with petrol or used as neat alcohol in vehicle engines due to higher octane number and high heat of vaporization (Farrell et al., 2006).

The thermophilic moulds have the ability to degrade the lignocellulosic materials by secreting an array of enzymes, which convert the agro-residues into fermentable sugars that can be further fermented to ethanol. The biofuels from lignocellulose will generate low greenhouse gas thus reducing environmental impacts, particularly climate change. In comparison to petrol and diesel, ethanol contains oxygen which results in the improved combustion and lower emissions of unburnt hydrocarbons, CO and particulate matter (Hahn-Hagerdal et al., 2006). Therefore, the use of biofuels would be more eco-friendly than petrol and diesel, which release a large number of greenhouse gases and other pollutants in the environment.

Bioremediation of Heavy Metals and Industrial Effluents

The occurrence of heavy metals represents a significant environmental hazard. The fungal biomass has been utilized for biosorption of heavy metals. Biomass of a thermophilic mould *Talaromyces emersonii* CBS 814.70 showed high biosorption of uranium (Bengtsson et al., 1995). Some species of *Mucor* and *Rhizopus* are also useful in accumulation and removal of heavy metals from waste water and mining operations (Gadd, 1990).

Water and soil pollution, by industrial effluents, have become a matter of great concern. The waste water from citrus processing industry contains pectinaceous materials that are not decomposed by the microbes during activated sludge treatment. Thermophilic moulds are well known producers of alkaline pectinases. *Sporotrichum thermophile* is also known to produce pectinase in solid state as well as submerged fermentation using wheat bran and citrus peal as the substrates (Kaur and Satyanarayana, 2004). The enzyme treated fruit pulps yielded more juices with high amount of total and reducing sugars (Kaur et al., 2004). The pectinases are used in reducing the level of pectic substances in the processing of fruits and juice industries and in the reduction of pectic wastes in the effluents. The industrial effluent from oil industries is rich in oil, and therefore, lipase-producing thermophilic moulds could find application in the treatment of such effluents. Thermophilic strain of *Rhizopus arrhizus* accumulates an acidic lipase in culture fluid when grown in a medium containing groundnut oil, milk powder and inorganic salts (Kumar et al., 1993). Arima et al. (1968) purified an extracellular lipase from *Humicola lanuginosa* strain Y-38, isolated from compost in Japan. The mould produced lipase in a medium containing soybean oil, starch, and corn steep liquor. *Rhizomucor miehei* and *Thermomyces lanuginosus* are well known lipase producing thermophilic moulds (Noel and Combes, 2003).

Taha et al. (2014) studied the decolourizing abilities of a thermophilic fungus, *Thermomucor indicae-seudaticae* using azure B, congored, trypan blue and remazol brilliant blue R. Inactivated biomass was more effective in decolorization than living biomass of *T. indicae-seudaticae*, *A. fumigatus* and their combined co-culture. Inactivated *T. indicae-seudaticae* was a faster and more effective dye decolourizer in the temperature range of 30–55°C at 100, 500 and 1000 mg l^{-1} concentrations than either *A. fumigatus* or the combined culture; acidic pH favoured effective adsorption by *T. indicae-seudaticae*.

Conclusions

Thermophilic moulds degrade the organic matter efficiently as their mycelia penetrate deeper into the substrate and produce extracellular enzymes. The degraded organic matter is used by the moulds to generate energy. The enzymes from thermophilic moulds are highly thermostable and sturdy, and therefore, can be exploited for industrial applications. The thermophilic moulds also play an important role in the environmental management by degrading organic matter in the industrial effluents, heavy metal biosorption and in bioethanol production from lignocellulosic residues. Thus, these microorganisms can further be exploited for the betterment of human and environmental lives.

Acknowledgements

The authors gratefully acknowledge the financial assistance provided by University Grant Commission (UGC), New Delhi, India during the course of writing this chapter.

References

Adams, R. and Deploey, J.J. 1978. Enzymes produced by thermophilic fungi. Mycologia, 70: 906–910.

Aikawa, J., Yamashita, T., Nishiyama, M., Horinouchi, S. and Beppu, T. 1990. Effects of glycosylation on the secretion and enzyme activity of *Mucor* rennin, an aspartic proteinase of *Mucor pusillus*, produced by recombinant yeast. J. Biol. Chem., 265: 13955–13959.

Aquino, A.C., Jorge, J.A., Terenzi, H.F. and Polizeli, M.L. 2001. Thermostable glucose-tolerant glucoamylase produced by the thermophilic fungus *Scytalidium thermophilum*. Folia Microbiol., 46(1): 11–16.

Archana, A. and Satyanarayana, T. 2001. Biodiversity and potential applications of thermophilic moulds. pp. 70–84. *In*: Manoharachari, C., Bagyanarayana, G., Bhadraiah, B., Reddy, B.N., Satyaprasad, K. and Nagamani, A. (eds.). Frontiers in Fungal Biodiversity and Plant Pathogen Relations. Allied Publishers Limited, India.

Arifolua, N. and Ögel, Z.B. 2000. Avicel-adsorbable endoglucanase production by the thermophilic fungus *Scytalidium thermophilum* type culture *Torula thermophila*. Enzyme Microb. Technol., 27(8): 560–569.

Arima, K., Iwasaki, S. and Tamura, G. 1968. Milk clotting enzymes from microorganisms. V. Purification and crystallization of *Mucor* rennin from *Mucor pusillus* Lindt. Appl. Microbiol., 16: 1727–1733.

Arnesen, S., Eriksen, S.H., Olsen, J.O. and Jensen, B. 1998. Increased production of α-amylase from *Thermomyces lanuginosus* by the addition of Tween 80. Enzyme Microb. Technol., 23(3-4): 249–252.

Baraznenoka, V.A., Becker, E.G., Ankudimova, N.V. and Okunev, N.N. 1999. Characterization of neutral xylanases from *Chaetomium cellulolyticum* and their biobleaching effect on eucalyptus pulp. Enzyme Microb. Technol., 25(8-9): 651–659.

Bengtsson, L., Johansson, B., Hackett, T.J. and McHale, A.P. 1995. Studies on the biosorption of uranium by *Talaromyces emersonii* CBS 814.70 biomass. Appl. Microbiol. Biotechnol., 42: 807–811.

Berka, R.M., Grigoriev, I.V., Otillar, R., Salamov, A., Grimwood, J., Reid, I., Ishmael, N., John, T., Darmond, C., Moisan, M.C., Henrissat, B., Coutinho, P.M., Lombard, V., Natvig, D.O.,

Lindquist, E., Schmutz, J., Lucas, S., Harris, P., Powlowski, J., Bellemare, A., Taylor, D., Butler, G., de Vries, R.P., Allijn, I.E., van den Brink, J., Ushinsky, S., Storms, R., Powell, A.J., Paulsen, I.T., Elbourne, L.D., Baker, S.E., Magnuson, J., Laboissiere, S., Clutterbuck, A.J., Martinez, D., Wogulis, M., de Leon, A.L., Rey, M.W. and Tsang, A. 2011. Comparative genomic analysis of the thermophilic biomass-degrading fungi *Myceliophthora thermophila* and *Thielavia terrestris*. Nat. Biotechnol., 29(10): 922–1007.

Berka, R.M., Rey, M.W., Brown, K.M., Byun, T. and Klotz, A.V. 1998. Molecular characterization and expression of a phytase gene from the thermophilic fungus *Thermomyces lanuginosus*. Appl. Environ. Microbiol., 64: 4423–4427.

Bhat, K.M. and Maheshwari, R. 1987. *Sporotrichum thermophile*: growth, cellulose degradation, and cellulase activity. Appl. Environ. Microbiol., 53: 2175–2182.

Bilai, T.I., Chernyagina, T.B., Dorokhov, V.V., Poedinok, N.L., Zakharchenko, V.A., Ellanskaya, I.A. and Lozhkina, G.A. 1985. Phosphatase activity of different species of thermophilic and mesophilic fungi. Mikrobiologia, 47: 53–56.

Campos, L. and Felix, C.R. 1995. Purification and characterization of a glucoamylase from *Humicola grisea*. Appl. Environ. Microbiol., 2436–2438.

Chadha, B.S., Gulati, H., Minhas, M., Saini, H.S. and Singh, N. 2004. Phytase production by the thermophilic fungus *Rhizomucor pusillus*. World J. Microbiol. Biotechnol., 20: 105–109.

Chadha, B.S., Singh, S., Vohra, G. and Saini, H.S. 1997. Shake culture studies for the production of amylases by *Thermomyces lanuginosus*. Acta Microbiol. et Immunol. Hung., 44(2): 181–185.

Charavgi, M.D., Dimarogona, M., Topakas, E., Christakopoulos, P. and Chrysina, E.D. 2013. The structure of a novel glucuronoyl esterase from *Myceliophthora thermophila* gives new insights into its role as a potential biocatalyst. Acta Crystallogr. D-Biol. Cryst., 69(Pt 1): 63–73.

Cooney, D.G. and Emerson, R. 1964. Thermophilic Fungi: An Account of their Biology, Activities and Classification, W.H. Freeman and Co., San Francisco, USA.

Craveri, R., Craveri, A. and Guicciardi, A. 1967. Research on the properties and activities of enzymes of eumycete thermophilic isolates of soil. Anal. Microbiol. Enzymol., 17: 1–30.

Deploey, J.J. 1976. Pectin utilization by thermophilic fungi. Proc. Pa. Acad. Sci., 50: 179–181.

dos Santos, E., Piovan, T., Roberto, I.C. and Milagres, A.M. 2003. Kinetics of the solid state fermentation of sugarcane bagasse by *Thermoascus aurantiacus* for the production of xylanase. Biotechnol. Lett., 25(1): 13–16.

Etoh, Y., Shoun, H., Beppu, T. and Arima, K. 1979. Physicochemical and immunochemical studies on similarities of acid proteases *Mucor pusillus* rennin and *Mucor miehei* rennin. Agric. Biol. Chem., 43: 209–215.

Farrell, A.E., Plevin, R.J., Turner, B.T., Jones, A.D., O'Hare, M. and Kammen, D.M. 2006. Ethanol can contribute to energy and environmental goals. Science, 311(5760): 506–508.

Gadd, G.M. 1990. Metal tolerance. pp. 178–210. *In*: Edwards, C. (ed.). Microbiology of Extreme Environments. McGraw-Hill Publishing Company. New York.

Ganju, R.K., Murthy, S.K. and Vithayathil, P.J. 1989. Purification and characterization of two cellobiohydrolases from *Chaetomium thermophile* var. *coprophile*. Biochim. et Biophys. Acta, 993: 266–274.

Gomes, D.J., Gomes, J. and Steiner, W. 1994. Production of highly thermostable xylanase by a wild strain of thermophilic fungus *Thermoascus aurantiacus* and partial characterization of the enzyme. J. Biotechnol., 37(1): 11–22.

Gomes, I., Gomes, J., Gomes, D.J. and Steiner, W. 2000. Simultaneous production of high activities of thermostable endoglucanase and β-glucosidase by the wild thermophilic fungus *Thermoascus aurantiacus*. Appl. Microbiol. Biotechnol., 53(4): 461–468.

Gool, M.P., van Muiswinkel, G.C., Hinz, S.W., Schols, H.A., Sinitsyn, A.P. and Gruppen, H. 2013. Two novel GH11 endo-xylanases from *Myceliophthora thermophila* C1 act differently toward soluble and insoluble xylans. Enzyme Microb. Technol., 53(1): 25–32.

Guimarães, L.H.S., Terenzi, H.F., Jorge, J.A. and Polizeli, M.L.T.M. 2001. Thermostable conidial and mycelial alkaline phosphatases from the thermophilic fungus *Scytalidium thermophilum*. J. Indust. Microbiol. Biotechnol., 27(4): 265–270.

Gupta, A.K. and Gautam, S.P. 1993. Purification and properties of an extracellular α-glucosidase from thermophilic fungus *Malbranchea sulfurea*. J. Gen. Microbiol., 139: 963–967.

Haarhoff, J., Moes, C.J., Cerff, C., van Wyk, W.J., Gerischer, G. and Janse, B.J.H. 1999. Characterization and biobleaching effect of hemicellulases produced by thermophilic fungi. Biotechnol. Lett., 21(5): 415–420.

Hahn-Hägerdal, B., Galbe, M., Gorwa-Grauslund, M.F., Lidén, G. and Zacchi, G. 2006. Bio-ethanol-the fuel of tomorrow from the residues of today. Trend. Biotechnol., 24(12): 549–556.

Han, P., Zhou, P., Hu, S., Yang, S., Yan, Q. and Jiang, Z. 2013. A novel multifunctional α-amylase from the thermophilic fungus *Malbranchea cinnamomea*: biochemical characterization and three-dimensional structure. Appl. Microbiol. Biotechnol., 170(2): 420–435.

Hasnain, S., Adeli, K. and Storer, A.C. 1992. Purification and characterization of an extracellular thiol-containing serine proteinase from *Thermomyces lanuginosus*. Biochem. Cell Biol., 70: 117–122.

Hirvonen, M. and Papageorgiou, A.C. 2003. Crystal structure of a family 45 endoglucanase from *Melanocarpus albomyces*: mechanistic implications based on the free and cellobiose-bound forms. J. Mol. Biol., 329(3): 403–410.

Hong, J., Tamaki, H., Yamamoto, K. and Kumagai, H. 2003. Cloning of a gene encoding a thermo-stable endo-beta-1,4-glucanase from *Thermoascus aurantiacus* and its expression in yeast. Biotechnol. Lett., 25(8): 657–661.

Huge-Jensen, B., Andreasen, F., Christensen, T., Christensen, M., Thim, L. and Boel, E. 1989. *Rhizomucor miehei* triglyceride lipase is processed and secreted from transformed *Aspergillus oryzae*. Lipids, 24: 781–785.

Ifrij, H. and Ögel, Z.B. 2002. Production of neutral and alkaline extracellular proteases by the thermophilic fungus, *Scytalidium thermophilum*, grown on microcrystalline cellulose. Biotechnol. Lett., 24(13): 1107–1110.

Jensen, B., Nebelong, P., Olsen, J. and Reeslev, M. 2002. Enzyme production in continuous cultivation by the thermophilic fungus, *Thermomyces lanuginosus*. Biotechnol. Lett., 24(1): 41–45.

Johri, B.N., Satyanarayana, T. and Olsen, J. 1999. Thermophilic Moulds in Biotechnology. Kluwer Academic Publishers, UK.

Kalogeris, E., Christakopoulos, P., Katapodis, P., Alexiou, A., Vlachou, S., Kekos, D. and Macris, B.J. 2003. Production and characterization of cellulolytic enzymes from the thermophilic fungus *Thermoascus aurantiacus* under solid state cultivation of agricultural wastes. Process Biochem., 38(7): 1099–1104.

Kalogeris, E., Christakopoulos, P., Kekos, D. and Macris, B.J. 1998. Studies on the solid state production of thermostable endoxylanases from *Thermoascus aurantiacus*, characterization of two isozymes. J. Biotechnol., 60: 155–163.

Kamra, P. and Satyanarayana, T. 2004. Xylanase production by the thermophilic mold *Humicola lanuginosa* in solid-state fermentation. Appl. Biochem. Biotechnol., 119(2): 145–157.

Karnaouri, A., Topakas, E., Paschos, T., Taouki, I. and Christakopoulos, P. 2013. Cloning, expression and characterization of an ethanol tolerant GH3 β-glucosidase from *Myceliophthora thermophila*. Peer J., 1: e46.

Karnaouri, A.C., Topakas, E. and Christakopoulos, P. 2014. Cloning, expression, and characterization of a thermostable GH7 endoglucanase from *Myceliophthora thermophila* capable of high-consistency enzymatic liquefaction. Appl. Microbiol. Biotechnol., 98(1): 231–242.

Katapodis, P., Vrsanska, M., Kekos, D., Nerinckx, W., Biely, P., Claeyssens, M., Macris, B.J. and Christakopoulos, P. 2003. Biochemical and catalytic properties of an endoxylanase purified from the culture filtrate of *Sporotrichum thermophile*. Carbohydr. Res., 338(18): 1881–1890.

Kaur, G. and Satyanarayana, T. 2004. Production of extracellular pectinolytic, cellulolytic and xylanolytic enzymes by a thermophilic mould *Sporotrichum thermophile* Apinis in solid state fermentation. Indian J. Biotechnol., 3: 552–557.

Kaur, G., Kumar, S. and Satyanarayana, T. 2004. Production, characterization and application of a thermostable polygalacturonase of a thermophilic mould *Sporotrichum thermophile* Apinis. Bioresour. Technol., 94(3): 239–243.

Kumar, K.K., Deshpande, B.S. and Ambedkar, S.S. 1993. Production of extracellular acidic lipase by *Rhizopus arrhizus* as a function of culture conditions. Hindustan Antibiot. Bull., 35(1-2): 33–42.

Kumar, S. and Satyanarayana, T. 2003. Purification and kinetics of a raw starch-hydrolyzing, thermostable and neutral glucoamylase of a thermophilic mold *Thermomucor indicae-seudaticae*. Biotechnol. Prog., 19(3): 936–944.

Kumar, S. and Satyanarayana, T. 2004. Statistical optimization of a thermostable and neutral glucoamylase production by a thermophilic mold *Thermomucor indicae-seudaticae* in solid-state fermentation. World J. Microbiol. Biotechnol., 20(9): 895–902.

Kvesitadze, E., Gomarteli, M., Bezborodov, A.M. and Adeishvili, E. 1998. Thermostable endoxylanases of thermophilic fungi. Appl. Biochem. Microbiol., 5: 469–472.

Li, A.N. and Li, D.C. 2009. Cloning, expression and characterization of the serine protease gene from *Chaetomium thermophilum*. J. Appl. Microbiol., 106(2): 369–380.

Li, A.N., Xie, C., Zhang, J., Zhang, J. and Li, D.C. 2011. Cloning, expression, and characterization of serine protease from thermophilic fungus *Thermoascus aurantiacus* var. *levisporus*. J. Microbiol., Microbiology, 49(1): 121–129.

Lin, J., Pillay, B. and Singh, S. 1999. Purification and biochemical characteristics of β-D-glucosidase from a thermophilic fungus, *Thermomyces lanuginosus*-SSBP. Biotechnol. Appl. Biochem., 30(Pt 1): 81–87.

Maheshwari, R. 2012. Fungi: Experimental Methods in Biology, 2nd Edition CRC Press, Boca Raton, London, New York, UK.

Maheshwari, R., Bharadwaj, G. and Bhat, M.K. 2000. Thermophilic fungi: Their physiology and enzymes, Microbiol. Mol. Biol. Rev., 64: 461–488.

Mishra, R. and Maheshwari, R. 1996. Amylases of the thermophilic fungus *Thermomyces lanuginosus*, their purification, properties, action on starch and response to heat. J. Biosci., 21: 653–672.

Mitchell, D.B., Vogel, K., Weimann, B.J., Pasamontes, L. and van Loon, A.P.G.M. 1997. The phytase subfamily of histidine acid phosphatase: isolation of genes for two novel phytases from the *Aspergillus terreus* and *Myceliophthora thermophila*. Microbiology, 143: 245–252.

Moriya, T., Watanabe, M., Sumida, N., Okakura, K. and Murakami, T. 2003. Cloning and overexpression of the avi2 gene encoding a major cellulase produced by *Humicola insolens* FERM BP-5977. Biosci. Biotechnol. Biochem., 67(6): 1434–1437.

Murray, P.G., Collins, C.M., Grassick, A. and Tuohy, M.G. 2003. Molecular cloning, transcriptional, and expression analysis of the first cellulase gene (cbh2), encoding cellobiohydrolase II, from the moderately thermophilic fungus *Talaromyces emersonii* and structure prediction of the gene product. Biochem. Biophys. Res. Commun., 301(2): 280–286.

Nampoothiri, K.M., Tomes, G.J., Roopesh, K., Szakacs, G., Nagy, V., Soccol, C.R. and Pandey, A. 2004. Thermostable phytase production by *Thermoascus aurantiacus* in submerged fermentation. Appl. Biochem. Biotechnol., 118(1-3): 205–214.

Narang, S., Sahai, V. and Bisaria, V.S. 2001. Optimization of xylanase production by *Melanocarpus albomyces* IIS68 in solid state fermentation using response surface methodology. J. Biosci. Bioeng., 91(4): 425–427.

Nguyena, Q.D., Judit, M., Claeyssens, R.M., Stals, I. and Hoschke, A. 2002. Purification and characterisation of amylolytic enzymes from thermophilic fungus *Thermomyces lanuginosus* strain ATCC 34626. Enzyme Microb. Technol., 31(3): 345–352.

Noel, M. and Combes, D. 2003. Effects of temperature and pressure on *Rhizomucor miehei* lipase stability. J. Biotechnol., 102(1): 23–32.

Omar, I.C., Nishio, N. and Nagai, S. 1987. Production of a thermostable lipase by *Humicola lanuginosa* grown on sorbitol-corn steep liquor medium. Agricult. Biol. Chem., 51: 2145–2151.

Ottesen, M. and Rickert, W. 1970. The isolation and partial characterization of an acid protease produced by *Mucor miehei*. C. R. Trav. Lab. Carlsberg, 37: 301–325.

Pandey, V. 2003. Pectinolytic Enzymes of a Thermophilic Mould *Sporotrichum thermophile* Apinis. M.Sc. Dissertation. University of Delhi South Campus, New Delhi, India.

Parry, N.J., Beever, D.E., Owen, E., Vandenberghe, I., Van Beeumen, J. and Bhat, M.K. 2001. Biochemical characterization and mechanism of action of a thermostable β-glucosidase purified from *Thermoascus aurantiacus*. Biochem. J., 353: 117–127.

Pasamontes, L., Haiker, M., Wyss, M., Tessier, M. and van Loon, A.P.G.M. 1997a. Gene cloning, purification, and characterization of a heat-stable phytase from the fungus *Aspergillus fumigatus*. Appl. Environ. Microbiol., 63: 1696–1700.

Pasamontes, L., Haiker, M., Wyss, M., Tessier, M. and van Loon, A.P.G.M. 1997b. Gene cloning, purification and characterization of a heat stable phytase from the fungus *Aspergillus fumigatus*. Applied and Environmental Microbiology, 63(5): 1696–1700.

Petrova, S.D., Ilieva, S.Z., Bakalova, N.G., Atev, A.P., Bhat, K. and Kolev, D.N. 2000. Production and characterization of extracellular α-amylases from the thermophilic fungus *Thermomyces lanuginosus* (wild and mutant strains). Biotechnol. Lett., 22(20): 1619–1624.

Prabhu, K.A. and Maheshwari, R. 1999. Biochemical properties of xylanases from a thermophilic fungus, *Melanocarpus albomyces*, and their action on plant cell walls. J. Biosci., 24(4): 461–470.

Prasad, G.S., Girisham, S. and Reddy, S.M. 2011. Potential of thermophilic fungus *Rhizomucor pusillus* NRRL28626 in biotransformation of antihelmintic drug albendazole. Appl. Biochem. Biotechnol., 165: 1120–1128.

Rao, P. and Divakar, S. 2002. Response surface methodological approach for *Rhizomucor miehei* lipase-mediated esterification of α-terpineol with propionic acid and acetic anhydride. World J. Microbiol. Biotechnol., 18(4): 345–349.

Roy, I., Gupta, A., Khare, S.K., Bisaria, V.S. and Gupta, M.N. 2003. Immobilization of xylan-degrading enzymes from *Melanocarpus albomyces* IIS 68 on the smart polymer Eudragit L-100. Appl. Microbiol. Biotechnol., 61(4): 309–313.

Roy, I., Sastry, M.S.R., Johri, B.N. and Gupta, M.N. 2000. Purification of α-amylase isoenzymes from *Scytalidium thermophilum* on a fluidized bed of alginate beads followed by concanavalin a–agarose column chromatography. Prot. Expr. Purif., 20(2): 162–168.

Sadaf, A. and Khare, S.K. 2014. Production of *Sporotrichum thermophile* xylanase by solid state fermentation utilizing deoiled *Jatropha curcas* seed cake and its application in xylooligosachharide synthesis. Bioresource Technology, 153: 126–130.

Satyanarayana, T. and Singh, B. 2004. Thermophilic moulds, diversity and potential biotechnological applications. pp. 87–110. *In*: Gautam, S.P., Sharma, A., Sandhu, S.S. and Pandey, A.K. (eds.). Microbial Diversity: Opportunities and Challenges. Shree Publishers and Distributors, New Delhi, India.

Satyanarayana, T., Chavant, L. and Montant, C. 1985. Applicability of API ZYM for screening enzyme activity of thermophilic moulds. Trans. Brit. Mycol. Soc., 85: 727–730.

Satyanarayana, T., Noorwez, S.M., Kumar, S., Rao, J.L.U.M., Ezhilvannan, M. and Kaur, P. 2004. Development of an ideal starch saccharification process using amylolytic enzymes from thermophiles. Biochem. Soc. Trans., 32: 276–278.

Satyanarayana, T., Jain, S. and Johri, B.N. 1988. Cellulases and xylanases of thermophilic moulds. pp. 24–60. *In*: Agnihotri, V.P., Sarbhoy, A.K. and Kumar, D. (eds.). Perspectives in Mycology and Plant Pathology. Malhotra Publishing House, New Delhi, India.

Satyanarayana, T., Johri, B.N. and Klein, J. 1992. Biotechnological potential of thermophilic fungi. pp. 729–762. *In*: Arora, D.K., Elander, R.P. and Mukerji, K.G. (eds.). Hand Book of Applied Mycology. Vol. 4 Fungal Biotechnology, Marcel Dekker Inc., New York, USA.

Sharma, H.S.S. and Johri, B.N. 1992. The role of thermophilic fungi in agriculture. pp. 707–728. *In*: Arora, D.K., Elander, R.P. and Mukerji, K.G. (eds.). Hand Book of Applied Mycology. Vol. 4 Fungal Biotechnology, Marcel Dekker Inc., New York, USA.

Shenolikar, S. and Stevenson, K.J. 1982. Purification and partial characterization of a thiol proteinase from the thermophilic fungus *Humicola lanuginosa*. Biochem. J., 205: 147–152.

Singh, B. 2014. *Myceliophthora thermophila* syn. *Sporotrichum thermophile*: a thermophilic mould of biotechnological potential. Critical Reviews in Biotechnology, doi: 10.3109/07388551.2014.923985.

Singh, B. and Satyanarayana, T. 2006a. Phytase production by thermophilic mold *Sporotrichum thermophile* in solid-state fermentation and its application in dephytinization of sesame oil cake. Appl. Biochem. Biotechnol., 133: 239–250.

Singh, B. and Satyanarayana, T. 2006b. A marked enhancement in phytase production by a thermophilic mould *Sporotrichum thermophile* using statistical designs in a cost-effective cane molasses medium. J. Appl. Microbiol., 101: 344–352.

Singh, B. and Satyanarayana, T. 2008a. Improved phytase production by a thermophilic mould *Sporotrichum thermophile* in submerged fermentation due to statistical optimization. Bioresour. Technol., 99: 824–830.

Singh, B. and Satyanarayana, T. 2008b. Phytase production by a thermophilic mould *Sporotrichum thermophile* in solid state fermentation and its potential applications. Bioresour. Technol., 99: 2824–2830.

Singh, B. and Satyanarayana, T. 2008c. Phytase production by *Sporotrichum thermophile* in a cost-effective cane molasses medium in submerged fermentation and its application in bread. J. Appl. Microbiol., 105: 1858–1865.

Singh, B. and Satyanarayana, T. 2009a. Characterization of a HAP-phytase from a thermophilic mould *Sporotrichum thermophile*. Bioresour. Technol., 100: 2046–2051.

Singh, B. and Satyanarayana, T. 2009b. Thermophilic moulds in environmental management. pp. 352–375. *In*: Misra, J.K. and Deshmukh, S.K. (eds.). Progress in Mycological Research. Vol. I Fungi from Different Environments. Environmental Mycology. Science Publishers, USA.

Singh, B. and Satyanarayana, T. 2010. Plant growth promotion by an extracellular HAP-phytase of a thermophilic mold *Sporotrichum thermophile*. Appl. Biochem. Biotechnol., 160(5): 1267–1276.

Singh, B. and Satyanarayana, T. 2011. Phytases from thermophilic molds: Their production, characteristics and multifarious applications. Process Biochem., 46(7): 1391–1398.

Singh, B., Kunze, G. and Satyanarayana, T. 2011. Developments in biochemical aspects and biotechnological applications of microbial phytases. Biotechnol. Mol. Biol. Rev., 6(3): 69–87.

Singh, S., Pillay, B. and Prior, B.A. 2000. Thermal stability of β-xylanases produced by different *Thermomyces lanuginosus* strains. Enzyme Microb. Technol., 26: 502–508.

Song, S., Tang, Y., Yang, S., Yan, Q., Zhou, P. and Jiang, Z. 2013. Characterization of two novel family 12 xyloglucanases from the thermophilic *Rhizomucor miehei*. Applied Microbiology and Biotechnology, 97(23): 10013–10024.

Stevenson, K.J. and Gaucher, G.M. 1975. The substrate specificity of thermomycolase, an extracellular serine proteinase from the thermophilic fungus *Malbranchea pulchella* var. *sulfurea*. Biochem. J., 151: 527–542.

Straatsma, G., Samson, R.A., Olijnsma, T.W., Camp, H.J.M.O.D., Gerrits, J.P.G. and Griensven, L.J.L.D.V. 1994. Ecology of thermophilic fungi in mushroom compost, with emphasis on *Scytalidium thermophilum* and growth stimulation of *Agaricus bisporus* mycelium. Appl. Environ. Microbiol., 60(2): 454–458.

Taha, M., Adetutu, E.M., Shahsavari, E., Smith, A.T. and Ball, A.S. 2014. Azo and anthraquinone dye mixture decolourization at elevated temperature and concentration by a newly isolated thermophilic fungus, *Thermomucor indicae-seudaticae*. Journal of Environmental Chemical Engineering, 2(1): 415–423.

Takai, K., Nakamura, K., Toki, T., Tsunogoi, U., Miyazaki, M., Miyazaki, J., Hirayama, H., Nakagawa, S., Nanoura, T. and Horikoshi, K. 2008. Cell proliferation at 122°C and isotopically heavy CH$_4$ production by a hyperthermophilic methanogen under high pressure cultivation. Proc. Natl. Acad. Sci. (USA), 105: 10949–10954.

Tansey, M.R. and Brock, T.D. 1978. Microbial life at high temperatures: Ecological aspects. pp. 159–195. *In*: Kushner, D.J. (ed.). Microbial Life in Extreme Environments, Academic Press, London.

Thakur, M.S., Karanth, N.G. and Krishnanand, G. 1993. Production of fungal rennet by *Mucor miehei* using solid state fermentation. Appl. Microbiol. Biotechnol., 32: 409–413.

Tong, C.C. and Cole, A.L. 1975. Physiological studies in the thermophilic fungus *Talaromyces thermophilus* Stolk. Mauri Ora, 3: 37–43.

Tong, C.C., Cole, A.L. and Shepherd, M.G. 1980. Purification and properties of the cellulases from the thermophilic fungus *Thermoascus aurantiacus*. Biochem. J., 191: 83–94.

Tosi, L.R.O., Terenzi, H.F. and Jorge, J.A. 1993. Purification and characterization of an extracellular glucoamylase from thermophilic fungus *Humicola grisea* var. *thermoidea*. Can. J. Microbiol., 39: 846–852.

Tuohy, M.G., Buckley, R.J., Griffin, T.O., Connelly, I.C., Shanley, N.A., Ximenes, E., Filho, F., Hughes, M.M., Gorgan, P. and Coughlan, M.P. 1989. Enzyme production by solid state cultures of aerobic fungi on lignocellulosic substrates. pp. 293–312. *In*: Coughlan, M.P. (ed.). Enzyme System for Lignocellulose Degradation, Elsevier Applied Science, London and New York, UK.

Venturi, L.L., Polizeli, L.M., Terenzi, H.F., Furriel, R.P. and Jorge, J.A. 2002. Extracellular β-D-glucosidase from *Chaetomium thermophilum* var. *coprophilum*: production, purification and some biochemical properties. J. Basic Microbiol., 42(1): 55–66.

Voordouw, G., Gaucher, G.M. and Roche, R.S. 1974. Physicochemical properties of thermomycolase, the thermostable, extracellular, serine protease of the fungus *Malbranchea pulchella*. Can. J. Biochem., 52: 981–990.

Whitehead, E.A. and Smith, S.N. 1989. Fungal extracellular enzyme activity associated with breakdown of plant cell biomass. Enzyme Microb. Technol., 11: 736–743.

Yadav, H. and Jaitly, A.K. 2011. Effect of salt on the production of xylanase in some thermophilic fungi. J. Phytol., 3: 12–14.

Yang, S., Hua, C., Yan, Q., Li, Y. and Jiang, Z. 2013. Biochemical properties of a novel glycoside hydrolase family 1 β-glucosidase (PtBglu1) from *Paecilomyces thermophila* expressed in *Pichia pastoris*. Carbohydrate Polymer, 92: 784–791.

CHAPTER 12

Wood-Inhabiting Fungi: Applied Aspects

Daniel Țura,[2,]* *Solomon P. Wasser*[1] *and Ivan V. Zmitrovich*[3]

ABSTRACT

The present chapter is a review of our modern knowledge on wood-inhabiting fungi emphasizing their role and importance as pathogens, their ability to decay timber, their usage in mycoremediation, medicine, cultivation, structural purposes or artwork. Symptoms, economic impact and control strategies are discussed for some of the most feared pathogenic and timber decay fungi. Wood-fungi involved in mycoremediation processes and their abilities to decompose various environmental pollutants such as the polycyclic aromatic hydrocarbons, chlorinated hydrocarbons or synthetic dyes are also presented. The second part of this chapter highlights the therapeutic effects of 33 wood-inhabiting fungi and some of the unsolved problems in the medicinal mushrooms science. A close attention is paid to the commercially important cultivated wood-decomposing fungi and discusses some of the problems involved in their growing process together with growing techniques and factors influencing their overall biological efficiencies. A new branch of mycology is also discussed here: the use of wood-inhabiting fungi for structural purposes and artwork.

[1] Department of Evolutionary and Environmental Biology, Institute of Evolution, Faculty of Natural Sciences, University of Haifa, Mt Carmel Haifa 31905, Israel.
[2] Aloha Medicinals Inc., 2300 Arrowhead Dr, Carson City, Nevada 89706, USA.
[3] V.L. Komarov Botanical Institute, Russian Academy of Sciences 2, Professor Popov St., St. Petersburg, 197376, Russia.
* Corresponding author: turadaniel21@yahoo.com

Introduction

Wood-inhabiting fungi represent an ecologically and economically important group of higher basidiomycetes that comprise various forms of macroscopic fungi in which the hymenophores are even, folded, toothed, poroid, or lamellate, and with tough or fleshy basidiocarps. Many species of wood-inhabiting fungi are saprobic wood-decayers; thus, these fungi are most often found on logs, stumps, or other dead wood and play a crucial role in nutrient cycling in forest ecosystems. According to their enzymatic systems, these fungi were grouped into two categories, namely, white-rot and brown-rot fungi. Wood consists mainly of lignin, cellulose, and hemicelluloses. Lignin is comprised of inter-unit carbon-carbon and ether bonds in polyphenolic construction that demand oxidative rather than hydrolytic degradative mechanisms (Cullen, 2002). Because of their ability to degrade lignin, white-rot fungi were extensively studied, and results revealed that three kinds of extracellular phenoloxidases, namely, lignin peroxidase, manganese peroxidase, and laccase are responsible for the depolymerization of lignin (Kirk and Farrell, 1987; Eriksson et al., 1990). Many types of organisms deteriorate wood, but brown-rot fungi are the most potent. With several exceptions, brown-rot fungi are principally associated with conifers, while white-rot fungi are more common on hardwoods.

Several species grow on living trees and cause decay of the non-functional heart wood, or invade plant tissue, as pathogens; only a few species are mycorrhizal. The current status of tree pathogenic and timber decaying fungi is rather obscure since no reliable methods of protection are available today. The scientific community is puzzled when applying protection strategies against certain fungal species affecting vast forest ecosystems or decaying timber used in construction. The economic impact is huge; therefore, some scientists are always preoccupied in finding eco-friendly solutions of treating infected trees or finding efficient prevention strategies against timber destroying fungi. The first part of this chapter discusses some of the most important wood-fungi as parasites and the relationships with their hosts (host symptoms, economic impact, and disease control). It highlights several wood-fungal species decaying timber together with various attempts of their prevention and control. The second part is devoted to medicinal fungi, fungi involved in mycoremediation processes and various biotechnological applications. Wood-decaying fungi are subject of an increasing attention due to their diverse applications in biotechnology. Many fungal species are currently screened in scientific labs worldwide for potential properties valuable from a scientific and applicational point of view. Some of these fungi are already well known in the Oriental traditional medicine from ancient times [e.g., *Ganoderma lucidum* (W. Curt.) P. Karst. known as "Ling Zhi" or "Reishi", *Lentinula edodes* (Berk.) Peglerknown as "Shiitake" or *Auricularia auricula-judae* (Bull.) Quél. known as "Kikurage"] and are currently subject of thousands of publications and sources of powerful pharmacological products available in the market today. Other wood-fungal species are subject to various ongoing experiments in order to obtain high quality enzymes with diverse industrial applications; some other fungi are involved in mycoremediation processes. Here we briefly discuss the role and importance of some wood-fungi in biotechnology.

Several publications on efficient ways to cultivate edible and medicinal fungi are treated in the chapter highlighting the importance of their biological efficiency (BE) as reflected in good practice to cultivate these fungi. Farmers and market suppliers worldwide must be familiar with the methods to obtain high BE. This will enable them to improve cultivation.

Applied Ecology of Wood-Inhabiting Fungi

Pathogenic Significance

Causal Agents

Because of their ability to destroy living wood tissue, pathogenic fungi received utmost attention not only by the scientists, but also by the people who are directly affected by them. Aggressive fungi attack the heartwood of living trees without invading the living tissues (except for a few species), causing the so-called "heart-rots", or attacking the sapwood of the roots resulting in the death of the host (the so-called "root-rot" fungi). However, serious pathogens are those that are able to invade living wood tissue at a faster rate, to cause the death of the host. Such pathogens are known from temperate as well as most tropical regions of the world, and have a very wide host range, including both, living broadleaf and coniferous trees, shrubs, and herbaceous plants.

The heart-rot fungi affect the strength and volume of wood, thus being able to cause great economic losses. The heart-rot is often encountered in old trees or particularly in trees affected by other stressful environmental factors such as drought, lack of nutrients, poor sunlight availability, wind intensity, or biotic impact. The overall stress lowers host resistance and stimulates development of the fungus. Human impact is positively related to the sickness of urban trees and plants especially due to improper tree pruning habits. Spores of pathogenic fungal species land on freshly trimmed trees and with conditions of optimal moisture and temperature begin to germinate. The resulting fungal hyphae invade the tree wood mass. Safety measures such as the trimming methods, right trimming season, and sealing processes of the freshly trimmed branches are needed in order to prevent the infection of healthy trees with aggressive wood-inhabiting pathogenic fungi. By contrast when wood infection occurs, invasion of wood by fungal hyphae takes time and mainly depends on the above mentioned environmental factors. In such cases, insects (e.g., beetles, wood-termites, wasps, etc.) play a crucial role by accelerating the wood decay process. Heart-rotten old and weakened trees are often predisposed to wind-throw. An important heart-rot causing fungus is *Inonotus rickii* (Pat.) D.A. Reid (Fig. 1a) that attacks living hardwoods of tropical and subtropical areas. This fungus not only has an ability to cause

Figure 1. Living tree pathogenic fungi: (a) *Inonotus rickii* parasiting *Delonix regia*; (b) *Chondrostereum purpureum* basidiomes; (c) *Ganoderma australe* growing on hardwood tree base; (d) *Armillaria tabescens* infecting roots of deciduous trees.

decay of heartwood, resulting in structural damage to the tree, but it also can parasitize the sapwood and cambium resulting in a progressive crown dieback (Ţura et al., 2011). Another example is *Phellinus tremulae* (Bondartsev) Bondartsev & P.N. Borisov that occurs almost everywhere aspen species grow and is reported to be able to spread through the sapwood. Among the Hymenochaetales, species causing the greatest losses to forestry are *Ph. pini* (Brot.) Bondartsev & Singer growing on pines and *Ph. igniarius* s.l. (L.) Quél. growing on various hardwoods. Both species destroy the heartwood (Larsson, 2007). Reports from United States (Covey et al., 1981; Bergdahl and French, 1985) suggest that *Trametes versicolor* (L.) Lloyd a common saprophyte is also capable of acting as an aggressive parasite that kills sapwood and cambial tissues of apple trees. Some other worth considering heart-rot and canker causing species include *Cerrena unicolor* (Bull.) Murrill, *Fomes fomentarius* (L.)

Fr., *Inonotus andersonii* (Ellis & Everh.), *I. hispidus* (Bull.) P. Karst., *I. dryophilus* (Berk.) Murrill, and *Phellinus punctatus* (Fr.) Pilát, etc.

True pathogens are considerably those species of fungi that attack the sapwood in the roots resulting in the death of the host. According to Hood (2006), in warm climate zones, important root diseases of plantation forests are caused by the species such as *Rigidoporus microporus* (Sw.) Overeem, *Junghuhnia vincta* (Berk.) Hood & M.A. Dick, *Phellinus noxius* (Corner) G. Cunn., and species of *Ganoderma*. Recently numerous projects were initiated to identify the genetic contents of several *Ganoderma* species. In 2013 thirteen whole genome sequencing projects were registered on *Ganoderma* species in databases. The results are meant to aid the identification of genes in biological pathways that are important in understanding the mechanism of pathogenesis, the biosynthesis of active compounds, life cycle or cellular development (Kües et al., 2015). A common species in European coniferous forests is *Heterobasidion annosum* (Fr.) Bref. s.l., a dangerous root-rot pathogen causing trunk rots, and finally the death of the host; fructifications continue to grow on fallen trees. *Inonotus obliquus* (Ach. ex Pers.) Pilát produces sterile conks on living trees. First, the invading species kills and decays the sapwood, after the tree dies, the fungus fruits under the bark. Some species of *Ganoderma* are saprobes, causing heart-rots in living trees and decaying fallen wood [e.g., *G. australe* (Fr.) Pat.—Fig. 1c], while others are harmful parasites [e.g., *G. applanatum* (Pers.) Pat., *G. lucidum*, or *G. resinacum* Boud.]. *Phellinus weirii* (Murrill) Gilb. is another serious pathogen infecting a wide range of conifer species throughout the north-western United States, British Columbia, and Canada. This fungus causes laminated root-rot leading to direct death of the host by reducing the amount of functioning root system or indirectly by increasing tree susceptibility to wind-borne fungal spores. It is considered the most important natural agent of disturbance that forest managers must deal with (Thies and Nelson, 1996). The fungus can survive saprophytically in the lower bole and roots of dead trees and serve as a source of infection for decades (Hansen, 1979; Harrington and Thies, 2007).

Some corticioid fungi are parasitic causing a variety of diseases in tree roots and leaves. Species in several corticioid genera form *Rhizoctonia* and *Fibulo rhizoctonia* anamorphs, e.g., *Athelia* and *Botryobasidium* (Kataria and Hoffmann, 1988).

Among corticioid fungi causing major diseases in trees, there are species such as *Chondrostereum purpureum* (Pers.) Pouzar, *Coniophora puteana* (Schumach.) P. Karst., and *Stereum sanguinolentum* (Alb. & Schwein.) Fr. *Chondrostereum purpureum*, occurring on a wide range of coniferous and hardwood trees (Fig. 1b), causes silver leaf disease. The fungus produces toxins causing characteristic foliar symptoms and ultimately killing branches or the entire tree (Ginns and Lefebvre, 1993). When associated with a white-rot, *Stereum sanguinolentum*, produces a sapwood rot of living coniferous trees, while *Coniophora puteana* causes a brown cubical heart-rot and more rarely a root-rot especially of living conifers.

The concern over the movement of plant pathogens across the world is very high. Dispersal of these fungi, most likely, occurs by transport of infected logs from one region to another. Many countries cannot efficiently control the spread of aggressive pathogenic fungal species. For example, Coetzee et al. (2001) reported the occurrence of the root-infecting fungus *Armillaria mellea* (Vahl.) P. Kumm. in central Cape Town (South Africa), that got into the South Africa by the early Dutch settlers. This fungus is a serious pathogen that has a wide host range including broadleaf and coniferous trees. Moreover, it attacks also shrubs and herbaceous plants. After the plant dies, *Armillaria* [e.g., *A. tabescens* (Scop.) Emel—Fig. 1d] persists as a saprophyte on the infected portions of the tree. *Armillaria* can also colonize orchids *Galeola* and *Gastrodia* but, in this case, the fungus is the host and the plant is the parasite. Similar to its contrasting relationships with plants, *Armillaria* acts as either host or parasite in its interactions with other fungi (Baumgartner et al., 2011). Many species of *Armillaria* can live for decades in suitable live host material, stumps and root fragments, and can disperse naturally through the spread of rhyzomorphs in the soil (Williams et al., 1986).

The heterobasidiomycetous fungi are usually considered as mycoparasitic and/or saprophytic (Roberts, 1993; Wells, 1994; Worrall et al., 1997; Wells and Bandoni, 2001), although little data is available on their trophic behavior in nature. In addition, some species are able to colonize living tissues, since these types of fungi include several orchid symbionts belonging to the genus *Rhizoctonia*. Heterobasidiomycetous species of *Sebacinaceae*, previously considered as saprophytes or parasites, were shown to form ectomycorrhizas on temperate forest trees (Selosse et al., 2002; Weiss et al., 2004).

Symptoms

Wood pathogenic fungi cause symptoms that can be classified into external and internal symptoms. Usually external symptoms include a general decline in vigor of the plant characterized by leaf discoloration and premature defoliation, resinosis (in conifers), dieback, basidiomes developing at the tree base, boles or major branches, etc. Internal symptoms usually show typical heart-rot, canker, or root-rot characterized by dark-stained, soft and crumbly wood, etc. Such symptoms may appear earlier or later, mostly depending on the species of the fungus, host species, its age, and stress inducing factors. Usually seedlings are very much affected by fungal infections and may die within 1–2 years, while in older trees symptoms often occur over several years. For example, the aggressive heart-rot fungus *Inonotus rickii* is usually found in urban areas attacking especially ornamental trees found in public gardens and along boulevards. According to Ramos et al. (2008) this fungus, when infecting *Celtis australis*, often develops external symptoms: dead branches in the upper part of the crown, sparse foliage, dark jelly exudations, swellings and cracks in the bark of the trunk. The internal symptoms show decayed wood with white to yellow-brown color separated from sound wood by zones of dark-stained wood. In some other pathogenic fungi symptoms are rather variable; for example, according to Baumgartner et al. (2011), the most apparent indicator of *Armillaria* root-rot disease is the disease centre displaying varying degrees of severity of stunted shoots, dwarfed foliage, wilting, premature defoliation, resinosis in the case of conifers and dwarfed fruit in the case of fruit and nut crops. The most common course of symptom development is a gradual multilayer reduction in shoot growth and yield leading to death of the host. Whatever symptoms may be, they are very important in identifying the pathogenic fungi. Often fungus identification is based on clear tree damage evidence when the disease already has spread in the wood mass. However, identification may be done in the presence of basidiome samples that usually appear on the roots, at the base, on boles, or on tree branches. In some other cases disease may be observed in incipient development stages, in such cases samples are taken and laboratory analyses are implemented.

Economic Impact

Fungal decay of wood is the most common problem in forest and timber industry. Wood rotting fungi damage forest wood even more than the insects, animals, or bacteria (Rauel and Barnoud, 1985). Affected wood tissue is not suitable for wood products. In the EU, annual losses attributed to growth reduction and degradation of wood is estimated at € 790 million (Woodward et al., 1998). Based on a rough estimate, the annual economic losses due to root- and butt-rot caused by *Heterobasidion* only in Finland are reported to be around € 35 million (Bendz-Hellgren et al., 1998). Laminated root-rot of *Pseudotsuga menziesii* (Mirb.) Franco (Douglas-fir) caused by *Phellinus weirii* annually reduces wood production in western North America by about 4.4 million m^3 (Nelson et al., 1981), while several *Armillaria* species act as primary pathogen and can cause significant economic losses by damaging timber (Bendel et al., 2006). In some cases it is more profitable to keep valuable tree species even if they show symptoms of infection with pathogenic fungi than to change them with other tree species (Piri, 2003).

Disease Control

Disease control is primarily a matter of prevention, which currently represents the most reliable strategy. Taking into consideration the major environmental factors responsible for rising stress levels in plants and the application of an efficient management strategy that involves proper pruning methods together with disease dispersal control could be an effective solution—at least on a local scale. Another important and necessary prevention measure is to avoid making wounds on trunks or branches of trees. Wounded trees are susceptible to infection by pathogenic fungi on a higher rate.

When symptoms of disease are present in trees, controlling strategies should be implemented. For example, recommended control practices in case of infection caused by *Inonotus rickii* include the prompt removal of inoculum sources and sanitation, a careful choice of a suitable tree species when replacing a tree, and the disinfection of pruning tools (Annesi et al., 2010). Post infection strategy to reduce the effect of pathogenic fungi causing root-rot is fumigation as a means of reducing the amount of inoculum. The injection of a fumigant to the soil as well as directly to the bole has been tested and discussed several times in literature. For example, chemical fumigants applied directly to Douglas-fir stumps and living trunks colonized by *Phellinus weirii* were found to eliminate the fungi without causing significant tree growth reduction (Thies and Nelson, 1982, 1987, 1996; Harrington and Thies, 2007). Recent research on post-infection controls has revealed that chemical fumigants in spite of their ability to eradicate root decaying fungi are now being regulated more seriously or banned outrightly because of their negative effects on the environment (Baumgartner et al., 2011). A common habit adopted by many farmers is to cut the infected trees and to clean the place by removing of the stumps and wood fragments. However, this method has proven to be rather inefficient since some of the most feared pathogenic fungi, e.g., *Armillaria*, *Heterobasidion annosum*, and *Phellinus weirii* are able to persist for decades in stumps, wood parts, and roots. These wood remnants may serve as inoculum for future plantations. Results are visible especially when using the same tree species which is obviously vulnerable to such pathogens. In such cases, resistant tree species to such pathogens proved to be crucial.

Another promising strategy showing effectiveness against root disease pathogens is biological control. Inoculation of the soil with antagonistic fungi such as *Trichoderma* spp. has been shown to be an effective parasite of *Armillaria* and *Phellinus weirii* (Goldfarb et al., 1989; Raziq and Fox, 2005). The positive point here is that *Trichoderma* spp. acts against a wide range of pathogenic fungi especially soil-borne fungal pathogens and are environmental friendly (Ha, 2010). According to Kumar and Gupta (2006), a certain species of actinomycetes, such as *Streptomyces*, known to produce not only antibiotics but also extracellular enzymes active in fungal cell wall degradation, are also good biocontrol agents (Trejo-Estrada et al., 1998b). For example, *Streptomyces violaceusniger* was shown to exhibit a strong antagonism towards white- and brown-rot fungi: *Postia placenta* (Fr.) M.J. Larsen & Lombard, *Phanerochaete chrysosporium* Burds., *Trametes versicolor*, and *Gloeophyllum trabeum* (Pers.) Murrill (Trejo-Estrada et al., 1998a,b; Kumar and Gupta, 2004).

Timber Decaying Fungi

Key Producers

Wood-decaying fungi are the most feared among wood-degrading organisms (e.g., beetles, marine borers, termites, and carpenter ants). Under favorable moisture and temperature conditions, wood-degrading fungi are able to decay timber, logs, or miscellaneous wood-products and can also cause serious economical losses to wood, construction, or paper industries. It has been estimated that the cost of repairing fungal damage of timber in construction in 1977 amounted to £3 million per week in Britain (Rayner and Boddy, 1988). Among wood-rotting fungi, the basidiomycetes are considered the major cause of losses in fiber yield used in paper industry (Nilsson, 1973).

Decaying fungi may appear on wood surface as variable sized patches, in some species rhizomorphs (strands) may be seen, or sometimes basidiomes may appear. Most wood-decaying fungi attack hardwoods, therefore, this type of wood is more susceptible to wood decay than conifers and should be well analysed before storing. Conifer wood is often used as a structural part in buildings of the temperate and northern hemisphere. Indoor wood-decay fungi cause considerable damage. Most of the structural damage to the indoors of buildings in Europe and North America is caused by brown-rot fungi that degrade conifer wood (Schmidt, 2007). These fungi selectively utilize the cellulose and hemicellulose components of wood. Brown-rot fungi adversely affect several strength properties of wood before significant weight loss is detected. The attacked wood undergoes drastic shrinkage and cracks across the grain; in advanced stages of decay, wood becomes crumbly with brown cubical chunks composed of largely or slightly modified lignin (Highley and Lutz, 1970; Wilcox, 1978; Ryvarden and Gilbertson, 1993). According to Zmitrovich et al. (2014), under xerophylic conditions (exposed fallen logs, and decorticated stands) the brown-rot fungi demonstrate rather weak activity causing the so-called "dry-rot". The typical dry-rot producers are representatives of the genera *Dacrymyces* and *Gloeophyllum*. However, when the process of water evaporation outside the wood is hampered by forest shade, ground conditions or the abundance of fallen wet logs, the rot stays very active, and in many cases it is accompanied by self-moisturizing of wood due to metabolic water. Such an active – wet brown-rot is characteristic of *Serpula*, *Tapinella panuoides* (Batsch) E.-J. Gilbert, and many species of *Antrodia* and *Fomitopsis*.

The indoor wood-decay fungi, *Serpula lacrymans* (Wulfen) J. Schröt. and *Coniophora puteana* were noted as the most harmful fungi occurring on indoor wood in temperate regions. An investigation of 3050 buildings in Poland revealed that the wood-rotting fungal population consists of *Serpula lacrymans* (54%), *Coniophora puteana* (22%), and *Fibroporia vaillantii* (DC.) Parmasto (11%) (Ważny and Czajnik, 1963). Emphasis, therefore, is placed on *Serpula lacrymans*, which is able to transport water through mycelial strands from the source of moisture into dry wood, and is a common destructive fungus reported in timber constructions (Ritschkoff, 1996). According to Schmidt (2007) and Schmidt and Huckfeldt (2011) the conventional wisdom is that *S. lacrymans* is the only fungus that can infect dry timber (at least 21% moisture) and brickwork (at least 6% water content) and is widely spread by mycelium and its rhizomorphs, thereby growing over and through various types of materials (Coggins, 1991). Moreover, experiments have shown that other fungal species are also able to cause dry-rot, to colonize wood and to grow through walls if the alkalinity of old mortar is sufficiently reduced (Table 1). Through their rhizomorphs these fungi may cross inert surfaces and penetrate deep within masonry. However, *S. lacrymans* seems to be the only fungus that has all of the important abilities to colonize a building, but considering more capabilities, other fungi are more powerful.

Table 1. Important characteristics of indoor wood-decaying basidiomycetes for spreading and damage in buildings (after Schmidt and Huckfeldt, 2011).

Indoor wood rotting basidiomycetes	Strand formation	Growth through brickwork	Spreading from brickwork	Colonization of dry wood[1]	Retarding substrate drying by surface mycelium
Serpula lacrymans	yes	yes	yes	yes	Well
Coniophora puteana	yes	yes	unknown	yes	Little
Donkioporia expansa	no	no	no	yes	very well
Fibroporia vaillantii	yes	rarely	unknown	yes	Moderately
Serpula himantioides	yes	yes	unknown	yes	Little
Hydnomerulius pinastri	yes	yes	doubtful	no	Little

[1]Colonization of dry wood if a close moisture source is available.

Gloeophyllum sepiarium, Amyloporia xantha (Fr.) Bondartsev & Singer, *A. sinuosa* (Fr.) Rajchenb., Gorjon & Pildain, *Antrodia serialis* (Fr.) Donkare probably the most important polypores that cause decay in houses. *Gloeophyllum sepiarium* is common on wooden roofs due to its high-temperature tolerance (Ryvarden and Gilbertson, 1993). Another species belonging to this genus, *G. trabeum* (Pers.) Murrill, is frequently found on dry wood used in construction on which it causes a brown cubical rot (Ţura et al., 2011). This fungus has been shown to persist in dry wood with 12% moisture content for 10 years (Esser and Bennett, 2002). This fungus is thermo-tolerant and grows well at high temperatures from 25 to 46°C. It can resist temperatures as high as 60–80°C for several hours (Viitanen, 1994). In conditions of constant humidity and temperature of mines, basements, and cellars, the species of *Antrodia* frequently develop. Another important species that produces decay of building timbers is the brown-rot causing fungus *Serpula incrassata* (Berk. & M.A. Curtis) Donk. This fungus unlike *S. lacrymans* uses the same survival strategy by transporting water through mycelial strands into dry wood. Therefore, wood protection from rainfall is not a solution against such type of fungi.

With several exceptions, brown-rot fungi are principally associated with conifers, while white-rot fungi are more common on hardwoods. The brown-rot fungi are more efficient in obtaining energy from wood for growth and are better suitable to survive in colder and drier habitats (Roy and De, 1996). Because brown-rots rapidly and drastically reduce wood strength early in the decay process, it is important to develop methods which can detect wood decay very early at an incipient stage prior to the occurrence of significant loss in strength. In the absence of fructifications, some of these fungi are hard to identify using classical methods. Techniques which have been used to detect incipient decay include: isolation and culturing of fungi, chemical staining, nuclear magnetic resonance, electrical resistance, biochemical, and immunological analyses (Cowling, 1961; Highley, 1987; Clausen, 1997).

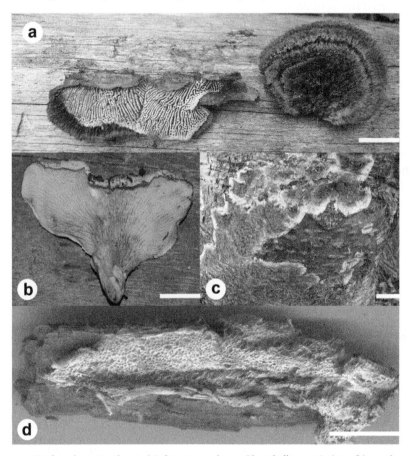

Figure 2. Timber decaying fungi: (a) dry-rot producer *Gloeophyllum sepiarium*; (b) wet brown-rot producer *Tapinella panuoides*; (c) dry-rot producer *Coniophora puteana*; (d) dry-rot producer *Fibroporia vaillantii*. Scale bar = 1 cm.

The DNA sequences of the ITS1 and ITS2 are highly variable; this feature can be exploited to generate restriction fragment length polymorphism (RFLP) patterns to identify wood decay fungi or to design taxon-specific primers (Jasalavich et al., 2000). For example, the report of Adair et al. (2002) was focused on early detection of wood-decaying fungi on wood chips used to supply high quality fiber in the pulp and paper industry. Since wood chips are stored outside 3–26 weeks before being processed into pulp, the authors developed a PCR–RFLP method that differentiated basidiomycetes from ascomycetes and detected the presence of basidiomycetes responsible for decay of wood chips at an incipient stage. Experiments on using fourier transform infrared microscopy (FTIR) were found to detect and to discriminate the two decaying fungi *Trametes versicolor* and *Schizophyllum commune* Fr. in experimentally infected beech blocks (Naumann et al., 2005). Sequencing of the internal transcribed spacers (ITS) of the rDNA is considered to be the best molecular tool (Schmidt, 2007) in early detection of wood-decaying fungi.

Control

There has been a tremendous interest to explore and to understand deep physiological mechanisms on how wood-decaying fungi act, and to discover efficient methods against their damaging effects. However, as previously stated, prevention measures must be adopted against timber decaying fungi. Information on how frequently occurring wood-destroying fungi act on particular types of wood, storing measures, decay influencing factors, and solid wood decay control strategies are essential. Moreover, wood can be treated with a preservative that improves service life. Such preserved wood is used most often for outdoor wood-products such as houses, bridges, fences, decks, railroad ties and so on. Some of the most important decay influencing factors are wood moisture and temperature (Table 2). Wood samples with low moisture content can be degraded and high temperatures as an alternative control measure do not kill mycelia (Schmidt, 2007).

Table 2. Moisture requirements of dry rot and cellar rot (Viitanen and Paajanen, 1988).

Fungus	Temperature °C/(F)			Timber moisture content (%)		
	Min.	Opt.	Max.	Min.	Opt.	Max.
Serpula lacrymans	−5 to +5 (23 to 41)	15–22 (59–71.6)	30–40 (86–104)	17–25	20–55	55–90
Coniophora puteana	0 to +5 (32 to 41)	20–25 (68–77)	40–46 (104–114.8)	15–25	30–70	60–80

In terms of control, various chemical fungicides were found to eradicate wood-decaying fungi. Most of these chemicals are water soluble and their application to the timber implies spraying or injection methods. In the latter method, wood walls need to be drilled first and then injected with chemical fungicides. However, such methods were found to be more or less effective and depending of a complex of factors such as: wood size, penetration rate of the chemical within wood, moisture levels in wood, temperature and so on Copper based wood preservatives were commonly used for pressure treatment of wood used for building construction. Each copper-based preservative type proved to be effective against one or several species of decaying fungi, but in many cases some fungi seemed to be tolerant to such chemicals. For example, in the study of Hastrup et al. (2005) 11 of 12 isolates of *Serpula lacrymans* were shown to be tolerant towards copper citrate. The report of Clausen and Yang (2007) suggests the use of biocides intended for indoor applications that must be non-toxic, non-volatile, odorless, hypoallergenic, and should be able to provide long-term protection under conditions of high humidity. Such biocides (borate-based multi-component biocide) were found to be effective against moulds, wood-decaying fungi (*Postia placenta*, *Gloeophyllum trabeum*, and *Trametes versicolor*) and termites. According to Ridout (2000) preservatives based on organic solvents perform well and do not have the disadvantages of some water-based systems. They do not appear to affect the strength characteristics of the timber, are non-deliquescent, do not cause metal corrosion and are not leached by water movement. Other preservative types are based on liquid petroleum gases. These are able to produce a drier and more evenly treated product, but are hazardous to use without extreme care.

The reports of Wang et al. (2005) and Cheng et al. (2006) found out that *Cinnamomum osmophloeum* Kaneh (*Lauraceae*), named as "Indigenous cinnamon tree", is one of the hardwood species that possesses significant antifungal activity. Leaf essential oils of *C. osmophloeum* were proved to be strong inhibitory for fungal growth. Results from the antifungal tests demonstrated that the leaf essential oils of cinnamaldehyde type and cinnamaldehyde/cinnamyl acetate type had an excellent inhibitory effect against white-rot fungi, *Trametes versicolor* and *Lenzites betulina* (L.) Fr. and brown-rot fungus *Laetiporus sulphureus* (Bull.) Murrill. The antifungal indices of leaf essential oils from these two chemotypes at 200 lg/ml against the above mentioned fungi were all hundred per cent.

In terms of biological control, species of *Trichoderma* were reported to eradicate the dry rot fungus *Serpula lacrymans* (Score and Palfreyman, 1994), but its usage is limited because *Trichoderma* has a negative impact upon human health.

Limiting fungal spread in buildings without the use of toxic pesticides may be done also by preventing amino acid translocation through mycelia cords. According to Watkinson and Tlalka (2008) the amino acid analogue AIB (α–aminoisobutyric acid) competitively excludes utilizable amino acids from the mycelium, and from the nutrient supply network of mycelia cords that enables basidiomycetes fungi including *Serpula lacrymans*, *Coniophora puteana*, *Gloeophyllum trabeum*, *Neolentinus lepideus* (Fr.) Redhead & Ginns and *Trametes versicolor* to spread through buildings.

Biotechnology of Wood-Inhabiting Fungi

Pharmaceutical Importance

Wood-decaying basidiomycetes include therapeutically important species of great interest. Several wood-decaying fungi such as *Ganoderma lucidum*, *Lentinula edodes* (Berk.) Pegler, *Tremella fuciformis* Berk.,

Auricularia auricula-judae (Bull.) Quél., and so on are well known in the Oriental traditional medicine from historical times due to their results in treating various types of ailments. Nowadays, in modern medicines also these fungi are greatly used. Technological development and scientific research have provided a better understanding of their anti-cancer, immunostimulating, anti-inflammatory or anti-bacterial activity. The market value of medicinal mushrooms and their derivative dietary supplements worldwide was ~US $1.2 billion in 1991, and was estimated to be US $6 billion in 1999 (Chang, 1996; Wasser et al., 2000). In the last decades, several compounds were discovered to have therapeutic value whereas limited information is available on physical, chemical, and pharmacodynamic properties of the active principles present in these extracts (Chu et al., 2002; Gao et al., 2006). Many forms of diseases (chronic diseases, cardiovascular disfunction or tumors) can, in part, be attributed to the diet. Now awareness of the relationship between the diet and a disease has evolved the concept of "functional foods" (Sadler and Saltmarsh, 1998). Some of the highly prized wood-decaying medicinal mushrooms are edible [e.g., *Lentinula edodes*, *Pleurotus ostreatus* (Jacq.) P. Kumm., *P. citrinopileatus* Singer, *P. pulmonarius* (Fr.) Quél., *P. eryngii* (DC.) Quél., *Auricularia auricula-judae*, *Laetiporus sulphureus*, *Hypsizygus marmoreus* (Peck) H.E. Bigelow, *Grifola frondosa*, *Armillaria mellea*, *Flammulina velutipes* (Curtis) Singer, *Tremella* spp., and *Polyporus umbellatus* (Pers.) Fr.–see Table 3] and they may be considered functional foods or "medicinal foods". Most of them are culturable and are available in the market today as fresh, dry, frozen, or in canned form.

According to Wasser (2011), biologically active mushroom compounds of therapeutic interest include: polysaccharides, secondary metabolites (lecitines, lactones, terpenoids, alkaloids, antibiotics, and metal chelating agents), enzymes (laccase, superoxide dismutase, glucose oxidase, and peroxidase), etc. Immunoceuticals isolated from more than 30 medicinal mushrooms have demonstrated antitumor activity in animal treatments. However, only a few have been tested for their anticancer potential in humans (β-D-glucans or β-D-glucans linked to proteins). Such immunoceuticals act mainly by elevating the host immune system due to activation of dendritic cells, NK cells, T cells, macrophages, and production of cytotokines. Several medicinal mushroom products, mainly polysaccharides and especially β-glucans, were developed with clinical and commercial purposes: Krestin (PSK) and polysaccharide peptide from *Trametes versicolor*; Lentinan, isolated from *Lentinus edodes*; Schizophyllan (Sonifilan, Sizofiran, or SPG), from *Schizophyllum commune*; Befungin from *Inonotus obliquus*; D-fraction from *Grifola frondosa*, *Ganoderma lucidum* polysaccharides fraction from *G. lucidum*; active hexose correlated compound; and many others.

Biologically active compounds isolated from several medicinally important species are available as medicinal products (health promoting dietary supplements, tea-bags, tinctures, tonic drinks, injections, etc.) mostly found in the pharmacies of Japan, China, Korea, United States, or European Union. Hot water extracts of medicinal fungi were traditionally used since

Table 3. Therapeutical effects of some wood-decaying mushroom species (adopted from Wasser and Weis, 1999a, and Stamets, 2005).

Species	Therapeutic effects																
	A	B	C	D	E	F	G	H	I	J	K	L	M	N	O	P	Q
Armillaria mellea (Vahl) P. Kumm.							+	+				+				+	
Auricularia auricula-judae (Bull.) Quél.		+					+	+	+						+		
A. polytricha (Mont.) Sacc.	+		+			+											
Flammulina velutipes (Curtis) Singer		+	+	+	+											+	
Fomes fomentarius (L.) J.J. Kickx		+			+	+											
Fomitopsis officinalis (Vill.) Bondartsev and Singer					+	+	+										
F. pinicola (Sw.) P. Karst.					+												
Ganoderma applanatum (Pers.) Pat.		+	+	+	+	+									+		
G. lucidum (Curtis) P. Karst.	+	+	+	+	+	+	+	+	+	+	+	+	+	+	+	+	+
G. oregonense Murrill		+	+		+		+						+		+		
G. tsugae Murrill	+	+	+														
Grifola frondosa (Dicks.) Gray		+	+	+	+	+	+		+		+		+		+	+	
Hericium erinaceus (Bull.) Pers.		+	+	+		+							+		+	+	
Hypsizygus marmoreus (Peck) H.E. Bigelow	+	+															
Inonotus hispidus (Bull.) P. Karst.	+				+												
I. obliquus (Ach. ex Pers.) Pilát	+	+	+	+	+	+	+				+					+	
Laetiporus sulphureus (Bull.) Murrill		+				+	+									+	
Lentinula edodes (Berk.) Pegler		+	+	+	+	+	+		+	+	+	+		+		+	+
Lenzites betulina (L.) Fr.		+					+										
Phellinus baumii Pilát	+	+		+		+											
Ph. igniarius (L.) Quél.	+	+															
Ph. linteus (Berk. & M.A. Curtis) Teng	+	+		+	+	+	+										
Piptoporus betulinus (Bull.) P. Karst.		+	+	+	+	+										+	
Pleurotus citrinopileatus Singer		+	+				+										

Table 3. contd....

Table 3. contd.

Species	Therapeutic effects																
	A	B	C	D	E	F	G	H	I	J	K	L	M	N	O	P	Q
P. eryngii (DC.) Quél.	+																
P. ostreatus (Jacq.) P. Kumm.		+			+	+		+		+			+				
P. pulmonarius (Fr.) Quél		+		+				+								+	
Polyporus umbellatus (Pers.) Fr.	+	+	+	+	+	+					+	+		+			
Schizophyllum commune Fr.		+	+	+	+	+					+					+	
Sparassiscrispa (Wulfen) Fr.		+	+														
Trametes versicolor (L.) Lloyd	+	+	+		+	+	+				+	+				+	
Tremella fuciformis Berk.		+	+	+			+			+	+			+			
T. mesenterica Retz.		+			+		+	+	+	+	+	+		+			

Note: antioxidant = **A**; antitumor = **B**; immunomodulating = **C**; anti-inflammatory = **D**; antiviral = **E**; antibacterial = **F**; antidiabetic = **G**; cardiovascular disorders = **H**; blood pressure regulator = **I**; cholesterol reducer = **J**; hepatoprotective = **K**; kidney tonic = **L**; nerve tonic = **M**; sexual potentiator = **N**; respiratory system disorders = **O**; antifungal = **P**; antistress = **Q**.

historical times, but modern practices are focusing rather on concentrated extracts, granular powders, mixtures of active compounds, etc. with the main purpose to obtain highly efficient drugs to cure various diseases. For example, bioactive compounds isolated from shiitake are used as LEM tablets (*Lentinula edodes* mycelium extract), tinctures, or as injectable lentinan of which the most effective is the LEM product (Kuhn and Winston, 2001). The precipitate obtained from a water solution of the mycelium by adding four volumes of ethanol was named LAP. Both LEM and LAP demonstrated strong antitumor activities orally and by injection to animal and humans (Wasser, 2005a). According to Lindequist et al. (2005) between 80 and 85% of all medicinal mushroom products are derived from the fruiting bodies. However, some disadvantages of this approach include: the long time required for the cultivation of fungal fruitbodies and additional attention and expenses necessary for obtaining optimal crop yields. Only ~ 15% of all products are based on extracts from mycelia. Examples are: PSK and PSP from *Trametes versicolor* and tremellastin from *Tremella mesenterica* Retz. A small percentage of mushroom products are obtained from culture filtrates, e.g., schizophyllan from *Schizophyllum commune*. Superior procedures in obtaining mushroom dietary supplements and pharmaceuticals include mycelia cultivation on solid substrate or in liquid or solid substrate fermentors. According to Smith et al. (2002) such an approach offers several advantages such as: speed of growth with reduction in production time, optimization of culture medium composition and physic-chemical conditions to allow regulation of mushroom metabolism, improved yield of specific products and possible designed variation in product types.

Medicinal products derived from mushrooms are often available in various types of mixtures (e.g., *Ganoderma lucidum* and *Lentinus edodes* bioactive compound mixture) and processed under patented methods with the main purpose to offer a more potent effect against target diseases. Since last decades interest of the world about medicinal mushrooms has increased and that has resulted into a variety of medicinal products currently available in the pharmacies and markets across the world. For example, shiitake is sold in some markets as fresh, dry, frozen, tea, wine, beer or as tonic drink, while reishi medicinal products are available as supplements, tinctures, tonic drinks (coffee and tea), toothpaste, soap, cream, etc. However, it is not known whether bioactive effects are caused by a single component or are the result of a synergistic impact of several ingredients. There is insufficient data whether mycelia powders are more potent than hot water, alcoholic, or hydro-alcoholic extracts (Chang and Wasser, 2012).

Ganoderma lucidum fruit body and mycelium contain approximately 400 different bioactive compounds, which mainly include steroids, lactones, alkaloids, polysaccharides, and triterpenes.

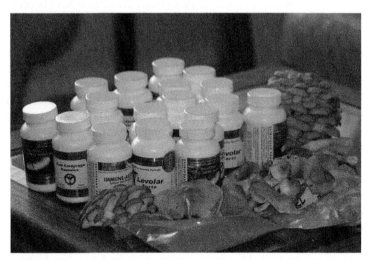

Figure 3. Different kinds of medicinal mushrooms dietary supplements. Photo courtesy: Darine Haddad.

According to Wasser (2005b) the key active constituents of this mushroom have shown pharmacological effects including immunomodulating, antiatherosclerotic, anti-inflammatory, analgesic, chemopreventive, antitumor, radioprotective, sleep promoting, antibacterial, antiviral (including anti-HIV), hypolipidemic, antifibrotic, hepatoprotective, diabetic, antioxidative and radical-scavenging, anti-aging, hypoglycemic, and anti-ulcer properties

(Wasser and Weis, 1997, 1999a; Chang and Buswell, 1999; Jong and Birmingham, 1992; Hobbs, 1995; McKeena et al., 2002; Wasser, 2002; Smith et al., 2002; Gao et al., 2006). *Ganoderma lucidum* is one of the most important mushrooms in Oriental medicine. Since thousands of years, in folk medicine, this fungus was used for a wide range of ailments including heart disease, diabetes, high altitude sickness, sexual impotency, chronic fatigue syndrome, hepatopathy, chronic hepatitis, nephritis, hypertension, arthritis, neurasthenia, insomnia, bronchitis, asthma, and gastric ulcers (Stamets, 2000; Wasser, 2005b) and hence this fungus has earned its reputation as "Ling Zhi" (Mushroom of herb and immortality – in China) or "Reishi" or "Mannentake" ("10,000 year mushroom" – in Japan). In traditional Chinese medicine the quality of Reishi mushroom is determined by color, shape and size: large fruitbodies, deep red and with a swirled ram's horn pattern are generally considered to have the highest quality (Upton et al., 2000).

Clinical evidence on humans, animals, and *in vitro* reports regarding the therapeutic effects of *G. lucidum* are tabulated (Table 3). Furthermore, Wang and Ng (2006) reported that protein 15-kDa isolated from *G. lucidum* show antifungal properties against *Botrytis cinerea*, *Fusarium oxysporum*, and *Physalospara pyricola*. The antitumor effect has also been stressed in many publications. For example, polysaccharides from *G. lucidum* have been found to have anticancer effects against HL-60 and U937 leukemic cell lines, lymphoma and multiple myeloma cells (Wang et al., 1997; Müller et al., 2006). Some of the triterpenes isolated from this fungus were found to exhibit cytotoxic activity against mouse sarcoma and mouse lung carcinoma *in vitro* (Min et al., 2000). Moreover, *G. lucidum* suppresses cell migration of highly invasive breast and prostate cancer cells and induces apoptosis in human prostate cancer cells PC-3 (Jiang et al., 2004; Mahajna et al., 2008). Tang et al. (2006) found the ganoderic acid T, which is a lanostane triterpenoid purified from methanol extract of *G. lucidum*, to show t cytotoxicity on various human carcinoma cell lines in a dose dependent manner and shown the major importance of using the ganoderic acid T as potential chemotherapeutic agent.

However, according to Wasser (2002) most of the clinical evidences for antitumor activity come from the commercial polysaccharides lentinan, PSK (Krestin), and schizophyllan, but polysaccharides of some other promising medicinal mushroom species (e.g., *Phellinus linteus* (Berk. & M.A. Curtis) Teng, *Flammulina velutipes*, *Hypsizygus marmoreus*, and *Agaricus blazei* Murrill) also show good results. The polysaccharides from mushrooms do not attack cancer cells directly, but produce their antitumour effects by activating different immune response in the host. Their activity is especially beneficial in clinics when used in conjunction with chemotherapy. The bioactive compound lentinan was isolated from Shiitake mushrooms (*Lentinus edodes*—Fig. 4a) by Chihara et al. (1970) inhibited the growth of Sarcoma 180 implanted in mice inducing almost complete regression of tumors at doses of 1 mg/kg x 10 with no signs of toxicity. Since then, several reports concluded lentinan to be a powerful polysaccharide demonstrating major antitumor effects including

Figure 4. Wood-decaying cultivated mushrooms: (a) *Lentinus edodes* growing on *Carpinus* spp. logs (Photo courtesy: Molnar D.); (b) *Pleurotus florida* growing on cardboard paper; (c) *Pleurotus* spp. growing on cardboard paper; (d) *P. citrinopileatus* growing on poplar sawdust.

the human colon carcinoma development (Ng and Yap, 2002), human MCF-7 breast carcinoma cells (Zhang and Cheung, 2002), or gastric tumours (Nakano et al., 1999). The mushroom polysaccharides appear to be well tolerated and compatible with chemotherapy and radiation therapy (Daba and Ezeronye, 2003).

Other novel active compounds found in Shiitake were noticed to decrease IL-1 production and apoptosis in human neutrophils (Sia and Candlish, 1999). Further experimental results proved Shiitake's bioactive compounds to have hypoglycemic effects (Yang et al., 2002; Kim et al., 2001), and to treat chronic persistent hepatitis and viral hepatitis B (Mizuno, 1996; Wasser and Weis, 1999a; Hobs, 2000), or to inhibit the activity of the human immunodeficiency virus (HIV) *in vitro* (Tochikura et al., 1988). Recent experiments on Shiitake extract showed an increased antimicrobial activity against 85% of the tested organisms, including 50% of the yeast and mould species demonstrating its potential in antimicrobial activity (Hearst et al., 2009), while the earlier report of Bender et al. (2003) stated its antimicrobial activity against *Staphylococcus aureus* when grown in malt extract broth concluding that oxalic acid was responsible for the antimicrobial effect.

Hot water extracts of *Trametes versicolor* (Yun Zhi "cloud-like mushroom" in Chinese) have been used in traditional Chinese medicine for a long time.

The active principle derived from *T. versicolor* belongs to a new class of elements called "biological response modifiers" (BRM) which are defined as agents capable of stimulating the immune system (Cheng and Leung, 2008). Modern studies have identified two important compounds, polysaccharide-K (PSK) or "Krestin", a water soluble protein-bound polysaccharide and PSP, a polysaccharide-peptide derived from mycelial cultures. These compounds have shown significant antitumor, anti-viral, and immuno-modulating activities (Țura and Nevo, 2007). The immune system boosting PSK is used to supplement the chemotherapy and radiotherapy of several cancers (gastric, esophageal, colorectal, breast and lung) and various infectious diseases (Cui and Chisti, 2003; Cheng and Leung, 2008; Wasser, 2002; Fisher and Yang, 2002). Evidence on PSK suggests that this polysaccharide can suppress pulmonary metastasis of methylcholanthrene-induced sarcomas, human prostate cancer DU145M, and lymphatic metastasis of mouse leukemia P388, and has prolonged the survival period in spontaneous metastasis models. It has also shown suppression of metastasis of rat hepatoma AH60C, mouse colon cancer colon 26 and mouse leukemia RL male 1 in artificial metastasis models (Kobayashi et al., 1995). Another study indicated that PSK has a direct cytotoxic activity *in vitro*, inhibiting tumour cell proliferation being independent of its previously described immunomodulatory activity in NK cells (Jiménez-Medina et al., 2008). Moreover, studies by Sakagami and Takeda (1993) indicated PSK to show a potent antimicrobial activity against *Escherichia coli* in ICR mice (Sakagami et al., 1990) and against other micro-organisms such as *Pseudomonas aeruginosa, Staphylococcus aureus, Candida albicans, Klebsiella pneumoniae* (Mayer and Drews, 1980), *Mycobacterium leprae, Listeria monocytogenes, Serratia marcescens, Streptococcus pneumoniae, Bacteroides fragilis, Cryptococcus neoformans*, and *Aspergillus fumigatus*. The second important bioactive compound—PSP, isolated from *Trametes versicolor*, shows strong immunomodulatory regulation rather than by direct cytotoxicity. It was reported that PSP in different concentrations promoted the proliferation of T-lymphocytes both in human peripheral blood and mouse splenocytes (Li, 1999). In addition, PSP has also been reported to have antitumor effects on patients suffering from non-small cell lung cancer (Hayakawa et al., 1993) and have been recommended as adjuvant treatment on the prognosis and after radiotherapy. Somewhat later, its clinical efficacy has been evaluated by Tsang et al. (2003) on 34 patients who had completed conventional treatment for advanced non-small cell lung cancer. The results have shown that after 28 days of treatment there was a significant improvement in blood leukocyte and neutrophil counts serum IgG and IgM, and percent of body fat among the PSP patients, but not the with control patients. Although the evaluable PSP patients did not improve in non-small cell lung cancer-related symptoms, there were significantly less PSP patients withdrawn due to disease progression. In China, *T. versicolor* is also considered useful in the treatment of hepatitis (Cheng and Leung, 2008). Data, based on accumulated evidence, suggest that Yun Zhi polysaccharides are non-toxic even when administered in higher dosages and over an extended period of time (Ng and Chan, 1997).

Several wood-decaying fungi belonging to the genus *Inonotus* s.l. are highly prized for their therapeutic properties, and are worth considering them here, namely, *I. levis* P. Karst., *I. obliquus*, *I. hispidus* (Bull.) P. Karst., *Phellinus baumii* Pilát, and *Ph. linteus*. The medicinal species *Inonotus obliquus* has been used for more than four centuries in folk remedies in Russia and Eastern Europe (Zheng et al., 2007; Balandaykin and Zmitrovich, 2015). Currently many studies claim the presence of valuable biologically active compounds (over 250) that have demonstrated hypoglycemic, hepatoprotective, anti-fungal, antitumor, anti-oxidative, and immuno-stimulating effects (Țura et al., 2009). Extracts of this fungus have shown definite improvement in the immune system and have also shown good results in treating chronic gastritis, psoriasis, and stomach complaints, including ulcers. Inonotodiol was found to be the most active ingredient and is able to destroy 100% of Walker 256 carcinosarcoma cells and MCF-7 human adenocarcinoma mammary cells. Several reports confirmed that *I. obliquus* has the ability to inhibit human cervical tumour cells—HeLa, whereas other reports demonstrated antiviral properties, and anti-inflammatory and anti-nociceptive effects (Kahlos et al., 1987; Kahlos and Tikka, 1994; Mizuno et al., 1999; Wasser and Weis, 1999b; Kim et al., 2005, 2006; Pietka and Grzywacz, 2006; Rogers, 2006; Lee et al., 2007). *Inonotus levis* has the capability of producing large amounts of exopolysaccharides (Vinogradov and Wasser, 2005), whereas studies relating to *Phellinus linteus* and its therapeutic potential are numerous, indicating that the species has polysaccharides with anti-cancer, anti-inflammatory, anti-bacterial, and anti-oxidant activity (Han et al., 1999; Song et al., 2003; Kim et al., 2003; Hur et al., 2004; Kim et al., 2007). The antitumor activity of the polysaccharides from the fruit body of this mushroom was first reported in 1968 when it was discovered that the inhibition rate of a hot water extract of *Ph. linteus* basidiocarps against the growth of sarcoma 180 in white mice was 96.7%, which was confirmed later by many investigators (Hwang et al., 2003). Rogers (2006) cited several works, mostly from researches in Korea, China, and Japan. It was found that water extracts of this fungus have protective effects on BNL cl.2 cells, indicating a liver cell sparing mechanism, and, in addition, the immune-stimulating effect was 71%. Recent work in Korea (see Kamei et al., 2003) looked at *Ph. linteus* and Adriamycin, a popular chemotherapy agent, and the ability to inhibit tumors. In mice that were given only the polypore extract, tumor growth was inhibited, metastases reduced, and they had the highest survival rate. *Ph. baumii*, another species present in our study, is considered to be effective in reducing various severe side effects of antitumor drugs. Both antitumor effects and an apparent decrease of side effects can be expected when using *Ph. baumii* for postoperative cancer patients. Additionally, other workers have demonstrated antitumor, hypoglycemic, anti-inflammatory, and anti-oxidant effects of this species (Dai and Xu, 1998; Hwang et al., 2005; Shon et al., 2003; Jong et al., 2004). Ethanolic extracts (hispidin and hispolon) prepared from fruit bodies and mycelial cultures of *Inonotus hispidus* showed considerable anti-viral activity against influenza viruses, type A and B, inhibiting the chemiluminescence response of human mononuclear blood cells and the mitogen-induced proliferation of spleen lymphocytes of mice (Ali et al., 1996, 2003). Its inhibition rate against sarcoma 180 and Ehrlich carcinoma is 80% and 70%, respectively. It was also found that *I. hispidus* and *I. rheades* (Pers.) Bondartsev & Singer stop hemorrhages, counteract noxious heat, relieve pain and cure mixed hemorrhoids, prolapsed rectum, and bleeding of piles (Jianzhe et al., 1987).

According to the traditional Chinese medicine, *Tremella mesenterica* has a well known healing capacity. It nourishes the lung, stomach, and kidney, strengthens bones, helps maintain ideal weight, and provides proper moisture to the skin. Nowadays, research has shown that *T. mesenterica*'s healing power is found in polysaccharide glucuronoxylomannan, which stimulates vascular endothelial cells, has antitumor activity, posseses pronounced anti-radiating effects, stimulates hematogenesis, has a hepatoprotective effect and anti-allergic, anti-inflammatory, and hypocholesterolemic activities (Ţura and Nevo, 2007). Glucuronoxylomannan is used also to produce skin products; some properties exhibit excellent skin moisture retention, skin protection effect, flexibility and flattening effects.

Wood-decaying fungi play a significant role in human health and hence have greatly attracted both scientific and public interest worldwide. Consequently, screening of new fungal material of medicinal importance and their detailed analyses of the bioactive substances, mechanisms of action, toxicity, and side effects are underway in many laboratories worldwide. Wasser (2011) has discussed in detail the future trends and unsolved problems of medicinal mushrooms which include: taxonomy and nomenclature, the study of culinary-medicinal mushrooms in pure culture, problems related to dietary supplements, medicinal mushrooms as natural products, unsolved problems in the study of structural characteristics, isolation process, receptor-mediated mechanism, and antitumor activity of medicinal mushrooms β-D-glucans.

Mycoremediation Aspects

The accumulation of agricultural wastes in large quantities has a negative environmental impact. Therefore, their bioconversion into eco-friendly valuable products, such as production of ethanol, for paper manufacturing, compost making for cultivation of edible fungi, biomass fuel production, and human and animal feed and so on, is essential (Sánchez, 2009). The bioconversion of lignocelluloses into bioproducts usually is a process requiring several steps which include: 1) pretreatment (mechanical, chemical or biological); 2) hydrolysis of polymers to produce readily metabolizable molecules (e.g., hexose or pentose sugars); 3) use of these molecules to support microbial growth or to produce chemical products; separation and purification

(Smith et al., 1987; Sun and Cheng, 2005). Wood-decaying fungi because of their lignolytic enzymes are able to degrade a large variety of such agricultural wastes and thus play an important role. For example, the biodegradation of agricultural wastes by 'oyster mushrooms' (*Pleurotus* spp.) has an increased applicability in the rural areas where various types of cellulosic substrates are readily available, e.g., cotton plant stalks are available in vast quantities in all cotton-growing areas. The material poses agrotechnical problems since the stalks have a fibrous structure resembling that of a hardwood and contain 46% cellulose, 20% hemicellulose and 21% lignin. *Pleurotus* spp. is able to degrade the lignocellulosic component of cotton stalks (Hadar et al., 1993). In addition, *Pleurotus* spp. is used for the bioremediation of organo-pollutants, xenobiotics and industrial contaminants (Cohen et al., 2002). Since the most utilized plastic is polyethylene, any reduction in the polyethylene waste would have a major impact on the overall reduction of the plastic waste in the environment. Recently, the study of Sivan (2011) points out that *Pleurotus ostreatus* and *Trametes versicolor* are potential candidates involved in direct biodegradation of polyethylene through their extracellular enzymes. And because laccase was found to be involved in polyethylene degradation, research is focusing on how to increase the laccase activity in these wood-decaying species. Laccase (benzenediol:oxygen oxidoreductases, EC 1.10. 3.2) is currently the focus of much attention because of its diverse applications, such as delignification of lignocellulosics, crosslinking of polysaccharides, bioremediation applications, such as waste detoxification, and textile dye transformation, food technologic uses, personal and medical care applications, and biosensor and analytical applications (Madhavi and Lele, 2009). The simple requirements of laccase catalysis (presence of substrate and O_2), as well as its apparent stability and lack of inhibition (as has been observed with H_2O_2 for peroxidase), make this enzyme both suitable and attractive for biotechnological applications (Majeau et al., 2010; Baldrian, 2006; Giardina et al., 2010). According to Majeau et al. (2010), the most important issue remains the high cost of enzyme production which could be reduced through the use of zero or negative cost substrates, such as tertiary matter, agricultural and food wastes or wastewater from the food or pulp and paper industries. This recent approach on laccase research of the last two decades is, however, still in experimental stage and more has to be done by testing the enzyme and its production on large-scale bioreactors for industrial applications. Currently, scientists across the world are working on the laccase optimization process which is influenced by many parameters such as: species, strain, type of cultivation, agitation (stationary or agitated culture mode), aeration and cultivation time and so on. However, the most critical factors are the glucose and nitrogen sources, their concentration and their ratio and the nature and concentration of the inducer. Among the most studied wood-decaying fungi responsible for high laccase activity are: *Trametes* spp. [*T. versicolor*, *T. pubescens* (Schumach.) Pilát, *T. hirsuta* (Wulfen) Lloyd and *Funalia. trogii* (Berk.) Bondartsev & Singer], *Pleurotus* spp. [*P. ostreatus*, *P. dryinus* (Pers.) P. Kumm., *P. pulmonarius*, and *P. sajor-caju* (Fr.)

Singer], *Pycnoporus* spp. [*P. cinnabarinus* (Jacq.) P. Karst. and *P. sanguineus* (L.) Murrill], *Ganoderma* spp. (*G. lucidum* and *G. australe*), and *Phellinus robustus* (P. Karst.) Bourdot & Galzin. Such fungi are able to degrade a variety of persistent environmental pollutants, such as chlorinated aromatic compounds, heterocyclic aromatic hydrocarbons, various dyes and synthetic high polymers (Bennett et al., 2002). Polycyclic aromatic hydrocarbons (PAHs) are one of the important recalcitrant pollutants that are present in our environment, and perhaps were the first recognized environmental carcinogens (Arun and Eyini, 2011). They are dangerous due to their carcinogenicity, mutagenicity, and teratogenicity as well long-term persistence in the environment (Easton et al., 2002; Miller et al., 2004; Ramsay et al., 2003; Jacques et al., 2005; Cajthaml et al., 2006). For example, anthracene is a PAH that has the ability to bioaccumulate and persist in soil and may lead to contamination of human food products such as vegetables and it can accumulate in other foods of the chains (Ping et al., 2008; Iwamoto and Nasu, 2001). A positive correlation between soil pollution with anthracene and biodegradation of it by *Pleurotus ostreatus* has been observed (Zebulum et al., 2011). Several important reports have demonstrated the abilities of wood-decaying fungi to degrade such hydrocarbons, among them the report of Mori et al. (2003) that *Phlebia lindtneri* (Pilát) Parmasto has ability to degrade chloronaphthalene and PAHs; the white-rot fungus *Anthracophyllum discolor* (Mont.) Singer has a high removal capability for PAHs such as phenanthrene (62%), anthracene (73%), fluoranthene (54%), pyrene (60%) and benzo(*a*)pyrene (75%) Acevedo et al. (2011); while according to Steffen et al. (2002) *Stropharia rugosoannulata* Farl. Ex Murrill is the most efficient PAH degrader, removing or transforming benzo(a)pyrene almost completely and about 95% of anthracene and 85% of pyrene, in cultures when supplemented with 200 mu M Mn, within 6 weeks. Some other wood-decaying fungi able to degrade PAHs are: *Phanerochaete chrysosporium* Burds, *Coriolopsis* spp., *Laetiporus sulphureus*, *Phellinus* sp., *Bjerkandera adusta* (Willd.) P. Karst., *Irpex lacteus* (Fr.) Fr., *Lentinus tigrinus* (Bull.) Fr., *Fibroporia vaillantii*, *Stropharia coronilla* (Bull. ex DC.) Quél., *Agrocybe praecox* (Pers.) Fayod, etc. (Arun and Eyini, 2011; Valentin et al., 2006; Steffen et al., 2002; Tekere et al., 2005; Tortella et al., 2005).

Chlorinated hydrocarbons are an important group of organopollutants widely found in soil, groundwater, and river sediments due to improper disposal practices and leakage from store tanks. Their chemical structural features preclude or retard biological attack and result in low levels of biodegradation and increased environmental persistence (Marco-Urrea et al., 2009). An interesting report of Dejong et al. (1994) claims, that common wood- and forest litter-degrading fungi produce chlorinated anisyl metabolites which are structurally related to xenobiotic chloroaromatics. Such compounds occur at high concentrations of approximately 75 mg of chlorinated anisyl metabolites per kg of wood or litter in the environment. Knowledge on natural occurrence of such compounds in forest ecosystems reveals the need of reducing their quantity due to anthropogenic activities by implementing efficient bio-control measures. In terms of biocontrol, scientific reports demonstrate that white-rot fungi are able to degrade chlorinated aromatic compounds reducing pollution caused by such compounds due to their enzymatic systems (e.g., lignin peroxidise, manganese peroxidase, or laccase) which are responsible for the depolymerisation of lignin. Twenty white-rot fungi mostly belonging to the genus *Phlebia–Phlebia acanthocystis* Gilb. & Nakasone, *Ph. brevispora* Nakasone, and *Ph. aurea* (Fr.) Nakasone, for example, when tested to degrade dieldrine and heptachlor (compounds of the most persistent organochlorine pesticides) showed significant ability to degrade dieldrine which was transformed to hydroxyl dieldrine and several unidentified metabolites. Heptachlor was degraded rapidly (compared to dieldrine) by almost all fungi belonging to the genus *Phlebia* (Kondo et al., 2010). Furthermore, the study of Marco-Urrea et al. (2009) reporting induction of hydroxyl radical production by *Trametes versicolor* which is able to degrade recalcitrant chlorinated hydrocarbons such as trichloroethylene, perchloroethylene, 1,2,4- and 1,3,5-trichlorobenzene and so on in the treatment of chlorinated hydrocarbons-contaminated wastewaters with promising results.

Synthetic dyes are used worldwide used in food, pharmaceutical, cosmetic, printing, textile and leather industries. In the textile and dyeing industries these dyes are used because of their ease and cost-effectiveness in synthesis, firmness, high stability to light, temperature, detergent and microbial attack and their of variety color as compared with natural dyes (Couto, 2008). Wastewaters from such industries are highly polluted showing a negative environmental impact. Consequently, environmental-friendly, reliable, and cost-effective decolorization and detoxification methods are needed and, therefore, catching the attention of the scientific community. White-rot fungi have proved to be very efficient in degrading synthetic dyes (Singh, 2006; Couto, 2008). Glenn and Gold (1983) discovered the white-rot fungus *Phanerochaete chrysosporium* to degrade polymeric dyes. Since then, about 50% of the known dye decolorization and degradation is achieved through white-rot fungi. Among them *Trametes versicolor*, *Bjerkandera adusta* and *Pleurotus ostreatus* have been found to be superior to *P. chrysosporium* (Singh, 2006). The study of Anastasi et al. (2010) involving 12 white-rot fungi in batch experiments to test their ability to decolorize and detoxify four model wastewaters from textile and tannery industries indicated that a strain of *Bjerkandera adusta* was able to completely degrade most of the dyes and to decolorize and detoxify three simulated wastewaters. The result opens up a new perspective for bioremediation of industrial effluents by using white-rot fungi. Mycoremediation by using white-decay fungi has been already patented and a few companies (EarthFax Development Corporation in Utah and Gebruder Huber Bodenrecycling in Germany) employ these fungi for bioremediation (Singh, 2006).

Accumulation of heavy metals (e.g., As, Cd, or Pb) by fungi has been known for a few decades now, and has an increased interest in the scientific community. Practically, wood-decaying fungi are able to take up heavy metals by deposition of particles from the atmosphere and absorption from the substrate (Gabriel et al., 1997). Because of their ability to grow and repeatedly fructify also in industrial areas with high concentrations of sulphur and nitrogen oxides, they can be used as a sensitive indicator for monitoring of the atmospheric pollution by heavy metals (Čurdová et al., 2004). Gabriel et al. (1994) studied the accumulation of aluminum, cadmium, lead and calcium in the wood-rotting fungi *Daedalea quercina* (L.) Pers., *Ganoderma applanatum*, *Stereum hirsutum* (Willd.) Pers. and *Schizophyllum commune* pointing out that lead content was high in the mycelium of *Stereum hirsutum* (90.6 mmol/g) while the mycelium of *Ganoderma applanatum* contained maximal values of cadmium (272 mmol/g), aluminum (600 mmol/g) and calcium (602 mmol/g) when cultured in liquid media in the presence of Al, Cd, Ph and Ca salts. Furthermore, the ability of white-rot fungi to adsorb and accumulate metals together with the excellent mechanical properties of fungal mycelial pellets provide an opportunity for application of fungal mycelia in selective sorption of individual heavy metal ions from polluted water (Baldrian, 2002). Another study indicates the ability of the edible wood-rotting fungus *Pleurotus citrinopileatus* to accumulate long-lived radioisotopes of cesium (^{134}Cs and ^{137}Cs). The ^{137}Cs can be even found far from the site of nuclear accident due to its high mobility in the environment. The fungus reaction after ^{137}Cs accumulation is to change color from white or yellow to black, hence, this black mushroom species may play an important role as bio-indicator of radioactive pollution caused by nuclear fallout containing ^{134}Cs or ^{137}Cs (Mukhopadhyay et al., 2007).

Disposal of decommissioned timber treated with preservatives based on chromium (Cr), copper (Cu) and arsenic (As) is of increasing concern because of the potential public health and ecological risks associated with the release of the toxic inorganic compounds (Kartal et al., 2004). In the United States and Canada alone, approximately 3–4 million m^3 of chromated-copper arsenate (CCA) preservative-treated wood are currently being removed from service annually, and it is estimated that this amount will increase to 16 million m^3 by 2020 (Cooper, 2003). Brown-rot fungi were used to remove CCA-treated wood, and the key element seems to be the oxalic acid produced by these fungi. Kartal et al. (2004) tested fungi *Fomitopsis palustris* (Berk. & M.A. Curtis) Gilb. & Ryvarden, *Laetiporus sulphureus* and *Coniophora puteana* and noted their CCA removing ability. *Fomitopsis palustris* and *Laetiporus sulphureus* exposed to CCA-treated sawdust for 10 days showed a decrease in arsenic by 100% and 85%, respectively; however, *Coniophora puteana* remediation removed only 18% arsenic from CCA-treated sawdust. Likewise, chromium removal by *Fomitopsis palustris* and *Laetiporus sulphureus* remediation processes was higher than those for *C. puteana*. Another such study indicates *Daedalea dickinsii* Yasuda, to achieve the most efficient removal of copper (82%) compared to *Fomitopsis palustris* and some other polypore species. However, the unknown polypore has shown a reduction in arsenic and chromium by 98% and 91%, respectively (Kim et al., 2009).

Other Biotechnological Applications of Wood-Inhabiting Fungi

Biotechnological approaches in the field of the wood and paper industry play a significant role by decreasing energy and costs involved in the industrial processes. Paper mainly consists of wood fibers or cellulose fibers which are conventionally obtained by mechanical or chemical pulping (Mai et al., 2004). Biological pulping has the potential to improve the quality of pulp, properties of paper as well as to reduce the costs of energy and environmental impact as compared to the traditional pulping operations. Other benefits include improved burst strength and tear indices of the product and reduced pitch deposition during the production process (Breen and Singleton, 1999). However, according to Maijala et al. (2008) for industrial application, biopulping is still considered too slow and technically demanding, and the direct application of enzymes on wood chips has become a more attractive alternative. By comparing the process of paper making aided by *Physisporinus rivulosus* (Berk. & M.A. Curtis) Ryvarden and its reference paper refined at similar specific energy consumption level, the authors observed that fungal treated chips significantly improved the paper strength and optical properties. Biobleaching reduces the usage of conventional bleaching by chemicals and improves pulp and eventually paper quality by improving brightness, breaking length, burst index, tear index, and manufacturing yield (Jiménez et al., 1997). For example, lignocelluloses-degrading enzymes obtained from *Phanerochaete chrysosporium* increases tensile index up to 45%, the tear index up to 35% and the burst index up to 9%, while the quality of chemical-mechanical pulp could reach 40–50% of the total fiber material. This results into an overall improvement in the quality of packaging paper (Boeva-Spiranova et al., 2007). Among *Ph. chrysosporium*, some other selected fungi used in the optimization process of paper quality include: *Physisporinus rivulosus* (Hatakka et al., 2003), *Ceriporiopsis subvermispora* (Pilát) Gilb. & Ryvarden (Akhtar et al., 2000), and *Phlebiopsis gigantea* (Fr.) Jülich (Behrendt and Blanchette, 1997).

Biotechnological application of wood decay fungi in the forest industry was intensively investigated during the last decades. The study of Schwarze and Schubert (2011) points out a positive correlation between the white-rot fungus *Physisporinus vitreus* (Pers.) P. Karst. and the permeability of preservatives in the so-called difficult-to-treat (refractory) wood type belonging to species such as *Picea abies* (Norway spruce) and *Abies alba* (European silver fir). *Physisporinus vitreus* has shown to have an extraordinary capacity to induce substantial permeability changes in the heartwood of *P. abies* without causing significant loss of impact on bending strength. Moreover, this fungus species in conditions of controlled use of its degrading pattern may produce the so-called "mycowood" with improvement in acoustic properties to overcome the shortage of natural wood with the superior tonal qualities desired by traditional musical instrument makers.

Investigations on alternative energy sources recently have focused on the production of fuel ethanol from lignocellulosic biomass (Krishna et al.,

1998; Krishna and Cowdary, 2000; Sreenath and Jeffries, 2000; Nigam, 2001). Currently, this aspect has a tremendous interest in the scientific world, and methods of developing a reliable technology to produce biofuel are ongoing. Agricultural residues are a major source of lignocellulosic biomass that can be transformed into biofuels through the mediation of wood-decaying fungi. Moreover, lignocellulosic biomass is an important resource for the production of biofuels because it is abundant in nature, is inexpensive, and production of such resources is eco-friendly (Khalil et al., 2011). By using *Ceriporiopsis subvermispora* and a new white-rot isolate, *Phellinus* sp. SKM2102, Baba et al. (2011) provided valuable data on pretreatment of Japanese cedar wood by ethanolysis with and without fungal treatment for enzymatic saccharification pointing out that combined pretreatment with the white-rot fungus *Ceriporiopsis subvermispora* and ethanolysis increased the sugar yields by 7 times than that of ethanolysis without fungal treatment. As we may notice, future prospects on using wood-decaying fungi for biofuel production are promising; however, new experiments and fungal taxa screening are still needed for developing an energy and cost efficient technology.

Cultivation Techniques

Wood-decaying fungi comprise some of the most popular edible mushrooms of the world. Through their enzymatic systems such fungi are able to depolymerize wood or cellulose into quality food and valuable medicinal products. The world production of mushrooms according to FAOSTAT (2008) was ~3.4×10^6 tonnes in 2008, China producing the highest 1.5×10^6 tonnes, followed by the USA producing 0.38×10^6 tonnes. Worldwide production of shiitake mushrooms being ranked as the second after the white button mushroom [*Agaricus bisporus* (J.E. Lange) Imbach]. China is the major producer of Shiitake, accounting for 85.1% of the total world production in 1997 (Chang, 1999; Chang and Miles, 2004; USDA, 2008). White-rot decaying fungi of the genus *Pleurotus* spp. (e.g., *P. ostreatus*, *P. eryngii*, *P. columbinus* Quél., *P. djamor* (Rumph. ex Fr.) Boedijn, *P. sajor-caju*, etc.) are the third largest commercially cultivated mushrooms in the world (Kües and Liu, 2000), while some other well-known edible and medicinally important ones that are widely cultivated are wood-decaying species belonging to the genera: *Grifola*, *Flammulina*, *Hypsizygus*, *Auricularia*, and *Hericium*.

Available literature indicates that ancient China is the first to successfully cultivate many popular species such as: *Auricularia auricula-judae* (600 AD), *Flammulina velutipes* (800–900 AD), *Lentinus edodes* (1000–1100 AD), *Volvariella volvacea* (Bull.) Singer (1700 AD), and *Tremella fuciformis* (1800 AD). However, during the last 50 years technological development has significantly improved the methods of cultivation of these fungi (Chang and Miles, 1989; Chang and Wasser, 2012). According to Stamets (1993), spawn making technology was revolutionized by Sinden's discovery of grain as a spawn carrier medium.

Somewhat later, Stoller (1962) significantly contributed to the technology of mushroom cultivation through a series of practical advances in using bags, collars, and filters. Since then, many other authors across the world described various mushroom cultivation procedures. The standard commercial mushroom cultivation consists of two phases: spawn technology and mushroom production technology. Spawn technology begins with the isolation of fungus species from their habitat followed by an intensive selection and breeding through classical and molecular genetics, as wild strains are normally not suitable for commercial cultivation. Genetic improvement is focused on high-yielding strains having additional characteristics, such as disease/ chemical resistance, earliness, tolerance to low or elevated temperatures, as well as shape, taste, and color of the fruit bodies (Martínez-Carrera et al., 2000).

Mushroom production technology refers to their outdoor and indoor cultivation. Outdoor cultivation is usually done using various types of logs as substrata according to the species of interest. It involves a simple procedure. However, despite its simplicity, large-scale cultivation on logs is not often used due to long incubation periods, low yields and environment dependence (Gregori et al., 2007). Indoor mushroom cultivation is rather connected to a large-scale mushroom production and involves equipments and efficient cultivation methods. It has short incubation periods and high yields. Literature based data revealed several distinct methods of mushroom culture aimed to increase overall yield. Enhanced yield was noticed to be dependent on a complex of factors including species, strain, substrate type, culture methodology, biological factors, and environmental conditions. The main parameter used to evaluate mushroom yield is the so-called "biological efficiency" (BE). According to Stamets (1993), this formula was originally developed by the White Button mushroom industry, and states that 1(lb)kg of fresh mushrooms grown from 1(lb)kg of dry substrate is 100% biological efficiency. This is also equivalent to growing 1(lb)kg of fresh mushrooms for every 4(lbs)kg-s of moist substrate, a 25% conversion of wet substrate mass to fresh mushrooms or achieving a 10% conversion of dry substrate mass into dry mushrooms.

Stamets (1993) used the BE formula to express oyster mushroom (*Pleurotus* spp.) yield observing that the combination of a good mushroom culture technique and a vigorous strain achieves substantial yields. The author observed that biological efficiency in oyster mushrooms often exceeds 100% and is some of the greatest in the world of cultivated mushrooms. He concluded that the formula is greatly affected by the stage at which the mushrooms are harvested. Studies on Shiitake yields revealed some other yield measuring formulas. For example, San Antonio (1981) used fresh weight of mushrooms (g) as a function of fresh weight of the log substrate (kg) to measure Shiitake production, while others have expressed Shiitake's yield as mushroom weight per volume of substrate logs (Bruhn et al., 2009). In terms of yield, the nutritional substrate types, on which the fungus is grown, and the supplementation with nitrogen-rich sources demonstrated significant increase in BE values. For example, Stamets (1993) highly recommended a substrate formula designed for maximizing yields of several wood decomposing species (*Lentinula edodes, Flammulina velutipes, Grifola frondosa, Hypholoma sublateritium* (Schaeff.) Quél., *Hericium erinaceus, Agrocybe aegerita* (V. Brig.) Singer, and *Pholiota nameko* (T. Itô) S. Ito & S. Imai. The formula consists of fast decomposing hardwoods (poplar, alder, and cotton wood) and a nitrogen-rich supplement: hardwood sawdust (100 pounds), ½–4 inches wood chips (50 pounds), oat, wheat or rice bran (40 pounds), gypsum (5–7 pounds) and water until 65–75% moisture is achieved. Several reports regarding substrate type, supplementation and BE are discussed below.

1. Shiitake (*Lentinula edodes*)—(Fig. 4a). A more natural cultivation of Shiitake is through outdoor log cultivation. However, this procedure is labor intensive, and slow in comparison with growing mushrooms on sterilized sawdust (Stamets, 1993). Log cultivation of shiitake for commercial purposes decreased in the USA in favor of synthetic logs able to offer increased yields, decreased time to complete a crop cycle, and a consistent market supply (Royce, 1996). In addition, Shiitake grown on sawdust show a higher nutritional composition (protein content, free amino acids, and sweet, umami, and bitter components) than those harvested from logs (Tabata et al., 2006). Most Shiitake are grown on synthetic substrates composed of oak sawdust (ca. 50%) and nutrient supplements (white millet, rye and wheat bran); however, other alternative raw material includes wheat straw, but this substrate has not been widely accepted due to infestation with *Trichoderma* spp. and low yields (Royse and Sanchez, 2007). Alternative fruiting substrates used in Shiitake cultivation include alder or oak sawdust supplemented with bran 4:1 or rye grass straw (Stamets and Chilton, 1983). In order to significantly boosts yields by 20% or more the final substrate may be supplemented with rice bran (20%), rye flour (20%), soybean meal (5%), molasses (3–5%), or sugar (1% sucrose) (Stamets, 1993). Subsequent reports suggest incorporating millet as a significant ingredient in Shiitake substrates (Royse, 1996). However, Royse (1985) reported BE and basidiome size to be positively correlated to the addition of combined wheat bran (10%) and millet (10%) rather than either millet (20%) or wheat bran (20%) used alone. The addition of 34% millet to a supplemented (wheat bran; rye grain; sucrose) sawdust substrate was observed to stimulate mushroom yield by 68% compared to a millet addition of 17%. In this case BE increased (ca. 13.5%) for each additional 6% increase in the amount of millet supplementation but average mushroom size decreased from 23 g at 17% millet to 13 g per mushroom at 34% millet supplementation. The experiments of Moonmoon et al. (2011) regarding Shiitake cultivation on sawdust offered two options supplementation dose dependent: yield and quality. The highest number of basidiomes (34.8/500 g packet), biological yield (153.3/500 g packet) and BE (76.6%) was observed when sawdust was supplemented with 25% wheat bran, whereas, yield of the best quality mushrooms was observed on sawdust with 40% wheat bran supplementation.

Several reports have discussed the yield potential of log cultivation of shiitake. According to Chen et al. (2000) there are three types of Shiitake strains, in part dependent upon the expected range of ambient temperatures throughout the productive life of the log and are categorized according to fruiting temperature requirements: low temperature (less than 10°C/50 F), mid-temperature (10–18°C/50–64.4 F), warm temperature (greater than 20°C/68 F) and wide range (5–35°C/41–95 F). As observed by Bruhn and Mihail (2009) and Bruhn et al. (2009) the use of a Shiitake strain selected for its propensity to fruit over a wide temperature range was more productive than strains which fruit best during warm or cold weather. The authors performed several outdoor Shiitake cultivation experiments in response to the temperature of forcing water, inoculum strain, substrate host species and physical orientation of the log during fruiting. It was also noticed that logs inoculated with traditional sawdust spawn outperformed those inoculated with dowel or thimble spawn. Several host species were compared, white oak (*Quercus alba*), shingle oak (*Q. imbricaria*) and black oak (*Q. velutina*) and sugar maple (*Acer saccharum*). Sugar maple had the highest yield under forced fruiting conditions with water as cool as 10–12°C/50–53.6 F. The 'forced fruiting' is a part of the Shiitake log cultivation process and it is employed by many growers across the world. Several reports are available to indicate that this practice enhances the overall yield. For example, Chang and Miles (2004) have reported that mushroom production may be stimulated by the episodic occurrence of sufficient rainfall or by the deliberate submersion of inoculated logs in water for 1–3 days depending on season and other variables. The report of (Shen et al., 2008) on outdoor and indoor Shiitake cultivation identified at least two variables that are important for increasing mushroom yield, i.e., substrate moisture content and log weight. Comparing moisture content in logs, the authors observed that yield distribution pattern in first and second break on logs with 50% substrate moisture content were more evenly distributed compared to high- and medium-moisture content logs and BE was greatest on 55% moisture logs. However, log weight had little effect on BE, but had a significant effect on mushroom yield/log.

Some of the 'spent' Shiitake substrates are recycled as animal feed or may be re-inoculated with other mushroom species (Stamets, 1993; Royse and Sanchez, 2007).

Oyster mushrooms (*Pleurotus* spp.)—(Fig. 4b,c,d). Mushrooms belonging to *Pleurotus* spp. are some of the highly appreciated culinary and medicinal mushrooms across the world. According to Martinez-Carerra (1998) empirical cultivation of *Pleurotus* started around 1917 in Germany, using natural spawn for inoculation of wood logs and stumps, and the first large-scale cultivation on logs was undertaken in Hungary in 1969. Later, other substrates were also found to be good for growing Oyster mushrooms and several other species were brought into cultivation such as *P. ostreatus*, *P. sajor-caju*, *P. cystidiosus*

O.K. Mill., *P. florida* nomen nudum, *P. citrinopileatus*, *P. flabellatus* Sacc., and *P. sapidus* Sacc.

Since sawdust causes reduction of wooded areas, straw and grass plants are popularly used worldwide to cultivate *Pleurotus* mushrooms (Liang et al., 2009). However, substrate formulas are dependent on local availability of lignocellulosic materials. A common procedure adopted by the European Oyster growers is to mix fully colonized pasteurized straw into ten times more pasteurized straw, thus attaining a tremendous amount of mycelial mileage (Stamets, 1993). The experiments of Fanadzo et al. (2010) found wheat straw to be superior over maize stover (*Zea mays*) and thatch grass (*Hyparrhenia filipendula*) when cultivating *P. sajor-caju*, concluding that maize stover was more suitable for *P. ostreatus* cultivation. Moreso, the addition of cotton seed hull improved yields when cultivating *P. ostreatus* using wheat straw. Hernández et al. (2003) suggested a simple procedure for preparing substrate for *P. ostreatus* cultivation. They used wooden crates for composting a mixture of 70% grass (*Digitaria decumbens*), and 30% coffee pulp, combined with 2% $Ca(OH)_2$. It has also been noticed by them that the crate composting considerably modifies the temperature in the substrate as compared to pile composting. The temperatures were lower and less homogenously distributed while the BE varied between 59.79% and 93% in the two harvests. In addition, it has also been observed that it is possible to produce *P. ostreatus* on a lignocellulosic non-composted, non-pasteurized substrate with an initial pH of 8.7 whereas composting for two to three days improved BE. Liang et al. (2009) compared the BE of *P. citrinopileatus* when grown on stalks of several grass plants such as *Panicum repens*, *Pennisetum purpureum*, and *Zea mays* and found 65.4% BE when *P. citrinopileatus* was grown on 45% *Zea mays* stalks combined with 45% hardwood sawdust supplemented with 9% rice bran, and 1% $CaCO_3$. However, according to Stamets (1993), yields of 132% BE for *Pleurotus* spp. are reported by Martinez-Carerra et al. (1985). They cultivated Oyster mushrooms on coffee pulp that was fermented for 5 days, pasteurized, and inoculated with wheat grain spawn. In 1987 Martinez-Carerra, validated the results with yields in excess of 100% biological efficiency on the same substrate and presented the first model for utilizing this abundant waste product. According to Martínez-Carrera et al. (2000), after mushroom cultivation about 27% of the original substrate remains unutilized. The chemical composition of spent coffee pulp, after *Pleurotus* cultivation, mainly contains carbohydrates (29.9%), crude protein (21.5%), crude fat (1.8%), and crude fiber (31.4%). Thus, spent coffee pulp substrate can be composted, either by aerobic composting or vermin-composting to produce an organic fertilizer or soil conditioner for crops. The experiments of Mandeel et al. (2005) with *P. columbinus*, *P. sajor-caju*, and *P. ostreatus* grown on untreated organic wastes including chopped office papers, cardboard, sawdust and plant fibers, indicated a high BE of *P. columbinus* grown on cardboard (134.5%) and paper (100.8%). Other investigated taxa such as *P. ostreatus* showed maximum yield when grown on cardboard (117.5%) and paper (112.4%). In farms, where environmental controls are rudimentary or lacking, growers use a casing overlay that minimises the loss of substrate moisture and allows more than one break (Oei, 2006; Rodriguez Estrada and Royse, 2008; Tan et al., 2005). The experiments of Rodriguez Estrada et al. (2009) about the yield improvement and BE of *P. eryngii* var. *eryngii* (DC.) Quél. pointed out that supplementation with a delay-release nutrient (Remo's, corn and soybean based) and the use of a casing layer significantly affected the yield by 141% in comparison to non-cased substrates. Moreover, when casing and supplementation were combined, yield increased by 179% over non-cased and non-supplemented substrates. The casing layer has some important properties such as water retention, structure and microflora and so on. It has the power to release and to absorb water, allow gas exchange and support primordial formation due to the presence of beneficial bacteria (Stamets and Chilton, 1983). As noticed by Stamets (1993), yield potential in *P. ostreatus* and *P. pulmonarius* (Fr.) Quél is between 75–200% BE, and is greatly affected by the size of the fruitbodies harvested, and the number of flushes orchestrated. Basidiome aspect and nutritional content proved to be dependent on substrate formula used for mushroom culture. Increased protein, aminoacid and lipid concentrations and a significant decrease in fiber, free sugar and carbohydrates was obtained by Shashirekha et al. (2005) by supplementing rice straw with cotton seed in *P. florida* cultivation. Rodriguez Estrada et al. (2009) also noticed darker color and more robust aspect of *P. eryngii* basidiomes grown in cased substrates.

Other commercially important cultivated wood decomposing fungi: *Grifola frondosa, Flammulina velutipes, Auricularia auricula-judae, Hypsizygus marmoreus,* and *Hericium erinaceus* are also cultivated commercially. However, data related to BE of these taxa grown on variable lignocellulosic substrates is rather scarce as compared to Shiitake and Oyster mushrooms. Japanese growers of *Flammulina velutipes, Auricularia auricula-judae* and allies, and *Pleurotus ostreatus* have a standard substrate formula consisting of 4 parts of sawdust and 1 part of bran (Stamets and Chilton, 1983), whereas synthetic substrates such as oak, beech, or larch are some of the most used culture substrates for Maitake (*Grifola frondosa*) cultivation and are often supplemented with cereal grains such as rice, wheat, oats, and corn. Shen and Royse (2001) used mixed sawdust (primarily red oak) supplemented with a combination of 10% wheat bran, 10% millet and 10% rye and obtained consistent yields (BE 44%) and best quality mushrooms. The report of Montoya et al. (2008) on *G. frondosa* culture pointed out that combination of oak sawdust and corn bran as substrate gave a high BE (35.3%) and best quality mushrooms whereas coffee spent-ground was found not suitable for the culture. The experiments of Hu et al. (2008) on *Hericium erinaceus* and *H. laciniatum* (Leers) Banker grown on sawdust, partially mixed with four different agro wastes (rice hull, rice straw, sugarcane bagasse, and soybean dregs) gave best results showing a BE of 80.4% for 500 g kg(–1) (dry wt) when sawdust was mixed with sugarcane bagasse. When *H. erinaceus* was grown on other substrata (e.g., fine and crude sawdust of ash and beech, mixed with wheat bran), the BE was best on fine sawdust from beech with wheat bran (73.6%) which is equivalent to an average result of 232.4 g/kg (Ehlers and Schnitzler, 1998). Regarding Bunashimenji (*Hypsizygus marmoreus*) cultivation, Akavia et al. (2009) reported a BE of 85.6% suggesting that the combination of corn cob, bran, and olive press cake is the most efficient substrate for the cultivation of this mushroom. Leifa et al. (2001) evaluated the feasibility of using coffee husk and spent-ground as substrates for the production of edible mushroom *Flammulina velutipes* under different conditions of moisture and spawn rate. It was found that even without supplementation, the BE reached 78% on spent-ground at a spawn rate of 25% and an optimal moisture of 50–60%. Coffee husk gave somewhat lower BE (56%) under the similar moisture conditions and spawn rate. Commercially, *F. velutipes* is usually cultivated in plastic jars containing 80–90% hardwood sawdust supplemented with 10–20% rice bran. Average yields are 160–220 g/800 ml bottle, whereas maximum yield is about 600 g/800 mm bottle (Stamets and Chilton, 1983). Supplementation of rice straw substrate with 30% maize powder and wheat bran was effective on the growth of the white milky mushroom (*Calocybe indica* Purkay. & A. Chandra) showing the maximum biological (459.30 g) and economic yield (457 g), highest BE (91.9%), and maximum pileus diameter (Alam et al., 2010).

Zervakis et al. (2001) investigated the influence of environmental parameters on mycelia linear growth of several wood-decaying taxa: *Pleurotus ostreatus* and *P. pulmonarius* (30°C/86 F), *P. eryngii* (25°C/77 F), *Agrocybe aegerita* (25–30°C/77–86 F), *Lentinus edodes* (20–30°C/68–86 F), and *Auricularia auricula-judae* (20–25°C/68–77 F) on seven substrata (wheat straw, cotton gin-trash, peanut shells, poplar sawdust, oak sawdust, corn cobs and olive press-cake). They observed highest mycelium extention on cotton gin-trash, peanut shells, and poplar sawdust for *Pleurotus* spp. and *A. aegerita* and wheat straw was found to be the most suitable substrate for *L. edodes* and *A. auricula-judae*.

The Use of Wood-inhabiting Fungi for Structural Purposes and Artwork

Several species of wood-inhabiting fungi present excellent binding abilities of adjacent substrate granules that are currently exploited in developing new products such as construction material or packaging. The new packaging type build out of fungus mycelium proved to be superior to the non-biodegradable synthetic foams and other products currently available on the market. With eco-friendly properties such materials are resistant to fire and present firmer texture. Companies such as Ecovative Design LLC (USA) did perform experiments on several species with binding abilities such as: *Ganoderma resinaceum*, *G. applanatum*, *Pleurotus ostreatus* or *Fomes fomentarius* grown on various substrates (rice hulls, cottonseed hulls, etc.) and evaluated their growth rate, thermal characteristics, aerobic and anaerobic characteristics, structural strength, and fungal resistance. Several structural insulation panels have been grown by colonizing agricultural wastes with such wood-inhabiting fungi and are currently used in some residential and commercial applications. Ecovative's biocomposites can replace polypropylene foams used in cars or synthetic foams found in bumpers, doors, roofs, engine bays, trunk liners, dashboards and seats because the material used have the same or better ability to absorb impact, insulate, dampen sound, and provide lightweight structure within an automobile (Zeller and Zocher, 2012). Another company, Mycoworks LLC (USA) is currently working with wood-inhabiting fungi mycelium in order to fabric high performance core materials for cabinetry, interiors, and custom design that are built to customer specifications. Several wood inhabiting fungi are currently launched in experimental work for binding or strength abilities. An important sign to start with when choosing fungi for structural work is the addressed mycelial mats on solid agar media or the fungus fruitbody tissues. Species belonging to genera such as *Ganoderma, Trametes, Phellinus*, or *Schizophyllum* proved to be some of the best candidates for structural work. Fungi species such as *Phellinus linteus, Ph. robustus, Bjerkandera fumosa, Cerrena unicolor*, or *Pycnoporus cinnabarinus* could be examined not only for technical characteristics but also for color.

Wood fungus mycelium is gaining popularity in artwork. Several artist names working with mycelium are currently popular in the US: Phil Ross, Ethan Levesque, Liora Yuklea, etc. Phill Rosses' mycotecture is one of the major attractions that inspire several worldwide artists. The artist launched a limited edition set of furniture from fungal tissue colonizing local agricultural waste (see Fig. 5).

Figure 5. Furniture build up of fungus mycelium colonized agricultural waste.

Artist Ethan Levesques experiments with wood fungus mycelium colonizing paper mache while Liora Yuklea launched 'A fine line' dining set highlighting the perfect and bland artificial food instead of natural, rich in look and taste.

As we may observe wood-inhabiting fungi follow a new path combining mycology with current society needs and even art.

Concluding Remarks

The progress achieved during the past decades in research on wood-inhabiting fungi highlights both their importance in nature and their high potential use in biotechnology. In nature, wood-inhabiting fungi are key agents involved in the decomposition of wood, soil humus formation and nutrient recycling. In terms of evolution, various wood-inhabiting fungi species belong to different groupings. Some are restricted to colonizing one type of substrate and follow the distribution range of the substrate that they prefer [e.g., *Inonotus tamaricis* (Pat.) Maire], while other fungi evolved differently. For example, species able to colonize wood at a faster rate occur on a wide range of substrata including both living and dead hardwoods and conifers and are able to spread over boreal, temperate as well as most tropical regions of the world and are represented by fungi well adapted to environmental fluctuations. This group of fungi have evolved by developing various survival strategies and are able to cause serious damages to forest and urban ecosystems (e.g., some of the most feared wood destroying fungi are able to kill living trees and decompose their wood structure and remain in soil on root fragments for several decades until new seedlings are planted: *Armillaria mellea, A. tabescens, Heterobasidion annosum, Ganoderma* spp., etc.). Some other fungi spread over wood surfaces poor in moisture content [dry-rot fungi, e.g., *Serpula lacrymans, S. himantioides* (Fr.) P. Karst., *Coniophora puteana, Fibroporia vaillantii,* etc.] causing serious damage to material used in construction or wood-made historical artifacts. In terms of disease control the most reliable strategy is the prevention by carefully selecting the type of plant when planting new tree species and adopting 'healthy' pruning habits. For infected plants correct fungus identification by observing both mycelia and fruitbody characters, disease symptoms are necessary before implementing any control strategies. Making people aware of the economical impact of some serious destructive fungi will also ensure a higher degree of prevention. Tree disease control and focus on adopting stronger regulations to check the dispersal of alien aggressive pathogenic species throughout borders would also be useful. Directing research towards finding practical solutions focused on eco-friendly biological control seems to be a much better option compared to using pesticides in order to control tree infections with fungi. *Cinnamomum osmophloeum* is one of the hardwood species known to show significant antifungal activity and therefore the use of such wood types would be a better option for construction purposes.

Several biotechnological methods have been adopted to exploit wood-inhabiting fungi. They are reservoir of therapeutically valuable by-products and enzymes that are widely required for diverse industrial applications and are in demand all over the world. Some of these products are commercially available while others are valuable in biotechnological applications. However, there is a huge gap between the use and knowledge about such fungi. Many people across the world are completely unaware of the health benefits of medicinal mushrooms. This aspect of medicinal mushrooms needs to be popularized for public benefits (Wasser, 2010).

We also need to understand that fungi are unique not only at a species level but also at strain level and this is reflected in both nature, applied biotechnology or cultivation. Some strains of a particular species are more vigurous than others. Therefore the concept according to which "everything is strain related" is strongly connected not only to quality and quantity of enzyme production but this concept is also followed by farmers growing edible and medicinal fungi. However, fungi cultivation obtaining high biological efficiencies is not only the result of using vigurous fungal strains in the cultivation process but taking into account a complex of other factors such as substrate type and size, nutrient additives, cultivation method, climate conditions, grow-room hygiene, etc.

We briefly underline here some of the current issues in the cultivation process of wood-inhabiting fungi:

1. Bioconversion of agricultural wastes would highly benefit especially poor and developing countries. Increasing popularity on how to grow mushrooms in such areas might have a positive economic impact in local communities.
2. More attention must be paid to developing methods and encouraging fungi cultivation as an easy procedure without misleading the general public that this could be done only in highly equipped facilities. Finding novel cultivation methods that offer high biological efficiencies without the use of sophisticated equipments would be of interest to people leaving in poor areas.
3. The public is still unaware of the heavy metals accumulation potential and other undesirable pollutants in some fungi fruitbodies while some mushroom farmers are still using various possible carcinogenic substances in the cultivation process. For example, in order to avoid contamination some books published 20–30 years ago give instructions on how to grow mushrooms by using formaldehyde (currently known as a carcinogenic substance). In addition, some professional growers kept their habits and still encourage the use of chemicals in the mushroom growing process.
4. Little attention is paid on finding potential candidates of wood-inhabiting fungi for cultivation purposes.

The use of wood-inhabiting fungi mycelium for structural purposes and artwork is a new path in applied mycology. Experiments should be employed in order to offer useful advice in this direction.

References

Acevedo, F., Pizzul, L., Castillo, M.P., Cuevas, R. and Diez, M.C. 2011. Degradation of polycyclic aromatic hydrocarbons by the Chilean white-rot fungus *Anthracophyllum discolor*. J. Hazard. Mater., 185: 212–219.

Adair, S., Kim, S.H. and Breuil, C. 2002. A molecular approach for early monitoring of decay basidiomycetes in wood chips. FEMS Microbiol. Lett., 211: 117–122.

Akavia, E., Beharav, A., Wasser, S.P. and Nevo, E. 2009. Disposal of agro-industrial by products by organic cultivation of the culinary and medicinal mushroom *Hypsizygus marmoreus*. Waste Manag., 29: 1622–1627.

Akhtar, M., Scott, G.M., Swaney, R.E. and Shipley, D.F. 2000. Biomechanical pulping a mill-scale evaluation. Resour. Conserv. Recycl., 28: 241–252.

Alam, N., Amin, R., Khair, A. and Lee, T.S. 2010. Influence of different supplements on the commercial cultivation of milky white mushroom. Microbiol., 38: 184–188.

Ali, N.A., Pilgrim, H., Lüdke, J. and Lindequist, U. 1996. Inhibition of chemiluminescence response of human mononuclear cells and suppression of mitogen-induced proliferation of spleen lymphocytes of mice by hispolon and hispidin. Pharmazie., 51: 667–670.

Ali, N.A., Mothana, R.A., Lesnau, A., Pilgrim, H. and Lindequist, U. 2003. Antiviral activity of *Inonotus hispidus*. Fitoterapia., 74: 483–512.

Anastasi, A., Spina, F., Prigione, V., Tigini, V. and Varese, C.G. 2010. Scale-up of a bioprocess for textile wastewater treatment using *Bjerkandera adusta*. Bioresource Technol., 101: 3067–75.

Annesi, T., D'amico, L., Bressanin, D., Motta, E. and Mazza, G. 2010. Characterization of Italian isolates of *Inonotus rickii*. Phytopatol. Mediterr., 49: 301–308.

Arun, A. and Eyini, M. 2011. Comparative studies on lignin and polycyclic aromatic hydrocarbons degradation by basidiomycetes fungi Bioresource. Technol., 102: 8063–8070.

Baba, Y., Tanabe, T., Shirai, N., Watanabe, T., Honda, Y. and Watanabe, T. 2011. Pretreatment of Japanese cedar wood by white-rot fungi and ethanolysis for bioethanol production. Biomass and Bioenergy, 35: 320–324.

Balandaykin, M.E. and Zmitrovich, I.V. 2015. Review on Chaga medicinal mushroom, *Inonotus obliquus* (higher basidiomycetes): realm of medicinal applications and approaches on estimating its resource potential. International Journal of Medicinal Mushrooms, 17(2): 95–104.

Baldrian, P. 2002. Interactions of heavy metals with white-rot fungi. Enz. and Microb. Technol., 32: 78–91.

Baldrian, P. 2006. Fungal laccases—occurrence and properties. FEMS Microb. Rev., 30(2): 215–242.

Baumgartner, K., Coetzee, M.P.A. and Hoffmeister, D. 2011. Secrets of the subterranean pathosystem of *Armillaria*. Molec. Pl. Pathol., 12: 515–534.

Behrendt, C.J. and Blanchette, R.A. 1997. Biological processing of pine logs for pulp and paper production with *Phlebiopsis gigantea*. Appl. Environ. Microbiol., 63: 1995–2000.

Bendel, M., Kienast, F., Baumgartner, K. and Rigling, D. 2006. Incidence and distribution of *Heterobasidion* and *Armillaria* and their influence on canopy gap formation in unmanaged mountain pine forests in the Swiss Alps. European J. Pl. Pathol., 116(2): 85–93.

Bender, S., Dumitrache, C.N., Backhaus, J., Christie, G., Cross, R.F. and Lonergan, G.T. 2003. A case for caution in assessing the antibiotic activity of extracts of culinary-medicinal Shiitake mushroom [*Lentinus edodes* (Berk.) Singer] (Agaricomycetidae). Int. J. Med. Mushrooms, 5: 31–35.

Bendz-Hellgren, M., Lipponen, K., Solheim, H. and Thomsen, I.M. 1998. Impact, control and management of *Heterobasidion annosum* root and butt rot in Europe and North America. The Nordic Countries. pp. 333–345. *In*: Woodward, S., Stenlid, J., Karjalainen, R. and Hüttermann,

A. (eds.). *Heterobasidion annosum*: Biology, Ecology, Impact and Control. CAB International, Wallingford, UK.

Bennett, J., Wunch, K. and Faison, B. 2002. Use of fungi biodegradation. pp. 960–971. *In*: Hurst, Ch. (ed.). Environmental Microbiology. ASM Press Washington, D.C.

Bergdahl, D.R. and French, D.W. 1985. Association of wood decay fungi with decline and mortality of apple trees in Minnesota. Plant Disease, 69: 887–890.

Boeva-Spiranova, R., Petkova, E., Georgieva, N., Yotova, L. and Spiridonov, I. 2007. Utilization of Chemical-mechanical pulp with improved properties from poplar wood in the composition of packing papers. BioResources, 2(1): 34–40.

Breen, A. and Singleton, F.L. 1999. Fungi in lignocellulose breakdown and biopulping. Current Opinion in Biotechnology, 10(3): 252–258.

Bruhn, J.N. and Mihail, J.D. 2009. Forest farming of shiitake mushrooms: Aspects of forced fruiting. Bioresource Technology Bioresource Technol., 100: 5973–5978.

Bruhn, J.N., Mihail, J.D. and Pickens, J.B. 2009. Forest farming of shiitake mushrooms: An integrated evaluation of management practices. Bioresource Technol., 100: 6472–6480.

Cajthaml, T., Erbanová, P., Šašek, V. and Moeder, M. 2006. Breakdown products on metabolic pathway of degradation of benz[a]anthracene by a ligninolytic fungus. Chemosphere, 64: 560–564.

Chang, S.T. and Miles, P.G. 1989. Edible Mushrooms and Their Cultivation, CRC Press, Boca Raton, USA.

Chang, S.T. 1996. Mushroom research and development - equality and mutual benefit. Mushroom Biology Mushroom Products, 2: 1–10.

Chang, S.T. 1999. Global impact of edible and medicinal mushrooms on human welfare in the 21st century: non-green revolution. Int. J. Med. Mushrooms, 1: 1–7.

Chang, S.T. and Buswell, J.A. 1999. *Ganoderma lucidum* (Curt.: Fr.) P. Karst. (Aphyllophoromycetideae)—a mushrooming medicinal mushroom. Int. J. Med. Mushrooms, 1(2): 139–146.

Chang, S.T. and Miles, P.G. 2004. Mushrooms: Cultivation, Nutritional Value, Medicinal Effect, and Environmental Impact, second ed. CRC Press, New York.

Chang, S.T. and Wasser, S.P. 2012. The role of culinary-medicinal mushrooms on human welfare with a pyramid model for human health. Int. J. Med. Mushrooms, 14(2): 95–134.

Chen, A.W., Arrold, N. and Stamets, P. 2000. Shiitake cultivation systems. pp. 771–778. *In*: Griensven, V. (ed.). Science and Cultivation of Edible Fungi. Balkema, Rotterdam.

Cheng, K.F. and Leung, P.C. 2008. General review of polysaccharopeptides (PSP) from *C. versicolor*: pharmacological and clinical studies. Cancer Therapy, 6: 117–130.

Cheng, S.S., Liu, J.Y., Hsui, Y.R. and Chang, S.T. 2006. Chemical polymorphism and antifungal activity of essential oils from leaves of different provenances of indigenous cinnamon (*Cinnamomum osmophloeum*). Bioresour Technol., 97(2): 306–12.

Chihara, G., Hamuro, J., Meada, Y., Arai, Y. and Fukuoka, F. 1970. Fractionation and purification of polysaccharides with marked antitumour activity, especially Lentinan from *Lentinus edodes* (Bark) Sing, an edible mushroom. Cancer Res., 30: 2776–2781.

Chu, K.K., Ho, S.S. and Chow, A.H. 2002. *Coriolus versicolor*: a medicinal mushroom with promising immunotherapeutic values. J. Clin. Pharmacol., 42: 976–984.

Clausen, C.A. 1997. Immunological detection of wood decay fungi—an overview of techniques development from 1986 to the present. Int. Biodet. and Biodeg., 39: 133–143.

Clausen, C.A. and Yang, V. 2007. Protecting wood from mould, decay, and termites with multicomponent biocide systems. Int. Biodet. and Biodeg., 59(1): 20–24.

Coetzee, M.P.A., Wingfield, B.D., Harrington, T.C., Steimel, J., Coutinho, T.A. and Wingfield, M.J. 2001. The root-rot fungus *Armillaria mellea* introduced into South Africa by early Dutch settlers. Mol. Ecol., 10: 387–396.

Coggins, C.R. 1991. Growth characteristics in a building. pp. 81–93. *In*: Jennings, D.H. and Bravery, A.F. (eds.). *Serpula lacrymans*. Wiley, Chichester, UK.

Cohen, R., Persky, L. and Hadar, Y. 2002. Biotechnological applications and potential of wood-degrading mushrooms of the genus *Pleurotus*. Appl. Microbiol. Biotechnol., 58: 582–594.

Cooper, P.A. 2003. A review of issues and technical options for managing spent CCA treated wood Proceedings of the American Wood Preservation Association 1999, Granbury, TX, pp. 1–23.

Couto, S.R. 2008. Dye removal by immobilised fungi. Biotechnol. Advances, 27(3): 227–235.

Covey, R.P., Larsen, H.J., Fitzgerald, T.J. and Dilley, M.A. 1981. *Coriolus versicolor* infection of young apple trees in Washington State. Plant Disease, 65: 280.

Cowling, E.G. 1961. Comparative biochemistry of the decay of sweetgum sapwood by white-rot and brown-rot fungi. US Dept. Agric. Tech. Bull., 258: 1–75.

Cui, J. and Chisti, Y. 2003. Polysaccharopeptides of *Coriolus versicolor*: physiological activity, uses, and production. Biotechnol. Advances, 21(2): 109–122.

Cullen, D. 2002. Molecular genetics of lignin-degrading fungi and their applications in organopollutant degradation. pp. 231. *In*: Esser, K. and Kempken, F. (eds.). The Mycota. Vol. 11. Agricultural Applications.

Čurdová, E., Vavrušková, L., Suchánek, M., Baldrian, P. and Gabriel, J. 2004. ICP-MS determination of heavy metals in submerged cultures of wood-rotting fungi. Talanta, 62(3): 483–487.

Daba, A.S. and Ezeronye, O.U. 2003. Anti-cancer effect of polysaccharides isolated from higher basidiomycetes mushrooms. African J. Biotech., 2(12): 672–678.

Dai, Y.C. and Xu, M.Q. 1998. Studies on the medicinal polypore *Phellinus baumii* and its kin, *Ph. linteus*. Mycotaxon, 67: 191–200.

Dejong, E., Field, J.A., Spinnler, H.E., Wijnberg, J. and Debont, J. 1994. Significant biogenesis of chlorinated aromatics by fungi in natural environments applied and environmental microbiology. Appl. Environ. Microbiol., 60(1): 264–270.

Easton, M.D.L., Luszniak, D. and Geest, E.V. 2002. Preliminary examination of contaminant loadings in farmed salmon, wild salmon and commercial salmon feed. Chemosphere, 46: 1053–1074.

Ehlers, S. and Schnitzler, W.H. 1998. Cultivation of the Basidiomycete *Hericium erinaceus* (Bull ex Fr) Pers. J. Appl. Bot.-Angewandte-Botanik, 72(1-2): 43–47.

Eriksson, K.E.L., Blanchette, R.A. and Ander, P. 1990. Microbial and enzymatic degradation of wood and wood components. Berlin: Springer-Verlag, 802 pp.

Esser, K. and Bennett, J.W. 2002. The Mycota. Industrial applications, Vol. 10. Springer Verlag, Berlin, 414 pp.

Fanadzo, M., Zireva, D.T., Dube, E. and Mashingaidze, A.B. 2010. Evaluation of various substrates and supplements for biological efficiency of *Pleurotus sajor-caju* and *Pleurotus ostreatus*. African J. Biotechnol., 9(19): 2756–2761.

FAOSTAT. 2008. Statistics. Food and Agriculture Organization of the United Nations. Rome, Italy [http://faostat.fao.org/site/567/DesktopDefault.aspx?PageID=567#ancor].

Fisher, M. and Yang, L.X. 2002. Anticancer effects and mechanisms of polysaccharide-K (PSK): implications of cancer immunotherapy. Anticancer Research, 22: 1737–1754.

Gabriel, J., Mokrejs, M., Bily, J. and Rychlovsky, P. 1994. Accumulation of heavy-metals by some wood-rotting fungi. Folia Microbiol., 39(2): 115–118.

Gabriel, J., Baldrian, P., Rychlovský, P. and Krenželok, M. 1997. Heavy metal content in wood-decaying fungi collected in Prague and in the national park Šumavain the Czech Republic. Bull. Environ. Contam. Toxicol., 59: 595–602.

Gao, J.J., Hirakawa, A., Nakamura, N. and Hattori, M. 2006. *In vivo* antitumor effects of bitter principles from the antlered form of fruiting bodies of *Ganoderma lucidum*. J. Nat. Med., 60: 42–48.

Giardina, P., Pezella, C. and Vanhulle, S. 2010. Laccases: a never-ending story. Cell. Mol. Life Sci., 67: 369–385.

Ginns, J.H. and Lefebvre, M.N.L. 1993. Lignicolous corticioid fungi of North America (systematics, distribution, and ecology). The American Phytopath. Society, Minnesota, pp. 1–247.

Glenn, J.K. and Gold, M.H. 1983. Decolorization of several polymeric dyes by the lignin-degrading basidiomycete *Phanerochaete chrysosporium*. Appl. Environ. Microbiol., 45(6): 1741–1747.

Goldfarb, B., Earl, E.N. and Everett, M.H. 1989. *Trichoderma* spp.: Growth rates and Antagonism to *Phellinus weirii in vitro*. Mycologia, 81(3): 375–381.

Gregori, A., Švagelj, M. and Pohleven, J. 2007. Cultivation techniques and medicinal properties of *Pleurotus* spp. Food Technol. Biotechnol., 45(3): 236–247.

Ha, T.N. 2010. Using *Trichoderma* species for biological control of plant pathogens in Viet Nam. J. ISSAAS, 16(1): 17–21.

Hadar, Y., Kerem, Z. and Gorodecki, B. 1993. Biodegradation of lignocellulosic agricultural wastes by *Pleurotus ostreatus*. J. Biotech., 30(1): 133–139.

Han, S.B., Lee, C.W., Jeon, Y.J., Hong, N.D., Yoo, I.D., Yang, K.H. and Kim, H.M. 1999. The inhibitory effect of polysaccharides isolated from *Phellinus linteus* on tumor growth and metastasis. Immunopharmacology, 41: 157–64.

Hansen, E.M. 1979. Survival of *Phellinus weirii* in Douglas-fir stumps after logging. Can. J. For. Res., 9: 484–488.

Harrington, C.A. and Thies, W.G. 2007. Laminated root rot and fumigant injection affect survival and growth of Douglas-Fir. West. J. Appl. For., 22(3): 220–227.

Hastrup, A.C.S., Green, F.I., Clausen, C.C. and Jensen, B. 2005. Tolerance of *Serpula lacrymans* to copper-wood preservatives. Int. Biodet. and Biodeg., 56(3): 173–177.

Hatakka, A., Maijala, P., Hakala, T.K., Hauhio, L. and Ellme´n, J. 2003. Novel white-rot fungus and use thereof in wood pretreatment. International patent application WO 03/080812.

Hayakawa, K., Mitsuhashi, N., Saito, Y., Takahashi, M., Katano, S., Shiojima, K., Furuta, M. and Niibe, H. 1993. Effect of Krestin (PSK) as adjuvant treatment on the prognosis after radiotherapy in patients with non-small cell lung cancer. Anticancer Research, 13: 1815–20.

Hearst, R., Nelson, D., McCollum, G., Millar, B.C., Maeda, Y., Goldsmith, C.E., Rooney, P.J., Lounghrey, A., Rao, J.R. and Moore, J.E. 2009. An examination of antibacterial and antifungal properties of constituents of Shiitake (*Lentinula edodes*) and Oyster (*Pleurotus ostreatus*) mushrooms. Complem. Therap. In Clin. Pract., 15(1): 5–7.

Hernández, D., Sánchezb, J.E. and Yamasakia, K. 2003. A simple procedure for preparing substrate for *Pleurotus ostreatus* cultivation. Bioresource Technol., 90(2): 145–150.

Highley, T.L. and Lutz, J.F. 1970. Progress in understanding how brown-rot fungi degrade cellulose. Biodet. Abstracts, 5: 231–244.

Highley, T.L. 1987. Change in chemical components of hardwood and softwood by brown-rot fungi. Mat u. Org., 22: 39–45.

Hobbs, C.H. 1995. Medicinal Mushrooms: An Exploration of Tradition, Healing, and Culture, 2nd Ed.; Botanica Press, Inc.: Santa Cruz, CA, USA.

Hobbs, C.H. 2000. Medicinal value of *Lentinus edodes* (Berk.) Sing. A literature review. Int. J. Med. Mushrooms, 2: 287–302.

Hood, I.A. 2006. The mycology of the Basidiomycetes. pp. 34–45. *In*: Potter, K., Rimbawanto, A. and Beadle, C. (eds.). Heart Rot and Root Rot in Tropical *Acacia* Plantations. Proceedings of a Workshop Held in Yogyakarta, Indonesia, 7–9 February 2006. ACIAR Proceedings No. 124. Published by Australian Centre for International Agricultural Research, Canberra.

Hu, S.H., Wang, J.C., Wu, C.Y., Hsieh, S.L., Chen, K.S., Chang, S.J. and Liang, Z.C. 2008. Bioconversion of agro wastes for the cultivation of the culinary-medicinal lion's mane mushrooms *Hericium erinaceus* (Bull.: Fr.) Pers. and *H. laciniatum* (Leers) Banker (Aphyllophoromycetideae) in Taiwan. Int. J. Med. Mushrooms, 10(4): 358–398.

Hur, J.M., Yang, C.H., Han, S.H., Lee, S.H., You, Y.O., Park, J.C. and Kim, K.J. 2004. Antibacterial effect of *Phellinus linteus* against methicillin-resistant *Staphylococcus aureus*. Fitoterapia, 75: 603–605.

Hwang, H.J., Kim, S.W., Choi, J.W. and Yun, J.W. 2003. Production and characterization of exopolysaccharides from submerged culture of *Phellinus linteus* KCTC 6190. Enz. Microb. Technol., 33: 309–319.

Hwang, H.J., Kim, S.W., Lim, J.M., Joo, J.H., Kim, H.O., Kim, H.M. and Yun, J.W. 2005. Hypoglycemic effect of crude exopolysaccharides produced by a medicinal mushroom *Phellinus baumii* in streptozotocin-induced diabetic rats. Life Sci., 76: 3069–3080.

Iwamoto, T. and Nasu, M. 2001. Current bioremediation practice and perspective. J. Biosci. Bioeng., 92: 1–8.

Jacques, R.J.S., Santos, E.C., Bento, F.M., Peralba, M.C.R., Selbach, P.A., Sa,´ E.L.S. and Camargo, F.A.O. 2005. Anthracene biodegradation by *Pseudomonas* sp. isolated from a petrochemical sludge land farming. Int. Biodeg. and Biodet., 56: 143–156.

Jasalavich, C.A., Ostrofsky, A. and Jellison, J. 2000. Detection and identification of decay fungi in spruce wood by restriction fragment length polymorphism analysis of amplified genes encoding rRNA. Appl. Env. Microbiol., 66: 4725–4734.

Jiang, J., Slivova, V., Valachovicova, T., Harvey, K. and Sliva, D. 2004. *Ganoderma lucidum* inhibits proliferation and induces apoptosis in human prostate cancer cells PC-3. Int. J. of Oncology, 24: 1093–1099.

Jianzhe, Y., Xiaolan, M., Qiming, M.A., Yichen, Z. and Huaan, W. 1987. Icons of medicinal fungi from China. Beijing (China): Sci. Press, pp. 235.

Jiménez, L., Martinez, C., Pérez, I. and López, F. 1997. Biobleaching procedures for pulp from agricultural residues using *Phanerochaete chrysosporium* and enzymes. Process Biochem., 32: 297–304.

Jiménez-Medina, E., Berruguilla, E., Romero, I., Algarra, I., Collado, A., Garrido, F. and Garcia-Lora, A. 2008. The immunomodulator PSK induces *in vitro* cytotoxic activity in tumour cell lines via arrest of cell cycle and induction of apoptosis. BMC Cancer, 8: 78.

Jong, S.C. and Birmingham, J.M. 1992. Medicinal benefits of the mushroom *Ganoderma*. Adv. Appl. Microbiol., 37: 101–134.

Jong, B.S., Kim, J.C., Bae, J.S., Rhee, M.H., Jang, K.H., Song, J.K., Kwon, O.D. and Park, S.C. 2004. Extracts of *Phellinus gilvus* and *Ph. baumii* inhibit pulmonary inflammation induced by lipopolysaccharides in rats. Biotechnol. Lett., 26: 31–3.

Kahlos, K., Kangas, L. and Hiltunen, R. 1987. Antitumor activity of some compounds and fractions from an n-hexane extract of *Inonotus obliquus*. Acta Pharmaceut. Fennica., 96: 33–40.

Kahlos, K. and Tikka, V.H. 1994. Antifungal activity of cysteine, its effect on C-21 oxygenated lanosterol derivatives and other lipids in *Inonotus obliquus, in vitro*. Appl. Microbiol. Biotechnol., 42: 385–390.

Kamei, T., Hoshino, G., Murata, K. and Toriumi, Y. 2003. Antitumour effect of *Phellinus baumii*: three post-operative cases with cancer. Focus Altern. Complement. Ther., 8: 505.

Kartal, S.N., Munir, E., Kakitani, T. and Imamura, Y. 2004. Bioremediation of CCA-treated wood by brown-rot fungi *Fomitopsis palustris, Coniophora puteana*, and *Laetiporus sulphureus*. J. Wood Sci., 50: 182–188.

Kataria, H.R. and Hoffmann, G.M. 1988. A critical review of plant pathogenic species of *Ceratobasidium* Rogers. Z. Pflanzenkrankh. Pflanzenschutz., 95: 81–107.

Khalil, M.I., Hoque, M.M., Basunia, M.A., Alam, N. and Khan, M.A. 2011. Production of cellulase by *Pleurotus ostreatus* and *Pleurotus sajor-caju* in solid state fermentation of lignocellulosic biomass. Turk J. Agric. For., 35: 333–341.

Kim, D.H., Yang, B.K., Jeong, S.C., Hur, N.J., Das, S., Yun, J.W., Choi, J.W., Lee, Y.S. and Song, C.H. 2001. A preliminary study on the hypoglycemic effect of the exo-polymers produced by five different medicinal mushrooms. J. Microbiol. Biotechnol., 11: 167–171.

Kim, G.Y., Kim, S.H., Hwang, S.Y., Kim, H.J., Park, Y.M., Park, S.K., Lee, M.K., Lee, S.H., Lee, T.H. and Lee, J.D. 2003. Oral administration of proteoglycan isolated from *Phellinus linteus* in the prevention and treatment of collagen-induced arthritis in mice. Biol. Pharm. Bull., 26: 823–31.

Kim, Y.O., Han, S.B., Lee, H.W., Ahn, H.J., Yoon, Y.D., Jung, J.K., Kim, H.M. and Shin, C.S. 2005. Immunostimulating effect of the endopolysaccharide produced by submerged culture of *Inonotus obliquus*. Life Sci., 77: 2438–56.

Kim, Y.O., Park, H.W., Kim, J.H., Lee, J.Y., Moon, S.H. and Shin, C.S. 2006. Anticancer effect and structural characterization of endopolysaccharide from cultivated mycelia of *Inonotus obliquus*. Life Sci., 79: 72–80.

Kim, H.G., Yoon, D.H., Lee, W.H., Han, S.K., Shrestha, B., Kim, C.H., Lim, M.H., Chang, W., Lim, S., Choi, S., Song, W.O., Sung, J.M., Hwang, K.C. and Kim, T.W. 2007. *Phellinus linteus* inhibits inflammatory mediators by suppressing redox-based NF-κB and MAPKs activation in lipopolysaccharide induced RAW 264.7 macrophage. J. Ethnopharmacol., 114: 307–315.

Kim, H.G., Choi, Y.S. and Kim, J.J. 2009. Improving the efficiency of metal removal from CCA-treated wood using brown-rot fungi. Environm. Technol., 30(7): 673–679.

Kirk, T.K. and Farrell, R.L. 1987. Enzymatic combustion: the microbial degradation of lignin. Ann. Rev. Microbiol., 41: 465–505.

Kobayashi, H., Matsunaga, K. and Oguchi, Y. 1995. Antimetastatic effects of PSK (Krestin), a protein-bound polysaccharide obtained from Basidiomycetes: an overview. Cancer Epid. Biomarkers and Prev., 4: 275–281.

Kondo, R., Pengfei, X., Kamei, I. and Mori, T. 2010. Bioconversion of organochlorine pesticides by wood-rotting fungi. J. Biotechnol., 150S: S1–S576.

Krishna, S.H., Prasanthi, K., Chowdary, G.V. and Ayyanna, C. 1998. Simultaneous saccharification fermentation of pretreated sugar cane leaves to ethanol. Process Biochem., 33: 825–830.

Krishna, S.H. and Chowdary, G.V. 2000. Optimization of simultaneous saccharification and fermentation for the production of ethanol from lignocellulosic biomass. J. Agric. Food. Chem., 48: 1971–1976.

Kües, U. and Liu, Y. 2000. Fruiting body production in basidiomycetes. App. Microbiol. and Biotechnol., 54: 141–152.

Kües, U., Nelson, D.R., Liu, C., Yu, G.-J., Zhang, J., Li, J., Wang, X.-C. and Sun, H. 2015. Genome analysis of medicinal *Ganoderma* spp. With plant-pathogenic and saprotrophic life-styles. Phytochemistry [in press].

Kuhn, M. and Winston, D. 2001. Herbal therapy and supplements: a scientific and traditional approach. Philadelphia: JB Lippincott.

Kumar, D. and Gupta, R.K. 2004. Antagonistic *Streptomyces violaceusniger* to control wood-rotting fungi, in Biodiversity and natural products: Chemistry and medical applications, Conference, held on 26–31 Jan, 2004 (IUPAC Int. Conf. New Delhi, India), 161 pp.

Kumar, D. and Gupta, R.K. 2006. Biocontrol of wood rotting fungi. Indian J. Biotechnol., 5: 20–25.

Larsson, K.H. 2007. Re-thinking the classification of corticioid fungi. Mycol. Res., 111: 1040–1063.

Lee, I.K., Kim, Y.S., Jang, Y.W., Jung, J.Y. and Yun, B.S. 2007. New antioxidant polyphenols from the medicinal mushroom *Inonotus obliquus*. Bioorg. Med. Chem. Lett., 17: 6678–6681.

Leifa, F., Pandey, A. and Soccol, C.R. 2001. Production of *Flammulina velutipes* on coffee husk and coffee spent-ground. Brazilian Arch. Biol. and Technol., 44(2): 205–212.

Li, X.Y. 1999. Advances in immunomodulating studies of PSP. In: Advanced Research in PSP. (Yang, Q.Y. ed.). Published by the Hong Kong Association for Health Care Ltd., pp. 39–46.

Liang, Z., Wu, C., Shieh, Z. and Cheng, S. 2009. Utilisation of grass plants for cultivation of *Pleurotus citrinopeleatus*. Int. Biodet. and Biodeg., 63: 509–514.

Lindequist, U., Niedermeyer, T.H.J. and Jülich, W.-D. 2005. The pharmacological potential of mushrooms. Evid. Based Complement. Alternat. Med., 2(3): 285–299.

Madhavi, V. and Lele, S.S. 2009. Laccase properties and applications. Bioresources, 4(4): 1694–1717.

Mahajna, J., Dotan, N., Zaidman, B.Z., Petrova, D.R. and Wasser, S.P. 2008. Pharmacological values of medicinal mushrooms for prostate cancer therapy: the case of *Ganoderma lucidum*. Nutrition and Cancer., 61(1): 16–26.

Mai, C., Kües, U. and Militz, H. 2004. Biotechnology in the wood industry. Appl. Microbiol. Biotechnol., 63: 477–494.

Maijala, P., Kleenb, M., Westinb, C., Poppius-Levlinb, K., Herranenc, K., Lehtoc, J.H., Reponend, P., Mäentaustae, O., Mettäläa, A. and Hatakkaa, A. 2008. Biomechanical pulping of softwood with enzymes and white-rot fungus *Physisporinus rivulosus*. Enz. and Microb. Technol., 43(2): 169–177.

Majeau, J.A., Satinder, K.B. and Rajeshwar, D.T. 2010. Laccases for removal of recalcitrant and emerging pollutants Bioresource Technol., 101(7): 2331–2350.

Mandeel, Q., Al-Laith, A. and Mohamed, S. 2005. Cultivation of oyster mushrooms (*Pleurotus* spp.) on various lignocellulosic wastes. World J. Microbiol. and Biotechnol., 21(4): 601–607.

Marco-Urreaa, E., Arandac, E., Caminalb, G. and Guillénc, F. 2009. Induction of hydroxyl radical production in *Trametes versicolor* to degrade recalcitrant chlorinated hydrocarbons. Bioresource Technol., 100(23): 5757–5762.

Martínez-Carrera, D., Guzmán, G. and Soto, C. 1985. The effect of fermentation of coffee pulp in the cultivation of *Pleurotus ostreatus* in Mexico. Mushroom Newsletter for the Tropics, 6: 21–28.

Martínez-Carrera, D. 1987. Design of a mushroom farm for growing *Pleurotus* on coffee pulp. Mushroom J. of the Tropics, (7): 13–23.

Martínez-Carrera, D. 1998. Oyster mushrooms. pp. 242–245. In: Licker, M.D. (ed.). McGraw-Hill Yearbook of Science and Technology, New York.

Martínez-Carrera, D., Aguilar, A., Martínez, W., Bonilla, M., Morales, P. and Sobal, M. 2000. Commercial production and marketing of edible mushrooms cultivated on coffee pulp in Mexico. Chapter 45. pp. 471–488. In: Sera, T.C., Soccol, A., Pandey, S. and Roussos, S. (eds.). Coffee Biotechnology and Quality. Kluwer Academic Publishers, Dordrecht, The Netherlands.

Mayer, J. and Drews, J. 1980. The effect of protein-bound polysaccharide from *Coriolus versicolor* on immunological parameters and experimental infections in mice. Infection, 8: 13–21.

McKenna, D.J., Jones, K. and Hughes, K. 2002. Reishi Botanical Medicines. The Desk reference for Major Herbal Supplements, 2nd Ed.; The Haworth Herbal Press: New York, London, Oxford, pp. 825–855.

Miller, K.P., Borgeest, C., Greenfield, C., Tomic, D. and Flaws, J.A. 2004. In utero effects of chemicals on reproductive tissues in females. Toxicol. Appl. Pharmacol., 198: 111–131.

Min, B.S., Gao, J.J., Nakamura, N. and Hattori, M. 2000. Triterpenes from the spores of *Ganoderma lucidum* and heir cytotoxicity against meth-A and LLC tumor cells. Chem. Pharm. Bull., 48: 1026–1033.

Mizuno, T. 1996. A development of antitumor polysaccharides from mushroom fungi. Food and Food Ingred. J. (Japan), 167: 69–85.

Mizuno, T., Zhuang, C., Abe, K., Okamot, H., Kiho, T., Ukai, S., Leclerc, S. and Meijer, L. 1999. Antitumor and hypo glycemic activities of polysaccharides from the sclerotia and mycelia of *Inonotus obliquus* (Pers.: Fr.) Pil. (Aphyllophoromycetideae). Int. J. Med. Mushrooms, 1: 301–316.

Montoya, B.S., Varón, L.M. and Levin, L. 2008. Effect of culture parameters on the production of the edible mushroom *Grifola frondosa* (maitake) in tropical weathers. World J. Microbiol. Biotechnol., 24: 1361–1366.

Moonmoon, M., Shelly, N.J., Khan, M.A., Uddin, M.N., Hossain, K., Tania, M. and Ahmed, S. 2011. Effects of different levels of wheat bran, rice bran and maize powder supplementation with saw dust on the production of shiitake mushroom (*Lentinus edodes* (Berk.) Singer). Saudi J. Bio. Sci., 18(4): 323–328.

Mori, T., Kitano, S. and Kondo, R. 2003. Biodegradation of chloronaphthalenes and polycyclic aromatic hydrocarbons by the white-rot fungus *Phlebia lindtneri*. Appl. Microbiol. Biotechnol., 61: 380–383.

Müller, C.I., Kumagai, T., O'Kelly, J., Seeram, N.P., Heber, D. and Koeffler, H.P. 2006. *Ganoderma lucidum* causes apoptosis in leukemia, lymphoma and multiple myeloma cells. Leukemia Res., 30(7): 841–848.

Mukhopadhyay, B., Nag, M., Laskar, S. and Lahiri, S. 2007. Accumulation of radiocesium by *Pleurotus citrinopileatus* species of edible mushroom. J. Radioanalyt. and Nuc. Chem., 273(2): 415–418.

Nakano, H., Namatame, K., Nemoto, H., Motohashi, H., Nishiyama, K. and Kumada, K. 1999. A multi-institutional prospective study of lentinan in advanced gastric cancer patients with unresectable and recurrent diseases: effect on prolongation of survival and improvement of quality of life. Kanagawa Lentinan Research Group. Hepatogastroenterology, 46(28): 2662–8.

Naumann, A., Gonzales, N.M., Peddireddi, S., Kües, U. and Polle, A. 2005. Fourier transform infrared microscopy and imaging: Detection of fungi in wood. Fungal Gen. and Bio., 42(10): 829–835.

Nelson, E.E., Martin, N.E. and Williams, R.E. 1981. Laminated root rot of western conifers. USDA For. Serv. For. Insect and Dis. Leafl., 159 pp.

Ng, T.B. and Chan, W.Y. 1997. Polysaccharopeptide from the mushroom *Coriolus versicolor* possesses analgesic activity but does not produce adverse effects on female reproductive or embryonic development in mice. Gen. Pharmacol., 29: 269–73.

Ng, M.L. and Yap, A.T. 2002. Inhibition of human colon carcinoma development by lentinan from shiitake mushrooms (*Lentinus edodes*). J. Altern. Complement. Med., 8(5): 581–589.

Nigam, J.N. 2001. Ethanol production from wheat straw hemicellulose hydrolysate by *Pichia stipitis*. J. Biotechnol., 87: 17–27.

Nilsson, T. 1973. Microorganisms in Chip Piles. Stockholm Skogshogskolan Inst. Virkeslara Ra, pp. R83.

Oei, P. 2006. Italy: Halfway Holland and China. Mushroom Business, 16: 10–11.

Pietka, J. and Grzywacz, A. 2006. Attempts at active protection of *Inonotus obliquus* by inoculating birches with its mycelium. Acta Mycol., 41: 305–12.

Ping, W., Ke-Zhao, D., Ya-Xian, Z. and Yong, Z. 2008. A novel analytical approach for investigation of anthracene adsorption onto mangrove leaves. Talanta, 76: 1177–1182.

Piri, T. 2003. Silvicultural control of *Heterobasidion* root rot in Norway spruce forests in southern Finland. The Finish For. Res. Inst. Res. Papers 898 pp.

Ramos, A.P., Caetano, M.F. and Melo, I. 2008. *Inonotus rickii* (Pat.) Reid: An important lignicolous Basidiomycete in urban trees. Revista de Ciências Agràrias, 31: 159–167.

Ramsay, J.A., Hao, L., Brown, R.S. and Ramsay, B.A. 2003. Naphthalene and anthracene mineralization linked to oxygen, nitrate, Fe(II) and sulphate reduction in a mixed microbial population. Biodeg., 14: 321–329.

Rauel, K. and Barnoud, F. 1985. Degradation of wood by microorganisms in Biosynthesis and biodegradation of wood components. Academic Press Inc. Florida, 441 pp.

Rayner, A.D.M. and Boddy, L. 1988. Fungal decomposition of wood. John Wiley, Chichester, 587 pp.

Raziq, F. and Fox, R.T.V. 2005. Combination of fungal antagonists for biological control of *Armillaria* root-rot of strawberry plants. Biol. Agric. Hortic., 23: 45–57.

Ridout, B. 2000. Timber decay in buildings: the conservation approach to treatment. Taylor and Francis, 232 pp.

Ritschkoff, A.C. 1996. Decay mechanisms of brown-rot fungi. Technical Res. Center of Finland, VTT Publications, 268: 67–138.

Roberts, P. 1993. Exidiopsis species from Devon, including the new segregate genera *Ceratosebacina*, *Endoperplexa*, *Microsebacina* and *Serendipita*. Mycol. Res., 97: 467–478.

Rodriguez-Estrada, A.E. and Royse, D.J. 2008. *Pleurotus eryngii* and *P. nebrodensis*: from the wild to commercial production. Mush. News, 56(2): 4–11.

Rodriguez-Estrada, A.E., Jimenez-Gasco, M. and Royse, D.J. 2009. Improvement of yield of *Pleurotus eryngii* var. *eryngii* by substrate supplementation and use of a casing overlay. Bioresource Technol., 100(21): 5270–5276.

Rogers, R. 2006. The Fungal Pharmacy: Medicinal Mushrooms of Western Canada. Edmonton, Alberta (Canada): Prairie Deva Press, 234 pp.

Roy, A. and De, A.B. 1996. Polyporaceae of India. R.P. Singh Gahlot, Dahra Dun, pp. 287.

Royse, D.J. 1985. Effect of spawn run time and substrate nutrition on yield and size of the shiitake mushroom. Mycologia, 77: 756–762.

Royse, D.J. 1996. Yield stimulation of shiitake by millet supplementation of wood chip substrate. pp. 277–283. *In*: Royse, D.J. (ed.). Mushroom Biology and Mushroom Products. Penn State University, USA.

Royse, D.J. and Sanchez, J.E. 2007. Ground wheat straw as a substitute for portions of oak wood chips used in shiitake (*Lentinula edodes*) substrate formulae. Bioresource Technol., 98: 2137–2141.

Ryvarden, L. and Gilbertson, R.L. 1993. European polypores. Part 1. *Abortiporus-Lindtneria*. Fungiflora, Oslo, pp. 1–387.

Sadler, M. and Saltmarsh, M. (eds.). 1998. Functional Foods: The Consumer, the Products and the Evidence. Royal Society of Chemistry, Cambridge.

Sakagami, H., Konno, K., Kurakata, Y., Takeda, M., Sato, T., Harada, H., Ohsawa, N., Fujimaki, M. and Komatsu, N. 1990. Effects of pretreatment with PSK, a protein-bound polysaccharide, on *Escherichia coli* infection in mice. Showa Univ. J. Med. Sci., 2: 7–10.

Sakagami, H. and Takeda, M. 1993. Diverse biological activity of PSK (Krestin), a protein-bound polysaccharide from *Coriolus versicolor* (Fr.) Quel (Review). Proceedings of the First International Conference on Mushroom Biology and Mushroom Products, pp. 237–245, Hong Kong.

San Antonio, J.P. 1981. Cultivation of the shiitake mushroom. Hortic. Sci., 16: 151–156.

Sánchez, C. 2009. Lignocellulosic residues: Biodegradation and bioconversion by fungi. Biotechnol. Advances, 27(2): 185–194.

Score, A.J. and Palfreyman, J.W. 1994. Biological control of the dry rot fungus *Serpula lacrymans* by *Trichoderma* species. Int. Biodet. and Biodeg., 33(2): 115–128.

Schmidt, O. 2007. Indoor wood-decay basidiomycetes: damage, causal fungi, physiology, identification and characterization, prevention and control. Mycol. Progress, 6: 261–279.

Schmidt, O. and Huckfeldt, T. 2011. Characteristics and identification of indoor wood-decaying basidiomycetes. pp. 117–180. *In*: Olaf, C.G.A. and Samson, R.A. (eds.). Fundamentals of Mold Growth in Indoor Environments and Strategies for Healthy Living. Part 2.

Schwarze, F.W. and Schubert, M. 2011. *Physisporinus vitreus*: a versatile white-rot fungus for engineering value-added wood products. Appl. Microbiol. Biotechnol., 92: 431–440.

Selosse, M.A., Bauer, R. and Moyersoen, B. 2002. Basal Hymenomycetes belonging to Sebacinaceae are ectomycorrhizal on temperate deciduous trees. New Phytol., 155: 183–195.

Shashirekha, M.N., Rajarathnam, S. and Bano, Z. 2005. Effects of supplementing rice straw growth substrate with cotton seeds on the analytical characteristics of the mushroom, *Pleurotus florida* (Block and Tsao). Food Chem., 92(2): 255–259.

Shen, Q. and Royse, D. 2001. Effects of nutrient supplements on biological efficiency, quality and crop cycle time of maitake (*Grifola frondosa*). Appl. Microbiol. Biotechnol., 57: 74–78.

Shen, Q., Liu, P., Wang, X. and Royse, D.J. 2008. Effects of substrate moisture content, log weight and filter porosity on shiitake (*Lentinula edodes*) yield. Bioresource Technol., 99: 8212–8216.

Shon, M.Y., Kim, T.H. and Sung, N.J. 2003. Antioxidants and free radical scavenging activity of *Phellinus baumii* (*Phellinus* of Hymenochaetaceae) extracts. Food Chem., 82: 593–7.

Sia, G.M. and Candlish, J.K. 1999. Effects of shiitake (*Lentinus edodes*) extract on human neutrophils and the U937 monocytic cell line. Phytother Res., 13(2): 133–137.

Singh, H. 2006. Mycoremediation: Fungal Bioremediation. Jonh Wiley and Sons, Inc. Hoboken, New Jersey, 592 pp.

Sivan, A. 2011. New perspectives in plastic biodegradation. Curr. Opin. Biotechnol., 22(3): 422–426.

Smith, J.E., Anderson, J.G., Senior, E.K. and Aiido, K. 1987. Bioprocessing of lignocellulose. Phil. Trans. R. Soc. Lond., A, 321: 507–21.

Smith, J., Rowan, N. and Sullivan, R. 2002. Medicinal Mushrooms. Their Therapeutic Properties and Current Medical Usage with Special Emphasis on Cancer Treatment; Special Report Commissioned by Cancer Research UK; The Univ. of Strathclyde in Glasgow, 256 pp.

Song, Y.S., Kim, S.H., Sa, J.H., Jin, C., Lim, C.J. and Park, E.H. 2003. Anti-angiogenic, antioxidant and xanthine oxidase inhibition activities of the mushroom *Phellinus linteus*. J. Ethnopharmacol., 88: 113–116.

Sreenath, H.K. and Jeffries, T.W. 2000. Production of ethanol from wood hydrolyzate by yeasts. Bioresource Technol., 72: 253–260.

Stamets, P. and Chilton, J. 1983. The Mushroom Cultivator. Olympia: Agaricon Press.

Stamets, P. 1993. Growing Gourmet and Medicinal Mushrooms. Berkeley, CA: Ten Speed Press.

Stamets, P. 2000. Growing Gourmet and Medicinal Mushrooms, 3rd edn. Berkeley, CA: Ten Speed Press.

Stamets, P. 2005. Mycelium Running: How Mushrooms can Help save the World. Berkeley, CA: Ten Speed Press.

Steffen, K.T., Hatakka, A. and Hofrichter, M. 2002. Removal and mineralization of polycyclic aromatic hydrocarbons by litter-decomposing basidiomycetous fungi. Appl. Microbiol. Biotechnol., 60: 212–217.

Stoller, B.B. 1962. Some practical aspects of making mushroom spawn. Mushroom Sci., 5: 170–184.

Sun, Y. and Cheng, J.J. 2005. Dilute acid pretreatment of rye straw and Bermuda grass for ethanol production. Bioresource Technol., 96: 1599–606.

Tabata, T., Tomioka, K., Iwasaka, Y., Shinohara, H. and Ogura, T. 2006. Comparison of chemical compositions of shiitake (*Lentinus edodes* (Berk.) Sing.) cultivated on logs and sawdust substrate. Food Sci. and Technol. Res., 12(4): 252–255.

Tan, Q., Wang, Z., Cheng, J., Guo, Q. and Guo, L. 2005. Cultivation of *Pleurotus* spp. in China. pp. 338–342. *In*: Proceedings of the Fifth International Conference on Mushroom Biology and Mushroom Products, Shanghai, China. Acta Edulis Fungi (Suppl.), April 8–12.

Tang, W., Liu, J.W., Zhao, W.M., Wei, D.Z. and Zhong, J.J. 2006. Ganoderic acid T from *Ganoderma lucidum* mycelia induces mitochondria mediated apoptosis in lung cancer cells. Life Sci., 80(3): 205–211.

Tekere, M., Read, J.S. and Mattiasson, B. 2005. Polycyclic aromatic hydrocarbon biodegradation in extracellular fluids and static batch cultures and selected sub-tropical white-rot fungi. J. Biotechnol., 115: 367–377.

Thies, W.G. and Nelson, E.E. 1982. Control of *Phellinus weirii* in Douglas-fir stumps by the fumigants chloropicrin, allyl alcohol, Vapam, or Vorlex. Can. J. For. Res., 12(3): 528–532.

Thies, W.G. and Nelson, E.E. 1987. Reduction of *Phellinus weirii* inoculum in Douglas-fir stumps by the fumigants chloropicrin, Vorlex or methylisothiocyanate. For. Sci., 33(2): 316–329.

Thies, W.G. and Nelson, E.E. 1996. Reducing *Phellinus weirii* inoculums by applying fumigants to living Douglas-fir. Can. J. For. Res., 26: 1158–1165.

Tochikura, T.S., Nakashima, H., Ohashi, Y. and Yamamoto, N. 1988. Inhibition (*in vitro*) of replication and of the cytopathic effect of human immunodeficiency virus by an extract of the culture medium of *Lentinus edodes* mycelia. Med. Microbiol. Immunol., 177: 235–244.

Tortella, G.R., Diez, M.C. and Duran, N. 2005. Fungal diversity and use in decomposition of environmental pollutants. Critical Rev. in Microbiol., 3: 197–212.

Trejo-Estrada, S.R., Sepulveda, I.R. and Crawford, D.L. 1998a. *In vitro* and *in vivo* antagonism of *Streptomyces violaceusniger* YCED9 against fungal pathogens of turf grass. World J. Microbiol. Biotechnol., 14: 865–872.

Trejo-Estrada, S.R., Paszczynski, A. and Crawford, D.L. 1998b. Antibiotics and enzymes produced by the biocontrol agent *Streptomyces violaceusniger* YCED 9. J. Ind. Microbiol. Biotechnol., 21: 81–90.

Tsang, K.W., Lam, C.L., Yan, C., Mak, J.C., Ooi, G.C., Ho, J.C., Lam, B., Man, R., Sham, J.S. and Lam, W.K. 2003. *Coriolus versicolor* polysaccharide peptide slows progression of advanced non-small cell lung cancer. Respir. Med., 97: 618–24.

Ţura, D. and Nevo, E. 2007. Medicinal Aphyllophorales mushrooms of Israeli mycobiota. Int. J. Med. Mushrooms, 9: 360–361.

Ţura, D., Zmitrovich, I.V., Wasser, S.P. and Nevo, E. 2009. Medicinal species from genera *Inonotus* and *Phellinus* (Aphyllophoromycetideae): Cultural-morphological peculiarities, growth characteristics, and qualitative enzymatic activity tests. Int. J. Med. Mushrooms, 11: 309–328.

Ţura, D., Zmitrovich, I.V., Wasser, S.P., Spirin, W.A. and Nevo, E. 2011. Biodiversity of Heterobasidiomycetes and non-gilled Hymenomycetes (former Aphyllophorales) of Israel. A.R.A. Gantner Verlag K.-G., Ruggell, 566 pp.

Upton, R., Petrone, C., Graff, A., Swisher, D., McGuffin, M. and Pizzorno, J. 2000. Reishi mushroom—*Ganoderma lucidum*: Standards of analysis, quality control, and therapeutics. American Herbal Pharmacopoeia, Santa Cruz, CA.

USDA (United States Department of Agriculture). 2008. Mushroom Production. Keystone Ag Digest, United States Department of Agriculture, National Agricultural Statistics Service, Early September, Vol. 7, No. 17 (see "Pennsylvania Publications" at www.nass.usda.gov/pa).

Valentin, L., Feijoo, G., Moreira, M.T. and Lema, J.M. 2006. Biodegradation of polycyclic aromatic hydrocarbons in forest and salt marsh soils by white-rot fungi. Int. Biodet. and Biodeg., 58(1): 15–21.

Viitanen, H. and Paajanen, L. 1988. The critical moisture and temperature conditions for the growth of some mould fungi and the brown-rot fungus *Coniophora puteana* on wood. Stockholm. Internat. Res. Group on Wood Pres. Doc. N:o IRG/WP 1369.

Viitanen, H. 1994. Factors affecting the development of biodeterioration in wooden constructions. Mater. Struct., 27: 483–493.

Vinogradov, E. and Wasser, S.P. 2005. The structure of polysaccharide isolated from *Inonotus levis* P. Karst. mushroom (Heterobasidiomycetes). Carbohydr. Res., 340: 2821–5.

Wang, H. and Ng, T.B. 2006. Ganodermin, an antifungal protein from fruiting bodies of the medicinal mushroom *Ganoderma lucidum*. Peptides, 27(1): 27–30.

Wang, S.Y., Hsu, M.L., Hsu, H.C., Lee, S.S., Shiao, M.S. and Ho, C.K. 1997. The anti-tumor effect of *Ganoderma lucidum* is mediated by cytokines released from activated macrophages and T lymphocytes. Int. J. Cancer, 70: 699–705.

Wang, S.Y., Chen, P.F. and Chang, S.T. 2005. Antifungal activities of essential oils and their constituents from indigenous cinnamon (*Cinnamomum osmophloeum*) leaves against wood decay fungi. Biores. Technol., 96(7): 813–818.

Wasser, S.P. and Weis, A.L. 1997. Medicinal Mushrooms. *Ganoderma lucidum* (Curtis: Fr.) P. Karst; Nevo, E. Eds.; Peledfus Publ. House: Haifa, Israel.

Wasser, S.P. and Weis, A.L. 1999a. Medicinal properties of substances occurring in higher Basidiomycetes mushrooms: current perspectives (review). Int. J. Med. Mushrooms, 1(1): 31–62.

Wasser, S.P. and Weis, A.L. 1999b. Therapeutic effects of substances occurring in higher Basidiomycetes mushrooms: a modern perspective. Crit. Rev. Immunol., 19: 65–96.

Wasser, S.P., Nevo, E., Sokolov, D., Reshetnikov, S. and Timot-Tismenetsky, M. 2000. Dietary supplements from medicinal mushrooms: diversity of types and variety of regulations. Int. J. Med. Mushrooms, 2: 1–19.

Wasser, S.P. 2002. Medicinal mushrooms as a source of antitumor and immunomodulating polysaccharides. Appl. Microbiol. Biotechnol., 60(3): 258–274.

Wasser, S.P. 2005a. Shiitake (*Lentinus edodes*). pp. 653–664. *In*: Encyclopedia of Dietary Supplements. DOI: 10.1081/E-EDS-120024880. Marcel Dekker: New York.

Wasser, S.P. 2005b. 'Reishi or Ling Zhi (*Ganoderma lucidum*)', Encyclopedia of Dietary Supplements, pp. 603–622.

Wasser, S.P. 2010. Medicinal mushroom science: history, current status, future trends, and unsolved problems. Int. J. of Med. Mush., 12(1): 1–16.

Wasser, S.P. 2011. Current findings, future trends, and unsolved problems in studies of medicinal mushrooms. Appl. Microbiol. Biotechnol., 89: 1323–1332.

Watkinson, S. and Tlalka, M. 2008. Limiting fungal spread in buildings without the use of toxic biocides. Cost Action E37 Final Conf. in Bordeaux 2008, pp. 41–46.

Ważny, H. and Czajnik, M. 1963. On the occurrence of indoor wood-decay fungi in Poland. Fol. For. Polon., 5: 5–17 (in Polish).

Weiss, M., Selosse, M.A., Rexer, K.H., Urban, A. and Oberwinkler, F. 2004. Sebacinales: a hitherto overlooked cosm of Heterobasidiomycetes with a broad mycorrhizal potential. Mycol. Res., 108: 1003–1010.

Wells, K. 1994. Jelly fungi, then and now! Mycologia, 86: 18–48.

Wells, K. and Bandoni, R.J. 2001. Heterobasidiomycetes. pp. 85–120. *In*: McLaughlin, D.J., Mclaughlin, E.G. and Lemke, P.A. (eds.). The Mycota VII. Systematics and Evolution. Part B. Springer Verlag, Berlin.

Wilcox, W.W. 1978. Review of literature on the effect of early stages of decay on wood strength. Wood and Fiber, 9: 252–257.

Williams, R.E., Shaw, C.G., Wargo, P.M. and Sites, W.H. 1986. *Armillaria* root disease. Forest Insect and Disease Leaflet 78, USDA Forest Service.

Woodward, S., Stenlid, J., Karjalainen, R. and Hüttermann, A. (eds.). 1998. *Hetembasidion annosum*: Biology, Ecology, Impact and Control. CAB International, Wallingford, UK.

Worrall, J.J., Anagost, S.E. and Zabel, R.A. 1997. Comparison of wood decay among diverse lignicolous fungi. Mycologia, 89: 199–219.

Yang, B.K., Kim, D.H., Jeong, S.C., Das, S., Choi, Y.S., Shik, J.S., Lee, S.C. and Song, C.H. 2002. Hypoglycemic effect of a *Lentinus edodes* exo-polymer produced from a submerged mycelial culture. Biosci. Biotechnol. Biochem., 66: 937–942.

Zebulum, H.O., Isikhuemhen, O.S. and Inyang, H. 2011. Decontamination of anthracene-polluted soil through white-rot fungus-induced biodegradation. Environmentalist, 31: 11–19.

Zeller, P. and Zocher, D. 2012. Ecovative's breackthrough biomaterials. Fungi, 5(1): 51–56.

Zervakis, G., Philippoussis, A., Ioannidou, S. and Diamantopoulou, P. 2001. Mycelium growth kinetics and optimal temperature conditions for the cultivation of edible mushroom species on lignocellulosic substrates. Folia Microbiol., 46(3): 231–234.

Zhang, P. and Cheung, P.C. 2002. Evaluation of sulfated *Lentinus edodes* α-(1→3)-D-Glucan as a potential antitumor agent. Biosci. Biotech. and Biochem., 66(5): 1052–1056.

Zheng, W., Liu, T., Xiao, X. and Qi, G. 2007. Sterol composition in field-grown and cultured mycelia of *Inonotus obliquus*. Acta Pharmaceut. Sinica., 42: 750–6.

Zmitrovich, I.V., Wasser, S.P. and Ţura, D. 2014. Wood-inhabiting fungi. pp. 17–74. *In*: Misra, J.K., Tewari, J.P., Deshmukh, S.K. and Vágvölgyi, C. (eds.). Fungi From Different Substrates. CRC Press.

CHAPTER 13

Fusarium Toxins in Cereals in Northern Europe and Asia

T. Yli-Mattila[1],* and T. Yu. Gagkaeva[2]

ABSTRACT

Several toxigenic *Fusarium* species are involved in *Fusarium* head blight which reduces both, crop yield and quality of cereals. *Fusarium* species have also caused several outbreaks in cereal-producing areas. Traditionally, *Fusarium* isolates have been grouped to species and sections according to their morphological and cultural characteristics. But now researchers have started to use molecular data and phylogenetic analyses for classifying the Fusaria, because often it is not possible to recognize new phylogenetic species by morphological characters. Therefore, the best way to identify and classify the isolates of Fusaria is now the polyphasic approach by using all available characters. Trichothecene-producing *Fusarium* species form a well-supported clade which is closely related to the *F. avenaceum*/*F. arthrosporioides*/*F. tricinctum*/*F. acuminatum*/ *F. torulosum* species complex that does not produce trichothecenes. Within the trichothecene-producing clade, the most important species in northern Europe and Asia belong to the deoxynivalenol (DON)— producing *F. graminearum* species complex and to the T-2/HT-2-toxin producing *F. sporotrichioides-F. langsethiae* species complex. New DON— (*F. ussurianum* and *F. vorosii*) and T-2/HT-2 toxin-producing (*F. langsethiae* and *F. sibiricum*) *Fusarium* species have been found recently in northern Europe and Asia. *F. langsethiae* is mainly distributed in Europe, while

---

[1] Molecular Plant Biology, Department of Biochemistry, University of Turku, FI-20014 Turku, Finland.

[2] Laboratory of Mycology and Phytopathology, All-Russian Institute of Plant Protection (VIZR), Podbelskogo 3, St. Petersburg-Pushkin, 196608, Russia.
   Email: t.gagkaeva@yahoo.com

* Corresponding author: tymat@utu.fi

F. *sibiricum* is distributed in Siberia and Russian Far East with two single isolates from Norway and Iran. So, it is probable that the actual distribution of *F. sibiricum* will be much larger than the present known distribution.

Introduction

Fusarium species are the most important phytopathogenic and toxigenic fungi in Nordic countries and globally (Parry et al., 1995; O'Donnell, 1996; McMullen et al., 1997; Bottalico and Perrone, 2002). Several *Fusarium* species are involved in *Fusarium* head blight (FHB), which reduces both crop yield and the quality of cereals. The most important mycotoxins produced by them in northern Europe and Asia are trichothecenes, zearalenone (ZON), moniliformin (MON) and enniatins (ENNs), including beauvericin (BEA) (Yli-Mattila, 2010). Morphological description has been the basis for *Fusarium* taxonomy and identification of *Fusarium* species. There have been a lot of problems in species identification and the number of *Fusarium* species has gone up from nine (Snyder and Hansen, 1945) to around one hundred (Gerlach and Nirenberg, 1982; Leslie and Summerell, 2006). Hence, biochemical, biological, molecular and phylogenetic methods have been applied to *Fusarium* to solve these problems. This chapter aims to present a brief review of the researches dealing with new emerging species of Fusaria that produce toxins and outbreaks of FHB in northern Europe and Asia.

Trichothecenes

Trichothecenes are a major group of secondary metabolites produced by several species of *Fusarium*. The historical background of trichothecene problems and discoveries of different trichothecenes have been described by Ueno (1983) and Desjardins (2006). The first trichothecene "trichothecin" was isolated from *Trichothecium roseum* as an antifungal metabolite (Freeman and Morrison, 1949). Trichothecenes are produced by *Myrothecium*, *Trichothecium*, *Cephalosporium*, *Verticimonosporium* and *Stachybotrys*, but *Fusarium* species are the most important producers of trichothecenes in cereals and other food and feed.

Chemically, trichothecenes, which are produced by *Fusarium* species, can be divided into two groups: type A (e.g., T-2 and HT-2 toxins and diacetoxyscirpenol) and B (e.g., deoxynivalenol (DON) and nivalenol (NIV)) trichothecenes and their mono- and di-acetylated derivatives (e.g., 3ADON, 15ADON and fusarenon X, Miller et al., 1991, Fig. 1). Type A trichothecenes are highly toxic; T-2 toxin has been reported to be roughly ten times more toxic in mammals than DON (Ueno, 1983). The proposed biosynthetic pathway for trichothecenes is shown in Fig. 2. *TRI7* and *TRI13* genes are involved in NIV synthesis where as for T-2 toxin synthesis, addition to these two genes, *TRI16* is also needed. DON is produced by *F. graminearum* instead of NIV, if *TRI7*

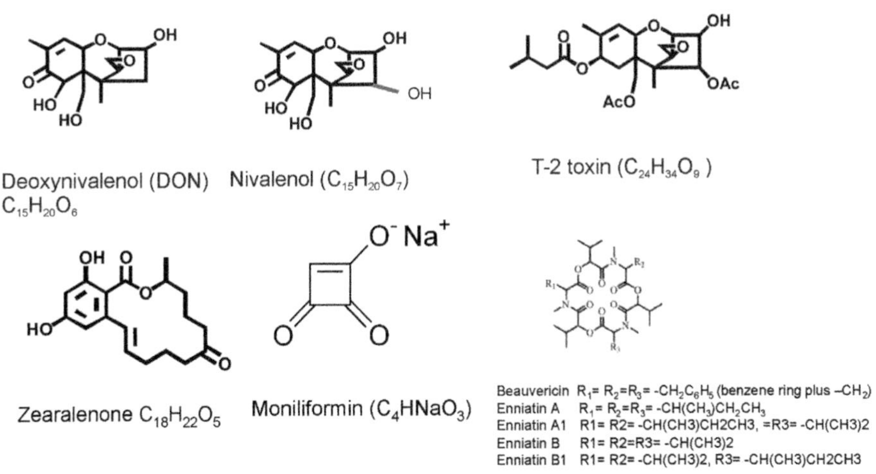

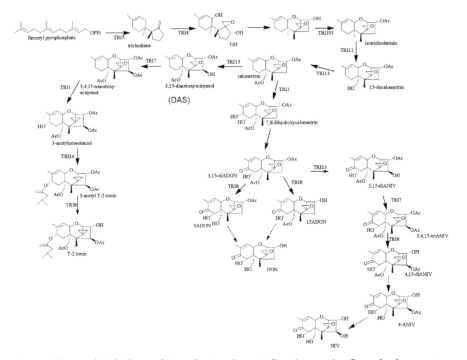

Deoxynivalenol (DON) $C_{15}H_{20}O_6$

Nivalenol ($C_{15}H_{20}O_7$)

T-2 toxin ($C_{24}H_{34}O_9$)

Zearalenone $C_{18}H_{22}O_5$

Moniliformin (C_4HNaO_3)

Beauvericin $R_1= R_2=R_3= -CH_2C_6H_5$ (benzene ring plus $-CH_2$)
Enniatin A $R_1= R_2=R_3= -CH(CH_3)CH_2CH_3$
Enniatin A1   R1= R2= -CH(CH3)CH2CH3, =R3= -CH(CH3)2
Enniatin B   R1= R2=R3= -CH(CH3)2
Enniatin B1   R1= R2= -CH(CH3)2, R3= -CH(CH3)CH2CH3

Figure 1. Chemical structure of deoxynivalenol, nivalenol, T-2 toxin, zearalenone, moniliformin, beauvericin and four enniatins. Nivalenol has one oxygen atom more than DON. The structures of HT-2 ($C_{22}H_{32}O_8$) and T-2 toxins differ only in the functional group at the C-4 position. The basic structure of beauvericin is similar to enniatins.

Figure 2. Proposed trichothecene biosynthetic pathway in *Fusarium* species. Genes for the enzymes are identified for each step, if they are known (modified from McCormick et al., 2011).

and *TRI13* genes are nonfunctional (Alexander et al., 2009; McCormick et al., 2011). Differences in the activity of the esterase encoded by *TRI8*, which are due to differences in the DNA sequences of the *TRI8* gene, determine whether 3ADON or 15ADON is produced by the isolates (Alexander et al., 2011).

T-2 toxin was first isolated in 1968 from *F. sporotrichioides* strain T-2 (originally identified as *F. tricinctum*), which was isolated from maize in France (Bamburg et al., 1968). DON was first isolated in 1973 in Japan and the USA from *F. graminearum* (Ueno, 1983; Desjardins, 2006), while NIV was first found and isolated in 1968 from a Japanese isolate, which was first identified as *F. nivale* (Tatsuno et al., 1968; Ueno, 1977) and reclassified as a new species, *F. kyushuense* (Aoki and O'Donnell, 1998). Later *F. kyushuense* was also found in China (Zhao and Lu, 2007). It is related to *F. poae*, which is the main NIV-producer in northern Europe (Pettersson, 1991; Yli-Mattila et al., 2004a, 2008) and northern Japan (Sugiura et al., 1993). NIV unlike DON occurs more frequently after dry and warm growing seasons (Pettersson et al., 1995). NIV is often detected in Europe and Asia, but incidence and mean level of this mycotoxin are usually lower than for DON even in these regions.

DON-contaminated feed and food can cause vomiting in animals and humans (D'Mello et al., 1999; Bhat et al., 1989). Toxicity of trichothecenes is due to the inhibition of protein synthesis, which is followed by problems in tissues with rapidly growing and dividing cells such as the gastrointestinal tract, skin, lymphoid and erythroid cells. T-2 toxin can also act through the skin (Sarkisov, 1954; Joffe, 1986). The effects of T-2 observed include non-specific symptoms like weight loss, feed refusal, dermatitis, vomiting, diarrhoea, haemorrhages and necrosis of the epithelium of stomach and intestine, bone marrow, spleen, testis and ovary (European Commission, 2001).

For DON, the highest allowed level in European Union (EU) for human food is 1750 ppb in oats, durum wheat and maize and 1250 ppb in most other cereals (EC, 2006a). For baby food the levels are much lower while for animal feed they are higher. For feed, the maximum limit of DON in EU is 900–12000 ppb (EC, 2006b). European regulations for NIV, T-2 and HT-2 toxins are under evaluation. The present levels for the sum of T-2 and HT-2 toxins in unprocessed cereals are 50–200 ppb for wheat, barley and corn and 500–1000 ppb for oats. The maximum limit of DON in Russia and China is 1000 ppb for several cereals used for human food. In Japan the provisional limit for DON in wheat is 1100 ppb. In Russia the maximal limit of T-2 toxin in cereals is 100 ppb. Mycotoxin NIV is not regulated anywhere (based on the information in www.mycotoxins.com).

In EU, the Scientific Committee on Food (SCF) has set a temporary Tolerable Daily Intake (tTDI) limit of 1, 0.7 and 0.06 µg/kg body weight/day for DON, NIV and sum of T2 and HT-2 toxins, respectively (Pronk et al., 2002). The tTDI limit for the sum of T-2 and HT-2 toxins is more than ten times lower as compared to those of DON and NIV. This is because *in vitro* T-2 is rapidly metabolised to HT-2 and the toxic effects of these metabolites cannot be differentiated.

Zearalenone, Enniatins and Moniliformin

ZON, is a mycotoxin, which has been associated with estrogenic syndromes in humans and animals. It was first isolated from corn contaminated by *F. graminearum* (Stob et al., 1962) and structurally characterized and named by Urry et al. (1966, Fig. 1). High ZON levels are usually associated with high DON levels, because ZON is mainly produced by the same species that produces DON (Desjardins, 2006; Fredlund et al., 2008; Jestoi et al., 2008, Table 1). For ZON the highest allowed level in European Union (EU) for human food is 200 ppb in maize and 100 ppb in other cereals (EC, 2006a). For feed, the maximum level of ZON in EU is 100–3000 ppb (EC, 2006b). The tTDI for ZON is 0.2 µg/kg bw/day (Pronk et al., 2002).

ENNs including BEA are cyclic hexapeptides produced by several *Fusarium* species (Table 1, Fig. 2). BEA was first isolated from entomopathogenic fungus (*Beauveria bassiana*) and it is toxic to insects (Desjardins, 2006). According to Logrieco et al. (2002), *F. avenaceum* isolated from rye from Finland (in the original paper the name of the crop was wheat by accident) is effective in

Table 1. Production of mycotoxins (ppm) by Finnish strains of seven *Fusarium* species after four weeks at 25°C with water activity of 0.997 on autoclaved rice (Jestoi et al., 2004, 2008). Ergosterol concentration, which is measuring fungal biomass, was 36–440 ppm in infected samples and < 3 ppm in control samples. n.d. = not detected. n = number of isolates studied. * = trace amounts. FX = fusarenon X. DAS = diacetoxyscirpenol.

Species	Mycotoxins									
	DON	3ADON	NIV	FX	DAS	HT-2/T-2	ZON	BEA	ENNs	MON
F. poae (n = 3)	n.d.	n.d.	11.8 ± 11.6	4.0 ± 3.9	3.7 ± 3.6	n.d.	n.d..	0.97 ± 0.77	220 ± 180	*
F. culmorum (n = 3)	0.60 ± 0.50	4.4 ± 4.2	*	n.d.	n.d.	n.d.	11.5 ± 10.9	n.d.	*	n.d.
F. sporotrichioides (n = 2)	n.d.	n.d.	n.d.	n.d.	7.1 (6.6–7.6)	232 (207–257)	n.d.	0.52 (0.30–0.73)	n.d.	n.d.
F. graminearum (n = 3)	5.9 ± 5.2	19.7 + 18.8	*	n.d.	n.d.	n.d.	334 ± 243	n.d.	*	n.d.
F. langsethiae (n = 1)	n.d.	n.d.	n.d.	n.d.	60.7	60.0	n.d.	n.d.	n.d.	n.d.
F. avenaceum (n = 3)	n.d.	n.d.	n.d.	n.d.	n.d.	n.d.	*	n.d.	105. ± 83	6.4 ± 3.1
F. arthrosporioides (n = 3)	n.d.	n.d.	n.d.	n.d.	n.d.	n.d.	n.d.	n.d.	514 ± 510	25.0 ± 13.8
F. tricinctum (n = 2)	n.d.	n.d.	n.d.	n.d.	n.d.	n.d.	n.d.	n.d.	797 (455–1140)	8.4 (2.1–14.7)
control (n = 4)	n.d.	n.d.	n.d.	n.d.	n.d.	n.d.	n.d.	n.d.	*	n.d.

producing BEA, but according to Jestoi et al. (2004, 2008, Table 1). *F. avenaceum, F. arthrosporioides* and *F. tricinctum* strains do not produce BEA in greater quantity. This report is in conformity with the results obtained by Uhlig (2005) from Norway and by Vogelgsang et al. (2008) from Switzerland. MON is a small cyclic compound, which is also produced by *F. avenaceum, F. arthrosporioides* and *F. tricinctum* strains (Jestoi et al., 2008, Table 1).

Detection Methods

Thin-layer chromatography (TLC), which is simple and cost effective, was one of the first methods used for mycotoxin analyses. However, it is not as precise as gas chromatographic (GC) and high performance liquid chromatographic (HPLC) techniques are, but TLC can be used for screening (Krska et al., 2001; Eskola et al., 2002). HPLC is usually coupled with different types of mass spectrometry (MS) detectors in order to identify and quantify mycotoxins (Malik et al., 2010). Immunochemical methods like enzyme linked immunosorbent assay (ELISA) are also widely used for screening of the mycotoxins because such techniques can rapidly detect them with minimal sample preparation and inexpensive equipment. GC and HPLC methods are expensive and slow, while often there are specificity problems with antibody-based detection methods. Molecular methods, like quantitative PCR (qPCR), with species-, mycotoxin- and chemotype-specific primers and probes can also be used for screening grain samples in order to predict the risk for mycotoxins (Yli-Mattila, 2010).

There is a good correlation between *F. graminearum/F. culmorum* DNA and DON levels in wheat, barley, oats and maize (Waalwijk et al., 2004; Sarlin et al., 2006; Yli-Mattila et al., 2008, 2011b; Fredlund et al., 2008; Nicolaisen et al., 2009; Demeke et al., 2010). A significant correlation has also been found between *F. poae* DNA and NIV levels in barley and oats (Yli-Mattila et al., 2004a, 2008, 2009a), between *F. sporotrichioides/F. langsethiae* DNA and T-2/HT-2 levels in oats and between ENNs/MON and *F. avenaceum* (in some cases also *F. tricinctum*) DNA levels in barley, wheat and oats (Yli-Mattila et al., 2006, 2008, 2009a; Fredlund et al., 2010, 2013; Edwards et al., 2012; Lindblad et al., 2013).

Mycotoxin Production *In Vitro* in Different Culture Conditions

The growth and mycotoxin production of fungi are affected by environmental factors, such as temperature, nutrients, gas composition and water activity and so on (Doohan et al., 2003; Shwab and Keller, 2008). In order to investigate mycotoxin production and the effect of environmental factors, the fungal isolates are grown on different media, usually on autoclaved grains (Jestoi et al., 2004, 2008; Kokkonen et al., 2010; Magan et al., 2011). In these investigations, it has been found that lower water activity or temperature can significantly increase DON and ZON production in *F. graminearum* and *F. culmorum* isolates, while higher temperature increases BEA and ENNs production in those *Fusarium* species, which produce these mycotoxins (Jestoi et al., 2004; Kokkonen et al., 2010). According to Jestoi et al. (2004), the highest T-2/HT-2 toxin production in *F. sporotrichioides* takes place at high temperature and water activity, while according to Kokkonen et al. (2010) the optimal conditions for these toxins in *F. langsethiae* and *F. sporotrichioides* are higher water activity and lower temperature. Medina and Magan (2011) studied T-2/HT-2 production in detail for *F. langsethiae* and found that the optimum water activity and temperature conditions for T-2 and HT-2 production were between 0.98–0.995, and 20–30°C, respectively. Generally, maximum toxin production seems to occur where mycelia growth is either under a slight water activity or temperature stresses (Magan et al., 2002).

Fusarium Head Blight and Mycotoxin Levels in Cereals

Fusarium head blight was first described in England in 1884 according to Goswami and Kistler (2004). The severity of FHBs of wheat, barley and oats and maize ear rot varies from year to year. *F. graminearum* is the most common cause of FHB in wheat, barley and other small grain cereals in most parts of the world, which means that FHB is usually closely connected to DON. In Finland "red mouldy oats" infected by *Fusarium* caused the first epidemic among farm animals in 1930–1931 (Rainio, 1932; Hintikka, 1983).

There have been several surveys of T-2 and HT-2 toxins in Norway, Finland, Denmark, Sweden and Baltic countries during the last 25 years (Hietaniemi and Kumpulainen, 1991; Langseth and Rundberget, 1999; Yli-Mattila et al., 2008, 2009a; Edwards et al., 2009; Nielsen et al., 2011; Pettersson et al., 2011; Suproniene et al., 2011; Cerveg database/Veli Hietaniemi; European Commission, 2007). In these surveys high T-2/HT-2 levels have most often been found in oats. Incidences of these toxins were low during most years, but higher levels were found in Norway in 1996, 1999 and 2000 and in Finland 2003, 2004, 2005, 2011 and 2012 especially in oats. Yli-Mattila et al. (2008, 2009a) have shown that DON and T-2/HT-2 levels were highest in oats as compared to other cereals.

In most surveys, the mean levels for the sum of T2/HT-2 toxins were below 100 ppb, but a few samples with higher toxin levels were found in some years. Incidences of the toxins seem to have increased after the year 2000. Some oats samples from UK were also found to contain very high T-2 and HT-2 toxin levels (6260–9990 ppb) (Edwards, 2009). T-2 and HT-2 toxin levels in oats were even higher in UK, Norway and Sweden in 2005–2007 than in Finland (Edwards et al., 2009). High levels of T-2 and HT-2 toxins in Norway have been observed in warm and dry summers (Langseth and Rundberget, 1999). In Finland, the exceptionally dry summer after flowering in 2006 resulted high levels of T-2/HT-2-producing *Fusarium* species in Finnish oats and barley, especially in plots with direct drilling (Yli-Mattila et al., 2009a).

Also in European part of Russia the occurrence of T-2/H-2 toxins is on the increase since the last few years (Gagkaeva et al., 2011).

In Denmark, highest DON levels were found in wheat in 1997–2000 and 2004 (Nielsen et al., 2011). In Sweden and Finland, the quality of the oat yield was poor in 2011 and elevated mycotoxin levels were found in oats, when harvesting was delayed due to the rains. The highest DON levels were found in Finland in 2000, 2005, 2007, 2011 and 2012 (Cerveg database/Veli Hietaniemi; Yli-Mattila et al., 2009a). In 2012, the exceptionally high DON levels in oats were due to the exceptionally high precipitation in Finland. In 2000, high NIV levels (200–3700 ppb) were detected in barley in Finland and in north-western Russia (Yli-Mattila et al., 2002). High NIV levels were found in south-western Finland in 2002 due to the natural high infection of *F. poae*. It is also possible to induce high NIV and *F. poae* levels in experimental plots of barley by artificial inoculation of other *Fusarium* species (Yli-Mattila et al., 2008, Table 2).

DON and NIV are the main mycotoxins in Japan but high levels of T-2 toxin have not been found there. In Korea, Taiwan and Vietnam, however, high levels of DON and NIV have been found (Desjardins, 2006). In South Korea FHB epidemics in rice seem to be mainly due to NIV-producing *F. asiaticum* isolates (Lee et al., 2009), while DON is the predominant mycotoxin in maize and NIV in barley (Sohn et al., 1999; Desjardins, 2006).

Table 2. NIV, *F. poae* contamination % and *F. poae* DNA levels in barley grains from Finland and Russia in 2002. Samples 1–4 and 12–13 were from naturally infected fields. Sample 5 was sprayed with water (water control) and samples 6–11 were artificially inoculated by different *Fusarium* species as described by Jestoi et al. (2008) in Marttila. *F. poae* DNA was measured by qPCR as described by Yli-Mattila et al. (2008). - = not analyzed.

Sample no.	Cultivar	Geographical origin and artifical inoculation	NIV (ppb)	*F. poae* (%)	TMpoaef/r/p (ng ng^{-1} total DNA)
1	Saana	Southwestern Finland, Marttila	1260	0	2.5×10^{-3}
2	Extract	Southwestern Finland, Vahto	1390	0.5	-
3	Extract	Southwestern Finland, Paattinen	110	0	-
4	Cellar	Southwestern Finland, Paattinen	490	0	10^{-5}
5	Scarlett	Southwestern Finland water control	2890	3	-
6	Scarlett	Southwestern Finland, *F. arthrosporioides*	4790	1	1.1×10^{-2}
7	Scarlett	*F. avenaceum*	5510	8	1.4×10^{-2}
8	Scarlett	*F. poae*	5460	5	2.9×10^{-3}
9	Scarlett	*F. sporotrichioides*	6200	3	9×10^{-3}
10	Scarlett	*F. culmorum*	1320	2	-
11	Scarlett	*F. graminearum*	4970	1	10^{-2}
12	Suzdalets	Northwestern Russia	140	5	-
13	Crinichniy	Northwestern Russia	740	7	5×10^{-3}

High ZON levels are usually connected to high DON and *F. graminearum* levels in what, barley and oats, while high MON and ENN levels are connected to high *F. avenaceum/F. arthrosporioides/F. tricinctum* levels in the same cereals (Yli-Mattila et al., 2006, 2008, 2009a,b; Jestoi et al., 2008).

High infestation of *F. graminearum* and DON levels in seeds have caused serious loss in germination at many instances, e.g., in Norway (Kimen Seed Laboratory, 2009), which may be connected to climate change (van der Fels-Klerx et al., 2012). This has led the seed industries to import seeds.

Fusarium Outbreaks

In Japan, contaminated wheat, barley and rice caused the red mould disease, akakabi-disease (Ueno et al., 1973; Ueno, 1977). Outbreaks of the red mould disease occurred several times since 1890. The main species of *F. graminearum* species complex in Japan like in Korea and China are *F. graminearum* and *F. asiaticum*, which mainly produce DON. Both, DON and NIV can cause red mould disease symptoms in animals.

In China there have been numerous outbreaks of FHB of wheat and barley and *Fusarium* ear rot of maize, especially after the year 1952, along the Yangtze river from Shanghai to Sichuan province (Yang et al., 2008; Choo, 2009).

In Russian Far East, FHB was first officially reported in 1882, but poisoning of animals and local people due to the diseased "pink grain" occurred long before that year (Naumov, 1916). The weather in the cereal producing region of Russian Far East is typically very damp and warm during the summer due to the Sea of Japan and Pacific Ocean. FHB was known as a problem of cereals in this area during the 19th century (Voronin, 1890; Jaczewski, 1904). From 1882 until about 1914, FHB epidemics occurred almost every year in this region. The use of seeds and straw contaminated with mycotoxins produced by various *Fusarium* species caused numerous cases of food poisoning in humans and animals. Initially, the symptoms after eating resembled alcohol intoxication and they were often referred to as "intoxicating bread" syndrome. The researches of several Russian mycologists revealed that the fungus *Gibberella saubinetii* Sacc. (now *G. zeae*, anamorph *F. graminearum*) was the principal causal organism of FHB (Palchevsky, 1891; Voronin, 1890; Jaczewski, 1904). FHB was a persistent problem in the Russian Far East during the first half of the 20th century (Jaczewski, 1904; Naumov, 1913) and it continues to be so even today (Gagkaeva et al., 2011).

The greatest tragedy due to the *Fusarium* toxins in Europe took place in the former Soviet Union before and during World War II, when harvesting was delayed due to the lack of farm labour and overwintered mouldy grains were consumed (Fig. 3). The alimentary toxic aleukia (ATA) outbreaks in Russia were probably due to T-2 toxin-producing *Fusarium* species and similar symptoms were obtained by treatment of extracts from ATA-associated grain samples or pure cultures of *F. sporotrichioides* (Sarkisov, 1954; Joffe, 1986). In the Orenburg

Figure 3. Women harvesting barley under the snow in East Karelia during World War II in 1942 (Rosen, 1998).

region near Ural, more than 10% of people were reported to be affected by the disease and mortality was high, because cereal products were the basic food. High levels of T-2 and HT-2 toxins have also been found during the last few years in oats in north-western region of Russia, Norway, Sweden, Finland and UK (Langseth and Rundberget, 1999; Yli-Mattila et al., 2008; Edwards et al., 2009; Gavrilova and Gagkaeva, 2010; Fredlund et al., 2010). T-2 and HT-2 toxins were associated in Europe with the presence of *F. sporotrichioides* (Desjardins, 2006) and later also with the presence of *F. langsethiae* (Torp and Langseth, 1999; Torp and Nirenberg, 2004; Yli-Mattila et al., 2004a; Knutsen et al., 2004; Thrane et al., 2004; Wilson et al., 2004).

It was suggested that NIV-producing *F. poae* and *F. cerealis* strains are responsible for the natural contamination with NIV found in the northernmost area of Japan (Ueno, 1983). NIV is also common in Europe due to NIV-producing *F. graminearum* and *F. culmorum* isolates in central Europe (Waalwijk et al., 2003; Jennings et al., 2004) and due to *F. poae* in northern Europe. However, *F. cerealis* is not very common in Europe.

Taxonomy and Identification of Trichothecene-producing *Fusarium* Species

The genus *Fusarium* was established by Link in 1809. The genus is having septate, branched mycelium with micro and macroconidia. Macroconida are fusoid, curved (banana-shaped), hyaline and septate, often gathered in slimy masses (sporodochia) on branched conidiophores (Fig. 4). The taxonomy of *Fusarium* species has been the subject of considerable debate for 200 years. One of first systematic and landmark publication is—*Die Fusarien* by Wollenweber and Reinking (1935). The existing classifications, with widely differing species concepts, are based on morphological and cultural characteristics. The heterogeneous features of *Fusarium* species has resulted in several classification systems. After Wollenweber and Reinking, the most considerable taxonomical schemes, based on morphological characters, have been made by Booth (1971), Gerlach and Nirenberg (1982), Nelson et al. (1983) and Marasas et al. (1984). Marasas et al. (1984) and Leslie and Summerell (2006) have also included the toxin production ability in the species description.

The ability to produce macro- and micro-conidia is a very important character in the taxonomy of *Fusarium*. As a rule DON-producing species are able to produce only macroconidia (FGSC, *F. culmorum*), while the T-2/HT-2 toxins producing species form a large number of microconidia (*F. sporotrichioides*, *F. langsethiae* and *F. sibiricum*). The main NIV-producing species (*F. poae*) forms many microconidia and only very rarely a few macroconidia, whereas other NIV-producing species (*F. cerealis* and

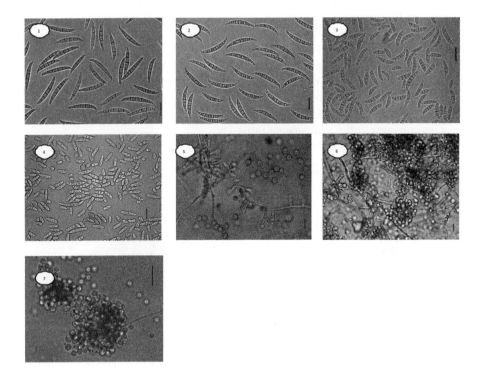

Figure 4. Variously shaped conidia of *Fusarium* spp.: 1. *F. graminearum*, 2. *F. cerealis*, 3. *F. culmorum*, 4. *F. sporotrichioides*, 5. *F. poae*, 6. *F. langsethiae*, 7. *F. sibiricum* (Scale bars 1–3, 6–7 = 20 μm; 4, 5 = 10 μm).

NIV-producing chemotypes of *F. graminearum* and *F. culmorum*) form only macroconidia. Macroconidia-forming species are also more aggressive pathogens than those species, which mainly produce microconidia (Jestoi et al., 2008; Imathiu et al., 2010; Divon et al., 2012).

Currently, researchers have started to use alternative ways for species identification and classification based on molecular data and phylogenetic analyses. Most of the phylogenetic characters are based on DNA sequences, which has dramatically increased the number of characters to be used for the identification and revised the systematics. It has been found that for example, *F. graminearum*, *F. sporotrichioides* and *F. avenaceum* actually represent species complexes based on phylogenetic analyses (O'Donnell et al., 2000, 2004; Yli-Mattila et al., 2002, 2004b, 2006, 2011a).

The use of species-specific primers (Doohan et al., 1998; Waalwijk et al., 2004; Yli-Mattila et al., 2004b; Jurado et al., 2006), microarrays and SNP (single nucleotide polymorphisms) analysis have made it possible to quickly detect, identify and quantify isolates of *Fusarium* species based on molecular data (Kristensen et al., 2006, 2007). A new phylogenetic species of *F. graminearum* species complex, *F. ussurianum*, has recently been found in Russian Far East (Yli-Mattila et al., 2009b). *F. vorosii* was also found in the same area. The main species of *F. graminearum* species complex in China is *F. asiaticum* (Qu et al., 2008; Yang et al., 2008), which together with *F. vorosii* and *F. ussurianum* form an Asian clade.

Each *Fusarium* species has a species-specific mycotoxin profile, which may be practically identical in closely related species (Jestoi et al., 2004, 2008, Table 1). DON, which is the most important trichothecene (Bottalico and Perrone, 2002; Desjardins, 2006; Fig. 5), is produced by species of the *F. graminearum* species complex and *F. culmorum* (Table 1).

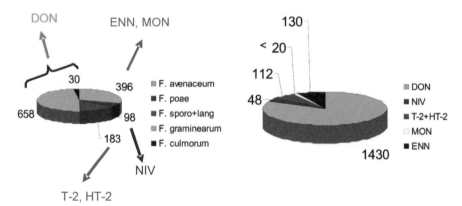

Figure 5. *Fusarium* DNA levels (10^{-6} ng ng^{-1} total DNA, left) as compared to *Fusarium* toxin levels (ppb, right) in Finnish oats samples in 2005 (Yli-Mattila et al., 2009a).

Phylogeny and Chemotypes of Trichothecene-producing *Fusarium* Species

According to the latest publications (Gräfenhan et al., 2011; O'Donnell et al., 2012), trichothecene-producing *Fusarium* species form a well-supported clade which is more closely related to trichothecene-nonproducing *F. avenaceum/F. arthrosporioides/F. tricinctum/F. acuminatum/F. torulosum* species complex (Yli-Mattila et al., 2006) rather than that to fumonisin-producing *Fusarium* species. Within the trichothecene-producing clade the most important species complexes are DON/NIV-producing *F. graminearum* species complex and T-2/HT-2-toxin producing *F. sporotrichioides-F. langsethiae* species complex.

Genealogical Concordance Phylogenetic Species Recognition analysis based on DNA sequences formally divided *F. graminearum* into 15 phylogenetically distinct species (O'Donnell et al., 2000, 2004, 2008; Yli-Mattila et al., 2009b; Sarver et al., 2011). *F. graminearum sensu stricto* is actually one of the phylogenetic species of the species complex of *F. graminearum* (O'Donnell et al., 2000, 2004, 2008). The proposed species names for the other lineages are: *F. austroamericanum* (lineage 1, South America), *F. meridionale* (lineage 2, originally from South America), *F. boothii* (lineage 3, originally from South America), *F. mesoamericanum* (lineage 4, originally from Central America), *F. acaciae-mearnsii* (lineage 5), *F. asiaticum* (lineage 6), *F. cortaderiae*, *F. brasilicum* (O'Donnell et al., 2000, 2004), *F. vorosii* (obtained from Japan, Russian Far East and Hungary), *F. gerlachii* (obtained from the midwestern United States) (Starkey et al., 2007), *F. aethiopicum* (from Ethiopia, O'Donnell et al., 2008), *F. ussurianum* (from Russian Far East and Siberia, Yli-Mattila et al., 2009a,b) *F. nepalense* and *F. louisianense* (Sarver et al., 2011). Some differences in toxin producing ability of these phylogenic species were found, but it is not possible to identify these species by morphological characters.

All *F. graminearum* lineages, as well *F. pseudograminearum*, produce type B trichothecene mycotoxins, such as DON and NIV. DON-producing isolates of *F. graminearum* species complex can be divided into 3ADON and 15ADON chemotypes (Miller et al., 1991). Three chemotypes have been recognized among the isolates: producers of DON and 3ADON (chemotype 3ADON), producers of DON and 15ADON (chemotype 15ADON), and producers of NIV and 4ANIV (chemotype NIV). Many lineages/species contain isolates of all chemotypes, whereas others, like *F. meridionale*, are all NIV chemotype (Ward et al., 2002). Multilocus genotyping and chemotype-specific primers can be used for screening isolates belonging to different chemotypes. *F. graminearum* and *F. culmorum* produce type B trichothecenes, but only two chemotypes are known in this species: NIV and 3ADON (Chandler et al., 2003). *F. cerealis* has only NIV chemotype (Chandler et al., 2003; Stepien et al., 2008; Yli-Mattila and Gagkaeva, 2010).

3ADON chemotype is dominating in most northern areas, such as Finland, Norway, Sweden, north-western Russia, Russian Far East, northern Japan and some parts of Canada (Yli-Mattila, 2010; Yli-Mattila and Gagkaeva, 2010;

Yli-Mattila et al., 2013). 3ADON isolates of *F. graminearum* and *F. culmorum* greatly inhibit the growth of wheat seedlings and cause more necrotic lesions than 15ADON isolates. This is in agreement with the higher DON producing 3ADON isolates (Ward et al., 2002; Yli-Mattila and Gagkaeva, 2010). 3ADON chemotype also grows quickly and produces more conidia than the 15ADON chemotype, which might explain, why the 3ADON chemotype is spreading so rapidly in Russian Far East and North America.

Recently, *F. graminearum* has been spreading northward in Europe (Waalwijk et al., 2003; Nicholson et al., 2003; Stepien et al., 2008; Yli-Mattila and Gagkaeva, 2010; Nielsen et al., 2011) and replacing the closely related *F. culmorum*. This change may be due to the changed agricultural practices and increased maize cropping. In Finland, *F. graminearum* was reported in cereals in 1932 (Rainio, 1932) and in the 1960s (Uoti and Ylimäki, 1974; Doohan et al., 1998), and it has been present in Finland since then (Ylimäki, 1981; Ylimäki et al., 1979; Rizzo, 1993; Eskola et al., 2001; Yli-Mattila, 2010), while in north-western Russia *F. graminearum* was found only a few years ago (Yli-Mattila et al., 2009b; Yli-Mattila and Gagkaeva, 2010). *F. graminearum* is also present in Norway, Sweden and Estonia, but not yet in Lithuania (Suproniene et al., 2010; Yli-Mattila et al., 2011b).

Type A trichothecene-producing species can be divided into two main groups, which are visible in phylogenetic trees (Proctor et al., 2009; Yli-Mattila et al., 2011a). The first group consists of species (*F. sporotrichioides, F. sibiricum, F. langsethiae* and *F. armeniacum*), which are able to produce large amounts of T-2 and HT-2 toxins. The rest of the type A trichothecene-producing species (*F. poae, F. kyushuense, F. venenatum* and *F. sambucinum*) are not able to produce large amounts of T-2 and HT-2 toxins.

Torp and Langseth (1999) found in Europe, T-2-producing *F. poae* isolates, which had less aerial hyphae than typical *F. poae* isolates. In 1998 COST action 835 started to investigate the international collection of *F. sporotrichioides, F. kyushuense, F. poae* and "powdery poae" isolates in order to characterize the new species morphologically, phylogenetically and based on metabolites. In these investigations it was found that the European T-2-producing "powdery poae" isolates belonged to a new species (*F. langsethiae*, Torp et al., 2000; Torp and Nirenberg, 2004; Yli-Mattila et al., 2004b). Although two subgroups of *F. langsethiae* could be separated by molecular analyses, based on IGS and elongation factor 1 α sequences, they could not be separated morphologically or geographically (Yli-Mattila et al., 2004b, 2011a; Konstantinova and Yli-Mattila, 2004; Figs. 6, 7).

Among the isolates originating from Siberia and morphologically resembling *F. poae* isolates producing high amounts of T-2 toxin were identified (Burkin et al., 2008). After detailed investigations these isolates were described as a new species, *F. sibiricum* (Yli-Mattila et al., 2011a; Fig. 8), which produces high levels of T-2 and HT-2 toxins. The identification is based on molecular, morphological and metabolite characters, but it is difficult to identify these three species based on only morphological characters.

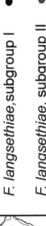

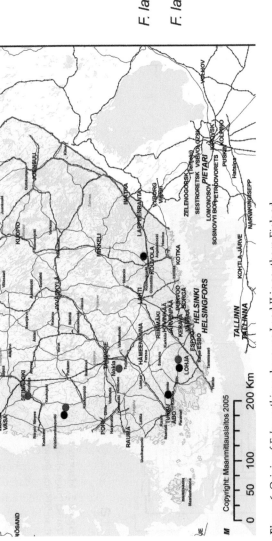

Figure 6. Origin of *F. langsethiae* subgroups I and II in southern Finland.

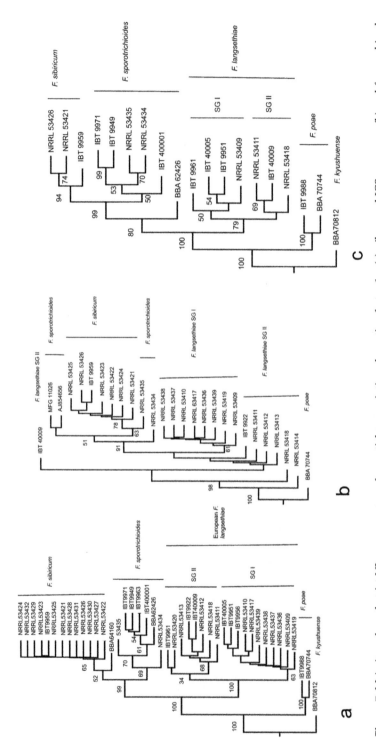

Figure 7. Majority rule and strict NJ consensus trees for partial transcription elongation factor 1 α (a), ribosomal IGS sequences (b) and for combined transcription elongation factor 1 α, IGS and beta-tubulin sequences (c, Yli-Mattila et al., 2004a,b,c, 2011a). Only branches with a bootstrap value higher than 50% are shown.

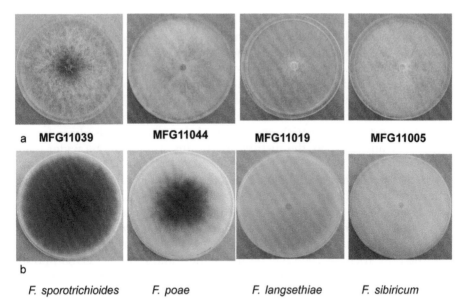

Figure 8. (a) *Fusarium* cultures on PSA after 7 days at 24°C in darkness. (b) Reverse of the same cultures on PSA after 7 days at 24°C in darkness.

Moreover, the Norwegian isolate IBT 9959 which is intermediate between *F. sporotrichioides* and *F. langsethiae* (Torp and Nirenberg, 2004; Yli-Mattila et al., 2004a; Knutsen et al., 2004) and another isolate from Iran (Kachuei et al., 2009; Yli-Mattila et al., 2015) were also identified as *F. sibiricum*. So, it is probable that the actual distribution of *F. sibiricum* is much wider than the presently known. Previously, Joffe (1986) and some others suggested that *F. poae* is an important T-2 producer, but these conclusions are probably due to unreliable morphological identification of the isolates, since *F. poae* isolates lack the functional gene *TRI16*, which is necessary for T-2 production (Proctor et al., 2009).

Future Prospects

The permissible limit of mycotoxins in food and feed should be uniform for the whole Europe, better if this is for the whole world. This will help international food and feed trade and protect public and animal health. Thus, further research dealing with mycotoxin chemistry, synthesis, production and detection is necessary.

Conclusions

Incidences of the *Fusarium* toxins seem to have increased after the year 2000. There are three main types of mycotoxin-producing *Fusarium* species in northern Europe and Asia. Type A (T-2 and HT-2 toxins) trichothecene-producers, type B (DON and NIV) trichothecene-producers and trichothecene-nonproducers. In Northern Europe, the main type A trichothecene-producers are *F. langsethiae* and *F. sporotrichioides*, the main DON-producers are *F. graminearum* and *F. culmorum*, while the main NIV-producer is *F. poae*. In most parts of northern Europe no NIV chemotype of *F. graminearum* and *F. culmorum* has been found. In Russian Far East and Siberia, *F. sibiricum* is also an important type A trichothecene producer, but it has also been found in Norway and Iran. So, the distribution of this species can be wider than suggested earlier.

In northern Europe the 3ADON chemotype of *F. graminearum* is dominating, while in the rest of Europe the 15ADON chemotype is the dominant one. In Russian Far East both 3ADON and 15ADON chemotypes of *F. graminearum* are common together with 3ADON-producing *F. ussurianum* and 15ADON-producing *F. vorosii*.

The only trichothecene having the highest allowed level (1750 ppb in oats, durum wheat and maize and 1250 ppb in most other cereals) in EU for human food is DON. Other trichothecenes, MON, ENNs and BEA are not yet regulated in EU.

Acknowledgements

The visits of Dr. T. Yli-Mattila to the All-Russian Institute of Plant Protection and the visits of Dr. T. Gagkaeva to the University of Turku were supported financially by the Academy of Finland (no. 126917 and 131957) and the Nordic network project New Emerging Mycotoxins and Secondary Metabolites in Toxigenic Fungi of Northern Europe (project 090014), which was funded by the Nordic Research Board. The investigation was supported by the grant No. 14-26-00067 of the Russian Science Foundation (RSF). We are grateful for the comments of Robert Proctor, Susan McCormick and Taha Abo-Dalam on the manuscript.

References

Alexander, N.J., Proctor, R.H. and McCormick, S.P. 2009. Genes, gene clusters, and biosynthesis of trichothecenes and fumonisins in *Fusarium*. Toxin Rev., 28: 198–215.

Alexander, N.J., McCormick, S.P., Waalwijk, C., van der Lee, T. and Proctor, R.H. 2011. The genetic basis for 3-ADON and 15-ADON trichothecene chemotypes in *Fusarium*. Fungal Genet. Biol., 38: 485–495.

Aoki, T. and O'Donnell, K. 1998. *Fusarium kyushuense* sp. nov. from Japan. Mycosci., 39: 1–6.

Bamburg, J.R., Riggs, N.V. and Strong, F.M. 1968. The structure of toxins from two strains of *Fusarium tricinctum*. Tetrahedron, 24: 3329–3336.

Bhat, R.V., Beedu, S.R., Ramakrishna, Y. and Munshi, K.L. 1989. Outbreak of trichothecene mycotoxicosis associated with consumption of mould damaged wheat products in Kashmir valley, India. Lancet, 1: 35–37.

Booth, C. 1971. The genus *Fusarium*. Commonw. Mycol. Inst., England.

Bottalico, A. and Perrone, G. 2002. Toxigenic *Fusarium* species and mycotoxins associated with head blight in small-cereals in Europe. Eur. J. Plant Pathol., 108: 611–624.

Burkin, A.A., Soboleva, N.A. and Kononenko, G.P. 2008. T-2 toxin production by *Fusarium poae* from cereal grain in Siberia and Far East regions. Mycol. Phytopathol., 42: 354–358 (in Russian).

Cerveg database/Veli Hietaniemi. https://portal.mtt.fi/portal/page/portal/kasper/pelto/peltopalvelut/cerveg (in Finnish).

Chandler, E.A., Simpson, D.R., Sthonsett, M.A. and Nicholson, P. 2003. Development of PCR assays to *Tri7* and *Tri13* trichothecene biosynthetic genes, and characterisation of chemotypes of *Fusarium graminearum* and *Fusarium cerealis*. Phys. Mol. Plant Pathol., 62: 355–367.

Choo, T.M. 2009. *Fusarium* head blight on barley in China. Can J. Plant Pathol., 31: 3–15.

Demeke, T., Gräfenhan, T., Clear, R.M., Phan, A., Ratnayaka, I., Chapados, J., Patrick, S.K., Gaba, D., Levasque, A. and Seifert, K.A. 2010. Development of a specific TaqMan® real-time PCR assay for quantification of *Fusarium graminearum* clade 7 and comparison of fungal biomass determined by PCR with deoxynivalenol content in wheat and barley. Intl. J. Food Microbiol., 141: 45–50.

Desjardins, A.E. 2006. *Fusarium* Mycotoxins: Chemistry, Genetics and Biology. The American Phytopathological Society, St. Paul, Minnesota, USA.

Divon, H.H., Razzaghian, J., Udnes-Aamot, H. and Klemsdal, S.S. 2012. *Fusarium langsethiae* (Torp and Nirenberg), investigation of alternative infection routes in oats. Eur. J. Plant Pathol., 132: 147–161.

D'Mello, J.P.F., Placinta, C.M. and Macdonald, A.M.C. 1999. *Fusarium* mycotoxins: a review of global implications for animal health, welfare and productivity. Anim. Feed Sci. Technol., 80: 183–205.

Doohan, F.M., Brennan, J. and Cooke, B.M. 2003. Influence of climatic factors on *Fusarium* species pathogenic to cereals. Eur. J. Plant Pathol., 109: 755–768.

Doohan, F.M., Parry, D.W., Jenkinson, P. and Nicholson, P. 1998. The use of species-specific PCR-based assays to analyse *Fusarium* ear blight of wheat. Plant Pathol., 47: 197–205.

Edwards, S.G. 2009. *Fusarium* mycotoxin content of UK organic and conventional oats. Food Add. Cont., 26: 1063–1069.

Edwards, S.G., Barrter-Guillot, B., Clasen, P.E., Hietaniemi, V. and Pettersson, H. 2009. Emerging issues of HT-2 and T-2 toxins in European cereal production. World Mycotox., J., 2: 173–179.

Edwards, S.G., Imathiu, S.M., Ray, R.V., Back, M. and Hare, M.C. 2012. Molecular studies to identify the Fusarium species responsible for HT-2 and T-2 mycotoxins in UK oats. Int. J. Food Microbiol., 156: 168–175.

Eskola, M. 2002. Study on Trichothecenes, Zearalenone and Ochratoxin A in Finnish cereals: Occurrence and Analytical Techniques. Ph.D. thesis, University of Helsinki.

Eskola, M., Parikka, P. and Rizzo, A. 2001. Trichothecenes, ochratoxin A and zearalenone contamination and *Fusarium* infection in Finnish cereal samples in 1998. Food Add. Cont., 18: 707–718.

European Commission (EC). 2001. Opinion on *Fusarium* Toxins—Part 5: T-2 toxin and HT-2 toxin (adopted by the SCF on 30 May 2001). http://ec.europa.eu/food/fs/sc/scf/out88_en.pdf.

European Commission (EC). 2006a. Commission Regulation (EC) No. 856/2005 amending Regulation (EC) No. 466/2001 as regards *Fusarium* toxins. Official Journal of the European Union, L 143: 3–8.

European Commission (EC). 2006b. Commission Recommendation of 17 August 2006 on the presence of deoxynivalenol, zearalenone, ochratoxin A, T-2 and HT-2 and fumonisins in products intended for animal feeding, L 229: 7–8.

European Commission (EC). 2007. Scoop Task 3.2.10. Collection of occurrence data of *Fusarium* toxins in food and assessment of dietary intake by the population of EU member states. http://ec.europa.eu/food/fs/scoop/task3210.pdf.

Fredlund, E., Gidlund, A., Olsen, M., Börjesson, T., Spliid, N.H.H. and Simonsson, M. 2008. Method evaluation of Fusarium DNA extraction from mycelia and wheat for down-stream real-time PCR quantification and correlation to mycotoxin levels. J. Microbiol. Meth., 73: 33–40.

Fredlund, F., Gidlund, A., Pettersson, H., Olsen, M. and Björnesson, T. 2010. Real time PCR detection of *Fusarium* species in Swedish oats and correlation to T-2 and HT-2 toxin content. World Mycotox. J., 3: 77–88.

Fredlund, E., Gidlund, A., Sulyok, M., Börjesson, T., Krska, R., Olsen, M. and Lindblad, M. 2013. Deoxynivalenol and other selected *Fusarium* toxins in Swedish oats—Occurrence and correlation to specific *Fusarium* species. Int. J. Food Microbiol., 167: 276–283.

Freeman, G.G. and Morrison, R.I. 1949. The isolation and chemical properties of trichothecin, an antifungal substance from *Trichothecium roseum*. Biochem. J., 44: 1–5.

Gagkaeva, T.Yu., Gavrilova, O.P., Levitin, M.M. and Novozhilov, K.V. 2011. *Fusarium* head blight. Plant Prot. Quarant., 5: 69–120 (in Russian).

Gavrilova, O. and Gagkaeva, T. 2010. *Fusarium* on all small-grain crops grown in North non-chernozem and Kaliningrad area in 2007–2008. Plant Prot. Quar., 2: 23–25 (in Russian).

Gerlach, W. and Nirenberg, H. 1982. The genus *Fusarium*—a pictorial atlas. P. Parey, Berlin, Germany.

Goswami, R.S. and Kistler, C. 2004. Heading for disaster: *Fusarium graminearum* on cereal crops. Mol. Plant Pathol., 5: 515–525.

Gräfenhan, T., Schroers, H.-J., Nirenberg, H.I. and Seifert, K.A. 2011. An overview of the taxonomy, phylogeny and typification of nectriaceous fungi in *Cosmospora, Acremonium, Fusarium, Stilbella* and *Volutella*. Stud. Mycol., 68: 79–113.

Hietaniemi, V. and Kumpulainen, J. 1991. Contents of *Fusarium* toxins in Finnish and imported grains and feeds. Food Add. Cont., 8: 171–182.

Hintikka, E.L. 1983. Toxicosis and natural occurrence of trichothecenes in Finland. pp. 221–228. *In*: Ueno, Y. (ed.). Trichothecenes—Chemical, Biological and Toxicological Aspects. Elsevier, Amsterdam, The Netherlands.

Imathiu, S.M., Hare, M.C., Ray, R.V., Back, M. and Edwards, S.G. 2010. Evaluation of pathogenicity and aggressiveness of *F. langsethiae* on oat and wheat seedlings relative to known seedling blight pathogens. Eur. J. Plant Pathol., 126: 203–216.

Jaczewski, A.A. 1904. About tempulent corn. Sheet Inf. Dis., cont., 11: 89–92 (in Russian).

Jennings, P., Coates, M.E., Walsh, K., Turner, J.A. and Nicholson, P. 2004. Determination of deoxynivalenol- and nivalenol-producing chemotypes of *Fusarium graminearum* isolates from wheat crops in England and Wales. Plant Pathol., 53: 643–652.

Jestoi, M., Paavanen-Huhtala, S., Uhlig, S., Rizzo, A. and Yli-Mattila, T. 2004. Mycotoxins and cytotoxicity of Finnish *Fusarium* strains grown on rice cultures. Proc. 2nd International Symposium on *Fusarium* Head Blight; incorporating the 8th European *Fusarium* seminar; 2004, 11–15; Orlando, Fl, USA, S.M. Canty et al. eds., Michigan State University, East Lansing, MI, USA, pp. 405–409.

Jestoi, M., Paavanen-Huhtala, S., Parikka, S. and Yli-Mattila, T. 2008. *In vitro* and *in vivo* mycotoxin production of *Fusarium* species isolated from Finnish grains. Arch. Phytopathol. Plant Prot., 41: 545–558.

Joffe, A.Z. 1986. *Fusarium* Species: Their Biology and Toxicology. J. Wiley & Sons, New York, USA.

Jurado, M., Vazquez, C., Marin, S., Patino, B. and Gonzales-Jaen, M.T. 2006. PCR detection assays for the trichothecene-producing species *Fusarium graminearum, Fusarium culmorum, Fusarium poae, Fusarium equiseti* and *Fusarium sporotrichioides*. Syst. Appl. Microbiol., 28: 562–568.

Kachuei, R., Yadegari, M.H., Rezaie, S., Allameh, A., Safaie, N., Zaini, F. and Yazd, F.K. 2009. Investigation of stored mycoflora, reporting the *Fusarium* cf. *langsethiae* in three provinces of Iran during 2007. Ann. Microbiol., 59: 383–390.

Kimen Seed Laboratory. 2009. http://www.kimen.no, in Norwegian.

Knutsen, A.K., Torp, M. and Holst-Jensen, A. 2004. Phylogenetic analyses of the *Fusarium poae, Fusarium sporotrichioides* and *Fusarium langsethiae* species complex based on partial sequences of the translation elongation factor-1 alpha gene. Int. J. Food Microbiol., 95: 287–295.

Kokkonen, M., Ojala, L., Parikka, P. and Jestoi, M. 2010. Mycotoxin production of selected *Fusarium* species at different culture conditions. Int. J. Food Microbiol., 143: 17–25.

Konstantinova, P. and Yli-Mattila, T. 2004. IGS-RFLP analysis and development of molecular markers for identification of *F. poae, F. langsethiae, F. sporotrichioides* and *F. kyushuense*. Int. J. Food Microbiol., 95: 321–331.

Kristensen, R., Gauthier, G., Berdal, K.G., Hamels, S., Remacle, J. and Holst-Jensen, A. 2006. DNA microarray to detect and identify trichothecene- and moniliformin-producing *Fusarium* species. J. Appl. Microbiol., 102: 1060–1070.

Kristensen, R., Berdal, K.G. and Holst-Jensen, A. 2007. Simulateneous detection and identification of trichothecene- and moniliformin-producing DNA microarray to detect and identify trichothecene- and moniliformin-producing *Fusarium* species based on multiplex SNP analysis. J. Appl. Microbiol., 102: 1071–1081.

Krska, R., Baumgartner, S. and Josephs, R. 2001. The state-of-the-art in the analysis of type-A and –B trichothecene mycotoxins in cereals. Fresenius J. Anal. Chem., 371: 285–299.

Langseth, W. and Rundberget, T. 1999. The occurrence of HT-2 toxin and other trichothecenes in Norwegian cereals. Mycopathol., 147: 157–165.

Lee, J., Chang, I.Y., Kim, H., Yun, S.H., Leslie, J.F. and Lee, Y.W. 2009. Genetic diversity and fitness of *Fusarium graminearum* populations from rice in Korea. Appl. Env. Microbiol., 75: 3289–3295.

Leslie, J.F. and Summerell, B.A. 2006. The Fusarium Laboratory Manual. Blackwell Publishing, USA.

Lindblad, M., Gidlund, A., Sulyok, M., Börjesson, T., Krska, R., Olsen, M. and Fredlund, E. 2013. Deoxynivalenol and other selected *Fusarium* toxins in Swedish wheat—Occurrence and correlation to specific *Fusarium* species. Int. J. Food Microbiol., 167: 284–291.

Logrieco, A., Rizzo, A., Ferracane, R. and Ritieni, A. 2002. Occurrence of beauvericin and enniatins in wheat affected by *Fusarium avenaceum* head blight. Appl. Env. Microbiol., 68: 82–85.

Magan, N., Hope, R., Colleate, A. and Baxter, E.S. 2002. Relationship between growth and mycotoxin production by *Fusarium* species, biocides and environment. Eur. J. Plant Pathol., 108: 685–690.

Magan, N., Medina, A. and Aldred, D. 2011. Possible climate-change effects on mycotoxin contamination of food crops pre- and postharvest. Plant Pathology, 60: 150–163.

Malik, A.K., Blasco, C. and Pico, Y. 2010. Liquid chromatography-mass spectrometry in food safety. J. Chrom., A 1217: 4018–4040.

Marasas, W.F.O., Nelson, P.E. and Toussoun, T.A. 1984. Toxigenic *Fusarium* species. Identity and mycotoxicology. Pennsylvania State University Press, University Park, USA.

McCormick, S., Stanley, A.M., Stover, Ni.A. and Alexander, N.J. 2011. Trichothecenes: From simple to complex mycotoxins. Toxins, 3: 802–814.

McMullen, M., Jones, R. and Gallenberg, D. 1997. Scab of wheat and barley: A re-emerging disease of devastating impact. Plant Disease, 81: 1340–1346.

Medina, A. and Magan, N. 2011. Water availability and temperature affects production of T-2 and HT-2 by *Fusarium langsethiae* strains from north European countries. Food Microbiol., 28: 392–398.

Miller, J.D., Greenhalgh, R., Wang, Y. and Lu, M. 1991. Trichothecene chemotypes of three *Fusarium* species. Mycologia, 83: 121–130.

Naumov, N.A. 1916. Tempulent corn. Observations under some species of genus *Fusarium*, SPb, Russia (in Russian).

Naumov, N.A. 1913. Observations and researches under tempulent corn in 1912. Seaside landlord. Nikoljsk-Ussuriisk, Russia (in Russian).

Nelson, P.E., Tousson, T.A. and Marasas, W.F.O. 1983. *Fusarium* species. An Illustrated Manual for Identification. Pennsylvania State University Press, University Park, USA.

Nicholson, P., Chandler, E., Draeger, R.C., Gosman, N.E., Simpson, D.R., Thomsett, M. and Wilson, A.H. 2003. Molecular tools to study epidemiology and toxicology of *Fusarium* head blight of cereals. Eur. J. Plant Pathol., 109: 691–703.

Nicolaisen, M., Supronienė, S., Nielsen, L.N., Lazzaro, I., Spliid, N.H. and Justesen, A.F. 2009. Real-time PCR for quantification of eleven individual *Fusarium* species in cereals. J. Microbiol. Meth., 76: 234–240.

Nielsen, L.K., Jensen, J.D., Nielsen, G.C., Jensen, J.E., Spliid, N.H., Thomsen, I.K., Justesen, A.F., Collinge, D.B. and Jorgensen, L.N. 2011. Fusarium head blight of cereals in Denmark: Species complex and related mycotoxins. Phytopathol., 101: 960–969.

O'Donnell, K. 1996. Progress towards a phylogenetic classification of *Fusarium*. Sydowia, 48: 57–70.

O'Donnell, K., Kistler, H.C., Tacke, B.K. and Casper, H.H. 2000. Gene genealogies reveal global phylogeographic structure and reproductive isolation among lineages of *Fusarium graminearum*, the fungus causing wheat scab. Proc. Natl. Acad. Sci. USA, 97: 7905–7910.

O'Donnell, K., Ward, T.J., Geiser, D.M., Kistler, H.C. and Aoki, T. 2004. Genealogical concordance between the mating type locus and seven other nuclear genes supports formal recognition of nine phylogenetically distinct species within the *Fusarium graminearum* clade. Fungal. Genet. Biol., 41: 600–623.

O'Donnell, K., Ward, T.J., Aberra, D., Kistler, H.C., Aoki, T., Orwig, N., Kimura, M., Bjørnstad, Å. and Klemsdal, S.S. 2008. Multilocus genotyping and molecular phylogenetics resolve a novel head blight pathogen within the *Fusarium graminearum* species complex from Ethiopia. Fungal. Genet. Biol., 45: 1514–1522.

O'Donnell, K., Humber, R.A., Geiser, D.M., Kang, S., Park, B., Robert, V.A.R.G., Crous, P.W., Johnston, 1 P.R., Aoki, T., Rooney, A. and Rehner, S. 2012. Phylogenetic diversity of insecticolous fusaria inferred from multilocus DNA sequence data and their molecular identification via FUSARIUM-ID and Fusarium MLST. Mycologia, 104: 427–445.

Palchevsky, N.A. 1891. Disease of cereal grain in the South-Ussuriiskiy region. St. Petersburg, Russia (in Russian).

Parry, D.W., Jenkins, P. and McLeod, L. 1995. *Fusarium* ear blight (scab) in small grain cereals—a review. Plant Pathol., 44: 207–238.

Pettersson, H. 1991. Nivalenol production by *Fusarium poae*. Mycotox Res., 7A: 26–30.

Pettersson, H., Hedman, R., Engstrom, B., Elwinger, K. and Fossum, O. 1995. Nivalenol in Swedish cereals—occurrence, production and toxicity towards chickens. Food Add. Cont., 12: 373–376.

Pettersson, H., Brown, C., Hauk, J., Hoth, S., Meyer, J. and Wessels, D. 2011. Survey of T-2 and HT-2 toxins by LC–MS/MS in oats and oat products from European oat mills in 2005–2009. Food Add. Cont. Part B., 4: 110–115.

Proctor, R., McCormick, S., Alexander, N. and Desjardins, A. 2009. Evidence that a secondary metabolic biosynthetic gene cluster has grown by gene relocation during evolution of the filamentous fungus *Fusarium*. Mol. Microbiol., 74: 1128–1142.

Pronk, M.E.J., Schothorst, R.C. and van Egmond, H.P. 2002. RIVM Report 388802024/2002, Toxicology and occurrence of nivalenol, fusarenon X, diacetoxyscirpenol, neosolaniol and 3- and 15-acetyldeoxynivalenol: a review of six trichothecenes. http://rivm.openrepository.com/rivm/bitstream/10029/9184/1/388802024.pdf.

Qu, B., Li, H.P., Zhang, J.B., Xu, Y.B., Huang, T., Wu, A.B., Zhao, C.S., Carter, J., Nicholson, P. and Liao, Y.C. 2008. Geographic distribution and genetic diversity of *Fusarium graminearum* and *F. asiaticum* on wheat spikes throughout China. Plant Pathol., 57: 15–24.

Rainio, A.J. 1932. Punahome *Fusarium* roseum Link. - *Gibberella saubinetii* (Mont.) *Sacc.* ja sen aiheuttamat myrkytykset kaurassa. Valt. Maatal. Koetoim. Julk., 50: 1–45 (in Finnish, abstract in German).

Rizzo, A.F. 1993. Determination of major naturally occurring *Fusarium* toxins in Finnish grains and feeds. The haemolytic activity of DON and T-2 toxin, and the lipid peroxidation induced in experimental animals. Ph.D. thesis, University of Helsinki. ISBN 952-90-4534-4, J-Paino Ky.

Rosén, G. 1998. Suomalaisina Itä-Karjalassa. Suomen Historiallinen Seura, Helsinki (in Finnish).

Sarkisov, A.K. 1954. Mycotoxicoses. Agricultural State Publishing House, Moscow, Russia (in Russian).

Sarlin, T., Yli-Mattila, T., Jestoi, M., Rizzo, A., Paavanen-Huhtala, S. and Haikara, A. 2006. Real-time PCR for quantification of toxigenic *Fusarium* species in barley and malt. Eur. J. Plant Pathol., 114: 371–380.

Sarver, B.A.J., Ward, T.J., Gale, L.R., Broz, K., Kistler, H.C., Aoki, T., Nicholson, P., Carter, J. and O'Donnell, K. 2011. Novel *Fusarium* head blight pathogens from Nepal and Louisiana revealed by multilocus genealogical concordance. Fungal. Genet. Biol., 48: 1096–1107.

Shwab, E.K. and Keller, N.P. 2008. Regulation of secondary metabolite production in filamentous ascomycetes. Mycological Research, 112: 225–230.

Snyder, W.C. and Hansen, H.N. 1945. The species concept in *Fusarium*. Amer. J. Bot., 1940: 64–67.

Sohn, H.B., Seo, J.A. and Lee, Y.W. 1999. Co-occurrence of *Fusarium* mycotoxins in mouldy and healthy corn from Korea. Food Add. Cont., 16: 153–158.

Starkey, D.E., Ward, T.J., Aoki, T., Gale, L.R., Kistler, H.C., Geiser, D.M., Suga, H., Toth, B., Varga, J. and O'Donnell, K. 2007. Global molecular surveillance reveals novel fusarium head blight species and trichothecene toxin diversity. Fungal Genet Biol., 44: 1191–1204.

Stepien, L., Popiel, D., Koczyk, G. and Chelkowsky, J. 2008. Wheat-infecting *Fusarium* species in Poland—their chemotypes and frequencies revealed by PCR assay. J. Appl. Genet., 49: 433–441.

Stob, M., Baldwin, R.S., Tuite, J., Andrews, F.N. and Gillette, K.G. 1962. Isolation of an anabolic, uterotrophic compound from corn infected with *Gibberella zeae*. Nature, 196: 1318.

Suproniene, S., Justesen, A.F., Nicolaisen, M., Mankeviciene, A., Dabkevicius, Z., Semaskiene, R. and Leistrumait, A. 2010. Distribution of trichothecene and zearalenone producing *Fusarium* species in grain of different cereal species and cultivars grown under organic farming conditions in Lithuania Ann. Agric. Environ. Med., 17: 73–80.

Sugiura, Y., Fukasaku, K., Tanaka, T., Matsui, Y. and Ueno, Y. 1993. *Fusarium poae* and *Fusarium crookwellense*, fungi responsible for the natural occurrence of nivalenol in Hokkaido. Appl. Environm. Microb., 59(10): 3334–3338.

Tatsuno, T., Saito, M., Enomoto, M. and Tsunoda, H. 1968. Nivalenol, a toxic principle of *Fusarium nivale*. Chem. Pharm. Bull., 16: 2519–2520.

Thrane, U., Adler, A., Clasen, P., Langseth, W. and Nielsen, K.F. 2004. Chemical diversity of *Fusarium poae, F. sporotrichioides* and *F. langsethiae* sp. Nov. Int. J. Food Microbiol., 95: 257–266.

Torp, M. and Langseth, W. 1999. Production of T-2 toxin by a *Fusarium* resembling *Fusarium poae*. Mycopathol., 147: 89–96.

Torp, M. and Nirenberg, H.I. 2004. Description of *Fusarium langsethiae* sp. nov. Int. J. Food Microbiol., 95: 89–96.

Torp, M., Langseth, W., Yli-Mattila, T., Klemsdal, S.S., Mach, R.I. and Nirenberg, H.I. 2000. Section *Sporotrichiella*—identification of a new *Fuarium* species, or a polyphasic approach to its taxonomy. pp. 19. *In*: Nirenberg, H.I. (ed.). 6th European *Fusarium* seminar and Third COST 835 Workshop of Agriculturally Important Toxigenic Fungi, Book of Abstracts, Berlin Germany.

Ueno, Y. 1977. Mode of action of trichothecenes. Pure Appl. Chem., 49: 1737–1745.

Ueno, Y. 1983. Trichothecenes: Chemical, Biological and Toxicological Aspects. Elsevier, Amsterdam, The Netherlands.

Ueno, Y., Sato, N., Ishi, K., Sakai, K., Tsunoda, H. and Enomono, N. 1973. Biological and chemical detection of trichothecene mycotoxins of *Fusarium* species. Appl. Microbiol., 25: 699–704.

Uhlig, S. 2005. *Fusarium avenaceum* Toxic metabolites, their occurrence and biological effects. Ph.D. thesis, The Norwegian School of Veterinary Science, Oslo.

Uoti, J. and Ylimäki, A. 1974. The occurrence of *Fusarium* species in cereal grain in Finland. Ann. Agric. Fenniae, 13: 5–17.

Urry, W.H., Wehrmeister, H.L., Hodge, E.B. and Hidy, P.H. 1966. The structure of zearalenone. Tetrahedron Lett., 27: 685–693.

Van der Fels-Klerx, H.J., Klemsdal, S., Hietaniemi, V., Lindblad, M., Iannou-Kakouri, E. and Van Asselt, E.D. 2012. Mycotoxin contamination of cereal grain commodities in relation to climate in North West Europe. Food Add. Cont. Part A, 29: 1581–1592.

Vogelgsang, S., Sulyok, M., Hecker, A., Jenny, E., Krska, R., Schuhmacher, R. and Forrer, H.-R. 2008. Toxigenicity and pathogenicity of *Fusarium poae* and *Fusarium avenaceum* on wheat. Eur. J. Plant Pathol., 122: 265–276.

Voronin, M.C. 1890. About tempulent corn in the South-Ussuriiskij region. Botanical notes. SPb, Russia, 3(1): 13–21 (in Russian).

Waalwijk, C., Kastelein, P., de Vries, I., Kerenyi, Z., van der Lee, T., Hesselink, T., Kohl, J. and Kema, G. 2003. Major changes in *Fusarium* spp. in the Netherlands. Eur. J. Plant Pathol., 109: 743–754.

Waalwijk, C., van der Heide, R., de Vries, I., van der Lee, T., Schoen, C., Costrel-de Corainville, G., Häuser-Hahn, I., Kastelein, P., Köhl, J., Lonnet, P., Demarquet, T. and Kema, G.H.J. 2004. Quantitative detection of *Fusarium* species in wheat using TaqMan. Eur. J. Plant Pathol., 110: 481–494.

Ward, T.J., Bielawski, J.P., Kistler, H.C., Sullivan, E. and O'Donnell, K. 2002. Ancestral polymorphism and adaptive evolution in the trichothecene mycotoxin gene cluster of phytopathogenic *Fusarium* Proc. Natl. Acad. Sci. USA, 99: 9278–9283.

Ward, T.J., Clear, R.M., Rooney, A.P., O'Donnell, K., Gaba, D., Patrick, S., Starkey, D.E., Gilbert, J., Geiser, D.M. and Nowicki, T.W. 2008. An adaptive evolutionary shift in fusarium head blight pathogen populations is driving the rapid spread of more toxigenic *Fusarium graminearum* in North America. Fungal Genet. Biol., 45: 473–484.

Wilson, A., Simpson, D., Chandler, E., Jennings, P. and Nicholson, P. 2004. Development of PCR assays for the detection and differentiation of *Fusarium sporotrichioides* and *Fusarium langsethiae* FEMS Microbiol. Lett., 233: 69–76.

Wollenweber, H.W. and Reinking, O.A. 1935. Die *Fusarium*, ihre Beschreiburg, Schadwirkung und Bekampfung. Paul Parey, Berlin, Germany.

Yang, L., van der Lee, T., Yang, X., Yu, D. and Waalwijk, C. 2008. *Fusarium* populations on Chinese barley show a dramatic gradient in mycotoxin profiles. Phytopathol., 98: 719–727.

Ylimäki, A. 1981. The Mycoflora of cereal seeds and some feedstuffs. Ann. Agric. Fenniae, 20: 74–78.

Ylimäki, A., Koponen, H., Hintikka, E.-L., Nummi, M., Niku-Paavola, M.-L., Ilus, T. and Enari, T.M. 1979. Mycoflora and occurrence of *Fusarium* toxins in Finnish grain. Technical Research Centre of Finland, Materials and Processing Technology publication 21, Valtion Painatuskeskus, Helsinki, 1–28.

Yli-Mattila, T. 2010. Ecology and evolution of toxigenic *Fusarium* species in cereals in northern Europe and Asia. J. Plant Pathol., 92: 7–18.

Yli-Mattila, T. and Gagkaeva, T. 2010. Molecular chemotyping of *Fusarium graminearum, F. culmorum* and *F. cerealis* isolates from Finland and Russia. Molecular identification of fungi. Gherbawy, Y. and Voigt, K. (eds.). Springer Verlag, Berlin, Heidelberg, New York, pp. 159–177.

Yli-Mattila, T., Paavanen-Huhtala, S., Parikka, P., Konstantinova, P., Gagkaeva, T., Eskola, M. and Rizzo, A. 2002. Occurrence of *Fusarium* fungi and their toxins in Finnish cereals in 1998 and 2000. J. Appl. Genetics, 43A: 207–214.

Yli-Mattila, T., Paavanen-Huhtala, S., Parikka, P., Hietaniemi, V., Jestoi, M. and Rizzo, A. 2004a. Real-time PCR detection and quantification of *Fusarium poae* as compared to mycotoxin production in grains in Finland. Proceedings book of the 2nd International Symposium on *Fusarium* Head Blight, December 11–15, 2004, Orlando, Florida, USA, pp. 422–425.

Yli-Mattila, T., Mach, R., Alekhina, I.A., Bulat, S.A., Koskinen, S., Kullnig-Gradinger, C.M., Kubicek, C. and Klemsdal, S.S. 2004b. Phylogenetic relationship of *Fusarium langsethiae* to *Fusarium poae* and *F. sporotrichioides* as inferred by IGS, ITS, β-tubulin sequence and UP-PCR hybridization analysis. Int. J. Food Microbiol., 95: 267–285.

Yli-Mattila, T., Paavanen-Huhtala, S., Parikka, P., Konstantinova, P. and Gagkaeva, T. 2004c. Molecular and morphological diversity of *Fusarium* species in Finland and northwestern Russia. Eur. J. Plant Pathol., 110: 573–585.

Yli-Mattila, T., Paavanen-Huhtala, S., Parikka, P., Jestoi, M., Klemsdal, S. and Rizzo, A. 2006. Genetic variation, real-time PCR, metabolites and mycotoxins of *Fusarium avenaceum* and related species. Mycotox. Res., 22: 79–86.

Yli-Mattila, T., Paavanen-Huhtala, S., Parikka, P., Hietaniemi, V., Jestoi, M., Gagkaeva, T., Sarlin, T., Haikara, A., Laaksonen, S. and Rizzo, A. 2008. Real-time PCR detection and quantification of *Fusarium poae, F. graminearum, F. sporotrichioides* and *F. langsethiae* as compared to mycotoxin production in grains in Finland and Russia. Arch. Phytopathol. Plant Prot., 41: 243–260.

Yli-Mattila, T., Parikka, P., Lahtinen, T., Ramo, S., Kokkonen, M., Rizzo, A., Jestoi, M. and Hietaniemi, V. 2009a. *Fusarium* DNA levels in Finnish cereal grains. pp. 107–138. *In*: Gherbawy, Y., Mach, R.L. and Rai, M. (eds.). Current Advances in Molecular Mycology, Nova Science Publishers, Inc., New York, USA.

Yli-Mattila, T., Gagkaeva, T., Ward, T.J., Aoki, T., Kistler, H.C. and O'Donnell, K. 2009b. A novel Asian clade within the *Fusarium graminearum* species complex includes a newly discovered cereal head blight pathogen from the Far East of Russia. Mycologia, 101: 841–852.

Yli-Mattila, T., Ward, T., O'Donnell, K., Proctor, R.H., Burkin, A., Kononenko, G., Gavrilova, O., Aoki, T., McCormick, S.P. and Gagkaeva, T. 2011a. *F. sibiricum* sp. nov; a novel type A trichothecene-producing *Fusarium* from northern Asia closely related to *F. sporotrichioides* and *F. langsethiae*. Int. J. Food Microbiol., 147: 58–68.

Yli-Mattila, T., Rämö, S., Tanner, R., Loiveke, H. and Hietaniemi, V. 2011b. *Fusarium* DNA levels as compared to mycotoxin levels in Finnish and Estonian grain samples. Plant Breed. Seed Sci., 64: 131–140.

Yli-Mattila, T., Rämö, S., Hussien, T., Carlobos-Lopez, A.L. and Cumagun, C.J.R. 2013. Molecular quantification and genetic diversity of toxigenic *Fusarium* species in northern Europe as compared to those in southern Europe. Microorg., 1: 162–174.

Yli-Mattila, T., Gavrilova, O., Hussien, T. and Gagkaeva, T. 2015. Identification of the first *Fusarium sibiricum* isolate from Iran and *F. langsethiae* isolate from Siberia by morphology and species-specific primers. J. Plant Pathol., 98: 183–187.

Zhao, Z.H. and Lu, G.Z. 2007. *Fusarium kyushuense*, a newly recorded species from China. Mycotax., 102: 119–126.

CHAPTER 14

Mycotoxins: Fungal Secondary Metabolites with Toxic Properties

P.M. Cano,[a] O. Puel[b] and I.P. Oswald*

ABSTRACT

Fungi are frequent contaminant of foodstuffs worldwide. According to environmental conditions and their own physiological particularities, they can develop on living plants in the field or during pre-harvest period as well as during storage in case of drying failure or re-moistening. During their evolution, the filamentous fungi have become one of the most important sources of secondary metabolites. Many of these secondary metabolites display properties interesting for pharmaceutical or food industry. But other metabolites, named mycotoxins, have several detriment effects on both human and animals. The variety of biological activities of fungal secondary metabolites is directly related to the large diversity of these metabolites in terms of chemical structures. This chemodiversity is largely due to the variety of not only "backbone" enzymes like polyketide synthases (PKS), isoprenyl disphosphate synthases, non ribosomal peptide synthetases (NRPS) and dimethylallyltryptophane synthases (DMATS) but also the tailoring enzymes involved in their biosynthesis. Mycotoxins can thus be classified into four main categories: polyketides, terpenes, non ribosomal peptides and hybrids.

INRA, UMR 1331, Toxalim, Research Center in Food Toxicology, FR 31027 Toulouse, France; and Université de Toulouse, INP, UMR 1331, Toxalim, F-31027 Toulouse, France.
[a] Email: paty.canog@gmail.com
[b] Email: opuel@toulouse.inra.fr
* Corresponding author: ioswald@toulouse.inra.fr

This chapter presents a detailed overview of some of the most representative mycotoxins of each of these categories with a special attention to mycotoxins that are regulated in a great number of countries: aflatoxins, fumonisins, ochratoxins, patulin, zearalenone and trichothecenes. For the non ribosomal peptides category, gliotoxin and ergot alkaloids have been chosen for their relevance in terms of occurrence and toxicity. This chapter includes information on the current understanding of the mechanisms of their toxinogenesis and summarizes their toxicological effects.

Introduction

The fungal kingdom plays an important role in natural ecosystems, as fungi are essential decomposers of dead organic matter, thus restoring carbon, nitrogen, phosphorous and mineral levels in the biosphere. In addition to that, filamentous fungi constitute one of the most important sources of secondary metabolites. Many of these secondary metabolites present industrially interesting properties that have been beneficial for humans in the fields of medicine, food industry, cosmetics, energy and construction. However, other metabolites, known as mycotoxins, have been shown to have deleterious effects resulting in human, animal and phyto-pathologies. For instance, aflatoxin B1, one of the most potent carcinogenic and genotoxic natural substances is of fungal origin (Bennett and Klich, 2003). The most important fungal genera that produce mycotoxins are *Aspergillus*, *Penicillium* and *Fusarium*. *Aspergillus* species are probably the most representative of spoilage fungi given their capacity to grow under a large array of environmental conditions. In fact, there are very few food commodities and raw materials in which *Aspergillus* cannot develop (Pitt and Hocking, 2009). *Penicillium* species are also ubiquitous saprophytes that are encountered mainly in soil and only accidentally in food/feed commodities, although some species are specifically used in the food production industry such as *P. roqueforti* (Pitt and Hocking, 2009). Contrarily to *Aspergillus* and *Penicillium* species, *Fusarium* species are plant pathogens that are widely distributed in soils and particularly in cultivated soils. Since development of *Fusarium* species requires high humidity levels, these species are most frequently encountered on the fields pre-harvest or at early stages of storage (Leslie and Summerell, 2006).

So far, over 1000 fungal secondary metabolites have been identified, among which, around 30 have proven toxicity and are considered as mycotoxins and only 6 are legally regulated (aflatoxins, zearalenone, deoxynivalenol, fumonisins, ochratoxin A and patulin). A list of the major mycotoxins, classified based on their biosynthetic origin is given in Table 1. As it can be seen, there is a large variety of metabolites with very different toxicities. However, in most cases, there is not enough data to assess the exposure risks, which makes it difficult to establish official regulations. For instance, toxicity of the so-called emerging mycotoxins, such as beauvericin, enniatins and moniliformin was acknowledged quite recently as compared to other mycotoxins. Therefore,

Table 1. Mycotoxins produced most frequently, worldwide (non-exhaustive).

Biosynthetic pathway	Mycotoxin	Producing fungal genera	Most frequent substrate	Most notorious effects	Regulation
Polyketides	Aflatoxins	*Aspergillus*	Corn, Peanuts, Tree nuts	Carcinogenicity (group 1*) Hepatotoxicity	Yes
	Citrinin	*Aspergillus Penicillium Monascus*	Grains, Cheese	Hepatotoxicity, Nephrotoxicity	No
	Fumonisins	*Fusarium*	Cereals (Maize)	Hepatotoxicity Disruption of lipid metabolism Carcinogenicity (group 2B)	Yes
	Moniliformin	*Fusarium*	Cereals	Cardiotoxicity DNA damage	No
	Mycophenolic acid	*Penicillium Byssochlamys*	Blue cheeses	Immunosuppressive	No
	Ochratoxins	*Aspergillus Penicillium*	Coffee, cocoa, spices, dried fruits, grapes	Nephrotoxicity, Carcinogenicity (group 2B)	Yes
	Patulin	*Aspergillus Penicillium Byssochlamys*	Fruits	Cytotoxicity DNA damage	Yes
	Penicillic acid	*Aspergillus Penicillium*	Coffee, Cereals	Synergism with Ochratoxin A	No
	Sterigmatocystin	*Aspergillus*	Cereals, Cheese	Hepatotoxicity, Carcinogenicity (group 2B)	No
	Zearalenone	*Fusarium*	Cereals	Hyperoestrogenism	Yes
Terpenes	Trichothecenes	*Fusarium*	Cereals	Haematotoxicity, Immunosuppressive	Yes
Non-ribosomal peptides	Beauvericin	*Fusarium*	Cereals, Fruits	Cytotoxic, ionophoric, Antibiotic and insecticidal activity	No
	Enniatins	*Fusarium*	Cereals, Fruits	Cytotoxic, ionophoric, Antibiotic and insecticidal activity	No
	Ergot Alkaloids	*Claviceps*	Rye, Wheat	Neurotoxicity	Yes
	Gliotoxin	*Aspergillus*	Animal feeds	Immunosuppressive	No
	Roquefortine C	*Penicillium*	Blue cheeses (rare)	Neurotoxicity	No
Hybrids	Cyclopiazonic acid	*Aspergillus Penicillium*	Oil seeds, nuts, peanuts, cereals, dried fruits	Inhibition of Ca^{2+}-ATPase (necrosis)	No
	Fusarin C	*Fusarium*	Maize	Carcinogenicity (group 2B)	No

*Carcinogenicity groups according to the International Agency for Research on Cancer (IARC).

there is still not enough data for a proper assessment of the risks related to exposure to these mycotoxins, and as yet no regulations have been established (Jestoi, 2008).

Mycotoxin-producing fungal species may develop all around the world in solid or liquid substrates as long as they are in presence of sufficient nutriments and humidity ($a_w > 0.6$). Therefore, food and feed commodities are particularly affected. A worldwide survey conducted in 2012 analyzed the presence of five of the main mycotoxins in 4,023 samples of food/feed commodities such as cereals, soybean and straw originating from North and South America, Europe, Africa, Middle East, Asia and Oceania (Fig. 1) (Schatzmayr and Streit, 2013). The results of this study show a worldwide occurrence of these mycotoxins but a heterogeneous distribution, mainly due to regional endemism of some species. Annual meteorological variations may also affect fungal development and the distribution of mycotoxins around the world.

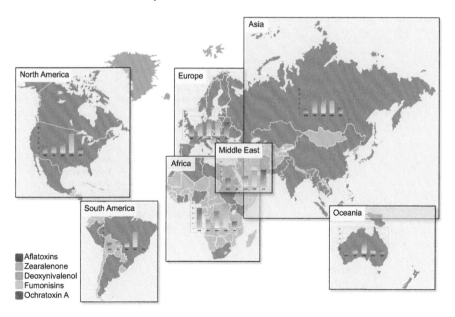

Figure 1. Worldwide occurrence of the 5 major mycotoxins: aflatoxins, zearalenone, deoxynivalenol, fumonisins and ochratoxin A. The number of samples analyzed for each mycotoxin in each of the regions was different depending on availabilities. However, for the purpose of clarity, in this figure, the results are expressed as the percentage of positive samples, independently of the total number of analyzed samples. (Adapted from Schatzmayr and Streit, 2013).

Mycotoxins are produced during fungal development on the fields or during storage and therefore, in some cases, they can reach human and animal alimentary chains. Human exposure can also occur by ingestion of contaminated animal products. In addition to ingestion, which is the most common route of exposure, mycotoxins may also enter the organism through dermal, respiratory and parental routes (Omar, 2013). Although cases of severe acute toxicity have already been reported for aflatoxin B1, and T-2 toxin, most of the concern related to mycotoxins exposure is due to chronic exposure. In any case, toxicity is dependent on the route, level and duration of exposure; on the age, health status, sex, genetics and dietary status of the exposed individuals and also on the possible interactions with other toxins (Omar, 2013). At a molecular level, mycotoxins can affect the DNA template, the RNA polymerase, thus hindering transcription, the ribosomes and polysomes and thereby hindering translation or different metabolic reactions. At a cellular level, these interactions may induce mutagenicity, carcinogenicity or teratogenicity. All of this may disrupt the function of the different organs of the exposed individuals consequently leading to hepatotoxicity, nephrotoxicity, neurotoxicity, immunotoxicity and so forth (Fig. 2) (Kiessling, 1986; Omar, 2013).

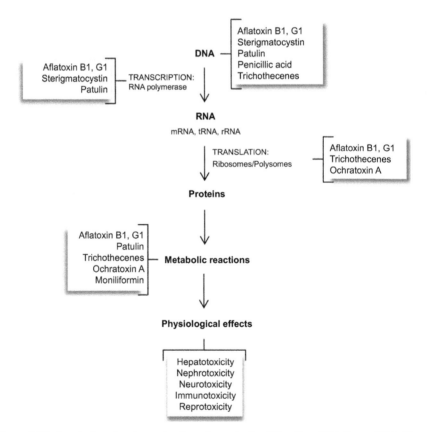

Figure 2. Biochemical mechanisms of action of mycotoxins (Kiessling, 1986; Omar, 2013). Given the large diversity of chemical structures of mycotoxins, they can cause a wide panel of toxic effects induced at a molecular level.

The variety of toxic effects induced by mycotoxins is directly related to the large diversity of these metabolites in terms of chemical structures. However, in spite of this diversity, mycotoxins are synthesized from simple building blocks, all of which are initially metabolized from glucose: (1) acetyl-CoA units for polyketides and terpenes and (2) aromatic amino acids for non-ribosomal peptides (Fig. 3). In contrast to bacterial toxins or toxins from higher fungi, which are usually ribosomal peptides, biosynthesis of mycotoxins require successive metabolic reactions that are orchestrated by a cascade of enzymes encoded by different genes that are generally gathered in clusters (Keller and Hohn, 1997). The structural diversity of mycotoxins results from the

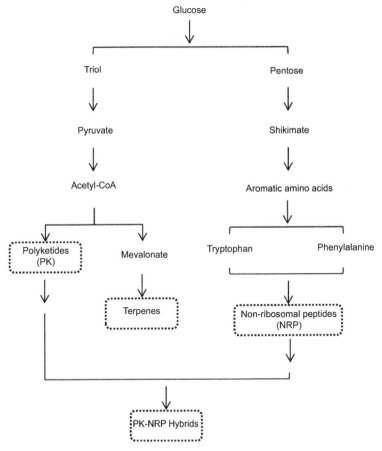

Figure 3. Overview of the different biosynthetic pathways involved in mycotoxin production. Mycotoxins can be classified into four main categories (polyketides, terpenes, non-ribosomal peptides and hybrids) that are synthesized from simple building blocks, all of which are initially metabolized from glucose.

variety of sub-units and the large array of processing reactions (cyclization, aromatization, alkylation, glycosylation, hydroxylation and epoxidation) involved in their biosynthesis (Boettger and Hertweck, 2012). Additional diversity is further obtained by combination of biosynthetic pathways such as the fusion of polyketide and non-ribosomal peptide biosynthetic pathways. Although quite wide-spread among microorganisms, this marriage remains enigmatic for the scientific community, especially concerning the biosynthetic programming and activity of the hybrid polyketide synthases (PKS) and the hybrid non-ribosomal peptide synthases (NRPS) (Boettger and Hertweck, 2012). Mycotoxins can thus be classified into four main categories: (1) polyketides, (2) terpenes, (3) non-ribosomal peptides and (4) hybrids (Fig. 3).

This chapter presents a detailed overview of some of the most representative mycotoxins of each of these categories. Special attention is paid to the mycotoxins that are regulated (aflatoxins, fumonisins, ochratoxins, patulin, zearalenone and trichothecenes). Gliotoxin and ergot alkaloids have been chosen for their relevance in terms of occurrence and toxicity.

Polyketides Mycotoxins

This group of secondary metabolites is characterized by the consecutive polymerization of ketide groups $(CH_2\text{-}CO)_n$, derived from the metabolization of acetate. It is a large and very diverse group of metabolites that includes polyphenols, polyenes and macrolides. Given the central role of acetyl-coenzyme A (acetyl-CoA) in polyketides biosynthesis, a relationship can be drawn with the biosynthesis of fatty acids. As a matter of fact, these primary metabolites were initially also classified as polyketides (Bentley and Bennett, 1999). However, two main differences exist between polyketides and fatty acids. Firstly, some of the polyketides synthases (PKS) involved in the biosynthesis of secondary metabolites require a different carboxylic acid than acetyl-CoA, which is not the case for the PKS involved in fatty acids biosynthesis, also known as fatty acid synthases (FAS). Secondly, unlike fatty acids, the oxidation state of polyketides is very variable. According to Nicholson et al. (2001), polyketides can be classified into three groups: non-reduced (like norsolorinic acid, an aflatoxin precursor), partially reduced (like 6-methylsalicylic acid, a patulin precursor) and highly reduced (like fumonisins). This classification further correlates with the structure and biosynthetic origin of each metabolite. Fungal polyketides can have a wide variety of biological activities, which is not surprising given the large chemical and structural diversity of these metabolites. Some of these activities present industrial interests such as the well-known immunosuppressant properties of mycophenolic acid, which is produced by fungi belonging to the genera *Penicillium*. This metabolite has been widely used to prevent rejection in organ transplantation (Lipsky, 1996) and also as a potent inhibitor of viral RNA synthesis *in vitro* (Gong et al., 1999). Another industrially interesting polyketide is lovastatin, which is mainly produced by *Aspergillus terreus*, and is used in the pharmaceutical industry as a cholesterol lowering drug (Seenivasan et al., 2008). However, many polyketides can also induce a large panel of toxic effects such as carcinogenic effects (aflatoxin, fumonisin B), hepatotoxicity (aflatoxin), nephrotoxicity (ochratoxin A), genotoxicity (patulin) and oestrogenic disruption (zearalenone). Interestingly, the impact of polyketides on public health is such that out of the 6 groups of mycotoxins that are regulated by the European Union, only the group of the trichothecenes does not belong to the polyketide family (European Commission, 2006). More details will be given concerning the effects of these mycotoxins in the following pages.

Aflatoxins

This family of mycotoxins was one of the first to be discovered, over 40 years ago, as the cause of the Turkey X disease which triggered acute hepatotoxicity and death in poultry in Great Britain (Nesbitt et al., 1962). Since then, aflatoxins have been shown to have immunosuppressive, mutagenic, teratogenic and hepatocarcinogenic effects in both humans and experimental animals (Eaton and Groopman, 1994). They are now considered as the most carcinogenic and genotoxic substances of natural origin (Bennett and Klich, 2003; CAST, 2003). Contamination by aflatoxins generally occurs before harvest in corn, peanuts, cotton seeds and tree nuts and also due to improper storage of food commodities, generating important economical losses (CAST, 2003). At a global level, contamination by aflatoxins affects over 5 billion people in the world through food and polluted air, most commonly in warm and humid places, such as Asia, Africa and Central America, where it is associated with liver/ lung cancer, HIV/AIDS, malaria, growth stunting, pediatric malnutrition and increased risk of adverse birth (Gong et al., 2003; Liu and Wu, 2010; Shuaib et al., 2010; Khlangwiset et al., 2011; Jolly et al., 2013).

Aflatoxins are mainly produced by 13 *Aspergillus* species, among which *A. flavus*, *A. ochraceoroseus* and *A. pseudotamarii* produce type B aflatoxins and *A. parasiticus*, *A. bombycis* and *A. nomius* produce both type B and G aflatoxins (Klich et al., 2003; Varga et al., 2011). *A. flavus* and *A. parasiticus* are the two main producing species. New *Aspergillus* species have been identified as producers of aflatoxins: *A. novoparasiticus*, *A. arachidicola*, *A. minisclerotigenes*, *A. pseudocaelatus* and *A. pseudonomius* (Pildain et al., 2008; Varga et al., 2011; Goncalves et al., 2012; Adjovi et al., 2014). In addition to *Aspergillus* species, aflatoxins or aflatoxin-related metabolites such as sterigmatocystin have also been isolated from culture extracts of *Fusarium kyusbuene* (Schmidt-Heydt et al., 2009), *Podospora anserina* (Slot and Rokas, 2011), *Dothistroma septosporum* and *D. pini* (Bradshaw et al., 2006). Recently, a sterigmatocystin genes cluster was reported in *Aspergillus ustus* genome although ability of this last species to produce this toxin is controversial (Pi et al., 2015; Rank et al., 2011).

The aflatoxins family encompasses 16 structurally related furanocoumarins (Bhatnagar et al., 2003), AFB1, AFB2, AFG1 and AFG2 being the four most abundant and naturally occurring (Huffman et al., 2010). AFB1 and AFB2 are characterized by a cyclopentane E-ring whereas AFG1 and AFG2 contain a xanthone ring instead (Huffman et al., 2010; Zeng et al., 2011). These differences result in Blue fluorescence under long wave UV light for AFB1 and AFB2 and Green fluorescence for AFG1 and AFG2. AFB2 and AFG2 differ from AFB1 and AFG1 in that their bisfuranyl ring is saturated (Fig. 4). In addition, AFM1 and AFM2 are the major metabolites of AFB1 and AFB2, respectively and they are produced in the liver of exposed animals by hydroxylation of the parent compounds (Gallagher et al., 1994). AFM1 and AFM2 represent an important threat to public health because they are excreted in milk, linked to casein proteins, and they persist in dairy products and human breast milk (Galvano et al., 2008; Keskin et al., 2009; Motawee and McMahon, 2009; Prandini et al., 2009). Recently, several studies have reported the presence of AFM1 in dairy products and, in some cases, above the maximum limit (50 ng/kg) established by the European regulation 1881/2006 (Boudra et al., 2007; Delialioglu et al., 2010; Fallah, 2010).

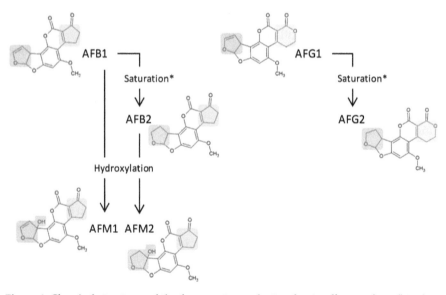

Figure 4. Chemical structures of the four most prevelant and naturally occuring aflatoxins: AFB1, AFB2, AFG1 and AFG2. AFB1 and AFB2 are characterized by a cyclopentane E-ring ( ) and blue fluorescence under long wave UV light whereas AFG1 and AFG2 contain a xanthone ring ( ) instead and display green fluorescence. AFB2 and AFG2 differ from AFB1 and AFG1 in that their bisfuranyl ring ( ) is saturated ( ). AFM1 and AFM2 are the major metabolites of AFB1 and AFB2, respectively and they are produced in the liver of exposed animals by hydroxylation ( ) of the parent compounds.

* Saturation of the Bisfuranyl Ring

Biosynthesis of aflatoxins is probably one of the most studied biosynthetic pathways of fungal secondary metabolites, starting with the cloning of the gene *nor-1* in *A. parasiticus* in the early 90's (Chang et al., 1992). The complete aflatoxin biosynthetic pathway gene cluster is 70 kb long and contains 25 genes (Trail et al., 1995a,b; Huffman et al., 2010). This biosynthetic pathway starts with the formation of hexanoate from acetate by the action of two fatty acid synthases (FASs). Hexanoate is then cyclisized by the action of a PKS into norsolorinic acid anthrone, which is the first stable intermediate. There are other stable intermediates like averantin, versicolorin A and B and sterigmatocystin. The production of versicolorins A and B is a branching point, which determines whether AFB1 and AFG1 or AFB2 and AFG2 are synthesized. Versicolorin A is synthesized by desaturation of versicolorin B and, therefore, it is the precursor of the desaturated AFB1 and AFG1 whereas, versicolorin B is the precursor of the saturated AFB2 and AFG2. Sterigmatocystin is the intermediate product in the versicolorin A pathway and it corresponds to dihydrosterigmatocystin in the versicolorin B pathway (Bhatnagar et al., 2003; Yu et al., 2004a; Yu et al., 2004b; Huffman et al., 2010; Roze et al., 2013). Notably, the last steps of the biosynthesis of aflatoxin take place in vesicules which play a key role in sterigmatocystin transformation and aflatoxin storage and export (Chanda et al., 2009). Interestingly, over 20 species, like *A. nidulans* and *A. versicolor*, produce sterigmatocystin as a final product instead of aflatoxin. The enzymatic pathway involved in the biosynthesis of sterigmatocystin in *A. nidulans* is the same than in the biosynthesis of aflatoxin by *A. parasiticus* or *A. flavus* (Keller and Adams, 1995; Brown et al., 1996; Keller and Hohn, 1997). The main differences lie on different gene distribution in the genetic clusters and also, the presence of several additional genes in the aflatoxin cluster of *A. parasiticus* and *A. flavus* (Keller and Adams, 1995; Brown et al., 1996; Yu et al., 2004a). That is why *A. nidulans* has been widely used as a model to study the fungal genetics and more precisely the biosynthetic pathway of sterigmatocystin. It has also been used to study the regulation of secondary metabolism in fungi and that is how the velvet protein VeA, which plays a central role in the regulation of primary and secondary metabolism in ascomycetes and basidiomycetes was first described (Kafer, 1965; Bayram and Braus, 2012).

Toxicity of aflatoxins is mostly related to the genotoxic and hepatocarcigenic AFB1, which has been classified in the group 1 (i.e., sufficient evidence of carcinogenicity in humans) in the evaluation of carcinogenic risks to humans by the *International Agency for Research on Cancer* (IARC) of the World Health Organization (WHO) (IARC, 2002). Toxicity of AFB1 follows its biotransformation mainly by cytochromes p450 in the liver (Fig. 5A) (Gallagher et al., 1994). The most reactive metabolite of AFB1 is the epoxydated exo-AFB-8, 9-epoxide form, also known as AFBO (Gross-Steinmeyer and Eaton, 2012; Hamid et al., 2013; Kim et al., 2013; Roze et al., 2013). This metabolite binds to proteins like albumin, which may result in hepatotoxicity (Sabbioni et al., 1987). It is also able to form DNA adducts which are responsible for hepatocarcinogenicity, and in humans, this is related to up to 50% of

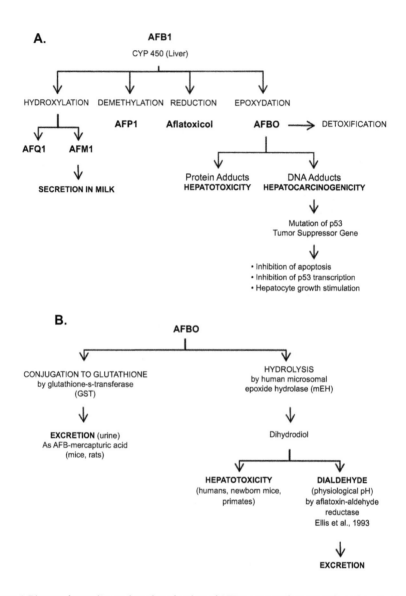

Figure 5. Biotransformation and modes of action of AFB1: a potent hepatotoxic and carcinogenic.
Bioactivation of AFB1 takes place in the liver under the action of cytochromes P450 and results in the formation of the epoxydated exo-AFB-8,9-epoxide (AFBO), which is the most reactive metabolite of AFB1 and the responsible of carcinogenic effects.

hepatocarcinomas (Bennett et al., 1981; Gross-Steinmeyer and Eaton, 2012). Most generally, DNA adducts formed by AFBO lead to DNA base and oxidative damage, which may result in a mutation of p53 Tumor Suppressor Gene. Consequently, there is an inhibition of apoptosis and p53 transcription as well as a stimulation of hepatocyte growth, which all together may ultimately lead to cancer (Fig. 5A). There are two main pathways of AFBO detoxification that are more or less effective depending on the species (Fig. 5B). In mice, AFBO is conjugated to glutathione by a glutathione-S-transferase (GST), and the resulting AFB-mercapturic acid is easily excreted through urine (Nayak et al., 2009). However, this is not very effective in humans where AFBO is hydrolyzed by human microsomal epoxide hydrolases (mEH) (Slone et al., 1995). This results in the production of a dihydrodiol of AFB that can contribute indirectly to hepatotoxicity by forming protein adducts (Eaton et al., 2001). It is the subsequent reduction of the dihydrodiol into a dialdehyde by aflatoxin-aldehyde reductases that finally detoxifies AFBO (Ellis et al., 1993; Kelly et al., 2002). There are other metabolites of AFB1 that come from demethylation (AFP1), hydroxylation (AFQ1 and AFM1) and reduction (aflatoxicol) reactions, but contrary to AFBO, these metabolites have a significantly lower toxicity as compared to their parent compound (Slone et al., 1995; Eaton et al., 2001; Gross-Steinmeyer and Eaton, 2012; Hamid et al., 2013; Kim et al., 2013) (Fig. 5A). However, AFM1 is also of great concern because of its presence in milk, and although, it is less toxic than AFB1, it has been classified by IARC in the group 2B (i.e., possibly carcinogenic to humans) (Hsieh et al., 1984; IARC, 2002). In addition to carcinogenic and genotoxic effects, AFB1 has also been reported to be immunotoxic. *In vitro* and *in vivo* studies have shown that it can reduce the number of circulating lymphocytes (Hinton et al., 2003), inhibit blastogenesis by lymphocytes (Meissonnier et al., 2008; Wada et al., 2008), alter the activity of natural killer cells and cytokine signaling (Methenitou et al., 2001), affect macrophage functions (Meissonnier et al., 2008; Bianco et al., 2012; Bruneau et al., 2012), and also impact on haematopoietic progenitors (Roda et al., 2010). Susceptibility to aflatoxins is very variable interspecies and interindividually mainly due to great differences in AFB1 biotransformation (Gross-Steinmeyer and Eaton, 2012; Hamid et al., 2013; Kim et al., 2013; Roze et al., 2013). For instance, the LD_{50} (Lethal Dose, 50%) for ducklings is 0.3 mg/kg.bw whereas, it is 9 mg/kg.bw for mice (Patterson and Allcroft, 1970). The order of decreasing toxicity has been described as follows: AFB1 > AFM1 > AFG1 > AFB2 > AFG2.

Fumonisins

Fumonisins constitute a large family of mycotoxins mainly produced by *Fusarium verticillioides, F. proliferatum* and other *Fusarium* species (Gelderblom et al., 1988; Marasas, 1996). These fungal species are important pathogens of corn responsible for "*Fusarium* kernel rot" or "pink ear rot" (Bullerman, 1996; Marasas et al., 2004). Fumonisins have also been isolated from cultures of *Alternaria alternata* and *Aspergillus niger* (Chen et al., 1992; Frisvad et al., 2007; Mansson et al., 2010; Mogensen et al., 2010). Contamination can occur at different stages like crop growth, harvesting or storage, depending on temperature and humidity. However, development of *Fusarium* species mainly occurs prior to harvest or in the early periods of storage when humidity levels are still high ($a_W > 0.9$) (Dutton, 2009). Human exposure to fumonisins is of greater importance in South American, Asian and African countries due to weather and storage conditions and also due to higher consumption levels of maize-based products (Dutton, 2009).

The fumonisins family encompasses 28 structurally related metabolites that can be classified into 4 series: A, B, C and P (Rheeder et al., 2002). All of them share a long carbon chain backbone with an amine, but have different hydroxyl, methyl and tricarboxylic acid side chains (Fig. 6). A series have an acetylated terminal amine group in C2 position whereas B and C series have free amino groups and P series have a 3-hydroxypiridinium moiety at this position. C series differ from the rest in that their acyl chain is condensated with glycine instead of an alanine residue so they have a C19 backbone chain instead of C20 (Huffman et al., 2010). One particular feature of fumonisins structure is the presence of two tricarballylic esters on C14 and C15, which is rare in natural products. Type B fumonisins are the most abundant in naturally contaminated commodities. Esterification of the backbone structure gives FB4 and further oxidation gives FB2, FB3 and FB1 (Huffman et al., 2010). Among type B fumonisins, fumonisin B1 (FB1) represents 70% of total fumonisins content and FB2 generally accounts for up to 25%. Concerning the other series, only C fumonisins have been detected in corn samples but at much lower concentrations (Thiel et al., 1991).

The biosynthesis of fumonisins has essentially been studied in the genome of *Fusarium* species (Proctor et al., 1999, 2003, 2008). The gene cluster of this biosynthetic pathway is concentrated in a 75 kb DNA region. Fumonisins biosynthetic pathway starts with the formation of a linear C18 backbone polyketide with two methyl groups and a terminal carbonyl, followed by condensation with alanine to form the C20 chain. Over 15 genes have been identified so far in the fumonisins biosynthetic cluster. However, the function of many of them remains poorly known (Brown et al., 2005). The first gene that was characterized was *FUM1* which assembles the 18 carbon backbone chain

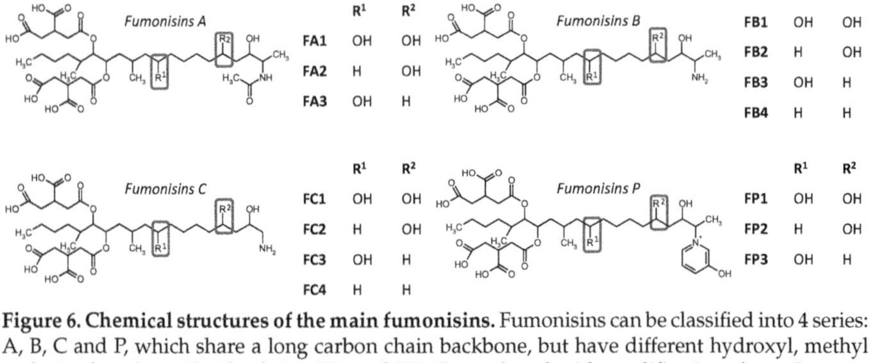

Figure 6. Chemical structures of the main fumonisins. Fumonisins can be classified into 4 series: A, B, C and P, which share a long carbon chain backbone, but have different hydroxyl, methyl and tricarboxylic acid side chains (R1 and R2). Reproduced with modifications from (Lazzaro et al., 2012).

from a molecule of acetyl-CoA and two molecules of S-adenosyl methionine (Huffman et al., 2010). *FUM8*, which encodes for an homologous of class II α-aminotransferases, is the next gene in the cluster and it is responsible for the condensation of the backbone chain (Huffman et al., 2010). In addition, *FUM21*, which encodes for a Zn (II)-2Cys6 DNA-binding transcription factor, is a specific pathway regulator (Brown et al., 2007). It positively regulates the expression of FUM genes and is, therefore, necessary for fumonisins production.

Absorption of fumonisins and particularly of FB1 after oral ingestion is very low and remains under 4% (Martinez-Larranaga et al., 1999). However, it is rapidly distributed to all organs and tissues, particularly to the liver and kidneys, which are the main target organs of this mycotoxin (Voss et al., 1993; Rotter et al., 1996b). Its elimination occurs through feces, biliary and urinary secretions as the parent compound or after microbial de-esterification, which results in partial or total hydrolization of FB1 into HFB1 (Shephard et al., 1994a,b,c; Martinez-Larranaga et al., 1999). There is no proof of FB1 biotransformation due to P450 cytochromes activities (Merrill et al., 1999). However, this mycotoxin is able to inhibit CYP2C11 and CYP1A2 activity by suppression of proteinase K (Spotti et al., 2000).

Toxicity of fumonisins was first reported for FB1 in farm animals in the late 80's with leukoencephalomalacia in horses (Marasas et al., 1988; Kellerman et al., 1990) and pulmonary oedemas in pigs (Harrison et al., 1990), these two species being the most sensitive to FB1. Later, laboratory studies showed hepatotoxic, nephrotoxic, carcinogenic, immunotoxic, developmental and reprotoxic effects (Riley et al., 1994; Voss et al., 1995; Gelderblom et al., 2001; Marasas et al., 2004; Bracarense et al., 2012). In addition, embryotoxicity and embryolethality caused by fumonisins is strongly suspected in poultry (Javed et al., 1993; Rauber et al., 2012). In humans, there is evidence supporting a role of FB1 in neural tube defects (Marasas et al., 2004) and in liver and oesophagus cancer (Sun et al., 2007; Alizadeh et al., 2012). FB1 has hence been classified in the group 2B by IARC as "possible carcinogenic to humans" (IARC, 2002).

Although FB1 increases micronucleus frequency and DNA strand breaks *in vitro* and *in vivo*, it is not considered as a genotoxic carcinogen since it lacks reactivity to DNA (Müller et al., 2012). Carcinogenicity of FB1 is rather explained by the induction of oxidative stress, which is related to the increase of reactive oxygen species (ROS) and the impairment of antioxidant defence mechanisms (Fig. 7A) (JECFA, 2001; Müller et al., 2012). These result in lipid peroxidation and oxidative DNA damage. However, the mechanisms behind the induction of oxidative stress by FB1 are still unknown.

Another possible mode of action for FB1 carcinogenicity is related to the induction of apoptosis, which has been shown *in vitro* and *in vivo* (Fig. 7A) (Müller et al., 2012). Induction of apoptosis triggers the activation of defence mechanisms such as cell proliferation, possibly leading to tumorigenesis. Again, the mechanisms by which FB1 induces apoptosis are not fully understood. Some studies provided evidence that apoptosis induction by FB1

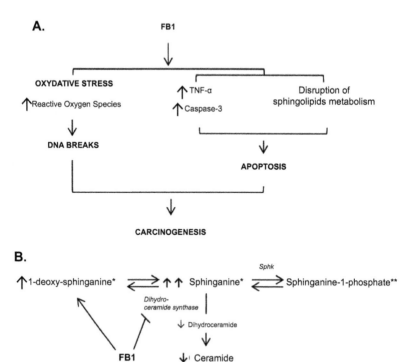

Figure 7. Modes of action of the carcinogenic FB1. The mode of action by which FB1 is carcinogenic is not very well understood so far. Different mechanisms are possible: (A) induction of oxidative stress and also (A) induction of apoptosis, both possibly leading to tumorigenesis. Apoptosis can occur by (A) an increase of TNF-α, which initiates apoptosis through cell surface-mediated mechanisms, such as the caspase-3 pathway or (B) by the increase of pro-apoptotic sphingoid bases (sphinganine, Sa and sphingosine, So) and the loss of complex sphingolipids (ceramide).

* Pro-apoptotic metabolites

** Anti-apoptotic metabolites

is directly related to the increase of TNF-α, which initiates apoptosis through cell surface-mediated mechanisms, like the caspase-3 pathway (Jones et al., 2001; Sharma et al., 2001; Bhandari and Sharma, 2002).

Apoptosis could also be induced by the increase of pro-apoptotic sphingoid bases and the loss of complex sphingolipids, which regulate cell adhesion. FB1 is indeed a well-known sphingolipid disruptor since it inhibits ceramide synthases (Soriano et al., 2005) due to similarities with sphinganine

(Sa) and sphingosine (So), the two ceramide precursors (Fig. 7B) (Soriano et al., 2005). This results in a dose-dependent increase of Sa, and a lower increase of So, before toxicity symptoms appear (Riley et al., 1993). Thus, the Sa/So ratio is commonly used as a biomarker of fumonisin exposure. In order to counterbalance the apoptotic effects of Sa and So, cells induce the phosphorylation of these sphingoid bases by Sphk1 into sphinganine 1-phosphate and sphingosine 1-phosphate, which are antiapoptotic (Müller et al., 2012). In addition, it has been shown that FB1 also increases the production of 1-deoxy-sphinganine *in vivo*, which is also a potent pro-apoptotic that cannot be phosphorylated like Sa and So to reduce its toxicity due to the lack of the hydroxyl group (Zitomer et al., 2009). Other studies have shown that when FB1 or HFB1 bind to ceramide synthase, they are transformed into their N-acetyl derivatives, which are more toxic than the parent compounds (Harrer et al., 2013).

Concerning the toxicity of other fumonisins, FC1 and FC2, which are structurally equivalent to FB1 and FB2 are highly phytotoxic, and FC3 and FC4, which are structurally similar to FB3 and FB4 have moderate phytotoxicity (Abbas et al., 1998).

Ochratoxins

Ochratoxins (OTs) belong to a large family of mycotoxins that encompasses more than 20 different metabolites among which ochratoxin A (OTA) is the most frequently occurring and most toxic. Ochratoxins are produced by *Aspergillus* species and mainly by *A. ochraceus* and *A. westerdijkiae*, which are usually found on coffee, cocoa, spices and dried fruits in tropical regions of the globe. Other OT-producing *Aspergillus* species are *A. alliaceus*, which has been detected on figs and nuts, and *A. carbonarius*, which is responsible for the decay of grapes and the presence of OTs in wine. Many *Penicillium* species also produce OTs like *P. verrucosum* that develops on cereals during storage in temperate regions of the world and *P. nordicum*, which is found on cheese and fermented meat (Reddy and Bhoola, 2010). As most mycotoxins, OTs are not degraded during most food-processing steps such as fermentation and cooking and they are, therefore, frequently encountered in food and feed (Huffman et al., 2010). Cereals, wine, beer and pork meat are, in this order of importance, the main sources of human exposure to OTs (Huffman et al., 2010). In addition, another important feature of OTs is that they readily bind to proteins and, therefore, they have also been detected in maternal milk, consequently in infant babies (Biasucci et al., 2011). Ochratoxins are usually co-detected with other mycotoxins like citrinin, aristolochic acid and fumonisins, with a resulting synergistic combination and an increased risk of genotoxicity. Nevertheless, very few studies have addressed the effects of co-exposure to these mycotoxins, and more data is needed (Reddy and Bhoola, 2010).

Ochratoxins are hybrid molecules composed of a pentaketide dihydroisocoumarin moiety linked to the amino acid phenylalanine by an amide bond at C7. There is a chlorine atom on one of the hydroisocoumarin rings that strongly contributes to its toxicity. That is why, OTB, which is the dechlorinated form of OTA, is at least 10 times less toxic than OTA (Huffman et al., 2010). Other important metabolites are: OTα and OTβ, in which there is a cleavage of the amide bond of OTA or OTB, respectively; OTC, which is the ethyl ester form of OTA and the hydroxylated OTA. OTA and OTB also exist in open lactone form (O'Callaghan et al., 2013).

Compared to other mycotoxins, little is known about the biosynthesis of OTs. Most of the studies have used *Penicillium* species and thus, a putative genetic cluster for OTA production has been identified in *P. nordicum* (Karolewiez and Geisen, 2005; Geisen et al., 2006). Given the chemical structure of OTA, three crucial steps for its biosynthesis have been defined: (1) the biosynthesis of the polyketide isocoumarin group, which is attributed to a PKS, (2) the ligation to phenylalanine, which is conducted by a NRPS and (3) the chlorination step, which is performed by a halogenase. Accordingly, sequences coding for a PKS (*otapksPN*), for a NRPS (*otanpsPN*) and for an enzyme homologous to a chlorinating enzyme (*otachlPN*) have been identified in the OTA genetic cluster in *P. nordicum*. In addition, other sequences have also been identified in the cluster coding for transporter proteins involved in OTA export (*otatraPN*) (Gallo et al., 2012). Homologous PKS genes have been identified in the genomes of *A. carbonarius*, *A. niger*, *A. ochraceus*, *A. westerdijkiae* and *P. nordicum* and homologous NRPS genes have been identified in *A. carbonarius* (Gallo et al., 2013).

Little is known on the order of the different steps leading to OTA production. The first biosynthetic pathway was proposed by Huff and Hamilton in 1979 (Huffman et al., 2010) (Fig. 8A). They suggested that condensation of 1 acetyl-CoA and 4 malonyl-CoA by a PKS resulted in the formation of mellein, which was oxidized into OTβ before chlorination into OTα. Then, ligation with phenylalanine would take places by the action of a NRPS, and would lead to the synthesis of OTC, which would finally be transformed into OTA by the action of an esterase. In 2001, Harris and Mantle suggested that chlorination took place in the final steps rather than before ligation to phenylalanine (Harris and Mantle, 2001). Finally, in 2012, Gallo et al. observed, in *A. carbonarius*, that when the NRPS gene in OTA cluster (*ACOTAnrps*) was disrupted, the mutant did not produce OTA, as expected, but neither did it produce OTC, OTB nor OTα (Gallo et al., 2012). Only OTβ and phenylalanine were detected. They also observed that the mutant kept the ability to produce OTα after addition of exogenous OTA. Given these results, they proposed a new version of OTA biosynthesis in which OTβ is ligated to phenylalanine by the NRPS leading to OTB production (Fig. 8B). OTB is then chlorinated to give OTA which can be further metabolized to form OTα as well as other metabolites. This last step can be performed by the fungi itself or other microorganisms such as rumen microbiota (Gallo et al., 2012).

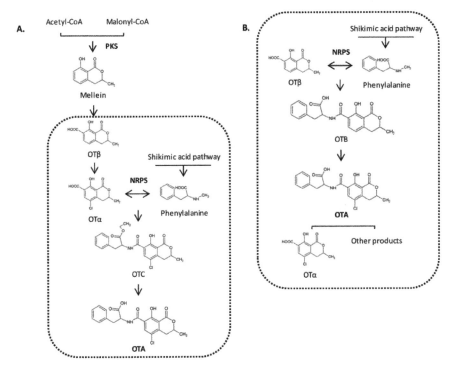

Figure 8. Different hypothesis for ochratoxin A (OTA) biosynthesis. (A) According to Huff and Hamilton (1979), (B) as suggested by Gallo et al. (2012).

Historically, exposure to ochratoxins has been associated with Balkan Endemic Nephropathy (BEN), a human fatal disease affecting southeastern European countries and characterized by disruption of renal function (el Khoury and Atoui, 2010). This pathology has also been related to exposure to aristolochic acid, which is generally found together with OTA in food samples and has similar effects. Kidneys are actually the main target organs of OTs and OTA has been shown to be nephrotoxic to all single-stomach animals tested including mammals and birds but not to adult ruminants (el Khoury and Atoui, 2010). Porcine nephropathies are the most severe conditions induced by OTA, causing atrophy of the proximal tubules, interstitial cortical fibrosis and sclerotized glomeruli. Impairment of tubular function is characterized by the reduction of the tubular excretion of para-aminohippurate for insulin clearance and an increase in glucose excretion (Reddy and Bhoola, 2010). OTA also causes lesions in the gastro-intestinal tract, the spleen, brain, liver, kidney, heart and lymphoid tissues in hamsters, rats, mice and chickens (el Khoury and Atoui, 2010). Embryotoxicity, teratogenicity and immunosuppression have also been reported after exposure to OTA but only at higher doses than those causing nephropathies (> 200–4000 μg/kg feed) (el Khoury and Atoui, 2010; Reddy and Bhoola, 2010; Vettorazzi et al., 2013). Finally, OTA has also been described as a carcinogen and it has been classified in the group 2B by IARC as "Possibly carcinogenic to humans" (IARC, 1993). Nephrocarcinogenicity is the most relevant carcinogenic effect of OTA and it has been proven in rats and mice so far, with rats being more sensitive than mice and males being more sensitive than females (Reddy and Bhoola, 2010). However, the mode of action by which OTA induces carcinogenicity is not well understood yet. Most studies support an indirect DNA non-reactive genotoxic mechanism involving different epigenetic factors such as oxidative stress, compensatory cell proliferation and disruption of cell signaling and cell division. Additional direct mechanisms like DNA adduct formation are also possibly involved and may influence epigenetic factors (Vettorazzi et al., 2013). Sex and species differences are not well understood. Excretion of OTA takes place by binding to carrier proteins that transport the mycotoxin through blood circulation to the liver where it is excreted in the bile or to the kidneys for urinary excretion (Reddy and Bhoola, 2010).

Patulin

Patulin was isolated in 1943 from *Penicillium griseofulvum*, *P. expansum* and later from *Aspergillus clavatus* (Puel et al., 2010). At that time, there was a general effort to find new active metabolites with the massive use of penicillins during World War II. Patulin was initially identified as a new antibiotic against Gram+ and Gram– bacteria; a British company sold it under the name "Tercinin". However, its use stopped a few years later due to strong neurotoxic effects to humans and animals.

Patulin is produced by several genera of filamentous fungi among which the most important are: *Penicillium* (13 species such as *P. expansum* and *P. griseofulvum* and so on), *Aspergillus* (4 species like *A. clavatus* and *A. giganteus*, etc.), *Byssochlamys* (*B. nivea*) and *Paecilomyces* (*P. saturatus*). However, only *B. nivea* and *P. expansum* are of great concern in terms of economical losses. *B. nivea* causes problems in silages, which may result in livestock exposure to patulin and *P. expansum* is responsible for decay of pomaceaous fruits (mostly apples), which is characterized by rapid soft rot and blue pustules (McKinley and Carlton, 1991). Other fruits such as cherries, apricots and peaches are subject to contamination by *P. expansum* and are thus suspected to contain patulin. This mycotoxin is degraded during fermentation processes and, therefore, it is not found in alcoholic cider (Puel et al., 2010).

Patulin is an unsaturated heterocyclic lactone. The study of its biosynthesis started in the 1950s. It was actually the first polyketide biosynthesis to be analyzed (Puel et al., 2010). The genetic cluster involved in patulin biosynthesis consists of 15 genes clustered in a 40 kb DNA region (Artigot et al., 2009). Previously isolated from *Aspergillus clavatus*, these fifteen genes cluster were reported very recently from several *Penicillium expansum* strains (Tannous et al., 2014; Ballester et al., 2015; Li et al., 2015). However only 5 of these genes have been fully characterized so far (Puel et al., 2010). *PatK, PatG, PatH,* and *PatI* are involved in the four first steps of the patulin biosynthesis pathway. They encodes for 6-methylsalicylic acid synthase (6MSAS), 6-methylsalicylic acid decarboxylase, *m*-cresol hydroxylase and *m*-hydroxybenzyl alcohol, respectively (Puel et al., 2010; Snini et al., 2014). PatN is involved in the transformation of isoepoxydon into phyllostin. The function of some of the other genes has not been established experimentally, but it can be attributed based on their sequence. For instance, *PatM, PatC* and *PatA* encode for ABC transporter, MFS transporter and acetate transporter, respectively (Puel et al., 2010). Biosynthesis of patulin consists of 10 steps, which starts with the formation of 6MSA with a molecule of acetyl-CoA and 3 molecules of malonyl-CoA by 6MSAS. Decarboxylation of 6MSA results in the production of *m*-cresol. The methyl group of *m*-cresol is then hydroxylated and oxidized into an aldehyde group to form gentisaldehyde (Puel et al., 2010). The conversion of this metabolite into the two ring structure of patulin occurs by successive formation of four post-aromatic precursors: isoepoxydon, phyllostine, neopatuline and ascladiol. However, it is still not clear whether the substrate of these reactions is gentisaldehyde or gentisyl alcohol, as shown in Fig. 9.

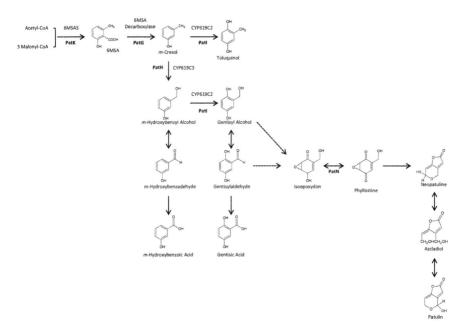

Figure 9. Biosynthesis of patulin. Biosynthesis of patulin consists of 10 steps, and out of the 15 genes involved in this pathway, only 5 have been fully characterized (*PatK, PatI, PatH* and *PatN*). The function of the other genes has been attributed putatively but has not been proved experimentally and thus, they are not represented in this figure. It is still not clear whether the substrate leading to the final reactions is gentisaldehyde or gentisyl alcohol (dotted arrows). Adapted from (Artigot et al., 2009; Puel et al., 2010; Snini et al., 2014).

Patulin is poorly absorbed after oral administration. Less than 3% of it is taken up by the gastric tissues and less than 20% reaches the systemic circulation. Excretion mainly takes place during the first 24 h by urine and feces. Most of the absorbed patulin is conjugated with glutathione intracellularly, which causes glutathione depletion in the cells (Rychlik et al., 2004).

Symptoms of acute toxicity encompass agitation, convulsions, dyspnea, pulmonary congestion, edema and ulceration, hyperemia and distention of the intestinal tract. Rodents seem to be the most sensitive species whereas, poultry seem less affected by patulin exposure (Puel et al., 2010). Sub-acute toxicity has been mainly studied in rats, and it results in weight loss, gastric and intestinal changes, neurotoxicity and alterations of renal functions at higher doses. Additionally, inhibition of several enzymes (ATPases) has also been observed in the brain and in the intestine with consequences particularly on lipid metabolism. Other studies confirmed these results in mice, hamsters and chickens (Puel et al., 2010). There are only a few and old studies concerning chronic exposures to patulin. The results of these studies concluded that the intestinal tract and the renal functions are essentially affected by chronic exposure to this mycotoxin (Puel et al., 2010). Concerning reproductive and developmental effects, in rats, 2 mg/kg body weight of patulin induced abortion of all embryos and in mice, the same dose provoked death of all new borns due to hemorrhages in brain, lung and skin. Embryotoxicity and teratogenicity have also been proven in eggs (Puel et al., 2010). Finally, patulin is also related to a diversity of immunomodulatory effects. It is responsible for the inhibition of macrophage function (inhibition of protein synthesis and alteration of membrane functions), and also for the reduction of IL-10, IFN-γ and IL-4 production in human macrophages, blood mononuclear cells and T-cells (Puel et al., 2010). Data on genotoxicity, mutagenicity and carcinogenicity are not very consistent between different studies. That is probably why IARC classified patulin in the group 3 as "not classifiable as to its carcinogenicity to humans" (IARC, 1986). However, the JECFA considers this mycotoxin as a genotoxic (JECFA, 1995).

The general mechanism of toxicity of patulin lies in its ability to bind to different side chains of cysteine, lysine and hystidine as well as to α-amino acid groups of proteins, which might result in biological inactivation of several enzymes. In addition, cytotoxicity and DNA-damages are caused by covalent bonding with sulfurydryl groups (thiols) like the cysteine-containing tripeptide gluthatione (GSH) (Fliege and Metzler, 2000). DNA damages induced by patulin cause micronuclei and chromosomal aberrations (clastogenic effects) (Puel et al., 2010).

Zearalenones

Zearalenones are a group of mycotoxins produced by several *Fusarium* species such as *F. graminearum*, *F. culmorum*, *F. crookwellense*, *F. equiseti*,

F. cerealis and *F. semitectum* and so on. In fact, they are produced by most of the toxigenic *Fusarium* species and, therefore, are usually detected in combination with other fusariotoxines like trichothecenes and fumonisins (SCOOP, 2003; Zinedine et al., 2007). As they are produced by *Fusarium* species, zearalenones are mostly encountered in temperate and warm countries on cereal crops, especially on corn but also on wheat, oats and soybean. Production of these mycotoxins mainly takes place in the field but also during improper cereal storage. Human exposure to zearalenones is essentially caused by consumption of contaminated raw maize, and to a lesser extent corn and wheat-derived products as well as beer (SCOOP, 2003). Absorption, metabolization and excretion of zearalenones are generally fast and, therefore, human exposure through meat and other animal products like milk is of little significance. However, the main concern related to exposure to zearalenones is due to the fact that they are the one of the most potent families of non-steroidal naturally occurring oestrogens (Zinedine et al., 2007).

In terms of chemical structure, zearalenones have a resorcinol (m-benzenediol) moiety fused to a 14-member macrocyclic lactone (Huffman et al., 2010). There are six major members of this family: Zearalenone (ZEN), α-zearalenol (α-ZEL), β-zearalenol (β-ZEL), α-zearalanol (α-ZAL), β-zearalanol (β-ZAL), zearalanone (ZAN). Until very recently, different abbreviations were used to refer to each of the metabolite. For the purpose of clarity, in this chapter, abbreviations will be used as described by Metzler (Metzler et al., 2010). As depicted in Fig. 10, α-ZEL, β-ZEL, α-ZAL and β-ZAL contain a hydroxyl group instead of a keto group at C-6 as compared to ZEN and ZAN. Also, ZAN, α-ZAL and β-ZAL have a saturated bond between C-1 and C-2 contrarily to ZEN, α-ZEL and β-ZEL.

Compared to other polyketide mycotoxins, biosynthesis of zearalenones has received far less attention. Generally, most clusters associated with polyketide biosynthesis contain a single PKS and several other genes encoding transformation enzymes or regulatory genes (Gaffoor and Trail, 2006). However, two PKS have been identified in the biosynthetic pathway of zearalenones. PKS13, which encodes for the protein ZEA1p, is a non-reduced resorcylic acid macrolactone megasynthase, and PKS4 which encodes for ZEA2p, on the contrary, is a highly reduced polyketide synthase (Huffman et al., 2010). ZEA2p is the starter enzyme in the biosynthesis of zearalenones. It is responsible for the formation of the highly reduced hexatide chain with a single acetyl-CoA and 8 malonyl-CoA molecules (Gaffoor and Trail, 2006). During biosynthesis, the polyketide chain can be partially reduced forming an alkene, which will further lead to ZEN or, the keto groups can be fully reduced forming an alkane, leading to ZAN production (Gaffoor and Trail, 2006). Then ZEA1p is in charge of the iterative elongation of the chain by the addition of unreduced ketones and is also in charge of the aromatic cyclization (Gaffoor and Trail, 2006; Huffman et al., 2010) (Fig. 10). In addition to PKS13 and PKS4, ZEB1 and ZEB2 have also been described in the biosynthetic pathway of zearalenones. ZEB1 is a putative isoamyl alcohol oxidase that oxydates α-ZEL

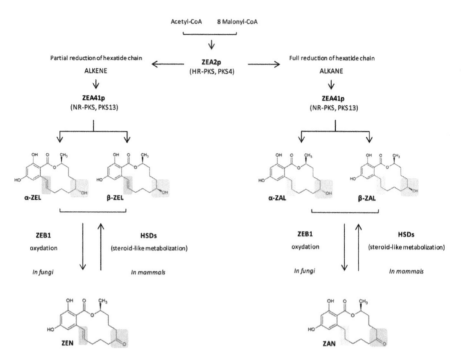

Figure 10. Biosynthesis of zearalenones. This biosynthetic pathway starts with the formation of the highly reduced hexatide chain with a single acetyl-CoA and 8 malonyl-CoA molecules by the action of ZEA2p, which is a highly reduced polyketide synthase encoded by the gene PKS4. Then, the polyketide chain can be partially reduced forming an alkene (■), which will further lead to ZEN or, the keto groups can be fully reduced forming an alkane ( ), leading to ZAN production. ZEA1p, which is a non-reduced resorcylic acid macrolactone megasynthase encoded by the gene PKS13, is in charge of the iterative elongation of the chain by the addition of unreduced ketones and is also in charge of the aromatic cyclization. Finally, ZEB1 is a putative isoamyl alcohol oxidase that oxydates α-ZEL and β-ZEL ( ) into ZEN (■), and α-ZAL and β-ZAL ( ) into ZAN ( ).

and β-ZEL into ZEN, and α-ZAL and β-ZAL into ZAN (Huffman et al., 2010) (Fig. 10). ZEB2 is a leucine zipper domain-containing regulator that controls the expression of the other genes of the cluster (Huffman et al., 2010).

Interestingly, ZEN can be metabolized back to α-ZEL and β-ZEL and ZAN back to α-ZAL and β-ZAL in mammalian cells (Metzler et al., 2010). Actually, ZEN can undergo both phase 1 and phase 2 metabolization in fungal, plant and mammalian cells. Phase 1 metabolism includes reduction of the keto groups and hydroxylation of aliphatic or aromatic groups. It has been shown recently that fungi produce a large variety of phase 1 metabolites like 13-formyl ZEN, 5,6-dehydroxy ZEN, 5-hydroxy ZEN and 10-hydroxy ZEN (Metzler et al., 2010). Mammalian cells with 3α- and 3β-hydroxysteroid dehydrogenases (HSDs) activities are able to reduce ZEN back to α-ZEL and β-ZEL (in rats, ruminants, swine, turkeys, hens and probably humans too) and ZAN back to α-ZAL and β-ZAL (in cattle, sheeps, horses, goats and deers) (Minervini and Dell'Aquila, 2008; Metzler et al., 2010) (Fig. 10). HSDs are involved in the synthesis of steroid hormones like E2. Liver is the major organ where this metabolism takes place but it also occurs in other organs with HSDs activity like kidneys, testis, prostate, hypothalamus, ovaries and intestine. Additionally, hydroxylation of the aromatic group by human CYP1A2 forms 13- and 15-hydroxyl ZEN, two highly unstable catechols, which are further oxidized into quinone-type electrophiles that can cause DNA-adducts (Metzler et al., 2010). Phase 2 metabolism consists of the conjugation of ZEN with glucuronides or sulfates. The resulting metabolites are readily excreted mainly through urines but also remain in blood and bile. From the bile, glucuronides-metabolites can be reabsorbed back in the intestine and hydrolyzed again (Metzler et al., 2010). In mammalian cells, conjugation is effected by uridine diphosphate glucuronyl transferases (UDPGT) and forms monoglucuronides of ZEN (Metzler et al., 2010). There is little evidence concerning conjugation of ZEN with sulfates but it has been soundly postulated (Metzler et al., 2010). In plants, conjugation occurs with polar molecules present in the host cells. The resulting metabolites are often difficult to detect because their polarity and solubility is different from the parent compound for which the detection method was designed. Therefore, many of the metabolites produced in plants are considered as masked mycotoxins.

Zearalenone was first detected after appearance of hyperestrogenism symptoms in swine fed moldy maize. These symptoms include prolonged estrus, anestrus, changes in libido, infertility, increased incidence of pseudo pregnancy, increased udder or mammary gland development, and abnormal lactation (Metzler et al., 2010). It is also related to other secondary complications namely stillbirths, abortions, mastitis, vulvovaginitis and rectal or vaginal prolapses (Metzler et al., 2010). Endocrine disruption related to zearalenone is attributed to both genomic and non-genomic effects. First, zearalenone competes effectively with endogenous steroid estrogens like 17β-estradiol (E2) for the specific binding sites of estrogen receptors (ER), with no preference between the two isomers ER-α and ER-β (Metzler et al., 2010) (Fig. 11). ER receptors are found in many different organs such as the uterus, the mammary glands, the bones, the liver and the brain, where zearalenone can, therefore, have an impact. Relative binding affinities of the main metabolites are as follows: α-ZAL > α-ZEL > β-ZAL > ZEN > β-ZEL. The α-isomers are about 10 fold more active than the rest because of the position of the hydroxyl group. Differences in production of α or β isomers explain differences in the sensitivity between species. For instance, pigs produce more α-ZEL whereas chickens produce more β-ZEL. Given its high affinity for the ER, α-ZAL has been used in cattle industry in the USA as a growth promoter under the name RalGro®, it is nevertheless forbidden in Europe (Metzler et al., 2010). Non-genomic endocrine disruption is caused by the ability of ZEN to disrupt the biosynthesis of steroid hormones by binding to HSDs and more particularly by inhibiting 3α-reduction of steroidal hormones (Fig. 11). This results in an accumulation of some active molecules that can induce reproductive and

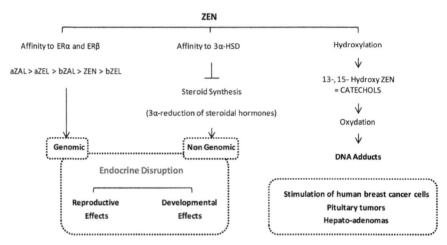

Figure 11. Mechanisms of toxicity of zearalenone (ZEN). Zearalenone is mainly known for its hyperoestrogenic effects, which are attributed to both genomic and non-genomic effects. Zearalenone is also related to carcinogenesis.

developmental toxicity in both females and males (Minervini and Dell'Aquila, 2008). Disruption of the reproductive tract of laboratory and domestic animals is highly dependent on the reproductive status (Minervini and Dell'Aquila, 2008). In female mice reproductive impairment has been related to fibrosis in uterus cystic ducts, in mammary glands and myelofibrosis (Minervini and Dell'Aquila, 2008). In male pigs, zearalenone induces a decrease in serum testosterone, weight of testes and spermatogenesis and also feminization and suppression of libido (Minervini and Dell'Aquila, 2008). In male rats, it has been shown that zearalenone causes inflammation of the prostate gland, testicular atrophy and cysts or cystic ducts in mammary glands (Minervini and Dell'Aquila, 2008). Developmental toxicity is mostly due to the fact that ZEN and its metabolites are able to cross the placental barrier and reach the fetus with the resulting exposure in early stages of life (Metzler et al., 2010) (Fig. 11). As a consequence, there is a reduction of embryonic survival probably caused by alterations of the uterus due to a decrease of the secretion of lutheinizing hormone (LH) and progesterone and to morphological impairments of uterine tissues. No teratogenic effects have been reported for zearalenones (JECFA, 2000).

In addition to these effects, exposure to zearalenone is also associated with stimulation of breast cancer cells in humans (Ahamed et al., 2001), pituitary tumors and hepato-adenomas in mice and rats (Zinedine et al., 2007). Also, symptoms of zearalenone poisoning due to scabby grains are generally similar to deoxynivalenol poisoning (described in the next section) that is nausea, vomiting and diarrhoea. Zearalenone has also been shown to have immunosuppressive effects in pigs *in vivo* (Marin et al., 2013; Pistol et al., 2013). However zearalenone has a low acute toxicity LD_{50} value after oral administration in all animal species tested: pigs and sheep being the most sensitive (Metzler et al., 2010).

Nevertheless, the mechanisms behind these pathologies are still not well understood. Some studies have shown that oxidation of catechols can induce DNA-adduct formation, and there is also evidence of DNA fragmentation and micronuclei production (Zinedine et al., 2007; Metzler et al., 2010) (Fig. 11). Based on current knowledge, zearalenone and its metabolites have been classified in the group 3 by IARC as: "not classifiable as to their carcinogenicity in humans" (IARC, 1993).

Terpene Mycotoxins

Terpenes compounds are derived from the combination of isopentenyl diphosphate (IPP) and dimethyl pyrophosphate (DMAPP) by the action of prenyl transferases, also known as isoprenyl diphosphate synthases. Geranyl diphosphate (GPP) is the result of one molecule of IPP and one molecule of DMAPP, addition of a supplementary IPP forms farnesyl diphosphate (FPP), which is an important precursor of fungal secondary metabolites. Terpenes are not only produced by fungi but they are also important secondary metabolites of bacteria and plants. In the latter, they play a role of pollinator attractors, chemical and physical barrier against herbivores and ovopositing insects (Chen et al., 2011). In addition, terpenes are of great interest in the pharmaceutical industry (taxol, used in treatments against lung, breast and ovary cancers), in the agroalimentary industry (menthol, used as a flavouring agent), in the cosmetics industry (santalols, used as fragances) and as biofuel precursors (farnesene) (Chen et al., 2011). Based on their structure, terpenes can be classified as monoterpenes, which are rare in fungi with the exception of chlorinated monoterpenes that have been isolated from mangrove endophytic fungus *Tryblidiopycnis* sp.; sesterpenes as neomangicols, which are produced by both plants and fungi; triterpenes, which are more frequently produced by plants and rarely reported from microorganisms; tetraterpenes or carotenoids which protect from intense sun radiation (reactive oxygen species and free radicals) (Ebel, 2010); and sesquiterpenes that encompass the family of trichothecenes that will be further described in this section.

Generalities on Trichothecenes

Trichothecenes are the most representative and best studied of terpene mycotoxins. They form a large family of over 150 secondary metabolites produced worldwide by fungi of different and taxonomically unrelated genera such as *Trichothecium*, *Fusarium*, *Stachybotrys*, *Myrothecium* and *Trichoderma* (Rocha et al., 2005). The first trichothecene described was trichothecin, an antifungal isolated from *Trichothecium roseum* cultures (Freeman and Morrison, 1948). However, *Fusarium* species are the most common of fungi trichothecenes-producing. These species are important plant pathogens that particularly contaminate cereals like wheat, barley, oats and maize and cause diseases viz., *Fusarium* head blight disease and stalk and cob rot, incurring heavy crop losses worldwide. *Fusarium* species can be classified in order of decreasing virulence, as follows: *F. graminearum*, *F. culmorum*, *F. sporotrichoides*, *F. langsethiae* and *F. poae*. Trichothecenes are readily encountered in food and feed because they are thermostable, and thus withstand food processing and autoclaving (Eriksen and Pettersson, 2004). After ingestion, they are not degraded in the stomach since they are also stable at neutral and acidic pH (Eriksen and Pettersson, 2004). Once in the intestine, trichothecenes are rapidly absorbed in their upper parts after 30 min of ingestion and are also rapidly excreted through feces and urine after 24 h of ingestion. Therefore, there is little risk for carry-over of the toxins from animals to human through food consumption (Eriksen and Alexander, 1998). Toxicity of trichothecenes has been reported for humans as well as for all animals species tested (Desjardins et al., 1993). The first alerts concerning human susceptibility to trichothecenes arose during the Second World War in the Soviet Union with outbreaks of alimentary toxic aleukia after consumption of grains infected with *F. sporotrichoides* (Desjardins et al., 1993; Bennett and Klich, 2003). *F. sporotrichoides* is a T-2 producing species and these outbreaks were thus associated with this mycotoxin, although there is still not enough evidence to prove it. T-2 toxin is also suspected of having been used as a biochemical weapon in Southeast Asia in the 1980s (Maddox, 1984).

All trichothecenes are sesquiterpenes, which share a common structural backbone made of a tricyclic nucleus, called trichothecene, and usually, a double bond at C-9, 10 and an epoxide at C-12, 13. This epoxide group is considered to be very important for toxicity (Desjardins et al., 1993). Based on differences in the position and number of hydroxylations and also the number and complexity of esterifications, trichothecenes can be classified into 4 groups (Ueno et al., 1973):

Type A: Contrary to the other trichothecenes, these trichothecenes lack a ketide group at C-8 position. The most representative mycotoxins of this group are T-2 toxin, its hydrolyzed derivate, HT-2 and diacetoxyscirpenol (DAS).

Type B: These trichothecenes have a ketide group at C-8 position. Deoxynivalenol (DON), nivalenol (NIV) and their acetylated derivatives (3-acetyldeoxynivalenol, 15-acetyldeoxynivalenol and 4-acetylnivalenol or fusarenon X) are the most occurring and best-characterized mycotoxins of this group.

Type C: In addition to the ketide group, these trichothecenes have an epoxyde group at C-7, 8 position. Crotocin is an example of type C trichothecenes. There is not much data on this type of trichothecenes, which will not be further addressed in this section.

Type D: These trichothecenes contain a macrocyclic ring between C-4 and C-15. Verrucarins and satratoxins are well-known type D trichothecenes.

Types A and B are the most occurring and they are generally produced by *Fusarium* species, whereas types C and D are generally produced by *Myrothecium, Trichothecium* and *Stachybotrys.* Among *Fusarium* species, *F. sporotrichoides, F. poae* and *F. langsethiae* are type A trichothecenes-producing species whereas, *F. graminearum* and *F. culmorum* are rather type B trichothecenes-producing species.

Trichothecenes Biosynthesis

Biosynthesis of trichothecenes has been widely studied since the 1980s, and initially on *F. sporotrichoides* and *F. sambucinum.* Fifteen genes have been identified as *Tri* genes that encode for the so-called TRI enzymes. Contrarily to most of the genes involved in the biosynthesis of mycotoxins, *Tri* genes are not in a cluster, rather they are located at three different loci on different chromosomes (Kimura et al., 2007; Merhej et al., 2011). There is a 12-gene core *Tri* cluster, but also a *Tri1, Tri6* locus and the single *Tri101* locus. Overall, structures and functions of *Tri* genes are well conserved among species. However, there exist some differences, such as gene inactivation, which explain why some species produce one or another trichothecene (Kimura et al., 2007). The biosynthetic pathway of T-2 and fusarenon X has mostly been used as models to study trichothecene biosynthesis and evolution of *Tri* genes because the production of other trichothecenes, such as deoxynivalenol or nivalenol, seem to have arisen due to the inactivation of some *Tri* genes involved in the biosynthesis of T-2 or fusarenon X (Kimura et al., 2007). The first step in the biosynthetic pathway is common to all trichothecenes and consists of cyclization of farnesyl pyrophosphate (FPP) into trichodiene under the action of an enzyme encoded by *Tri5* (Fig. 12). Then, 9 reactions orchestrated by TRI4, TRI101, TRI11 and TRI3 result in the production of calonectrin. From that point, the biosynthetic pathways of T-2 and NIV diverge from that of DON. The T-2 and NIV biosynthetic pathways share the same enzymatic steps until the synthesis of 3, 4, 15-triacetoxyscirpenol. Then, they differ with the action of species-specific genes *FsTri1, FsTri16* and *FsTri8* which lead to T-2 production in *F. sporotrichoides. FgTri1,* and *FgTri8* allow the synthesis of nivalenol in *F. graminearum.* Interestingly, FgTRI1 is able to perform a double hydroxylation on 3, 4, 15-triacetoxyscirpenol on the contrary of FsTRI1 (McCormick et al., 2004). FgTRI1 is also able to hydroxylate calonectrin directly to produce 3, 15-acetyldeoxynivalenol, which is then tranformed into 3-acetyldeoxynivalenol (3-ADON) or 15-acetyldeoxynivalenol (15-ADON) by the action of FgTRI8. Finally, these are then deacetylated into DON. Notably, FgTRI7 and FgTRI13 are not operative in DON-producing species (Brown et al., 2004). Production of 3-ADON or 15-ADON by FgTRI8 is strain specific and depends on DNA sequence polymorphisms of Fg*Tri8.* This polymorphism determines different chemotypes of type B-producing *Fusarium* species. Fg*Tri8* DNA sequence in 3-ADON chemotype is particularly different from the sequence in 15-ADON

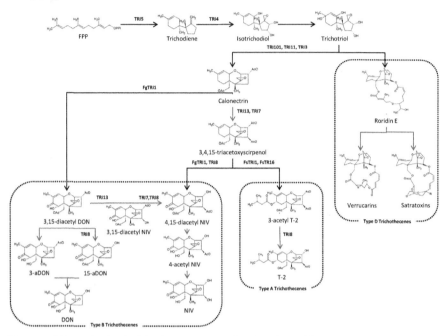

Figure 12. Biosynthesis of trichothecenes. Reproduced with modifications from (Nielsen, 2002; Alexander et al., 2011).

and NIV chemotypes (only 78% homologies), whereas 15-ADON and NIV chemotypes have a sequence homology of 85% (Alexander et al., 2011).

Biosynthesis of macrocyclic trichothecenes (type D) diverges from that of the other trichothecenes after trichotriol production (Nielsen, 2002) (Fig. 12). However, this biosynthetic pathway is less well known than those of types A and B, and more particularly there is little knowledge on the Tri genes involved in it. After trichotriol, different steps result in the synthesis of roridin E, which is the first macrocyclic metabolite (Nielsen, 2002). Verrucarins and satratoxins derive from roridin E.

Other *Tri* genes play a role in the regulation of trichothecenes biosynthesis. For example, TRI6, which is a Cys2His2 zinc finger transcription factor, is a central pathway specific regulator in both type A- and type B-producing *Fusarium* species (Merhej et al., 2011). TRI10 is also a pathway specific regulator which is upstream TRI6 but plays a more important role in *F. sporotrichoides* than in *F. graminearum*. On the contrary, Tri14, another regulator, reduces pathogenicity only in *F. graminearum* but not in *F. sporotrichoides*.

Toxicity and Mode of Action

Trichothecenes are primarily phytotoxins responsible for wilting, chlorosis and necrosis, which result in growth retardation, inhibition of seedling and inhibition of plant regeneration. It has been shown that production of trichothecenes favors fungal development on the plant, but it is not essential for the appearance of disease symptoms (Boenisch and Schafer, 2011). Trichothecenes have also been shown to be toxic to humans and to all tested animals. The mechanism of such toxicity lies on multiple inhibitory effects on the primary metabolism of eukaryotic cells and mainly on the synthesis of proteins (Rocha et al., 2005). Trichothecenes are also responsible for the inhibition of DNA and RNA synthesis as well as inhibition of mitochondrial activity and modulation of mitogen-activated protein kinases (MAPK) (Pestka et al., 2004). The inhibition of protein synthesis is due to the binding at the active site of a peptidyl transferase on the 60S ribosomal subunit. This results in, what is known as, ribotoxic stress response, which triggers most of the other inhibitory effects caused by trichothecenes. These effects at the molecular level generate alterations at the cellular level such as cell proliferation, cytokine production and modulation of the immune system, increased susceptibility to infectious diseases, growth retardation and reproductive disorders (Oswald et al., 2005; Rocha et al., 2005). The effects induced by trichothecenes are highly dependent on the dose and frequency of exposure. As an example, they can be immune-stimulatory at low doses and short exposure times by activating inflammatory genes; but, at high dose exposure, they can also trigger immunosuppression by inducing apoptosis of leukocytes (Pestka et al., 2004). As already mentioned, toxicity of trichothecenes is due to the presence of an epoxy group; therefore, de-epoxydation appears as an efficient detoxifying mechanism (Eriksen and Pettersson, 2004). Differences in the ability to de-epoxydate trichothecenes explain the differences in susceptibility to trichothecenes and also why the pig is one of the most sensitive animal species (Oswald et al., 2005).

In the last few years, there is an increasing number of studies showing the co-occurrence of trichothecenes in corn, wheat and barley (SCOOP, 2003). That is why, there is also an increasing number of studies that pay attention to the effects of multicontaminations, and more particularly of the most frequent DON and NIV co-occurrence (Alassane-Kpembi et al., 2013; Alassane-Kpembi et al., 2015).

A more detailed description of type A, B and D trichothecenes is given below.

Type A Trichothecenes

T-2 Toxin

T-2 toxin is considered to be the most toxic of the trichothecenes family produced by *Fusarium* species, although it is not very common in occurrence (Rocha et al., 2005; Li et al., 2011). It is produced mainly by species of the section *Sporotrichiella*, such as *F. sporotrichoides*, *F. poae* and *F. langsethiae*, which are mainly found in cold climate regions or during wet storage conditions.

As other trichothecenes, T-2 can induce a large variety of effects like weight loss, emesis, disruption of hematological parameters, immunosuppression, damages in the liver and stomach and ultimately death. These are because of inhibition of protein synthesis (Rocha et al., 2005; Li et al., 2011). This mycotoxin is also related to embryo and fetal toxicity in pigs (Rocha et al., 2005).

Among 20 metabolites related to T-2 identified so far (Wu et al., 2011), HT-2 is the major metabolite of T-2. It is both produced by *F. sporotrichoides, F. poae* and *F. langsethiae* but by mammalian cells from the metabolization of T2 toxin. It is produced mainly in the liver and in the gut by the hydrolysis of T-2 by a non-specific carboxyesterase. HT-2 is further metabolized into 3-hydroxy HT-2 and 4-hydroxy HT-2 by the action of a CYP450 in the liver. Toxicity of these hydroxylated metabolites has been shown to be greater than T-2 and HT-2. 3-hydroxy HT-2 is excreted free in urine or in its glucuronide form (Wu et al., 2011). HT-2 can also be hydrolyzed into T-2 tetraol, which is less toxic than T-2 and HT-2 and is easily excreted in urine. As for the other trichothecenes, de-epoxydation is a major detoxification mechanism for T-2. De-epoxy T-2 is especially produced in rodents, ruminants and swine (Wu et al., 2011). Although, T-2 metabolites are well known, knowledge on the toxicity of these compounds is rather limited and deserve more attention (Li et al., 2011).

Type B Trichothecenes

Type B trichothecenes are the most common contaminants of cereal grains in temperate regions of the world. A large scale data survey indicates that DON, 15-ADON, NIV, Fusarenon X (FX) and 3-ADON are present in 57, 20, 16, 10 and 8%, respectively, of food samples collected in the European Union (SCOOP, 2003). Type B-producing *Fusarium* species can be classified in different chemotypes according to the set of type B trichothecenes they produce.

Nivalenol and Fusarenon X

Nivalenol is mainly produced by *Fusarium* species of the section *Discolor* like *F. crookwellense* and *F. poae* and to a lesser extent by *F. graminearum* and *F. culmorum* (Eriksen and Alexander, 1998).

Phytopathogenicity associated to this mycotoxin is low and, therefore, apparently healthy cereal grains may contain high levels of nivalenol. Highest concentrations of nivalenol can be found in oats, maize, barley and wheat cereals and their by-products (EFSA, 2013).

Nivalenol is responsible for reduction of feed intake, growth retardation, different nephropathies, immunotoxicity and haematotoxicity (Rocha et al., 2005; Wu et al., 2011; EFSA, 2013). In addition, nivalenol is 10 times more toxic to lymphocytes than DON due to the presence of an hydroxyl group at the position C-4, that is absent in the structure of DON (Ueno, 1985). In mice, NIV can be transferred to fetus by placenta and to suckling mice through milk (Wu et al., 2011).

Fusarenon X (FX), which is produced by acetylation of NIV at C-4 position, is the major metabolite of this mycotoxin in mice, chicken and ducks (Wu et al., 2011). After ingestion, FX is rapidly metabolized back to NIV in liver and kidney, and it can be then excreted in urine (Wu et al., 2011). In ruminants, rats and swine, NIV is de-epoxydated in the rumen or in the lower parts of the gastrointestinal tract and its toxicity is thus drastically decreased (Eriksen and Pettersson, 2004).

Deoxynivalenol and its Acetylated Derivatives

Although, it is not one of the most toxic trichothecenes, deoxynivalenol is of great concern in terms of public health because it is one of the most occurring mycotoxins that can be found in cereals (Wu et al., 2011). It is mainly produced by *Fusarium* species of the section *Discolor* like *F. graminearum*, *F. culmorum*, *F. crookwellense* and *F. sambucinum*. Its acetylated derivatives, 3-ADON and 15-ADON, that are directly formed by fungi, are most generally found together with DON in contaminated samples, although the proportions in which they are present vary depending on the strain chemotype.

In plants, DON is considered as a virulence factor during the infection of wheat spikes, which enables fungal colonization of the wheat rachis after infection of the florets (Proctor et al., 1995; Desjardins et al., 1996; Boenisch and Schafer, 2011). In humans and animals, DON is also known as vomitoxin due to its emetic effects after acute exposures (Vesonder et al., 1973). In addition to that, symptoms caused by acute exposure are similar to those observed after ionization: abdominal pain, salivation, diarrhoea, vomiting, leucocytosis and gastro-intestinal haemorrhage (Pestka and Smolinski, 2005). Chronic exposure to DON causes feed refusal (Maresca, 2013), due to alterations of seritogenic activity of both the peripherous and central nervous system (Fig. 13A) (Razafimanjato et al., 2011). Other effects related to chronic ingestion of DON are weight loss, growth retardation, intestinal irritation, disruption of immune response and death (Rocha et al., 2005).

Toxicity of DON is caused by the same molecular mechanisms as the other trichothecenes (Rotter et al., 1996a). Immune modulation induced by DON has been extensively studied and it has been shown that inflammation arises from MAPK activation, followed by activation of the prostaglandin-endoperoxide synthase 2 (Cox-2) and a consecutive increase of prostaglandins which triggers the production of inflammatory cytokines (Moon and Pestka, 2002) (Fig. 13A). Recently, DON has been shown to maintain inflammation by favoring the Th-17-related immune response, which could be associated to chronic inflammatory intestinal diseases (Maresca and Fantini, 2010; Cano et al., 2013b). Low doses of this toxin can induce morphological and histological

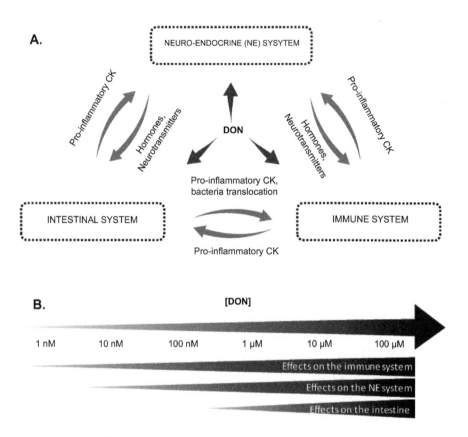

Figure 13. Effects of deoxynivalenol (DON). In humans and animals, DON affects the intestinal system, the immune system and the neuro-endocrine (NE) system. There is an important interplay between these three systems as depicted in the figure (A). For instance, feed refusal induced by this mycotoxin is caused by alterations of the seritogenic activity of both the peripherous and central nervous system. The production of pro-inflammatory cytokines (CK) is also modulated by the presence of DON and this has repercussions on the immune response, the intestine and the NE system. The three systems have different sensitivities to DON (B). The immune system appears to be the most sensitive followed by the NE system and finally by the intestinal system. Reproduced with modifications from (Maresca, 2013).

impairments of the intestinal epithelial cells, and also alter the intestinal permeability (Pinton et al., 2012). Toxicity of DON thus has repercussions on the neuro-endocrine system, the intestinal system and the immune system. Based on DON concentrations, the immune system appears to be the most sensitive, followed by the neuro-endocrine system and finally by the intestinal system (Fig. 13B).

Detoxification of DON is principally due to de-epoxydation processes that lead to de-epoxy DON, known as DOM-1. DON de-epoxydation is essentially performed by bacteria of the genus *Eubacterium*, in the gut of ruminants, rodents, poultry and swine (Volkl et al., 2004; Wu et al., 2011). Results of the BrdU bioassay, which detects the incorporation of 5-bromo-2'-deoxyuridine (BrdU) during DNA replication as an indicator of cell proliferation, showed that DOM-1 is 55 times less toxic than DON (Eriksen and Pettersson, 2004). De-epoxydation of DON has been estimated to be 95% efficient in ruminants, 80% in poultry but only 5% in swine, which explains the high susceptibility of this species to DON exposure (Turner et al., 2010; Yu et al., 2010).

Type D Trichothecenes

Satratoxins

Stachybotryotoxicosis was first described in the 1930s in Russia, and it became a serious issue during World War II because it decimated the horses of the Russian army (Nielsen, 2002). Satratoxins and macrocyclic trichothecenes stand among the most toxic mycotoxins of the trichothecenes family, being 10 times more cytotoxic than type A trichothecenes (Pestka and Forsell, 1988). They are mainly produced by the genus *Myrothecium* and *Stachybotrys* and more particularly, by *Stachybotrys chartarum* (Nielsen, 2002; Amuzie et al., 2010). They are essentially encountered in moldy straw in Eastern and Northern Europe and also in damp and moldy buildings, in which they are associated with damp building-related illnesses (Nielsen, 2002; Amuzie et al., 2010).

Stachybotryotoxicosis most typically occurs with inflammation of the stomach, and the mouth as well as necrosis of some tissues. This is followed by a decreasing number of thrombocytes and thus, a decrease in blood clotting and of leukocytes. Less frequently, satratoxins triggers nervous disorders, which result in loss of reflexes, elevated body temperatures, and altered cardiac action (Nielsen, 2002). Additionally, satratoxins are readily transported by spores and they can thus be taken up by the respiratory tract of exposed individuals. It has been shown that they are then distributed throughout the body, although a large part remains in the nasal cavity where they induce inflammation and apoptosis of nasal olfactory sensory neurons and from where they are slowly cleared (Amuzie et al., 2010).

Additionally, recent studies on satratoxin H showed that this mycotoxin induces chromosome breaks in the micronucleus assay (Nusuetrong et al., 2012). Furthermore, it was also observed that it increases the level of phosphorylation of histone H2A, indicating that it causes DNA double-stranded breaks in PC12 cells, and suggesting its genotoxicity (Nusuetrong et al., 2012).

Non-Ribosomal Peptide Mycotoxins

The production of peptides with a non-ribosomal origin was discovered in the 1960's by analysis of cell extracts in which peptide had been synthesized in the presence of RNases and ribosome inhibitors (Finking and Marahiel, 2004). Non-ribosomal peptides differ from ribosomal peptides in that they are not involved in the primary metabolism and, therefore, there are no proofreading mechanisms involved in their biosynthesis. Also, several hundred substrates, used as building blocks for non-ribosomal peptides, have been identified so far, whereas only 20 amino acids are known to date for ribosomal peptides biosynthesis. As a result, non-ribosomal peptides present a much larger variety of structures and thus, a broader range of biological activities (Finking and Marahiel, 2004). Microorganisms such as gram + bacteria of the genera *Actinomycetes* and *Bacilli* and filamentous fungi are the main producers of this type of peptides. Additionally, enzymes similar to non-ribosomal peptide synthases (NRPS) have also been identified in higher eukaryotes like the amino adipic acid semialdehyde dehydrogenase (U26) in mice (Finking and Marahiel, 2004). NRPS are organized in different modules that incorporate different amino acids into the polypeptide chain. Each module is sub-divided into domains that correspond to different enzymatic units. At least three domains are necessary to incorporate one amino acid: (1) the A-domain, which determines the amino acid that will be inserted in the polypeptide chain and activates its amino acid or hydroxyl acid; (2) the T or PCP domain, which is a thiolation or peptidyl-carrier protein that transports the activated unit between the active sites of the domains; and (3) the C-domain, which is a condensation domain that catalyses the formation of the peptide bond (C-N) between the elongated chain and the new amino acid (Gallo et al., 2013). Other domains may intervene in the biosynthesis of non-ribosomal peptides by adding different special side-features to the peptidic chain like fatty acids (in fengicin and surfactin), non proteinogenic amino acids (in cyclosporine A and gramicidin S), carboxy acids (in enterobactin and myxochelin A), heterocyclic rings (in epothilone and bleomycin A2) and modified amino acids (in thaxtornin A and anabaenopeptide 90-A) (Finking and Marahiel, 2004). A list of the so-called tailoring domains is shown in Table 2. NRPS domains share highly conserved sequences known as core-motifs that can be identified at the protein level and that allow to determine a putative function for unknown NRPS (Finking and Marahiel, 2004). In fungi and contrary to bacteria, only one NRPS is generally involved in the biosynthesis of one secondary metabolite and, therefore, fungal NRPS contain several catalytic domains that are usually larger enzymes (Finking and Marahiel, 2004). This section will introduce the two important and most studied families of toxic non-ribosomal peptides.

Table 2. Tailoring domains of non-ribosomal peptide synthases.

Tailoring domain		Function
Epimerization domain	E	Transformation of L-amino acid into D-amino acid
Dual/epimerization domain	E/C	Epimerization and condensation
Reductase	R	Reduction of final peptide
Methylation	MT	N-methylation of the amide nitrogen
Cyclization	Cy	Formation of heterocyclic rings
Oxydation	Ox	Oxydation of thiazoline ring, formation of aromatic thiazol at the C-terminus of A-domain or downstream of the PCP domain
Thioesterase domain	TE	Release of final peptide through cyclization or hydrolysis

Ergot Alkaloids

Ergot alkaloids (EA) constitute a very large family of indole-derived fungal metabolites that were first isolated from sclerotia of *Claviceps purpurea*. Sclerotia are dark structures corresponding to a resting stage of the fungus that replace rye grains essentially but also other cereals like wheat. Sclerotia can be directly eaten by grazing animals, harvested with the ripened crop and consequently contaminate food and feed, or they can fall on the ground where they form sexual spores. These spores can remain on the soil but they can also become airborne and infect flowers. Development of sclerotia on the flowers results in the formation of asexual spores that can either spread and infect other plants or form new sclerotia (Eadie, 2003). Sclerotia of *Claviceps purpurea* have been shown to contain many pharmacologically active and inactive substances among which are over 40 different ergot alkaloids.

Other species related to the family *Clavicipitaceae* like *Claviceps paspali* and *C. fusiformis* produce ergot alkaloids. Ergot alkaloids have also been isolated from other genera such as *Penicillium* and *Aspergillus* species and also from *Epichloe* and *Balansia*, that are grass endophytes responsible for severe livestock intoxications (Tudzynski et al., 2001). Nevertheless, the genus *Claviceps* has been the most generally studied and used for biotechnological purposes.

The pharmacological interest of ergot alkaloids lies on their structural homologies with different neurotransmitters and their subsequent effect on the central nervous system (Tudzynski et al., 2001). Historically, these mycotoxins were associated to St. Anthony's fire in the Middle Ages, but they have also been extensively used in obstetrics as anti haemorrhagics during childbirth or abortion (Tudzynski et al., 2001; Eadie, 2003). In the twentieth century, some of these active substances, like ergotamine, were purified, making their use in medicine safer. Additionally, new synthetic alkaloids have been produced by genetic engineering of their biosynthetic pathways by insertion of new enzymes with different substrate specificities or additional chemical conversions (Tudzynski et al., 2001).

Ergot alkaloids are 3,4-substituted indol derivatives with a tetracyclic ergoline ring (Tudzynski et al., 2001). Based on their structure, they can be classified as peptides (ergopeptides and ergopeptames), amides (alkylamides or small peptides) or clavines (clavine alkaloids) and can be synthesized from paspalic acid, L-isolysergic acid or D-lysergic acid (Eadie, 2003). Metabolites deriving from L-isolysergic acid (names ending with –inine) are not biologically active unless dissolved in an aqueous solution in which they isomerize with active derivatives of D-lysergic acid (names ending with –ine) (Eadie, 2003). D-lysergic acid derivatives are simple amino alcohol (like ergotamine) or short peptide chain (like ergometrine) attached to the ergoline nucleus in amide linkage via a carboxy group at C-8 (Tudzynski et al., 2001). D-lysergic acid derived EA have a wide variety of activities depending on the substituents attached to the carboxy group at C-8 of the ergoline ring. Different substituents will determine the affinity to a different neuro-receptors like the noradrenaline receptor, the dopamine receptor or the serotonin receptor from which they will give a different effect (Tudzynski et al., 2001). Structural alteration of the ergoline ring or addition of side chains at C-8 can have drastic effects on the affinity to neuro-receptors. For instance, hydrogenation of ergotamine leads to an important increase of its adrenolytic effect (Tudzynski et al., 2001).

Biosynthesis of ergot alkaloids starts with the formation of the ergoline ring, which has been mostly studied in *C. purpurea*, *C. fusiformis* and *C. paspali* (Tudzynski et al., 2001) (Fig. 14). The ergoline ring is formed with the condensation of a molecule of L-tryptophan with a molecule of 3.3-dimethylallyl pyrophosphate (DMAPP) by the action of the enzyme dimethylallyltryptophane (DMAT) synthase. This constitutes the carboline skeleton of the ergoline system. DMATs activity is positively regulated by tryptophan but is down regulated by phosphate. A cascade of enzymatic reactions following DMAT synthesis lead to the formation of the other two rings. However, apart from DMAT synthase, there is not much information concerning the other enzymes, due to their high instability (Tudzynski et al., 2001). The next step involves a methyltransferase which methylates the amino nitrogen of DMAT. Then, decarboxylation and closure of the ring C is performed by a chanoclavine-I cyclase. Other enzymes have been described as depicted in Fig. 14A. The formation of the ergoline ring ends after the biosynthesis of paspalic acid. However, some species stop the biosynthesis earlier due to natural mutations or absence of some genes; for instance, *C. fusiformis* stops after biosynthesis of chanoclavine I (Tudzynski et al., 2001). The second part of EA biosynthesis consists in the assembly of the ergot peptides and the L-isolysergic acid or D-lysergic acid amids (Tudzynski et al., 2001). Ergopeptines such as ergotamine, ergoxine and ergotoxine are derived from the binding of the D-lysergic acid to a bicyclic cyclol-lactam structure, which is made of three amino acids by D-lysergic peptide synthase (LPS) Fig. 14B. These amino acids differ among ergopeptines. LPS is formed by two subunits: LPS2, which contains the module for D-lysergic acid activation (Lsa)

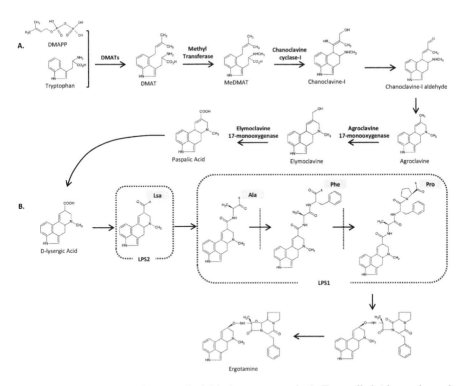

Figure 14. Biosynthesis of Ergot alkaloids (e.g., ergotamine). Ergot alkaloids are formed by an ergoline ring assembled to an amide group. (A) The next step involves a methylation, decarboxylation and closure of the ring C. (B) Then, in the case of ergotamine, the D-lysergic acid is bound to the two subunits of the D-lysergic peptide synthase (LPS): LPS2 which contains the module for D-lysergic acid activation (Lsa) and LPS1, which contains three modules that successively add one of the three amino acids (Alanine, Phenylalanine and Proline). Based on (Tudzynski et al., 2001).

and LPS1, which contains three modules that successively add one of the three amino acids of the bicyclic cyclol-lactam (Alanine, Phenylalanine and Proline, in the case of ergotamine) (Tudzynski et al., 2001).

Different pathologies are associated with the ingestion of ergot alkaloids. The most frequent forms of ergotism are due to repeated exposure to these metabolites. There are two main forms of chronic ergotism that differ geographically and symptomatically. Gangrenous ergotism, which has been more frequent in western regions of the Rhine, is associated with an overdose of ergotamine causing important ischemia in the limbs (leading to the burning sensation described as St. Anthony's fire). Further, vasoconstriction triggers dry gangrene with a mummified effect of the affected limbs. These symptoms are caused by the agonistic effect of ergotamine on the serotonin receptor 5-HT1B/D and 5-HT2, which results in arterial constriction and gangrene (Eadie, 2003). Gangrene is usually related to secondary infections, possibly leading to death (Eadie, 2003). East of the Rhine, the predominant form of ergotism has been convulsive ergotism, which is associated with chronic ingestion of ergoline. This type of ergotism is characterized by involuntary distortion of the trunk and the limbs and also by a delirious, lethargic, manic and hallucinogenic state of the patients. Epileptic episodes related with convulsive ergotism are usually fatal (Eadie, 2003). Convulsive ergotism may have been the cause of the Salem witch trials that happened in 1692 where several women were accused of bewitching young girls presenting symptoms of ergotism (Woolf, 2000). Although some arguments point at ergot contamination of rye grains as the cause of these events, there is still a lot of questioning surrounding these events. Concerning the effects of EA on the central nervous system, it is still not clear whether or not EA like ergotamine are able to cross the blood brain barrier. Different hypothesis have been suggested for this matter. Vitamin A deficiency would lead to an increase of brain permeability, allowing the entry of EA like ergotamine. Some EA like ergocriptine inhibit P-glycoprotein, which also increases brain permeability (Eadie, 2003).

Today, cereals are mechanically cleaned from sclerotia, and this has drastically decreased the human exposure to these mycotoxins (Tudzynski et al., 2001). Animals such as sheep, cows and goats are, however, still at risk, principally due to grazing. Animals are subjected to the similar pathologies as humans.

Gliotoxin

Gliotoxin belongs to the family of epipolythiodioxopiperazines (ETP), which is a class of fungal toxins characterized by the presence of an internal disulphide bridge across the piperazine ring, responsible for its toxicity (Gardiner and Howlett, 2005). Gliotoxin was first described as a metabolite of *Gliocladium fimbriatum*, from which its name is derived. However, production of this mycotoxin is mostly attributed to other fungal genera such as *Aspergillus*, *Penicillium* and *Trichoderma*. In *Aspergillus* species, such as *A. fumigatus*, gliotoxin has been shown to act as a potent virulence factor that promotes tissue colonization (Gardiner et al., 2005; Kwon-Chung and Sugui, 2009). *A. fumigatus* is an important human pathogen responsible for invasive aspergillosis that can be fatal in immunocompromised patients. Interestingly, gliotoxin has been more frequently detected in strains isolated in medical facilities than in environmental isolates; it could, thus, play an important role in infection of hospital patients (Kwon-Chung and Sugui, 2009). Apart from its toxinogenic effects, gliotoxin has been shown to protect *A. fumigatus* strains from peroxy-induced oxidative stress (Gallagher et al., 2012).

Although different studies have addressed gliotoxin biosynthesis, knowledge concerning the enzymes and the intermediary metabolites involved in the pathway is still not complete. In 2005, a putative 12 gene cluster responsible for gliotoxin biosynthesis was identified in *A. fumigatus* based on its homology with the gene cluster of sirodesmin another ETP produced by *Leptosphaeria maculans* (Gardiner and Howlett, 2005). Disruption of some of the genes included in the cluster of *A. fumigatus* confirmed that it was indeed responsible for gliotoxin production (Kwon-Chung and Sugui, 2009). This gene cluster has also been found in *A. terreus* (except for one gene, *GliN*) and *A. flavus* (Kwon-Chung and Sugui, 2009). A biosynthetic pathway has recently been suggested based on the knowledge on the gene cluster and the structure of gliotoxin (Scharf et al., 2012) (Fig. 15). However, some steps still remain unclear. The first step consists in the formation of diketopiperazine (DKP) by condensation of phenylalanine and serine by the action of the NRPS GliP. This NRPS contains the three basic domains of NRPS, adenylation, condensation and thiolation (Scharf et al., 2012). Interestingly, deletion of *GliP* decreases the expression of all the other genes in the cluster but addition of exogenous gliotoxin restores mycotoxin production indicating that gliotoxin is self-regulated (Kwon-Chung and Sugui, 2009; Scharf et al., 2012). The next step involves hydroxylation of DKP into C-hydroxylated DKP by the P450 cytochrome, GliC. Then, introduction of the sulphur atom takes place. This has remained the most intriguing step in gliotoxin biosynthesis until *GliG* was fully characterized in 2011 (Scharf et al., 2011). GliG is a gluthatione-S-transferase that mediates the insertion of the sulphur atom from cysteine or sodium sulphate. Then GliT, a pyridine dinucleotide-dependent oxidoreductase, forms the internal disulfide bridge. Oxygen serves as a co-substrate for reduction and reoxydation and flavin adenine dinucleotide (FAD) serves as a co-factor in the redox reactions (Scharf et al., 2012). GliT is able of keeping gliotoxin in the sulphur-bridge form, which does not induce the production of ROS species or protein conjugates in fungal cell. Consequently, this enzyme has a key role in self-resistance of *A. fumigatus* to this mycotoxin (Scharf et al., 2012). Also, the disulfide bridge may facilitate excretion from fungal cells (Scharf et al., 2012). All the enzymes that have been described so far in the gliotoxin cluster in *A. fumigatus* are listed in Table 3.

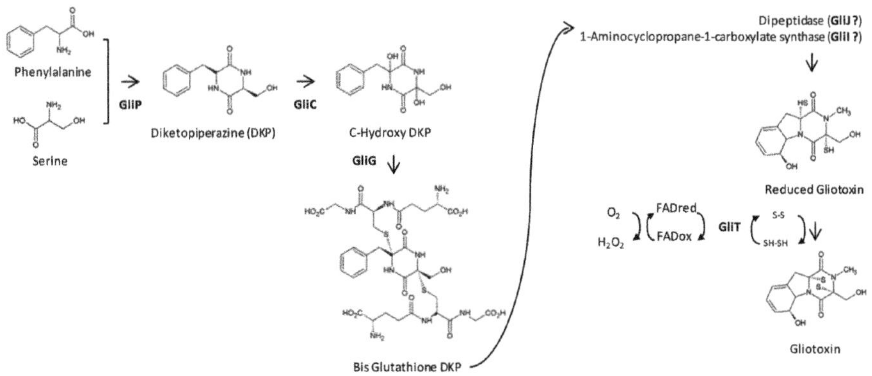

Figure 15. Biosynthesis of gliotoxin. This pathway has been suggested recently by Scharf et al. based on the knowledge on the gene cluster and the structure of gliotoxin (Scharf et al., 2012). Steps marked with a question mark (?) are still unclear.

Table 3. Enzymes involved in gliotoxin biosynthesis.

Enzyme	Comments and proposed function	Reference
GliP	Two-module non-ribosomal peptide synthetase (dioxopiperazine synthase) involved in synthesis of the dipeptide	(Gardiner et al., 2005)
GliT	Thioredoxin reductase (dithiolpiperazine oxidase); formation of disulphide bond	(Gardiner et al., 2005)
GliC, GliF	Cytochrome P450 monooxygenase	(Gardiner et al., 2005; Kwon-Chung and Sugui, 2009)
GliI	High similarity to amino cyclopropane carboxylate synthases	(Gardiner et al., 2005)
GliJ	Dipeptidase domain; unknown function	(Gardiner et al., 2005)
GliM	*O*-Methyl transferase domain; unknown function	(Gardiner et al., 2005)
GliN	Methyl transferase domain; unknown function	(Gardiner et al., 2005)
GliG	Glutathione *S*-transferase domain; encodes enzyme that makes/breaks C–S bonds	(Gardiner et al., 2005; Scharf et al., 2012)
GliZ	Zinc finger transcription factor	(Kwon-Chung and Sugui, 2009)
GliA	Transporter	(Kwon-Chung and Sugui, 2009)
GliL	Aminocyclo-propane carboxylic acid synthase	(Kwon-Chung and Sugui, 2009)
GliK	Hypothetical protein	(Kwon-Chung and Sugui, 2009)
GliH	Conserved hypothetical protein	(Scharf et al., 2012)

The mode of action of gliotoxin has been extensively described in animal cell cultures. It is directly related to the presence of the disulphide bridge which triggers: (1) conjugation to proteins with susceptible thiol residues and subsequent inactivation of the protein by catalysis of the internal sulphide bond, and (2) generation of reactive oxygen species (ROS) such as superoxide and hydrogen peroxide via redox cycling between the reduced (dithiol) and the oxidized (disulphide) forms of gliotoxin (Gardiner et al., 2005). All of this results in immunosuppression by inhibition of NF-kB, phagocytosis by macrophages and apoptosis of macrophages and monocytes (Kwon-Chung and Sugui, 2009). Apoptosis is triggered by activation of BAK, which is a member of Bcl-2 family involved in the initiation of apoptosis by induction of ROS production. Consecutively, cytochrome c and other mitochondrial apoptosis-inducing factors are released resulting in caspase induction and apoptosis (Kwon-Chung and Sugui, 2009). Apoptosis may turn to necrosis with a concentration of gliotoxin above 10 µM *in vitro* (Gardiner et al., 2005). Neutrophils are also important targets of gliotoxin although this population of immunologically active cells is resistant to the apoptosis induced by this mycotoxin (Kwon-Chung and Sugui, 2009). Additionally, gliotoxin is responsible for mitochondrial damage by inhibiting ATP synthesis, and thus triggering hyperpolarization of the mitochondrial membrane. This affects the electrophoretic mobility of the adenine nucleotide transporter (ATN), an important mediator of apoptosis. By affecting the mitochondrial function, gliotoxin may, therefore, induce apoptotic cell death (Gardiner et al., 2005). Mitochondrial calcium influx is also disrupted by gliotoxin and this has been shown to cause thymocyte necrosis (Gardiner et al., 2005). In cases of invasive aspergillosis, gliotoxin enhances colonization of the respiratory mucosa by *A. fumigatus* by damaging the ciliated respiratory epithelium (Gardiner et al., 2005). Finally, gliotoxin inhibits RNA replication by inhibiting the reverse transcriptase and it has hence been used as a therapeutic agent for diseases such as cancer (Gardiner et al., 2005).

Conclusions and Future Need

As is evident this review, there is great diversity in the biosynthesis, chemical structure and toxicity of mycotoxins. In spite of enough efforts put in by the researchers from all over the world, only a limited number of mycotoxins and fungal secondary metabolites are known today, especially as compared to the large variety of species and the knowledge on the richness of their genome.

The biological role of secondary metabolites and mycotoxins for fungi still remains unknown to fungal microbiologists. Most of the scientists agree that secondary metabolites play an important role in the pathogenicity of fungi (Calvo et al., 2002; Fox and Howlett, 2008; Hof, 2008). But, a difference has been suggested between metabolites that act as pathogenicity factors or destructive weapons in hostile environments and metabolites that act as virulence factors, which are produced in order to obtain direct benefits in terms of self defence and host invasion (Fox and Howlett, 2008). However, fungal secondary metabolites also act as protectants against other microorganisms (Reverberi et al., 2010). For instance, penicillic acid produced by *Penicillium* species inhibits quorum-sensing systems of bacteria and also, production of secondary metabolites by *A. nidulans* protects from fungivores (Rohlfs et al., 2007). A string of studies has showed the relationship between regulatory mechanisms of fungal secondary metabolism and resistance against fungivorous grazers (Rohlfs, 2015). Secondary metabolites, such as aflatoxins, also help maintain the oxidative state inside fungal cells, preventing harmful effects of endogenous oxidants. In addition, a close relationship has been established between the production of secondary metabolites and fungal development and more precisely, sporulation (Calvo et al., 2002). Production of secondary metabolites is highly dependent, not only on external signals from the environment like light, temperature and humidity, etc., but also on the signals from the host cells and rival microorganisms. Therefore, a better understanding of the interactions that take place between these signals and production of secondary metabolites, including mycotoxins, would help to understand the function of these compounds and to design new techniques to control and limit the their production.

While, aflatoxins and trichothecenes have been extensively studied, there is still a lot of information missing regarding the other mycotoxins. Indeed, biosynthetic pathways, modes of action and carcinogenicity would deserve more attention. Patulin biosynthesis is also not fully understood. A genetic cluster has clearly been identified and a putative function has been attributed to the different genes of the cluster but for most of them, this function has not been confirmed experimentally. More data should also be gathered concerning carcinogenicity of mycotoxins since in most cases, there is not enough evidence to conclude on their carcinogenicity as is the case of fumonisin B1, ochratoxin A, patulin and zearalenone.

Given their toxicological effects and occurrence in food and feed, assessment of the risks generated by mycotoxins for consumers should be taken very seriously. Proper risk assessment requires the effective and precise detection of these toxins. In most cases, only analyses targeting a limited number of known mycotoxins are performed, although great efforts are currently made to maximize the number of detected metabolites (Varga et al., 2013). Untargeted methods to analyze fungal metabolomes could lead to the discovery of new secondary metabolites potentially toxic for humans, animals or plants. One possibility for untargeted methods is to combine high resolution mass spectrometry and double isotopic labeling, as was done recently for the analysis of the metabolome of *A. fumigatus*, resulting in the identification of new secondary metabolites (Cano et al., 2013a). This kind of studies should be carried out on different fungal species, especially species belonging to the genera *Aspergillus*, *Penicillium* and *Fusarium*, which are the main producers of secondary metabolites.

However, many factors may hinder proper mycotoxin-related risk assessment. For instance, the majority of food and feed products undergo different transformation processes such as fermentation and baking before consumption. In some cases, these techniques degrade mycotoxins thus decreasing toxicity; but in other cases, the situation is worsened due to the metabolization of mycotoxins into more toxic metabolites or due to the release of mycotoxins trapped in the raw materials and that had not been detected in previous analysis. In all cases, the toxicity may be increased by metabolic processes taking place in the humans or animals after ingestion of contaminated commodities. Another important aspect of mycotoxin detection that has just started receiving attention is the issue of masked mycotoxins. There are two forms of masked mycotoxins: (1) mycotoxins covalently or non-covalently bound to the vegetal matrix and that are not extractable unless liberated and (2) conjugated mycotoxins, which can only be detected with appropriate methods (Berthiller et al., 2013). Masked mycotoxins have become an important challenge since their presence leads to an underestimation of the real mycotoxin content and their related risk for the consumer. Metabolization can occur as phase I reactions (hydrolysis or oxidation) in the case of lipophilic compounds or as phase II reactions (conjugation to glucose, malonic acid or glutathione) in the case of hydrophilic compounds. The final aim of these reactions is the detoxification and excretion of these toxic metabolites although, in some cases, the intermediary metabolites might be as toxic as or even more toxic than the parent compound. Trichothecenes are a good example of glucose conjugation resulting in decreased toxicity. DON-glucoside (D3G), FX-glucoside, NIV-glucoside and 3-O-glucosides of T-2 and HT-2 have been detected in contaminated cereal samples with relative proportions around 20%. Zearalenone also undergoes conjugation transformations and it has been detected in its glucoside form (Z14G) and in its sulfate form (Z14S). Fumonisins are good examples of mycotoxins covalently bound to the matrix by binding to hydroxyl groups of carbohydrates and to amino groups of amino acids. Despite a lower toxicity of the substances investigated so far, an increased risk for the consumer is not to be ignored due to an increased bioavailability and partial reactivation of masked mycotoxins in mammalian cells. To date, there is little information available concerning toxicokinetics and toxicodynamics of masked mycotoxins that would allow a proper risk assessment of their presence in food commodities (Berthiller et al., 2013).

Risk characterization should also take into account that: (1) the same fungal species may produce several mycotoxins, (2) the same product may be contaminated by different fungal species and (3) our diet is composed of a variety of food commodities. Therefore, co-exposure to several mycotoxins is the most frequent scenario. However, there is not much information concerning the toxicity of such interactions as reviewed by Grenier et al. (Grenier and Oswald, 2011). Studying toxicity of multi-contaminations is a complex matter in which the natures of the toxins, the concentrations and also the ratios in the food/feed commodities have to be considered. In addition, risk characterization should also consider the different sensitivity of each species as well as individual differences, related to the sex, the age and the targeted organ as well as on the dose and frequency of exposure.

Based on the previous considerations, different regulations have been enacted worldwide in order to establish maximal permissible contents of some mycotoxins in food/feed commodities as well as maximal tolerable daily intake (TDI) doses. These efforts to protect humans and livestock are not only lead by governmental agencies but also by intergovernmental agencies. Risk assessment is usually performed by intergovernmental agencies such as the Joint FAO/WHO Expert Committee on Food Additives (JECFA) and the European Food Safety Authority (EFSA) and so on. Then, based on these evaluations, risk management is lead by governmental structures like the European Commission. For instance, in Europe the Commission regulation 1881/2006 sets maximum levels in food for aflatoxins (B1+ B2+ G1+ G2), aflatoxin M1, DON, fumonisins (B1+ B2), OTA, patulin, T2, H-T2 and zearalenone. Other regulations have been implemented for animal feed such as the directive 2002/32/EC for AFB1 and rye ergots, the recommendation 2006/576/EC for DON, fumonisins (B1+ B2), OTA and zearalenone and the recommendation 2013/165/EU for T-2 and HT-2. Finally, the risks related to mycotoxins are different around the globe depending on the legislations that are applied in each country, the handling and storage conditions of food/feed commodities as well as the combination with other pathological conditions such as malnutrition, malaria and hepatitis (Omar, 2013).

In a nutshell, the issue of mycotoxins is quite complex and it has to be analyzed from several points of view including fungal biology, its environment (physico-chemical and biological), and then the transfer of mycotoxins to food/feed commodities and their possible transformations and interactions before reaching the consumers. Although the articles included in this chapter depict the great progress concerning each of the different aspects related to mycotoxins, a great deal still remain largely unknown.

References

Abbas, H.K., Shier, W.T., Seo, J.A., Lee, Y.W. and Musser, S.M. 1998. Phytotoxicity and cytotoxicity of the fumonisin C and P series of mycotoxins from *Fusarium* spp. fungi. Toxicon, 36: 2033–2037.

Adjovi, Y.C.S., Bailly, S., Gnonlonfin, B.J.G., Tadrist, S., Querin, A., Snanni, A., Oswald, I.P., Puel, O. and Bailly, J.D. 2014. Analysis of the contrast between natural occurrence of toxinogenic *Aspergillii* of the *Flavi* section and aflatoxin B1 in cassava. Food Microbiol., 38: 151–159.

Ahamed, S., Foster, J.S., Bukovsky, A. and Wimalasena, J. 2001. Signal transduction through the Ras/Erk pathway is essential for the mycoestrogen zearalenone-induced cell-cycle progression in MCF-7 cells. Mol. Carcinog., 30: 88–98.

Alassane-Kpembi, I., Kolf-Clauw, M., Gauthier, T., Abrami, R., Abiola, F.A., Oswald, I.P. and Puel, O. 2013. New insights into mycotoxin mixtures: The toxicity of low doses of Type B trichothecenes on intestinal epithelial cells is synergistic. Toxicol. Appl. Pharmacol., 272: 191–198.

Alassane-Kpembi, I., Puel, O. and Oswald, I.P. 2015. Toxicological interactions between the mycotoxins deoxynivalenol, nivalenol and their acetylated derivatives in intestinal epithelial cells. Arch. Toxicol., 89: 1337–1346.

Alexander, N.J., McCormick, S.P., Waalwijk, C., van der Lee, T. and Proctor, R.H. 2011. The genetic basis for 3-ADON and 15-ADON trichothecene chemotypes in *Fusarium*. Fungal Genet. Biol., 48: 485–495.

Alizadeh, A.M., Rohandel, G., Roudbarmohammadi, S., Roudbary, M., Sohanaki, H., Ghiasian, S.A., Taherkhani, A., Semnani, S. and Aghasi, M. 2012. Fumonisin B1 contamination of cereals and risk of esophageal cancer in a high risk area in northeastern Iran. Asian Pac. J. Cancer Prev., 13: 2625–2628.

Amuzie, C.J., Islam, Z., Kim, J.K., Seo, J.H. and Pestka, J.J. 2010. Kinetics of satratoxin G tissue distribution and excretion following intranasal exposure in the mouse. Toxicol. Sci., 116: 433–440.

Artigot, M.P., Loiseau, N., Laffitte, J., Mas-Reguieg, L., Tadrist, S., Oswald, I.P. and Puel, O. 2009. Molecular cloning and functional characterization of two CYP619 cytochrome P450s involved in biosynthesis of patulin in *Aspergillus clavatus*. Microbiology, 155: 1738–1747.

Ballester, A.R., Marcet-Houben, M., Levin, E., Sela, N., Selma-Lázaro, C., Carmona, L., Wisniewski, M., Droby, S., González-Candelas, L. and Gabaldón, T. 2015. Genome, transcriptome, and functional analyses of *Penicillium expansum* provide new insights into secondary metabolism and pathogenicity. Mol. Plant Microbe Interact., 28: 232–248.

Bayram, O. and Braus, G.H. 2012. Coordination of secondary metabolism and development in fungi: the velvet family of regulatory proteins. FEMS Microbiol. Rev., 36: 1–24.

Bennett, J.W. and Klich, M. 2003. Mycotoxins. Clin. Microbiol. Rev., 16: 497–516.

Bennett, R.A., Essigmann, J.M. and Wogan, G.N. 1981. Excretion of an aflatoxin-guanine adduct in the urine of aflatoxin B1-treated rats. Cancer Res., 41: 650–654.

Bentley, R. and Bennett, J.W. 1999. Constructing polyketides: From Collie to combinatorial biosynthesis. Annu. Rev. Microbiol., 53: 411–446.

Berthiller, F., Crews, C., Dall'Asta, C., Saeger, S.D., Haesaert, G., Karlovsky, P., Oswald, I.P., Seefelder, W., Speijers, G. and Stroka, J. 2013. Masked mycotoxins: A review. Mol. Nutr. Food Res., 57: 165–186.

Bhandari, N. and Sharma, R.P. 2002. Fumonisin B(1)-induced alterations in cytokine expression and apoptosis signaling genes in mouse liver and kidney after an acute exposure. Toxicology, 172: 81–92.

Bhatnagar, D., Ehrlich, K.C. and Cleveland, T.E. 2003. Molecular genetic analysis and regulation of aflatoxin biosynthesis. Appl. Microbiol. Biotechnol., 61: 83–93.

Bianco, G., Russo, R., Marzocco, S., Velotto, S., Autore, G. and Severino, L. 2012. Modulation of macrophage activity by aflatoxins B1 and B2 and their metabolites aflatoxins M1 and M2. Toxicon, 59: 644–650.

Biasucci, G., Calabrese, G., Di Giuseppe, R., Carrara, G., Colombo, F., Mandelli, B., Maj, M., Bertuzzi, T., Pietri, A.A. and Rossi, F. 2011. The presence of ochratoxin A in cord serum and in human milk and its correspondence with maternal dietary habits. Eur. J. Nutr., 50: 211–218.

Boenisch, M. and Schafer, W. 2011. *Fusarium graminearum* forms mycotoxin producing infection structures on wheat. BMC Plant Biology, 11: 110.

Boettger, D. and Hertweck, C. 2012. Molecular diversity sculpted by fungal PKS–NRPS hybrids. Chem. Biol. Chem., 14: 28–42.

Boudra, H., Barnouin, J., Dragacci, S. and Morgavi, D.P. 2007. Aflatoxin M1 and ochratoxin A in raw bulk milk from French dairy herds. J. Dairy Sci., 90: 3197–3201.

Bracarense, A.P., Lucioli, J., Grenier, B., Drociunas Pacheco, G., Moll, W.D., Schatzmayr, G. and Oswald, I.P. 2012. Chronic ingestion of deoxynivalenol and fumonisin, alone or in interaction, induces morphological and immunological changes in the intestine of piglets. Br. J. Nutr., 107: 1776–1786.

Bradshaw, R.E., Jin, H., Morgan, B.S., Schwelm, A., Teddy, O.R., Young, C.A. and Zhang, S. 2006. A polyketide synthase gene required for biosynthesis of the aflatoxin-like toxin, dothistromin. Mycopathologia, 161: 283–294.

Brown, D.W., Butchko, R.A., Busman, M. and Proctor, R.H. 2007. The *Fusarium verticillioides* FUM gene cluster encodes a Zn(II)2Cys6 protein that affects FUM gene expression and fumonisin production. Eukaryot. Cell, 6: 1210–1218.

Brown, D.W., Cheung, F., Proctor, R.H., Butchko, R.A., Zheng, L., Lee, Y., Utterback, T., Smith, S., Feldblyum, T., Glenn, A.E., Plattner, R.D., Kendra, D.F., Town, C.D. and Whitelaw, C.A. 2005. Comparative analysis of 87,000 expressed sequence tags from the fumonisin-producing fungus *Fusarium verticillioides*. Fungal Genet. Biol., 42: 848–861.

Brown, D.W., Dyer, R.B., McCormick, S.P., Kendra, D.F. and Plattner, R.D. 2004. Functional demarcation of the *Fusarium* core trichothecene gene cluster. Fungal Genet. Biol., 41: 454–462.

Brown, D.W., Yu, J.H., Kelkar, H.S., Fernandes, M., Nesbitt, T.C., Keller, N.P., Adams, T.H. and Leonard, T.J. 1996. Twenty-five coregulated transcripts define a sterigmatocystin gene cluster in *Aspergillus nidulans*. Proc. Natl. Acad. Sci. USA, 93: 1418–1422.

Bruneau, J.C., Stack, E., O'Kennedy, R. and Loscher, C.E. 2012. Aflatoxins B(1), B(2) and G(1) modulate cytokine secretion and cell surface marker expression in J774A.1 murine macrophages. Toxicol *In Vitro*, 26: 686–693.

Bullerman, L.B. 1996. Occurrence of *Fusarium* and fumonisins on food grains and in foods. Adv. Exp. Med. Biol., 392: 27–38.

Calvo, A.M., Wilson, R.A., Bok, J.W. and Keller, N.P. 2002. Relationship between secondary metabolism and fungal development. Microbiol. Mol. Biol. Rev., 66: 447–459.

Cano, P.M., Jamin, E.L., Tadrist, S., Bourdaud'hui, P., Pean, M., Debrauwer, L., Oswald, I.P., Delaforge, M. and Puel, O. 2013a. New untargeted metabolic profiling combining mass spectrometry and isotopic labeling: Application on *Aspergillus fumigatus* grown on wheat. Anal. Chem., 85: 8412–8420.

Cano, P.M., Seeboth, J., Meurens, F., Cognie, J., Abrami, R., Oswald, I.P. and Guzylack-Piriou, L. 2013b. Deoxynivalenol as a new factor in the persistence of intestinal inflammatory diseases: an emerging hypothesis through possible modulation of Th17-mediated response. PLoS One, 8: e53647.

CAST. 2003. Mycotoxins: risks in plant, animal and human systems. Potential economic costs of mycotoxins in United States Task Force Rep., 138: 136–142.

Chanda, A., Roze, L.V., Kang, S., Artymovich, K.A., Hicks, G.R., Raikhel, N.V., Calvo, A.M. and Linz, J.E. 2009. A key role for vesicles in fungal secondary metabolism. Proc. Natl. Acad. Sci. USA, 106: 19533–19538.

Chang, P.K., Skory, C.D. and Linz, J.E. 1992. Cloning of a gene associated with aflatoxin B1 biosynthesis in *Aspergillus parasiticus*. Curr Genet., 21: 231–233.

Chen, F., Tholl, D., Bohlmann, J. and Pichersky, E. 2011. The family of terpene synthases in plants: a mid-size family of genes for specialized metabolism that is highly diversified throughout the kingdom. Plant J., 66: 212–229.

Chen, J., Mirocha, C.J., Xie, W., Hogge, L. and Olson, D. 1992. Production of the mycotoxin fumonisin B(1) by *Alternaria alternata* f. sp. *lycopersici*. Appl. Environ. Microbiol., 58: 3928–3931.

Delialioglu, N., Otag, F., Ocal, N.D., Aslan, G. and Emekda, G. 2010. Investigation of aflatoxin M1 levels in raw and market milks in Mersin Province, Turkey. Mikrobiyol. Bul., 44: 87–91.

Desjardins, A.E., Hohn, T.M. and McCormick, S.P. 1993. Trichothecene biosynthesis in *Fusarium* species: chemistry, genetics, and significance. Microbiol. Mol. Biol. Rev., 57: 595–604.

Desjardins, A.E., Proctor, R.H., Bai, G.H., McCormick, S.P., Shaner, G., Buechley, G. and Hohn, T.M. 1996. Reduced virulence of trichothecene-nonproducing mutants of *Gibberella zeae* in wheat field tests. Mol. Plant Microbe Interact, 9: 775–781.

Dutton, M.F. 2009. The African *Fusarium*/maize disease. Mycotoxin Res., 25: 29–39.

Eadie, M.J. 2003. Convulsive ergotism: epidemics of the serotonin syndrome? Lancet Neurol., 2: 429–434.

Eaton, D.L. and Groopman, J.D. 1994. The Toxicology of Aflatoxins: Human Health, Veterinary, and Agricultural Significance, Academic Press, San Diego, CA, 544 pp.

Eaton, D.L., Bammler, T.K. and Kelly, E.J. 2001. Interindividual differences in response to chemoprotection against aflatoxin-induced hepatocarcinogenesis: implications for human biotransformation enzyme polymorphisms. Adv. Exp. Med. Biol., 500: 559–576.

Ebel, R. 2010. Terpenes from Marine-Derived Fungi. Marine Drugs, 8: 2340–2368.

EFSA. 2013. Scientific Opinion on risks for animal and public health related to the presence of nivalenol in food and feed. EFSA J., 11: 3262.

Ellis, E.M., Judah, D.J., Neal, G.E. and Hayes, J.D. 1993. An ethoxyquin-inducible aldehyde reductase from rat liver that metabolizes aflatoxin B1 defines a subfamily of aldo-keto reductases. Proc. Natl. Acad. Sci. USA, 90: 10350–10354.

Eriksen, G.S. and Alexander, N.J. 1998. *Fusarium* Toxins in Cereals: A Risk Assessment. Nordic Council of Ministers, Copenhagen, Danemark.

Eriksen, G.S. and Pettersson, H. 2004. Toxicological evaluation of trichothecenes in animal feed. Anim. Feed Sci. Technol., 114: 205–239.

European Commission. 2006. 1881/2006/EC: Maximum levels for certain contaminants in foodstuffs.

Fallah, A.A. 2010. Assessment of aflatoxin M1 contamination in pasteurized and UHT milk marketed in central part of Iran. Food. Chem. Toxicol., 48: 988–991.

Finking, R. and Marahiel, M.A. 2004. Biosynthesis of nonribosomal peptides 1. Annu. Rev. Microbiol., 58: 453–488.

Fliege, R. and Metzler, M. 2000. Electrophilic properties of patulin. Adduct structures and reaction pathways with 4-bromothiophenol and other model nucleophiles. Chem. Res. Toxicol., 13: 363–372.

Fox, E.M. and Howlett, B.J. 2008. Secondary metabolism: regulation and role in fungal biology. Curr. Opin. Microbiol., 11: 481–487.

Freeman, G.G. and Morrison, R.I. 1948. Trichothecin; an antifungal metabolic product of *Trichothecium roseum* Link. Nature, 162: 30.

Frisvad, J.C., Smedsgaard, J., Samson, R.A., Larsen, T.O. and Thrane, U. 2007. Fumonisin B2 production by *Aspergillus niger*. J. Agric. Food Chem., 55: 9727–9732.

Gaffoor, I. and Trail, F. 2006. Characterization of two polyketide synthase genes involved in zearalenone biosynthesis in *Gibberella zeae*. Appl. Environ. Microbiol., 72: 1793–1799.

Gallagher, E.P., Wienkers, L.C., Stapleton, P.L., Kunze, K.L. and Eaton, D.L. 1994. Role of human microsomal and human complementary DNA-expressed cytochromes P4501A2 and P4503A4 in the bioactivation of aflatoxin B1. Cancer Res., 54: 101–108.

Gallagher, L., Owens, R.A., Dolan, S.K., O'Keeffe, G., Schrettl, M., Kavanagh, K., Jones, G.W. and Doyle, S. 2012. The *Aspergillus fumigatus* protein *GliK* protects against oxidative stress and is essential for gliotoxin biosynthesis. Eukaryot. Cell. 11: 1226–1238.

Gallo, A., Bruno, K.S., Solfrizzo, M., Perrone, G., Mulè, G., Visconti, A. and Baker, S.E. 2012. New insight into the Ochratoxin A biosynthetic pathway through deletion of a nonribosomal peptide synthetase gene in *Aspergillus carbonarius*. Appl. Environ. Microbiol., 78: 8208–8218.

Gallo, A., Ferrara, M. and Perrone, G. 2013. Phylogenetic study of polyketide synthases and nonribosomal peptide synthetases involved in the biosynthesis of mycotoxins. Toxins, 5: 717–742.

Galvano, F., Pietri, A., Bertuzzi, T., Gagliardi, L., Ciotti, S., Luisi, S., Bognanno, M., La Fauci, L., Iacopino, A.M., Nigro, F., Li Volti, G., Vanella, L., Giammanco, G., Tina, G.L. and Gazzolo, D. 2008. Maternal dietary habits and mycotoxin occurrence in human mature milk. Mol. Nutr. Food Res., 52: 496–501.

Gardiner, D.M. and Howlett, B.J. 2005. Bioinformatic and expression analysis of the putative gliotoxin biosynthetic gene cluster of *Aspergillus fumigatus*. FEMS Microbiol. Lett., 248: 241–248.

Gardiner, D.M., Waring, P. and Howlett, B.J. 2005. The epipolythiodioxopiperazine (ETP) class of fungal toxins: distribution, mode of action, functions and biosynthesis. Microbiology, 151: 1021–1032.

Geisen, R., Schmidt-Heydt, M. and Karolewiez, A. 2006. A gene cluster of the ochratoxin A biosynthetic genes in *Penicillium*. Mycotoxin Res., 22: 134–141.

Gelderblom, W.C., Abel, S., Smuts, C.M., Marnewick, J., Marasas, W.F., Lemmer, E.R. and Ramljak, D. 2001. Fumonisin-induced hepatocarcinogenesis: mechanisms related to cancer initiation and promotion. Environ. Health Perspect., 109: (Suppl. 2) 291–300.

Gelderblom, W.C., Jaskiewicz, K., Marasas, W.F., Thiel, P.G., Horak, R.M., Vleggaar, R. and Kriek, N.P. 1988. Fumonisins-novel mycotoxins with cancer-promoting activity produced by *Fusarium moniliforme*. Appl. Environ. Microbiol., 54: 1806–1811.

Goncalves, S.S., Stchigel, A.M., Cano, J.F., Godoy-Martinez, P.C., Colombo, A.L. and Guarro, J. 2012. *Aspergillus novoparasiticus*: a new clinical species of the section *Flavi*. Med. Mycol., 50: 152–160.

Gong, Y.Y., Egal, S., Hounsa, A., Turner, P.C., Hall, A.J., Cardwell, K.F. and Wild, C.P. 2003. Determinants of aflatoxin exposure in young children from Benin and Togo, West Africa: the critical role of weaning. Int. J. Epidemiol., 32: 556–562.

Gong, Z.J., De Meyer, S., Clarysse, C., Verslype, C., Neyts, J., De Clercq, E. and Yap, S.H. 1999. Mycophenolic acid, an immunosuppressive agent, inhibits HBV replication *in vitro*. J. Viral. Hepat., 6: 229–236.

Grenier, B. and Oswald, I.P. 2011. Mycotoxin co-contamination of food and feed: meta-analysis of publications describing toxicological interactions. World Mycotox. J., 4: 285–313.

Gross-Steinmeyer, K. and Eaton, D.L. 2012. Dietary modulation of the biotransformation and genotoxicity of aflatoxin B1. Toxicology, 299: 69–79.

Hamid, A.S., Tesfamariam, I.G., Zhang, Y. and Zhang, Z.G. 2013. Aflatoxin B1-induced hepatocellular carcinoma in developing countries: Geographical distribution, mechanism of action and prevention. Oncol. Lett., 5: 1087–1092.

Harrer, H., Laviad, E.L., Humpf, H.U. and Futerman, A.H. 2013. Identification of N-acyl-fumonisin B1 as new cytotoxic metabolites of fumonisin mycotoxins. Mol. Nutr. Food Res., 57: 516–522.

Harris, J.P. and Mantle, P.G. 2001. Biosynthesis of ochratoxins by *Aspergillus ochraceus*. Phytochemistry, 58: 709–716.

Harrison, L.R., Colvin, B.M., Greene, J.T., Newman, L.E. and Cole, J.R., Jr. 1990. Pulmonary edema and hydrothorax in swine produced by fumonisin B1, a toxic metabolite of *Fusarium moniliforme*. J. Vet. Diagn. Invest., 2: 217–221.

Hinton, D.M., Myers, M.J., Raybourne, R.A., Francke-Carroll, S., Sotomayor, R.E., Shaddock, J., Warbritton, A. and Chou, M.W. 2003. Immunotoxicity of aflatoxin B1 in rats: effects on lymphocytes and the inflammatory response in a chronic intermittent dosing study. Toxicol. Sci., 73: 362–377.

Hof, H. 2008. Mycotoxins: pathogenicity factors or virulence factors? Mycoses, 51: 93–94.

Hsieh, D.P., Cullen, J.M. and Ruebner, B.H. 1984. Comparative hepatocarcinogenicity of aflatoxins B1 and M1 in the rat. Food Chem. Toxicol., 22: 1027–1028.

Huffman, J., Gerber, R. and Du, L. 2010. Recent advancements in the biosynthetic mechanisms for polyketide-derived mycotoxins. Biopolymers, 93: 764–776.

IARC. 1986. Some naturally occurring and synthetic food components, furocoumarins and ultraviolet radiation. IARC Monogr. Eval. Carcinog Risks Hum., 40: 83–98.

IARC. 1993. Some naturally occurring substances: Food items and constituents, heterocyclic aromatic amines and mycotoxins. IARC monographs on the evaluation if carcinogenic risks to humans, 56: 489.

IARC. 2002. Some traditional herbal medicines, some mycotoxins, naphthalene and styrene. IARC Monogr. Eval. Carcinog Risks Hum., 82: 1–556.

Javed, T., Richard, J.L., Bennett, G.A., Dombrink-Kurtzman, M.A., Bunte, R.M., Koelkebeck, K.W., Cote, L.M., Leeper, R.W. and Buck, W.B. 1993. Embryopathic and embryocidal effects of purified fumonisin B1 or *Fusarium proliferatum* culture material extract on chicken embryos. Mycopathologia, 123: 185–193.

JECFA. 1995. Evaluation of certain food additives and contaminants. WHO Technical report series, 859 (44).

JECFA. 1999. Evaluation of certain food additives and contaminants. WHO Technical report series, 896 (53).

JECFA. 2001. Safety evaluation of certain mycotoxins in food. WHO Food additives series, 47.

Jestoi, M. 2008. Emerging *Fusarium*-mycotoxins fusaproliferin, beauvericin, enniatins, and moniliformin: a review. Crit. Rev. Food Sci. Nutr., 48: 21–49.

Jolly, P.E., Inusah, S., Lu, B., Ellis, W.O., Nyarko, A., Phillips, T.D. and Williams, J.H. 2013. Association between high aflatoxin B-1 levels and high viral load in HIV-positive people. World Mycotoxin J., 6: 255–261.

Jones, C., Ciacci-Zanella, J.R., Zhang, Y., Henderson, G. and Dickman, M. 2001. Analysis of fumonisin B1-induced apoptosis. Environ. Health Perspect., 109: 315–320.

Kafer, E. 1965. Origins of translocations in *Aspergillus nidulans*. Genetics, 52: 217–232.

Karolewiez, A. and Geisen, R. 2005. Cloning a part of the ochratoxin A biosynthetic gene cluster of *Penicillium nordicum* and characterization of the ochratoxin polyketide synthase gene. Syst. Appl. Microbiol., 28: 588–595.

Keller, N.P. and Adams, T.H. 1995. Analysis of a mycotoxin gene cluster in *Aspergillus nidulans*. SAAS Bull. Biochem. Biotechnol., 8: 14–21.

Keller, N.P. and Hohn, T.M. 1997. Metabolic pathway gene clusters in filamentous fungi. Fungal Genet. Biol., 21: 17–29.

Kellerman, T.S., Marasas, W.F., Thiel, P.G., Gelderblom, W.C., Cawood, M. and Coetzer, J.A. 1990. Leukoencephalomalacia in two horses induced by oral dosing of fumonisin B1. Onderstepoort J. Vet. Res., 57: 269–275.

Kelly, E.J., Erickson, K.E., Sengstag, C. and Eaton, D.L. 2002. Expression of human microsomal epoxide hydrolase in *Saccharomyces cerevisiae* reveals a functional role in aflatoxin B1 detoxification. Toxicol. Sci., 65: 35–42.

Keskin, Y., Baskaya, R., Karsli, S., Yurdun, T. and Ozyaral, O. 2009. Detection of aflatoxin M1 in human breast milk and raw cow's milk in Istanbul, Turkey. J. Food Prot., 72: 885–889.

Khlangwiset, P., Shephard, G.S. and Wu, F. 2011. Aflatoxins and growth impairment: A review. Crit. Rev. Toxicol., 41: 740–755.

Klich, M.A., Cary, J.W., Beltz, S.B. and Bennett, C.A. 2003. Phylogenetic and morphological analysis of *Aspergillus ochraceoroseus*. Mycologia, 95: 1252–1260.

Kiessling, K.H. 1986. Biochemical mehcanisms of action of mycotoxins. Pure Appl. Chem., 58: 327–338.

Kim, J.E., Bunderson, B.R., Croasdell, A., Reed, K.M. and Coulombe, R.A., Jr. 2013. Alpha-Class Glutathione S-Transferases in Wild Turkeys (*Meleagris gallopavo*): Characterization and Role in Resistance to the Carcinogenic Mycotoxin Aflatoxin B1. PLoS One, 8: e60662.

Kimura, M., Tokai, T., Takahashi-Ando, N., Ohsato, S. and Fujimura, M. 2007. Molecular and genetic studies of *Fusarium* trichothecene biosynthesis: Pathways, genes, and evolution. Biosci. Biotech. Biochem., 71: 2105–2123.

el Khoury, A. and Atoui, A. 2010. Ochratoxin A: General overview and actual molecular status. Toxins, 2: 461–493.

Kwon-Chung, K.J. and Sugui, J.A. 2009. What do we know about the role of gliotoxin in the pathobiology of *Aspergillus fumigatus*? Med. Mycol., 47: (s1) S97–S103.

Lazzaro, I., Falavigna, C., Dall'Asta, C., Proctor, R.H., Galaverna, G. and Battilani, P. 2012. Fumonisins B, A and C profile and masking in *Fusarium verticillioides* strains on fumonisin-inducing and maize-based media. Int. J. Food Microbiol., 159: 93–100.

Leslie, J.F. and Summerell, B.A. 2006. The *Fusarium* Laboratory Manual. Blackwell Publishing, Oxford, UK.

Li, Y., Wang, Z., Beier, R.C., Shen, J., De Smet, D., De Saeger, S. and Zhang, S. 2011. T-2 toxin, a trichothecene mycotoxin: review of toxicity, metabolism, and analytical methods. J. Agric. Food Chem., 59: 3441–3453.

Li, B., Zong, Y., Du, Z., Chen, Y., Zhang, Z., Qin, G., Zhao, W. and Tian, S. 2015. Genomic characterization reveals insights into patulin biosynthesis and pathogenicity in *Penicillium* species. Mol. Plant Microbe Interact., 28: 635–647.

Lipsky, J.J. 1996. Mycophenolate mofetil. Lancet, 348: 1357–1359.

Liu, Y. and Wu, F. 2010. Global burden of aflatoxin-induced hepatocellular carcinoma: a risk assessment. Environ. Health Perspect., 118: 818–824.

Maddox, J. 1984. Natural history of yellow rain. Nature, 309: 207.

Mansson, M., Klejnstrup, M.L., Phipps, R.K., Nielsen, K.F., Frisvad, J.C., Gotfredsen, C.H. and Larsen, T.O. 2010. Isolation and NMR characterization of fumonisin B2 and a new fumonisin B6 from *Aspergillus niger*. J. Agric. Food Chem., 58: 949–953.

Marasas, W.F. 1996. Fumonisins: history, world-wide occurrence and impact. Adv. Exp. Med. Biol., 392: 1–17.

Marasas, W.F., Kellerman, T.S., Gelderblom, W.C., Coetzer, J.A., Thiel, P.G. and van der Lugt, J.J. 1988. Leukoencephalomalacia in a horse induced by fumonisin B1 isolated from *Fusarium moniliforme*. Onderstepoort J. Vet. Res., 55: 197–203.

Marasas, W.F., Riley, R.T., Hendricks, K.A., Stevens, V.L., Sadler, T.W., Gelineau-van Waes, J., Missmer, S.A., Cabrera, J., Torres, O., Gelderblom, W.C., Allegood, J., Martinez, C., Maddox, J., Miller, J.D., Starr, L., Sullards, M.C., Roman, A.V., Voss, K.A.,Wang, E. and Merrill, A.H., Jr. 2004. Fumonisins disrupt sphingolipid metabolism, folate transport, and neural tube development in embryo culture and *in vivo*: a potential risk factor for human neural tube defects among populations consuming fumonisin-contaminated maize. J. Nutr., 134: 711–716.

Maresca, M. 2013. From the gut to the brain: journey and pathophysiological effects of the food-associated trichothecene mycotoxin deoxynivalenol. Toxins, 5: 784–820.

Maresca, M. and Fantini, J. 2010. Some food-associated mycotoxins as potential risk factors in humans predisposed to chronic intestinal inflammatory diseases. Toxicon, 56: 282–294.

Marin, D.E., Pistol, G.C., Neagoe, I.V., Calin, L. and Taranu, I. 2013. Effects of zearalenone on oxidative stress and inflammation in weanling piglets. Food Chem. Toxicol., 58: 408–415.

Martinez-Larranaga, M.R., Anadon, A., Diaz, M.J., Fernandez-Cruz, M.L., Martinez, M.A., Frejo, M.T., Martinez, M., Fernandez, R., Anton, R.M., Morales, M.E. and Tafur, M. 1999. Toxicokinetics and oral bioavailability of fumonisin B1. Vet. Hum. Toxicol., 41: 357–362.

McCormick, S.P., Harris, L.J., Alexander, N.J., Ouellet, T., Saparno, A., Allard, S. and Desjardins, A.E. 2004. Tri1 in *Fusarium graminearum* encodes a P450 oxygenase. Appl. Environ. Microbiol., 70: 2044–2051.

McKinley, E.R. and Carlton, W.W. 1991. Patulin. pp. 191–236. *In*: Sharma, R.P. and Salunkhe, D.K. (eds.). Mycotoxins and Phytoalexins. CRC Press, Boca Raton, USA.

Meissonnier, G.M., Pinton, P., Laffitte, J., Cossalter, A.M., Gong, Y.Y., Wild, C.P., Bertin, G., Galtier, P. and Oswald, I.P. 2008. Immunotoxicity of aflatoxin B1: impairment of the cell-mediated response to vaccine antigen and modulation of cytokine expression. Toxicol. Appl. Pharmacol., 231: 142–149.

Merhej, J., Richard-Forget, F. and Barreau, C. 2011. Regulation of trichothecene biosynthesis in *Fusarium*: recent advances and new insights. Appl. Microbiol. Biotechnol., 91: 519–528.

Merrill, A.H., Jr., Morgan, E.T., Nikolova-Karakashian, M. and Stewart, J. 1999. Sphingomyelin hydrolysis and regulation of the expression of the gene for cytochrome P450. Biochem. Soc. Trans., 27: 383–387.

Methenitou, G., Maravelias, C., Athanaselis, S., Dona, A. and Koutselinis, A. 2001. Immunomodulative effects of aflatoxins and selenium on human natural killer cells. Vet. Hum. Toxicol., 43: 232–234.

Metzler, M., Pfeiffer, E. and Hildebrand, A. 2010. Zearalenone and its metabolites as endocrine disrupting chemicals. World Mycotoxin. J., 3: 385–401.

Minervini, F. and Dell'Aquila, M.E. 2008. Zearalenone and reproductive function in farm animals. Int. J. Mol. Sci., 9: 2570–2584.

Mogensen, J.M., Frisvad, J.C., Thrane, U. and Nielsen, K.F. 2010. Production of Fumonisin B2 and B4 by *Aspergillus niger* on grapes and raisins. J. Agric. Food Chem., 58: 954–958.

Moon, Y. and Pestka, J.J. 2002. Vomitoxin-induced cyclooxygenase-2 gene expression in macrophages mediated by activation of ERK and p38 but not JNK mitogen-activated protein kinases. Toxicol. Sci., 69: 373–382.

Motawee, M.M. and McMahon, D.J. 2009. Fate of aflatoxin M(1) during manufacture and storage of feta cheese. J. Food Sci., 74: T42–45.

Müller, S., Dekant, W. and Mally, A. 2012. Fumonisin B1 and the kidney: Modes of action for renal tumor formation by fumonisin B1 in rodents. Food Chem. Toxicol., 50: 3833–3846.

Nayak, S., Tanuja, P. and Sashidhar, R.B. 2009. Synthesis and characterization of mercapturic acid (N-acetyl-L-cysteine)-aflatoxin B1 adduct and its quantitation in rat urine by an enzyme immunoassay. J. AOAC Int., 92: 487–495.

Nesbitt, B.F., O'Kelly, J., Sargeant, K. and Sheridan, A. 1962. *Aspergillus flavus* and turkey X disease. Toxic metabolites of *Aspergillus flavus*. Nature, 195: 1062–1063.

Nicholson, T.P., Rudd, B.A.M., Dawson, M., Lazarus, C.M., Simpson, T.J. and Cox, R.J. 2001. Design and utility of oligonucleotide gene probes for fungal polyketide synthases. Chem. Biol., 8: 157–178.

Nielsen, K.F. 2002. Mould Growth on Building Materials: Secondary Metabolites, Mycotoxins and Biomarkers. Ph.D. Thesis, Danish Builduing and Urban Research, Horsolm, Danemark.

Nusuetrong, P., Saito, M., Kikuchi, H., Oshima, Y., Moriya, T. and Nakahata, N. 2012. Apoptotic effects of satratoxin H is mediated through DNA double-stranded break in PC12 cells. J. Toxicol, Sci., 37: 803–812.

O'Callaghan, J., Coghlan, A., Abbas, A., Garcia-Estrada, C., Martin, J.-F. and Dobson, A.D.W. 2013. Functional characterization of the polyketide synthase gene required for ochratoxin A biosynthesis in *Penicillium verrucosum*. Int. J. Food Microbiol., 161: 172–181.

Omar, H.M. 2013. Mycotoxins-induced oxidative stress and disease. *In:* Makun, H. (ed.). Mycotoxin and Food Safety in Developing Countries. InTech. Available from: http://www.intechopen.com/books/mycotoxin-and-food-safety-in-developing-countries/mycotoxins-induced-oxidative-stress-and-disease.

Oswald, I.P., Marin, D.E., Bouhet, S., Pinton, P., Taranu, I. and Accensi, F. 2005. Immunotoxicological risk of mycotoxins for domestic animals. Food Addit. Contam., 22: 354–360.

Patterson, D.S. and Allcroft, R. 1970. Metabolism of aflatoxin in susceptible and resistant animal species. Food Cosmet. Toxicol., 8: 43–53.

Pestka, J.J. and Forsell, J.H. 1988. Inhibition of human lymphocyte transformation by the macrocyclic trichothecenes roridin A and verrucarin A. Toxicol. Lett., 41: 215–222.

Pestka, J.J. and Smolinski, A.T. 2005. Deoxynivalenol: toxicology and potential effects on humans. J. Toxicol. Environ. Health B Crit. Rev., 8: 39–69.

Pestka, J.J., Zhou, H.R., Moon, Y. and Chung, Y.J. 2004. Cellular and molecular mechanisms for immune modulation by deoxynivalenol and other trichothecenes: unraveling a paradox. Toxicol. Lett., 153: 61–73.

Pi, B., Yu, D., Dai, F., Song, X., Zhu, C., Li, H. and Yu, Y. 2015. A Genomics Based Discovery of Secondary Metabolite Biosynthetic Gene Clusters in *Aspergillus ustus*. PLoS One, 10: e0116089.

Pildain, M.B., Frisvad, J.C., Vaamonde, G., Cabral, D., Varga, J. and Samson, R.A. 2008. Two novel aflatoxin-producing *Aspergillus* species from Argentinean peanuts. Int. J. Syst. Evol. Microbiol., 58: 725–735.

Pinton, P., Tsybulskyy, D., Lucioli, J., Laffitte, J., Callu, P., Lyazhri, F., Grosjean, F., Bracarense, A.P., Kolf-Clauw, M. and Oswald, I.P. 2012. Toxicity of deoxynivalenol and its acetylated derivatives on the intestine: differential effects on morphology, barrier function, tight junction proteins, and mitogen-activated protein kinases. Toxicol. Sci., 130: 180–190.

Pistol, G.C., Gras, M.A., Marin, D.E., Israel-Roming, F., Stancu, M. and Taranu, I. 2014. Natural feed contaminant zearalenone decreases the expressions of important pro- and anti-inflammatory mediators and mitogen-activated protein kinase/NF-kappaB signalling molecules in pigs. Br. J. Nutr., 111: 452–464.

Pitt, J.I. and Hocking, D.A. 2009. Fungi and Food Spoilage, 3rd Ed. Springer Science, New York, USA.

Prandini, A., Tansini, G., Sigolo, S., Filippi, L., Laporta, M. and Piva, G. 2009. On the occurrence of aflatoxin M1 in milk and dairy products. Food Chem. Toxicol., 47: 984–991.

Proctor, R.H., Brown, D.W., Plattner, R.D. and Desjardins, A.E. 2003. Co-expression of 15 contiguous genes delineates a fumonisin biosynthetic gene cluster in *Gibberella moniliformis*. Fungal Genet. Biol., 38: 237–249.

Proctor, R.H., Busman, M., Seo, J.A., Lee, Y.W. and Plattner, R.D. 2008. A fumonisin biosynthetic gene cluster in *Fusarium oxysporum* strain O-1890 and the genetic basis for B versus C fumonisin production. Fungal. Genet. Biol., 45: 1016–1026.

Proctor, R.H., Desjardins, A.E., Plattner, R.D. and Hohn, T.M. 1999. A polyketide synthase gene required for biosynthesis of fumonisin mycotoxins in *Gibberella fujikuroi* mating population A. Fungal Genet. Biol., 27: 100–112.

Proctor, R.H., Hohn, T.M. and McCormick, S.P. 1995. Reduced virulence of *Gibberella zeae* caused by disruption of a trichothecene toxin biosynthetic gene. Mol. Plant Microbe. Interact., 8: 593–601.

Puel, O., Galtier, P. and Oswald, I.P. 2010. Biosynthesis and toxicological effects of patulin. Toxins, 2: 613–631.

Rauber, R.H., Dilkin, P., Mallmann, A.O., Marchioro, A., Mallmann, C.A., Borsoi, A. and Nascimento, V.P. 2012. Individual and combined effects of *Salmonella typhimurium* lipopolysaccharide and fumonisin B1 in broiler chickens. Poult. Sci., 91: 2785–2791.

Rank, C., Nielsen, K.F., Larsen, T.O., Varga, J., Samson, R.A. and Frisvad, J.C. 2011. Distribution of sterigmatocystin in filamentous fungi. Fungal Biol., 115: 406–420.

Razafimanjato, H., Benzaria, A., Taieb, N., Guo, X.J., Vidal, N., Di Scala, C., Varini, K. and Maresca, M. 2011. The ribotoxin deoxynivalenol affects the viability and functions of glial cells. Glia, 59: 1672–1683.

Reddy, L. and Bhoola, K. 2010. Ochratoxins—food contaminants: impact on human health. Toxins, 2: 771–779.

Reverberi, M., Ricelli, A., Zjalic, S., Fabbri, A. and Fanelli, C. 2010. Natural functions of mycotoxins and control of their biosynthesis in fungi. Appl. Microbiol. Biotechnol., 87: 899–911.

Rheeder, J.P., Marasas, W.F. and Vismer, H.F. 2002. Production of fumonisin analogs by *Fusarium* species. Appl. Environ. Microbiol., 68: 2101–2105.

Riley, R.T., An, N.H., Showker, J.L., Yoo, H.S., Norred, W.P., Chamberlain, W.J., Wang, E., Merrill, A.H., Jr., Motelin, G., Beasley, V.R. and Haschek, W.M. 1993. Alteration of tissue and serum sphinganine to sphingosine ratio: an early biomarker of exposure to fumonisin-containing feeds in pigs. Toxicol. Appl. Pharmacol., 118: 105–112.

Riley, R.T., Hinton, D.M., Chamberlain, W.J., Bacon, C.W., Wang, E., Merrill, A.H., Jr. and Voss, K.A. 1994. Dietary fumonisin B1 induces disruption of sphingolipid metabolism in Sprague-Dawley rats: a new mechanism of nephrotoxicity. J. Nutr., 124: 594–603.

Rocha, O., Ansari, K. and Doohan, F.M. 2005. Effects of trichothecene mycotoxins on eukaryotic cells: A review. Food Addit. Contam., 22: 369–378.

Roda, E., Coccini, T., Acerbi, D., Castoldi, A.F. and Manzo, L. 2010. Comparative *in vitro* and *ex-vivo* myelotoxicity of aflatoxins B1 and M1 on haematopoietic progenitors (BFU-E, CFU-E, and CFU-GM): species-related susceptibility. Toxicol *In Vitro*, 24: 217–223.

Rohlfs, M., Albert, M., Keller, N.P. and Kempken, F. 2007. Secondary chemicals protect mould from fungivory. Biology Letters, 3: 523–525.

Rohlfs, M. 2015. Fungal secondary metabolite dynamics in fungus-grazer interactions: novel insights and unanswered questions. Front Microbiol., 5: 788.

Rotter, B.A., Prelusky, D.B. and Pestka, J.J. 1996a. Toxicology of deoxynivalenol (vomitoxin). J. Toxicol. Environ. Health, 48: 1–34.

Rotter, B.A., Thompson, B.K., Prelusky, D.B., Trenholm, H.L., Stewart, B., Miller, J.D. and Savard, M.E. 1996b. Response of growing swine to dietary exposure to pure fumonisin B1 during an eight-week period: growth and clinical parameters. Nat. Toxins, 4: 42–50.

Roze, L.V., Hong, S.-Y. and Linz, J.E. 2013. Aflatoxin Biosynthesis: Current Frontiers. Annual Review of Food Science and Technology, 4: 293–311.

Rychlik, M., Kircher, F., Schusdziarra, V. and Lippl, F. 2004. Absorption of the mycotoxin patulin from the rat stomach. Food Chem. Toxicol., 42: 729–735.

Sabbioni, G., Skipper, P.L., Buchi, G. and Tannenbaum, S.R. 1987. Isolation and characterization of the major serum albumin adduct formed by aflatoxin B1 *in vivo* in rats. Carcinogenesis, 8: 819–824.

Scharf, D.H., Heinekamp, T., Remme, N., Hortschansky, P., Brakhage, A.A. and Hertweck, C. 2012. Biosynthesis and function of gliotoxin in *Aspergillus fumigatus*. Appl. Microbiol. Biotechnol., 93: 467–472.

Scharf, D.H., Remme, N., Habel, A., Chankhamjon, P., Scherlach, K., Heinekamp, T., Hortschansky, P., Brakhage, A.A. and Hertweck, C. 2011. A dedicated glutathione S-transferase mediates carbon-sulfur bond formation in gliotoxin biosynthesis. J. Am. Chem. Soc., 133: 12322–12325.

Schatzmayr, G. and Streit, E. 2013. Global occurrence of mycotoxins in the food and feed chain: facts and figures. World Mycotoxin J., 6: 213–222.

Schmidt-Heydt, M., Hackel, S., Rufer, C.E. and Geisen, R. 2009. A strain of *Fusarium kyushuense* is able to produce aflatoxin B1 and G 1. Mycotoxin Res., 25: 141–147.

SCOOP. 2003. Collection of occurrence data of *Fusarium* toxins in food and assessment of dietary intake by the population of EU member states. Toxicol. Lett., 153: 133–143.

Seenivasan, A., Aravindan, R., Subhagar, S. and Viruthagiri, T. 2008. Microbial production and biomedical applications of lovastatin. Indian J. Pharm. Sci. 70: 701–709.

Sharma, R.P., Bhandari, N., He, Q., Riley, R.T. and Voss, K.A. 2001. Decreased fumonisin hepatotoxicity in mice with a targeted deletion of tumor necrosis factor receptor 1. Toxicology, 159: 69–79.

Shephard, G.S., Thiel, P.G., Sydenham, E.W. and Alberts, J.F. 1994a. Biliary excretion of the mycotoxin fumonisin B1 in rats. Food Chem. Toxicol., 32: 489–491.

Shephard, G.S., Thiel, P.G., Sydenham, E.W., Alberts, J.F. and Cawood, M.E. 1994b. Distribution and excretion of a single dose of the mycotoxin fumonisin B1 in a non-human primate. Toxicon, 32: 735–741.

Shephard, G.S., Thiel, P.G., Sydenham, E.W., Vleggaar, R. and Alberts, J.F. 1994c. Determination of the mycotoxin fumonisin B1 and identification of its partially hydrolysed metabolites in the faeces of non-human primates. Food Chem. Toxicol., 32: 23–29.

Shuaib, F.M.B., Ehiri, J., Abdullahi, A., Williams, J.H. and Jolly, P.E. 2010. Reproductive health effects of aflatoxins: A review of the literature. Repr. Toxicol., 29: 262–270.

Slone, D.H., Gallagher, E.P., Ramsdell, H.S., Rettie, A.E., Stapleton, P.L., Berlad, L.G. and Eaton, D.L. 1995. Human variability in hepatic glutathione S-transferase-mediated conjugation of aflatoxin B1-epoxide and other substrates. Pharmacogenetics, 5: 224–233.

Slot, J.C. and Rokas, A. 2011. Horizontal transfer of a large and highly toxic secondary metabolic gene cluster between fungi. Current Biology, 21: 134–139.

Snini, S.P., Tadrist, S., Lafitte, J., Jamin, E.L., Oswald, I.P. and Puel, O. 2014. The gene *PatG*, involved in biosynthesis pathway of patulin, a food-borne mycotoxin, encodes a 6-methylsalicylic acid decarboxylase. Int. J. Food. Microbiol., 171: 77–83.

Soriano, J.M., Gonzalez, L. and Catala, A.I. 2005. Mechanism of action of sphingolipids and their metabolites in the toxicity of fumonisin B1. Prog. Lipid Res., 44: 345–356.

Spotti, M., Maas, R.F., de Nijs, C.M. and Fink-Gremmels, J. 2000. Effect of fumonisin B(1) on rat hepatic P450 system. Environ. Toxicol. Pharmacol., 8: 197–204.

Sun, G., Wang, S., Hu, X., Su, J., Huang, T., Yu, J., Tang, L., Gao, W. and Wang, J.S. 2007. Fumonisin B1 contamination of home-grown corn in high-risk areas for esophageal and liver cancer in China. Food Addit. Contam., 24: 181–185.

Tannous, J., El Khoury, R., Snini, S.P., Lippi, Y., El Khoury, A., Atoui, A., Lteif, R., Oswald, I.P. and Puel, O. 2014. Sequencing, physical organization and kinetic expression of the patulin biosynthetic gene cluster from *Penicillium expansum*. Int. J. Food Microbiol., 189: 51–60.

Thiel, P.G., Marasas, W.F., Sydenham, E.W., Shephard, G.S., Gelderblom, W.C. and Nieuwenhuis, J.J. 1991. Survey of fumonisin production by *Fusarium* species. Appl. Environ. Microbiol., 57: 1089–1093.

Trail, F., Mahanti, N. and Linz, J. 1995a. Molecular biology of aflatoxin biosynthesis. Microbiology, 141: 755–765.

Trail, F., Mahanti, N., Rarick, M., Mehigh, R., Liang, S.H., Zhou, R. and Linz, J.E. 1995b. Physical and transcriptional map of an aflatoxin gene cluster in *Aspergillus parasiticus* and functional disruption of a gene involved early in the aflatoxin pathway. Appl. Environ. Microbiol., 61: 2665–2673.

Tudzynski, P., Correia, T. and Keller, U. 2001. Biotechnology and genetics of ergot alkaloids. Appl. Microbiol. Biotechnol., 57: 593–605.

Turner, P.C., Hopton, R.P., Lecluse, Y., White, K.L., Fisher, J. and Lebailly, P. 2010. Determinants of urinary deoxynivalenol and de-epoxy deoxynivalenol in male farmers from Normandy, France. J. Agric. Food Chem., 58: 5206–5212.

Ueno, Y. 1985. The toxicology of mycotoxins. Crit. Rev. Toxicol., 14: 99–132.

Ueno, Y., Sato, N., Ishii, K., Sakai, K. and Tsunoda, H. 1973. Biological and chemical detection of trichothecene mycotoxins of *Fusarium* species. Appl. Microbiol., 25: 699–704.

Varga, E., Glauner, T., Berthiller, F., Krska, R., Schuhmacher, R. and Sulyok, M. 2013. Development and validation of a (semi-) quantitative UHPLC-MS/MS method for the determination of 191 mycotoxins and other fungal metabolites in almonds, hazelnuts, peanuts and pistachios. Anal. and Bioanal. Chem., 405: 5087–5104.

Varga, J., Frisvad, J.C. and Samson, R.A. 2011. Two new aflatoxin producing species, and an overview of *Aspergillus* section *Flavi*. Stud. Mycol., 69: 57–80.

Vesonder, R.F., Ciegler, A. and Jensen, A.H. 1973. Isolation of the emetic principle from *Fusarium*-infected corn. Appl. Microbiol., 26: 1008–1010.

Vettorazzi, A., van Delft, J. and Lopez de Cerain, A. 2013. A review on ochratoxin A transcriptomic studies. Food Chem. Toxicol., 59: 766–783.

Volkl, A., Vogler, B., Schollenberger, M. and Karlovsky, P. 2004. Microbial detoxification of mycotoxin deoxynivalenol. J. Basic Microbiol., 44: 147–156.

Voss, K.A., Chamberlain, W.J., Bacon, C.W. and Norred, W.P. 1993. A preliminary investigation on renal and hepatic toxicity in rats fed purified fumonisin B1. Nat. Toxins, 1: 222–228.

Voss, K.A., Chamberlain, W.J., Bacon, C.W., Riley, R.T. and Norred, W.P. 1995. Subchronic toxicity of fumonisin B1 to male and female rats. Food Addit. Contam., 12: 473–478.

Wada, K., Hashiba, Y., Ohtsuka, H., Kohiruimaki, M., Masui, M., Kawamura, S., Endo, H. and Ogata, Y. 2008. Effects of mycotoxins on mitogen-stimulated proliferation of bovine peripheral blood mononuclear cells. J. Vet. Med. Sci., 70: 193–196.

Woolf, A. 2000. Witchcraft or mycotoxin? The Salem witch trials. J. Toxicol. Clin. Toxicol., 38: 457–460.

Wu, Q., Dohnal, V., Huang, L., Kuca, K. and Yuan, Z. 2011. Metabolic pathways of trichothecenes. Drug Metab. Rev., 42: 250–267.

Yu, H., Zhou, T., Gong, J., Young, C., Su, X., Li, X.Z., Zhu, H., Tsao, R. and Yang, R. 2010. Isolation of deoxynivalenol-transforming bacteria from the chicken intestines using the approach of PCR-DGGE guided microbial selection. BMC Microbiol., 10: 182.

Yu, J., Bhatnagar, D. and Cleveland, T.E. 2004a. Completed sequence of aflatoxin pathway gene cluster in *Aspergillus parasiticus*. FEBS Lett., 564: 126–130.

Yu, J., Chang, P.K., Ehrlich, K.C., Cary, J.W., Bhatnagar, D., Cleveland, T.E., Payne, G.A., Linz, J.E., Woloshuk, C.P. and Bennett, J.W. 2004b. Clustered pathway genes in aflatoxin biosynthesis. Appl. Environ. Microbiol., 70: 1253–1262.

Zeng, H., Hatabayashi, H., Nakagawa, H., Cai, J., Suzuki, R., Sakuno, E., Tanaka, T., Ito, Y., Ehrlich, K.C., Nakajima, H. and Yabe, K. 2011. Conversion of 11-hydroxy-O-methylsterigmatocystin to aflatoxin G1 in *Aspergillus parasiticus*. Appl. Microbiol. Biotechnol., 90: 635–650.

Zinedine, A., Soriano, J.M., Molto, J.C. and Manes, J. 2007. Review on the toxicity, occurrence, metabolism, detoxification, regulations and intake of zearalenone: An oestrogenic mycotoxin. Food Chem. Toxicol., 45: 1–18.

Zitomer, N.C., Mitchell, T., Voss, K.A., Bondy, G.S., Pruett, S.T., Garnier-Amblard, E.C., Liebeskind, L.S., Park, H., Wang, E., Sullards, M.C., Merrill, A.H., Jr. and Riley, R.T. 2009. Ceramide synthase inhibition by fumonisin B1 causes accumulation of 1-deoxysphinganine: a novel category of bioactive 1-deoxysphingoid bases and 1-deoxydihydroceramides biosynthesized by mammalian cell lines and animals. J. Biol. Chem., 284: 4786–4795.

CHAPTER 15

Nutritional Profile of Wild Edible Mushrooms of North India

N.S. Atri,[1],* Babita Kumari,[1] Sapan Kumar,[1] R.C. Upadhyay,[2] Ashu Gulati,[3] Lata[1] and Arvind Gulati[3]

ABSTRACT

The chapter embodies nutritional profile of 20 wild edible mushrooms of North India, which include *Lentinus cladopus* Lèv, *L. connatus* Berk., *L. sajor-caju* (Fr.) Fr., *L. squarrosulus* Mont., *L. torulosus* (Pers.: Fr.) Lloyd, *Termitomyces badius* Otieno, *T. heimii* Natarajan, *T. mammiformis* Heim, *T. medius* Heim & Gasse, *T. microcarpus* (Berk. & Br.) Heim, *T. radicatus* Natarajan, *T. striatus* (Beeli) Heim, *Macrolepiota dolichaula* (Berk. & Broome) Pegler & Rayner, *M. procera* (Scop.: Fr.) Sing., *M. rhacodes* (Vitt.) Sing., *Pleurotus cystidiosus* O.K. Miller, *P. floridanus* Singer, *P. pulmonarius* (Fr.) Quél, *P. sajor-caju* (Fr.) Sing. and *P. sapidus* Quél., with respect to their carbohydrates, crude fat, proteins, fibers, ash, minerals, heavy metals and vitamins contents. Carbohydrate constitutes the largest fraction of mushroom dry matter ranging from 33.3% in *Termitomyces medius* to 89.10% in *Lentinus cladopus*. Maximum percentage of crude fat (3.4%) has been documented in *Macrolepiota procera*, while minimum amount (0.48%) was observed in *Lentinus connatus*. Protein percentage is maximum in *Termitomyces medius* (46.2) and minimum in *Lentinus connatus* (0.525). The percentage of fiber ranged from 1.83 in *L. squarrosulus* to 8.0 in *Termitomyces mammiformis*. The ash content was

---

[1] Department of Botany, Punjabi University, Patiala-147002, (Pb.), India.
[2] Directorate of Mushroom Research (ICAR), Chambaghat, Solan-173230 (H.P.), India.
[3] CSIR Institute of Himalayan Bioresource Technology, Palampur (H.P.), India.
* Corresponding author: narinderatri04@gmail.com

within 1.03% (*Pleurotus pulmonarius*) to 12.13% (*Termitomyces striatus*) range. Moisture percentage ranged from 4.168% in *Lentinus cladopus* to 8.8% in *Macrolepiota procera*. There is no set pattern in the presence and the amount of different minerals documented. Macro and micro elements which have been evaluated include Ca, Cu, Fe, K, Mg, Mn, Na, Se and Zn. Amongst the evaluated minerals K and Na were not detected in species of *Macrolepiota* Sing. and *Termitomyces* R. Heim while Mn and Se were not detected in the species of *Lentinus* Fr. and *Pleurotus* (Fr.) P. Kumm. Amongst the heavy metals As, Hg, and Cd was detected in the species of *Macrolepiota* and *Termitomyces*, although the amount was within the permissible limit for human consumption. None of the species of *Lentinus* and *Pleurotus* tested positive quantitively for As, Pb, Ag, Hg and Sn. As far as vitamins are concerned in some species of *Macrolepiota* and *Termitomyces*, the presence of retinol, thiamine and riboflavin has been documented while substantial amounts of ascorbic acid has been reported in the species of *Macrolepiota*, *Termitomyces*, *Lentinus* and *Pleurotus*.

Introduction

Mushroom is a general layman term commonly used to describe the sporocarps of Ascomycetous and Basidiomycetous fungi (Tsujikawa et al., 2003; Adejumo and Awosanya, 2005; Atri et al., 2010). These are mostly saprophytic; however, many of them form mycorrhizal association with plants and a few of them including *Armillaria mellea* (Vahl.) P. Kumm. and *Ganoderma lucidum* (Curtis) P. Karsten are parasitic as well. Out of more than 41,000 species of mushrooms reported, approximately 850 species are documented from India (Deshmukh, 2004). Some of them including *Agaricus bisporus* (J.E. Lange) Imbach, *Lentinula edodes* (Berk.) Pegler, various species of *Pleurotus* and *Volvariella* has attained commercial status (Lindequest et al., 2005; Tewari, 2005). Man has been hunting for wild mushrooms since times immemorial due to their palatability, nutritional value, medicinal utility and unique flavour (Chang and Miles, 2004; Adejumo and Awosanya, 2005; Manimozhi and Kaviyarasan, 2013). There are references about their use as food supplement in various cultures. Their therapeutic value has also been recognized in early civilizations including Greek, Aryan, Egyptian, Roman, Chinese and Mexican. Greeks regarded mushrooms as strength food for warriors, while Romans considered them as "Food for God", and Chinese treated them as "Elixir of Life" (Tewari, 2005). The famous "Somrus" of Hindu Rig-Veda is considered as the decoction of mushrooms (Wasson, 1969). Besides their traditional uses, currently researches are underway to work out their chemical profile with a view to explore them for bioactive compounds for use in medicine. Proximate composition of large number of mushrooms have been documented by Crisan and Sands (1978), Purkayastha and Chandra (1985), Bano and Rajarathnam (1982), Chang and Miles (2004), Rai and Arumuganathan (2005), Manimozhi and Kaviyarasan (2013) and Kumari and Atri (2014). Investigations, to work out nutraceutical profile and bioactive properties of the components present therein and their therapeutic importance, have opened up altogether new horizon for the mushroom Mycologists (Lindequest et al., 2005). Some of the noteworthy contributions in this regard are those of Barros et al. (2008), Johnsy et al. (2011), Atri et al. (2012) and Sharma et al. (2014). Thus, mushrooms are known to have all essential components of a balanced food such as proteins, vitamins, minerals. They are poor in fats and carbohydrates and therefore, are low in calories. Besides, they also contain biologically active polysaccharides, proteins, and score of other biochemicals having antitumor, immunomodulating and antioxidant properties (Wasser and Weis, 1999; Tewari, 2005).

Besides traditionally cultivated mushrooms, many of their wild relatives are hunted regularly for consumption by local inhabitants of North India (Atri et al., 2010). These are mostly species of *Morchella* Dill. ex Pers., *Tuber* P. Micheli ex F. H. Wigg., *Pleurotus* (Fr.) P. Kumm., *Termitomyces* R. Heim, *Volvariella* Speg., *Macrolepiota* Sing., *Coprinus* Pers., *Russula* Pers., *Lactarius* Pers., *Boletus* L., *Lycoperdon* Pers. and *Podaxis* Desv., etc. The chapter presents nutritional attributes of five species each of *Lentinus* and *Pleurotus*, three species of *Macrolepiota* and seven species of *Termitomyces*. All these species are being consumed by the local inhabitants from the wild in North India.

Material and Methods

Twenty wild edible mushrooms were collected from different localities of North India. These include five species of *Lentinus* [*L. cladopus* Lèv, *L. connatus* Berk., *L. sajor-caju* (Fr.) Fr., *L. squarrosulus* Mont., *L. torulosus* (Pers.: Fr.) Lloyd], seven species of *Termitomyces* [*T. badius* Otieno, *T. heimii* Natarajan, *T. mammiformis* Heim, *T. medius* Heim & Gasse, *T. microcarpus* (Berk. & Br.) Heim, *T. radicatus* Natarajan, *T. striatus* (Beeli) Heim], three species of *Macrolepiota* [*M. dolichaula* (Berk. & Broome) Pegler & Rayner, *M. procera* (Scop.: Fr.) Sing., *M. rhacodes* (Vitt.) Sing.] and five species of *Pleurotus* [*P. cystidiosus* O.K. Miller, *P. floridanus* Singer, *P. pulmonarius* (Fr.) Quél., *P. sajor-caju* (Fr.) Sing. and *P. sapidus* Quél.].

Standard techniques of analysis as published by AOAC (1995) for determining the composition of mushrooms (carbohydrates, crude fat, proteins, fibers, ash, moisture content, minerals, heavy metals and vitamins) were employed.

The per cent carbohydrate content present in mushroom samples was estimated by subtracting total components excepting carbohydrates from 100 g of mushroom sample as per following relation after Crisan and Sands (1978):

Carbohydrate % = 100 − (Protein + Fat + Fiber + Ash content + Moisture content)

So as to determine fat content in wild samples of edible mushrooms, solvent extraction method as outlined by AOAC (1995) was followed. The per cent crude fat in mushroom was determined by employing the following relation:

$$\text{Crude Fat \%} = \frac{\text{Weight of Ether Soluble Material}}{\text{Weight of the Sample}} \times 100$$

For calculation of crude protein content in the form of per cent nitrogen, following relation as given by AOAC (1990) was followed:

$$\text{Nitrogen \%} = \frac{\text{Titrate Volume} \times 0.00014 \times \text{Volume made}}{\text{Aliquot taken} \times \text{Weight of the Sample}} \times 100$$

The fiber estimation was done according to the method given by Maynard (1970):

$$\text{Crude Fiber \%} = \frac{\text{Loss of Weight on Ignition}}{\text{Weight of the Sample}} \times 100$$

The ash content was determined as residue after incineration, following the protocol given by AOAC (2003):

$$\text{Crude Ash \%} = \frac{\text{Weight of Ash (g)}}{\text{Weight of the Sample}} \times 100$$

The moisture content estimation was done according to the method given in AOAC (2003):

$$\text{Moisture Content \%} = \frac{\text{Loss of Weight (g)}}{\text{Weight of the Sample}} \times 100$$

Quantitative estimation of minerals and heavy metals was done by the method of Jackson (1967) with the use of an atomic absorption spectrophotometer (Perkin Elmer Analyst 400).

Vitamins were estimated by the methods given in Indian Pharmacopoeia (1996).

Results and Discussion

The results of the proximate nutritional composition (expressed on dry weight basis) of the 20 wild edible mushrooms collected from different localities of North India are given in Tables 1–6.

Table 1. Proximate composition of wild edible mushrooms from North India on dry wt. basis.

S. No.	Name of species	Carbohydrate (%)	Crude fat (%)	Proteins (%)	Fibers (%)	Ash (%)	Moisture (%)
1	*Lentinus cladopus*	89.10 ± 0.6	0.80 ± 0.01	2.362 ± 0.01	1.91 ± 0.4	1.66 ± 0.4	4.168 ± 0.2
2	*L. connatus*	88.32 ± 0.2	0.48 ± 0.04	0.525 ± 0.03	2.02 ± 0.01	2.11 ± 0.01	6.545 ± 0.03
3	*L. sajor-caju*	85.82 ± 0.05	0.80 ± 0.02	1.05 ± 0.02	3.99 ± 0.01	1.91 ± 0.03	6.43 ± 0.2
4	*L. squarrosulus*	87.42 ± 0.04	0.62 ± 0.08	1.712 ± 0.05	1.83 ± 0.06	2.21 ± 0.06	6.208 ± 0.01
5	*L. torulosus*	87.33 ± 0.04	0.62 ± 0.02	2.45 ± 0.034	2.11 ± 0.02	1.52 ± 0.02	5.97 ± 0.5
6	*Macrolepiota dolichaula*	56.2 ± 0.10	3.2 ± 0.20	19.95 ± 1.35	4.85 ± 0.18	7.3 ± 0.15	8.5 ± 1.47
7	*M. procera*	60.82 ± 0.11	3.4 ± 0.08	19.95 ± 1.06	5.1 ± 0.22	1.93 ± 0.06	8.8 ± 0.62
8	*M. rhacodes*	68.19 ± 0.17	2.9 ± 0.11	16.45 ± 0.54	2.5 ± 0.01	2.16 ± 0.14	7.8 ± 0.62
9	*Pleurotus cystidiosus*	85.86 ± 0.029	0.80 ± 0.02	3.10 ± 0.04	3.12 ± 0.021	2.0 ± 0.07	5.12 ± 0.02
10	*P. floridanus*	86.63 ± 0.034	0.84 ± 0.1	2.27 ± 0.02	3.00 ± 0.021	2.20 ± 0.011	5.06 ± 0.04
11	*P. pulmonarius*	88.38 ± 0.02	0.79 ± 0.015	1.4 ± 0.12	2.98 ± 0.01	1.03 ± 0.03	5.42 ± 0.03
12	*P. sajor-caju*	87.15 ± 0.013	0.62 ± 0.04	2.36 ± 0.01	2.76 ± 0.04	1.91 ± 0.02	5.2 ± 0.03
13	*P. sapidus*	86.73 ± 0.024	0.72 ± 0.013	2.362 ± 0.03	2.97 ± 0.023	1.28 ± 0.7	5.938 ± 0.01
14	*Termitomyces badius*	39.0 ± 0.17	2.2 ± 0.10	44.00 ± 0.10	2.5 ± 0.01	6.6 ± 0.03	5.7 ± 0.22
15	*T. heimii*	36.2 ± 0.72	1.65 ± 0.19	40.95 ± 0.84	5.0 ± 0.11	8.6 ± 0.05	7.6 ± 0.23
16	*T. mammiformis*	47.65 ± 0.02	3.3 ± 0.17	23.45 ± 0.04	8.0 ± 0.26	9.9 ± 0.09	7.7 ± 0.06
17	*T. medius*	33.3 ± 0.37	2.0 ± 0.05	46.2 ± 0.02	7.5 ± 0.04	5.0 ± 0.20	6.0 ± 0.17
18	*T. microcarpous*	33.55 ± 0.11	2.5 ± 0.01	37.45 ± 0.45	5.0 ± 0.11	15.6 ± 0.06	5.9 ± 0.24
19	*T. radicatus*	41.07 ± 0.03	1.8 ± 0.12	40.00 ± 0.20	4.8 ± 0.04	6.3 ± 0.08	6.03 ± 0.02
20	*T. striatus*	60.27 ± 0.20	3.25 ± 0.06	12.95 ± 0.05	4.1 ± 0.15	12.13 ± 0.33	7.3 ± 0.18

Table 2. Macro and Micro elements in twenty wild species (mg/100 g of dry sample).

S. No.	Name of species	Ca	Cu	Fe	K	Mg	Mn	Na	Se	Zn
1	*Lentinus cladopus*	5.66 ± 0.02	2.33 ± 0.2	37 ± 0.03	0.057 ± 0.004	1.260 ± 0.08	ND	1.150 ± 0.03	ND	4 ± 0.02
2	*L. connatus*	221.3 ± 0.6	1 ± 0.01	18.6 ± 0.02	0.018 ± 0.00	1.722 ± 0.06	ND	0.708 ± 0.05	ND	3 ± 0.024
3	*L. sajor-caju*	97.3 ± 0.04	1.33 ± .02	14 ± 0.01	0.027 ± 0.05	1.028 ± 0.07	ND	0.726 ± 0.05	ND	4 ± 0.04
4	*L. squarrosulus*	3.33 ± 0.026	0.75 ± .03	6.4 ± 0.02	0.053 ± 0.00	1.200 ± 0.03	ND	2.392 ± 0.02	ND	6.10 ± 0.02
5	*L. torulosus*	6 ± 0.03	3.33 ± .03	11 ± 0.02	0.019 ± 0.00	807 ± 0.07	ND	0.837 ± 0.05	ND	3.33 ± 0.02
6	*Macrolepiota dolichaula*	5 ± 0.72	5 ± 0.92	241 ± 1.73	ND*	143 ± 1.00	1 ± 0.23	ND	0.10 ± 0.05	0.08 ± 0.01
7	*M. procera*	14 ± 0.6	9 ± 0.32	276 ± 0.87	ND	254 ± 2.00	5 ± 0.30	ND	0.08 ± 0.03	0.06 ± 0.03
8	*M. rhacodes*	28 ± 1.40	217 ± 0.50	248 ± 1.74	ND	217 ± 0.50	3 ± 0.53	ND	0.06 ± 0.03	0.09 ± 0.02
9	*Pleurotus cystidiosus*	3.4 ± 0.01	2 ± 0.04	20.24 ± 0.012	0.08 ± 0.00	1.28 ± 0.03	ND	1.12 ± 0.01	ND	6.21 ± 0.023
10	*P. floridanus*	7.3 ± 0.01	1 ± 0.03	12.33 ± 0.027	0.02 ± 0.01	1.26 ± 0.03	ND	1.18 ± 0.01	ND	5.33 ± 0.024
11	*P. pulmonarius*	6 ± 0.021	1 ± 0.00	17.33 ± 0.12	0.05 ± 0.00	1.68 ± 0.04	ND	0.95 ± 0.05	ND	8.33 ± 0.011
12	*P. sajor-caju*	4.2 ± 0.012	1 ± 0.01	22.31 ± 0.03	0.07 ± 0.00	1.22 ± 0.02	ND	0.99 ± 0.017	ND	4.39 ± 0.06
13	*P. sapidus*	2.3 ± 0.034	1 ± 0.06	21.33 ± 0.18	0.09 ± 0.0	1.24 ± 0.05	ND	0.94 ± 0.027	ND	4.33 ± 0.002
14	*Termitomyces badius*	24 ± 1.42	7 ± 1.48	144 ± 0.90	ND	205 ± 1.05	3 ± 0.32	ND	0.08 ± 0.02	0.06 ± 0.02
15	*T. heimii*	28 ± 1.35	6 ± 0.09	388 ± 0.74	ND	287 ± 0.50	5 ± 0.27	ND	0.1 ± 0.02	0.07 ± 0.01

Table 2. contd....

Table 2. contd.

S. No.	Name of species	Ca	Cu	Fe	K	Mg	Mn	Na	Se	Zn
16	*T. mammiformis*	30 ± 1.00	4 ± 0.46	673 ± 1.00	ND	277 ± 0.50	2 ± 0.53	ND	0.07 ± 0.01	0.06 ± 0.01
17	*T. medius*	204 ± 0.50	7 ± 1.84	454 ± 1.00	ND	330 ± 1.00	13 ± 0.79	ND	0.07 ± 0.01	0.06 ± 0.01
18	*T. microcarpus*	24 ± 1.42	6 ± 0.30	86 ± 1.00	ND	6 ± 0.20	3 ± 0.53	ND	0.12 ± 0.02	0.08 ± 0.02
19	*T. radicatus*	109 ± 1.00	9 ± 0.32	482 ± 1.50	ND	272 ± 1.98	10 ± 0.15	ND	0.09 ± 0.01	0.04 ± 0.01
20	*T. striatus*	15 ± 0.22	11 ± 1.08	82 ± 0.82	ND	191 ± 1.00	2 ± 0.28	ND	0.05 ± 0.02	0.07 ± 0.0

ND* – Not detected.

Table 3. Qualitative analysis for heavy metals in *Lentinus* and *Pleurotus* species.

Copper foil colour	Heavy metals	Results
Black or Brown	Arsenic, Lead	No colour change
Silver Shining	Silver, Mercury	No colour change
Bluish Black	Antimony	No colour change

Table 4. Quantitative analysis for heavy metals in *Macrolepiota* and *Termitomyces* Species (mg/100 g of the sample).

S. No.	Name of species	As	Pb	Ag	Hg	Sb	Cr	Cd
1	*Macrolepiota dolichaula*	ND*	ND	ND	0.062	ND	ND	0.0019
2	*M. procera*	ND	ND	ND	0.087	ND	ND	0.0019
3	*M. rhacodes*	0.0074	ND	ND	0.069	ND	ND	0.0014
4	*Termitomyces badius*	0.00002	ND	ND	0.096	ND	ND	0.0045
5	*T. heimii*	ND	ND	ND	0.018	ND	ND	0.0040
6	*T. mammiformis*	0.0037	ND	ND	0.043	ND	ND	0.0027
7	*T. medius*	0.00010	ND	ND	0.10	ND	ND	0.0039
8	*T. microcarpous*	ND	ND	ND	0.094	ND	ND	0.00488
9	*T. radicatus*	ND	ND	ND	0.016	ND	ND	0.0022
10	*T. striatus*	0.0185	ND	ND	0.099	ND	ND	0.0017

ND* – Not detected.

Table 5. Vitamins in lepiotoid and termitophilous mushrooms (mg/100 g of dry sample).

S. No.	Name of species	Vitamin A (Retinol)	Vitamin B1 (Thiamine)	Vitamin B2 (Riboflavin)	Vitamin C (Ascorbic Acid)
1	*Macrolepiota dolichaula*	0.07 ± 0.01	0.75 ± 0.05	0.13 ± 0.02	0.48 ± 0.02
2	*M. rhacodes*	0.09 ± 0.00	0.80 ± 0.04	0.13 ± 0.03	0.36 ± 0.01
3	*Termitomyces heimii*	0.12 ± 0.01	0.21 ± 0.01	0.25 ± 0.01	0.24 ± 0.01
4	*T. mammiformis*	0.11 ± 0.01	0.28 ± 0.01	0.21 ± 0.01	0.30 ± 0.0
5	*T. radicatus*	0.10 ± 0.02	0.42 ± 0.04	0.20 ± 0.01	0.96 ± 0.03
6	*T. reticulatus*	0.01 ± 0.00	0.26 ± 0.02	0.23 ± 0.01	1.45 ± 0.10

Carbohydrates

As is evident from the nutritional profile of different edible mushrooms in Table 1 that carbohydrates as in other edible mushrooms constitute the largest fraction of mushrooms per 100 g of dry matter. It has been documented to range from 33.3% in *Termitomyces medius* to 89.10% in *Lentinus cladopus*. The results of the present study reveal that the carbohydrate content on dry weight

Table 6. Ascorbic acid composition of *Lentinus* and *Pleurotus* species (mg/100 g of dry sample).

S. No.	Name of species	Ascorbic acid
1	*Lentinus cladopus*	0.46 ± 0.03
2	*L. connatus*	0.45 ± 0.01
3	*L. squarrosulus*	0.48 ± 0.02
4	*L. sajor-caju*	0.42 ± 0.03
5	*L. torulosus*	0.49 ± 0.03
6	*Pleurotus cystidiosus*	0.49 ± 0.03
7	*P. floridanus*	0.48 ± 0.013
8	*P. pulmonarius*	0.47 ± 0.09
9	*P. sapidus*	0.46 ± 0.02
10	*P. sajor-caju*	0.46 ± 0.04

basis was highest in *L. cladopus* (89.10%) followed by *Pleurotus pulmonarius* (88.38%) and *Lentinus connatus* (88.32%). Amongst various species of *Lentinus, Pleurotus, Macrolepiota* and *Termitomyces* minimum carbohydrate percentage was documented in the species of *Termitomyces*, in which it ranged from 33.3% in *T. medius* to 60.27% in *T. striatus*. In case of *Termitomyces* species evaluated by Mukiibi (1973) and Zakia et al. (1964) carbohydrate content has been reported to range from 54.2–62%. In comparison, in all other species of *Termitomyces* evaluated presently except for *T. striatus* (60.27%), the carbohydrate percentage is substantially low (33.3–47.65%). Amongst various species of *Lentinus* (85.82–89.10%) and *Pleurotus* (85.86–88.38%) evaluated the percentage of carbohydrate is quite high in comparison to carbohydrate percentage in the species of *Termitomyces* (33.3–60.27%) and *Macrolepiota* (56.2–68.19%). Adriano and Cruz (1933) reported substantially high carbohydrate percentage ranging from 75.1–85.2% in *Lentinus* which is almost comparable to the results obtained during the present study. While working with *L. squarrosulus* Nawanze et al. (2006) documented 56.23% carbohydrate content in the pileus and 65.07% carbohydrate content in the stipe which is much less in comparison to the percentage (87.42%) of carbohydrate evaluated in the presently examined specimen. Similarly in *L. sajor-caju* (85.82% instead of 68.24%) and *L. torulosus* (87.33% instead of 64.95%) substantially high amount of carbohydrate was evaluated in comparison to the proportion of carbohydrate reported in these species by Singdevsachan et al. (2013).

In case of *Termitomyces* species evaluated the amount of carbohydrate documented ranged from 33.3% in *T. medius* to 60.27% in *T. striatus*, which is much less in comparison to 77.6% carbohydrate percentage evaluated in *T. robustus* by Obodai et al. (2014).

In case of wild species of *Pleurotus*, the percentage of carbohydrate ranged from 85.86% in *P. cystidiosus* to 88.38% in *P. pulmonarius*. This is substantially on the higher side in comparison to the amount of carbohydrate reported in

P. tuber-regium (79.10%), *P. ostreatus* (64.14%) and *P. sajor-caju* (68.24%) by Obodai et al. (2014). However Bano and Rajarathnam (1982) reported the carbohydrate content in *Pleurotus* species to range from 46.6–81.8%. Crisan and Sands (1978) also reported carbohydrate content almost in the same range (46.6–73.7%) in different *Pleurotus* species. Phan et al. (2014) while studying the proximate composition of wild and commercial *Pleurotus giganteus* presented the average carbohydrate content to range from 647–672 g/kg. The present observations largely conform to the higher range of the carbohydrate content (55.3–85.9%) reported in 4 different *Pleurotus* species by Valverde et al. (2015). As is the case of the presently studied species, major portion of carpophore of *Pleurotus florida* has been reported to be composed of carbohydrate (Pushpa and Purushothama, 2010).

Amongst three *Macrolepiota* species, *M. rhacodes* (68.19%) contained maximum percentage of carbohydrate while *M. dolichaula* (56.2%) contained the least amount. This is much less in comparison to the percentage of carbohydrate documented in different species of *Lentinus* and *Pleurotus* but is much more than the carbohydrate percentage in many of the species of *Termitomyces* evaluated presently except *T. striatus* (Table 1).

Crude Fat

As documented by Crisan and Sands (1978) crude fat content of mushrooms has been reported to range from less than 1% to as high as 15–20%. Contrary to the percentage of carbohydrates, the amount of crude fats in the studied species of *Macrolepiota* (2.9–3.4%) and *Termitomyces* (1.65–3.3%) was much higher in comparison to that of *Lentinus* (0.48–0.80%) and *Pleurotus* (0.62–0.84%). Maximum amount of crude fats was recorded in *Macrolepiota procera* (3.4%). The percentage was comparable in *M. dolichaula* (3.2%), *Termitomyces mammiformis* (3.3%) and *T. striatus* (3.25%). Similar trend has been documented by Mukiibi (1973) and Zakia et al. (1964) for *Termitomyces* (2.2–6%) as for crude fat proportion is concerned. Nabubuya et al. (2010) evaluated 2.32% lipids in *T. microcarpus*. Masamba and Kazombo-Mwale (2010) documented 0.7% fats in *T. le-testui* (Pat.) Heim. Valverde et al. (2015) evaluated 2.2% lipids in *Agaricus bisporus* which is almost comparable and near to the proportion of fat evaluated in the species of *Termitomyces* and *Macrolepiota* presently.

Minimum percentage of crude fat was determined in *Lentinus connatus* (0.48%) followed by *L. squarrosulus* (0.62%), *L. torulosus* (0.62%) and *Pleurotus sajor-caju* (0.62%). Adriano and Cruz (1933), while working with different species of *Lentinus*, documented 0.6–3.4% crude fat on dry weight basis. In comparison to the presently achieved values Singhdevsachan et al. (2013) documented substantially higher fat content in *L. sajor-caju* (2.42%) and *L. torulosus* (1.36%). Nawanze et al. (2006) depicted 6.01% fat content in stipe and 6.56% in pileus of *L. squarrosulus* which is substantially high in comparison to the percentage of crude fat (0.62%) evaluated in the Indian sample of

L. squarrosulus (Table 1). Lipid content has also been reported to vary between 4.3–5.4% in Shiitake and Oyster mushrooms (Furlani, 2004) which is much more than the percentage of crude fats in the species of *Lentinus, Pleurotus* and *Macrolepiota* evaluated presently. Bano and Rajarathnam (1982) reported crude fat content in the range of 1.08–9.4% with an average of 2.85% in *Pleurotus* species. Rai et al. (1988) reported fat content ranging between 0.09% in *P. florida* to 0.18% in *P. sajor-caju* which is much less in comparison to the amount of crude fat (0.62–0.84%) determined in the presently evaluated species of *Pleurotus* (Table 1). Jandaik and Kapoor (1975) and Sivaprakasam (1983) documented 2.26 g and 0.25 g fat content in the dried samples of *P. sajor-caju*, respectively. In comparison substantially low amount of crude fat percentage ranging between 0.62–0.84% has been estimated in the presently studied five wild *Pleurotus* species. As compared in case of *P. opuntiae* (Duriev & Lév.) Sacc high crude fat content (2.4%) has been documented by Adriano and Cruz (1933). Even in case of *P. ostreatus* (Jacq.) P. Kumm. (1.6%) by Mendel (1898) and *P. florida* Sing. (1.54%) by Pushpa and Purushothama (2010) comparatively higher amount of crude fat has been documented. Phan et al. (2014) documented 37 g/kg fat in commercial *Pleurotus giganteus* and 31 g/kg in wild *P. giganteus*. Valverde et al. (2015) reported crude fat content in the range of 1–4.3% in *Pleurotus* species as compared to 0.62–0.84% crude fat evaluated presently (Table 1).

Proteins

Mushrooms are rich source of quality proteins and these are reported to rank well above common vegetables and fruits in quantity (Rai and Arumuganathan, 2005). During present study amongst species of *Lentinus, Macrolepiota, Pleurotus* and *Termitomyces* evaluated, the species of *Termitomyces* contained highest amount of proteins which ranged from 12.95% in *T. striatus* to 46.2% in *T. medius*. Mukiibi (1973) and Zakia et al. (1964) documented protein content of *Termitomyces* species to vary between 27.4–33%. The net protein percentage in the presently evaluated lepiotoid and termitophilous mushrooms (12.95–46.2%) is much more than the total protein proportion documented in FAO/WHO (1989) data given for some common fruits and vegetables including apple (0.3%), carrots (4%), onion (1.4%) orange (1.0%), peas (5.38%) and potatoes (1.6%). Botha and Eicker (1992) reported that protein of *T. umkowaani* (Cooke & Massee) D.A. Reid has very high sulphur containing amino acids which exceeds the essential amino acid content of the reference protein by 15%. Adejumo and Awosanya (2005) documented high protein content in *T. mammiformis* (36.8%) followed by *Lactarius trivialis* (Fr.) Fr. and other mushroom species. Oboh and Shodehinde (2009) while working with *T. mammiformis* from Nigeria documented 28.6% protein in pileus and 24.6% protein in the stipe which is almost comparable to the protein percentage documented in the Indian sample (23.45%) but is much less in comparison to the report of Adejumo and Awosanya (2005) in this regard (36.8%). Masamba and Kazombo-Mwale (2010) evaluated 3.9% protein while working with *T. le-testui*. Nabubuya et al. (2010) studied the nutritional properties of *T. microcarpus* and reported 25.48% protein content which is much less in comparison to the net amount (37.45%) of protein evaluated in the Indian sample of *T. microcarpus* (Table 1).

Even the presently studied species of *Macrolepiota* exhibited better protein percentage which ranged from 16.45% in *M. rhacodes* to 19.95% in *M. dolichaula* and *M. procera*. Amongst the five *Lentinus* species presently investigated the amount of protein has been recorded to be between 0.525–2.45%. Maximum amount of protein was documented in *L. torulosus* (2.45%) followed by *L. cladopus* (2.362%), while minimum percentage of protein was recorded in *L. connatus* (0.525%). As compared, the range of 6.1–13.1% protein content was documented by Adriano and Cruz (1933) in *Lentinus* species. Sugimori et al. (1971) reported 17.5% protein in *Lentinula edodes*. On fresh weight basis 1 g protein content per 100 g was recorded in *Lentinus polychorus* Lév by Natarajan and Manjula (1978). Gbolagade et al. (2006) worked out the proximate composition of *L. subnudus* Berk. and reported 5.1% protein. Nawanze et al. (2006), in *L. squarrosulus* reported 7.64% protein while in mycelial extract of *L. squarrosulus* 57.6 mg protein per 100 g of samples was reported by Omar et al. (2010). Manjunathan and Kaviyarasan (2011) documented 18.07% protein from the wild fruit bodies and 25% protein from the cultivated fruit bodies of *L. tuber-regium* (Fr.) Fr. Zhang et al. (2015) depicted 363.70–393.12 mg/g protein content of water extracts of *Lentinus edodes* and *Auricularia auricula*. Singdevsachan et al. (2013) reported 28.36 g/100 g protein in *Lentinus sajor-caju* and 27.31 g/100 g in *L. torulosus*. As compared to all the above reports by different workers much less percentage of protein (0.525–2.45%) has been documented in the five species of *Lentinus* including *L. connatus*, *L. cladopus*, *L. sajor-caju*, *L. squarrosulus* and *L. torulosus* during the present study (Table 1). The protein content has been reported to range between 12.51–28.40 g/100 g in *Pleurotus tuber-regium*, *P. ostreatus* EM-1, *P. sajor-caju* PScW, *Lentinus squarrosulus* SQW, *L. squarrosulus* LSF and *Termitomyces robustus* by Obodai et al. (2014).

In the present study the amount of protein in all species of *Pleurotus* except *P. pulmonarius* (1.4%), was higher which ranged from 2.36% in *P. sapidus* and *P. sajor-caju* to 3.10% in *P. cystidiosus*. For *Pleurotus* species Bano and Rajarathnam (1982) reported protein content to range between 8.9–38.7% on dry weight basis. Rai et al. (1988) reported protein content evaluated by employing different methods in seven species of *Pleurotus* to range from 1.57–3.47%. These results are in conformity with the range of protein (1.4–3.10%) percentage documented while working with five species of *Pleurotus* including *P. floridanus*, *P. pulmonarius*, *P. sajor-caju*, and *P. sapidus* during the present study. Desai et al. (1991) evaluated 33.68% crude proteins in case of pink Oyster. Investigation of composition of *P. sajor-caju* by Kattan et al. (1999) showed 34.80% crude protein which is substantially high in comparison to the protein percentage (2.36%) obtained in the Indian sample of

P. sajor-caju. The protein content has been reported to vary between 20–25.5% in two *Pleurotus* species (Alam et al., 2007). Similarly Pushpa and Purushothama (2010) also reported high (23.18%) amount of protein in *P. florida*. Phan et al. (2014) while studying the proximate composition of commercial and wild *Pleurotus giganteus* presented protein content to be ranging from 154–192 g/kg. As compared, the range of 7.0–37.4% protein content was documented by Valverde et al. (2015) in *Pleurotus* species. All these values are much higher in comparison (Table 1).

Fibers

Mushrooms are rich in fiber content. The fiber percentage on dry weight basis in termitophilous mushrooms range from 2.5–8% which is substantially high in comparison to the percentage of fiber in pleurotoid (2.76–3.12%) and lentinoid (1.83–3.99%) mushrooms evaluated in the present study. Amongst species of *Macrolepiota*, *M. procera* has been evaluated to contain 5.1% fibers in comparison to 4.85% in *M. dolichaula* and 2.5% in *M. rhacodes*. Mukiibi (1973) and Nabubuya et al. (2010) reported 2.2% and 11.21% fiber content in *Termitomyces microcarpus*, respectively. In case of *T. mammiformis*, Adejumo and Awosanya (2005) documented 7.2% fiber content while amongst the presently evaluated species, *T. mammiformis* (8%) is quite rich in this regard followed by *T. medius* (7.5%) and minimum percentage has been documented in *T. badius* (2.5%). Masamba and Kazombo-Mwale (2010) documented 0.5% fiber content in *T. le-testui*.

During the present investigations, amongst *Lentinus* species, minimum percentage of fiber has been determined in *L. squarrosulus* (1.83%), followed by *L. cladopus* (1.91%). Adriano and Cruz (1933) reported 4.2–5.3% fiber content in *Lentinus* species. According to Rai and Sohi (1988) and Rai et al. (1988), fiber content is high in *Lentinus* species. Adejumo and Awosanya (2005) documented 7.8% crude fiber content in *L. tigrinus* (Bull.) Fr. Nawanze et al. (2006) reported 6.8% crude fiber in the stipe of *L. squarrosulus* and 8.48% in the pileus. In case of *Lentinula edodes*, Regula and Siwulski (2007) evaluated about 1.95% soluble fibers and 44.2% insoluble fibers. In the mycelial extract of *L. squarrosulus* 0.1 g fiber content per 100 g of sample has been reported by Omar et al. (2010). High fiber content in cultivated species as compared to wild species of *L. tuber-regium* was documented by Manjunathan and Kaviyarasan (2011). Amongst *Pleurotus* species, *P. cystidiosus* contained maximum fiber percentage (3.12%), followed by *P. floridanus* (3%) while minimum percentage of fibers was analyzed in *P. sajor-caju* (2.76%). Adriano and Cruz (1933) worked on *P. limpidus* (Fr.) Gillet and *P. opuntiae* and reported 7.5% fiber content in both the species. Mendel (1898) estimated 8.7%, Kaul and Janardhanan (1970) estimated 1.2% and FAO (1972) report documented 8.7% fiber content in *P. ostreatus*. Bano and Rajarathnam (1982) reported 7.5–27.6% fiber content on fresh weight basis in *Pleurotus* species. Fiber content in *Pleurotus* species on fresh weight basis has been documented to range between 0.7–1.3% and highest in *P. membranaceus* Massee (Rai et al., 1988). In case of *P. ostreatus*, Regula and Siwulski (2007) reported 2.01% soluble dietary fibers and 39.8% insoluble fibers. Alam et al. (2007) documented 22–24% crude fibers in *P. florida* and *P. sajor-caju*. Patil et al. (2010) evaluated the proximate composition of *P. ostreatus* grown on various substrates and concluded that fiber content varied from 7.15–7.70% in fruit bodies obtained from different substrates. The fiber content of 5 wild species, namely *Calocybe indica* Purkayastha and Chandra (13.20%), *Agaricus bisporus* (18.23%), *Pleurotus florida* (23.18%), *Lyophyllum decastes* (Fr.) Sing. (29.02%) and *Russula delica* Fr. (15.42%), was documented by Pushpa and Purushothama (2010). Nile and Park (2014) reported 32 g/100 g, 37 g/100 g and 27 g/100 g fiber content in *Lentinus squarrosulus*, *Pleurotus djamor* and *P. sajor-caju*, respectively.

Ash

Amongst various edible species of *Lentinus, Pleurotus, Macrolepiota* and *Termitomyces* evaluated in the present study for ash content on dry weight basis, termitophilous mushrooms contained substantially high percentage of ash content which ranged from 5.0% in *T. medius* to 15.6% in *T. microcarpus*. Mukiibi (1973) and Zakia et al. (1964) documented high percentage (6.8–15.6%) of dry ash in different species of *Termitomyces* including *T. microcarpus* (14.1%). As compared, in the species of *Lentinus*, it ranged from 1.52% (*L. torulosus*) to 2.21% (*L. squarrosulus*). Adriano and Cruz (1933) reported 4.2–11.2% ash content for *Lentinus* species. Relatively much less ash percentage (1.52–2.21%) was detected in the presently examined species of *Lentinus* (Table 1). Adejumo and Awosanya (2005) documented 5–8% ash content in *L. tigrinus*. Nawanze et al. (2006) reported 6.62% ash in the stipe and 8.42% ash in the pileus of *L. squarrosulus*. On dry weight basis 6.73% ash content was documented by Regula and Siwulski (2007) in *Lentinula edodes*. Adedayo et al. (2011) reported 0.90% ash content in *Lentinus subnudus*. Manjunathan and Kaviyarasan (2011) documented high ash content in the cultivated fruit bodies of *L. tuber-regium*. Singdevsachan et al. (2013) documented 4.88 g/100 g ash content in *Lentinus sajor-caju* and 13.16 g/100 g in *L. torulosus*.

As is evident from the data given in Table 1, in species of *Pleurotus* the ash content ranged from 1.03% (*P. pulmonarius*) to 2.20% (*P. floridanus*). Rai and Arumuganathan (2005) documented 8–10% ash content in *Agaricus bisporus* and 5–15% in *Pleurotus* species. Mendel (1898) documented high amount (7.5%) of ash per 100 g on dry weight basis in *P. ostreatus* whereas 8.3% ash content was documented by Adriano and Cruz (1933) and 8.7% in FAO (1972) data. Furlani (2004) while working on Oyster mushrooms and Shiitake in Brazil documented 7.0–12% ash content in the carpophores. Patil et al. (2010), in cultivated *P. ostreatus*, documented 5.90–6.70% ash content which is very high in comparison to the ash percentage documented in the evaluated species of

Pleurotus. Adedayo et al. (2011) evaluated 0.85% ash content in *P. ostreatus* which is much less from the amount of ash content (1.03–2.20%) obtained in the presently evaluated samples of *Pleurotus.* Obodai et al. (2014) studied the nutritional composition of 6 edible mushrooms and reported ash content range between 5.60–6.38 g/100 g. Amongst *Macrolepiota* species investigated, the maximum ash percentage has been documented in *M. dolichaula* (7.3%) while minimum in *M. procera* (1.93%).

Moisture

Moisture is an integral part of the proximate composition of various mushrooms. It is normally determined to have the exact estimate of various nutritional components, including carbohydrates, fat, proteins, fibers and ash. The exact value of moisture content varies substantially from species to species as is evident from the data given in Table 1.

Crisan and Sands (1978) documented an average mushroom to contain approximately 90% moisture when fresh and 10–12% when air-dried. During the present investigation the moisture percentage in air-dried samples ranged between 4.168% in *Lentinus cladopus* to 8.8% in *Macrolepiota procera.* The moisture percentage ranges between 4.168–6.545% in dried samples of five species of *Lentinus* evaluated presently (Table 1). Adriano and Cruz (1933) documented 19.7% moisture content in dried samples of *Lentinus.* Cuptapun et al. (2010) worked out four edible mushrooms (*L. lepideus* (Fr.: Fr.) Fr., *Pleurotus ostreatus, P. sajor-caju* and *Lentinula edodes*) and documented 7.21–7.5% moisture content on dry weight basis. Johnsy et al. (2011) studied the nutritional composition of 10 naturally growing edible mushrooms and reported moisture content between 87.3–95.17%. Singdevsachan et al. (2013) evaluated 80.29% moisture content in *Lentinus sajor-caju* and 80.97% in *L. torulosus* on fresh weight basis.

For *Pleurotus ostreatus,* Adriano and Cruz (1933) evaluated the moisture percentage at 10.7% when air-dried. As compared in the air-dried samples of presently examined species of *Pleurotus,* minimum percentage of moisture was documented in *P. floridanus* (5.06%) while maximum percentage in *P. sapidus* (5.938%). On fresh weight basis 86%, 92% and 85% moisture content was reported by Nile and Park (2014) in *Lentinus squarrosulus, Pleurotus djamor* and *P. sajor-caju,* respectively. As reported by Mukiibi (1973), the air-dried samples of *Termitomyces* possess moisture content between 8–12%. Meghalatha et al. (2014) estimated 26.2% moisture in *Microporus xanthopus.* In the presently evaluated species, maximum moisture percentage was documented in the air dried samples of *T. mammiformis* (7.7%) while minimum moisture content was evaluated in *T. badius* (5.7%). Amongst the species of *Macrolepiota,* *M. procera* (8.8%) possessed higher moisture percentage on dry weight basis in comparison to other two species evaluated (Table 1).

Although moisture has no nutritional relevance but it needs to be taken into consideration so as to have the correct value of nutritionally and nutraceutically important constituents. As documented by Crisan and Sands (1978) not evaluating the moisture percentage will normally result in inflated estimates of various nutritional components present in the samples.

Minerals

Mushrooms are an excellent source of important minerals. About 56–70% of the total ash has been reported to be normally composed of different minerals such as Ca, Mg, Na, K, P, etc. (Li and Chang, 1982). The fruit body mineral composition of mushrooms is reported to be largely influenced by uptake of minerals by growing mycelium from the substrate (Chang and Hayes, 1978; Chang and Miles, 2004). In most of the cases, K, Na and P are reported to be present in significant amount, followed by Ca and Fe being in poor amount (Chang and Hayes, 1978). The results of the present study to determine the proximate mineral content in dried samples of wild edible species of *Lentinus*, *Pleurotus*, *Macrolepiota* and *Termitomyces* have been summarized in Table 2. As is evident from the data, there is no set pattern in the presence and the amount of different minerals in various species. Amongst the evaluated taxa, *Lentinus connatus* (221.3 mg) contained maximum amount of Ca. As compared *Macrolepiota rhacodes* (217 mg) possessed highest amount of Cu, *Termitomyces mammiformis* (673 mg) contained maximum Fe content, *Pleurotus sapidus* (0.09 mg) showed highest level of K. The amount of Mg was the maximum in *L. torulosus* (807 mg) and Mn was the maximum in *Termitomyces medius* (13 mg). Sodium (Na) content was found to be maximum in *L. squarrosulus* (2.392 mg) while Se level was highest in *Termitomyces microcarpus* (0.12 mg) and Zn content in *Pleurotus pulmonarius* (8.33 mg), respectively.

Amongst five species of *Lentinus*, *L. connatus* (221.3 mg) is richest in Ca followed by *L. sajor-caju* (97.3 mg). As compared, the remaining three species are poorer in this regard as the amount of Ca evaluated in them is substantially low. *Lentinus torulosus* (3.33 mg) contained highest amount of Cu while maximum amount of Fe was documented in *L. cladopus* (37 mg). Even the highest level of K (0.057 mg) was also recorded in this species. Besides Ca, *L. connatus* also possessed highest amount of Mg (1.722 mg). The other two elements, Na (2.392 mg) and Zn (6.10 mg), were the maximum in *L. squarrosulus*. Adriano and Cruz (1933) documented the presence of Ca (132 mg), P (622 mg), Fe (27 mg), Na (831 mg), K (1290 mg) and Mg (60 mg) per 100 g in fresh samples of *Lentinus exilis* Klotzsch: Fr. Manjunathan and Kaviyarsan (2011) while working on cultivated and wild samples of *L. tuber-regium* documented K (90.8 mg and 7.53 mg), Ca (87 mg and 2.66 mg), Na (37.3 mg and 1.2 mg), Mg (30.4 mg and 2.45 mg), Cu (1 mg and 0.11 mg) and Mn (1.7 mg and 0.08 mg) per 100 g of the respective samples. Singdevsachan et al. (2013) while working with *Lentinus sajor-caju* and *L. torulosus* documented P (0.10 g/100 g and 0.24 g/100 g), K (0.14 g/100 g and 0.85 g/100 g), Mn (0.12 mg/kg and 0.05 mg/kg), Ni (0.05 mg/kg and 0.04 mg/kg) and Fe (2.37 mg/kg and 2.94 mg/kg) in the respective mushrooms. During evaluation Mn and Se were not detected in any of the *Lentinus* species evaluated for macro and micromineral profile.

Amongst species of *Macrolepiota* evaluated, maximum amounts of Ca (28 mg), Cu (217 mg) and Zn (0.09 mg) were documented in *M. rhacodes,* Fe (276 mg), Mg (254 mg) and Mn (5 mg) in *M. procera,* and Se (0.10 mg) in *M. dolichaula.* Olfati et al. (2009) documented the presence of P (8.8 mg), K (23.9 mg), Na (1.1 mg), Ca (1.4 mg) and Mg (1.8 mg) in *M. procera.* During the present study K and Na were not detected in any *Macrolepiota* species. Meghalatha et al. (2014) documented the presence of Na (1.01 ppm), K (0.57 ppm), P (0.33 ppm), Ca (0.18 ppm), Mg (0.10 ppm), Zn (11.45 ppm), Cu (26.25 ppm) and Mn (32.89 ppm) in *Microporus xanthopus.*

Amongst the species of *Pleurotus,* Ca (7.3 mg) and Na (1.18 mg) was the maximum in *P. floridanus,* Cu (2.0 mg) in *P. cystidiosus,* Fe (22.31 mg) in *P. sajor-caju,* K (0.09 mg) in *P. sapidus,* Mg (1.68 mg) and Zn (8.33 mg) in *P. pulmonarius.* Mn and Se were not detected in any of the samples evaluated. FAO (1972) report documented the presence of Ca (33 mg), P (1348 mg), Fe (15.2 mg), Na (83 mg) and K (3793 mg) in the fresh samples of *P. ostreatus.* Jandaik and Kapoor (1975) while evaluating fresh specimens of *Pleurotus sajor-caju* for different mineral contents, documented the presence of Ca (0.04%), P (1.62%), Fe (0.01%), Na (0.20%), K (2.5%) and Cu (0.02%). Bano et al. (1981) reported a range of variation in Cu (12.2–21.9 ppm), Ca (0.3–0.5 ppm) and Pb (1.5–3.2 ppm) contents in different species of *Pleurotus.* Phan et al. (2014) reported a range of variation in K (11.71–13.46 g/kg), P (4.01–5.27 g/kg), Mg (0.65–0.67 g/kg), Ca (0.058–0.087 g/kg), Na (0.047–0.058 g/kg), Fe (0.014–0.019 g/kg), Zn (0.027–0.042 g/kg), Mn (0.041–0.043 g/kg) and Cu (0.228–0.60 mg/kg) in commercial and wild strain of *Pleurotus giganteus.*

In species of *Termitomyces,* maximum level of Ca (204 mg), Mn (13 mg) and Mg (330 mg) was documented in *T. medius.* Copper (Cu) level (11 mg) was highest in *T. striatus.* In case of *T. mammiformis* the amount of Fe (673 mg) was substantially high. The level of Se (0.12 mg) and Zn (0.08 mg) was the maximum in *T. microcarpus.* Potasium (K) and Na were not detected in any samples of *Termitomyces.* There are reports of occurrence of minerals in variable concentrations in wild and cultivated mushrooms. Crisan and Sands (1978) documented higher mineral content in canned and dried mushrooms as compared to fresh mushrooms.

Sanmee et al. (2003) documented presence of Ca, P, Fe, Mn, Cu and Zn in popular wild edible mushrooms from Northern Thailand. As also observed in the present studies, Vetter (2003) reported low concentration of Na in mushrooms. Manimozhi and Kaviyarasan (2013) evaluated *Agaricus heterocystis* Heinem. & Gooss-Font. for the presence of minerals and documented K (422 mg), Mg (39 mg), Ca (81 mg), Fe (39 mg), Cu (3.72 mg), Mn (0.8 mg) and Zn (1.9 mg) per 100 g of the sample. Ravikrishan et al. (2015) documented presence of Ca (90.60 mg/100 g), P (31 mg/100 g), Fe (3 mg/100 g),

K (104 mg/100 g), Na (37.70 mg/100 g) and Mg (177 mg/100 g) in *Lentinus polychorus* from Western Ghats, Southern India.

Heavy Metals

All 20 samples of wild edible mushrooms were also evaluated for the presence of heavy metals (As, Pb, Ag, Hg, Sb, Cr and Cd). The species of *Lentinus* and *Pleurotus* were evaluated on qualitative basis only as these were not directly growing on the soil, instead were collected from the wooden stumps. However, no heavy metals were detected (Table 3). The species of *Macrolepiota* and *Termitomyces* that were in direct contact with the ground were found to possess heavy metals in varying quantity (Table 4). Arsenic (As) content was the maximum in *Termitomyces striatus* (0.0185 mg), followed by *Macrolepiota rhacodes* (0.0074 mg), *Termitomyces mammiformis* (0.0037 mg), *T. medius* (0.00010 mg) and *T. badius* (0.00002 mg), but it was not detected in the remaining species of *Macrolepiota* and *Termitomyces*. Other heavy metals—Pb, Ag, Sb and Cr were not detected in *Termitomyces* and *Macrolepiota*. However, presence of Hg was detected that ranged from 0.062–0.087 mg in the dry samples of *M. dolichaula, M. procera, M. rhacodes,* and 0.016–0.1 mg in all seven species of *Termitomyces* evaluated. Cadmium (Cd) was another heavy metal which ranged from 0.0014–0.0019 mg in all the three species of *Macrolepiota* and 0.0017–0.00488 mg in seven species of *Termitomyces* investigated. However, the amount of different heavy metals present in the species of *Macrolepiota* and *Termitomyces* is well within the range of permissible limits for human consumption (FAO/WHO, 1976). Pb and Cd were not detected in *Microporus xanthopus* by Meghalatha et al. (2014); however, in comparison all the species of *Macrolepiota* and *Termitomyces* evaluated presently did contain some proportion of Cd, but within the permissible limit for human consumption (Table 4). Singdevsachan et al. (2013) stated the absence of Cd in *Lentinus sajor-caju* and *L. torulosus*. Phan et al. (2014) reported the absence of Se in the commercial and wild strains of *Pleurotus giganteus*.

According to different workers mushrooms are reported to possess very effective mechanism to readily absorb heavy metals from their substrates (Cibulka et al., 1996; Isiloglu et al., 2001; Turkekul et al., 2004). There are reports of high concentration of heavy metals in the fruit bodies of different mushrooms collected adjacent to heavy metal smelters and oil polluted areas (Kalač et al., 1991; Isiloglu et al., 2001; Elekes et al., 2010). In a study on *Pleurotus ostreatus, Boletus* and *Agaricus bisporus* species, Demirbas (2001) documented accumulation of high concentration of Cd and Hg in *A. bisporus*. Although there, are several reports about the accumulation of heavy metals in mushrooms (Higgs et al., 1972; Thomas et al., 1972; Leh, 1975), however, during the present investigation no heavy metals were detected in *Lentinus* and *Pleurotus* species. This is probably due to their lignicolous nature. The species of *Macrolepiota* and *Termitomyces* were found to have some traces of heavy metals that were within the permissible limits of human consumption.

Vitamins

An important factor in the overall nutritional value of food is its vitamin content. Mushrooms are rich in vitamin B-complex (Breene, 1990; Mattila et al., 1994, 2001; Chang and Buswell, 1996) while poor in fat soluble vitamins (A, D, E and K). During the present investigation wild edible lentinoid and pleurotoid mushrooms were evaluated for ascorbic acid (vitamin C) and some of the lepiotoid and termitophilous mushrooms were evaluated for retinol (vitamin A), thiamine (vitamin B_1), riboflavin (vitamin B_2) and ascorbic acid (vitamin C) (Tables 5 and 6). The results obtained in present study are in agreement with those of Barros et al. (2007, 2008); Heleno et al. (2009); Ouzouni et al. (2009), and Grangeia et al. (2011). Comparative account of vitamin in edible mushrooms has been reported by number of workers including Esselen and Fellers (1946), Litchfield (1964) and Manning (1985).

Amongst the evaluated taxa *T. reticulatus* (1.45 mg) shows the maximum amount of ascorbic acid (Table 5), followed by *Lentinus torulosus* and *Pleurotus cystidiosus* (0.49 mg) (Table 6). Bano (1976) reported 13–14.7 mg of ascorbic acid/100 g in different mushrooms. Sapers et al. (1999) and Mattila et al. (2001) reported that mushrooms contain ascorbic acid in small amount. Vitamin C is one of the major contributors to the antioxidant activity of mushrooms (Caglarirmak et al., 2002). As for other vitamins are concerned, *Macrolepiota rhacodes* possessed highest amount of thiamine (0.80 mg), *Termitomyces heimii* contained maximum content of retinol (0.12 mg) and riboflavin (0.25 mg). Furlani and Godoy (2008) analyzed vitamins B_1 and B_2 contents in cultivated mushrooms (*Agaricus bisporus, Lentinula edodes,* and *Pleurotus* species) and the results showed that the amount of thiamine (vitamin B_1) ranged from 0.004–0.08 mg/100 g in comparison to riboflavin (vitamin B_2) that ranged from 0.04–0.3 mg/100 g. Amongst the five species of *Lentinus* evaluated, *L. torulosus* is the richest in ascorbic acid (0.49 mg), followed by *L. squarrosulus* (0.48 mg), while *L. sajor-caju* contained the least (0.42 mg) amount. Mycelial extract of *L. squarrosulus* has been reported to contain appreciable amount of vitamin A, E, B_1 and B_3 (Omar et al., 2010). Phan et al. (2014) depicted variable amount of vitamin B_1, vitamin B_2, vitamin B_3, and vitamin C in commercial and wild strains of *Pleurotus giganteus*. In different species of *Pleurotus* examined, maximum level of ascorbic acid (0.49 mg) was documented in *P. cystidiosus* followed by *P. floridanus* (0.48 mg). In case of *P. ostreatus*, 4.8 mg thiamine, 4.7 mg riboflavin and 108.7 mg of niacin have been documented in FAO (1970) report from 100 g of mushrooms on dry weight basis. Li and Chang (1985) assessed vitamin C content of some mushrooms and documented 2–3 mg in *P. ostreatus* and 4–5 mg in *P. sajor-caju*. Rai et al. (1988) documented vitamin C that ranged from 2.2–4.9 mg in *Pleurotus* species on fresh weight basis. In case of *P. sajor-caju*, Jandaik and Kapoor (1975) evaluated 0.06 mg of ascorbic acid per 100 g of the dry sample which is comparable to the present results. The ascorbic acid content of cultivated *P. ostreatus* has been reported to range from 12.52–15.80 mg (Patil et al., 2010). Okwulehie et al. (2007) reported 0.490 mg ascorbic acid in *Schizophyllum commune* Fr. and 0.370 mg ascorbic acid in *Polyporus* P. Micheli: Adams. sample per 100 g on dry weight basis. These results are in conformity with the present observations on wild samples of *Lentinus* and *Pleurotus*. The amount of vitamin C in the presently investigated mushrooms is higher in comparison to the amount of vitamin C reported in *Lycoperdon perlatum* Pers., *Clavaria vermicularis* Sw., *Marasmius oreades* (Bulton) Fr., *Russula delica*, *Morchella conica* Pers. and *Pleurotus pulmonarius* (Aziz et al., 2007; Türkoğlu et al., 2009; Ramesh and Pattar, 2010).

Conclusion

Nutritional evaluation of wild edible species of *Lentinus, Pleurotus, Termitomyces* and *Macrolepiota* revealed the presence of macronutrients, minerals and vitamins in substantial amount in all the species. About 33–89% of macronutrient composition is constituted by carbohydrates, 0.48–3.4% by fat, 0.5–46% by proteins, 1–8% by fibers, 1–16% by ash and remaining proportion is moisture on dry weight basis. The investigated mushrooms are also rich sources of macro and micro elements, including Ca, Cu, Fe, K, Mg, Mn, Na, Se and Zn as well as vitamin A (retinol), vitamin B_1 (thiamine), vitamin B_2 (riboflavin) and vitamin C (ascorbic acid) which is extremely important in enhancing the functioning of immune system. The presence of substantial amounts of nutritional components in the wild edible species accounts for their nutritional credentials. In view of their proximate composition profile these mushrooms provide equally potent culinary option as is provided by mushrooms like *Agaricus bisporus, Lentinula edodes*, etc.

References

Adedayo, J., Majekodumni, A. and Rachel, M. 2011. Proximate analysis of four edible mushrooms. J. Appl. Sci. Environ. Manage., 15: 9–11.

Adejumo, T.O. and Awosanya, O.B. 2005. Proximate and mineral composition of four edible mushroom species from South Western Nigeria. Afr. J. Biotechnol., 4: 1084–1088.

Adriano, F.T. and Cruz, R.A. 1933. The chemical composition of Philippine mushrooms. Philipp. J. Agaric, 4: 1–11.

Alam, N., Khan, A., Hossain, S., Ruhul, A.S.M. and Liakot, A.K. 2007. Nutritional analysis of dietary mushroom *Pleurotus florida* Singer and *Pleurotus sajor-caju* (Fr.) Singer. Bangladesh J. Mush., 1: 1–7.

AOAC (Association of Official Analytical Chemistry). 1990. Official methods of analysis of the Association of Official Analytical Chemists. 15th ed. Arlington, 1105–1106.

AOAC International. 1995. Official methods of analysis of AOAC International. 2 vols. 16th ed. Arlington, VA, USA, Association of Analytical Communities.

AOAC International. 2003. Official methods of analysis of AOAC International. 17th ed. 2nd revision. Gaithersburg, MD, USA, Association of Analytical Communities.

Atri, N.S., Sharma, S.K., Joshi, R., Gulati, A. and Gulati, A. 2012. Amino acid composition of five wild *Pleurotus* species chosen from North West India. Europ. J. Bio. Sci., 4: 31–34.

Atri, N.S., Saini, M.K., Gupta, A.K., Kaur, A., Kour, H. and Saini, S.S. 2010. Documentation of wild edible mushrooms and their seasonal availability in Punjab. pp. 161–169. *In*: Mukerji, K.G. and Manoharachary, C. (eds.). Taxonomy and Ecology of Indian Fungi. I. K. International Publishing House, Pvt. Ltd., New Delhi.

Aziz, T., Mehmet, E.D. and Nazime, M. 2007. Antioxidant and antimicrobial activity of *Russula delica* Fr: An edible wild mushroom. Eurasian J. Analyt. Chem., 2: 54–63.

Bano, Z. 1976. Nutritive value of Indian mushrooms and medicinal practices. Eco. Bot., 31: 367–371.

Bano, Z. and Rajarathnam, S. 1982. *Pleurotus* mushroom as a nutritious food. *In*: Chang, S.T. and Quimio, T.H. (eds.). Tropical Mushrooms Biological Nature and Cultivation Methods. The Chinese University Press.

Bano, Z., Bhagya, S. and Srinivasan, K.S. 1981. Essential amino acid composition and proximate analysis of the mushroom *Pleurotus eous* and *P. florida*. Mushroom News Lett. Tropics., 1: 6–10.

Barros, L., Baptista, P., Correia, D.M., Casal, S., Oliveira, B. and Ferreira, I.C.F.R. 2007. Fatty acid and sugar compositions and nutritional value of five wild edible mushrooms from north east Portugal. Food Chem., 105: 140–145.

Barros, L., Cruz, T., Baptista, P., Letícia, M.E. and Ferreira, I.C.F.R. 2008. Wild and commercial mushrooms as source of nutrients and nutraceuticals. Food and Chem. Toxicol., 46: 2742–2747.

Botha, W.J. and Eicker, A. 1992. Nutritional values of *Termitomyces* mycelial protein and growth of mycelium on natural substrate. Mycol. Res., 96: 350–354.

Breene, W.M. 1990. Nutritional and medicinal value of specialty mushrooms. J. Food Protect., 53: 883–894.

Caglarirmak, N., Otles, S. and Unal, K. 2002. Nutritional value of edible wild mushrooms collected from the Black Sea region of Turkey. Micol. Apl. Int., 14: 1–5.

Chang, S.T. and Buswell, J.A. 1996. Mushroom nutriceuticals. World J. Microbiol. Biotechnol., 12: 473–476.

Chang, S.T. and Hayes, W.A. 1978. The Biology and Cultivation of Edible Mushrooms. Academic Press Inc., London, pp. 819.

Chang, S.T. and Miles, G. 2004. Mushroom: cultivation, nutritional value, medicinal effects and environmental impact (2nd Ed.) CRC Press, Boca Raton, Fla, pp. 455.

Cibulka, J., Sisak, L., Pulkrab, K., Miholova, D., Szakova, J. and Fucikova, A. 1996. Cadmium, lead and caesium levels in wild mushrooms and forest berries from different localities of the Czech Republic. Sci. Agric. Biochem., 27: 113–129.

Crisan, E.V. and Sands, A. 1978. Nutritional value. pp. 251–293. *In*: Chang, S.T. and Hayes, W.A. (eds.). The Biology and Cultivation of Edible Mushrooms. New York, Academic Press.

Cuptapun, Y., Hengsawadi, D., Mesomya, W. and Sompoch, Yaieiam. 2010. Quality and quantity of protein in certain kinds of edible mushrooms in Thailand. Kasetsart J. (Nat. Sci.), 44: 664–670.

Demirbas, B. 2001. Concentration of 21 metals in 18 species of mushrooms growing in the East Black Sea region. Food Chem., 75: 453–457.

Desai, A.V.P., Earanna, N. and Shetty, K.S. 1991. Pink *Pleurotus*—a new edible oyster mushroom. Adv. Mushroom Sci., pp. 27.

Deshmukh, S.K. 2004. Biodiversity of tropical basidiomycetes as a source of novel secondary metabolites. pp. 121–140. *In*: Jain, P.C. (ed.). Microbiology and Biotechnology for Sustainable Development. CBS Publishing & Distribution New Delhi.

Elekes, C.C., Busuioc, G. and Ionita, G. 2010. The bioaccumulation of some heavy metals in the fruiting body of wild growing mushrooms. Not. Bot. Hort. Agrobot. Cluj., 38: 147–151.

Esselen, W.B. and Fellers, C.R. 1946. Mushrooms for food and flavor. Bull. Mass. Agric. Exp. Sta., pp. 434.

FAO. 1970. Food and Agricultural Organization (No. 12) FAO, Rome.

FAO. 1972. Food Composition Table for Use in East Asia. Food Policy and Nutrition. Div., Food Agric Org. U.N. Rome.

FAO/WHO Standards. 1976. List of maximum levels recommended for contaminants by the Joint FAO/WHO Codex Alimentarius.

FAO/WHO. 1989. Protein quality evaluation. Report of the joint FAO/WHO expert consultation. Food and Nutrition Paper no. 51. Food and Agriculture Organizations and the World Health Organization, Rome, Italy.

Furlani, R.P.Z. 2004. Valor nutricional de cogumelos cultivados no Brasil (Thesis) Campinas. SP (Brasil), Universidade Estadual de Campinas (Faculdade de Engenharia de Alimentos).

Furlani, R.P.Z. and Godoy, H.T. 2008. Vitamins B_1 and B_2 contents in cultivated mushrooms. Food Chem., 106: 816–819.

Gbolagade, J.S., Fasidi, I.O., Ajayi, E.J., Wiley, J. and Sobowale, A.A. 2006. Effect of physicochemical factors and semi synthetic media on vegetative growth of *Lentinus subnudus* Berk. an edible mushroom from Nigeria. Food Chem., 99: 742–747.

Grangeia, C., Heleno, S.A., Barros, L., Martins, A. and Ferreira, I.C.F.R. 2011. Effects of trophism on nutritional and nutraceutical potential of wild edible mushrooms. Food Res. Inter. doi: 10.1016/j.foodres.2011.03.006.

Heleno, S.A., Barros, L., Sousa, M.J., Martins, A. and Ferreira, I.C.F.R. 2009. Study and characterization of selected nutrients in wild mushrooms from Portugal by gas chromatography and high performance liquid chromatography. Microchem. J., 93: 195–199.

Higgs, D.J., Morris, V.C. and Levander, O.A. 1972. Effect of cooking on selenium content of food. J. Agric. Food Chem., 20: 678–680.

Indian Pharmacopoeia. 1996. Addendun 2000. The Controller of Publications, Govt. of India, Ministry of Health and Family Welfare, Delhi, India.

Isiloglu, M., Merdivan, M. and Yilmaz, F. 2001. Heavy metal contents in some macrofungi collected in the North western part of Turkey. Enr. Cont. Toxi., 41: 1–7.

Jackson, M.L. 1967. Soil Chemical Analysis. New Delhi: Prentice Hall of India Ltd.

Jandaik, C. and Kapoor, J. 1975. Nutritive value of mushroom *Pleurotus sajor-caju*. The Mushroom Journ., 36: 408–410.

Johnsy, G., Sargunam, S.D., Dinesh, M.G. and Kaviyarasan, V. 2011. Nutritive value of edible wild mushrooms collected from the western ghats of Kanyakumari district. Bot. Res. Int., 4: 69–74.

Kalač, P., Burda, J. and Staskova, I. 1991. Concentration of lead, cadmium, mercury and copper in mushroom in the vicinity of a lead smelter. Sci. Tot. Env., 105: 109–119.

Kattan, E.I., Helmy, M.H., Abdel, Z.A.E.I.H., Leithy, M. and Abdelkawi, K.A. 1999. Studies on cultivation techniques and chemical composition of oyster mushrooms. Mushroom J. Tropics., 11: 59–66.

Kaul, T. and Janardhanan, K. 1970. Experimental cultivation of *Pleurotus ostreatus* white strain. Indian Phytopath., 23: 578.

Kumari, B. and Atri, N.S. 2014. Nutritional and nutraceutical potential of wild edible macrolepiotoid mushrooms of north India. Int. J. Pharm. Pharmac. Sci., 6(2): 200–204.

Leh, H.O. 1975. Bleigehalte in Pilzen. Zeitschrift für Lebensmittel-Untersuchung und –Forschung, 157: 141–142.

Li, G.S.F. and Chang, S.T. 1982. Nutritive value of *Volvariella volvacea*. pp. 199–219. *In*: Chang, S.T. and Quimio, T.H. (eds.). Tropical Mushroom—Biological Nature and Cultivation Methods. The Chinese University Press, Hong Kong.

Li, G.S.F. and Chang, S.T. 1985. Determination of vitamin C (ascorbic acid) in some mushrooms by differential pulse polarography. Mushroom J. Tropics., 5: 11–16.

Lindequist, U., Niedermeyer, T.H.S. and Julich, W.D. 2005. The pharmacological potential of mushroom. ECAM., 2: 285–299.

Litchfield, J.H. 1964. Nutrient content of morel mushroom mycelium: B vitamin composition. J. Food Sci., 29: 690–691.

Manimozhi, M. and Kaviyarasan, V. 2013. Nutritional composition and antibacterial activity of indigenous edible mushroom *Agaricus heterocystis*. IJABR, 4: 78–84.

Manjunathan, J. and Kaviyarasan, V. 2011. Nutrient composition in wild and cultivated edible mushroom, *Lentinus tuber-regium* (Fr.) Fr. Tamil Nadu, India. Int. Food Res. J., 18: 784–786.

Manning, K. 1985. Food value and chemical composition. pp. 221–230. *In*: Flegg, P.B., Spencer, D.M. and Wood, D.A. (eds.). The Biology and Technology of the Cultivated Mushroom. John Willey and sons, New York.

Masamba, K.G. and Kazombo-Mwale, R. 2010. Determination and comparison of nutrient and mineral contents between cultivated and indigenous edible mushrooms in Central Malawi. Afri. J. Food. Sci., 4: 176–179.

Mattila, P., Konko, K., Eurola, M., Pihlava, J.M., Astola, J., Vahteristo, L., Hietaniemi, V., Kumpulainen, J., Valtonen, M. and Piironen, V. 2001. Content of vitamins, mineral elements, and some phenolic compounds in cultivated mushrooms. J. Agric. Food Chem., 49: 2343–2348.

Mattila, P.H., Piironen, V.I., Uusi, R. and Koivistoinen, P.E. 1994. Vitamin D contents in edible mush. J. Agric. Food Chem., 42: 2449–2453.

Maynard, A.J. 1970. Extraction methods and separation processes. pp. 141–155. *In*: Joslyan, A.M. (ed.). Methods of Food Analysis. 2nd edn. Academic Press. New York.

Meghalatha, R., Ashok, C., Natraja, S. and Krishnappa, M. 2014. Studies on chemical composition and proximate analysis of wild mushrooms. World J. Pharm. Sci., 2(4): 357–363.

Mendel, L.B. 1898. The chemical composition and nutritive value of some edible American fungi. Am. J. Physio., 1: 225–238.

Mukiibi, J. 1973. The nutritional value of some Uganda mushrooms. Acta Hortic., 33: 171–175.

Nabubuya, A., Muyonga, J.H. and Kabasa, J.D. 2010. Nutritional and hypocholesterolemic properties of *Termitomyces microcarpus* mushrooms. Afri. J. Food Agri. Nutri. Dev., 10: 2235–2257.

Natarajan, K. and Manjula, M. 1978. Studies on *Lentinus polychorus* Lév., *Lentinus squarrosulus* Mont. Indian Mush. Sci., 1: 451–453.

Nawanze, P.I., Khan, A.U., Ameh, J.B. and Umoh, V.J. 2006. Nutritional studies with *Lentinus squarrosulus* (Mont.) Singer and *Psathyrella atroumbonata* Pegler: Animal assay. Afric. J. Biotechnol., 5: 457–460.

Nile, S.H. and Park, S.W. 2014. Toatl, soluble and insoluble dietary fibre contents of wild growing edible mushrooms. Crech J. Food Sci., 32(3): 302–307.

Obodai, M., Ferreira, I.C.F.R., Fernandes, A., Barros, L., Mensah, D.L.N., Dzomeku, M., Urben, A.F., Prempeh, J. and Takli, R.K. 2014. Evaluation of the chemical and antioxidant properties of wild and cultivated mushrooms of Ghana. Molecules, 19: 19532–19548.

Oboh, G. and Shodehinde, S.A. 2009. Distribution of nutrients, polyphenols and antioxidant activities in the pileus and stipes of some commonly consumed edible mushrooms in Nigeria. Bull. Chem. Soc. Ethiop., 23: 391–398.

Okwulehie, I.C., Nwosu, C.P. and Okoroafor, C.J. 2007. Pharmaceutical and nutritional prospects of two wild macro-fungi found in Nigeria. Res. Journ. of Appl. Sci., 2: 715–720.

Olfati, J., Peyvast, G. and Mami, Y. 2009. Identification and chemical properties of popular wild edible mushrooms from Northern Iran. J. Hort. Forest., 1: 48–51.

Omar, N.A.M., Abdullah, N., Rani, U., Kuppusamy, M., Abdulla, A. and Sabaratnam, V. 2010. Nutritional composition, antioxidant activities, and antiulcer potential of *Lentinus squarrosulus* Mont. mycelia extract. Hindawi Publishing Corporation Evidence-Based Complementary and Alternative Medicine. Vol. 2011, Article ID 539356. doi: 10.1155/2011/539356.

Ouzouni, P.K., Petridis, D., Koller, W.D. and Riganakos, K.A. 2009. Nutritional value and metal content of wild edible mushrooms collected from West Macedonia and Epirus, Greece. Food Chem., 115: 1575–1580.

Patil, S.S., Ahmed, S.A., Telang, M.S. and Baig, M.M.V. 2010. The nutritional value of *Pleurotus ostreatus* (Jacq. Fr.) Kumm. cultivated on different lignocellulosic agrowastes. Innovat. Rom. Food Biotechnol., 7: 66–76.

Phan, C., David, P., Tan, Y., Naidu, M., Wong, K., Kuppusamy, U.R. and Sabaratnam, V. 2014. Intrastrain comparison of the chemical composition and antioxidant activity of an edible mushroom, *Pleurotus giganteus*, and its potent neuritogenic properties. The Sci. World J., 2014: 1–10. Article ID 378651, http://dx.doi.org/10.1155/2014/378651.

Purkayastha, R.P. and Chandra, A. 1985. Mannual of Indian edible mushrooms. Jagmander Book Agency, New Delhi, India, pp. 267.

Pushpa, H. and Purushothama, K.B. 2010. Nutritional analysis of wild and cultivated medicinal mushrooms. World J. Dairy Food Sci., 5: 140–144.

Rai, R.D. and Arumuganathan, T. 2005. Mushroom, their role in nature and society. pp. 27–36. *In*: Rai, R.D., Upadhyay, R.C. and Sharma, S.R. (eds.). Frontiers in Mushroom Biotechnology, NRCM, Chambaghat, Solan.

Rai, R.D. and Sohi, H.S. 1988. How protein rich are mushrooms. Indian Hort., 33: 2–3.

Rai, R.D., Saxena, S., Upadhyay, R.C. and Sohi, H.S. 1988. Comparative nutritional value of various *Pleurotus* species grown under identical conditions. Mushroom J. Tropics., 8: 93–98.

Ramesh, C.H. and Pattar, M.G. 2010. Antimicrobial properties, antioxidant activity and bioactive compounds from six wild edible mushrooms of Western Ghats of Karnataka, India. Pharmacognosy Res., 2: 107–111.

Ravikrishan, V., Naik, P., Ganesh, S. and Rajashekhar, M. 2015. Amino acid, fatty acid and mineral profile of mushrooms *Lentinus polychrous* Lev, from Western Ghats, Southern India. Int. J. Plant, Animal & Environ. Sci., 5(1): 278–281.

Regula, J. and Siwulski, M. 2007. Dried shiitake (*Lentinus edodes*) and oyster (*Pleurotus ostreatus*) mushrooms as a good source of nutrient. Acta Sci. Pol. Technol. Alim., 6: 135–142.

Sanmee, R., Dell, B., Lumyong, P., Izumori, K. and Lumyong, S. 2003. Nutritive value of popular wild edible mushroom from Northern Thailand. Food Chem., 82: 527–532.

Sapers, G.M., Miller, R.L., Choi, S.W. and Cooke, P.H. 1999. Structure and composition of mushrooms as affected by hydrogen peroxide wash. J. Food Sci., 64: 889–892.

Sharma, S.K., Atri, N.S., Thakur, R. and Gulati, A. 2014. Taxonomy and compositional analysis of two new for science medicinal mushroom taxa from India. Int. J. Med. Mush., 16(6): 593–603.

Singdevsachan, S.K., Patra, J.K. and Thatoi, H. 2013. Nutritional and bioactive potential of two wild edible mushrooms (*Lentinus sajor-caju* and *Lentinus torulosus*) from Similipal Biosphere Reserve, India. Food Sci. Biotechnol., 22(1): 137–145.

Sivaprakasam, K. 1983. Nutritive Value of Sporophores of *Pleurotus sajor–caju*. Silver Jubilee Symposium on "Science and cultivation technology of edible fungi". Srinagar (Abstract).

Sugimori, T., Oyama, Y. and Omichi, T. 1971. Studies on Basidiomycetes. Production of mycelium and fruiting body from non carbohydrate organic substances. J. Fermnt. Technol., 49: 435–446.

Tewari, R.P. 2005. Mushroom, their role in nature and society. pp. 1–8. *In*: Rai, R.D., Upadhyay, R.C. and Sharma, S.R. (eds.). Frontiers in Mushroom Biotechnology NRCM, Chambaghat, Solan.

Thomas, B., Roughan, J.A. and Walter, E.D. 1972. Lead and cadmium content of some vegetables food stuffs. J. Sci. Food Agric., 23: 1493–1498.

Tsujikawa, K., Kanamori, T., Iwata, Y., Ohmae, Y., Sugita, R., Inoue, H. and Kishi, T. 2003. Morphological and chemical analysis of magic mushrooms in Japan. Forensic Sci. Int., 138: 85–90.

Turkekul, I., Elmastas, M. and Tuzen, M. 2004. Determination of iron, copper, manganese, zinc, lead and cadmium in mushroom samples from Tokat, Turkey. Food Chem., 84: 389–392.

Türkoğlu, A., Duru, M.E. and Mercan, N. 2009. Antioxidant and antimicrobial activity of *Russula delica* Fr: An edible wild mushroom. Eurasian J. Analytical Chem., 2: 54–63.

Valverde, M.E., Perez, T.H. and Lopez, O.P. 2015. Edible mushrooms: improving human health and promoting quality life. Int. J. Micro., Volume 2015, Article ID 376387, 14 pages http://dx.doi.org/10.1155/2015/376387.

Vetter, J. 2003. Data on sodium content of common edible mushrooms. Food Chem., 81: 589–593.

Wasser, S.P. and Weis, A.L. 1999. Medicinal properties of substances occurring in higher basidiomycetes mushrooms: current perspectives (review). Int. J. Med. Mush., 1: 31–62.

Wasson, G.R. 1969. Soma-Divine Mushroom of Immortality XIII, Hew Court Brace & world Inc., New York, 318.

Zakia, B., Ahmed, R. and Srivastva, H.C. 1964. Amino acids of edible mushrooms, *Lepiota* and *Termitomyces* species. Indian J. Chem., 2: 380–391.

Zhang, N., Chen, H., Zhang, Y., Xing, L., Li, S., Wang, X. and Zhen, S. 2015. Chemical composition and antioxidant properties of five edible hymenomycetes mushrooms. Int. J. Food Sci. and Tech., 50: 465–471.

CHAPTER 16

Strategies for the Management of Rice Pathogenic Fungi

*Truong H. Xuan** and *Evelyn B. Gergon*[a]

ABSTRACT

Diversified population structures of major filamentous fungi evolved in specific rice intensive production systems. They are speedily analysed and accurately identified by molecular detection tools. If unchecked, pathogenic fungi may initiate disease epidemics that may pose a serious threat to the food security of over 800 million people across the tropical and sub-tropical climates. Several strategies, that are adopted worldwide, are integrated to develop an effective disease management system, are discussed. Crop losses can be minimized by appropriate crop husbandry practices such as exclusion of the pathogenic inoculum, stabilization of pathogenic virulence mechanisms, durable host plant resistance and recycle sources. These management strategies are vital keys to ensure production efficiency, stability of agricultural ecosystems, avoid fungicide resistance and advance research development and extension services for the future.

Rice blast caused by the ascomyceteous *Magnaporthe oryzae* and sheath blight caused by the necrotrophic soil-borne *Rhizoctonia solani* are the two most destructive diseases in rain-fed lowlands and dry upland ecologies where predisposing factors favour the development of epidemics. The dynamic evolution and adaptation of these pathogens have been challenging the effectiveness of deployed resistance genes in commercial cultivars and fungicidal compounds. Development of an effective and durable resistance requires a comprehensive understanding of the

Philippine Rice Research Institute, Science City of Muñoz, Nueva Ecija, The Philippines 3119.
[a] Email: egergon@yahoo.com
* Corresponding author: truongxuan893@yahoo.com

host-pathogen interaction, population structure and diversity, aggressiveness, pathogenesis, and pathogenecity of pathogen complex in different pathosystems. Durable resistance is attributed to a combination of complete and partial resistance and provides a broad spectrum of resistance. Detection of pathogen diversity and seed transmission of *Bipolaris oryzae* and *Sarocladium oryzae* have been addressed by PCR-based techniques. The long time debate on the pathogen infection process of false smut caused by *Ustilaginoidea virens* has been validated by cytological and molecular-based techniques. The view undergoing sexual reproduction in the field and a gene responsible for the production of mycotoxins by *Fusarium fujikuroi* and *F. graminearum* were detected by real time RT-PCR, which is an essential tool for assessing risk factors and crop losses and predicting the disease epidemics in order to develop a reliable disease management to produce high quality seeds. The management of other rice diseases such as sheath blight, brown leaf spot, narrow leaf spot, foot rot, stem rot and sheath rot are not discounted but crop husbandry practices and biological control are presented.

Introduction

Rice is an important staple food for about half of the world's population. It is exposed to several fungal diseases and ten of them are economically important in rice production. Lack of durable host resistance to the pathogen diversity and virulence can result in disease epidemic and pose a serious threat to world's rice economy. The catastrophic widespread of brown leaf spot, *Cochliobolus miyabeanus* that caused Great Bengal Famine of 1943 implies that in some regions of the world, a large proportion of the population is dependent on a single crop or a few crops as the principal staple food and that population could be devastated at an instance when the crop fails due to a ravaging disease. Records show that more than 800 million people do not have adequate food amidst 10% or more reduction in global food production owing to plant diseases (Christou and Twyman, 2004; James, 1998).

The risk is particularly greater in developing poor countries where populations are growing fastest and are dependent on locally produced staples. The infrastructure of research, development, and extension are often poor resource, although modern rice varieties have achieved greater productivity and suppressed the epidemic potential of many important diseases. There are still other diseases in tropical rice ecosystems that cause chronic yield losses and destabilize annual production (Evenson and Gollin, 2003). As cultural practices and cropping intensity have changed over time, some diseases that were previously minor have become major problems (Evenson, 1998). Assessment of current rice disease problems across all production situations in South and Southeast Asia have shown that sheath blight and brown leaf spot have accounted for the highest yield loss with 6% and 5%, respectively

(Savary et al., 2000a,b). Yield losses by severe epidemics of rice blast across Asia were estimated from 0.3% to 5%. Sheath blight and brown spot have become prominent due to limitation of effective resistance in the germplasm or changing weather patterns. The problem owing to plant diseases, particularly in developing countries, is aggravated by lack of resources to quantify the effects of plant diseases and avoid crop failure. Strategies developed worldwide are usually applicable but there is a need for a holistic, integrated, and objective approach to manage fungal diseases effectively in rice production to ensure global food security. The aforesaid aspects of the rice disease management have been dealt with in this chapter.

Identification of Inoculum Sources

Once the identity of an organism is reliably established, it becomes easy to evaluate the risk and potential effects of the pathogen, provide sufficient information on inoculum sources and dissemination, and control the disease (Crop Protection Compendium, 2004). The exclusion of the inoculum sources, through plant quarantine to avoid or predict the pathogen invasion, is the first line of defence and, therefore, deserves more resources to guarantee global food security (Sutherst and Maywald, 1985; Lanoiselet et al., 2001).

Development of disease management strategies requires knowledge on pathogen population at species and races levels, sources of inoculum, wide host range of the pathogen, and availability of sources of durable resistance. The risk assessments of tropical climates that allow continuous cropping and consequent build-up of inoculum are critical for improving the crop husbandry. Understanding of virulence mechanisms is crucial to exploit the gene pool for breeding programs in stabilising pathogen genetic flux using reliable molecular detection tools. A single variety, planted in large areas repeatedly, often results in a serious threat of varietal breakdown because of a multitude of pathogenic races. To address appropriately such a challenge, research on host plant resistance must not only aim to respond to significant diseases but also to use multiple genetic resources in managing them. A combination of molecular biology and biological information with the traditional techniques have to be employed to analyse and optimize the genetic diversity of rice fields, sustain the resistant resources, and balance the ecosystems. Polymerase chain reaction (PCR) techniques based on transcribed spacer (ITS) regions of ribosomal RNA target unknown isolates for amplification and rapid identification rather than using the long process of morphological analysis (Bridge et al., 1997). Real-time (RT)-PCR is a very sensitive and reliable detection tool that can produce results on-site to contain and control pathogens (Qi and Yang, 2002). Suitable marker-assisted selection has also been explored to determine plant-pathogen interaction in developing host resistance with elite traits.

Development of Crop Husbandry Practices

Crop husbandry practices are based on the historical disease problem and existing pathogen structure in order to design intercropping and crop rotation schemes, maintain genetic diversity in crop plants, enhance soil flora and fauna for biological control and plant growth promoting organisms, judicious use of pesticides, and seed health management to effectively exclude pathogen population and limit the impact of diseases.

Risk and Its Assessment

The rice plant is a principal source of human nutrition and an organic source for the pathogens. Fungi may cause catastrophic plant diseases for the following reasons: (1) they can prolifically produce conidial inoculum, which may widespread within few days; (2) they may produce high-density inoculum which can disperse long distances by wind or water and become established in a favourable microclimatic environment; (3) they may produce mycotoxins that can destroy the plant's structure; (4) they may affect the plant's nutrient uptake and its growth regulators such as cytokinins; and (5) they may reduce crop productivity (Strange et al., 2005).

Among the fungal diseases infecting rice, blast induced by an ascomycete, *Pyricularia oryzae* (teleomorph *Magnaporthe grisea*) is considered the most economically important. In 1995, due to blast epidemic, 700 ha of rice of diverse genotypes with varying levels of resistance in Bhutan were destroyed (Thinlay et al., 2000a,b). At large, blast infection can result in annual crop losses of 10% to 30% (Talbot, 2003). Seedborne pathogens, *Fusarium graminearum* and *Gibberella fujikuroi*, both causing head blight and foot rot diseases, do not only reduce plant biomass but also produce mycotoxins that can affect human and animals. Considering the seriousness of fungal diseases, it is essential to establish the pathogen population, identify the sources of inoculum, to know the disease spread and dispersal, and to estimate the affected crop production.

Measurement of disease occurrence and assessment of crop losses remain difficult. There is usually no simple relationship between measures of symptoms and the failure to reach achievable yields. The relation of disease severity to yield may not be obvious and requires empirical measurement. Various models—critical point, multiple-point, and area under the disease progress curve (AUDPC) have been used in crop modelling for risk assessment, estimation and prediction of crop loss at large area. This is usually done to come up with comparative values in order to determine the economic impact of a certain disease (Gaunt, 1995; Madden and Nutter, 1995; Maywald et al., 1997; Kaundal et al., 2006; Mousanejad et al., 2010).

Host-pathogen Interaction

Plant pathogen has a state of mutability or capability to overcome the resistance of its host plant. Understanding the nature of pathogenicity is essential and relevant in determining the host-pathogen interaction. For an organism to be a pathogen, it must have a number of virulence functions such as the production of degradative enzymes (Song and Goodman, 2001). Upon ingress of the pathogen, the plant tissues may produce and accumulate phytoalexins which are low-molecular weight antimicrobial compounds. One of the phytoalexins produced in rice plant is sakuranetin, which is synthesized in response to UV irradiation or blast infection (Kodama et al., 1992). Other several defence-related genes of rice are activated by the fungal elicitor (Kim et al., 2000). The plant cells also activate a variety of early signal transduction defence such as TNP2-like gene Rim2 upon initial recognition of the avr-proteins or elicitors of *M. oryzae* (He et al., 2000). Once chitinase and glucanase genes in plant were activated with elicitors, plant cells produce chitinases and b-1,3-glucanases to hydrolyse chitin and b-1,3-glucan, respectively, which are major constituents of fungal cell walls (Schweizer et al., 1997; He et al., 1998; Kim et al., 1998). Hydrolysis of the fungal cell wall constituents leads to the inhibition of the growth of several fungi *in vitro* (Fujikawa et al., 2012). Rice plants secrete α-1, 3-glucanase (AGL-rice) to show a strong resistance not only to *Magnaporthe oryzae* but also to its distantly relative, *C. miyabeanus* and *Rhizoctonia solani*. Treatment with α-1, 3-glucanase *in vitro* activates both chitinase and glucanase and causes fragmentation of infectious hyphae of *R. solani* indicating that α-1,3-glucan is also involved in maintaining infectious structures in some fungi (Anarutha et al., 1996).

Rice plant can evoke its inherent immunity against *M. oryzae* challenges upon recognition of the pathogen-associated molecular patterns (PAMPs) such as fungal cell wall chitin (Fujikawa et al., 2012). However, pathogens may bypass the host PAMP-triggered immunity. The pathogen can also mask cell wall surfaces with α-1,3-glucan during invasion as discussed earlier. The surface α-1,3-glucan is indispensable for the successful infection of the fungus by interfering with the plant's defence mechanisms. The α-1, 3-glucan synthase gene *MgAGS1* is not essential for infectious structure development but is required for infection in *M. oryzae*. The protective role of α-1, 3-glucan of the pathogen against plants' antifungal chitinase occurs during infection. Many infection-responsive genes in the pathogen and host plant which have a wide dynamic range for expression and have been characterized using the emerging RNA-Sequence technique (Yoshihiro et al., 2012). It was shown that when the primary infection hyphae penetrate leaf epidermal cells, the gene expression of both the host plant and the pathogen occurs simultaneously in the same infected leaf blade in natural infection conditions at 24 h after inoculation. The regulation of transcript encoding glycosyl hydrolases, cutinases and LysM domain-containing proteins occur in the blast fungus, whereas pathogenesis-related and phytoalexin biosynthetic genes are regulated in rice plant. The transcriptome analysis is useful for the simultaneous elucidation of the tactics of host plant defence and pathogen infection.

Rice blast disease caused by *M. grisea* remains as one of the most devastating diseases of rice worldwide despite of the availability of several major resistance genes (R-genes) known as *Pi* genes (Yamada et al., 1976; Zeigler et al., 1994a; Tsunematsu et al., 2000). Use of differential varieties (DV) is also a useful tool to characterize the pathogenicity of the pathogen population. A set of DV consisting of 31 monogenic lines, each carrying a single gene, were inoculated with 20 blast isolates from the Philippines. Those DV having *Pi1*, *Pi9(t)*, *Pia*, *Pib*, and *Pik-s* were susceptible to most pathogen races, while other DV lines harboured R-genes *Pi5(t)*, *Pi9(t)*, *Piz-5*, *Pit*, and *Pish* which confer a wide spectrum of resistance from moderate resistance to resistance to majority of isolates (Fukuta et al., 2004).

It is well known that the use of R-resistant cultivars is essential to control blast disease. However, some resistance mediated *Pi* genes are not durable and are typically overcome by new races under field conditions soon after the variety is commercially released. Hundreds of pathogenic races are identified based on their infection spectra on differential rice cultivars. DNA fingerprinting analyses showed that rice blast fungal populations are asexual and consist of several discrete lineages having apparently predictable responses toward specific R-genes (Levy et al., 1993; Zeigler et al., 1994b).

The rice blast system is a classical gene-for-gene system (Flor, 1971) in which avirulent genes (avr-genes) in the pathogens play special functions correspondent to the particular R-genes in rice crop (Silué et al., 1992; Zeigler et al., 1994b). Avr-genes of the pathogen encode molecules that are recognized by the corresponding R-gene product encoding enzymes. Recognition of the pathogen, triggers host resistance and stops the infection. Field efficacy of any R-gene depends on the biology of the interaction of avr-gene that may provide for the pathogen in triggering R-gene-mediated resistance. The avr-gene mutation helps the pathogen to encounter the R-gene in the rice plant in order to survive and compete under field conditions.

The R-gene such as *Pi-ta* in the resistant rice variety receives pathogen elicitors encoded by the corresponding avr-gene that activates the plant defence system. *Pi-ta* encodes a predicted 928-amino acid cytoplasmic receptor with a centrally localized nucleotide binding site that is linked to the centromere of chromosome 12 in rice (Bryan et al., 2000), while avr-*Pita*, encoding a predicted 223-aminoneutral zinc metalloprotease that are tightly linked to the telomere on chromosome 3 in *M. oryzae* (Orbach et al., 2000). A direct interaction of *Pi-ta* protein and avr-*Pita* protein triggers a complete resistance to *M. oryzae* (Jia et al., 2000, 2004). Diverse mutations in avr-*Pita* which occur by point mutations, insertions, and deletions (Singh et al., 2006) help the pathogen to avoid triggering resistance responses mediated by *Pi-ta* and subsequently abolish the avirulence function of *Pi-ta* (Zhou et al., 2005a). Such a host-plant interaction indicates that rice cultivars such as Katy, 'Drew', 'Kaybonnet',

'Cybonnet', and 'Ahrent' containing *Pi-ta* gene became susceptible to the pathogen variant (Dean et al., 2005; Zhou et al., 2005b).

Rice Blast Management

Rice (*Oryza sativa*), the second largest cultivated crop worldwide and principal staple food of 90% of the population in Asian region, is considerably affected by various biotic and abiotic stresses. Among the biotic stresses, *M. grisea* is the most serious constraint that can limit rice productivity. Major blast epidemics occur annually resulting in 11% to 30% crop losses estimated at about 157 M metric tonnes (Talbot, 2003). One of the most economical options to manage blast disease is the use of resistant varieties. But, the resistance of varieties to blast usually remains effective for only 2 to 3 years because of highly variable nature of the pathogen. Pyramiding of two or more genes for blast resistance has been used in overcoming this problem. Another helpful strategy is to develop a broad spectrum race-non-specific resistance using molecular biology approaches. In farming practices, management of rice blast requires an integration of many strategies contributed by the rice growers and intensive research efforts worldwide. Right choice of an appropriate strategic option can save the rice crop from this important disease.

Exclusion and Reduction of Inoculum

The most fundamental approach to disease management is to ensure that the inoculum is excluded through quarantine (Ebbels, 2003; Lanoiselet et al., 2001). Regional plant protection organizations, the European and Mediterranean Plant Protection Organization (EPPO), play a significant role in coordinating the phytosanitary activities across a region (Smith et al., 1996), and the Secretariat of the International Plant Protection Convention has a global responsibility in this field. There is enormous scope for simply strengthening the resources and capacity of plant protection services in developing countries, with the goal of improving food security (Asher, 1996). In the Philippines, the Plant Quarantine Service of the Bureau of Plant Industry under the Department of Agriculture (PQS-BPI-DA) assumes the responsibility of protecting the country from the unwanted entry of new pathogens and pests and for coordinating programs to eradicate those that have recently arrived in the country. Plant materials are tested for specific quarantine pathogens by a number of rapid and sensitive techniques that are available (Schaad et al., 2002). If a pathogen is endemic in a region, the infected material should be destroyed to eliminate the amount of inoculum. Seeds and other materials of propagation are sometimes treated with biocides while infected plants in containment fields are rogued and incinerated (Ou, 1985).

Use of Healthy Seeds

Palay-Check (2007), a dynamic rice crop management system that presents the best key technology and management system, presents eight key checks that have impact on the growth, yield, and grain quality of rice and on the sustainability of the environment. Its first key check is the use of high quality seeds that are of the same variety, clean, full and uniform in size, and have at least 85% germination rate. High quality seeds are expected to produce healthy seedlings that grow fast and uniformly and can resist pests better. All factors combined brought by high quality seeds give an increase of 5–20% in rice yield (Mew et al., 2003).

Seed is one of the least expensive inputs in crop production and yet the most important factor influencing yield potential (Brick, 2004). It is essential that seeds that will be used for planting especially for seed purposes must be of high quality and do not carry organism within or on the surface that may affect the germinating seedlings. Poor seeds may give rise to weak seedlings or reduce seed germination. Fungi that are seed transmitted may also cause disease on the subsequent crop when they are sown in the field. Percentage of seed transmission of a disease is high for some fungi like *F. fujikuroi* but it could also be low as in the case of *P. grisea*. Aside from seed quality, the seed lot must not also contain weed seeds, insects, and sclerotia that may be disseminated through irrigation water.

Crop Husbandry

There are other ways by which farmers can take action to reduce disease in the field. Adjusting planting time, practicing fallow period, ploughing after harvest, flooding during and after land preparation, and crop rotation using non-host plants are the most simple and economical but vital operations to reduce inoculum from crop residues. These simple practices are better alternatives than the application of pesticides when an infection is severe.

Host Resistance and Pathogen Structure

The rice blast disease, caused by *P. grisea*, is the most prevalent disease of rice in both the tropics and temperate zones (Zeigler et al., 1994a). Its pathogen populations have been analysed without doing pathogenicity of the constituent isolates using a neutral repetitive DNA sequence of *M. grisea* repeat (MGR) which was discovered in the late 1980s, it is a useful tool for analysing populations without pathogenicity of the constituent isolates (Correa-Victoria et al., 1995). The similarity of the MGR "fingerprints" generated by DNA analyses of the different isolates permitted an estimate of their relatedness. The use of combination of MGR and DV has revealed a direct relationship between fingerprint types or lineages and pathogenic races (Hamer et al., 1989).

Their combined use is a vital strategy in developing host plant resistances and durable resistance. This finding suggests that the rice breeding can focus on selecting cultivars containing a combination of R-genes that are effective against avirulence spectrum of all lineages in a target population. This breeding strategy is referred to as lineage-exclusion. An important assumption of this strategy is that *P. grisea* populations are comprised of a few number of distinct lineages and that these lineages have different and stable virulence spectra without gene flow across or genetic recombination among lineages (Zeigler et al., 1994b). This assumption was tested in two populations from blast resistance screening nurseries in the Philippines. The analysis of lineage virulence revealed that a combination of resistance genes with complementary resistance spectra can increase the utility of major gene resistance (Chen et al., 1995; Zeigler et al., 1995). In the Philippine blast population, rice plants with a combination of resistance genes *Pi-1* and *Pi Z⁵* (*Pi-2*) are effectively resistant across all 10 lineages (Zeigler et al., 1995). Chen et al. (1996, 1999, 2005) gave other combination of *Pi2*(t) and *Pi44*(t) together with other genes conditioning qualitative and quantitative resistance that would provide a broad spectrum and possibly durable resistance. This strategy for obtaining durable blast resistance was effective in the different test areas in the Philippines where modern varieties have been grown. The durable resistance to rice blast was attributed to the early deployment of blast resistant varieties such as IR36, IR60, IR64, IR72, and IR74 in irrigated lowland in the Philippines (Babujee et al., 2000; Khush and Virk, 2002) resulting in simpler pathogen population than those populations in other rice-growing regions in Asia. However, the lineage exclusion may not be suitable for its adoption where pathogen populations are very complex, and the virulence spectra of all lineages cannot be practically characterized. In this environment, the pathogen has the potential to evolve to new pathotypes from a single asexual spore (Mekwatanakarn et al., 1999). Once the resistant varieties are released for only a single pathotype and exposed to an array of pathogenic variation in the fields, a rapid breakdown of blast resistance is expected to occur due to disease pressure from different virulent races existing in the field. Zeigler and Correa (2000) noted that the proponents of "hypervariability" are usually encountered from the populations in Asia which is the center of rice origin. The *P. grisea* from a traditional rice-growing area of Northeast Thailand is composed of a very complex population with 49 lineages represented by only one or a few isolates (Mekwatanakarn et al., 2000). There was no obvious relationship between lineages and DV or cultivars with known resistance. Nevertheless, the complexity of the Northeast Thailand population revealed the same complementary effectiveness of resistance genes Pi1 and *Pi z⁵* (Zeigler et al., 1995). A high lineage diversity was also observed in the Indian Himalayas (Kumar et al., 1999) and in the Himalayan Kingdom of Bhutan (Thinlay et al., 2000a,b). It is impossible to determine the virulence spectrum of lineages comprising these populations because many lineages are represented by only one isolate. Besides, sexually fertile field isolates were reported from India (Kumar et al., 1999; Zeigler, 1998), and Thailand (Mekwatanakarn et al., 1999).

In contrast, there are *P. grisea* populations that are relatively simple and stable in large rice growing areas. The proponents of stability may be attributed to the recent introduction of rice in the areas that have been planted with a few varieties carrying several R-genes (Latterell, 1971; Latterell and Rossi, 1986; Zeigler and Correa, 2000).

Deployment of the Major Host Plant Resistance

Rice varieties with some elite traits that harbour multiple major genes for blast resistance (*Pita, Pi20, Pik*) have been developed by different international and national research institutions and approved annually for release to seed growers by the National Seed Industry Council in the Philippines (NSIC) (Table 1). Rotation of resistant varieties is a temporal measure of genetic diversity (Wang et al., 1998). Among the released resistant varieties, PSB Rc4 (*Pita* and *Pib*), Rc10 (*Pi20, Pita, Pib,* and *Pi3*), Rc160 (with parental lines harbour *Pi20, Pita, Pia,* and *Pib*) have sustained their resistance in rain-fed lowland in the Philippines over 5 to 20 years (Table 1). Sequential release of varieties over time and space is one form of diversifying varieties or resistance genes in the field. This system, however, should be supported by good race prediction and survey data. Combination of multitude R-genes and quantitative trait loci (QTL) into one variety is considered very effective and durable against a broad spectrum of pathogens (Bonman et al., 1991; Wang et al., 1994; Manosalva et al., 2009). This strategy using horizontal resistance has currently been applied in the development of host resistance to reduce the selection pressure on pathogens.

Deployment of Multilines

Practice of rice monoculture is convenient and remains the norm in intensive rice production systems. However, the exploration of genetic diversity definitely offers a great stability of performance and sustainability of inherent resistance to the rice blast virulence (Zhu et al., 2003; Liu et al., 2003). A success story of rice blast management achieved through deployment of multiline systems was performed over a large area in Yunnan Province of China. Growing of two traditional rice and susceptible varieties in single rows and interspersed between four to six rows of resistant hybrid varieties resulted in a significant reduction of blast disease severity by 94% as compared with monocultures (Zhu et al., 2000). The study considered that there will be no component in the mixture that are susceptible to the same pathogen and that the pathogen does not have a wide host range. Geographic distribution of pathogen structure and genetic were pre-assessed. The physical appearances of the multilines were also properly documented (Zhu et al., 2000). The design of multilines was effectively achieved in BaoXiu, Shiping County, China. The strategy decreased

Table 1. List of varieties resistant to rice blast identified by the National Collaborative Trial and approved for release by the Philippine Seed Board (PSB)/National Seed Industry Council (NSIC)[a].

Item#	Entry	NCT[a]	VT[b]	Line designation/Resistant genes[c]	Year of release
1	IR42	NO	I/R	IR2071-586-5-6-3 Pita, Pib	1977
2	IR65	R	R	IR21015-196-3-1-3 Pita, Pib	1985
3	IR72	NO	I	IR35366-90-3-2-1-2 Pita, Pib	1988
4	PSB Rc4	R	R	IR41985-111-3-2-2 Pita, Pib	1991
4	IR70	NO	R	IR28228-12-3-1-1-2 Pita, Pik-t	1988
6	PSB Rc28	R	R	IR56381-139-2-2 Pita, Pib, Pi20	1995
7	PSB Rc82	I	I	IR64683-87-2-2-3-3 Pita, Pib, Pi20	2000
8	PSB Rc10	R	R	IR50404-57-2-2-3 Pita, Pib, Pi20, Pi3	1992
9	PSB Rc18	I	R	IR51672-62-2-1-1-2-3 Pita, Pib, Pi20, Pi3	1994
10	PSB Rc20	I	R	IR57301-195-3-3 Pita, Pib, Pi20, Pi3	1994
11	IR64	No	R	IR18348-36-3-3 Pita, Pib, Pi20, Piz-t	1985
12	NSIC Rc160	I	R/I	PR30536-B-48-2 Pita, Pib, Pi20, Pia	2007
13	IR60	NO	R	IR13429-299-2-1-3 Pita, Pib, Pik-s, Piz-t	1983
14	IR74	NO	R	IR32453-20-3-2-2 Pia, Pib, Pi20, Pik-t	1988
15	IR20	NO	R	IR57301-195-3-3 Pib, Pik-s	1969
16	IR66	NO	I	IR32307-107-3-2-2 Pib, Pik-s	1987
17	PSB Rc2	I	NO	IR32809-26-3-3 Pib, Pik-s, Pi20, Piz-t	1991
18	PSB Rc30	I	R	IR58099-41-2-3 Pib, Pik-s, Pi20	1996
10	NSIC Rc128	S	S	PR26645-B-7A	2004
39	NSIC Rc130	I	S	PR30244-AC-9-1	2004
19	NSIC Rc23	I/R	R	IR79913-B-176-B-4	2011

Table 1. contd....

Table 1. contd.

Item#	Entry	NCT[a]	VT[b]	Line designation/Resistant genes[c]	Year of release
20	NSIC Rc122	R	S/I	IR61979-138-1-3-2-3	2003
21	NSIC Rc144	I	R	C6053-B-1-2-1-2	2006
22	NSIC Rc150	I	R	PR27842-P-127	2007
23	NSIC Rc156	S	S	C6624-B-1-2	2007
24	NSIC Rc158	I	I	IR77166-122-2-2-3	2007
25	NSIC Rc170	R	R	IR68333-R-R-B-22	2008
26	NSIC Rc214	I	R	IR78568-1-2-1-2	2009
27	NSIC Rc216	S	S	PR34141-38-1-J2	2009
28	NSIC Rc224	I	R	PR31091-17-3-1	2010
29	NSIC Rc226	S	S/I	PR33373-10-1-1-B	2012
30	NSIC Rc238	NO	S	IR78555-68-3-3-3	2011
31	NSIC Rc240	NO	R	PR31132-B-1-1-1-3=3	2011
32	NSIC Rc272	S/I	R	PR34363-4 Pokkali/AC-45-M_5R-19	2011
33	NSIC Rc274	I/R	S	IR81412-B-B-82-1	2011
34	NSIC Rc276	S/I	I	C8108-B-10-2-2-1	2011
35	NSIC Rc278	S/I	R	IR81023-B-116-1-2	2011
36	NSIC Rc280	S/I	R	IR72667-116-1-2	2011
37	NSIC Rc282	S/I	R	C8231-B-1-1	2011
38	NSIC Rc284	S/I/R	R	IR74963-262-5-1-3-3	2011
39	NSIC Rc286	S/I/R	S	C6393-2B-3-3-1-2	2011
40	NSIC Rc288	I/R	R	PR25769-B-9-1	2011
41	NSIC Rc296	S/I	S	IR71896-3R-8-3-1	2011
42	NSIC Rc298	S/I	I	PR34159-13-1	2012
43	NSIC Rc300	S/I/R	R	PR31379-2B-10-1-2-1-2	2012

[a] NCT: National Collaborative Testing across rice ecosystem for selecting lines resistant to rice blast disease. Reaction at the time of released: R = resistant, I = intermediate or moderately resistant, and S = susceptible.

[b] VT: Verified trials were conducted recently for at least two trials in rainfed lowland.

[c] Resistant genes were estimated based on the reactions of isolines to 14 blast isolated from the Philippines (Fukuta et al., 2007).

the incidence of rice blast below 4.78% for successive three years (Wang et al., 1998). The reduction of blast severity by 75% in the multiline set-up was comparable to the level of control achieved by fungicide treatments (Koizumi, 2001). Based on a simulation model of pathogen evolution, the use of varietal mixture is the best deployment strategy for rice blast as compared with gene pyramids and gene rotation (Winterer et al., 1994). The emergence of complex races was not observed in fields using varietal mixtures (Chin and Husin, 1982). To ensure the success of multiline trials, the differences in genetic backgrounds of the multiline to the blast pathogen, the spatial distribution of conidia and agro-economical characteristics of rice varieties were pre-determined (Sun et al., 2002). The pathogen, *P. grisea,* and the microclimate conditions in the mixture field environment were also considered. With the use of resistance gene analogue technique, varieties with higher degree of polymorphism showed better control of the disease when they were planted in multiline (Liu et al., 2003; Zhu et al., 2003). The pathogenicity test also showed an excellent resistance of the hybrid varieties with virulence frequency of 13.8%, while susceptible traditional varieties showed resistance with virulence frequency of 86.2%. There was no obvious effect of disease control when varieties with low genetic differences were mixed.

Nutrient Management

Nutrition greatly influences all of the interacting components affecting disease severity (Huber and Haneklaus, 2010). The interaction of nutrition in these components is dynamic and all essential nutrients are reported to influence the incidence or severity of some diseases. A particular element may decrease the severity of some diseases, but increase others, and some have an opposite effect in different environments. Rice blast for instance, is highly favoured by excessive nitrogen fertilization coupled with aerobic soils and drought stress. High nitrogen rates and nitrate nitrogen also increase rice susceptibility to the disease. Extended drain periods that encourage the conversion of ammonium to nitrate also favour blast incidence (Webster and Greer, 2004).

The use of right nutrient elements is one of the keys to reduce blast severity in most rice environments. Excess of nitrogen fertilizers increases the disease severity resulting in crop loss. A balanced NPK rate (90-40-40 NPK kgha^{-1}) is generally recommended for rain-fed lowland in the Philippines. Typical fertilizer recovery efficiency averages are 0.40 kg/kg^{-1} N, 0.2 kg/kg^{-1} P and 0.35 kg/kg^{-1} K (Dobermann and Fairhurst, 2000). After harvest, turning-over rice stubbles and crop residues, sun drying, and practicing fallow periods for at least two weeks can eliminate the inoculum source while returning various nutrient elements into the soil such as silicon (Si). Recycled rice straw and rice husks contribute Si in irrigated lowland rice with good crop management and grain yield of 5–7 t ha^{-1}. Si positively contributes in the development of strong leaves, stems, and roots and plays a vital role in reducing the susceptibility of rice crop to fungal diseases. Si is present in plant tissue (8–10%), rice leaves (71%), hulls (13%), roots (10%), and stems (6%) (Chen, 1990; Epstein, 1991), rice straw (5–6%), and rice husks (10%). Inorganic source of Si fertilizer for rice are $CaSiO_3$, $MgSiO_3$ and $MnSiO_3$.

Proper amendment of various organic and inorganic silicon sources can reduce 50% incidence of rice blast (Aleshin et al., 1987). Fine ground calcium silicate slag was more effective in reducing the neck blast severity than the coarse grades (Datnoff et al., 1992). Application of Si at a rate of 100 kg ha^{-1} was as effective as edifenphos and tricyclazole for controlling leaf and neck blast at low severity levels (Seebold et al., 2004).

Water Management

Rain-fed and upland conditions favour blast development than irrigated lowland. Growing seedlings in extended dry seedbed exposes the crop to early disease development. Disease usually becomes severe when there is long duration of high leaf wetness after water deficit. To reduce blast development, regulation of intermittent water irrigation (2.5–3.0 cm water depth) during seedling stage is effective. Reducing drought stress in upland rice could be a cost-effective component of an integrated blast management in this ecosystem.

Biological Control

The biological method of plant disease management is an alternative to chemical fungicides in managing the blast disease. The biocontrol products at present contribute less than 1% of the fungicide market (Morton and Staub, 2008). A diverse group of BCAs exists in nature. Among them, bacterial antagonists are considered ideal candidates because they have the same habitats (rhizosphere, episphere, and endophytic) with the pathogens, grow rapidly, easy to handle, and aggressively colonize population. Deployment of potential BCA is an eco-friendly and cost effective strategy which can be used in integration with other strategies for a greater level of protection with sustained rice yields. Sustainable disease management requires a genuine integration of technologies with sound cultural and sanitation measures, balance use of host plant resistance and BCA, and restrained use of chemical component.

Mycofungicides

Antagonists—*Trichoderma harzianum* (CPO-80) and *Chaetomium globosum* (N76-1) are two of the popular mycofungicides. They could inhibit the mycelial growth and conidial germination of *P. oryzae* by 70–88%. Ferimzone (TF-164), a new fungicide with biological properties, was also found to inhibit the mycelial growth of *M. grisea* by more than 90% at a concentration of 5 µg ml^{-1} but it has no effect on conidial germination (Gouramanis, 1994).

Bacterial Antagonists

Bacillus subtilis subsp. *subtilis,* a gram-positive bacterium, produces endospores that make it tolerant to heat and desiccation. This characteristic is a good feature of *B. subtilis* which is required for efficient field application of BCA in tropical and humid regions (Table 2). *Serratia marcescens* isolated from shoots and roots is an ideal agent for the control of *P. oryzae* (Jaiganesh et al., 2007). It can produce chitinolytic enzymes that degrade the fungal cell walls and induce plant defence and certain element with antifungal property (Someya et al., 2000). In talc-formulation, *S. marcescens* survived and increased its population in the phyllosphere after being sprayed on rice variety, IR50, and effectively controlled blast inoculum. The negative pseudomonads–rod-shaped bacteria

Table 2. List of fungicides with their Mode of Action Code, Resistance Action Committee Code (FRAC), and Chemical Group for Monitoring Fungicide Resistance to *P. oryzae*.

Common name*	MOA code**	FRAC code***	Chemical group
Oxystrobin	C3:QoI	11	Acrylates
Bacillus subtilis strain QST713 *B. subtilis* strain FZB24	F6	44	Microbial disrupters of cell membranes
Benomyl	B1:MBC	1	Benzimidazole
Blasticidin-S	D2	23	Enopyranuronic acid
Carbendazim	B1:MBC	1	Benzimidazole
Carpropamid Dicyclomet	I2:MBI-D	16.2	Cyclopropanecarboxamide
Edifenphos	F2	6	Phosphorothiolates
Fenoxanil	I2:MBI-D	16.2	Propionamide
Fermions	Unknown	U14	Pyrimidinone hydrazones
Fthalide	I1:MBI-R	16.1	Isobenzofuranone
Iprobenfos	F2	6	Phosphorothidates
Isoprothiolane	F2	6	Dithiolane
Kasugamycin	D3:antibiotic	24	Hexopyranosyl
Mancozeb, Maneb, Zineb	Multi-site:M3		Dithiocarbamates
Probenazole	P2	P	Benzisothiazole
Pyroquilon	I1:MBI-R	16.1	Pyrroloquinolone
Tricyclazole	I1:MBI-R	16.1	Triazolobenzothiazole
Thiophanate-methyl	B1:MBC	1	Thiophanate

* The Common Names, MOA Code, FRAC Code and Chemical Group names included in this list are those used in the FRAC Code List and the Associated List of Plant Pathogenic Organisms Resistant to Disease Control Agents. Last update: December 2008.
** **Mode of Action Code (MOA Code)**
Different letters (A to I, with added numbers) are used to distinguish fungicide groups according to their mode of action (MOA) in the biosynthetic pathways of plant pathogens. The grouping was made according to processes in the metabolism starting from: A = nucleic acids synthesis to secondary metabolism as follows: I = melanin synthesis (I) at the end of the list, P = host plant defense inducers (P), U = unknown mode of action and unknown resistance risk, and M = multi-site inhibitors (M).
*** **FRAC Code** Numbers are used to distinguish the fungicide groups according to their cross resistance behaviour. The numbers were assigned primarily according to the time of product introduction to the market (numbers 1 to 44, status 2008).

are excellent colonizers and widely prevalent in rice rhizosphere. Several antagonists associated with upland and lowland rice rhizosphere have also been found effective *in vitro* against *P. oryzae*. Strains of *Pseudomonas fluorescens* (Pf7-14) produce antifungal antibiotics that inhibit germination of conidia of the blast pathogen (Vasudevan et al., 2002). Formulation of Pf7-14 applied as seed and multiple foliar sprays suppress 60% and 72% leaf and neck blast infection, respectively.

Chemical Control

Chemical methods of disease control still play a big role in managing blast using several classes. Field monitoring from crop establishment until the booting stage for presence of leaf lesions can help decide whether to apply or not to apply fungicide. If blast lesions are present and increasing just before the booting stage, a fungicidal treatment may be justified. Right decision of treatment should consider disease progress curve, crop growth stages, environmental conditions, and rice varieties. Fungicide treatments for managing rice blast can be applied as seed treatment usually in the form of slurry in order to prevent seedling infection after germination. Fungicide can also be sprayed on rice seedlings at 10–14 day after sowing. Since rice blast has a multiple disease cycle, a preventive fungicide measure of leaf blast early in the season is generally ineffective in reducing the incidence of neck blast and yield losses, especially in susceptible varieties. An application for blast should be done from late booting stage to 10% heading. If weather conditions that favour blast development persisted, then an additional application is also recommended about after 7 days (when 50% of panicles have emerged). The development of highly effective systemic fungicides against blast has opened new strategies for managing the disease under the disease-conducive environments. Small quantity of tricyclazole (Froyd et al., 1976) is needed for seed treatment that is effective and economical against leaf blast. However, fungicide resistance in *P. grisea* poses a threat to rice blast control measures. Naturally occurring resistant strains caused rice crop failure in paddies treated solely with kasugamycin (Ohmori, 1967; Uesugi et al., 1969; Miura et al., 1975; Sakurai et al., 1975; Miura et al., 1976; Ito and Yamaguchi, 1977). Other fungicide resistances were gradually revealed including blasticidin-S (Nakamura and Sakurai, 1968; Hwuang and Chung, 1977); IBP, kitazin P (S-benzyl diisopropyl phosphorothiolate (Katagiri and Uesugi, 1977); EDP, kitazin, EDDP, and edifenphos (ethyl SS-diphenylphos-phorodithiolate) (Uesugi et al., 1969); and isoprothiolane (diisopropyl 1,3-dithiolane-2-ylidenemalonate) (Katagiri and Uesugi, 1977). Taga et al. (1979) identified three loci (kas-1, kas-2, and kas-3) for kasugamycin resistance in *P. oryzae*. The mutation at the kas-3 locus was expected to be responsible for resistance to blasticidin-S.

Over application of a single fungicide on a large scale rice production can result in the vulnerability of a variety within few years after release (Suzuki et al., 2010). Melanin biosynthesis inhibitor targeting scytalone dehydratase (MBI-D) fungicides (Table 2) were introduced in 1998 as efficient and specific fungicides for control of rice blast. MBI-D is a tightly binding competitive inhibitor of SDH, a key enzyme in the biosynthesis of melanin. Nursery box technique was adapted for low cost application of MBI-D. The use of MBI-D fungicides spread quickly, reduced application frequency and quantity, and shortens labour. After large deployment of this fungicide for three years, MBI-D failed to control rice blast in Saga Prefecture (Yamaguchi et al., 2002). Resistance of *P. oryzae* to MBI-D was associated with a single point mutation of the SDH gene that resulted in the replacement of valine by methionine (Takagaki et al., 2004). MBI-D-resistant strains soon spread throughout most of Kyushu during 2002–2003 (Sawada et al., 2004; Suzuki et al., 2007), and were subsequently reported in many other prefectures (Ishi, 2006; Sasaki et al., 2006). The resistant isolates possessed high genetic diversity, indicating that the resistance occurred in a multi-genetic background (Suzuki et al., 2007). An immediate countermeasure against the development of resistance was to discontinue the use of MBI-D fungicides in Saga in 2003 leading to decrease in the frequency of resistant isolates and became undetectable in 2007 (Suzuki et al., 2007). Other types of melanin biosynthesis inhibitors, targeting polyhydro-xynaphthalene reductase (MBI-R) and plant defence activators, were placed under MBI-D (Table 2). The influence of MBI-D switching revealed the fitness of MBI-D-resistant population structure of rice blast under practical conditions. Understanding the fitness of MBI-D-resistant strains would have significant benefits for rice blast management (Suzuki et al., 2007).

If field resistance is known to one member of the Group of fungicides, it is most likely but not exclusively valid that cross resistance to other group members is present. There is increasing evidence that the degree of cross resistance can differ between group members and pathogen species or even within species. Resistance mechanisms vary, but mainly involve modification of the primary site of action of the fungicide within the fungal pathogen (Keith and Derek, 2007). Thus, application must involve chemicals with different modes of action to prevent development of resistance to the chemical by the pathogen. This threat of fungicide resistance and the fact that cross-resistance often exists to related products from different manufacturers has led to a close collaboration between them and the Fungicide Resistance Action Committee (FRAC). FRAC has produced several monographs on various aspects of fungicide resistance and has grouped the available fungicides according to various criteria that facilitate the understanding of the resistance and cross-resistance risks (Table 2). The above rice blast scenario indicates that both host plant resistance and selective fungicides are prone to adaptation by the blast pathogen virulence. The balance between the genetic and chemical control should be continuously assessed to decide a timely management option. During the endemic seasons of 2008–2010, resistant PSB Rc82 amended with two foliar applications of edifenphos at 45 and 60 days after transplanting effectively managed the leaf and panicle blast.

However, their effectiveness has been seriously affected in some situations by the development of resistance in target fungi. An ability to determine the risk of resistance arising would help greatly both the selection of candidate chemicals for development and the establishment of strategies to ensure their durability. Monitoring against large pathogen populations over years can readily detect the early stages of polygenic resistance spread enough to cause field problems (Takagaki et al., 2004; Suzuki et al., 2007).

Past experiences of fungicide resistance indicates clearly that the risk of resistance development depends greatly upon the chemical class to which a fungicide belongs. Each chemical class is characterised by a typical resistance behaviour pattern. Thus, certain major classes of fungicide such as copper-based, phthalimides, and dithiocarbamates have never shown risk of resistance even after many years of use (Brent and Derek, 1998). By contrast, all fungicides belonging to benzimidazoles, phenylamides, and dicarboximides, met serious resistance problems within 2–10 years after the commercial release. In addition, several epidemiological risk factors, specific to *P. oryzae* target pathogen, affect the rate of resistance development. Short generation time, abundant sporulation and isolation of pathogen populations tend to increase resistance risk. And the interaction of inherent biochemical, genetic and epidemiological risk factors can be mitigated by an integration of different disease management approaches: reduce application frequency, rotation with other types of fungicide, and concurrent use of non-chemical disease-control. These measures tend to lower the risk of resistance development.

Sheath Blight Management

Cultural Management

High seeding rate and nitrogen fertilizer rate greatly influence the sheath blight (ShB) severity. The uniform standing crop, comprising 17–20 tillers per hill at 35 days after transplanting, might give consistent attainable yields at harvest and is ideal to manage ShB (Palay-Check System, 2007). High N rates favour severe sheath blight during the growing season as seen at the low-lying areas of the paddies. ShB grows rapidly upward on plants receiving too much nitrogen fertilizer of higher than 120 kg N ha^{-1} in dry season and higher than 90 kg N ha^{-1} in wet season in most irrigated lowland rice in the Philippines. Application of Si may also be beneficial to plant growth and helps plants to overcome abiotic and biotic stresses by preventing lodging and increasing resistance to pests and diseases as well as other stresses. Ma et al. (2006) identified low silicon rice 1 gene (Lsi1), which controls silicon accumulation in rice, a typical silicon-accumulating plant. This gene belongs to the aquaporin family and is constitutively expressed in the roots. On the other hand, Lsi2 functions as an efflux silicon transporter and belongs to a putative anion transporter family without any similarity to Lsi1. Si uptake mediated by

Lsi1 and Lsi2 shows different pathways between rice and other plant species (Ma et al., 2011). The identification of a silicon transporter provides both an insight into the silicon uptake system in plants, and a new strategy for producing crops with high resistance to multiple stresses by genetic modification of the root's silicon uptake capacity.

Use of Host Resistance

Host plant resistance is considered an effective approach for managing ShB since host plant resistance is limited (Bonmann et al., 1992). Tall plants and late maturing varieties may concede to disease escape (Srinivasachary et al., 2011). Recently, ShB-QTL that are independent of morphological traits have been identified in a population derived by Jasmine85/Lemont population (Zou et al., 2000); qSB-9TQ from an Indica cultivar, Teqing (Pinson et al., 2005; Tan et al., 2005; Yin et al., 2009), and qSBR11-1 from a HP2216/Tetep population (Channamallikarjuna et al., 2010).

To reduce the inevitable loss, prudent adoption of tall varieties that tend to have a more open canopy for more sunlight penetration and lower humidity than the shorter ones is recommended. In this practice, the lesion height of ShB hardly reaches over 30% of plant height of NSIC Rc282 (Truong et al., unpublished data). The serious damage is mostly due to the rotten flag leaf and panicle of susceptible short varieties like NSIC Rc214. Farmers who recognized the value of planting NSIC Rc160 having tall and thick culms for only one rice season a year in rain-fed lowland, have consistently sustained their target yield, without fungicide application.

Biological Control Agent

Many BCAs can effectively suppress *R. solani* particularly those that are incorporated with crop residues into the soil. This practice enhances BCAs diversity and number in the soil over the soil-borne pathogen. BCAs become intimately associated with nutrients constantly released by the root system into the rhizosphere and eventually, BCAs colonize and compete with pathogens for ecological niche, inhibitory allelopathy, and induce a systemic resistance in host plants (Haas and Défago, 2005; Alabouvette et al., 2006). BCA-treated rice seeds with selected strains were found to protect the rice seedlings from *R. solani* infection and promote the growth of the rice plants (Mew and Rosales, 1986; Rosales and Mew, 1997; Mew et al., 2004a). Several BCAs can produce a variety of plant hormone-like compounds including auxins, gibberellins, and cytokinins, especially under abiotic stress conditions that can promote plant growth (Lugtenberg and Kamilova, 2009; Francis et al., 2010). Besides, two promising *Pseudomonas fluorescens* strains (Pf1 and FP7) isolated from the rhizosphere induce systemic resistance in ShB-susceptible rice variety IR 50. Post-inoculated plants with BCA increase chitinase activity significantly. Western blot analysis of chitinase indicated the expression of 28–38 kDa proteins in rice sheaths against *R. solani* (Radjacommare et al., 2004). Application of talc formulations of *P. fluorescens* also significantly reduced ShB incidence and increased grain yield comparable to the treatment with a systemic fungicide, carbendazim (Mathivanan et al., 2005). Antagonistic bacterial strains Pf7-14 and PpV14I formulated with combination of methyl cellulose and talc at 1:4 and $CaCO_3$ were most satisfactory. Formulated PpV14I applied as seed treatment, root dip or foliar sprays could suppress ShB up to 60% (Mew et al., 1998; Chen et al., 2001; Nilpanit et al., 2001; Vasudevan et al., 2002). The sustainable ShB management in direct seeded rice in Central Thailand and in transplanted rice in Jiangsu required a continuous application of BCA for nine crop cycles. Further trials of *B. subtilis* strain B916 B in Jiangsu, China showed that BCA applied twice during the booting stage suppressed ShB infection to the minimum when disease pressure was moderate (Chen et al., 2001). However, under conditions of high disease pressure, BCA alone was not adequate to suppress disease spread. Effectiveness of BCA can be increased by mixing B916 B strain with Jinggangmycin or Validamycin to improve the control value of BCA. In addition, Wiwattanapatapee et al. (2007) formulated endospores of *Bacillus megaterium* using lactose, polyvinyl pyrrolidone K-30, citric acid, tartaric acid and sodium bicarbonate to sustain its high density of 10^9 $CFUg^{-1}$, and a long shelf life of 12 months storage at room temperature. Application of this product by either broadcasting or foliar spraying reduced the incidence of ShB in the greenhouse.

Chemical Control

Fungicide is the last option for ShB management. The use of currently registered fungicides has seldom achieved its economic benefits due to several factors including the patchy nature of the disease in most fields. ShB is only able to cause yield loss, if it is able to destroy the upper two leaves of infected plants before grain filling is complete. It is essential to scout for a profitable fungicide before applying the recommended control measure (Cartwright and Lee, 2006).

Brown Leaf Spot Management

Molecular techniques are vital tools in detecting seed borne pathogens. Information on the mechanism of transmission is important in the development of management strategies such as varietal resistance, use of certified seeds, land preparation and maintenance of the fields, and establishment of proper crop cycles to avoid the disease epidemiology and subsequently optimize rice productivity. Direct assessment of genetic variation in a given fungal population is addressed by almost unlimited number of polymorphic loci (Bridge et al., 1997; Bridge et al., 2004).

Cultural Management

Since the pathogen, *Drechslera oryzae*, is an internal seedborne organism, the use of high-quality seeds for planting is the most practical approach for the management of brown leaf spot disease. The use of certified or pre-treated certified seeds for planting every season rather than using farmer's saved seeds is most feasible. Farmer's fields are not usually monitored for the occurrence of pests and diseases so seeds may come from diseased plants. Field sanitation, crop rotation, adjustment of planting date, and proper fertilization can also contribute to the decrease of brown spot infection. Severe brown leaf spot is often associated with weak plants grown under stressful environment such as high plant population.

Host Plant Resistance

Several rice cultivars suited in lowland ecosystem such as PSB Rc82 and NSIC Rc220 have been reported to be resistant to brown leaf spot, including Tetep (Ou, 1985). Improved varieties NSIC Rc9, Rc11, and Rc92, UPLRi5 and Ri7 and a few selected traditional cultivars that are well adapted in the upland ecology in the Philippines where brown leaf spot prevails, are tolerant to brown leaf spot disease (Truong et al., unpublished data). Varieties with partial resistance and three quantitative trait loci (QTL) for disease resistance have been identified (Sato et al., 2008). Since resistant or tolerant varieties are available, the use of host plant resistance would be the most economical mean of controlling brown leaf spot.

Nutrition and Water Management

Proper nutrition for optimum plant growth and prevention of water stress are the most important control measures of brown leaf spot. Improving soil fertility through regular monitoring of nutrients in the soil and the application of right amount of fertilizers are the simple ways for the management of the disease. Nitrogen and other nutrient elements can be easily monitored using leaf colour chart, minus-one nutrient element (detection technique), and soil test kits. A combined N and K rate may contribute in reducing the intensity of brown leaf spot in rice while improving plant development (Carvalho et al., 2010).

The application of calcium silicate slag before crop establishment has been recommended for soils that are low in silicon (Datnoff et al., 1992). Fine grade silicon fertilizer such as calcium silicate slag can reduce the disease to as much as 48%–80% and increases yields of rice by 20–26%. Dallagnol et al. (2011) reported high activities of peroxidase and chitinase activities in rice cultivar Oochikara and the *lsi1* mutant supplied with Si. However, polyphenoloxidase activity did not contribute to the resistance of the test cultivars, regardless of

Si supply. The finding indicates that the involvement of Si is a complex defence mechanism rather than a simple formation of a physical barrier of the rice plant to avoid or delay fungal penetration. In the location where Si fertilizers are not commercially available, regular return of rice stalk after harvest can contribute to Si amendment for long term and reduce the disease severity.

Physical Control

Hot water seed treatment at 53–54°C for 10–12 min with pre-soaking of seeds in cold water for 2 h protects the seedlings from primary infection from seedborne pathogens (Mew and Gonzales, 2002). Precautionary measures, however, should been taken into consideration when doing hot water treatment. Prolonged exposure to heat may kill the embryo. Hot water-treated seeds must also be sown immediately because germination rate of treated seeds decline faster than untreated seeds.

Biological Control

Information on the BCA against brown leaf spot is very limited. Harish et al. (2007) reported that *Cladosporium* spp., *Penicillium* spp. and *Aspergillus flavus* isolated from leaf surface were effective in inhibiting the mycelial growth and spore germination of *Helminthosporium oryzae*.

Chemical Control

In developing countries the use of inorganic pesticides is critically important for seed health, weed control, and insect pests. Pesticide use is particularly appropriate in systems of integrated pest management in which the objective is to generate a sustainable balance of complementary approaches (Maredia et al., 2003). Seed treatment with fungicides prior to sowing can reduce the initial inoculum and protect the growing seedlings from infection. Seed treatment is a common practice for vegetable seeds. Rice seeds for export are also treated with fungicide as a phyto-sanitary requirement of importing countries.

Airborne infections at seedling stage may also be reduced by foliar spraying with dithiocarbamates before heading. However, field application of the fungicidal compound may not always be feasible.

Sheath Rot Management

Host plant resistance to sheath rot has not been developed. The use of certified seeds as planting material is the most economical option to prevent the spread of the disease. Crop rotation and tillage can reduce disease pressure by eliminating the pathogen in the soil. Bacterization with *Pseudomonas fluorescens* was also found to reduce sheath rot severity by 20–42%, enhanced crop growth, and increased grain yield (Sakthivel and Gnanamanickam, 1987). The use of fungicide such as dithiocarbamate and benzimidazoleas a general disease control program can reduce sheath rot damage. However, fungicidal application should only be done when disease spread goes beyond control. Pesticide application should always be the last resort in disease management especially with the use of highly toxic chemicals.

Foot Rot (Bakanae) Management

Understanding the fungus mating types that interact with the host plants and microbial community, response to the various biotic and abiotic factors, and production of mycotoxins are important factors needed in developing and implementing management strategies against the bakanae disease (Desjardins, 2006; Leslie and Summerell, 2006; Jouany, 2007). This typical seed-borne disease is monocyclic (Mew, 1997; Mew et al., 2004b). As such, when seed infection is controlled, the problem concerned is solved. A rice seed lot may carry some microorganisms at one time, but the disease management strategy is focused only on the major pathogen.

Use of Certified Seed

The use of high-quality certified seeds for planting is a cost-effective disease management for saving labour, reducing seed-borne inoculum, seeding rate, producing a uniform standing crop, and sustainable crop productivity. This practice has recently been adapted widely in irrigated and rain-fed lowland in the Philippines. Farmers using own seeds saved from previous rice harvest for planting in the next seasons during 1990s could greatly reduce the disease in most locations in the tropical Asia (Mew et al., 2003). Promotion of planting high-quality certified seeds produced by the National Seed Network has been widely accepted by the farmers.

Biological Control

Successful biocontrol of a disease requires a better understanding of the diversity of the community of rice-associated antagonistic bacteria in the rhizosphere as it relates to the complex regulation of disease suppression. Some bacterial strains found in rice rhizosphere such as *Bacillus subtilis* subsp. *subtilis* ces332p, *B. pumilus* CES6 and *Paenibacilus polymyxa* CES6-4, and *Pseudomonas aeruginosa* NTAVCP and endophytic *Burkholderia pyrrocinia* ZAMSED have high potential as antagonists of *P. oryzae* and *R. solani* pathogens (Truong, unpublished data). The last three strains are plant growth promoting bacteria. The recent advent of high throughput sequencing may be helpful to establish the abundance, richness and function of the BCA community associated with rice agro-ecosystem. Yang et al. (2008) explored techniques such as amplified ribosomal DNA restriction analysis, BOX-PCR and 16S rRNA gene sequence analysis to reveal a significant diversity among the rice-associated bacteria isolated from different microenvironments Endorhiza associated with the pathogen had the highest antagonistic population with 58%; followed by rhizosphere with 26%; and phyllosphere, 7.9%. The most prominent genus in all microenvironments was *Bacillus*. Of these, the endophytic *Brevibacillus brevis* strain 1Pe2 and *Deinococcus aquaticus* strain 1Re14 are reported for the first time as rice-associated bacteria and have potential for field application. Among endorhizobia, 1.4% isolates were found antagonistic to bakanae. In addition, Niknejad and Anvary (2009) demonstrated that two phenazine C and D genes loci of *P. fluorescens* isolated from the rhizosphere have also been found to naturally suppress the soil-borne *F. verticillioides* causing collar and root rot of rice.

Chemical Control

Pre-test followed by seed health protocol and treatment with recommended fungicides for seeds intended for seed exchange is suggested. Imported seed lots are treated with fungicides and closely monitored during crop growth (Ou, 1985). Seed treatments with fungicides such as dithiocarbamates, and thiophanate-methylate at the rate of 1–2% by seed weight have been found effective and are still being used in seed health testing laboratories.

Narrow Brown Leaf Spot

Regular planting within the recommended planting period to meet all irrigation scheduling and nutrient element required help reduce the disease. The use of resistant varieties is the most economical option for managing the disease, but monitoring closely the presence of the pathogenic races to avoid any break down of resistance that may happen over time, is also important. Chemical treatment may not be feasible since the disease occurs late at crop maturity (Groth and Hollier, 2010).

Leaf Scald Management

Cultural Method

Avoidance of excessive nitrogen application to optimize vegetative growth population and wider spacing between plants may limit the development of the disease (Mondal et al., 1986). Regulation of irrigation scheduling to supply adequate water and maintaining a 2–3 cm water depth starting from booting stage and gradually reducing at the end of the reproductive stage to avoid high relative humidity of > 85% and high temperature between 30°C and 33°C under the canopy are desirable. Practice of turning-over the rice stalks after harvest, for medium term, improves silicon in the paddy.

Chemical Control

Foliar applications of validamycin, ethylene bisdithiocarbamate, copper oxychloride, and benzimidazole have been found to reduce infection by *M. albescens*.

Stem Rot Management

An integrated disease management practice has proven effective to suppress stem rot. The sclerotia that serve as carry-over inoculum must be destroyed by ploughing under the soil immediately after harvesting. Stem rot infection is also affected by susceptibility of the host plant. Plants with shorter and thinner culms like PSB Rc52 are usually more susceptible to the diseases than those plants with larger culm and leaves like those of rice variety NSIC Rc160.

Basal and side-dress application of potassium fertilizer at the rate of 50–100 kg/ha depending on pre-test soil fertility in combination with 80–150 kg ha^{-1} (N), 30 kg ha^{-1} (P$_2$O$_5$), and 50 kg ha^{-1} (K$_2$O$_5$) effectively managed stem rot infestation (Collado et al., personal communication) (Fig. 1). A

Figure 1. High rate of nitrogen without potassium (210-30-0 kg ha^{-1}) induced a severe stem rot infestation on NSIC Rc160 (A). The disease was manageable by a combination of N-P-K at 100-30-50 kg ha^{-1} (B).

balanced N-P-K rate based on pre-test soil is the most effective practice for most rice varieties. Application of protectant fungicide such as azoxystrobin targeted at other diseases can also reduce stem rot severity although its application was not economically beneficial.

False Smut Management

Understanding of pathogen population and host plant interaction and cultural practices are cost-effective measures for managing this emerging disease. Use of certified seeds is better than thorough removal of infected seeds and plant debris after harvest (Singh and Khan, 1989; Rush et al., 2000; Xu et al., 2002). Cultural management practices such as adapting varieties that are resistant to false smut in specific location, and crop rotation (Anders and Yeater, 2009; Brooks et al., 2011) are some other ways of managing the disease.

Intermittent irrigation and cultivation in clay soil may reduce disease (Brooks et al., 2010). Crop applied with a low rate of N (Singh et al., 1987) and early transplanting (Narinder and Singh, 1989; Singh et al., 1987; Singh and Khan, 1989; Ahonsi et al., 2000) can escape from false smut infection as well as bad weather (Fujita et al., 1990).

Field scouting is not applicable in managing the disease since sign of the disease is visible after spikelet formation. Disease control using sprays is also ineffective at this stage. Rotation and foliar sprays of rice crop, that had a historical disease, with either copper-based and benzimidazole, dicarboximide, or dithiocarbamate compounds at the beginning of booting stage to early heading stage could significantly reduce the disease incidence (Giffin, 1981; Ou, 1985; Tsai et al., 1990; Chen et al., 1994; Ahonsi and Adeoti, 2003; Atia, 2004; Bagga and Kaur, 2006; Tsuda et al., 2006). Application of *B. subtilis* in solution of validamycin with 4.5 L/hm^2 at 6 days before heading reduced the floret infestation (Yan et al., 2014).

Fusarium Head Blight Management

Risk of the disease outbreaks can be traced to several factors: 1) large growing areas with highly susceptible varieties, 2) seed-borne and virulent pathogen and existence of inoculum in the crop residues and farming equipment, 3) distribution of mycotoxin-producing races, 4) presence of alternate host (corn), and 5) humid weather during flowering and flood favourable for infection. Head blight can reduce yield and the quality of grain. Contaminated grains with mycotoxins at certain threshold are unfit for human and livestock consumption and have no marketable value. Several chemical and physical methods have been studied to detoxify mycotoxins. Unfortunately, there is no easy, economical way to reduce the toxicity of the mycotoxin-contaminated kernels. Agricultural inspectors are responsible and have authority to enhance the standard for disease management within their own jurisdiction, to enforce pest control measures in their respective locality, to implement guidelines and update the management strategy intended to assist growers across the locality.

Crop Husbandry

The disease can be effectively managed by appropriate cultural operations such as avoidance of overlapping cropping system of susceptible hosts like corn and rice and non-exposure to the dispersal period of conidia in the humid tropic. Field sanitation after harvest and fallow period of one month also eliminate inoculum source on crop residues. Close field monitoring revealed that the landing of the conidia of the fungus occur at early heading stage. This indicate that gradually reducing the water supply to keep low humidity under the canopy (< 80%) and leaf wetness (< 20%) and temperature (> 32°C) until the mid of ripening stage could be a good practice to reduce infection.

Most cultural strategies for the control of head blight are based on avoiding or limiting the exposure of cereal spikes to spores during flowering and early grain filling. Following pathogenic colonization of susceptible plants, *F. graminearum* survives saprophytically on residues of corn, small grain cereals, and numerous other plant species and produces both macroconidia and ascospores on these substrates (Warren and Kommedahl, 1973; Windels and Kommedahl, 1974; Sutton, 1982; Pereyra and Dill-Macky, 2008).

Tillage operations that bury infested cereal residues below the soil surface may also be employed to reduce exposure to fungal spores. Sequential crop rotation, followed by a non-host crop of *F. graminearum* will also reduce exposure to inoculum produced on host residues (Pereyra and Dill-Macky, 2008).

Biological Control

Since management of the disease is best applied after the flowering stage where fungicidal application is inappropriate, BCAs play an important role in managing the disease. Spore-producing *B. subtilis* and yeast *C. flavescens* OH 182.9 have effectively reduced head blight and mycotoxin contamination (Schisler et al., 2002; Khan et al., 2004). A combination of these BCAs, salicylic acid, and isonicotinic acid, benzothidiazoles and b-amino-n-butyric acid has shown potential to activate resistance in many crops against the diseases (Ryals, 1994; Cohen, 2001; Oostendorp et al., 2001).

Chemical Control

Paul et al. (2008) confirmed that the combination of prothioconazole and tebuconazole is the most efficacious fungicide for suppressing head blight disease and mycotoxin level. However, fungicides are rarely used against head blight due to high cost, variable efficacy, and the erratic nature of disease epidemics.

Integration of Management Strategies

When environment highly favours disease infection, the use of a single management strategy often fails to control the disease and reduce mycotoxin to acceptable levels. An integration of several approaches that include crop rotation, cultivar resistance, use of BCAs, and right timing of right fungicide has benefited in the years or locations with low disease pressure. To keep the disease at manageable level, 75% reduction in head blight index must be obtained when disease pressure is high (Willyerd et al., 2012). With the advent of disease prediction models, growers can access to tools that help them quantify the risk of disease and the need for fungicide applications to reduce further the impact of a disease epidemic.

References

Ahonsi, M.O., Adeoti, A.A., Erinle, I.D., Alegbejo, T.A., Singh, B.N. and Sy, A.A. 2000. Effect of variety and sowing date on false smut incidence in upland rice in Edo State, Nigeria. IRRI Notes, 25(1): 14.

Ahonsi, M.O. and Adeoti, A.A. 2003. Evaluation of fungicides for the control of false smut of rice caused by *Ustilaginoidea virens* (Cooke) Tak. Moor J. Agric. Res., 4(1): 118–122.

Alabouvette, C., Olivain, C. and Steinberg, C. 2006. Biological control of plant diseases: the European situation. European J. Plant Pathol., 114: 329–341.

Aleshin, N.E., Avakyan, E.R., Dyakunchak, S.A., Aleskin, E.P., Baryahok, V.P. and Voronkov, M.G. 1987. Role of silicon in resistance of rice to blast. DOKL. Akad. Nauk. SSSR, 291: 217–219.

Anarutha, C.S., Zen, K.C., Cole, K.C., Mew, T.M. and Muthukrishnan, S. 1996. Induction of chitinase and b-1, 3-glucanase in *Rhizoctonia solani*-infected rice plants: isolation of an infection-related chitinase cDNA clone. Physiologica Plantarum, 97: 39–46.

Anders, M.M. and Yeater, K.M. 2009. Effect of cultural management practices on the severity of false smut and kernel smut of rice. Plant Dis., 93: 1202–1208.

Asher, R., Ben-Ze'ev, I.S. and Black, R. 1996. Plant clinics in developing countries. Annu. Rev. Phytopathol., 34: 51–66.

Atia, M.M.M. 2004. Rice false smut (*Ustilaginoidea virens*) in Egypt. J. Plant Dis. Prot., 111: 71–82.

Babujee, L. and Gnanamanickam, S.S. 2000. Molecular tools for characterization of rice blast pathogen (*Magnaporthe grisea*) population and molecular marker-assisted breeding for disease resistance. Curr. Sci., 78: 248–257.

Bagga, P.S. and Kaur, S. 2006. Evaluation of fungicides for controlling false smut (*Ustilaginoidea virens*) of rice. J. Indian Phytopathol., 59: 115–117.

Bonman, J.M., Estrada, B.A., Kim, C.K., Ra, D.S. and Lee, E.J. 1991. Assessment of blast disease and yield loss in susceptible and partially resistant rice cultivars in two irrigated lowland environments. Plant Dis., 75: 462–466.

Bonman, J.M., Khush, G.S. and Nelson, R.J. 1992. Breeding for resistance to pests. Ann. Rev. Phytopathol., 30: 507–523.

Brent, K.J. and Derek, W.H. 1998. Fungicide resistance: the assessment of risk. FRAC Monograph No. 2: 49 pp. Published by GCPF (Brussels), September 1998. Printed by Aimprint in the United Kingdom.

Brick, M.A. 2004. Improve yield with high quality seed. Colorado State University Extension No. 303. Dept. of Soil and Crop Sciences., pp. 1–2.

Bridge, P.D., Pearce, D.A., Rutherford, M.A. and Rivero, A. 1997. VNTR-derived oligonucleotides as PCR primers for population studies in filamentous fungi. Letters Appl. Microbiol., 24: 426–430.

Bridge, P.D., Singh, T. and Arora, D.K. 2004. The application of molecular markers in the epidemiology of plant pathogenic fungi. pp. 57–68. *In*: Arora, D.K. (ed.). Fungal Biotechnology in Agricultural, Food, and Environmental Applications. Mycology Series, ISBN 0-8247-4770-4.

Brooks, S.A., Anders, M.M. and Yeater, K.M. 2010. Effect of furrow irrigation on the severity of false smut in susceptible rice varieties. Plant Dis., 94: 570–574.

Brooks, S.A., Anders, M.M. and Yeater, K.M. 2011. Influences from long-term crop rotation, soil tillage, and fertility on the severity of rice grain smuts. Plant Dis., 95: 990–996.

Bryan, G.T., Wu, L.K., Farrall, Y., Jia, H.P., Hershey, S.A., McAdams, K.N., Faulk, G.K., Donaldson, R.T. and Valent, B. 2000. A single amino acid difference distinguishes resistant and susceptible alleles of rice blast resistance gene *Pi-ta*. Plant Cell, 12: 2033–2045.

Carvalho, M.P., Rodrigues, F.A., Silveira, P.R., Andrade, C.C.L., Baroni, J.C.P., Paye, H.S. and Loureiro, Jr., J.E. 2010. Rice resistance to brown spot mediated by nitrogen and potassium. J. Phytopathol., 158: 160–166.

Cartwright, R. and Lee, F. 2006. Management of Rice Diseases. Rice Production Handbook. Online: http://www.uaex.edu/Other-Areas/publications/PDF/MP192/chapter10.pd.

Channamallikarjuna, V., Sonah, H., Prasad, M., Rao, G.J.N., Chand, S., Upreti, H.C., Singh, N.K. and Sharma, T.R. 2010. Identification of major quantitative trait loci qSBR11-1 for sheath blight resistance in rice. Mol. Breed., 25: 155–166.

Chen, Y. 1990. Characteristics of silicon uptaking and accumulation in rice. J. Guizhou Agric. Sci., 6: 37–40.

Chen, Z.Y., Yin, S. and Chen, Y.L. 1994. Biological characteristics of *Ustilaginodea virens* and the screening *in vitro* of the causal pathogen of false smut of rice and fungicides. Chinese Rice Res. Newsl., 2(1): 4–6.

Chen, D., Zeigler, R.S., Leung, H. and Nelson, R.J. 1995. Population structure of *Pyricularia grisea* at two screening sites in the Philippines. Phytopathology, 85: 1011–1019.

Chen, D.H., Zeigler, R.S., Ahn, S.W. and Nelson, R.J. 1996. Phenotypic characterization of the rice blast resistance gene *Pi2(t)*. Plant Dis., 80: 52–56.

Chen, D.H., dela Viña, M., Inukai, T., Mackill, D.J., Ronald, P.C. and Nelson, R.J. 1999. Molecular mapping of the blast resistance gene, *Pi44(t)*, in a line derived from a durably resistant rice cultivar. Theor. Appl. Genet., 98: 1046–1053.

Chen, Z., Xu, Z., Gao, T., Ni, S., Yan, D., Lu, F. and Liu, Y. 2001. Biological control of rice diseases. pp. 61–63. No. 6, *In*: Mew, T.W. and Cottyn, B. (eds.). 1987. Proceedings of the Seed Health and Seed-Associated Microorganisms for Rice Disease Management. International Rice Research Institute, Los Banos, Philippines.

Chen, S., Wang, L., Que, Z., Pan, R. and Pan, Q. 2005. Genetic and physical mapping of *Pi37(t)*, a new gene conferring resistance to rice blast in the famous cultivar St. No. 1. Theor. Appl. Genet., 111: 1563–1570.

Chin, K.M. and Husin, A.N. 1982. Rice variety mixtures in disease control. Proceedings of International Conference of Plant Protection in the Tropics, pp. 241–246.

Christou, P. and Twyman, R.M. 2004. The potential of genetically enhanced plants to address food insecurity. Nutr. Res. Rev., 17: 23–42.

Cohen, Y. 2001. b-Aminobutyric acid-induced resistance against plant pathogens. Plant Dis., 86: 448–457.

Correa-Victoria, F.J. and Zeigler, R.S. 1995. Stability of partial and complete resistance in Rice to *Pyricularia grisea* under rain-fed upland conditions in eastern Colombia. Phytopathology, 85: 977–982.

Crop Protection Compendium. 2004. CD-Rom. Wallingford, UK: CAB Int. Online: http://www.cabi.org/compendia.

Dallagnol, L.J., Rodriguez, F.A., DaMatta, F.M., Mielli, M.V.B. and Pereira, S.C. 2011. Deficiency in silicon uptake affects cytological, physiological, and biochemical events in the rice *Bipolaris oryzae* interaction. Phytopath., 101(1): 92–104.

Datnoff, L.E., Snyder, G.H. and Deren, C.W. 1992. Influence of silicon fertilizer grades on blast and brown spot development and on rice yields. Plant Dis., 76: 1011–1013.

Dean, R.A., Talbot, N.J., Ebbole, D.J. and Farman, M.L. 2005. The genome sequence of the rice blast fungus *Magnaporthe grisea*. Nature, 434: 980–986.

Desjardins, A.E. 2006. *Fusarium* Mycotoxins: Chemistry, Genetics, and Biology. The Amer. Phytopathol. Soc. St. Paul, Minnesota. APS Press, 260 p.

Dobermann, A. and Fairhurst, T. 2000. Rice: nutrient disorders and nutrient management. Potash and Phosphate Institute, Singapore, and International Rice Research Institute, Los Baños, Philippines, 191 p.

Ebbels, D.L. 2003. Principles of Plant Health and Quarantine.Wallingford, UK: CAB Intl.

Epstein, E. 1991. The anomaly of silicon in plant biology. Proc. Nat'l. Acad. Sci., USA, 91: 11–170.

Evenson, R.E. 1998. The economic value of genetic improvement in rice. pp. 303–320. *In*: Dowling, N.G., Greenfield, S.M. and Fischer, K.S. (eds.). Sustainability of Rice in the Global Food System. International Rice Research Institute, Manila.

Evenson, R.E. and Gollin, D. 2003. Assessing the impact of the green revolution, 1960–2000. Science, 300: 758–762.

Flor, H.H. 1971. Current status of the gene-for-gene concept. Annu. Rev. Phytopathol., 9: 275–296.

Francis, I., Holsters, M. and Vereecke, D. 2010. The gram-positive side of plant-microbe `interactions. Environ. Microbiol., 12(1): 1–12.

Froyd, J.D., Paget, C.J., Guse, L.R., Dreiborn, B.A. and Pafford, J.L. 1976. Tricyclazole: A new systemic fungicide for control of *Pyricularia oryzae* on rice. Phytopathol., 66: 1135–1139.

Fujikawa, T., Sakaguchi, A., Nishizawa, Y., Kouzai, Y., Minami, E., Shigekazu, Y., Hironori, K. and Tetsuo, M. 2012. Surface α-1, 3-Glucan facilitates fungal sheath infection by interfering with innate immunity in plants. PLoS Pathol., 8(8): e1002882. doi: 10.1371/journal.ppat. 1002882.

Fujita, Y., Sonoda, R. and Yaegashi, H. 1990. Evaluation of the false smut resistance in rice. Ann. Rep. Plant Prot. North Jpn., 41: 205 (In Japanese with English abstract).

Fukuta, Y., Telebanco-Yanoria, M.J., Imbe, T., Tsunematsu, H., Kato, H., Ban, T., Ebron, L.A., Hayashi, N., Ando, I. and Khush, G.S. 2004. Monogenic lines as an international standard differential set for blast resistance in rice (*Oryza sativa* L.). B. Research Notes. Genetics of disease and insect resistance, 21(4): 70–72.

Fukuta, Y., Ebron, L.A. and Kobayashi, N. 2007. Genetic and breeding analysis of blast resistance in elite indica-type rice (*Oryza sativa* L.) bred in International Rice Research Institute. JARQ 4: 101–114.

Gaunt, R.E. 1995. The relationship between plant disease severity and yield. Annu. Rev. Phytopathol., 33: 119–144.

Giffin, D.H. 1981. Fungal physiology. John Wiley and Sons, New York, 383 p.

Gouramanis, G.D. 1994. Biological and chemical control of rice blast disease (*Pyricularia oryzae*) in Northern Greece. National Agricultural Research Foundation, Plant Protection Institute of Thessaloniki (Greece). CIHEAM—Options Mediterraneennes Cahiers Options Méditerranéennes, 15: 3.

Groth, D. and Hollier, C. 2010. Leaf scald of rice. Louisiana Plant Pathology Disease identification and Management Series. Online: http://www. Isuagcenter.com., November 21, 2012.

Haas, D. and Défago, G. 2005. Biological control of soil-borne pathogens by fluorescent pseudomonads. Nature Rev. Microbiol., 3: 307–216.

Hamer, J.E., Farral, L., Orbach, M.J., Valent, B. and Chumley, F.G. 1989. Host species-specific conservation of a family of repeated DNA sequences in the genome of a fungal plant pathogen. Proc. Natl. Acad. Sci. (USA), 86: 9981–9985.

Harish, S., Saravanakumar, D., Kamalakannan, A., Vivekananthan, R., Ebenezar, E.G. and Seetharaman, K. 2007. Phylloplane microorganisms as a potential biocontrol agent against *Helminthosporium oryzae* Breda de Haan, the incitant of rice brown spot. Arch. Phytopathol. Plant Prot., 40: 148–157.

He, D.-Y., Yazaki, Y., Nishizawa, Y., Takai, R., Yamada, K., Sakano, K., Shibuya, N. and Minami, E. 1998. Gene activation by cytoplasmic acidication in suspension—cultured rice cells in response to the potent elicitor, N-acetylchitoheptaose. Molecular Plant-Microbe Interactions, 11: 1167–1174.

He, Z.H., Dong, H.T., Dong, J.X., Li, D.B. and Ronald, P.C. 2000. The rice Rim2 transcript accumulates in response to *Magnaporthe grisea* and its predicted protein product shares similarity with TNP2-like proteins encoded by CACTA transposons. Molecular General Genetics, 264: 2–10.

Huber, D.M. and Haneklaus, S. 2010. Managing nutrition to control plant disease. www.ifao. com/pdfs/2010. huberhandout2.pdf.

Hwuang, B.K. and Chung, H.S. 1977. Acquired tolerance to Resistance blasticidin-S in *Pyricularia oryzae*. Phytopathol., 67: 421–424.

Ishi, H. 2006. Impact of fungicide resistance in plant pathogens on crop disease control and agricultural environment. Japan Agric. Res. Quarterly, 40: 205–211.

Ito, H. and Yamaguchi, T. 1977. Occurrence of kasugamycin resistant rice blast fungus influenced by the application of fungicides. Ann. Phytopathol. Soc. Jpn., 43: 301–303.

Jaiganesh, V., Eswaran, A., Balabaskar, P. and Kannan, C. 2007. Antagonistic activity of *Serratia marcescens* against *Pyricularia oryzae*. Not. Bot. Hort. Agrobot. Cluj., 35(2): 48–54.

James, C. 1998. Global food security. Abstr. Int. Congr. Plant Pathol, 7th Edinburgh, UK, Aug. No. 4.1GF. Online: http://www.bspp.org.uk/icpp98/4/1GF.htm.

Jia, Y., McAdams, S.A., Bryan, G.T., Hershey, H.P. and Valent, B. 2000. Direct interaction of resistance gene and avirulence gene products confers rice blast resistance. EMBO J., 19: 4004–4014.

Jia, Y., Wang, Z., Fjellstrom, R.G., Moldenhauer, K.A.K., Azam, M.A., Correll, J., Lee, F.N., Xia, Y. and Rutger, J.N. 2004. Rice *Pi-ta* gene confers resistance to the major pathotypes of rice blast fungus in the United States. Phytopathology, 94: 296–301.

Jouany, J.P. 2007. Methods for preventing, decontaminating and minimizing the toxicity of mycotoxins in feeds. pp. 342–362. *In*: Morgavi, D.P. and Riley, R.T. (eds.). *Fusarium* and Their Toxins: Mycology, Occurrence, Toxicity, Control and Economic Impact. Anim. Feed Sci. Technol., 137.

Katagiri, M. and Uesugi, Y. 1977. Similarities between the fungicidal action of isoprothiolane and organophosphorus thiolate fungicides. Phytopathol., 67: 1415–1417.

Kaundal, K.R., Kapoor, A.S. and Raghava, G.P.S. 2006. Machine learning techniques in disease forecasting: a case study on rice blast prediction. MC Bioinformatics, 7: 485. doi: 10.1186/1471-2105-7-485.

Keith, J.B. and Derek, W.H. 2007. Fungicide resistance in crop pathogens: how can it be managed? Second edition, Fungicide Resistance Action Committee 2007, FRAC Monograph No. 1, Croplife International, 56 pp. Online: www.frac.info.

Khan, N.I., Schisler, D.A., Boehm, M.J., Lipps, P.E. and Slininger, P.J. 2004. Field testing of antagonists of *Fusarium* head blight incited by *Gibberella zeae*. Biol. Control., 29: 245–255.

Khush, G.S. and Virk, P.S. 2002. Rice improvement: Past, present, and future. pp. 17–42. *In*: Kang, M.S. (ed.). Crop Improvement: Challenges in the Twenty-first Century. Food Products Press, New York.

Kim, C.Y., Gal, S.W., Choe, M.S., Jeong, S.Y., Lee, S.I., Cheong, Y.H., Lee, S.H., Choi, Y.J., Han, C.D., Kang, K.Y. and Cho, M.J. 1998. A new class II rice chitinase, Rcht2, whose induction by fungal elicitor is abolished by protein phosphatase 1 and 2A inhibitor. Plant Molecular Biology, 37: 523–534.

Kim, C.Y., Lee, S.H., Park, H.C., Bae, C.G., Cheong, Y.H., Choi, Y.J., Han, C.D., Lee, S.Y., Lim, C.O. and Cho, M.J. 2000. Identification of rice blast fungal elicitor-responsive genes by differential display analysis. Molecular Plant Microbe Interactions, 13: 470–474.

Kodama, O., Miyakawa, J., Akatsuka, T. and Kiyosawa, S. 1992. Sakuranetin, a flavanone phytoalexin from ultraviolet-irradiated rice leaves. Phytochemistry, 31: 3807–3809.

Koizumi, S. 2001. Rice blast control with multilines in Japan. pp. 1443–158. *In*: Mew, T.W., Borromeo, E. and Hardy, B. (eds.). Exploiting Biodiversity for Sustainable Pest Management. International Rice Research Institute, Manila, Philippines.

Kumar, J., Nelson, R.J. and Zeigler, R.S. 1999. Population structure and dynamics of *Magnaporthe grisea* in the Indian Himalayas. Genetics, 152: 971–984.

Lanoiselet, V., Cother, E.J. and Ash, G.J. 2001. Risk assessment of exotic plant diseases to the Australian rice industry, with emphasis on rice blast, 51 p. Online: http//prijipati.library. usyd.edu.au/bitstream/2123/149/1/PR2407FR09-01.pdf.

Latterell, F.M. 1971. Phenotypic stability of pathogenic races of *Pyricularia oryza*, and its implications for breeding of blast resistant rice varieties. pp. 9–234. *In*: Proceedings, Seminar on Horizontal Resistance to Blast Disease of Rice. Centro International de Agricultura Tropical (CIAT), Cali, Colombia.

Latterell, F.M. and Rossi, A.E. 1986. Longevity and pathogenic stability of *Pyricularia oryza*. Phytopathol., 76: 231–235.

Leslie, J.F. and Summerell, B.A. 2006. The *Fusarium* Laboratory Manual. Blackwell Publishing, Ames, Iowa, USA, 388 p.

Levy, M., Correa, F.J., Zeigler, R.S., Xu, S. and Hamer, J.E. 1993. Genetic diversity of the rice blast fungus in a disease nursery in Colombia. Phytopathol., 83: 1427–1433.

Liu, E.M., Zhu, Y.Y., Xiao, F.H., Luo, M. and Ye, H.Z. 2003. Using genetic diversity of rice varieties for sustainable control of rice blast disease. Scientia Agricultural Sinica, 36(2): 164–168.

Lugtenberg, B. and Kamilova, F. 2009. Plant-growth-promoting-rhizobacteria. Ann. Rev. Microbiol., 63: 541–556.

Ma, J.F., Tamai, K., Yamaji, N., Mitani, N., Konishi, S., Katsuhara, M., Ishiguro, M., Murata, M. and Yano, M. 2006. A silicon transporter in rice. Nature, 440(7084): 688–691.

Ma, J.F., Yamaj, N. and Mitani-Ueno, N. 2011. Transport of silicon from roots to panicles in plants. Proc. Jpn. Acad. Series B, Phys. Biol. Sci., 87: 377–385.

Madden, L.V. and Nutter, F.W. 1995. Modelling crop losses at the field scale. Can. J. Plant Pathol., 17: 124–35.

Manosalva, P.M., Davidson, R.M., Liu, B., Zhu, X., Hulbert, S.H., Leung, H. and Leach, J.E. 2009. A germin-like protein gene family functions as a complex quantitative trait locus conferring broad-spectrum disease resistance in rice. Plant Physiol., 149(1): 286–296. doi: 10.1104/ pp.108.128348 .PMCID: PMC2613727.

Maredia, K.M., Dakouo, D. and Mota-Sanchez, D. (eds.). 2003. Integrated Pest Management in the Global Arena. Wallingford, UK: CAB International, 512 p.

Mathivanan, N., Prabavathy, V.R. and Vijayanandraj, V.R. 2005. Application of talc formulations of *Pseudomonas fluorescens* Migula and *Trichoderma viridae* Pers. ex S.F. Gray decrease the sheath blight disease and enhance the plant growth and yield in rice. J. Phytopathol., 153: 697–701.

Maywald, G.F., Sutherst, R.W. and Zalucki, M.P. 1997. Generic modelling for integrated pest management. pp. 1115–1116. *In*: Proceedings of MODSIM 97, International Congress on Modelling and Simulation, Hobart, Australia.

Mekwatanakarn, P., Kositratana, W., Phromraksa, T. and Zeigler, R.S. 1999. Sexually fertile Magnaporthe grisea rice pathogens in Thailand. Plant Dis., 83: 939–943.

Mekwatanakarn, P., Kositratana, W., Levy, M. and Zeigler, R.S. 2000. Pathotype and avirulence gene diversity of *Pyricularia grisea* in Thailand as determined by rice lines near-isogenic for major resistance genes. Plant Dis., 84: 60–70.

Mew, T.W. and Rosales, A.M. 1986. Bacterization of rice plants for control of sheath blight caused by *Rhizoctonia solani*. Phytopathology, 76: 1260–1264.

Mew, T.W. 1997. Developments in rice seed health testing policy. pp. 129–138. *In*: Hutchins, J.D. and Reeves, J.C. (eds.). Seed Health Testing: Progress Towards the 21st Century. CAB International, Wallingford, UK.

Mew, T.W., Chen, Z.Y., Nilpanit, N. and Parkpian, A. 1998. Getting biological control technology to rice farmers. pp. 5.2.3S. *In*: Proceedings of the 7th International Congress of Plant Pathology, 9–16 August 1998. Edinburgh, Scotland.

Mew, T.W. and Gonzales, P. 2002. A Handbook of Rice Seedborne Fungi. International Rice Research Institute, Los Banos, Philippines, and Science Publishers, Inc., Enfield, NH, USA, 83 p.

Mew, T.W., Rickman, J., Bell, M., Balsubramanian, V. and Shires, D. 2003. Using good seed. Rice Fact Sheet. International Rice Research Institute, College, Laguna.

Mew, T.W., Cottyn, B., Pamplona, R., Barrios, H., Li, X., Chen, Z., Lu, F., Arunyanart, P., Nilpanit, N., Rasamee, D., Kim, P.V. and Du, P.V. 2004a. Applying rice seed-associated antagonistic bacteria to manage rice sheath blight in developing countries. Plant Dis., 88: 557–564.

Mew, T.W., Leung, H., Savary, S., Vera Cruz, C.M. and Leach, J.E. 2004b. Looking ahead in rice disease research and management. Crit. Rev. Plant Sci., 23(2): 103–127.

Miura, H., Ito, H. and Takahashi, S. 1975. *Pyricularia oryzae* to kasugamycin as a cause of the diminished fungicidal activities to rice blast. Ann. Phytopathol. Soc. Jpn., 41: 415–417 (In Japanese, with English summary).

Miura, H., Katagiri, M., Yamaguchi, T., Uesugi, Y. and Ito, H. 1976. Mode of occurrence of kasugamycin resistant rice blast fungus. Ann. Phytopathol. Soc. Jpn., 42: 117–123.

Mondal, A.S., Ahmed, H.U. and Miah, S.A. 1986. Effect of nitrogen on the development leaf scald disease of rice. Bangladesh J. Botany, 15: 213–215.

Morton, V. and Staub, T. 2008. A short history of fungicides. APS *net* Features. doi: 10.1094/APSnetFeature-2008-0308.

Mousanejad, S., Alizadeh, A. and Safaie, N. 2010. Assessment of yield loss due to rice blast disease in Iran. J. Agr. Sci. Tech., 12: 357–364.

Nakamura, H. and Sakurai, H. 1968. Tolerance of *Pyricularia oryzae* Cavara to blasticidin S. Bull. Agric. Chem. Inspec. Stn. Jpn., 8: 21–25 (In Japanese, with English summary).

Narinder, S. and Singh, M.S. 1989. Effect of different levels of nitrogen and dates of transplanting on the incidence of false smut of paddy in Punjab. Indian J. Ecol., 14(1): 164–167.

Niknejad, K.M. and Anvary, M. 2009. Cloning of phenazine carboxylic acid genes of antagonists bacteria in e dh5α. Agricultura Tropica et Subtropica, 42(4): 157.

Nilpanit, N., Arunyanart, P. and Mew, T.W. 2001. Sustaining biological control in farmers' fields. pp. 55–59. *In*: Mew, T.W. and Cottyn, B. (eds.). Limited Proceedings of the Seed Health and Seed-Associated Microorganisms for Rice Disease Management. International Rice Research Institute, Los Banos, Philippines.

Ohmori, K. 1967. Studies on characters of *Pyricularia oryzae* made resistant to kasugamycin. J. Antibiotics, A-20: 109–114.

Oostendorp, M., Kunz, W., Dietrich, B. and Staub, T. 2001. Induced disease resistance in plants by chemicals. Eur. J. Plant Pathol., 107: 19–28.

Orbach, M.J., Farrall, L., Sweigard, J.A., Chumley, F.G. and Valent, B. 2000. A telomeric avirulence gene determines efficacy for the rice blast resistance gene *Pi-ta*. Plant Cell, 12: 2019–2032. www.plantcell.org © 2000 American Society of Plant Physiologists. DOI 10.1105/tpc.12.11.2019.

Ou, S.H. 1985. Rice Diseases. 2nd Edition, Commonwealth Mycological Institute, Kew, UK, 380 p.

Palay-Check System for Irrigated Lowland Rice. 2007. PhilRice and FAO, 91 p.

Paul, P.A., Lipps, P.E., Hershman, D.E., McMullen, M.P., Draper, M.A. and Madden, L.V. 2008. Efficacy of triazole-based fungicides for *Fusarium* head blight and deoxynivalenol control in wheat: A multivariate meta-analysis. Phytopathology, 98: 999–1011.

Pereyra, S.A. and Dill-Macky, R. 2008. Colonization of the residues of diverse plant species by *Gibberella zeae* and their contribution to *Fusarium* head blight inoculum. Plant Dis., 92: 800–807.

Pinson, S.R.M., Capdevielle, F.M. and Oard, J.H. 2005. Confirming QTLs and finding additional loci conditioning sheath blight resistance in rice using recombinant inbred lines. Crop Sci., 45: 503–510.

Qi, M. and Yang, Y. 2002. Quantification of *Magnaporthe grisea* during infection of rice plants using real-time polymerase chain reaction and Northern blot/phosphor-imaging analyses. Phytopathol., 92: 870–876.

Radjacommare, R., Kandan, A., Nandakumar, R. and Samiyappan, R. 2004. Association of the hydrolytic enzyme chitinase against *Rhizoctonia solani* in rhizobacteria-treated rice plants. J. Phytopathol., 152: 365–370.

Rosales, A.M. and Mew, T.W. 1997. Suppression of *Fusarium moniliforme* in rice by rice-associated antagonistic bacteria. Plant Dis., 81: 49–52.

Rush, M.C., Shahjahan, A.K.M., Jones, J.P. and Groth, D.E. 2000. Outbreak of false smut of rice in Louisiana. Plant Dis., 84: 100.

Ryals, J. 1994. Induction of systemic acquired resistance in plants by chemicals. Ann. Rev. Phytopathol., 32: 439–459.

Sakthivel, N. and Gnanamanickam, S.S. 1987. Evaluation of *Pseudomonas fluorescens* for suppression of sheath rot disease and for enhancement of grain yields in rice (*Oryza sativa* L.). Appl. Environ. Microbiol., 53(9): 2056–2059.

Sakurai, H., Naito, H. and Yoshida, K. 1975. Studies on cross resistance to antifungal antibiotics in kasugamycin-resistant strains of *Pyricularia oryzae* Cavara. Bull. Agric. Chem. Inspec. Stn. Jpn., 15: 82–91 (In Japanese, with English summary).

Sasaki, N., Arai, M. and Suzuki, F. 2006. Rep-PCR fingerprinting analysis of *Pyricularia oryzae* isolates with decreased sensitivity to dehydratase inhibitors in melanin biosynthesis (MBI-D) in Iwate prefecture and their pathogenic races (In Japanese). Annu. Rep. Plant Prot. North Jpn., 57: 10–13.

Sato, H., I., Hirabayashi, H., Takeuchi, Y., Arase, S., Kihara, J., Kato, H., Imbe, T. and Nemoto, H. 2008. QTL analysis of brown spot resistance in rice (*Oryza sativa* L.). Breed. Sci., 58: 93–96.

Savary, S., Willocquet, L., Elazegui, F.A., Castilla, N.P. and Teng, P.S. 2000a. Rice pest constraints in tropical Asia: Quantification of yield losses due to rice pests in a range of production situations. Plant Dis., 84: 357–369.

Savary, S., Willocquet, L., Elazegui, F.A., Castilla, N.P., Teng, P.S., Du, P.V., Zhu, D., Tang, Q., Huang, S., Lin, X., Singh, H.M. and Srivastava, R.K. 2000b. Rice pest constraints in tropical Asia: Characterization of injury profiles in relation to production situations. Plant Dis., 84: 341–356.

Sawada, H., Sugihara, M., Takagaki, M. and Nagayama, K. 2004. Monitoring and characterization of *Magnaporthe grisea* isolates with decreased sensitivity to scytalone dehydratase inhibitors. Pest Mgt. Sci., 60: 777–785.

Schaad, N.W., Opgenorth, D. and Gaush, P. 2002. Real-time polymerase chain reaction for one-hour on-site diagnosis of Pierce's disease of grape in early season asymptomatic vines. Phytopathol., 92: 721–728.

Schisler, D.A., Khan, N.I., Boehm, M.J. and Slininger, P.J. 2002. Greenhouse and field evaluation of biological control of *Fusarium* head blight on durum wheat. Plant Dis., 86: 1350–1356.

Schweizer, P., Buchala, A. and Metraux, J.P. 1997. Gene-expression patterns and levels of jasmonic acid in rice treated with the resistance inducer 2, 6-dichloroiso-nicotinic acid. Plant Physiol., 115: 61–70.

Seebold, Jr., K.W., Datnoff, L.E., Correa-Victoria, F.J., Kucharek, T.A. and Snyder, G.H. 2004. Effects of silicon and fungicides on the control of leaf and neck blast in upland rice. Plant Dis., 88: 253–258.

Silué, D., Notteghem, J.L. and Tharreau, D. 1992. Evidence for a gene-for-gene relationship in the *Oryza sativa–Magnaporthe grisea* pathosystem. Phytopathology, 82: 577–580.

Singh, G.P., Singh, R.N. and Singh, A. 1987. Status of false smut (FS) of rice in eastern Uttar Pradesh, India. IRRI News Letter, 12(2): 28.

Singh, R.A. and Khan, A.T. 1989. Field resistance to false smut and narrow brown leaf spot in Eastern Uttar Pradesh, India. IRRI News Letter, 14: 16–17.

Singh, P., Jia, Y., Correll, J. and Lee, F.N. 2006. Developing molecular marker from *AVR-Pita* for surveillance of durable rice blast resistance conferred by *Pi-ta* in Arkansas. pp. 150–158. *In*: Norman, R.J., Meullenet, J.F. and Moldenhauer, K.A.K. (eds.). B.R. Wells Rice Research Studies 2005. University of Arkansas Agricultural Experiment Station Research Series, 540.

Smith, I.M., McNamara, D.G., Scott, P.R. and Holderness, M. 1996. Quarantine Pests for Europe, 2nd edition. CAB International, Wallingford, UK.

Someya, N., Kataoka, N., Komagata, T., Hibi, T. and Akutsu, K. 2000. Biological control of cyclamen soil borne diseases by *Serratia marcescens* strain B2. Plant Dis., 84: 334–340.

Song, F. and Goodman, R. 2001. Molecular biology of disease resistance in rice. Physiological and Mol. Plt. Pathol., 59: 1–11.

Srinivasachary, Willocquet, L. and Savary, S. 2011. Resistance to rice sheath blight (*Rhizoctonia solani* Kuhn) [(teleomorph: *Thanatephorus cucumeris* (A.B. Frank) Donk.] disease: current status and perspectives. Euphytica, 178: 1– 22.

Strange, R.N. and Scott, P.R. 2005. Plant disease: A threat to global food security. Annu. Rev. Phytopathol., 43: 83–116. Doi: 0.1146/annurev. phyto. 43.113004. 133839.

Sun, Y., Wang, Y.Y., He, Y.Q., Fan, J.H., Chen, J.B. and Zhu, Y.Y. 2002. Analysis of resistance gene analogue for rice cultivars in Yunnan province. Scientia Agricultural Sinica, 1(5): 502–507.

Sutherst, R.W. and Maywald, G.F. 1985. A computerised system for matching climates in ecology. Agriculture Ecosystems and Environment, 13: 281–299.

Sutton, J.C. 1982. Epidemiology of wheat head blight and maize ear rot caused by *Fusarium graminearum*. Can. J. Plant Pathol., 4: 195–209.

Suzuki, F., Arai, M. and Yamaguchi, J. 2007. Genetic analysis of *Pyricularia grisea* population by rep-PCR during development of resistance to scytalone dehydratase inhibitors of melanin biosynthesis. Plant Dis., 91: 176–184.

Suzuki, F., Yamaguchi, J., Koba, A., Nakajima, T. and Arai, M. 2010. Changes in fungicide resistance frequency and population structure of *Pyricularia oryzae* after discontinuance of MBI-D fungicides. Plant Dis., 94(3): 329–334.

Taga, M., Nakagawa, H., Tsuda, M. and Ueyama, A. 1979. Identification of three different loci controlling kasugamycin resistance in *Pyricularia oryzae*. Phytopathol., 69(5): 463–466.

Takagaki, M., Kaku, K., Watanabe, S., Kawai, K., Shimizu, T., Sawada, H., Kumakura, K. and Nagayama, K. 2004. Mechanism of resistance to carpropamid in *Magnaporthe grisea*. Pest Mgt. Sci., 60: 921–926.

Talbot, N.J. 2003. On the trail of a cereal killer: exploring the biology of *Magnaporthe grisea*. Annu. Rev. Microbiol., 57: 177–202.

Tan, C.X., Ji, X.M., Yang, Y., Pan, X.Y., Zuo, S.M., Zhang, Y.F., Zou, J.H., Chen, Z.X., Zhu, L.H. and Pan, X.B. 2005. Identification and marker-assisted selection of two major quantitative genes controlling rice sheath blight resistance in backcross generations. Acta Gen. Sin., 32: 6.

Thinlay, X., Finckh, M.R., Bordeos, A.C. and Zeigler, R.S. 2000a. Effects and possible causes of an unprecedented rice blast epidemic on the traditional farming system of Bhutan. Agric. Ecosyst. Environ., 78: 237–248.

Thinlay, X., Zeigler, R.S. and Finckh, M.R. 2000b. Pathogenic variability of *Pyricularia grisea* from the high- and mid-elevation zones of Bhutan. Phytopathology, 90: 621–628.

Tsai, W.H., Chien, C.C. and Hwang, S.C. 1990. Ecology of rice false smut disease and its control. J. Agric. Res. China, 39: 102–112.

Tsuda, M., Sasahara, M., Ohara, T. and Kato, S. 2006. Optimal application timing of simeconazole granules for control of rice kernel smut and false smut. J. Gen. Plant Path., 72(5): 301–304.

Tsunematsu, H., Yanoria, M.J.T., Ebron, L.A., Hayashi, N., Ando, I., Kato, H., Imbe, T. and Khush, G.S. 2000. Development of monogenic lines of rice for rice blast resistance. Breed. Sci., 50: 229–234.

Uesugi, Y., Katagiri, M. and Fukunaga, K. 1969. Resistance in *Pyricularia oryzae* to antibiotics and organophosphorus fungicides. Bull. Nat. Inst. Agric. Sci. Tokyo., C-23: 93–112 (In Japanese, with English summary).

Vasudevan, P.S., Kavitha, V.B., Priyadarisini, V.B., Babujee, L. and Gnanamanickam, S.S. 2002. Biological control of rice diseases. pp. 11–32. *In*: Gnanamanickam, S.S. (ed.). Biological Control of Crop Diseases. Marcel Dekker Inc., New York.

Wang, G.-L., Mackill, D.J., Bonman, J.M., McCouch, S.R., Champoux, M.C. and Nelson, R.J. 1994. RFLP mapping of genes conferring complete and partial resistance to blast in a durably resistant rice cultivar. Genetics, 136: 1421–1434.

Wang, Y.Y., Fan, J.X., Zhao, J.J. and Zhu, Y.Y. 1998. The layout and rotation of rice varieties and blast control. Journal of China Agricultural University, 3 (supplement): 12–16.

Warren, H.L. and Kommedahl, T. 1973. Fertilization and wheat refuse effects on *Fusarium* species associated with wheat roots in Minnesota. Phytopathology, 63: 103–108.

Webster, R.K. and Greer, C.A. 2004. UC IPM Pest Management Guidelines: Rice. UC ANR Publication 3465. http://www.ipm.ucdavis.edu/PMG/r682100611.html.

Willyerd, K.T., Li, C., Madden, L.V., Bradley, C.A., Bergstrom, G.C., Sweets, L.E., McMullen, M., Ransom, J.K., Grybauskas, A., Osborne, L., Wegulo, S.N., Hershman, D.E., Wise, K., Bockus, W.W., Groth, D., Dill-Macky, R., Milus, E., Esker, P.D., Waxman, K.D., Adee, E.A., Ebelhar, S.E., Young, B.G. and Paul, P.A. 2012. Efficacy and stability of integrating fungicide and cultivar resistance to manage *Fusarium* head blight and deoxynivalenol in wheat. Plant Dis., 96: 957–967.

Windels, C.E. and Kommedahl, T. 1974. Population differences in indigenous *Fusarium* species by corn culture of prairie soil. Am. J. Bot., 61: 141–145.

Winterer, J., Klepetka, B., Banks, J. and Kareiva, P. 1994. Strategies for minimizing the vulnerability of rice pest epidemics. pp. 53–70. *In*: Teng, P.S., Heong, K.L. and Moody, K. (eds.). Rice Pest Science and Management. International Rice Research Institute, Manila.

Wiwattanapatapee, R., Chumthong, A., Pengnoo, A. and Kanianamaneesathian, M. 2007. Effervescent fast-disintegrating bacterial formulation for biological control of rice sheath blight. J. Control Release, 119: 229–235.

Xu, J.L., Xue, Q.Z., Luo, L.J. and Li, Z.K. 2002. Preliminary report on quantitative trait loci mapping of false smut resistance using near-isogenic introgression lines in rice. Acta Agric. Zhejiangensis, 14: 14–19.

Yamada, M., Kiyosawa, S., Yamaguchi, T., Hirano, T., Kobayashi, T., Kushibuchi, K. and Watanabe, S. 1976. Proposal of a new method for differentiating races of *Pyricularia oryzae* Cavara in Japan. Ann. Phytopathol. Soc. Jpn., 42: 216–219.

Yamaguchi, J., Kuchiki, F., Hirayae, K. and So, K. 2002. Decreased effect of carpropamid for rice blast control in the west north area of Saga Prefecture in 2001 (In Japanese) Jpn. J. Phytopathol., 68: 261.

Yan, L., Xue-mei, Z., De-qiang, L., Fu, H., Pei-song, H. and Yun-liang, P. 2014. Integrated approach to control false smut in hybrid rice in Sichuan province, China. Rice Science, 21(6): 354–360.

Yang, J.H., Liu, H.X., Zhu, G.M., Pan, Y.L., Xu, L.P. and Guo, J.H. 2008. Diversity analysis of antagonists from rice-associated bacteria and their application in biocontrol of rice diseases. J. Appl. Microbiol., 104: 91–104.

Yin, Y., Zuo, S., Wang, H., Chen, Z., Gu, S., Zhang, Y. and Pan, X. 2009. Evaluation of the effect of qSB-9Tq involved in quantitative resistance to rice sheath blight using near-isogenic lines. Can. J. Plant Science, 89: 731–737.

Yoshihiro, K., Oono, Y., Kanamori, H., Matsumoto, T., Itoh, T. and Minami, E. 2012. Simultaneous RNA-Seq analysis of a mixed transcriptome of rice and blast fungus interaction. PLoS One, 7(11): e49423. doi: 10.1371/journal.pone.0049423.

Zeigler, R.S., Teng, P.S. and Leong, S.A. 1994a. Rice Blast Disease. Commonwealth Agricultural Bureaux, Wallingford, UK, 626 p.

Zeigler, R.S., Tohme, J., Nelson, R.J., Levy, M. and Correa, F.J. 1994b. Lineage exclusion: A proposal for linking blast population analysis to resistance breeding. pp. 267–292. *In*: Zeigler, R.S., Teng, P.S. and Leong, S.A. (eds.). Rice Blast Disease. Commonwealth Agricultural Bureaux, Walllingford, U.K.

Zeigler, R.S., Cuoc, L.X., Scott, R.P., Bernardo, M.A., Chen, D.H., Valent, B. and Nelson, R.J. 1995. The relationship between lineage and virulence in *Pyricularia grisea* in the Philippines. Phytopathol., 85: 443–451.

Zeigler, R.S. 1998. Recombination in *Magnaporthe grisea*. Annual Review Phytopathol., 36: 249–276.

Zeigler, R.S. and Correa, F.J. 2000. Applying *Magnaporthe grisea* population analyses for durable rice blast resistance. APS*net* Features. Online: doi: 10.1094/APSnetFeature-2000-0700A.

Zhou, E., Jia, Y., Correll, J.C. and Lee, F.N. 2005a. Molecular mechanisms of the instability of avirulence gene *AVR-Pita* in rice blast fungus *Magnaporthe oryzae*. AAES Research Series, 540: 160–167.

Zhou, E., Jia, Y., Lee, F.N., Lin, M., Jia, M., Correll, J.C. and Cartwright, R.D. 2005b. Evidence of the instability of a telomeric *Magnaporthe grisea* avirulence gene *AVR-Pita* in the U.S. Phytopathology, 95(6): S117–S118.

Zhu, Y.Y., Chen, H.R., Fan, J.H., Wang, Y.Y., Li, Y., Fan, J.X., Yang, S.S., Ma, G.L., Chen, J.B., Li, Z.S. and Lu, B.R. 2003. The use of rice varietal diversity for rice blast control. Scientia Agricultural Sinica, 36(5): 521–527.

Zhu, Y.Y., Chen, H.R., Fan, J.H., Wang, Y.Y., Li, Y., Chen, J.B., Fan, J.X., Yang, S.S., Hu, L.P., Leung, H., Mew, T.W., Teng, P.S., Wang, Z.H. and Mundt, C.C. 2000. Genetic diversity and disease control in rice. Nature, 406: 718–722.

Zou, J.H., Pan, X.B., Chen, Z.X., Xu, J.Y., Lu, J.F., Zhai, W.X. and Zhu, L.H. 2000. Mapping quantitative trait loci controlling sheath blight resistance in two rice cultivars (*Oryza sativa* L.). Theor. Appl. Genet., 101: 569–573.

CHAPTER 17

Integrated Management of Fungal Diseases of Soybean [*Glycine max* (L.) Merrill] Occurring in India

S.K. Sharma,[1] *S.K. Srivastava*[2],* *and Moly Saxena*[3]

ABSTRACT

Soybean ranks first among the oilseeds in terms of area (111.27 m ha), production (276.4 m t) and productivity (2,484 kg/ha), in the world. It contributes nearly 25% of the world's total oil and fats production and is a primary source of protein and oil.

In India, exhibiting an unparallel growth, the area and production of soybean have increased to 12.03 m ha and 14.67 m t with productivity of 1350 kg/ha. Besides contributing about 25% to the national edible oil pool of the country and export earnings to the tune of approximately USD 1103 million per annum, it has improved the socio-economic conditions of the small, poor and marginal farmers of central India.

Besides many other factors, diseases have become one of the major constraints in harnessing the full productivity potential of the crop. Out of 29 fungal, 6 bacterial, 18 viral, 6 nematodes and 3 mycoplasma diseases

[1] Directorate of Soybean Research, Khandwa Road, Indore 452 001, Madhya Pradesh, India.
[2] Presently, Director Extension Services, Rajmata Vijayaraje Scindia Agricultural University, Gwalior M.P., India.
[3] Principal Scientist, Rajmata Vijayaraje Scindia Agricultural University, Campus: Sehore, M.P., India.
* Corresponding author: sksrivastava03@gmail.com

recorded on this crop, approximately 10 fungal pathogens are of regular appearance in different parts of the globe entailing into serious economic losses.

Amongst them, six pathogens (*Sclerotium rolfsii, Macrophomina phaseolina, Colletotrichum truncatum, Phakopsora pachyrhizi, Cercospora sojina* and *Cercospora kikuchii*) cause severe damage in India. Although, a number of measures have been adopted for their control, consolidated modules for their management are not yet available to the growers. In this chapter, how the effective and timely adoption of the disease management modules can minimise the losses of the crop due to fungal pathogens has been described and discussed besides the distribution and brief symptoms of the major diseases.

Introduction

The Soybean [*Glycine max* (L.) Merrill], a native of Eastern Asia, is a promising leguminous crop. It was domesticated by the farmers in the eastern part of Northern China during the Shang dynasty. For several thousand years people in Eastern Asia have used soybeans for food and animal feed. Today, soybean is grown, to some extent, in most parts of the world and is a primary source of protein and oil.

Soybean ranks first among the oilseeds in the world and contributes for nearly 25% of the world's total oil and fats production. Currently, the area under soybean in the world is 111.27 m h with a production of 276.4 m t and productivity of 2484 kg/ha (Table 1). The USA leads in terms of area and production of soybean, while India ranks fourth in area and fifth in production in the world. USA, Argentina, Brazil, China and India are the major producers of soybean accounting for 90% of world production. The productivity of soybean in India is less than half of the world's average.

Soybean is a major monsoon season crop in the rainfed agro-ecosystem of central and peninsular India. The region spreads between 15° and 25° N

Table 1. Comparative global coverage of soybean (Area in million ha, prod. in million tonnes and yield in kg/ha).

Country	2011–12			2012–13			2013–14		
	Area	Prod	Yield	Area	Prod	Yield	Area	Prod	Yield
Argentina	18.75	48.88	2607	17.58	40.10	2281	19.42	49.31	2539
Brazil	23.97	74.82	3121	24.98	65.85	2637	27.86	81.70	2932
China	7.89	14.49	1836	6.75	13.05	1933	6.60	12.50	1894
India	10.18	12.21	1200	10.84	14.67	1353	12.20	11.95	979
USA	29.86	84.19	2820	30.80	82.05	2664	30.70	89.48	2914
World	103.81	261.94	2523	104.92	241.14	2298	111.27	276.41	2484

Source: FAOSTAT.

latitude, including the states of Madhya Pradesh, Maharashtra, Rajasthan, Andhra Pradesh, Karnataka and Chhattisgarh. These contribute 98% of the total area under soybean in the country. Madhya Pradesh alone contributes 55% of the total soybean production of the country. The crop is predominantly grown on Vertisols and associated soils with an average crop season rainfall of about 900 mm, which varies greatly across locations and years. Introduction of soybean in these areas has led to a shift in the cropping system from the rain season fallow followed by post-rain season wheat or chickpea (fallow-wheat/chickpea) system to soybean followed by wheat or chickpea (soybean-wheat/chickpea) system. This has resulted in an enhancement in the cropping intensity and resultant increase in the profitability per unit land area.

The advent of commercial exploitation of soybean in India is nearly four decades old. In this short spell of time, the crop has shown unparallel growth in area and production. The area under the soybean has increased from a meager 0.03 million ha in 1970 to 12.03 million ha in 2013. Mean national productivity increased from 0.43 t/ha in 1970 to 1350 kg/ha in 2012. Since the beginning of commercial cultivation of soybean in India, the crop has contributed significantly to supplement the edible oil production of the country, and is responsible for improving the socio-economic conditions of the small, poor and marginal farmers of central India. Currently, soybean contributes about 25% to the national edible oil pool of the country. India, the largest importer of edible oil in the world, the supplementation of soybean oil has helped greatly in curtailing the expenditure on import. Moreover, the crop is playing an important role in national economy by way of earning foreign money to the tune of approximately USD 1103 million per annum.

A variety of pathogens, bacteria, viruses, nematodes, mycoplasma, including fungi, attack soybean. So far, 29 fungal, 6 bacterial, 18 viral, 6 nematodes and 3 mycoplasm diseases have been recorded on this crop (Sinclair and Dhingra, 1975; Sinclair and Shurtleff, 1975; Sinclair and Backman, 1989; Tisselli et al., 1980; Verma et al., 1988; Sharma, 1990). Out of them, only 35 pathogens are known to cause economic damage (Sinclair, 1983).

The first record of the soybean disease dates back to 1882, having a brief description of root gall symptoms due to *Meloidogyne* sp. in soybean plants grown in a green-house in Berlin (Frank, 1882). The first description of a fungal disease affecting aerial parts came about two decades later when Massalongo (1900) reported *Phyllosticta sojaecola* as a new species affecting soybean leaves. Similarly, Butler and Bisby (1931) listed the fungal diseases recorded in India. Soon after, in 1938 Mundkar brought the first supplement to Indian fungi and reported *Pernospora manshurica* from Kashmir. Much later description of soybean diseases common in the U.S.A. appeared in form of a farmer's Bulletin (Morse and Carter, 1939; Morse et al., 1949).

In the present chapter, major/minor fungal diseases of soybean reported in India, have been described and discussed with regards to their distribution, brief symptoms and management in Table 2.

Table 2. Major soybean diseases caused by fungi and their distribution.

Disease	Losses	Distribution	
		India	World
Collar rot	60–70%	Agarwal and Kotasthane, 1971	Worldwide (Aycock, 1966; Weber, 1931), Japan (Kurata, 1960)
Charcoal rot	1.234 million metric tonnes (Wrether et al., 1997). 30–50% in Missouri State (USA) (Wyllie, 1988)	(Likhite, 1936) Agarwal, 1973; Agarwal et al., 1973; Gangopadhyay et al., 1973; Madhyay Pradesh, Maharastra, Rajasthan and Delhi (Gupta and Chauhan, 2005)	Bermuda (Waterson, 1939), North and South America, Australia, Asia, Europe and Africa continents (McGee, 1991), North and South America, Australia, Asia, Europe and Africa continents (McGee, 1991)
Anthracnose *Colletotrichum truncatum* [(Schw.) Andrus and W.D. Moore]	50% in Thailand and 100% in India (Sinclair and Backman, 1989; Ploper and Backman, 1992; Manandhar and Hartman, 1999)	First time from India by Nene and Srivastava (1971), U.P. (Singh et al. (1973); Saxena and Sinha (1978); Singh and Shukla (1987)). A.P. (Saikia and Phukan, 1983), H. P. (Bhardwaj and Thakur, 1991), Karnataka (Banu et al., 1990); M.P. (Nicholson and Sinclair, 1973; Singh, 1993), Maharashtra (Rao et al., 1989), Rajasthan (Singh and Srivastava, 1989)	Korea (Nakata and Takimoto, 1934), North Borneo (Johnston, 1960), Java (Goot and Muller, 1932). North Carolina (Wolf and Lehman, 1924)
Rust *Phakopsora pachyrhizi*	20–100% losses in India (Gupta, 2004; Sharma and Gupta, 2006)	Low hills of UP, West Bengal and North East Region, Madras (Ramakrishnan, 1951), Pantnagar (Sarbhoy et al., 1972), U.P. and Kalyani in Bengal (Singh and Thapliyal, 1977), Meghalaya and Assam plain (Maiti et al., 1981; Sharma, 1990)	Asia, many countries of Europe, Africa, Australia and America. Japan (Yang, 1977), Eastern and Western Hemisphere (Sinclair, 1977)
Frog-eye spot *Cercospora sojina*	22% yield loss (Gupta, 2003, 2004)	India (Miura, 1930; Thirumalachar and Chupp, 1948), U.P. and rest of India (Mehta et al., 1950), H.P., M. P., Karnataka and Uttaranchal (Gupta, 2003, 2004)	Japan (Hara, 1915), Australia, USA and China (Adams, 1933; Lehman, 1928, 1934, 1942; Sun, 1958), USA, Europe and China (Mann's and Adams, 1934; Kornfield, 1935; Tai, 1936; Takasugi, 1936), Venezuela (Muller and Chupp, 1942), Canada (Conners and Savile, 1944), South Western Ontario (Koch and

Table 2. contd....

Table 2. contd.

Disease	Losses	Distribution	
		India	**World**
			Hildebrand, 1944), Virginia (USA) (Fenne, 1942, 1949), Chimaltenago, Guatemala (Muller and Chupp, 1950), Iowa and different states of USA (Weimer, 1947; Crall, 1952), Primorsk (Pecific Coastal) Region of Russia (Mikhalenko, 1965; Ovchinnikova, 1968), Western USSR (Shoshiashviii, 1940), Zambia (Javaid and Ashraf, 1978), China (Feng et al., 2004)
Purple seed stain *Cercospora kikuchii*		India (Nene and Srivastava, 1971), Nainital district of U.P. now in Uttaranchal Pradesh (Agarwal and Joshi, 1971), North-Eastern Hills of India (Maiti et al., 1983; Vishwadhar and Chaudhary, 1982; Verma et al., 1988; Sharma, 1990, 2003; Sharma et al., 1988, 1989, 1993, 2000)	Western Hemisphere (Matsumoto and Tomoyasu, 1925), Eastern Hemisphere, (Yoshii, 1927), in Japan (Hara, 1930), Manchuria, China (Miura, 1930) China (Takasugi, 1936; Nakata and Asyuama, 1941), Korea (Nakata and Takimoto, 1934), in Borneo (Johnston, 1960), in New Guinea (Johnston, 1961), Taiwan (Ling, 1948; Liu, 1948), Taiwan, United States (Gardner, 1924, 1927, 1928), Maryland and Iowa (Petty, 1943; Mc New, 1948). America New Jersey (Haenseler, 1946, 1947), Mississippi (Johnson and Kilpatrick, 1953; Kilpatrick, 1955), Columbia (Patino, 1967), Sao Paulo in Brazil (Do Amaral, 1951), Pecific side of Nicaragua (Litzenberger and Stevenson, 1957), Venezuela (Diaz, 1966), Yugoslavia (Lusin, 1960; Nagata, 1962). African- Tanzania (Riley, 1960)

COLLAR ROT

Collar rot of soybean caused by *Sclerotium rolfsii* Sacc. (teleomorph *Athelia rolfsii* (Curzi) Tu & Kimbrough) is a devastating soil-borne disease with a wide distribution (Aycock, 1966; Punja, 1988). It becomes a serious disease in most of the soybean growing areas. Under heavy soil moisture conditions, some times more than 60–70% crop has been found damaged due to this disease.

Symptom

Collar rot disease attacks mainly host stems, although it may infect any part of a plant under favorable environmental conditions including roots, fruits, petioles, leaves, and flowers. The first signs of infection, though usually undetectable, are dark-brown lesions on the stem at or just beneath the soil level. The first visible symptoms are progressive yellowing and wilting of the leaves. Following this, the fungus produces abundant white and fluffy mycelium on infected tissues. Characteristic disease symptom in seedling occurs in the form of damping-off. The collar region starts decaying resulting into the drooping and wilting of plants. Mature sclerotia resemble mustard seed. The fungus occasionally produces basidiospores under humid conditions at the margins of the lesions, though this form is not common. Seedlings are highly susceptible and die quickly once they become infected. However, older plants that have formed woody tissue are gradually girdled by lesions and eventually die.

Favourable Condition

S. rolfsii is favoured by hot weather conditions and prefers high temperature (28–30°C). It is able to survive within a wide range of environmental conditions. Growth is possible within a broad pH range, though it is best on acidic soils.

Collar rot of soybean is more common in sandy soils as compared to the clay soils. The pathogen perpetuates in the soil. Sclerotia produced during crop season remain viable in the soil and serve as a primary source of inoculum for the disease in the successive crop. These sclerotia germinate, when suitable moisture and temperatures are available and cause infection in young germinating soybean seedlings. Secondary infection is limited due to restricted movement of sclerotia and mycelium. Borkar (1992), while studying the influence of weather on the incidence of *Sclerotium* root rot, reported that 100% death of young plants occurred after a dry spell for three continuous days with soil temperatures between 29°C and 30°C followed by a shower of light rain.

High temperatures and moist conditions are associated with germination of sclerotia (Punja, 1985). High soil moisture, dense planting and frequent irrigation promote infection (Aycock, 1966; Clark and Moyer, 1988; Paola, 1933).

Integrated Management

Cultural practices: Collar rots can be managed by use of cultural practices rather than the application of any chemicals. Long crop rotation with non susceptible crop like wheat and paddy should be followed. Good field sanitation should be maintained by rouging out the infected plants.

Chemical Control

There are several reports where fungicides have been used for the control of soil-borne pathogens including *S. rolfsii* (Seoud et al., 1982; Ilieseu et al., 1985). Captan and Benomyl have been used successfully against several seed-borne fungi under laboratory and field conditions (Goulart, 1992). The high amount of Benzoldehyde (0.4 ml/kg of soil) and Velvet bean (100 g/kg) inhibited mycelial growth and sclerotial germination of *S. rolfsii*. The number of soybean (*Glycine max*) plants was higher and the disease was lower in amended soil as compared to the non-amended ones (Blum and Rodriguez, 2004).

Seed treatment: Seed treatment by fungicides is effective to some extent in reducing losses caused by *S. rolfsii* in crops, particularly at seedling stage. Seed treatment with Captan @ 4 g/kg seed significantly decreased root rot of groundnut caused by *S. rolfsii* (Muthamilan and Jayrajan, 1992). Thiophanate-methyl was found to be effective in controlling *S. rolfsii* in stevia (Hilal and Baiuomy, 2000).

Seed treatment of soybean with Hexaconazole and Propiconazole inhibited *S. rolfsii*. These fungicides were found to be absorbed by roots and translocated to the shoots and the leaves (Tajane et al., 2002). Torray et al. (2007) in *in vitro* evaluation found that Captan, Kavach and Thiram showed some reduction in pre-emergence mortality of chickpea. Similarly Thakur et al. (2002) reported reduction in collar rot infection of chickpea by Bavistin, Benomyl and Captan. Seed treatment with Thiram + Carbendazim was equally effective for the control of the disease (Shukla and Singh, 1992).

Khode and Raut (2010) reported that seed treatment and soil application of fungicides, bioagents and its combinations were effective in increasing seed germination and reducing pre and post emergence mortality. Minimum pre and post emergence mortality was obtained with application of Thiram + Carbendazim + Trichoderma @ 3+1+4 g/kg.

Soil treatment: In order of their efficacy, Benomyl, Thiophanate-methyl, Thiram, Thiabendazole, Triforine and Captan decreased viability of microsclerotia in soil and in soybean stem pieces (Ilyas et al., 1976). In inoculated soil, the fungicides did not affect after emergence, but number of microsclerotia was greatly reduced by Benomyl. Soil treatment with Thiram (2000 ppm) minimized pre- and post-emergence mortality of barley caused by *S. rolfsii*. They also evaluated the efficacy of Hexaconazole (0.1% and 0.2%), Carbendazim (0.2%) and Thiophanate-methyl (0.2%) under *in vitro* conditions against *S. rolfsii* of gram and sunflower. Hexaconazole was found to be highly effective followed by Carbendazim and Thiophanate-methyl.

Soil treatment with Propiconazole and Hexaconazole effectively reduced stem rot of groundnut caused by *S. rolfsii* (Charde et al., 2002). However, Gogoi et al. (2002) reported that soil-drenching with Captan (0.2%) reduced the collar rot disease of elephant's foot yam caused by *S. rolfsii*.

Pawar et al. (2014) observed that the fungicidal seed treatment with Captan, Carbendazim and their combination; soil application of organic amendments with Groundnut cake (314 g/plot), Safflower cake (314 g/plot), Sunflower cake (314 g/plot), Neem cake (314 g/plot), Cotton cake (314 g/plot), FYM (314 g/plot), and their interactions significantly reduced the pre- and post mortality induced by *S. rolfsii* in soybean Cv. MAUS-71. Of the six organic amendments tested, Neem cake recorded significantly low pre-emergence (05.63%) and post-emergence mortality (05.47%). Among the interactions, Carbendazim + Captan x Neem cake recorded significantly least pre-emergence mortality (01.66%). Post-emergence mortality was significantly reduced with treatment interactions of Captan x FYM (01.23%). Interactions of Carbendazim + Captan with Sunflower cake recorded significantly increased number of pods/plant (37.00). Significantly highest test weight was recorded with the interactions of Carbendazim + Captan x Neem cake (12.50 g). The treatment interactions of Carbendazim + Captan x Sunflower cake recorded significantly highest seed yield (2040.00 kg/ha). Soil drenching with Chloroneb @ 20 kg/ha is also recommended in soil heavily infested with sclerotia of *S. rolfsii*.

Bio-control: The potential for the use of fungal antagonist as biocontrol agent of plants diseases was suggested more than 70 years ago by Weinding (1932), who was the first to report the parasitic activity of *Trichoderma* spp. against *Rhizoctonia solani* and *S. rolfsii*. *T. harzianum* and *T. viride* were found to be effective in inhibiting the mycelial growth of *S. rolfsii* by several workers (Mathur and Sarbhay, 1979; Roberti et al., 1996; Pushpavati and Rao, 1998; Prasad and Rangeshwaran, 2001; Patil et al., 2004; Anju and Varma, 2007; Jadhav et al., 2008; Mandhare and Suryawanshi, 2008).

Seed treatment with bioagents like *Trichoderma harzianum*, *T. viride* and *Gliocladium virens* is also effective against *S. rolfsii*. B-183 isolate of *Pseudomonas fluorescens* was reported to be antagonistic against *S. rolfsii* (Cattelan, 1994).

Ansari (2005) studied the suitability of different biological agents to control the collar rot of soybean. He noticed highest seed germination in *P. fluorescens* treated seeds at 10 g/kg (51.11%), followed by *T. viride* treated seeds at 4 g/kg (47.82%) and Thiram + Carbendazim (2:1) treated at 3 g/kg (41.63%). Germination was lowest in the control (37.50%). Both biological control agents were effective in controlling collar rot incidence, and both increased the emergence (varied from 71 to 93%) and decreased the pre-emergence mortality. Treatment with biological control agents produced more study and vigorous plants than those treated with chemicals and untreated control.

Maurya et al. (2008) used _T. harzianum_ (104, 106 and 108 spore/ml) and two PGPRs (_P. fluorescens_ strains 4 and _P. aeruginosa_) as foliar spray against collar rot of chickpea. Foliar application of _T. harzianum_ (108 spore/ml) and _P. fluorescens_ strain 4 (108 cfu/ml) showed maximum efficacy in reducing plant mortality (15–25%) as compared to the control.

Singh and Upadhyay (2009) reported that in the absence of PCNB, the mutant _Th mu 6_ and _Th mu 19_ significantly controlled the disease up to 76.5% and 63.1%, respectively as compared to the inoculated control plants. The mutant _Th mu 11_ was found less efficient in controlling the disease as compared to other two mutants. The wild type strain showed lower efficiency in disease control (51.2%) as compared to all three mutant strains. Treatment combination of antagonists and PCNB showed drastic control of the disease as compared with the inoculated control. The mutant _Th mu 6_ significantly controlled the disease (90.8%) followed by the mutant _Th mu 19_ [(78.4%) (Singh and Upadhyay, 2009)].

Application of mutant strains of _T. harzianum_ 4572 as sodium alginate beads to the rhizosphere of soybean plants was highly effective both in the glasshouse and field experiments (Singh and Upadhyay, 2009). In the glass house, wild type and mutants of _T. harzianum_ 4572 were delivered through alginate beads at the rate of 1 g per 100 g of soil.

Lahre et al. (2012) conducted a field experiment to find efficacy of _Trichoderma_ by seed treatment (4 g/kg) along with soil application of Neem cake, Mustard cake and Karanj cake against _S. rolfsii_. The result revealed that seed treatment of bio-agent _Trichoderma_ with Neem cake as soil application was found to be the most effective, recording maximum seed germination (93.05%) and minimum total mortality (8.32%), followed by _Trichoderma_ with Mustard cake (85.03% and 9.55%) and _Trichoderma_ with Karanj cake (80.86% and 11.41%). Maximum reduction in collar rot (44.10%) was achieved in soil application of _Trichoderma_ @ 3 kg/50 kg FYM/ha and seed treatment @ 4 g/kg + soil application @ 2 kg/ha, respectively. Dual application, i.e., seed treatment and soil application 4 g + 2 kg/FYM and seed treatment with Thiram + Carbendazim + _Trichoderma_ (3 + 1 + 4 kg) proved to be effective in recording the higher grain yield with maximum reduction in diseases (Mane et al., 2013).

Host Resistance

Soybean disease survey was conducted in different locations of the USA, i.e., Alabama, Georgia, Louisiana, Mississippi and South Carolina, during 1944–1946 (Weimer, 1947). He reported that cultivars, _Roanoke, Volstate, F.C. 30261-1,_ and Woods Yellow showed resistance to _Sclerotium_ blight caused by _S. rolfsii._ Agarwal and Kotasthane (1971) tested 25 soybean cultivars in India and found cultivar IC 216, Taichung E-32, Improved Pelican, Monetta, Pb-1, Shelby, Palmetto, Hood and Jackson resistant to collar rot while, Hill, Scott,

Dare, Blenville, Pickett, Clark 63, Lee, Semms, Harasoy, Shih Shih, Dorman and Hardee were moderately resistant.

CHARCOAL ROT

Soybean is one of the most important hosts of *Macrophomina phaseolina* (Tassi) Goid. This pathogen causes charcoal rot, dry rot, and seedling blight disease (Reichert and Hellinger, 1947; Su et al., 2001). The fungus is primarily soil inhabiting but is also seed borne in many crops including soybean. It survives in the soil mainly as microsclerotia, which germinates repeatedly in the crop growing season.

Symptoms

The disease symptoms vary with the growth stages at which the soybean plant is infected. Usually, charcoal rot develops late in the season but it can cause seedling disease, since the roots of the plant can be infected at any time during the soybean-growing season.

On seedlings: Seedlings can be infected when soil temperature is continuously above 35°C for 2–3 weeks. After emergence, symptoms can be visible on cotyledonary leaves as brown to dark spots. Sometimes, the margins of the cotyledonary-leaves become brown to black and such leaves fall down at an early stage. From the unifoliate leaf stage onwards, the symptoms appear on emerging hypocotyls of infected seedlings as circular to oblong, reddish brown lesions that may turn dark brown to black after several days. This lesion may extend up to the stem.

On adult plant: The pathogen causes lesions on the roots, stem, pods and seeds. From ground level upward, superficial lesions, brown to grey in colour, infrequently appear on the stem. Microsclerotia are formed in the vascular tissue and in the pith, giving a grayish black appearance to the subepidermal tissues of the stem. The pathogen also produces ashy stem blight symptoms (Dhar and Sarbhoy, 1987). A reddish-brown discolouration of the vascular elements of roots and lower stem precedes the premature yellowing as the fungus spread upto the stem during the season. Mature dry pods are covered with locally or widely distributed black bodies (microsclerotia). Charcoal rot symptoms may easily be observed after death of the plant, which comprise of numerous, minute, pinhead-size microsclerotia. These can be seen readily when the epidermal tissue of the lower stems and roots are pealed off from the affected parts.

In India, epiphytotics occur in areas where temperature ranges from 36–40°C during cropping season. During 1997 season, charcoal rot had caused substantial loss to plant stand and yield in Guna district of Madhya Pradesh (Gupta and Chauhan, 2005).

Control Measures

Since the charcoal rot disease is associated with stressed plants, the disease incidence can be managed through proper fertilization, weed control, and irrigation. Therefore, crop rotation with a less susceptible crop, minimizing plant stress by avoiding excessive seeding rate, fertilizing when necessary and also irrigation to keep soil moisture high during pod development periods, as well as application of fungicides to seed and soil help in the management of the disease.

Currently, no single control measure is satisfactory for various reasons. The measures that form an integral part of the Integrated Disease Management (IDM) are discussed below:

Cultural Practices

Through crop management, the inoculum levels can be brought down and soil moisture can also be retained, which in turn can culminate in reduced incidence of charcoal rot as well as scarcity of microsclorotia. Crop rotation with non-host crops for two to three years is considered necessary to lower *M. phaseolina* infection levels in severely infected field (Wyllie, 1988).

There is a little information available on the direct role of fertility or plant nutrition on the management of charcoal rot in soybean (Todd et al., 1987). However, timely use of recommended dose of fertilizer and adequate phosphorus and potash nutrition is important for charcoal rot management in soybean (Gupta and Chauhan, 2005; Gupta et al., 2006). Increase in NPK supply is also known to reduce *M. phaseolina* infection in soybean (Csondes et al., 2008). Ansari (2010) observed, significant effects of seed treatment (ST) with *Bradyrhizobium japonicum, Trichoderma viride*, Vitavax power (Carboxin 37.5% + Thiram 37.5%), irrigation at the time of moisture stress, and soil application of Zn, B and Fe alone or in different combinations, on disease incidence, number of chaffy (empty) pods per plant, yield and 100 seed weight in variety NRC 7. The maximum chaffy pods were noticed in untreated control and minimum in combined soil application of Zn along with *B. japonicum* and *T. viride* followed by combinations of bio-agents with B + Fe and Zn + Fe. Boron, Zn and Fe individually and in their combinations reduced charcoal rot incidence. Seed treatment with *B. japonicum* and *T. viride* and soil application of Zn with B and Fe reduced chaffy pods as well as the disease incidence up to 75% over untreated control. Seed treatment with *Trichoderma* and irrigation at the time of moisture stress reduced the intensity of disease to about 50 per cent.

By growing soybean cultivars of full season maturity groups that escape the hottest and driest conditions in the post-flowering period, the disease can be controlled effectively (Bowen and Schapaugh, 1989).

Charcoal rot disease could be controlled by organic amendments such as Farm Yard Manure (FYM), Neem and Mustard cake (Rathore, 2000). Muthusamy and Mariappan (1992) investigated *in-vitro*, the mode of action of oil cakes on *M. phaseolina* microsclarotia from soybean. With the addition of Neem or Gingelly cake @ 3%, they observed 90 and 80% microsclarotia disintegration, respectively due to stimulation in germination and lysis as compared to the untreated control. Maximum germtube production and lysis were observed in Coconut cake extracts. Although, solarisation has been suggested as a possible method to manage soil born pathogens, but solarisation alone was not effective in controlling *M. phaseolina* (Grinstein et al., 1979; Katan et al., 1980; Pullman et al., 1981; Usmani and Ghaffar, 1982).

Host Resistance

Resistance has been identified in some varieties of bean against charcoal rot disease (Dhingra and Sinclair, 1978), which has been attributed in part to the variability of the pathogen (Wyllie, 1988). By growing soybean cultivars of full season maturity groups that escape the hottest and driest conditions in the post-flowering period, the disease can be controlled effectively (Bowen and Schapaugh, 1989). The cultivars that had long duration may escape stress because of low temperatures and rainfall less sporadic during critical time for infection (Pearson et al., 1944). Gangopadhyay et al. (1973) reported that the soybean cultivar Hill was least susceptible while Harosoy was most susceptible to the infection. In India, soybean varieties viz., NRC 2, NRC 37, JS 7105, LSB-01 and MACS13 have been identified as less susceptible to charcoal rot and are recommended for cultivation (Gupta and Chauhan, 2005).

Agarwal et al. (1973) reported Hill cultivars as least susceptible and Horsoy most susceptible to infection of *M. phaseolina*. Out of 126 elite soybean, ten genotypes viz., JS 335, G 213, Birsa Sova-1, GS 1, GC 175320, G 9, G-688, NRC 37, DSb 6-1 and RSC 14 were highly resistant (1.0% mortality), whereas 60 lines were moderately resistant (1.1–10.0% mortality) to charcoal rot (Ansari, 2007). Talukdar et al. (2009) screened 100 diverse genotypes for resistance by paper towel method and noticed 7 germplasm lines to be highly resistant while 23 germplasms were moderately resistance. Gopal and Jagadeeshwar (1997) reported 3 genotypes resistant, 1 moderately resistant, 9 moderately tolerant, 37 moderately susceptible and 20 susceptible to charcoal rot out of 70 genotypes tested.

Mengistu et al. (2011) evaluated reactions of 27 maturity group (MG) III, 29 Early MG IV, 34 Late MG IV, and 59 MG V genotypes for *M. phaseolina* in artificially infested soil for three years and identified six genotypes (one genotype in MG III, one in Late MG IV, and four in MG V) as moderately resistant. Some of the commercial and public genotypes were resistant to *M. phaseolina* at levels equal or greater than the standard DT97-4290, a moderately resistant cultivar.

Chemical Control

Carbendazim (Bavistin) and Propiconazole were found highly effective *in vitro* with 100 and 93.7% mycelial growth inhibition of *R. bataticola* [*M. phaseolina* (Prashanthi et al., 2000)]. Carbendazim showed 100% inhibition at all concentrations 250, 500 and 1000 micro g/ml. Among the non-systemic fungicides, Chlorothalonil SC at 3000 ppm was the most effective (82.45% inhibition of mycelial growth). Konde et al. (2008) revealed that Carbendazim + Thiram (0.1 + 0.2%), Penconazole (0.1%) and Thiophanate-M (0.1%) completely inhibited (100%) the growth of *R. bataticola*. Prajapati et al. (2002) also reported complete inhibition of the growth of *R. bataticola* by Carbendazim, Carbendazim + Thiram, Carboxin, and Thiophanate-methyl.

Salunke et al. (2008) reported complete growth inhibition of *R. bataticola* at 0.2% concentration of Campanion and Zineb for all isolates. All isolates showed tolerance to Copper oxychloride, thereby showing its least efficacy.

Ammajamma and Hegde (2009) reported 100% growth inbition of *R. bataticola* by Carboxin (0.05 and 0.1) and Hexaconazole, Metalaxyl and Traidemifon at 0.1 per cent. Among the non-systemic fungicides tested Thirum at 0.1, 0.2 and 0.3 per cent concentrations was found effective against *R. bataticola*. Kumar et al. (2011) also reported that Bavistin and Vitavax inhibit 100 per cent of mycelial growth at 50 ppm.

Soil treatment: In order of efficacy, Benomyl, Thiophanate methyl, Thiram, Thiabendazole triforine and Captan decreased viability of microsclerotia in soil, and in soybean stem pieces (Ilyas et al., 1976). In inoculated soil in the field, fungicides did not affect emergence, but microsclerotia numbers were greatly reduced by Benomyl and to a lesser extent by the other fungicides. Seedling infection was best controlled by Benomyl and Thiabendazole but both showed some phytotoxicity (Ilyas et al., 1975). Other studies indicated that the above fungicides along with Carbendazim, PCNB and Macozab reduced population of the pathogen under laboratory condition on soybean stem pieces (Ilyas et al., 1976; Dwivedi and Dubey, 1987).

Soil fumigation with Sodium methyldithiocarbamate reduced population of the fungus on soybean residues, and in the roots of plant grown in field plots (Gray, 1978; Kittle and Gray, 1982). Fumigation with Methyl bromide significantly increased yields and reduced number of viable microsclerotia and prevalence of *M. phaseolina* (Walters, 1961; Watanabe et al., 1970).

Seed treatment: Seed treatment by fungicides is effective to some extent in reducing losses caused by *M. phaseolina* in crops, which are particularly vulnerable at the seedling stage. Vir et al. (1972) reported that out of four fungicides namely, BAS 3191 F, Topsin, Benlate and Difolation and one antibiotic—Piomy, that were tested in the field each at 1000 ppm against *M. phaseolina* on soybean, Topsin and BAS 3191 F gave the best control. Seed treatment with Carbendazim 50 WP (2.09 g/kg seed) and Thiophanate methyl (1.0 g/kg seed) was effective in eliminating the pathogen (*M. phaseolina*) from the infected seeds of soybean (Kumar and Singh, 2000). Gupta and Chauhan (2005) also recommended seed treatment with Captan @ 3 g/kg or Thiram @ 3 g/kg or Thiram + Carbendazim (2:1) @ 3 g/kg or Thiram + Carboxin @ 2 g/kg seed.

Most commonly used fungicides for seed treatment for improvement in seedling emergence were: Carbendazim (Theradimani and Marimuthu, 1993), Captan and Macozeb (Singh et al., 1990), Thiram, Chlorothalonil, Iprodione and Captan (Raut and Somani, 1987), Quintozene (Chauhan, 1988), Thiophanate methyl (Ahmad et al., 1992), Benomyl and Carboxin (Abawi and Pastor-Corrales, 1990) and Dichlone (El-Ghany et al., 1975).

Biological control: The antagonistic activity of *Aspergillus, Penicillium, Trichoderma* spp., *Actinomycetes, Psuedomonas fluorescens, P. cepacia, Bacillus* spp., *Xanthomonas matophilia,* and a few strains of *Rhizobacteria* against *M. phaseolina* has been observed by several workers.

T. harzianum has been investigated extensively as a biological seed treatment for *M. phaseolina*. In India, use of *T. viride* or *T. harzianum* on a seed treatment @ 4–5 g/kg seed has been recommended for the management of charcoal rot in soybean (Gupta and Chauhan, 2005). Seed treatment with *P. aeruginosa* reduced infection of *M. phaseolina* to the extent of 14 to 100% depending on the strain of bacteria and variety of soybean used (Etheshamul-Haque et al., 2007). The inoculation of seeds or roots with *Rhizobium japonicum* reduced severity of charcoal rot disease in soybean on account of the fungitoxic action of rhizobitoxine (Chakraborty and Purkayastha, 1984; Pearson et al., 1984; Prasad, 1988; Prabarakan et al., 1989). The bacterium reduced growth of the pathogen *in vitro* and rhizobitoxine obtained from culture extracts was toxic to the fungus. This toxin was also recovered from the roots inoculated with *R. japoxicum* (Chakraborty and Purkayastha, 1984). Several antagonistic rhizosphere inhabiting fungi and soybean bacterial endophytes were also indentified during *in vitro* tests (Dubey and Dwivedi, 1988; Senthil Kumar et al., 2009).

Suriachandraselvan et al. (2005) reported that FYM @ 12.5 t/ha and Neem cake @ 0.25 t/ha were effective in reducing the inoculum levels of *M. phaseolina* and significantly reduced the incidence of charcoal rot to 13.3 and 15.0%, respectively as compared to 63.3% in the control. Konde et al. (2008) in *in vitro* experiment found that *T. viride* inhibited the growth of *R. bataticola* to the extent of 96.39 per cent. Similarly, Salunke et al. (2008) reported *T. viride* to be a most efficient bioagent which suppressed mycelial growth ranging from 71.11 to 75.92% followed by *T. harzianum* 67.29 to 74.93% and *P. fluorescens* 60.37 to 70.00 per cent.

Rajput et al. (2010) in *in vitro* experiment found that *T. harzianum* inhibited the maximum radial growth of *R. bataticola* (89.67%). Kumar et al. (2011) also reported *T. harzianum* exhibited the highest mycelial growth inhibition (58.9%) against *R. bataticola*.

ANTHRACNOSE

Colletotrichum truncatum [(Schw.) Andrus and W.D. Moore], is one of the most important seed-borne fungal pathogen of soybean that causes anthracnose disease (Sinclair and Backmam, 1989). This is the most common species associated with this disease, but several other *Colletotrichum* species have also been identified to be involved.

Symptom

Infected plant parts like leaves, stem and pods may not show any symptoms in early stages. At an advanced stage, black fruiting bodies (acervuli) with minute black spines (setae) are visible even with unassisted eyes. These are the diagnostic characteristics for its preliminary identification. Pre-mature defoliation may occur throughout the canopy when cankers girdle the leaf petiole. Morgan and Johnson (1964) gave a descriptive account of the cankers that girdled petioles and caused leaf blades to shed leaving only the shriveled petioles attached. Leaf rolling is also noticed.

Melhus (1942) reported that the infected plants were stunted at the apical region, tip of the stem was curled and brown upper leaves showed small necrotic lesions. The pulvinus part of most of the leaves showed a dark-brown water-soaked appearance. Rhoads (1944) and Parbery and Lee (1972) also observed the poor seed germination and cankers on cotyledons due to *C. dematium*.

Infection at an early stage may cause no seed formation, or if seed develop, may be smaller and fewer in number. Infected seeds are shriveled or moldy and can develop brown stained irregular, grey areas with black specks or may not show any symptoms. The fungus is confined at first to the seed coat. Such seeds may die during germination, or if they germinate, may produce infected seedlings. Infected seeds may lead the pre- and post-emergence damping-off. Cotyledons may have the dark-brown, sunken lesions which may gradually extend up to epicotyles and radical and may become water-soaked.

Seed-borne inoculum of *C. truncatum* (*C. dematium* f. *truncatum*) was responsible for three types of infection on soybean: (i) pre-emergence killing (ii) seedling blight and (iii) symptoms-less establishment of internal mycelium. At the host maturity under proper environmental conditions, the fungus may fruit abundantly on stem and pods (Tiffany, 1951).

Anthracnose of maturing plants causes serious losses, particularly during the rainy period when shaded lower branches and leaves are killed. Yield losses up to 50% in Thailand and 100% in India have been reported due to anthracnose (Sinclair and Backman, 1989; Ploper and Backman, 1992; Manandhar and Hartman, 1999).

Control Strategies

Cultural practices: Steps like crop rotation and tillage programs, promoting soybean residue decomposition before the next crop of soybeans is planted, help reduce diseases such as pod and stem blight and anthracnose.

The effect of time of fungicide spraying on the control of late season diseases of soybean including anthracnose caused by *C. truncatum* were studied in Brazil (Klingelfuss et al., 2001). Treatments included; 3 tillage practices (non-tillage, minimum tillage and conventional tillage); and difeconazole applied at R5.1, R5.2, R5.3, R5.4, R5.5, R6, R5.1 + R5.4, R5.2 + R5.5 and R5.3 + R6. Due to prevailing water deficit and high temperature during reproductive phases of soybeans, disease severity was low and no yield differences were detected. They also observed the significant differences when the fungicide was applied at 5.1 + R5.4. Among the soil management practices, conventional tillage showed greater values than non-tillage for all parameters (severity of late season diseases, percentage defoliation, and crop yield and 1000-seed weight).

Chemical control: Several fungicides reduced infection of *C. truncatum* on soybean seeds and improved their emergence. The most effective were Thiram, Difolatan, and Captan in India (Khare and Chacko, 1983). Benomyl + Thiram, Thiophanate methyl + Thiram, Thiabendazole, and Phenapronil gave effective control in Korea (Lee, 1984). In Portugal, Tolyfluenid, Pencycuron + Tolygluanid, Pencycuron, Captan, Thiram and Benomyl were tested as seed treatment to control seed-borne *C. truncatum* with 9% natural infection in the laboratory and in the field on soybean. All the treatments controlled the disease, but Tolyfluanid was the most effective, followed by Benomyl and Pencycuron + Tolyfluamid (Goulart, 1991).

Ahn and Chung (1970) investigated the effect of seed disinfectants for controlling the soybean anthracnose. All the chemicals tried (Arasan, Orthocide, Phygon-XL, PTAB, and Mercuron) reduced significantly the seedling infection as compared to the control. In regard to soil infection, the result was similar but percentage of infection was lower than in the seed inoculation. Thiram proved to be the best among all fungicides viz., Thiram (75%), Captan (75%), PCNB 75%, and a mixture (Thiram + Captan + PCNB in a 1:1:1 ratio) @ 4.5 g/kg of seed and with Aureofungin @ 25 ppm for seed treatment in improving the emergence (Nene et al., 1971).

Soybean seed treatment with Captan, Carboxin + Thiram (WP and EC), Thiram (WP and EC), Thiabendazole + Thiram and Thiabendazole + Quintozene eradicated at different levels the incidence of *C. truncatum* (Picinini and Fernandes, 1996). In Brazil, the most effective control of soybean seed-borne pathogens, including *C. truncatum* was obtained with Tolyflunid + Thiabendazole, Tolyfluanid + Thiabendazole, Tolyfluanid + Carbendazim, Thiabendazole + Thiram, Thiabendazole + Captan, Carbendazim + Thiram, Benomyl + Captan and Benomyl + Thiram (Goulart, 1998).

Goulart et al. (2000) investigated the efficiency of fungicide seed dressings on the control of soybean seed-borne fungi and incidence of *Phomopsis* spp., *F. semitectum* (*F. pallidoroseum*), *C. truncatum* and *C. kikuchii*. The most effective fungicides were Benzimidazoles (Benomyl, Carbendazim or Thiabendazole) and the protective ones were Captan, Thiram and Tolyfluamid. The residual effect and efficacy of fungicides (Benomyl, Carbendazim, Tebuconazole, Difenoconazole and Acibenzolar-s-methyl) were investigated by Klingelfuss and Yorinori (2000). They also observed that latent infection by *C. truncatum* was shown to have occurred prior to fungicide spray and *Cercospora kikuchii* infection was first recorded 7 days after fungicide spraying. Klingelfuss et al. (2001) observed significant differences in disease severity when fungicide Difeconazole was applied at R5.1 + R5.4. Among the soil management practices, conventional tillage was found of greater values than no-tillage.

C. truncatum was completely eliminated when soybean seeds were pre-treated with a dye like Methylene blue, Methyl red or Carmine and irradiated with a laser for 10 min. Seed germination was stimulated on exposure of the seed to 1 minute of irradiation. At this dose, most of the dyes were accelerators where as the higher doses were inhibitory to seed germination (Ouf and Abdel Hady, 1999).

Jagtap et al. (2013) reported that the Carbandazim spray @ 0.1% effectively reduced the disease intensity to 19.55% and pod infection to 9.63% as compared to 40.73% and 75.73% under unsprayed control and was followed by Mancozeb @ 0.1%.

Bio-control: Feykuy-Eaka, ginger (*Zingiber officinale*), garlic (*Allium sativum*) and neem (*Azadirachta indica*) extracts gave excellent control of seed-borne *C. dematium* var. *truncatum* when seeds were dipped for 30 minutes (Hossain et al., 1999).

The effect of two plant extracts, two antagonists and two chemicals on seed germination, root length, shoot length, vigour index and anthracnose disease (*C. truncatum*) of soybean cv. CO 2 were studied by Chandrasekaran et al. (2000) *in vitro* and in pot culture conditions. The treatment comprised of 10% leaf extracts of *Lawsonia inermis* and *Prosopis juliflora*; talc-based formulation of *T. viride* and *T. harzianum*; 0.1 and 1% alum; 0.1% dipotassium hydrogen phosphate and 0.1% Carbendazim. They observed that the seed treatment with alum recorded maximum seedling vigour index. *L. inermis* leaf extract was at par with *T. harzianum* in increasing seed germination, root and shoot length as compared to the control. The *L. inermis* leaf extract also reduced anthracnose disease incidence significantly as compared to other treatments.

Arora and Kaushik (2003) reported that seventeen plant extracts (aqueous) were effective against one or more pathogens. Dry hot water extracts of *Berberis aristate*, *Boenninghausenia albiflora* and *Lantana camara* were highly potent against *C. truncatum*. *Boenninghausenia albiflora*, *Polygonum glabrum*, *Origanum vulgare*, and *Rhododendron arboretum* (*R. arboretum*), significantly inhibited the growth of *C. truncatum*.

Seed treatment with alum (0.1%) recorded the greatest seed germination of 90.1%, compared to 68.0% in the untreated control. Seed treatment + a foliar spray with *Lawsonia inermis* leaf extract (1%) + alum (0.1%) recorded leaf anthracnose and pod blight incidences of 7.0 and 4.2%, respectively with a grain yield of 2191 kg/ha. Seed treatment + *L. inermis* (1%) + *Trichoderma viride* (0.4%) + alum (0.1%) registered 7.4% leaf anthracnose, 5.6% pod blight incidence and a yield of 2186 kg/ha (Chandrasekaran et al., 2000).

The individual and combined effects of plant extracts, biological control agents and chemicals on foliar anthracnose and pod blight caused by *C. truncatum* of soybean was studied by Chandrasekaran and Rajappan (2002). They observed that the combined application of leaf extracts of *Lawsonia inermis* (5%) with alum at 1 and 0.1% showed 100% reduction in pod blight infection under laboratory conditions. They also recorded that seed treatment + foliar spray of *L. inermis* (1%) + alum (0.1%) minimises leaf anthracnose and pod blight incidence upto 10 and 7%, respectively and it was at par with anthracnose and pod blight incidence recorded by seed treatment + foliar spray of *L. inermis* (1%) + *T. viride* 4 (0.4%) + alum (0.1%) and there was no significant difference between seed treatment recorded in those two treatments.

The anti fungal effects of some plants extracts namely, tobacco leaf, keora seed, keora, mahogoni, gaint Indian milky weed, garlic and ginger at different concentrations (30%, 40%, 50%, 60% and 70%) on the growth and development of *C. gloeosporioides* were evaluated. The growth inhibition increased with the increase of concentration of all the plant extracts. Highest mycelial growth inhibition (74.35%) was observed in case of garlic extracts at 70% concentration. Garlic extract at 50% and 60% concentration were also effective than other treatments (Mukherjee et al., 2011).

RUST

Soybean rust caused by *Phakopsora pachyrhizi*, is a most serious disease of economic importance. It is known to occur in Asia, Europe, Africa, Australia and America. In India, the disease was earlier considered as a minor one and was known to occur in the low hills of UP, West Bengal and North East Region. But after 1993, the rust scenario took a serious turn and now every year it appears in many areas. Therefore, it has become imperative to know more about the disease and its eco-friendly integrated management.

Host Range

Phakopsora pachyrhizi has extremely wide host range since it infects a large number of dicotyledonous plants such as the common bean (*Phaseolus vulgaris*), wild soybean (*Glycine usseriensis*) and so on, both in the field and in the laboratory (Yang, 1977). Keogh (1974) has made a rather thorough study of the host plants indigenous to Australia. A host list of *P. pachyrhizi* Syd. is compiled (Keogh, 1974; Kitani and Inowe, 1960; Lin, 1966). Nine isolates have been separated into six pathogenic groups, differing mainly in their reaction types, with and without sporulation, or no infection on *Vigna unguiculata*, *P. vulgaris*, and *Pachyrhizus erosus*. Burdon and Speer (1984) established a set of differential *Glycine* hosts for the identification of *P. pachyrhizi*.

Legume species identified as host of *P. pachyrhizi* in Thailand are: *Canavaria gladiata*, *Carjinus* sp., *Centrosema pubescens*, *P. rerosus*, *Vigna mungo*, *Dolichos lablab*, *Phaseolus aureus*, *P. lathyiodes*, *Pueraria thunbergiana*, *Pisum sativum*, *Vigna sinensis* (Poonpolgul and Surin, 1980).

Control Strategies

Soybean rust can be effectively managed through the timely application of appropriate fungicides, use of disease-resistant cultivars, and also through appropriate practices of good husbandry (Bromfield, 1984). However, currently the control of soybean rust relies primarily on the application of fungicides, but the efficiency depends largely on correct timing of spraying.

Cultural practices: Some cultural practices can be used in the control of soybean rust. For example, it has been reported that the destruction of host weeds and increased phosphorus levels in the fields reduce the incidence of soybean rust. Where the crop is grown under irrigation, water should be supplied in the middle of the day so that the leaves dry before dew sets in (Caldwell et al., 2002).

In Brazil, soybean is planted before the onset of the epidemic and harvested early to reduce the period of exposure to environmental conditions favorable to soybean rust (Yang et al., 1991). In recent years, there has been a trend towards planting cultivars from earlier maturity groups than those that were previously grown in various soybean production regions in Brazil. Implementation of a "vazio santitário", or host-free period, has been a highly successful regional rust management strategy in Brazil since 2006, and it was adopted in Paraguay in 2011.

Early planting and growing early maturing soybean cultivars may also limit disease progress and development. Ribeiro Do-Vale et al. (1985) observed that the crop sown in October remained free from rust up to the R 8 stage of growth and they concluded that the long cycle cultivars exhibited higher disease severity. In Nigeria, disease severity was higher on the medium-maturing varieties and those planted late in the season (Akinsanmi et al., 2001).

It has been observed that balanced and adequate crop nutrition reduce soybean rust disease severity, this might be due to the fact that adequate crop nutrition can increase soybean's ability to tolerate pathogen. Kitani et al. (1960a,b) studied that, when soybean plants in the field are sprayed with lime and sulphur, a gaseous substance evolves from the deposits of the fungicide. This substance affects the uredospores already formed by reducing their germination rate. It also depresses the infectivity of scattered spores.

The pre-disposing effects of host nutrition on rust intensity at 3 levels of NPK fertilizers was studied by Sharma (1990), Sharma and Verma (1995) and Sharma et al. (1996). The lowest rust intensity was recorded at 0:60:20, 0:30:0 and 40:30:40 NPK doses, while the highest intensity of the disease was observed at 60:0:20 followed by 60:0:40. Hung and Liu (1961) and Wang (1961) found best control of the rust by 5 weekly sprays of Dithane M-45 or Dithane T-78, wettable sulphur 0-318 B or Bordeaux mixture in Taiwan.

Chemical Control

Several studies on the efficacy of fungicides for the control of soybean rust have been conducted. Early research from Asia indicated that mancozeb was effective (Hartman et al., 1992). Additionally, fungicide trials in India (Patil and Anahosur, 1998) and Southern Africa (Levy et al., 2002) identified several Triazole compounds and Triazole mixes that could manage the disease. Other compounds that reduce disease severity have also been identified although their efficacy has been inconsistent. For example, studies in Zimbabwe and South Africa showed that fungicide application before flowering did not increase the yield of soybean (Miles et al., 2003). In Uganda, Kawuki (2002) evaluated three fungicides (Dithane M-45, Saprol and Folicur) under three different spray regimes for the management of soybean rust. The results demonstrated that Dithane M-45 offered the best protection with 5 sprays at a weekly interval from disease on-set while, Saprol and Folicur showed best protection with 2 and 3 sprays from disease on-set to full seed formation. Although, fungicide application is a strategy commonly used in soybean rust management, but it is costly (Miles et al., 2003; Pedro et al., 2008) and associated with environmental hazards (Bromfield, 1984).

The effect of different fungicidal chemicals on the germination of uredospores *in-vitro* as well as *in-vivo* was studied by Jan and Wu (1971). Sangawangse (1973) found that Plantvax and Plantvax+Benlate reduced the defoliation significantly but without any increases in the yield. Thapliyal and Singh (1974) on the other hand, reported best control of rust by Dithane M-45 and Dithane Z-78 followed by Benomyl and Plantvax that could increase yield. Quebral (1977) tested the efficiency of five fungicides in the Philippines and found that the Dithane M-45 was the most effective in controlling the disease as well as increasing the yield during both wet and dry seasons. Maiti et al. (1982) found Saprol and Detan more effective than Dithane M-45 and RH 124. They found that Detan and Saprol gave almost three-fold increase in grain yield. Later on, Maiti et al. (1983) reported Saprol as the most effective fungicide both in respect to the disease control as well as increase in yield. Junqueira et al. (1984) found that Triforine and Triadimefon had better curative effects than Benomyl, while the latter gave better protection. Among the non-systemic fungicides, Maneb gave the best control. Efficiency of nine fungicides and two modes of treatments were studied by Sharma (1990) to control rust. He observed that the seed treatment followed by two sprays of Tilt or seed-treatment followed by three sprays of Dithane M-45 were most effective against rust.

Ruengwiset and Poonpolgul (1999) observed that 2 spray of Chorothalonil (35 g/20 l water) at 35 and 42 days significantly reduced the rust infection that was followed by Mancozeb 80% WP (30 gm/20 l water). Two sprays (last at 63 DAP) of Hexaconazole (40 cc/20 l water) significantly decreased the rust severity which was followed by Mancozeb (Suraponchal and Poonpolgul, 2002). They also recommended that in Thailand the disease can be controlled by three applications of Mancozeb (80% WP) at 7 day interval when the first symptom is found on the unifoliate leaves at flowering stage or when there are continuous shower for 3 days or when there are dew on leaf surface at least 3–4 hours in the cool morning. However, the fungicides Tetraconazole and Tebuconazole were equally efficient in reducing rust severity and soybean productivity loss in Brazil (Blum et al., 2004).

Mueller et al. (2009) found that a single fungicide application at the R3 growth stage (early pod development) was an effective treatment to control soybean rust at most locations. At locations with favorable environmental conditions for an epidemic, vigorous monitoring is necessary for early detection of soybean rust at, or just before its onset, so that fungicide can be applied at the optimum time (Yang et al., 1991). However, fungicide applications increase production costs (Mueller et al., 2009).

Biological Control: Biological controls offer great potential as an ideal component of IPM. There are many reports of using biological agents experimentally to control rust of soybean. Some of the reported biological agents for soybean rust are: *Gliocladium roseum, Penicillium thomii* and *Tricothecium roseum* (Kumar and Jha, 2002), *Darluca filum, Tuberculina vinosa* and *Verticillium lecanii* (Blakeman and Fokkema, 1982).

Jahagirdar (2014) found that seed treatment with recommended seed dressing dosage of *T. harzianum* @ 6 g/kg with 106 cfu/g + Spray with Cow urine @ 10% + *T. harzianum* @ 0.5%, recorded minimum (35.1) Per cent Disease Index (PDI) followed by 37.4, 38.9 PDI in case of spray with Cow urine @ 10% + potassium phosphonate @ 0.3% and neem oil @ 1%, respectively. However, minimum PDI was recorded in Hexaconazole @ 1 ml/l (30.5) which are statistically at par with each other. The maximum incidence of 87.8 PDI was recorded in the control.

Rust Resistance Gene

Soybean cultivars with rust resistance may be a cost-effective way to control soybean rust, but breeding for resistance has not been easy and durable, due to pathogenic variability in *P. pachyrhizi* populations (Yorinori, 2008; Miles et al., 2011).

Specific resistance to *P. pachyrhizi* has been attributed to four single dominant genes (*Rpp1, Rpp2, Rpp3,* and *Rpp4*). Hartwig and Whitten (1995) summarized that the *Rpp1* is having immune reaction when inoculated with a few isolate, including India 73-1. Inoculation of most rust isolates on Rpp1 or the other genes produces a resistant red brown (RB) lesion with no or sparsely sporulating uredinia.

A large number of soybean accessions have been screened in Taiwan, Thailand, the Philippines, Indonesia and India (Tisselli et al., 1980). Lantican (1977) reported that the rust organism could continuously develop new strains leading to the breakdown of resistance rather quickly. He suggested for greater collaborative efforts to generate requisite information on: (i) prevalent strains of the rust organism in an area; (ii) identification of sources of genetic resistance against each strain and, (iii) a practical system for maintenance of each strain in pure culture.

In India, Singh and Thapliyal (1977) screened over 3,300 lines out of which six accessions including *PI 200465, PI 200466, PI 200477, PI 200490, PI 200492* and *PI 224268* were found to be resistant. Kumar and Verma (1985) and Verma (1985) have reported reactions of some indigenous and exotic (AVRDC) soybean cultivars. Five cultivars viz., *PK 73-75, DS 73-16, PK 73-94, PK 73-84* and *MGSB 75* were found to exhibit immune type reaction. Sharma (1990) and Sharma et al. (1989) screened 425 and 296 national, international, and local collections at mid and high altitudes of Meghalaya and reported three cultivars viz., *IC 18694, IC 89529* and *89530* at Barapani (mid altitude) and two cultivars viz., *JS 2* and *JS 76-188* at upper Shillong (high altitude) were completely free from rust infection. However, a large number of varieties exhibited variable reactions at the two locations. A few resistant cultivars viz., *IC 2511, JS 71-05* and *UPSM 534* and 29 moderately resistant cultivars exhibited stability in their reactions at the two locations. He also observed that the majority of the cultivars tested for three years also exhibited unstable reaction over the years. Yet, the investigation helped to identify several resistant and moderately resistant cultivars showing stability both over locations and years.

As per current information, the resistance against *P. pachyrhizi* is governed by four independent and dominant genes, which have been identified in *PI 200492, PI 230970, PI 462312* (*Ankur*) and *PI 459025* (Sinclair and Backman, 1989). Similarly, nine races have so far been identified in some wild *Glycine* sp. from Australia.

Khan et al. (2013) screened eighty-four genotypes of soybean under natural epiphytotic condition and found resistant reaction in only two genotypes EC-241778 and EC-241780.

FROG-EYE LEAF SPOT

Frog-eye leaf spot disease of soybean caused by *Cercospora sojina* Hara is world wide in distribution and causes substantial yield losses. Frog-eye leaf spot, as the name indicates, is primarily a foliar disease of soybean.

Symptoms

Infection occurs on leaf, stem, pods, and seeds. Hara (1915, 1930), for the first time, briefly described the disease as a leaf spot disease of soybean caused by *Cercospora sojina*. Minute, discrete spots that are reddish-brown and circular to sub-circular or angular, 0.5–5 mm in diameter first appear on the upper surface of the leaves. As the lesions enlarge and age, the central area becomes olive-gray or ash-gray and is surrounded by a narrow, dark reddish-brown border. There is no chlorotic zone surrounding the lesions. On the lower leaf surface, the spots are darker brown or gray. Dark conidiophores are borne singly or in very dense fascicles in the center of each lesion, mostly on the underside of the leaf. Older spots become very thin, often paper-white, and translucent. Several spots may coalesce to form larger, irregular spots. When the lesions are numerous, leaves wither and fall prematurely.

Infection on stem is conspicuous but not very common and appear later in the season. On stem, young lesions are deep red with a narrow, dark-brown to black margin. The central area is slightly sunken. As the lesions enlarge, the center becomes brown to light gray, usually with a narrow, dark-brown border. Infected seeds develop conspicuous light-to dark gray or brown areas, varying from minute specks to large blotches that can cover the entire seed coat. Usually, there is some cracking or flaking of the seed coat. Lehman (1934) recorded that the fungus was frequently observed on stems, pods and seeds of soybean. In stem, the fungus is chiefly confined to the cortex and the injury to phloem and cambium is usually due to diffusion of the toxic substance from the necrotic cortex. In pods the mycelium penetrates through the pod wall, entering the thin white membranes lining the pod and closely infecting the seeds.

Seed Borne Nature

C. sojina is seed borne. Infected seeds exhibit light to dark gray or brown discolourations as specks to large blotches and in some cases cover the whole seed surface. Alternate light to brown bands also develop on seed coat and cracks. Histopathology of infected soybean seeds exhibited the mycelium in aggregates in the hilar region, seed coat layers, in the space between seed coat and embryo. The cotyledons were not infected (Singh and Sinclair, 1985). The seeds get infected when the hyphae penetrate through pores and cracks in the seed coats as well as through hilar tracheids. Plant to seed transmission of

C. sojina is at the beginning of pod formation and at pod filling stage (Lin et al., 1991). Production of *Cercosporin* by *C. sojina* has been found to be inhibitory for other seed borne fungi to colonize soyabean seeds (Velicheti and Sinclair, 1992). Khare et al. (2000) has described the pathogen as seed borne, its method of detection and isolation from the seeds along with its management.

Control: Frog-eye leaf spot causes significant yield losses; hence its management practices are aimed at reducing the amount of inoculum available for infection in the field and protecting plants from infection. Fungicide seed treatment can reduce the risk of infection. Spray applications of fungicides after growth stage R 1 can reduce disease severity.

Cultural practices: Effect of NPK doses on the severity of the Frog-eye leaf spot disease was studied by Sharma (1990). The disease severity at different NPK doses exhibited similar variations over the years. He also observed that with increasing dose of N the disease severity increased significantly upto 60 ha^{-1}. However, the effect of increasing doses of P in relation to K did not show any definite trend (Sharma et al., 1996).

Chemical control: As early as 1928, Lehman suggested cultural practices viz., plowing the crop residues and 2–3 years rotation besides the use of early maturing and resistant cultivars for the control of Frog-eye leaf spot. However, he later realized that plowing in of infected stubbles was not practicable and the only means to reduce the disease was to use healthy seeds (Lehman, 1934). Studies by Adair et al. (1950) also suggested crop rotation and the use of resistant cultivars as the only measures to control the disease, besides seed treatment with Arasan, Spergon or improved Ceresan, particularly when the seed quality was poor. Gottlieb and Pate (1960), however, reported the efficacy of Tetrin, a non-toxic antifungal antibiotic against Frog-eye leaf spot.

Ovchinnikova (1968) studied the efficacy of fungicidal spray and found that 2–3 sprays of Bordeaux mixture were most effective followed by Ziram + Copper, Ziram and Dyrene. Almeida (1981) evaluated both preventive (Maneb) and curative (Benomyl, Thiophanate-methyl + Maneb) fungicides in an artificially inoculated experiment to control *C. sojina*. Estimation of diseased pods and seeds, defoliaton and seed weight indicated that the curative fungicides were equally effective if applied up to 6 days after inoculation. Maneb on the other hand was found to prevent the disease when applied even before inoculation.

In a study, application of Mancozeb reduced the infection and enhanced the yield followed by Captafol, Thiophanate-methyl and Carbendazim (El-Gantiry et al., 1990).

The efficacy of 9 fungicides and 2 modes of treatments were studied by Sharma (1990) to control Frog-eye leaf spot disease. Combinations of seed treatment + Saprol + Dithane M-45 spray were most effective showing curative effect also. Sharma (1990) and Sharma et al. (1997) while reporting the yield correlation with the incidence and severity of the 3 diseases indicated that yield was negatively correlated with all the 3 diseases. The Frog-eye leaf spot disease seems to exert maximum influence over yield reduction. An average 21.95% yield loss of soybean var. VL Soya 2 was assessed by Mittal (1996) and need-based sprays of Carbendazim (0.1%) and Macozeb (0.2%) were found to be effective against Athracnose and Frog-eye leaf spot disease. However, Gupta (2003) mentioned that the foliar spray of Ziram (0.2%) gives the best control of Frog-eye leaf spot disease. Two sprays of Carbendazim (0.05%) or Thiophanate-methyl (0.05%) control most of the foliar diseases including Frog-eye leaf spot disease of soybean.

Host Resistance

Although, the fungus is highly variable, yet attempts, to identify resistant cultivars against the disease have yielded considerable information. Athow and Probst (1952) studied the inheritance of resistance to this disease and found that a single major dominant Mendelian gene controls the resistance. This was further confirmed by their studies on soybean cultivars *Lincoln, Illini, Wabash* and the strain *C 1076* (Probst and Athow, 1958). Johnson (1960) has listed several varieties as resistant to Frog-eye leaf spot including *Hill, Hood, Lindarina, Merit* and *Shelby.* Athow et al. (1962) identified the cultivars *CNS, Dorman, Hood, Kanrich, Kent, Lee, Ogden* and *Roanoke* as resistant to both races 1 and 2. Hinson and Hartwig (1964) also found *Hardee* as resistant to *C. sojina.* Ovchinnikova (1968) found all the local cultivars grown in Pacific Coastal region of Russia as susceptible but the cultivar *Capital* was almost immune. Another race 5 of *C. sojina* was identified by Phillips and Boerma (1981) from Georgia infecting the cultivars *Corsoy 17* and *Bragg.* In both the cultivars, a single dominant gene controlled the resistance. Bisht and Sinclair (1985) reported that two soybean cultivars, *Corsoy 79* and *Wells,* were susceptible and moderately resistant, respectively to *C. sojina* and showed a negative correlation with yield and 1000 seed weght. Sharma (1990) screened 398 cultivars at Barapani and 327 at Upper Shillong, India, which indicated different levels of resistance and susceptibility. Out of them 26 and 10 were highly resistant at Barapani and Upper Shillong, respectively. Out of the 10 highly resistant cultivars at Upper Shillong only four viz., *PK 408, JS 75-46, JS 8 1-608* and *UPSM 534* exhibited the immune reaction at both the locations. Of the remaining six, *JS 76-20 1, JS 80-21* and *JS 80-291* behaved as resistant; *JS 2, JS 76-188* as, moderately resistant, and *PK 74-279* as moderately susceptible at Barapani. He also observed that most of the varieties exhibited variations on disease reaction. Only 8 resistant cultivars viz., *IC 150, IC 13754, IC 18753, EC 25676, HIMSO 773-A, P 3, P 81* and *Kandaghat* at Barapani and 2 viz., *EC 39494* and *EC 85949* at Upper Shillong were stable in their reaction over 3 years. Similarly, *IC 37182* gave consistently moderate resistant reaction during all the 3 years (Sharma et al., 1989; Sharma, 1990). Gupta (2003) listed *Bragg,*

Js 79-8 1, JS 80-21, KHSb 2, NRC 37, Pusa 37, and *VLS 21* as resistant cultivars against the Frog-eye leaf spot disease in India.

The high adaptability and variability of *C. sojina* was studied by Gao and Yang (2004). The resistance to this disease has been reported to be controlled both by major gene and polygene through fixed inheritance model and two-gene loci inheritance model. They observed that the genetic parameters indicated the additive dominance and epistasis effects which existed and played an important role in the resistance to *C. sojina.*

PURPLE SEED STAIN

Purple seed stain disease caused by *Cercospora kikuchii* (Matsomoto and Tomoyasu, 1925) M.W. Gardner, is one of the major diseases of soybean and can be found wherever soybeans are grown. The fungus causes leaf blight, purple discolouration of the seed coat and adversely affects the germination and quality of the seeds.

Symptom

The purple seed stain disease is also known as purple spot, purple blotch, purple speck or lavender spot. Symptoms of purple seed stain are quite conspicious and are easily distinguishable on the seeds. The seed discolouration varies from pink or a pale purple to a dark purple and ranges from minute specks to a large, irregular blotches, which may cover the entire surface of the seed coat. The pathogen (*Cercospora kikuchii*) also causes leaf blight, which is characterized by reddish-brown to purple, angular to irregular lesions, which expand irregularly. On the lower sides of the leaves irregular reddish-brown to light purple blotches appear which cause necrosis. Sometimes upper leaves appear light purple, leathery and dark. Blighting of younger upper leaves over large areas is the striking symptom of the disease. Veinal necrosis, defoilation along with early senescence of the plant is also very common.

Seed Borne Nature

In *C. kikuchii* infected seeds, mycelium can be seen abundantly in the aleurone and parenchyma cell layers of the seed coat. The mycelium concentration was high in the hilum region but sparse in the palisade layer (Ilyas et al., 1975). The infection spreads to young stem, unifoliate leaves, trifoliate leaves, branches, petioles and finally to the pods and seeds. Secondary infection of new tissues occurs simultaneously by the conidia produced from the infected tissues.

Control

Purple seed stain disease is a seed-borne disease, hence its management aims to eliminate the primary inoculum present in the seeds.

Cultural practices: Effect of NPK on the incidence and severity of purple seed stain disease has not been studied much. The maximum disease intensity in 1988 was recorded at 60:30:40 which was at par with 60:30:20 and 40:30:20. The least disease intensity was recorded at 60:60:40 and 40:60:40 NPK doses (Sharma, 1990). According to Ito et al. (1993) the number of purple stained seeds decreased with the increase of the rate of K.

Physical treatment: Echeverry et al. (1983) reported that the seed-borne *C. kikuchii* could be inactivated by hot water treatment of the seeds at 49°C for 5 minutes. However, water temperature above 55°C was found to significantly reduce seed germination and emergence.

Chemical treatment: Effectiveness of fungicides for controlling soybean seed pathogens including *C. kikuchii* was evaluated by Picinini et al. (2004), both *in vitro* and *in vivo*. The fungal incidence (%) in check treatments for Kaiabo and Pintado cultivars was 14.0 and 11.0, respectively. All fungicides evaluated *in vitro* reduced the incidence. In treatment with Difenconazole, the transmission of *C. kikuchii* was observed to be 1.0%.

Seed treatment (chemical): Johnson (1950) advocated seed treatment with Arasan and Spergon to get good crop stand from infected seeds and application of tribasic copper sulphate to the crop to reduce *C. kikuchii* infection in seeds. Simple seed treatment with 0.4 percent Thiram + Benomyl (1:1) reduced the seed infection (Agarwal et al., 1974). Singh and Agarwal (1984) found that seed treatment with Bavistin gave the maximum germination of seeds. They also concluded that the viability of the seeds and their germination depended on the degree of infection by *C. kikuchii* and the area of the seed coat discoloured. Later, they estimated that 100 percent purple stained seeds exhibited more than 30 percent loss in viability. Sharma (2003) recommended seed treatment with Dithane M-45 (maneb) before planting.

Several other fungicides have also been recommended by various workers for seed treatment to control seed borne *C. kikuchii* like Thiram (Agarwal and Joshi, 1971); Sodium hypochlorite solution 1.31% for 3 minutes (Chamberlain and Gray, 1974); Thiophanate methyl benomyl (Fujita, 1990); TBZ, Captan, Benlate (Vitti et al., 1993); TBZ (Yuyama and Henning, 1997); Tolyfluanid + TBZ (50 + 15 g ai/100 kg seed); Tolyfluanid + carbandazim (15 + 90 g/100 kg seed) Benomyl + Captan (30 + 90 g/100 kg seed); Benomyl + Thiram (30 + 70 g/100 kg seed) (Goulart, 1998); and Carbandazim, Difenconazole (Kingelfuss and Yorinori, 2000).

Chemical spray: Han (1959) reported the efficacy of Dithane M-45 and Phygon XL in reducing the purple seed stain disease incidence and increasing the yield.

Agarwal et al. (1973) tested the efficacy of the fungicides. They standardized the schedule of Benomyl spray, first at 25 days after planting followed by a second spray after another 30 days to control *C. kikuchii*.

Sasaki and Ogawa (1983) found Thiophanatemethyl as the most effective fungicide. It showed 95 percent efficacy in reducing the incidence of purple seed stain disease with a single application of the fungicide at 15–50 days after full bloom stage. Tekrony et al. (1985) found that a single application of Benomyl at growth stage R6 checks plants from seed infection. Suzuki (1985) has also emphasized that under high disease pressure, spray of Thiophanate-methyl and Benomyl are necessary to contain the disease.

Efficacy of 9 different fungicides and 2 modes of application in controlling the purple seed stain disease were studied by Sharma (1990). The maximum control of the disease, both in terms of incidence and severity, was achieved by seed treatment with Captafol + Topsin M during 1988 and by seed treatment with Dithane M-45 + Tilt spray during 1989.

Biocontrol: Seed treatment with culture of *Bacillus subtilis* or with its metabolites controlled seed borne *C. kikuchii*. Nonprotien amino acid 2-amino-3-cyclopropylbutanoic acid from *Amanita cokeri* is toxic to *C. kikuchii*. Dipping of seeds in ginger, garlic and neem extracts for five minutes checked seed borne *C. kikuchii* in soybean (Hossain et al., 1999).

Host Resistance

Hard seeded varieties with impermeable seed coat exhibited some degree of resistance to *C. kikuchii* infection and purple colouration in soybean seed (Fujita and Suzuki, 1988; Roy et al., 1994). Matsumoto and Tomoyasu (1925) reported that early maturing cultivars were more susceptible than the late maturing cultivars. According to Okabe et al. (1990) varieties with short stems were more resistant to the disease. Lehman (1950) indicated differential susceptibility among the soybean cultivars to the disease. Han (1959) listed the cultivars viz., *Pingtung Green Bean, Acadian, Pingtung-chu-tzu-tou, improved Pelican Bean, Kanto 13, Biofield, Chichibu musheachin, Dortchsoy Z gr 2, Hale, Ogdan gr 1, Mandarin, Seminole T,* and *M 3234* as resistant to *C. kikuchii*. Similarly, Chamberlain (1963) reported the cultivars viz., *Kent, Dorman, Lee, Ogdan* and *Roanoke* as resistant to purple seed stain disease. *Zane* (Mc Blain et al., 1989) and *TVC-G16* (Ricci and Devani, 1991) were moderately resistant to the disease. Vishwadhar and Chaudhary (1982) found 5 out of 15 varieties as resistant at two locations of Meghalaya in India. Sharma (1990) screened 237 cultivars at Barapani and 69 at Upper Shillong and indicated that maximum number of cultivars were resistant at both the locations. Two cultivars viz., *P 27-3* and *VLS 1* exhibited variability in their reactions over the years.

Sharma et al. (1993) observed *EC 93741* and *PK 74-93* as immune, 60 as resistant, 27 as moderately resistant; 11 as moderately susceptible and rest as susceptible out of 137 entries screened during 1986. While, out of 107 entries evaluated in 1987, *EC 93746, EC 93748, PK 74-289, PK 408* and *P 100* were immune, 62 were resistant, 25 were moderately resistant and rest entries were either moderately susceptible or susceptible.

Conclusion

Some of the reports indicate that the soybean diseases could be reduced by the application of organic manures and micronutrients, integrated rotations, soil fertility, sowing density and irrigation. For a disease to start and establish, environment plays a crucial role and, therefore, there is an urgent need to work out weather based forecasting models for prediction of disease outbreak. Selection of effective crops for crop rotation, use of composting and biocontrol agents including fungal antagonists, possible use of arbuscular mycorrhizal fungi, soil solarization coupled with organic amendments could be the other areas of research which can ensure the effective containment of the disease in the absence of high resistant varieties and very effective chemicals. Since, the reaction of genotypes and cultural practices to the disease varies with the regions, a development of region specific management strategy is a need.

Research has shown that with low manganese content of plants is related with susceptibility for the fungal pathogens. Boron (B) is another element that is correlated with the production of small fissures and cracks that might enhance the chances of fungal diseases. The nutrient uptake in "genetically modified" soybean is considered to be a novel area for disease control in the future.

References

Abawi, G.S. and Pastor-Corrales, M.A. 1990. Seed transmission and effect of fungicide seed treatments against *Macrophomina phaseolina* in dry edible beans. Turrialba, 40(3): 334–339.

Adams, J.F. 1933. Reports of the plant pathologist for 1932. Dalaware State Board Agr. Quart. Bull., 23: 3–16.

Adair, C.R., McClelland, C.R. and Cralley, E.M. 1950. Soybean research in Arkansas, 1936–48. Varietal test for seed and hay and studies in disease control. Arkansas Agr. Expt. Sta. Bull., 490: 62.

Agarwal, D.K., Gangopadhyay, S. and Sarbhoy, A.K. 1973. Effect of temperature on charcoal rot disease of soybean. Indian Phytopath., 26(3): 587–598.

Agarwal, D.K. 1973. Pathological and Physiological Studies of Some Root Infecting Fungi. Ph.D. Thesis. Agra University, Agra, India.

Agarwal, S.C. and Kotasthane, S.R. 1971. Resistance in some soybean varieties against *Sclerotium rolfsii* Sacc. Indian Phytopath., 24: 401–403.

Agarwal, V.K. and Joshi, A.R. 1971. A preliminary note on the purple stain disease of Soybean. Indian Phytopath., 24: 811–814.

Agarwal, V.K., Singh, O.V., Thapliyal, P.N. and Malhotra, R.K. 1973. Soybean (*Glycine max*) purple stain (*Cercospora kikuchii*). Fungicide and Nematicide Tests Results of 1972, Amer. Phytopath. Soc., 28: 100.

Agarwal, V.K., Singh, O.V., Thapliyal, P.N. and Malhotra, B.K. 1974. Control of purple stain disease of soybean. Indian J. Myco. Pl. Path., 4(1): 4.

Ahmad, Y., Hameed, A. and Aslam, M. 1992. Efficacy of different fungicides in controlling maize stock rot. Pakistan J. Phytopath., 4(1): 14–19.

Ahn, J.K. and Chung, B.K. 1970. Effects of seed disinfectants for controlling the soybean anthracnose. J. Pl. Prot. (Korea), 9: 21–24.

Almeida, A.M.R. 1981. Evaluation of the culture and preventive effect of fungicides on soybean [*Glycine max* (L.) Merril]. Fitopathol. Braseleira, 6(2): 173–178.

Ammajamma, R. and Hegde, Y.R. 2009. Efficacy of fungicides against *Rhizoctonia bataticola* causing wilt of *Coleus forskohlii* (Wild) Briq. Intl. J. Pl. Prot., 2(1): 31–32.

Akinsanmi, O.A., Ladipo, J.L. and Oyekan, P.O. 2001. First report of soybean rust (*Phakopsora pachyrhizi*) in Nigeria. Pl. Dis., 85: 97.

Ansari, M.M. 2005. Management of collar rot of soybean through biological agents. Pl. Dis. Res., 20(2): 171–173.

Ansari, M.M. 2007. Evaluation of soybean genotypes against *Macrophomina phaseolina* (*Rhizoctonia bataticola*) causing charcoal rot in soybean. Soybean Res., 5(3): 68–70.

Ansari, M.M. 2010. Integrated management of charcoal rot of soybean caused by *Macrophomina phaseolina* (Tassi) Goid. Soybean Res., 8: 39–47.

Anju, G. and Varma, R.K. 2007. Non-target effect of agro-chemicals and bio-controls agents on important soil-borne pathogen of soybean. JNKVV. Res. J., 4(1): 65–71.

Arora, C. and Kaushik, R.D. 2003. Fungicidal activity of plants extracts from Uttaranchal hills against soybean fungal pathogen. Allelopathy J., 11(2): 217–228.

Athow, K.L. and Probst, A.H. 1952. The Inheritance of resistance to frog-eye leaf spot of soybean. Phytopathol., 42: 660–662.

Athow, K.L., Probst, A.H., Kurtzman, C.P. and Lauiolette, F.A. 1962. A newly identified physiological race of *Cercosporin sojina* on soybean. Phytopathol., 52: 712–714.

Aycock, R. 1966. Stem rot and other diseases caused by *Sclerotium rolfsii*. NC Agric. Exp. Stn. Tech. Bull., 174. 202 p.

Banu, I.S.K.F., Shivanna, M.B. and Shetty, H.S. 1990. Seed-borne nature and transmission of *Colletotrichum truncatum* in chilli. Adv. Pl. Sci., 3(2): 200–206.

Bhardwaj, C.L. and Thakur, D.R. 1991. Efficacy and economics of fungicide spray schedules for control of leaf spots and pod blight in urdbean. Indian Phytopath., 44(4): 470–475.

Bisht, V.S. and Sinclair, J.B. 1985. Effect of *Cercospora sojina* and *Phomopsis sojae* alone or in combination on seed quality and yield of soybeans. Plant Dis., 69(5): 436–439.

Blakeman, J.P. and Fokkema, N.S. 1982. Potential for biological control of plant diseases on the phylloplane. Ann. Rev. Phytopathol., 20: 167–192.

Blum, Luiz E.B. and Rodriguez–Kabana, Rodrigo. 2004. Effect of organic ammendament of sclerotial germination, mycelia growth and *Sclerotium rolfsii*–induced diseases. Fitopathol. Bras., 29: 1–66.

Borkar, S.G. 1992. Influence of weather on *Sclerotium* root rot and wilt in soybean. Indian Mycol. Pl. Pathol., 22(2): 193–194.

Bowen, C.R. and Schapaugh, W.T. 1989. Relationship among charcoal rot infection, yield and stability estimates in soybean blends. Crop Sci., 29: 42–46.

Bromfield, K.R. 1984. Soybean Rust. Monograph, American Phytopathological Society, 11. St. Paul, MN.

Burdon, J.J. and Speer, S.S. 1984. A set of differential *Glycine* hosts for the identification of races of *Phakopsora pachyrhizi* Syd. Euphytica, 33: 891–896.

Butler, E.J. and Bisby, G.R. 1931. The fungi of India. Imp. Counc. Agr. Res., India, Sci. Mono, 1: 44 p.

Caldwell, P., Laing, M. and Ward, J. 2002. Soybean Rust. An Important New Diseases of Soybeans. Cambridge, UK.

Cattelan, A.J. 1994. Antagonism of fluorescent *Pseudomonas* to phytopathogenic fungi from soil and soybean seeds. Revista Brasileira de Ciencia do solo., 18(1): 37–42.

Chakraborty, B.N. and Purkayastha, R.P. 1984. Role of rhizobitoxine in protecting soybean roots from *Macrophomina phaseolina* infection. Can. J. Microbiol., 30(3): 285–289.

Chamberlain, D.W. 1963. Spot and stop soybean diseases. II. The leaves and seeds. Crops and Soils, 15(8): 16–20.

Chamerlain, D.W. and Gray, L.E. 1974. Germination, seed treatment, and microorganisms in soybean seed produced in Illinois. Plant Dis. Rep., 58: 50–54.

Chandrasekaran, A. and Rajappan, K. 2002. Effect of plant extracts, antagonists and chemicals (individual and combined) on foliar anthracnose and pod blight of soybean. J. Mycol. Pl. Path., 32(1): 25–27.

Chandrasekaran, A., Narasimhan, V. and Rajappan, K. 2000. Integrated management of anthracnose and pod blight of soybean. Ann. Pl. Prot. Sci., 8(2): 163–165.

Charde, J.D., Waghale, C.S. and Dhote, V.L. 2002. Management of stem rot of groundnut caused by *Sclerotium rolfsii*. Plant Dis. Res., 11: 220–221.

Chauhan, M.S. 1988. Relative efficiency of different methods for the control of seedling disease of cotton caused by *Rhizoctania bataticola*. Indian J. Mycol. Pl. Path., 18(1): 25–30.

Clark, C.A. and Moyer, J.W. 1988. Compendium of Sweet Potato Diseases. Amer. Phytopath. Soc., St. Paul, Minnesota.

Conners, I.L. and Savile, D.B.O. 1944. Twenty-third annual report of the Canadian Plant Disease Survey, 1943: 122.

Crall, J.M. 1952. Soybean diseases in Iowa in 1951. Plant Dis. Rep., 36: 302.

Csöndes, I., Balikó, K. and Dégenhardt, A. 2008. Effect of different nutrient levels on the resistance of soybean to *Macrophomina phaseolina* infection in field experiments. Acta Agron. Hungarica, 56(3): 357–362.

Dhar, V. and Sarbhoy, A.K. 1987. Ashy stem blight of soybean in India. Curr. Sci., 56(22): 1182–1183.

Dhingra, O.D. and Sinclair, J.B. 1978. Biology and Pathology of *Macrophomina phaseolina* (Univeridade Feederal de Vicosa: Minas Gerais, Brazil), 166 p.

Diaz, P.C. 1966. *Cercospora Kikuchii* on soybean, a new pathogen in Venezuela. Agr. Trop., 16: 213–221.

Do Amaral, J.E. 1951. Principal diseases of cultivated plants in the State of Sao Paulo and their control. Biologico, 17: 179–188.

Dubey, R.C. and Dwivedi, R.S. 1988. Antagonism between *Macrophomina phaseolina* and some rhizosphere microfungi of soybean. Intl. J. Trop. Pl. Dis., 6(1): 89–94.

Dwivedi, R.S. and Dubey, R.C. 1987. Effect of fungicides on survival of *Macrophomina phaseolina* in soil and in soybean stem in soil. Intl. J. Trop. Pl. Dis., 5(2): 147–152.

Echeverry, A.A., Rojas, M.M. and Zarater, R.D. 1983. Effect of soybean (*Glycine max.* (L.)) seed treatment with hot water on the control of purple discolouration due to *Cercospora kikuchii* (Matsumoto & Tomoyasu) Gardner. Acta Agrono., 33(3): 53–59.

El-Ghany, A.K.A., Seoud, M.B., Azab, M.W., El-Alfy, K.A. and El-Gawwad, M.A.A. 1975. Control of root rot and wilt diseases by seed treatment with fungicides. Agric. Res. Rev., 53(2): 79–83.

El-Gantiry, S.M., Abou-Zeid, N.M., Hassanein, A.M. and Mohamed, H.A. 1990. Occurrence of frog eye leaf spot of soybean in Egypt and cultural studies of the fungal pathogen. Agric. Res. Rev., 68(3): 455–462.

Etheshamul-Haque, S., Sultana,V., Ara, J. and Athat, M. 2007. Cultivar response against root-infecting fungi and efficacy of *Pseudomonas aeruginosa* in controlling soybean root rot. Pl. biosys., 141(1): 51–55.

Fenne, S.B. 1942. Two new records for the frog eye leaf spot: Virginia. Plant Dis. Reptr., 26: 383.

Fenne, S.B. 1949. Alfalfa and soybean diseases in Virginia, 1948. Plant Dis. Reptr., 33: 90–91.

Feng, F. and Wang, H.G. 2004. Soybean diseases report in China. Proc. VII World Soybean Research Cant Brazil, 29th Feb.–5th March' 2004, 331–334.

Frank, A.B. 1882. Gallen der *Angaillula radicicola* Greff and *Soja hispida, Medicago sativa, Lactuca sativa*, and *Pirus communnis*. Verb. Bot. Prov. Brandenburg., 23(2): 54–55.

Fujita, Y. 1990. Ecology and control of purple seed stain of soybean caused by *Cercospora kikuchii*. Bull. Tohoku Natl. Agric. Experi. Stn., 81: 51–109.

Fujita, Y. and Suzuki, H. 1988. Histological study on the resistance in soybean (*Glycine max*) and a wild soybean (*G. soja*) to purple seed stain caused by *Cercospora Kikuchii*. Ann. Phytopathol. Soc, Japan, 54: 151–157.

Gangopadhyay, S. and Wyllie, T.D. 1974. The role of peroxidase and polyphenol oxidase systems in relation to charcoal rot disease of soybean incited by *Macrophomina Phaseolina*. Indian J. Mycol. Pl. Path., 4(2): 132–137.

Gangopadhyay, S., Agarwal, D.K., Sarbhoy, A. and Wadhi, S.R. 1973. Charcoal rot disease of soybean in India. Indian Phytopath., 26(4): 730–732.

Gao, Y.P. and Yang, Q.K. 2004. Inheritance of resistance to *Cercospora sojina* Hara of many races in soybean. Proc. VII World Soybean Research Conf. Brazil, 29th Feb.–5th March 2004, 54 p.

Gardner, M.W. 1924. Indiana plant disease, 1921. Proc. Indiana Acad. Sci., 33: 163–201.

Gardner, M.W. 1927. Indiana plant diseases, 1925. Proc. Indiana Acad. Sci., 36: 231–247.

Gardner, M.W. 1928. Indiana plant diseases (1927). Proc. Indiana Acad. Sci., 38: 143–157.

Gogoi, N.K., Phokan, A.K. and Narzary, B.D. 2002. Management of collar rot of elephant's foot yam. Indian Phytopath., 55: 238–240.

Goot, P.V.D. and Muller, H.R.A. 1932. Pests and diseases of the soybean crop in Java. Preliminary report Landhuw Tijdschr. Vereen. Landb. Nederl.-Indie., 7: 683–704.

Gopal, K. and Jagadeeshwar, R. 1997. Reaction of soybean genotypes to charcoal rot (*Macrophomina phaseolina*). J. Mycol. Pl. Pathol., 27(1): 87–88.

Gottlieb, O. and Pate, H.L. 1960. Tetrin, an antifungal antibiotic. Phytopathol., 50: 817–822.

Goulart, A.C.P. 1991. Efficiency of chemical treatment of soybean seeds on the control of *Colletotrichum dematium* var. *truncate*. Revista Brasileira de Sementes, 13(1): 1–4.

Goulart, A.C.P. 1992. Effect of fungicides on the control of pathogens on cotton (*Gossypium hirsutum*) seeds. Summa Phytopathologica, 18: 173–177.

Goulart, A.C.P. 1998. Efficacy of seed treatment of soybeans with fungicides for the control of pathogens. Boletimde Pesquisa EMBRAPA Centro de Pesquisa Agropecuaria, 4: 20.

Goulart, A.C.P., Andrade, P.J.M. and Borges, E.P. 2000. Control of soybean seed borne pathogens by fungicide treatment and its effects on emergence and yield. Summa Phytopathologica, 26(3): 341–346.

Gray, L.E. 1978. Effect of soil fumigation on soybean disease and plant yield. Plant Dis. Reptr., 62(7): 613–615.

Grinstein, A., Katan, J., Abdul-Razik, A., Zeydan, O. and Elad, Y. 1979. Control of *Sclerotium rolfsii* and weeds in peanuts by solar heating of the soil. Plant Dis. Reptr., 63(11): 1056–1059.

Gupta, G.K. 2003. Integrated Disease Management of Soybean Diseases. Training Mannual DRS/02/2003. Dept. of Plant Breeding and Genetics, J.N. Krishi Vishwa Vidhyalaya. Jabalpur, M.P., 114–125.

Gupta, G.K. 2004. Soybean diseases and their management. pp. 145–168. *In*: Singh, N.B., Chauhan, G.S., Vyas, A.K. and Joshi, O.P. (eds.). Soybean Production and Improvement in India. National Research Centre for Soybean, Indore, M.P., India.

Gupta, G.K. and Chauhan, G.S. 2005. Symptoms, Identification and Management of Soybean Diseases. Technical Bulletin, 10. National Research centre for Soybean, Indore, India, 92 p.

Gupta, G.K., Sharma, A.N., Billore, S.D. and Joshi, O.P. 2006. Status and prospects of integrated pest management strategies in selected crops: Soybean. pp. 198–233. *In*: Singh, A., Sharma, O.P., Garg, D.K. (eds.). Integrated Pest Management—Principle and Application Volume 2: Application. CBS Publishers & Distributors, New Delhi, India.

Haenseler, C.M. 1946. Soybean diseases in New Jersey. New Jersey Agr. Expt. Sta. Pl. Dis. Notes, 23: 17–20.

Haenseler, C.M. 1947. Pathologist describes four principal diseases of fluid soybeans in Jersey. New Jersey Agr., 29(3): 4.

Hara, K. 1930. Pathologia Agriculturalis Planturam. Pathologia Agriculturalis Plantarum. Tokyo, 950 p.

Hara, K. 1915. Spot disease of soybean. Agr. Country, 9: 28.

Hartman, G.L., Wang, T.C. and Hymowitz, T. 1992. Sources of resistance to soybean rust in perennial *Glycine* species. Plant Dis., 76: 396–399.

Hartwig, E.E. and Whitten, J. 1995. Resistance to soybean rust. pp. 65–66. *In*: Sinclair, J.B. and Hartman, G.L. (eds.). Proceeding Soyabean Rust Workshop 9–11 Aug. 1995. College of Agricultural Consumer, & Environmental Sciences, National Soyabean Research Lab. Urbana, IL.

Hinson, K. and Hartwig, E.E. 1964. Bragg and Hardee soybean. Crap Sd., 4: 664.

Hilal, A.A. and Baiuomy, M.A. 2000. First record of fungal diseases of Stevia (*Stevia reboudiana* Bertoni.) in Egypt. Egyptian J. Agric. Res., 78: 1435–1448.

Hossain, I., Suratuzzaman, M. and Khalil, M.I. 1999. Seed health of soybean and control of seed borne fungi with botanicals. Bangladesh J. Train. Develo., 12(1-2): 99–105.

Hung, C.H. and Liu, K.C. 1961. Soybean spraying experiment for rust disease control. Agri. Res. (Taiwan), 10: 35–40.

Ilieseu, H., Sesan, T., Csep, N., Ionita, A., Stolca, V. and Cariciu, M. 1985. Seed treatment an important link in the prevention and control of some cryptogenic diseases of sunflower. Problems de Protectia Plantelor, 13: 173–188.

Ilyas, M.B., Ellis, M.A. and Sinclair, J.B. 1975. Evaluation of soil fungicides for control of charcoal rot of soybeans. Plant Dis. Reptr., 59(4): 360–364.

Ilyas, M.B., Ellis, M.A. and Sinclair, J.B. 1976. Effect of soil fungicides on *Macrophomina phaseolina* sclerotium vianility in soil and in soybean stem pieces. Phytopathol., 66(3): 355–359.

Ito, M.F., Truaka, M.A.S., Mascarenhas, H.A.A., Tanaka, R.T., Dudienas, C. and Gallo, P.B. 1993. Residual effect of soil liming and potassium fertilization on the *Cercospora* leaf blight (*Cercospora kikuchii*). Summa Phytopathologica, 19: 21–23.

Jadhav, P.B., Dake, G.N., Shete, M.H. and Barahate, B.G. 2008. Efficacy of bio-control agents and fungicides against *Sclerotium rolfsii* causing stem rot of tomato. J. Soil crops, 18(1): 135–137.

Jahagirdar, Shamarao. 2014. Bioformulations and Indigenous Plant Protection Measures in Enhancing the Vitalities of Bio-Control Agents for Induced Systemic Resistance Suppressing Asian Soybean Rust in India. International Conference on Biological, Civil and Environmental Engineering (BCEE-2014) March 17–18, 2014 Dubai (UAE), 123–126.

Jagtap, G.P., Gavate, D.S. and dey, Utpal. 2013. Management of *Colletotrichum truncatum* causing anthracnose/pod blight of soybean by fungicides. Indian Phytopath., 66(2): 177–181.

Javaid, I. and Ashraf, M. 1978. Some observations on soybean diseases in Zambia and occurrence of *Pyrenochaeta glycines* on certain varieties. Plant Dis. Reptr., 62(1): 46–47.

Jan, C. and Wu, L. 1971. Chemical control of soybean rust. Mem. Coll. Agr. Natl. Taiwan Univ., 12(1): 173–190.

Johnson, H.W. 1950. Pland disease research on forage crops in the bereau of plant industry, soils and agricultural engineering. Plant Dis. Reptr., 191: 42–59.

Johnston, A. 1960. A preliminary plant disease survey in North Borneo. Plant Prol. Prot. Div., Food and Agr. Org. Rome, 43 p.

Johnston, A. 1961. A preliminary plant disease survey in Netherlands, New Guinea. Netherlands New Guine Dienst Econ. Zaken, Meded. Landbourk. Ser., 4: 55 p.

Johnson, H.W. and Kilpatrick, R.A. 1953. Soybean diseases in Mississippi in 1951–1952. Plant Dis. Reptr., 37: 154–155.

Junqueira, N.T.V., Zambolim, L. and Chaves, G.M. 1984. Translocation of Benomyl and Triadimefon in soybean tissues and its effects on the infection with *Phakopsora pachyrhizi*. Fitopatol. Brasileira, 9(1): 119–127.

Katan, J., Rotem, I., Finkel, Y. and Daniel, J. 1980. Solar heating of the soil for control of Pink root and other soil borne diseases on Onions. Phytoparasitica, 8(1): 39–50.

Kawuki, R. 2002. Soybean Germplasm Reaction to Rust in Uganda, Associated Yield Loss and Rust Control Using Fungicides. M. Sc. Thesis. Makerere University.

Keogh, R.C. 1974. *Phakospora pachyrhizi* Syd. The causal agent of soybean rust. Australian Pl. Path. Soc. News Ltr., 3: 5.

Khan, M.H., Tyagi, S.D. and Dar, Z.A. 2013. Screening of soybean (*Glycine max* (L.) Merrill) genotypes for resistance to rust, yellow mosaic and pod shattering. *In:* Hany, A. and El-Shemy (eds.). Soybean—Pest Resistance. Agricultural and Biological Sciences.

Khare, M.N. and Chacko, S. 1983. Factors affecting seed infection and transmission of *Colletotrichum dematium* f. sp. *truncate* in soybean. Seed Sci. Tech., 11: 853–858.

Khare, M.N., Bhale, M.S. and Kumar, Kumud. 2000. Seed borne pathogens causing diseases in soybean, their detection, diagnosis and management. pp. 77–94. *In:* Narain, U., Kumar, K. and Srivastava, M. (eds.). Advances in Plant Disease Management. Creative Offset Press, New Delhi.

Khode, S.W. and Raut, B.T. 2010. Management of root rot/collar rot of soybean. India Phytopath., 63(3): 298–301.

Kilpatrick, R.A. 1955. Soybean diseases in the delta of Mississipi in 1954. Plant Dis. Reptr., 39: 578–579.

Kitani, K. and Inoue, Y. 1960. Studies on the soybean rust and its control measures (Part 1). Studies on the soybean rust. Shikoku Agr. Expt. Sta. (Zentsuji, Japan). Bull., 5: 319–342.

Kitani, K., Inoue, Y. and Natsume, T. 1960a. Ecological studies on the mobilization of lime-sulphur spraying. Efficacy to the wheat brown rust and the soybean rust. Shikoku Agr. Expt. Sta. (Zentsuji, Japan) Bull., 5: 225–306.

Kitani, K., Inoue, Y. and Natsume, T. 1960b. Studies on the Soybean rust and its control measure (Part 2). Studies on the control measure on the soybean rust. Shikoku Agr. Expt. Ste. (Zentsuji, Japan) Bull., 5: 343–358.

Kittle, D.R. and Gray, L.E. 1982. Response of soybeans and soybean pathogens to soil fumigation and foliar fungicide sprays. Plant Dis., 66(3): 213–215.

Klingelfuss, L.H. and Yorinori, J.T. 2000. Latent infection by *Colletotrichum truncatum* and *Cercospora kikuchii* and effect of fungicides on late season diseases of soybean. Summa Phytopathologica, 26(3): 356–361.

Klingelfuss, L.H., Yorinori, J.T., Ferreiro, L.P. and Pereira, J.E. 2001. Timing of fungicide sprays for the control of late season diseases of soybean *Glycine max* (L.) Merrill. Acta Scientiarum, 23(5): 1287–1292.

Koch, L.W. and Hilderbrand, A.A. 1944. Soybean diseases in South Western Ontario in 1943. Can. Pl. Dis. Surv. Ann. Report, 23: 29–32.

Konde, S.A., Raut, B.T. and Gade, M.R. 2008. Chemical and biological management of root rot (*Rhizoctonia bataticola*) of soybean. Ann. Plant Physiol., 22(2): 275–277.

Kornfield, A. 1935. Schadigungen and Krankheiten der Olbohne (Soja), Soweit sie bisher in Europe bekannt geworden sind. Zeitscher Pflanenkr., 45: 577–613.

Kumar, K. and Singh, J. 2000. Location, survival, transmission and management of seed-borne *Macrophomina phaseolina*, causing charcoal rot in soybean. Ann. Plant Prot. Sci., 8(1): 44–46.

Kumar, S. and Verma, R.N. 1985. Soybean rust in NEH hills of India: Further observations. Soybean Rust News Letter, 7: 17–19.

Kumar, S. and Jha, D.K. 2002. *Trichoderma roseum* a potential agent for the biological control of soybean rust. Indian Phytopath., 55: 232–234.

Kumar, S., Sharma, S., Pathak, D.V. and Beniwal, J. 2011. Integrated management of *Jatropha* root rot caused by *Rhizoctonia bataticola*. J. Trop. For. Sci., 23(1): 35–41.

Kurata, H. 1960. Studies on fungal diseases of soybean in Japan. Natl. Inst. Agr. Sci. (Tokyo) Bull. Ser. C., 12: 1–154.

Lahre, S.K., Khare, N., Lakpale, N. and Chaliganjewar, S.D. 2012. Efficacy of bio-agents and organic amendments against *Sclerotium rolfsii* in Chickpea. J. Plant Dis., 7(1): 32–34.

Lantican, R.M. 1977. Observations and theories on cultivar resistance of soybeans to rust. Rust of Soybean: The problem and research needs. INTSOY Ser. Univ. of Illinois, 54–57.

Lee, D.H. 1984. Fungi associated with soybean seed, their pathogenicity and seed treatment. Korean J. Mycol., 12(1): 27–33.

Lehman, S.C. 1928. Frog eye leaf spot of soybean caused by *Cercospora daizu* Miura. J. Agr. Res., 36: 811–833.

Lehman, S.G. 1934. Frog eye (*Cercospora daizu* Miura) on stems, pods, and seeds of soybean and the relation of these infections to recurrence of the disease. J. Agr. Res., 48: 131–147.

Lehman, S.C. 1942. Notes on plant diseases in North Carolina in 1941 Soybean. Plant Dis. Reptr., 26: 111.

Lehman, S.G. 1950. Purple stain of soybean seeds. North Carolina Agr. Expt. Stat. Bull., 369.

Likhite, V.N. 1936. Host range of the Gujrat cotton root rot. Proc. Assoc. Econ. Biol. Coimbatore, 3: 18–20.

Levy, C., Techagwa, J.S. and Tattersfield, J.R. 2002. The Status of Soybean Rust in Zimbabwe and South Africa. Paper read at Brazilian Soybean Congress, at Foz do Iguaçu, Parana, Brazil.

Lin, S. 1966. Studies on physiologic races of soybean rust fungus, *Phakopsora pachyrhizi* Syd. Taiwan Agr. Res., 51: 24–28.

Lin, P.L., Li, Y.P., Liu, J., Wu, B.Z. and Li, Y. 1991. Study on time of fungicide application for controlling soybean frog eye spot (*Cercospora sojina* Hara). Soybean Sci., 10: 135–138.

Ling, L. 1948. Host index of the parasitic fungi of Szechwan, China. Plant Dis. Reptr. Suppl., 173: 1–38.

Litzenberger, S.C. and Stevenson, J.A. 1957. A preliminary list of Nicaraguan plant diseases. Plant Dis. Rep. Suppl., 243: 1–19.

Liu, S.T. 1948. Seed brone diseases of soybean. Bot. Bull. Acad. Sinica, 2: 69–80.

Lusin, V. 1960. *Cercospora kikuchii*-Soybean disease. Savremena Poljoprivreda, 8: 601–604.

Maiti, S., Vishwadhar and Verma, R.N. 1981. Rust of Soybean in India—Are appraisal. Soybean Rust Newsletter, 4(1): 14–16.

Maiti, S., Vishwadhar and Verma, R.N. 1982. Bio-efficacy of fungicides for controlling soybean rust in India. Soybean Rust Newsletter, 6(1): 8–13.

Maiti, S., Kumar, S., Verma, R.N. and Vishwadhar. 1983. Current status of soybean diseases in North East India. Soybean Rust Newsletter, 6(1): 14–21.

Manandhar, J.B. and Hartman, G.L. 1999. Anthracnose. pp. 13–14. *In*: Hartman, G.L., Sinclair, J.B. and Rupe, J.C. (eds.). Compendium of Soybean Diseases, 4th edition. St. Paul, APS Press, USA.

Mandhare, V.K. and Suryawanshi, A.V. 2008. Efficacy of some botanicals and *Trichoderma* species against soil borne pathogens infecting chickpea. J. Food Legumes, 21(2): 122–124.

Mane, S.S., Khode, S.W., Ghawade, R.S. and Shrirao, A.V. 2013. Management of root and collar rot of soybean through bio-control agent *Trichoderma*. J. Pl. Dis. Sci., 8(1): 86–91.

Manns, T.F. and Adams, J.F. 1934. Department of Plant Pathology. Delaware Agr. Expt. Sta. Bull., 188: 3646.

Massalongo, C. 1900. De nonnullis speciebus novis micromycetum agri veronensis. Atti. R. Inst. Veneto Sci. Letl. Ed. Arti., 59: 683–690.

Mathur, S.B. and Sarbhay, A.K. 1979. Biological control of *Sclerotium* root rot of sugar beet. Indian Phytopath., 31(3): 365–367.

Matsumoto, T. and Tomoyasu, R. 1925. Studies on purple speck of soybean seed. Ann. Phytopathol. Soc. Japan, 1: 1–14.

Maurya, S., Singh, R., Singh, D.P., Singh, H.B., Singh, U. and Srivastava, J.S. 2008. Management of collar rot of chickpea (*Cicer arietinum*) by *Trichoderma harzianum* and plant growth promoting rhizobacteria. J. Pl. Prot. Res., 48(3): 347–354.

Mc Blain, B.A., Martin, S.T., Walker, S.K. and Schmitthener, A.F. 1989. Zane soybeans. Special Circular of Ohio Agricultural Research and Development Center, 134: 11.

Mc Gee, D.C. 1991. Soybean Diseases. A Reference Source for Seed Technologist. APS Press, St. Paul, Minnesota, 151 p.

Mc New, G.L. 1948. Study of soybean diseases and their control. Iowa Agr. Expt. Sta. Rpt. On Agr. Res. For the year ending June 30, 1948, 188–189 pp.

Mehta, P.R., Garg, D.N. and Mathur, S.C. 1950. Important diseases of food crops, their distribution in India and Uttar Pradesh (India). Dept. Agr. Tech. Bull., 2: 13.

Mejia, A.S. 1954. *Sclerotium* wilt of supa (*Sindora supa* Merr.). Phillipp. J. For., 9: 119–132.

Mengistu, A., Arelli, P.A., Bond, J.P., Shannon, G.J., Wrather, A.J., Rupe, J.B., Chen, P., Little, C.R., Canaday, C.H., Newman, M.A. and Pantalone, V.R. 2011. Evaluation of soybean genotypes for resistance to charcoal rot. Online: Plant Health Progress doi: 10.1094/PHP-2010-0926-01-RS.

Melhus, I.E. 1942. Soybean diseases in Iowa in 1942. Plant Dis. Rep., 26: 431–432.

Miles, M.R., Bonde, M.R., Nester, S.E., Berner, D.K., Frederick, R.D. and Hartman, G.L. 2011. Characterizing resistance to *Phakopsora pachyrhizi* in soybean. Plant Dis., 95: 577–581.

Miles, M.R., Hartman, G.L., Levy, C. and Morel, W. 2003. Current status of soybean rust control by fungicides. Pesticide Outlook.

Mikhalenko, A. 1965. Disease of legumes in the Primorsk region. Zashch. Rast. Vredit. Bolez., 10: 4143.

Mittal, R.K. 1996. Yield loss by foliar diseases in soybean. Indian J. Mycol. Pl. Pathol., 26(1): 107.

Miura, M. 1930. Diseases of the main agricultural crops in Manchuria. Manchuria Railway Co. Agr. Expt. Sta. Bull., 11(Rev. ed.) 56 p.

Morgan, F.L. and Johnson, H.W. 1964. Leaf symptoms of soybean anthracnose. Phytopathol., 54: 625.

Morse, W.J. and Carter, J.L. 1939. Soybean culture and varieties: U.S. Dept. Agr. Farmer Bull., 1520.

Morse, W.J., Carter, J.L. and Williams, L.F. 1949. Soybean: culture and varieties. U.S. Dept. Agr. Res. Serv. Farmers Bull., 1520(rev.) 38.

Mukherjee, A., Khandker, S., Islam, M.R. and Sonia B. Shahid. 2011. Efficacy of some plant extracts on the mycelial growth of *Colletotrichum gloeosporioides*. J. Bangladesh Agril. Univ., 9(1): 43–47.

Muller, A.S. 1941. Survey of diseases of cultivated plants in Venezuela, 1937–1941. Soc. Venezolana de Cienc. Natl. Bol., 7: 99–113.

Muller, A.S. 1950. A preliminary survey of plant diseases in Guatemala. Plant Dis. Rep., 34: 161–164.

Muller, A.S. and Chupp, C. 1942. Soc. Venezolana de Cience, Nail. Bol., 8: 35–59.

Muller, A.S. and Chupp, C. 1950. *Cercospora* in Guatemala. Cieba, 1: 171–178.

Mueller, T.A., Miles, M.R., Morel, W., Marois, J.J., Wright, D.L., Levy, R.C. and Hartman, G.L. 2009. Effect of fungicide and timing of application on Soybean rust severity and yield. Plant Dis., 93(3): 243–248.

Mundkar, B.B. 1938. Fungi of India. Supplement 1. India Impor. Counc. Agr. Res. Sci. Mono., 12: 54 p.

Muthusamy, S. and Mariappan, V. 1992. Disintegration of microslerotia of *Marcophomina phaseolina* (soybean isolate) by oil cake extracts. Indian Phytopath., 45(2): 271–273.

Muthamilan, M. and Jayrajan, R. 1992. Effect of antagonistic fungi on *Sclerotium rolfsii* causing root rot of groundnut. J. Biol. Control, 6: 88–92.

Nagata, T. 1962. Report to the Govt. of Yugoslavia on improvement of soybean cultivation. FAO Expanded Tech. Asstt. Prog. Fao Rept., 1465: 22 p.

Nakata, K. and Takimoto, K. 1934. A list of crop diseases in Korea. Agr. Expt. Sta. Govt. Central Chosen Res. Rpt., 15: 1–146.

Nakata, K. and Asyuama, H. 1941. Report on diseases on main agricultural crops. Manchuria— Bereau Indus. Manchuria Rpt., 32: 1–166.

Nene, Y.L. and Srivastava, S.S.L. 1971. Report on the newly recorded purple stain of soybean seed caused by *Cercospora kikuchii* and pod blight caused by *Colletotrichum damatium* var. *truncate*. Plant Protec. Bull. FAO, 19(3): 66–67.

Nicholson, J.F. and Sinclair, J.B. 1973. Effect of planting date, storage conditions and seedborne fungi on soybean seed quality. Plant Dis. Rep., 57: 770–774.

Okabe, A., Sasaki, K. and Kazunri, K. 1990. Varietal difference in resistance to purple seed stain of soybean and the method of selecting for the resistance. Japanese Agri. Res. Quarterly, 23: 163–169.

Ouf, S.A. and Abdel Hady, N.F. 1999. Influence of He-Ne laser irradiation of soybean seeds on seed mycoflora, growth, nodulation and resistance to *Fusarium solani*. Folia Microbiol., 44(4): 388–396.

Ovchinnikova, A.M. 1968. *Cercospora* on soybean in Primorsk (Pecific Coastal) region, Zashch Rast. Mosk., 13: 27–28.

Parbery, D.G. and Lee, C.K. 1972. Anthracnose of soybean. Australian Plant Path. Soc. Newslett., 1: 10–11.

Patil, P.V. and Anahosur, K.H. 1998. Control of soybean rust by fungicides. Indian Phytopath., 51: 265–268.

Patil, P.P., Raut, B.T., Shinde, V.B. and Ingole, M.N. 2004. Role of antagonists in inhibition of *Rhizoctonia bataticola* and *Sclerotium rolfsii* causing seed/root/stem rot of sunflower. Ann. Plant Physiol., 18(2): 195–197.

Patino, H.C. 1967. Diseases of Oleaginous Annuals in Colombia. Agr. Trop., 23: 532–539.

Pawar, B.K., Surywanshi, A.P., Gawade, D.B., Zagade, S.N. and Wadje, A.G. 2014. Effects of organic amendments and fungicides on the survival of collar rot fungus of soybean incited by *Sclerotium rolfsii*. African J. Agric. Res., 9(27): 2124–2131.

Pearson, C.A.S., Schwenk, F.W., Crowe, F.J. and Kelley, K. 1984. Colonization of soybean roots by *Macrophomina phaseolina*. Plant Dis., 68(12): 1086–1088.

Pearson, L.H. 1944. List of plant diseases observed during survey in Mississippi and Louisiana, August to November, 1943. Plant Dis. Rep., Suppl., 148: 280–283.

Pedro, H.B., Aliny, S.R., José, U.V.M., Larissa, C.L., Breno, F.R., Wilmar, F.L., Carlos, A.A.A., Marcelo, F.O. and José, F.F. 2008. New soybean (*Glycine max* Fabales, Fabaceae) sources of qualitative genetic resistance to Asian soybean rust caused by *Phakopsora pachyrhizi* (Uredinales, Phakopsoraceae). Physiology, 131: 872–877.

Petty, A.M. 1943. Soybean disease incidence in Maryland in 1942 and 1943. Plant Dis. Reptr., 27: 347–349.

Phillips, D.V. and Boerma, H.R. 1981. *Cercospora sojina* race 5: A threat to soybean in the South Eastern United States. Phytopathol., 71: 334–336.

Phillips, D.V. and Boerma, H.R. 1982. Two genes for resistance to race 5 of *Cercospora sojina* in soybeans. Phytopathol., 72: 764–766.

Picinini, E.C. and Fernandes, J.M.C. 1996. Fungicide efficiency in controlling soybean seed-borne pathogens. Fitipatol. Brasileira, 21(4): 492–495.

Picinini, E.C., Fernandes, J.M.C., Bargas, P.R. and Adernt, P. 2004. *In vitro* and *in vivo* fungicides efficacy for controlling soybean seed pathogens. Proc. VIIth World Soybean Research Conf., Brazil, 29th Feb.–5th March, 2004. 156 p.

Ploper, L.D. and Backman, P.A. 1992. Nature and management of fungal diseases affecting soybean stems, pods and seeds. pp. 174–184. *In*: L.G. Green, M.B. and Ress, R.T. (eds.). Pest Management in Soybean Copping. SCI: Elsevier Applied Science, London.

Paola, M.A. 1933. A *Sclerotium* seed rot and seedling stem rot of mango. Philippine J. Sci., 52: 232–261.

Poonpolgul, S. and Surin, P. 1980. Study on host range of soybean rust fungus in Thailand. Soybean Rust Newslett., AVRDC., 3: 30–31.

Prabarakan, J., Samiappan, R. and Arjunan, G. 1989. Effect of inoculation of *Rhizobium japonicum*, *Macrophomina phaseolina*, and K application on the growth and nodulation of soybean. pp. 500–505. *In*: Pascle, A.J. (ed.). World Soybean Research Conference IV Proceeding. Orientation Grafica Editora, Bruenos Aires, Argentina.

Prajapati, R.K., Gangwar, R.K., Srivastava, S.S.L. and Ahamad, Shahid. 2002. Efficacy of fungicides, non-target pesticides and bio-agents against the dry root rot of chickpea. Ann. Pl. Prot. Sci., 10(1): 154–155.

Prasad, K.V.V. 1988. Efficacy of fungicidal combinations for seed treatment of soybean. Indian Phytopath., 41(2): 300 (Abst.).

Prasad, R.D. and Rangeshwaran, R. 2001. Biological control of root and collar rot of chickpea caused by *Sclerotium rolfsii*. Ann. Pl. Prot. Sci., 9(2): 297–303.

Prashanthi, S.K., Kulkarani, S., Sangam, V.S. and Kulkarni, M.S. 2000. Chemical control of *Rhizoctonia bataticola* (Taub.) Butler, the causal agent of root rot of safflower. Plant Dis. Res., 15(2): 186–190.

Probst, A.H. and Athow, K.L. 1958. Additional studies on the inheritance of resistance to Frog-eye leaf spot of soybeans. Phytopathol., 48: 414–416.

Pullman, G.S., DeVay, J.E., Garber, R.H. and Weinhold, A.R. 1981. Soil solarization: Effects on *Verticillium dahliae*, *Pythium* spp. *Rhizoctonia solani*, and *Thielaviopsis basicola*. Phytopathol., 71(9): 954–959.

Punja, Z.K. 1985. The biology, ecology and control of *Sclerotium rolfsii*. Annual Rev. Phytopathol., 23: 97–127.

Punja, Z.K. 1988. *Sclerotium (Athelia) rolfsii* a pathogen of many plant species. pp. 523–534. *In*: Sidhu, G.S. (ed.). Advances in Plant Pathology. Vol. 6, Genetic of Plant Pathogenic Fungi. Academic Press, London.

Pushpavati, B. and Rao, K.C. 1998. Biological control of *Sclerotium rolfsii*, the inicitant of groundnut stem rot. Indian J. Pl. Prot., 26(2): 149–154.

Quebral, F.C. 1977. Chemical control of soybean rust in the Phillipines. Rust of Soybean: The Problem and Research Needs. INTSOY Ser.12, Univ. of Illinois, pp. 81–84.

Quebral, F.C. and Pua, D.R. 1976. Screening of soybean against *Sclerotium rolfsii*. Trop. Grain Leg. Bull., 6: 22–23.

Rao, V.G., Pande, Alaka and Patwardhan, P.G. 1989. Three new records of fungal diseases of economic plants in Maharashtra State. Biovigyanam, 15(1): 51–53.

Rajput, V.A., Konde, S.A. and Thakur, M.R. 2010. Evaluation of bioagents against chickpea wilt complex. J. Soils Crop., 20(1): 155–158.

Rathore, B.S. 2000. Effect of organic amendments on incidence of seed rot and seedling blight of mothbean. Plant Dis. Res., 15: 217–219.

Raut, B.T. and Somani, R.B. 1987. Efficacy of different fungicides. IV. Field trails on root rot of chickpea. PKV Res. J., 11(2): 182–184.

Ramakrishnan, T.S. 1951. Additions to Fungi of Madras XI. Indian Acad. Sci. Sect. B., 34: 157–164.

Reichert, I. and Hellinger, E. 1947. On the Occurrence, morphology and parasitism of *Sclerotium bataticola*. Palestine J. Bot. Rehovot Ser., 6: 107–147.

Rhoads, A.S. 1944. Summary of observations on plant diseases in Florida during the emergency plant disease prevention project surveys, July 25 to December 31, 1943. Plant Dis. Rep. (supp.), 148: 262–276.

Ribero Do Vale, F.X., Chaves, G.M. and Zambolim, L. 1985. Host range study of soybean rust in Brazil. Soybean Rust News Lett., 7: 7–9.

Ricci, O.R. and Devani, M.R. 1991. Soybean: the new cultivar *TUC G-16*. Advance Agroindustrial, 12: 5–7.

Riley, E.A. 1960. A Revised List of Plant Diseases in Tanganyika Territory. CMI., Mycol. Papers, 75: 42 p.

Roberti, R., Flor, P. and Pisi, A. 1996. Biological control of soil borne *Sclerotium rolfsii* infection by treatment of bean seed with species of *Trichoderma*. Petria, 6(2): 105–116.

Roy, K.W., Keith, B.C. and Andrews, C.H. 1994. Resistance of hard seeded soybean lines to infection by *Phomopsis*, other fungi and soybean mosaic virus. Can. J. Pl. Pathol., 16: 122–128.

Ruengwiset, K. and Poonpolgul, S. 1999. Efficacy of chlorothaionil 50 and SC to control rust of soybean cause by *Phakopsora pachyrhizi*. Technical papers Ladda Co. Ltd., 11 p.

Saikia, V.N. and Phukan, A.K. 1983. Occurrence of seedling blight of soybean in Assam. J. Res., Assam Agric. Univ., 4(2): 171–172.

Salunke, V.N., Armarkar Sarita and Ingle, R.W. 2008. Efficacy of fungicides and antagonistic effect of bioagent against *Rhizoctonia bataticola* isolates. Ann. Pl. Physiol., 22(1): 134–137.

Sangawangse, P. 1973. A preliminary report of the study on soybean rust. Thai J. Agric. Sci., 6(2): 165–169.

Sasaki, Y. and Ogawa, M. 1983. Chemical control of soybean seed disease caused by *Cercospora kikuchii* Matsumoto et Tomoyasu. Bull. Hiroshima Prefectural Agri. Expt. Sta., 46: 33–40.

Sarbhoy, A., Thapliyal, P.N. and Payak, M.M. 1972. *Phakospora pachyrhizi* on soybean in India. Sci. Cult., 38(4): 198.

Saxena, R.M. and Sinha, S. 1978. Seed-borne infection on *Vigna radiate* (L.) Wilczek var. *radiate* in Uttar Pradesh—new records. Sci. Cult., 44(8): 377–379.

Seoud, M.B., El-Dib, A.A., El-Wakel, A.A., El-Gawwed, M.A.A. and Thoma, A.T. 1982. Chemical control of root rot and wilt diseases of sesame in Egypt. Agric. Res. Rev., 60: 117–126.

Senthil Kumar, M., Swarnalakshmi, K., Govindasamy, V., Lee, Y.K. and Annapurna, K. 2009. Biocontrol potential of soybean bacterial endophytes against charcoal rot fungus, *Rhizoctonia bataticola*. Curr. Microbiol., 58(3): 288–293.

Shukla, A.K. and Singh, A.K. 1992. Studies on survey, surveillance, epidemiology and other biological aspect of major root and seed diseases of soybean. Annual Report NRCS, Indore, 58 p.

Sharma, S.K. 1990. Phytopathological Studies on Important Fungal Diseases of Soybean of NEH India with Special Reference to Soybean Rust and Purple Seed Stain. Ph.D. Thesis, Jiwaji University, Gwalior, M.P., India, 168 p.

Sharma, S.K. 2003. Field study on purple seed stain (*Cercospora kikuchii*) disease of soybean at mid altitude of Meghalaya. Crop Res., 26(3): 512–514.

Sharma, S.K. and Verma, R.N. 1995. Effect of NPK doses on rust of soybean from North-East India. J. Hill Res., 8(1): 113–119.

Sharma, S.K. and Gupta, G.K. 2006. Current status of Soybean rust (*Phakospora pachyrhizi*). A review. Agric. Rev., 27(2): 91–102.

Sharma, S.K., Verma, R.N. and Chauhan, S. 1988. Studies on the field reaction of soybean cultivar against purple seed stain disease (*Cercospora kikuchii*). Biol. Bull. India, 10(3): 117–121.

Sharma, S.K., Verma, R.N. and Chauhan, S. 1989. Seed borne microbes of purple stain infected seeds of soybean cultivars. Indian Jr. Hill Farm., 2(2): 109–111.

Sharma, S.K., Verma, R.N. and Chauhan, S. 1991. A new host Record of *Colletotrichum capsici* on soybean from India. Intl. J. Tropical Plant Dis., 10(2): 277.

Sharma, S.K., Verma, R.N. and Chauhan, S. 1993. Sources of resistance to *Cercospora kikuchii* incitant of purple stain disease of soybean. pp. 415–427. *In*: Current Trends in Life Sciences Vol. 29, Recent Trend in Plant Disease Control. Today and Tomorrow's Printer & Publsihers. New Delhi.

Sharma, S.K., Verma, R.N. and Chauhan, S. 1996. Effect of NPK doses on yield and severity of three major diseases of soybean at medium altitude of East Khasi Hills (Meghalaya). J. Hill Res., 9(2): 279–295.

Sharma, S.K., Verma, R.N. and Chauhan, S. 1997. Field evaluation of some fungicides against frog-eye leaf spot (*Cercospora sojina*) disease of soybean (*Glycine max*) in Meghalaya. Pestology XXIII, 10: 3540.

Sharma, S.K., Verma, R.N. and Chauhan, S. 2000. Distribution and magnitude of purple seed stain disease of soybean in North-East India. J. Hill Res., 13(2): 105–107.

Shoshiashvili, I. 1940. With reference to studies on soybean and groundnut diseases. Bull. Georgian Expt. Sta. Prot. Ser. A. Phytopathol., 2: 71–283.

Sinclair, J.B. 1977. Infections soybean diseases of world importance. PANS, 23(1): 49–57.

Sinclair, J.B. 1983. Fungicide sprays for control of internally seedborne fungi. Seed Sci. Tech., 11:

Sinclair, J.B. 1991. Latent infection of soybean plants and seeds by fungi. Plant Dis., 75: 220–224.

Sinclair, J.B. and Dhingra, O.D. 1975. An Annotated Bibliography of Soybean Diseases 1882–1974. Urbana, USA. INTSOY, Univ. Illinois, 280 p.

Sinclair, J.B. and Shurtleff, M.C. 1975. Compendium of Soybean Diseases. American Phytopathological Society, 68 p.

Sinclair, J.B. and Backman, P.A. 1989. Compendium of Soybean Diseases. 3rd edition. American Phytopathological Society, St. Paul, MN.

Singh, D.P. 1993. Relative susceptibility of soybean cultivars to pod blight caused by *Colletotrichum truncatum* (Schw.). Agric. Sci. Digest Karnal, 13(2): 90–92.

Singh, B.B. and Thapliyal, P.N. 1977. Breeding for resistance to soybean rust in India. Rust of Soybean: The problem and research needs. INTSOY Series, 12: 62–65.

Singh, D.P. and Agarwal, V.K. 1984. Effect of different levels of purple stain infection on viability and germination on soybean seed. Seed Res., 12(2): 44–48.

Singh, Tribhuwan and Sinclair, J.B. 1985. Histopathology of *Cercospora sojina* in soybean seeds. Phytopathol., 75(2): 185–189.

Singh, R.R. and Shukla, P. 1987. Amino acid changes in black gram leaves infected with *Colletotrichum truncatum*. Indian Phytopath., 40(2): 241–242.

Singh, S.K. and Srivastava, H.P. 1989. Some new fungal diseases of moth bean. Indian Phytopath., 42(1): 164–167.

Singh, B.K. and Upadhyay, R.S. 2009. Management of southern stem blight of soybean by PCNB resistant mutants of *Trichoderma harzianum* 4572 incited by *Sclerotium rolfsii*. J. Agric. Tech., 5(1): 85–98.

Singh, O.V., Agarwal, V.K. and Nene, Y.L. 1973. Seed health studies in soybean raised in the Nainital Tarai. Indian Phytopath., 26: 260–267.

Singh, S.N., Srivastava, S.K., Bhargava, P.K. and Khare, M.N. 1990. Chemical control of seedling mortality in cv. *JS 72-44* and bacterial pustule in cv. *Punjab-1* of soybean (*Glycine max.* (L.) Merrill). Legume Res., 13(1): 17–20.

Sun, S.D. 1958. Soybean. Moscow, 248 p.

Su, G., Suh, S.O., Schneider, R.W. and Russin, J.S. 2001. Host specialization in the charcoal rot fungus, *Macrophomina phaseolina*. Phytopathol., 91(2): 120–126.

Suranpongchal, P. and Poonpolgul, S. 2002. Efficacy of Forwavit 5 SC (hexaconazole 5% W/v SC) to control Soybean rust pathogen (*Phakopsora pachyrhizi*). Technical papers Forward (Thailand 1989) Co. Ltd., 11 p.

Suriachandraselvan, V., Salalrajan, F. and Aiyyanathan, K.A.E. 2005. Effect of organic soil amendments against charcoal rot in sunflower. Madras Agric. J., 92(10): 705–708.

Suzuki, H. 1985. Mode of occurrence and control of purple speck of soybean. J.A.R.Q., 19(1): 7–17.

Tai, F.L. 1936. Notes on Chinese fungi. VII. Chinese Bot. Soc. Bull., 2: 45–66.

Tajane, V.S., Behere, G.T. and Aage, V.E. 2002. Efficacy and translocation of fungicides against collar rot of soybean caused by *Sclerotium rolfsii*. Plant Dis. Res., 17: 196–197.

Takasugi, H. 1936. Division of Plant Pathology and Entomology. Control. Agr. Expt. Sta. Manchurian R.R., 1933, 583–739.

Talukdar, A., Verma, K., Gowda, D.S.S., Lai, S.K., Sapra, R.L., Singh, K.P., Singh, R. and Sinha, P. 2009. Molecular breeding for charcoal rot resistance in soybean. I. Screening and mapping population development. Indian J. Genet. Pl. Breed., 69(4): 367–370.

Tekrony, D.M., Egli, D.B., Stuckey, R.E. and Loeffler, T.M. 1985. Effect of Benomyl applications on soybean seed borne fungi, seed germination and yield. Plant Dis., 69(9): 763–765.

Thakur, K.S., Keshry, P.K., Tamrakar, D.K. and Sinha, A.K. 2002. Studies of management of collar rot disease (*Sclerotium rolfsii*) of chickpea by use of fungicides. PKV Res. J., 26(1/2): 51–52.

Thapliyal, P.N. and Singh, K.P. 1974. Soybean (*Glycine max*) rust *Phakospora pachyrhizi*. Amer. Phytopath. Soc. Fungicide and Nematicide Tests. Results of 1973, 29: 94.

Theradimani, M. and Marimuthu, T. 1993. Effect of decomposed coconut coirpith on damping-off of chillies and root rot of blackgram. Plant Dis. Res., 8(1): 1–5.

Thirumalachar, M.J. and Chupp, C. 1948. Notes on some *Cercospora* in India. Mycologia, 40: 352–362.

Tiffany, L.H. 1951. Delayed sporulation of *Colletotrichum* on soybean. Phytopathol., 41: 975–985.

Tisselli, O., Sinclair, J.B. and Hymowitx, T. 1980. Sources of resistance to selected fungal, bacterial, viral and nematode diseases of soybeans. INTSOY Series No. 18. Univ. of Illinois at Urbana, 134 p.

Todd, T.C., Pearson, C.A.S. and Schwenk, F.W. 1987. Effect of *Heterodera glycines* on charcoal rot severity in soybean cultivars resistant and susceptible to soybean cyst nematode. Ann. Appl. Nematol., 1(1): 35–40.

Toorray, N.K., Verma, K.P. and Sinha, A.K. 2007. Evaluation of fungicides and bio-agents against *Sclerotium rolfsii* Sacc. causing collar rot disease of chickpea in laboratory and field condition. Adv. Pl. Sci., 20(2): 439–442.

Usmani, S.M.H. and Ghffar, A. 1982. Polyethylene mulching of soil to reduce viability of *Sclerotium oryzae*. Soil Biol. Biochem., 14(3): 203–207.

Velicheti, R.K. and Sinclair, J.B. 1992. Reaction of seed borne soybean fungal pathogens to Cercosporin. Seed Sci. Technol., 20: 149–154.

Verma, R.N. 1985. Reaction of six AVRDC soybean accessions to rust in Shillong, India in 1984. Soybean Rust Newsletter, 7: 20.

Verma, R.N., Kumar, S., Chandra, S., Sharma, S.K. and Chauhan, S. 1988. Diseases of soybean in North-Eastern states of India. Nat. Symp. on Insect Pests and Diseases of Soybean, 30–31 (Abstr.).

Vir, D.S., Gangopadhyay, S. and Gaur, A. 1972. Evaluation os some systemic fungicides and antibiotics against *Macrophomina phaseolina*. Pesticides, 6(11): 25–26.

Vishwadhar and Chaudhary, D.N. 1982. Field reaction of some soybean varieties to purple stain disease. Indian J. Pl. Pathol., 12(3): 312–313.

Vitti, A.J., De Carvalho, M.L. and Menten, J.O.M. 1993. Effect of chemical treatment on the performance of soybean. Ecossistema, 18: 75–83.

Walters, H.J. 1961. A premature dying of soybeans in Arkansas. Phytopathol., 51: 646.

Wang, C. 1961. Chemical Control of Soybean Rust. Agr. Assoc. China (Taiwan). J., 35: 51–54.

Watanabe, T.R., Smith, S. and Snyder, W.C. 1970. Populations of *Macrophomina phaseolina* in soil as affected by fumigation and cropping. Phytopathol., 60(12): 1717–1719.

Waterson, J.M. 1939. Annotated List of Diseases of Cultivated Plants in Bermuda. Bermuda Dept. Agr. Bull., 18: 38 p.

Weber, G.F. 1931. Blight of carrots caused by *Sclerotium rolfsii*, with geographic distribution and host range of the fungus. Phytopathol., 21: 1129–1140.

Weimer, J.L. 1947. Disease survey of soybean nurseries in the South. Plant Dis. Reptr. Suppl., 168: 27–53.

Weinding, R. 1932. *Trichoderma lignorum* as a parasite of other soil fungi. Phytopathol., 22: 837–845.

Wolf, F.A. and Lehman, S.G. 1924. Report of division of plant pathology. North Carolina Agr. Ext. Sta. Ann. Rpt., 47 pp. 83–85.

Wrether, J.A., Anderson, T.R., Arysad, D.M., Gai, J., Ploper, L.D., Porta-Puglia, A., Ram, H.H. and Yorimori, J.T. 1997. Soybean disease loss estimates for the top ten soybean producing countries in 1994. Plant Dis., 81: 107–110.

Wyllie, T.D. 1988. Charcoal rot of soybean—Current status. pp. 106–113. *In*: Wyllie, T.D. and Scott, D.H. (eds.). Soybean Disease of the North Central Region. St. Paul., USA, APS Press.

Yang, C.Y. 1977. Soybean rust in the Eastern Hemisphere. Rust of Soybean: The problem and research needs. INTSOY Series No., 12: 22–33.

Yang, X.B., Dowler, W.M. and Tschanz, A.T. 1991. A simulation model for assessing soybean rust epidemics. J. Phytopathol., 133: 187–200.

Yoshii, H. 1927. Crop diseases in 1926. Ann. Agr. Expt. Sta. Chosan, 7: 21–34.

Yuyama, M.M. and Henning, A.A. 1997. Evaluation of thiabendazole and thiram on the control of the main pathogens in soybean seeds. Revista Brasileira-de-Sementes, 19(2): 267–270.

Index

Printed and bound by CPI Group (UK) Ltd, Croydon, CR0 4YY

01/11/2024

01782623-0015